Applied Hierarchical Modeling in Ecology

Analysis of distribution, abundance and species richness in R and BUGS

Applied Hierarchical Modeling in Ecology
Analysis of distribution, abundance and species richness in R and BUGS

Volume 1

Prelude and Static Models

Marc Kéry
Swiss Ornithological Institute,
Sempach, Switzerland

J. Andrew Royle
USGS Patuxent Wildlife Research Center,
Laurel MD, USA

AMSTERDAM • BOSTON • HEIDELBERG • LONDON • NEW YORK • OXFORD
PARIS • SAN DIEGO • SAN FRANCISCO • SINGAPORE • SYDNEY • TOKYO

Academic Press is an imprint of Elsevier

Academic Press is an imprint of Elsevier
125 London Wall, London EC2Y 5AS, UK
525 B Street, Suite 1800, San Diego, CA 92101-4495, USA
225 Wyman Street, Waltham, MA 02451, USA
The Boulevard, Langford Lane, Kidlington, Oxford OX5 1GB, UK

Copyright © 2016 Elsevier Inc. All rights reserved.

No part of this publication may be reproduced or transmitted in any form or by any means, electronic or mechanical, including photocopying, recording, or any information storage and retrieval system, without permission in writing from the publisher. Details on how to seek permission, further information about the Publisher's permissions policies and our arrangements with organizations such as the Copyright Clearance Center and the Copyright Licensing Agency, can be found at our website: www.elsevier.com/permissions.

This book and the individual contributions contained in it are protected under copyright by the Publisher (other than as may be noted herein).

Notices
Knowledge and best practice in this field are constantly changing. As new research and experience broaden our understanding, changes in research methods, professional practices, or medical treatment may become necessary.

Practitioners and researchers must always rely on their own experience and knowledge in evaluating and using any information, methods, compounds, or experiments described herein. In using such information or methods they should be mindful of their own safety and the safety of others, including parties for whom they have a professional responsibility.

To the fullest extent of the law, neither the Publisher nor the authors, contributors, or editors, assume any liability for any injury and/or damage to persons or property as a matter of products liability, negligence or otherwise, or from any use or operation of any methods, products, instructions, or ideas contained in the material herein.

ISBN: 978-0-12-801378-6

British Library Cataloguing-in-Publication Data
A catalogue record for this book is available from the British Library

Library of Congress Cataloging-in-Publication Data
A catalogue record for this book is available from the Library of Congress

For information on all Academic Press publications
visit our website at http://store.elsevier.com/

J. A. Royle was the principal author of chapters 2, 7, 8, 9. Use of product names does not constitute endorsement by the U.S. Government.

For Jim Nichols, who changed the way in which we think about Ecology

Contents

Foreword ix
Preface xiii
Acknowledgments xxiii

PART 1 PRELUDE

CHAPTER 1 Distribution, Abundance, and Species Richness in Ecology 3
CHAPTER 2 What Are Hierarchical Models and How Do We Analyze Them? 19
CHAPTER 3 Linear Models, Generalized Linear Models (GLMs), and Random Effects Models: The Components of Hierarchical Models 79
CHAPTER 4 Introduction to Data Simulation 123
CHAPTER 5 Fitting Models Using the Bayesian Modeling Software BUGS and JAGS 145

PART 2 MODELS FOR STATIC SYSTEMS

CHAPTER 6 Modeling Abundance with Counts of Unmarked Individuals in Closed Populations: Binomial N-Mixture Models 219
CHAPTER 7 Modeling Abundance Using Multinomial N-Mixture Models 313
CHAPTER 8 Modeling Abundance Using Hierarchical Distance Sampling 393
CHAPTER 9 Advanced Hierarchical Distance Sampling 463
CHAPTER 10 Modeling Static Occurrence and Species Distributions Using Site-Occupancy Models 551
CHAPTER 11 Hierarchical Models for Communities 631

Summary and Conclusion 729
References 737
Author Index 761
Subject Index 771

Additional Online Content: http://www.mbr-pwrc.usgs.gov/pubanalysis/keryroylebook/

Foreword

I started graduate school in 2003 working on a simple project to understand how forest management practices affect bird populations in the White Mountain National Forest in New Hampshire, USA. As with many projects of its kind, we collected point count data to characterize abundance at a collection of sites that had received different management actions. Knowing that we would fail to detect many of the individuals present at several sites, and that detection probability might covary with habitat variables, my adviser David King recommended that we survey each site multiple times and record the distance to each individual detected. When the field season came to an end, I had my data in hand and was ready to knock out a quick analysis. That is when the difficulties began. The first thing I tried was a multiple regression, but I was immediately stumped as to what the appropriate response variable was. Should it be the mean number of individuals detected at each site, the maximum number detected, or perhaps the median? Different authorities recommended different strategies, and to my dismay, the results differed with each approach. In addition to this problem, it was apparent that the model made no distinction between the explanatory variables that I had collected to describe variation in abundance and the variables that I had collected to explain variation in detection. As a result, I could use my model to predict the effect of management on observed counts, but not on abundance, the state variable that I was actually interested in.

I soon abandoned the regression approach and turned my attention to distance sampling methods. Here again, I was quickly surprised to find that although I could pool my data to account for the effect of distance on detection probability, I couldn't directly model the effects of management variables (some of which were continuous) on abundance. Even the two-stage approaches that were recommended at the time were not possible because of the sparseness of my data, which included many sites with no detections. Moreover, even if I had been able to correct for detection and then model my estimates, it would have been very difficult to properly account for the covariance of the estimates. As I searched for solutions to these problems, my despair continued to grow as I read several distance sampling papers proclaiming that the data I had worked so hard to collect might be impossible to use for my purpose. The only option seemed to be to study something else!

And then in 2004, Andy Royle and colleagues published two papers that seemed to have been written with exactly my problem in mind. Royle (2004b, *Biometrics*) demonstrated how repeated count data could be used to model spatial variation in abundance while accounting for various factors influencing detection probability, and Royle et al. (2004, *Ecology*) explained how spatial variation in abundance could be modeled using distance data collected using standard point count survey methods. Unlike all other methods I had seen, there was no need to do a two-stage analysis, and it was straightforward to model covariates of both abundance and detection. Here were hierarchical models that not only provided conceptual clarity, but allowed me to make full use of my data so that I could test the hypotheses I was interested in.

There was only one problem with the exciting models that I was reading about—I had no clue how to fit them to my data. I had taken a couple of statistics courses, but I had never heard of marginal likelihood, and I certainly had no idea how to write code to maximize an integrated likelihood function. The very next year, however, Marc Kéry, Andy Royle, and Hans Schmid published a paper in *Ecological Applications* that provided additional details about these models, and the paper included an appendix with a short R script that demonstrated how to obtain maximum likelihood estimates. It's no exaggeration to say that I learned more about statistics and programming by trying to understand this one bit of code than I did from any formal education up to that point. From there, I quickly began modifying the code to make it more general and to deal with other types of data that I had gathered. This eventually led me to develop some general functions, which were coming together just as I learned about an R package being developed by Ian Fiske to fit this class of models. I offered to help, and even though I never met Ian in person, we had a couple of good years of collaboration that resulted in the R package `unmarked`.

The reason for conveying this bit of personal history is not just to make the point that Marc and Andy have had a huge impact on my own career, but to convey a story that I know applied ecologists around the world can relate to. Simple practical problems turn out to pose serious challenges when we are unable to directly observe the processes of interest. Indices of abundance and population trends are nearly useless when we're concerned about absolutes like extinction risk or harvest limits. Applied ecologists have recognized these problems for a long time, and hierarchical models provide a solution by enabling researchers to directly model the ecological process, rather than some poorly defined index, while also modeling the observation process. Moreover, these models allow practitioners to solve problems in such a way that the results can be clearly communicated to managers and policy makers. But why did it take so long for these methods, the foundations of which were developed a long time ago, to make their way into the hands of practitioners? In my opinion, the power of hierarchical models would never have been realized had it not been for researchers like Marc and Andy, who have deliberately worked to make these tools accessible to those of us lacking advanced degrees in statistics. This book is a phenomenal synthesis of that effort. Unlike many books on statistical modeling that seem to have been written by statisticians for statisticians, the main audience of this volume is clearly the practicing ecologist. The writing style is clear and engaging, and for virtually every technical problem, worked examples and code are provided. This accomplishment, of distilling advanced modeling techniques into an accessible format, is in my view one of the two great contributions of this work to ecology and related environmental sciences.

The second great contribution of this work lies outside the context of applied research. The hierarchical modeling approach that has been developed and illustrated here is part of an emerging trend, one that is far more general and in some ways far more consequential for ecology as a science. The methods covered in this book provide a framework for advancing knowledge of ecological systems by narrowing the chasm between theoretical and statistical models. It has always struck me as troubling that the mathematical models covered in ecology textbooks often fall by the wayside when we get our hands on data. As soon as the data are facing us on the computer, the instinct arises to turn toward the latest development in statistics and leave theory behind. For example, we have >100 years of theory on the factors limiting species distributions—factors such as competition, Allee effects, and physiological constraints—yet we cram data, often collected for other purposes, into species distribution models that ignore demographic processes and biotic interactions, not to mention the observation processes that will cause severe bias if ignored. Or, we often have data from metapopulations that we run through a machine learning algorithm or to which we fit some sort of GAM with loads of random effects. The tools described in this book provide an alternative. They offer a framework that allows one to fit metapopulation models to metapopulation data, to estimate the strength of biotic interactions, and to test for effects of abiotic covariates on abundance, occurrence, or population growth rates. This is the power of hierarchical modeling: that we can tailor our statistical models to the scientific question at hand and not the other way around.

This is not to say that this book is full of theory. Rather, it provides the tools necessary to build hierarchical models based on theory instead of relying on purely phenomenological approaches. Why is this so important at this point in time, when focus is increasingly shifting to prediction? Why don't we just hire a team of Netflix data miners to forecast the future of ecological systems? In my view, prediction without mechanism falls well outside the realm of science. For instance, we know that we can develop good predictive models by simply modeling spatial and temporal autocorrelation. Abundance at one location can often be accurately predicted as a function of abundance at an adjacent location, just as voting habits in one county can be predicted from the behavior of neighboring counties, and just as the weather today often tells us something about the weather tomorrow. But what is learned about the underlying processes from fitting models lacking mechanism? Very little, in my estimation, which is why I'm happy to have this new book that presents such a powerful alternative.

As great as the hierarchical modeling framework is, I think it is important to emphasize that it is not meant to be an alternative to classical methods of experimental design and analysis. In fact, I would suggest that it is only via manipulative experiments that can we achieve the ultimate goal of causal inference. The problems we face in ecology, however, are that we often cannot bring our system into the lab, and we can't always manipulate one component while holding the others constant. To complicate matters further, processes that hold at one point in space and time may operate differently at another. We are therefore forced to combine experimental approaches with observational ones if we wish to advance knowledge and inform conservation efforts. Once again, the methods presented in this book provide a formal way of building mechanisms into our models so that we can unify the insights gained from experimental studies with the information contained in field data. It is this unified approach that I think offers the greatest promise for advancing our field, and it is exciting to be working in a time when we finally have the tools available for the task. So it is with great pleasure that I congratulate Marc and Andy on a fantastic book, one that, as large as it is, is just the beginning. I'll be looking forward to the next volume and all the excellent work that is sure to benefit from it.

Richard Chandler
University of Georgia

Preface

This is volume 1 of our new book on the applied hierarchical modeling of the three central quantities in ecology—abundance, or density, occurrence, and species richness—as well as of parameters governing their change over time, especially survival and recruitment. Hierarchical modeling is a growth industry in ecology. In the last 10 years there have been a dozen or more books focused on hierarchical modeling in ecology including Banerjee et al. (2004), Clark and Gelfand (2006), Clark (2007), Gelman and Hill (2007), McCarthy (2007), Royle and Dorazio (2008), King et al. (2009), Link and Barker (2010), Kéry and Schaub (2012), Hobbs and Hooten (2015), etc. How can we possibly add 700+ more pages (and perhaps 1500 if you count volume 2) to what is known on this topic? That's a good question!

In this book we cover several classes of models that have previously received only cursory or no treatment at all in the hierarchical modeling literature, and yet they are extremely important in ecology (e.g., distance sampling). Moreover, we give complete recipes for analyzing these models and all others covered in the book, using program R in general and the R package `unmarked` in particular, and very extensively using the generic Bayesian modeling software BUGS. The use of `unmarked` is completely novel compared to these other books. Some models that we cover here and especially in volume 2 were simply unimaginable a couple years ago, e.g., the open hierarchical distance sampling models, the metacommunity abundance models and models with explicit population dynamics (i.e., the famous model of Dail and Madsen, 2011), which can be fit to counts and related data, including even distance sampling data. Much of this material is extremely new, and some has only just appeared in the literature in the last year or so. Thus, this book represents a timely synthesis and extension of the state of hierarchical modeling in ecology that builds on previous efforts, but covers much new and important territory, and provides implementations using both the likelihood (`unmarked`) and Bayesian (BUGS) frameworks.

A BOOK OF MONOGRAPHS

In a sense, *Applied Hierarchical Modeling for Ecologists* (*AHM*) is a book of books, or a *book of monographs*. Volume 1 contains the first two parts, a prelude, which introduces the necessary concepts and techniques in five chapters, followed by six chapters that deal with static demographic models of distribution, abundance, and species richness and other descriptors of communities and metacommunities. Volume 2 will contain two further parts on dynamic models and on advanced demographic models for populations and communities; see below for more information on the division of topics between volumes 1 and 2 of *AHM* and on the content that we envision for volume 2.

Looking back at volume 1 now, at the time of writing of this preface, we feel as if we have packaged almost a dozen independent books into this one book. There are general, introductory "monographs" on the concepts of distribution, abundance, and species richness and their measurement and modeling in practice (Chapter 1), on hierarchical models and their analysis (Chapter 2), on linear, generalized linear, and mixed models (Chapter 3), on data simulation in R (Chapter 4), and on the celebrated BUGS language and software (Chapter 5). After that, there are six comprehensive monographs on demographic models for distribution, abundance, and species richness in the context of what we call a "meta-population design," that is, the extremely common situation where you measure something in a population or a community at more than a single point in space.

In the prelude, and following the introductory Chapter 1, we have one monograph to cover hierarchical models (HMs) and their Bayesian and frequentist analyses (Chapter 2). The next monograph (Chapter 3) provides a highly accessible review of that "heart" of applied statistics: linear models, generalized linear models (GLMs), and simple mixed models, all of them illustrated in the context of one extremely simple ecological data set. Data simulation is one of the defining features of this book because it provides such immense benefits for the work of ecologists (and also for statisticians). Hence, the next monograph (Chapter 4) is dedicated to this essential topic and walks you through the R code necessary for the generation of one simple type of data set that is fundamental to the classes of models covered in this book: the case where one goes out and counts birds (or any other species) at multiple places (e.g., 20, 100, or 267) and repeats these counts at each site multiple times (e.g., 2 or 3).

The BUGS model definition language is implemented in three currently used BUGS engines for Bayesian inference: WinBUGS (Lunn et al., 2000), OpenBUGS (Thomas et al., 2006), and JAGS (Plummer, 2003). It has also just been adopted in the exciting new R package NIMBLE (NIMBLE Development Team, 2015; de Valpine et al., in review), which is a general model fitting software that uses and extends the BUGS language for flexible specification of HMs and allows analysis of HMs with both maximum likelihood and Bayesian posterior inference.

Over the first 25 years of its existence, BUGS has been instrumental in the surge of Bayesian statistics in all kinds of sciences including ecology (Lunn et al., 2009). It has grown by far into the most important general, Bayesian modeling language, and its user population keeps growing at a rapid rate (and of course we hope to increase that rate even more with this book). BUGS is unique in giving you as a nonstatistician a modeling freedom that lets you develop, test, and fit models that you wouldn't even have dared to dream of in the pre-BUGS era of ecological modeling (which we might call the ecological Stone Age;...). Although there are now many useful introductory books on BUGS (e.g., McCarthy, 2007; Kéry, 2010; Lunn et al., 2013; Korner-Nievergelt et al., 2015), we have decided to write yet another practical BUGS introduction and package it into Chapter 5. It is our latest and best attempt at covering as much as possible on this topic and including some of the latest tricks in BUGS modeling in a mere 70 book pages, illustrating the use of all three BUGS engines and focusing on the models covered in Chapter 3, i.e., linear models, GLMs, and simple mixed models. By introducing BUGS for exactly the kinds of models that you are likely to be familiar with already, we hope to make it especially easy for you to grasp the Bayesian side of the analysis and the implementation of these essential models in the BUGS language.

In the second part of the book, we present six monographs that contain a comprehensive treatment of important classes of models for inference about distribution, abundance, and species richness, and related demographic population or community metrics in so-called "meta-population designs" (Royle, 2004a; Kéry and Royle, 2010), i.e., for the frequent case where you are interested in these things not at a single place but have studied them at multiple sites. Specifically, in Chapter 6 we cover binomial mixture, or N-mixture, models (Royle, 2004b), which are a unique type of model for count data on unmarked individuals (that is, you do not need to keep track of which individual is which across the repeated measurements of abundance at a site) and that contains an explicit measurement error model, which corrects your inferences for the biases that would otherwise be caused by undercounting due to imperfect detection probability. Chapter 7 covers a "sister-type" of model, the multinomial mixture model (Royle, 2004a; Dorazio et al., 2005), which only differs from binomial mixture models in the type of data to which it is fitted: typically you need individual recognition, that is, you have capture-recapture-type of data, but again collected not at a single site but at multiple places. Both types of mixture models have been around for about 10 years now and previously they have been featured in some of the above-cited hierarchical modeling books (mostly in Royle and Dorazio, 2008), but never before have they been covered in such detail and, especially, in a manner that makes them so accessible to you as an ecologist.

Chapters 8 and 9 are special in that they provide perhaps the first large, and yet practical and applied, synthesis in distance sampling more than 10 years after the two classics by Buckland et al. (2001, 2004a) were published. In our two distance sampling monographs, we provide a fresh, new look at distance sampling in the context of hierarchical models in an essentially book-length treatment. We hope that this will help to make this important type of model even more widely understood and used by ecologists. While we cover mainly static models in volume 1 and then cover dynamic models in more detail in volume 2 of *AHM*, we have deviated slightly from this rule in Chapters 8 and 9, where we have preferred topical unity over conceptual unity by keeping all of distance sampling (closed and open) together. Nevertheless, we plan to cover several more cutting-edge open and other novel extensions of hierarchical distance sampling (HDS) models in volume 2.

These two monographs, and more specifically the wealth of material on hierarchical distance sampling, are perhaps those with most novelty in our book. Though again invented just over 10 years ago (around 2004), HDS has recently experienced a boost with the widespread realization that this type of specification of distance sampling models enables extremely flexible modeling of spatially or temporally replicated distance sampling data or combined analyses of data sets collected under differing protocols ("integrated models"), which was thought impossible before or at least was never achieved. For instance, it is perfectly doable (or even trivial) to model population dynamics (Sollmann et al., 2015) or community size and composition (Sollmann et al., in press) from distance sampling data within the context of hierarchical models. And, the power of BUGS nowadays makes the implementation of such models possible even for ecologists, since really such models differ in only relatively minor ways from similar models for other data types (e.g., of the capture-recapture type).

Hence, we hope that we contribute to change your view of "capture-recapture" and "distance sampling" as being two widely separated fields to a new way of seeing them as really relatively minor variations on the overarching theme of hierarchical models, which have one model component for abundance, or density, and in another model component describe the measurement error that induces imperfect detection and therefore undercounting (Borchers et al., 2015). The only thing that changes when you move from a capture-recapture to a distance sampling model is the specific parameterization of the measurement error underlying the observed data and of course the type of data that you need to estimate the parameters of that measurement error model. This wonderful, unifying power of describing statistical models in a hierarchical way lets you much better grasp the similarities among large numbers of models that were often thought as totally distinct hitherto. It is one of the main themes of this book and one on which we will say much more throughout the book. For instance, we hope that you will recognize that there are really only quite minor differences between a binomial mixture model for counts of unmarked individuals, a multinomial mixture model for capture-recapture data, and a hierarchical distance sampling model—the only difference is again the measurement error model, while the state model, that is, the description of the essential biological quantity (abundance or density), is exactly the same in all three types of models.

The penultimate monograph (Chapter 10) is on occupancy modeling (MacKenzie et al., 2002; Tyre et al., 2003). This powerful type of model for occurrence or distribution comes with an explicit measurement error component model for both false-negatives and false-positives (models by Royle and Link, 2006; Aing et al., 2011; Miller et al., 2011, 2013b; Sutherland et al., 2013; Chambert et al.,

2015) or with a measurement error model for false negatives only (all other types of occupancy models). Occupancy models have become huge in ecology and have experienced a steep growth curve in both the number of papers that further develop the theory of these models and especially also in studies that apply this design and the associated models. (We have even heard rumors that the vigorous growth of the field has "scared" some ecology journal editors so that they put a cap on the number of occupancy papers they accept—a strange way of stifling progress one would think.) Occupancy models have received one book-length treatise so far (MacKenzie et al., 2006), with a second edition that is in preparation, and several customized software products that specialize in them, especially PRESENCE (Hines, 2006) and MARK (White and Burnham, 1999; Cooch and White, 2014). In this first *AHM* volume, we deal with single-species occupancy models in great detail and cover some topics (e.g., some models for data collected along space or time "transects") that haven't been covered in any book before. In volume 2, we will add several more monographs on a large variety of occupancy model types; see below.

The final monograph in volume 1 covers community models, that is, community or multispecies variants of all the previous models. Specifically, we cover the community variant of an occupancy model (Chapter 10) and the community variant of a binomial *N*-mixture model (Chapter 6). These powerful hierarchical models enable inferences at multiple scales, that of the individual species, that of a local community, and that of an entire metacommunity. As always in this book, both come with an explicit measurement error model for the desired state of inference, presence/absence, or abundance of each individual species at every site in the "meta-population." These models have experienced much increased attention in the very recent past (Iknayan et al., 2014; Yamaura et al., 2012, in press), and we provide a much needed, comprehensive and yet supremely practical monograph on both the abundance and on the occupancy-based community models.

Of course, apart from serving as a standalone introduction to this large range of powerful and useful hierarchical models, the material in volume 1 also lays the groundwork for more models and more advanced material in volume 2. See below for more about the division of content between the two volumes.

UNIFYING THEMES

AHM is not just a hodgepodge of models that have not previously been covered in detail or at all. Rather, our development and organization of these models has a number of unifying themes that we emphasize throughout the book:
- hierarchical modeling
- data simulation
- measurement error models
- dual inference paradigm approach (Bayesianism and frequentism)
- accessible and gentle style (including hierarchical likelihood construction and data simulation)
- "cookbook recipes"
- predictions

One, we advocate *hierarchical modeling* as a unifying concept and overarching principle in modeling and also conceptually; as we have emphasized before, when seen as HMs all these models almost look the same (or very similar) and it is quite trivial to move from one to another, e.g., from a capture-recapture model to a distance sampling model to an occupancy model or even to a community or metacommunity model. We dedicate an entire chapter to introduce and explain the crucial concept of HMs, which permeates every section of this book.

Two, we use *data simulation* throughout the book, because this is so tremendously important in practice, for statisticians, but much more so still for ecologists. This is done in hardly any other book we know of, except for two of our earlier books (Kéry, 2010; Kéry and Schaub, 2012). Though quite frequently done by statisticians and also by ecologists in many different modes, we believe that data simulation should be done *much* more widely still. We dedicate an entire chapter to data simulation (Chapter 4) and therein explain the major advantages for you when you start doing this routinely for your work. Among them, perhaps the two most important benefits of data simulation are, first, that it enforces on you a complete understanding of your model. If you don't understand your model, you will not be able to write R code to simulate data under that model—it's as simple as that. We would even go as far as saying that a data simulation algorithm provides a complete description of a statistical model. Indeed, throughout the book we use data simulation in R in a completely novel fashion *to explain a statistical model*!

The second major benefit of data simulation is that it serves an important role to validate both your MCMC algorithm (whether written by yourself or produced by an MCMC black box such as BUGS) and to validate your model code. For most model classes in the book we provide R functions to simulate data under various types of models. We hope that these will be widely used in the many different modes of data simulation (as per Chapter 4)…or perhaps sometimes simply to marvel at the pretty and highly variable graphical output they produce.

Three, one defining feature of all main classes of models in our book is the presence of a submodel that contains an explicit description of the *measurement error* process underlying all data on the distribution and abundance of individual species and even more perhaps when you study them in entire communities or metacommunities. Unlike the types of measurement error for continuous variables (such as body length) to which you may have been exposed, the measurement error for discrete measurements (e.g., counts and presence/absence measurements) is of a radically different nature and comes in exactly two types: false-positive and false-negative measurement error, with the complement of the latter typically being called detection or encounter probability, or detectability for short. These are very different types of measurement error, which you cannot expect to cancel out in the mean over several measurements. Hence, unless you account for them in your models for distribution, abundance, and species richness, badly biased inferences may result. Our *AHM* book is an "estimationist book" in line with a rapidly increasing number of previous works that emphasize the measurement error processes in ecological models for distribution and abundance in ecology, such as Otis et al. (1978), Seber (1982), Buckland et al. (2001), Borchers et al. (2002), Williams et al. (2002), Buckland et al. (2004a), Amstrup et al. (2005), MacKenzie et al. (2006), Royle and Dorazio (2008), King et al. (2009), Kéry and Schaub (2012), McCrea and Morgan (2014), and Royle et al. (2014).

Four, we are neither purebred Bayesians nor hardcore frequentists, rather, we are big fans of a *dual inference paradigm approach*, i.e., the use of Bayesianism *and* frequentism alongside, as it seems especially useful for the particular case. While perhaps both of us have a slight personal slant toward Bayesianism, there are advantages and disadvantages of both Bayesianism and frequentism, and these may come into play more or less for any given data set or scientific question (Little, 2006; de Valpine, 2009, 2011). In addition, the choice of whether a Bayesian or a frequentist analysis is most appropriate will also be affected by the availability of a well-trained analyst and/or a fast computer, with Bayesian solutions often requiring more statistical and programming experience and faster computers. Thus, we are firm believers in the value of a dual inference paradigm approach, and this is a pervasive theme of our book as well. The dual inference paradigm approach appears in all but one chapters of this book. This approach has been done a little bit in some previous books (especially in Royle and Dorazio, 2008) but never to our knowledge in such a completely integrated way as in this book. Every topical chapter in Part 2 covers a class of models using both inference paradigms and emphasizes things that are easier or harder to do one way or the other (the exception being Chapter 11, where it is very hard to do a non-Bayesian analysis of these parameter-rich models).

Five, we have striven to make this book *gentle and accessible in style and easy to read*. This means that we do of course present formulae and equations, but perhaps fewer than in many other comparable statistics books. Many ecologists cannot read even moderately complex likelihood expressions. This is perhaps not a good state of affairs, but it is a simple fact of life that is unlikely to change anytime soon. We believe that the hierarchical construction of the likelihood, as a series of conditional probability statements as in every topical chapter in this book (and as we naturally do when specifying these models in the BUGS language), is perhaps the *only* way in which a fairly large proportion of ecologists have any chance of being able to read and understand the likelihood of a somewhat complex model. In addition to algebra, we use especially data simulation (and hence R code) to describe our models throughout the book. We find that R code for data simulation is an extremely clear and transparent way of implicitly describing the likelihood of a model. This seems to be a completely novel idea that has never been expressed explicitly before.

Six, we illustrate analyses of each class of models using a complete set of steps that you would use in your work ("*cookbook recipes*"). This includes not just fitting the models but producing summary analyses such as response curves and prediction, and especially illustrating maps of abundance and occurrence, and also assessing the goodness-of-fit of models. We believe that providing cookbook recipes is frowned upon by many statisticians because there is a feeling that this encourages people to do things that they don't understand. We are convinced that this sentiment is mostly unfounded. First, and most importantly, for any but an extremely trivial analysis, the practitioner will still have to understand the model and the analysis in order to not make any of a myriad of trivial errors that will make the BUGS program crash. Second, without at least some understanding, he will probably not be able to describe the results in an intelligible way in the results section of his paper or to explain them to her supervisor, advisor, or colleague. On the other hand, even some of the most basic of statistical analyses, namely linear models with factor levels, are extremely widely misunderstood, i.e., people don't understand what the intercept means and what the treatment contrast parameters are. Hence, some abuse is to be expected with *any* kind of statistical model for which easy-to-use code is made widely available. In addition, in complex models in this and similar books, sometimes even very basic steps such as formatting the data into a three- or four-dimensional array can be a complete stumbling block to an R novice, even though he may have a decent conceptual grasp of a model. In this case, the availability of cookbook analysis code is essential. Finally, it is the experience of at least one of the authors that only fitting a model and looking at the estimates may sometimes really let one understand what these parameters mean. Of course, this latter effect may be magnified still when you fit the model to simulated data, where you know what you input into your data set and therefore what ballpark estimates you can expect. In summary, we believe that it is not evil to hand out cookbook recipes but rather that they *ought* to be given much more widely, and we do exactly this throughout our book.

Seven, and finally, one of the examples of us giving ample code recipes is for *prediction*, i.e., for the computation of the expected value of some quantity (e.g., the response or some parameter) for a range of values for one or more covariates. Forming such predictions is extremely important for you in two ways: to even understand what the model is telling you about the form of some covariate relationship when you have log, logistic, or similar link functions, polynomial terms, or interactions; and second to present the results of your analysis, e.g., in a figure in your paper. We emphasize prediction throughout the book, especially predictions in geographic space, leading to maps of species abundance and occurrence (the associated models are then called "species distribution models"); this is a very hot topic nowadays. The forming of predictions, and how to put these predictions on a map, is the focus of every single monograph in the second part of the book and also appears extensively in the prelude chapters.

Although this book is especially geared toward ecologists, it presents the cutting edge of the current state and understanding of all of the models presented. At several places, we were not shy to lay open our partial lack of understanding about some topics, in the hope to emphasize the need for further research; this includes goodness-of-fit in these models (and probably in many other classes of hierarchical models in general), the good fit versus bad prediction dilemma with some negative binomial N-mixture models in Chapter 6, or the use of spatial instead of temporal replication for obtaining information about measurement error in occupancy models in Chapter 10. Clearly, our aim in writing this book is not to show off how much we know but to help you to learn these models to understand and apply them. This includes a recognition of where their limits or the general limits of our understanding about them are and where you could therefore make a contribution to the progress in this field.

THE unmarked PACKAGE

Somebody once said that he (or she) did not trust any R package unless it has a book written about it. So now you can finally trust the R package unmarked (Fiske and Chandler, 2011) because this is also a book about unmarked. The unmarked package is fully general, and as part of the R programming environment, it allows you to embed your analyses seamlessly into your R programming. This is a great help when running simulations, for data processing and formatting, running analyses in batch mode (e.g., looping over many species, years, sites), documenting your data processing and analysis steps, and when analyzing results to produce plots, summary analyses, fit assessments, and model selection.

The unmarked package permits you to fit a large variety of closed and open hierarchical models and, to the best of our knowledge, it is the only package for likelihood estimation of (almost) all classes of models we cover in this book, although PRESENCE (Hines, 2006), MARK (White and Burnham, 1999; Cooch and White, 2014), and E-SURGE (Choquet et al., 2009b; Gimenez et al., 2014) fit occupancy models, and the former two also Royle-Nichols and binomial N-mixture models. One of the benefits of using unmarked for analyzing these various hierarchical models is that it streamlines and standardizes the work flow across models. An analysis of any class of hierarchical model in unmarked has a few basic steps, which include: (1) processing and packaging the data into an "unmarked frame" using standard constructor functions that ensure data are in the proper format; (2) utilization of a standard model fitting function that produces parameter estimates, standard errors, AIC, and other summary statistics; (3) summary analyses that include producing model selection tables, goodness-of-fit analyses (e.g., using parametric bootstrapping), and plotting predictions or fitted values. Each of these summary analyses is supported by standard functions that are part of the unmarked package.

The unmarked package is supported by an active and most of the time very friendly e-mail user group (groups.google.com/forum/#!forum/unmarked), which you can subscribe to for following developments and bug reports, or for requesting assistance. Finally, unmarked is an open source software development project. The source code is readily available and can be easily modified and extended by anyone. We encourage you to participate in the unmarked community.

COMPUTING[1]

In a sense this is a book about ecological computing. While we emphasize the formulation and analysis of models, a vast majority of the effort to do so requires programming in the R language and running various functions in unmarked and in WinBUGS or JAGS. For Bayesian analysis we adopt the implementations of the BUGS language using WinBUGS and JAGS (and could equally well have used OpenBUGS or NIMBLE). These are used almost equivalently with the help of the R packages R2WinBUGS and jagsUI, and there are only a very small number of minor differences between the JAGS and WinBUGS implementations of the BUGS language (see the JAGS manual, available on the Internet, and Lunn et al., 2013).

Interestingly, the use of BUGS is often frowned upon by statisticians as some kind of inferior approach to things, as compared to writing your own MCMC algorithm, and reliance on BUGS is readily criticized in reviews of papers and conference presentations (especially those that are widely attended by statisticians). The academic statistician's view is often that you should be writing your own MCMC because then you understand what's going on under the hood. We disagree with this view. Now, and we think even 20 years into the future, the vast majority of ecologists will not be able to write their own MCMC nor even will most ecologists want to do that. Indeed, many statisticians can't do that either. On the other hand, BUGS makes accessible to ecologists the extremely convenient and useful technique of MCMC and consequently the ability to describe models and analyze them without having to have a PhD statistician helping them out. We therefore strongly advocate for the use of the BUGS language in whatever implementation is convenient (WinBUGS, JAGS, OpenBUGS, NIMBLE, or some future implementation). To be sure, custom MCMC algorithms may be *much* more efficient for any particular model or application. However, the time to produce custom algorithms really renders that approach impractical for most situations and for most people. Moreover, perhaps the greatest thing about the BUGS programs is the BUGS model definition language. This has proved to be supremely easy to understand for statisticians and nonstatisticians alike in their attempts to formulate, with confidence, even very complex statistical and simulation models. Therefore, we feel that the BUGS *language* is here to stay. There may be some inefficiency to the current implementations, but who's to say that a more efficient implementation won't be invented in the future? And, of course computing power is always improving and will continue to do so. In particular, computers will certainly have many more cores, and therefore multicore processing will improve the runtime of many models.

As to the diffuse fears of some when using a computational MCMC black box such as BUGS, we have argued before that data simulation has an important role to play in modern ecological modeling. Analyzing simulated data can give you much confidence about the good or bad behavior of a computational procedure—of course not for every single particular case (but you can't have

---

[1]Use of product names does not imply endorsement by the US government.

this anyway, e.g., your likelihood maximization algorithm may always get stuck at a local maximum or along some flat ridge), but on average, and that is what really counts. For instance, over the years we have used BUGS to fit models to literally many thousands of simulated data sets for the types of models presented in this book. And we have only exceedingly rarely experienced cases where the algorithm converged to a place in parameter space that was not close to or right at the correct value, i.e., the value used to simulate the data set. Thus, we are not at all made nervous by the occasional claims heard about how terribly difficult it is to achieve chain convergence in an MCMC analysis.

ORGANIZATION

When we started *AHM* we did not think of it as comprising two volumes. But then we realized the wealth of material we had at our hands, and so now *AHM* comes in two volumes. As planned now, there are 25 chapters that are grouped in four parts, with two parts per volume (see Table 1). As already explained, the split of the whole *AHM* project into two volumes has introductory material including basic concepts of statistical modeling and inference and data simulation (Part I) and then single- and multispecies models of abundance and occurrence in static systems (Part II) in volume 1 (with the slight exception mentioned for distance sampling in Chapter 9). Volume 2 of the book will focus on dynamic and spatial models and other "advanced" topics.

WHO SHOULD READ THIS BOOK?

This book has two target audiences: first, ecologists and scientists and managers in related disciplines, where the demographic analysis of populations, meta-populations, communities, and metacommunities is a focus of interest. And second, statisticians, especially those hitherto unacquainted with these classes of HMs, which are hardly ever taught in standard methodology classes or in typical classical applied statistics texts. For the former group, the book represents a practical how-to guide for each class of models and thus should be accessible to anyone with basic R programming knowledge. Use of the BUGS language is needed also, but we hope you can gather the requisite skills by reading the earlier chapters of the book (3–5) or else you may consult an introductory BUGS book such as McCarthy (2007), Kéry (2010), Lunn et al. (2013), or Korner-Nievergelt et al. (2015). Because R programming is the standard now in many university curricula, we think the book should be ideal for a graduate level class on quantitative methods, either as a complete semester long course or part of a course covering specific models such as hierarchical modeling of abundance using N-mixture models, on occupancy models, and on hierarchical distance sampling.

BOOK WEB SITE AND USER GROUP E-MAIL LIST

For every analysis in the book we provide the complete instructions for organizing the data, fitting the model, and summarizing the results. Most of the commands are given directly in the book, although our companion Web site (http://www.mbr-pwrc.usgs.gov/pubanalysis/keryroylebook/) also provides the scripts for ready download. In addition, you find other information there, notably the solution to exercises, a list of errata as we find them (or, more likely, you detect and report them to us), etc.

Table 1 Outline of volume 1 and volume 2 of the Book *Applied Hierarchical Modeling in Ecology* (*AHM*).

AHM volume 1: Prelude and static models

Preface

Part 1: Prelude

1. Distribution, abundance, and species richness in ecology
2. What are hierarchical models and how do we analyze them?
3. Linear models, generalized linear models (GLMs), and random effects: the components of hierarchical models
4. Introduction to data simulation
5. The Bayesian modeling software BUGS and JAGS

Part 2: Models for static systems

6. Modeling abundance using binomial N-mixture models
7. Modeling abundance using multinomial N-mixture models
8. Modeling abundance using hierarchical distance sampling
9. Advanced hierarchical distance sampling
10. Modeling distribution and occurrence using site-occupancy models
11. Community models

AHM volume 2: Dynamic and advanced models

Part 3: Models for dynamic systems

12. Modeling population dynamics with Poisson generalized linear mixed models (GLMMs) and some extensions
13. Modeling population dynamics with replicate counts within a season
14. Modeling population dynamics with distance sampling data
15. Hierarchical models of survival
16. Modeling species distribution and range dynamics using dynamic occupancy models
17. Modeling metacommunity dynamics using dynamic community models

Part 4: Advanced models

18. Multistate occupancy models
19. Modeling false-positives
20. Models for species interactions
21. Spatial models I
22. Spatial models II
23. Combination approaches/Integrated models
24. Spatial distance sampling and spatial capture-recapture
25. Conclusions

The *AHM* book (or indeed our larger "hierarchical modeling enterprise") has an associated e-mail user group (http://groups.google.com/forum/?hl=en#!forum/hmecology), which you can subscribe to for following developments and bug reports, or for requesting assistance. There is some overlap with the `unmarked` e-mail user group, but the hierarchical modeling user group is more general and, in particular, is the only one specifically for questions about BUGS software in ecological modeling. We would again encourage you to become an active member of that community.

Acknowledgments

We would like to thank James D. Nichols (USGS Patuxent Wildlife Research Center, Laurel, MD) for teaching us to "think hierarchically" in ecology and for emphasizing so forcefully the need for accommodating in our ecological models the ever-present measurement errors, especially detection probability, which in ecological studies afflicts the assessment of demographic quantities such as distribution and abundance. Jim's sharp brain, his never-flagging enthusiasm, his interest in other people's projects, his willingness to help, and at the same time his legendary modesty, have been a huge influence for us both. We would like to dedicate this book to you, Jim—thank you so much!

Then, we owe super-special thanks to Ian Fiske and Richard Chandler for creating the `unmarked` package, which along with BUGS is the main software that we use in this book. Had it not been for his new position at the University of Georgia/Athens, Richard would have been a coauthor of this book. Now, we are very grateful that he wrote the foreword and thus is still associated with the book project. Ken Kellner created the `jagsUI` package, which is our favorite R/JAGS interface. Many times, Ken has been incredibly quick at replying to our queries about things that were problems to us or things that we thought would be nice to add to `jagsUI`, and we are extremely grateful for Ken's time. Marc Mazerolle's `AICcmodavg` package has grown to be extremely useful for many types of models, and it contains a large suite of functions specifically for models fit with `unmarked`. Marc has been particularly helpful as well in answering our queries and adding functionality to the package as needs were perceived.

Next, we would like to warmly thank the developers of WinBUGS (Gilks et al., 1994; Lunn et al., 2000, 2009, 2013) for inventing the wonderful BUGS language and also to thank them and the developers of OpenBUGS (Thomas et al., 2006) and JAGS (Plummer, 2003) for giving these unbelievably powerful and user-friendly programs to the world. You have changed our (scientific) lives! The BUGS model definition language and the three BUGS engines (or four, with NIMBLE) have been revolutionizing the way in which especially nonstatisticians can fit complex and customized models to their complicated data. We think that the service of the BUGS developers and maintainers to ecology, and many sciences beyond, can hardly be exaggerated.

Many people have reviewed parts of the book, sometimes under extreme time constraints, and we are extremely grateful to them, including (in alphabetical order) Courtney Amundson, Evan Cooch, Tara Crewe, Nathan Crum, Francisco Dénes, Emily Dennis, Gurutzeta Guillera-Arroita, Jose Lahoz-Monfort, Abby Lawson, Dan Linden, Mike Meredith (we always appreciate your openness, Mike), Dana Janine Morin, Danielle Rappaport, Benedikt Schmidt, Rahel Sollmann, Nicolas Strebel, Chris Sutherland, and Yuichi Yamaura. Several people have been generously sharing data or R or BUGS code, including Courtney Amundson (code in Chapter 9), Scott Sillett (ISSJ data, graphics, photos), Wolf Theunissen from the Dutch Centre for Field Ornithology Sovon (Dutch wagtail data in Chapter 9), and Rahel Sollmann (material of Chapter 9 related to open HDS models). We also thank all the photographers who offered us their great photos, sometimes for free, to illustrate the fascinating animals behind the numbers that we crunch. A warm thanks goes to Bert Orr for allowing us to use his wonderful dragonfly art on the covers of the two volumes of *AHM*.

We owe a special thank you to our colleague Hans Schmid, who is the father and manager of the Swiss breeding bird survey MHB (Monitoring Häufige Brutvögel), and to the hundreds of volunteers who annually survey the 267 1-km^2 quadrats in a largely mountainous Switzerland. In many respects, the MHB is an exemplary biological survey. We are privileged to have ready access to the high-quality data produced by it, and we have made ample use of its beautiful data; see the analyses in Chapters 6, 7, 10, and 11. And, had it not been for the MHB and the first papers that MK and JAR wrote together back in 2005, perhaps this book would never have been written at all.

Other people who have contributed to the book in various ways include Jérôme Guélat, Guido Häfliger, Fränzi Korner Nievergelt, Michael Schaub, Benedikt Schmidt, Richard Schuster, and Nicolas Strebel.

Finally, here are some special thanks from the two of us.

MK: Most of all I thank my coauthor Andy for being my colleague and friend. The publication of this book marks a little more than the 10th anniversary of what is the most important and productive collaboration in my professional life. It would be hard to exaggerate Andy's role in my development as a quantitative ecologist. I have always been astonished by your generosity to contribute your brains and your power as a statistical modeler to my projects. It is a tremendous honor and a huge pleasure for me to coauthor this book with you, Andy. Thank you very much! Next, I would like to thank my employers, the Swiss Ornithological Institute and especially my former boss Niklaus Zbinden (now retired), for granting me the much-appreciated academic freedom in my research, which is required for conducting a big book project. And lastly, but especially importantly, I am indebted to my family, Susana and Gabriel, for tolerating so much investment of my time and energy in this project, which sometimes appeared to be growing over our heads.

JAR: I extend equally effusive and heartfelt thanks and gratitude to my friend and colleague Marc Kéry. Without your persistence in pushing forward our collaboration 10 years ago, none of this would have happened. I owe you so much, my friend! I would also like to thank the Patuxent community, my colleagues, and past and present postdoctoral researchers with whom I've had the pleasure to work. Nothing is more satisfying in science than pushing forward new ideas with enthusiastic young researchers who are at the start of their careers.

MK and JAR, Lima/Peru and Laurel/MD
July 21, 2015

PART 1

PRELUDE

CHAPTER 1

DISTRIBUTION, ABUNDANCE, AND SPECIES RICHNESS IN ECOLOGY

CHAPTER OUTLINE

- 1.1 Point Processes, Distribution, Abundance, and Species Richness 3
- 1.2 Meta-population Designs 10
- 1.3 State and Rate Parameters 12
- 1.4 Measurement Error Models in Ecology 13
- 1.5 Hierarchical Models for Distribution, Abundance, and Species Richness 16
- 1.6 Summary and Outlook 16
- Exercises 17

1.1 POINT PROCESSES, DISTRIBUTION, ABUNDANCE, AND SPECIES RICHNESS

Distribution and abundance are the two fundamental state variables in ecology (Begon et al., 1986; Krebs, 2009) and species richness is the most widely used measure for biodiversity (Purvis and Hector, 2000; Balmford et al., 2003). All three are the focus of a preponderance of both theoretical ecological studies and especially of studies focused on specific management or conservation problems involving rare or endangered species, game animals, and invasive species. Interestingly, though, all three are only derived quantities, i.e., summaries of a more fundamental quantity: *point patterns*. Point patterns are the outcome of stochastic processes known as point processes, and, not surprisingly, statistical models describing them are called point process models (PPMs; Illian et al., 2008; Wiegand and Moloney, 2014). PPMs treat both the number *and* the locations of discrete points as random quantities governed by an underlying, continuous intensity field. The intensity is the expected number of points (e.g., animals or plants) per unit area in some study area and is the modeled parameter.

Both distribution and abundance are simple areal summaries of spatial point patterns for a single animal or plant species, that is, aggregations of a point pattern over some area. To develop a basic understanding of the relationships between a point pattern and abundance and occurrence, we jump right in and run our first simple data simulation in program R. Thus, consistent with how we often approach the understanding of a new model in the rest of this book, *we here use simulation to explain and to understand* a model, such as a PPM. Function `sim.fn` lets you experiment with the relationship between a point pattern and abundance and occurrence as a function of the intensity of the pattern (which is something that you cannot control and is the result of the biology you're interested in) and of the grid size, or more specifically, the size of the cells making up that grid; this is something that you *can* control—or somebody else has done it for you (for instance, the people who designed the monitoring program that produces the data you are analyzing). The default settings of the function are:

```
sim.fn(quad.size = 10, cell.size = 1, intensity = 1)
```

The function simulates animal or plant locations in a grid of cells forming a quadrat with total length (in arbitrary units) equal to `quad.size`, according to a Poisson process where individuals are randomly distributed in space. This process is characterized by a constant `intensity`, which is the average number of animals or plants ("points") per unit area. The resulting point pattern is then discretized by overlaying a grid with quadratic cells of length `cell.size`. It is only this discretization of space that lets one define abundance in the first place and then presence/absence, or occurrence, in a second step. Species richness, the third crucial quantity in the title of this chapter, is the sum of the species occurring at a site, hence, a summary of the point patterns not for a single species but for all species (or for some set of species) that occur at a site.

As usual for the data simulation functions in this book, execution of the function produces both numerical output (data that you can save and do things with after) and informative plots that visualize the simulated process and the resulting data set (Figure 1.1). We will use many such functions in this book; also note that we have a whole chapter on the simulation of data in R (Chapter 4).

To appreciate the randomness inherent in the stochastic process defined by this function, we encourage you to call the function repeatedly without random number seeds, or with different seeds, and with changed function arguments. There is nothing like data simulation to help you realize what stochasticity really means—that lack of exact reproducibility of a process, which can therefore only be predicted in some average sense. Therefore, we urge you to play! Play with this data simulation function and also with all other data simulation functions in this book. You will see that this book gives you much to play with. "To play" means that you vary the function arguments and observe the changes in the output from the process represented by the data simulation function. We are convinced that this can be a huge help for your understanding of the process represented by the function. In addition, our hierarchical models directly represent the processes underlying the observed data, hence, if you understand the data simulation process that serves to *assemble* a data set, you will also understand the model that serves to *disassemble* the data set in the analysis, where disassembly means to "break the data apart" into coefficients of covariates, random effects variances etc. (Kéry, 2010).

For now, we execute the function once, with a specific random number seed, so you get the same results as we do. Afterwards, you can do `str(tmp)` to see the objects created by the function and saved in the object `tmp`, but we simply focus on the graphical output for now. This is all that we need to make our point about the one-way deterministic relationship between a point pattern, abundance and distribution (remember that "distribution" is simply a certain spatial pattern of presence/absence).

```
set.seed(82)
tmp <- sim.fn(quad.size = 16, cell.size = 2, intensity = 0.5)
```

This relationship among the three quantities is visualized in the first three panels of Figure 1.1. Without spatial discretization, neither abundance nor occurrence (or presence/absence) is defined; both necessarily require discretization of continuous space into what you can think of as one or more "sites." In this simulation, a "site" is represented by one cell in the entire grid. You can perhaps think of the entire grid as a region wherein your study takes place. Only once we have established that discretization of space is abundance (which we like to denote as N) or occurrence (presence/absence,

1.1 POINT PROCESSES, DISTRIBUTION, ABUNDANCE

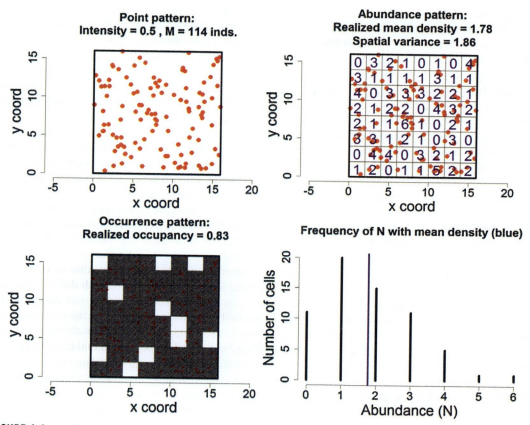

FIGURE 1.1

Relationship among three fundamental quantities in ecology: a *point pattern* of individual animals or plants (top left), a map of *abundance* with the local abundance (N) in each cell shown in blue (top right), and a species distribution map showing binary *presence/absence or occurrence* (bottom left), with occupied cells shown in gray and unoccupied cells in white. At the bottom right is the distribution of cell-based abundance (which is Poisson in this simulation), along with the mean shown in blue (which estimates the Poisson mean lambda). This figure is the graphical output from running function `sim.fn`.

which we like to denote as z) defined. Then, abundance N is simply the number of points falling into each "site" (i.e., cell)—if there is no point in a cell, abundance is zero; if there is one point, abundance is one; and so on. Furthermore, presence/absence (z) simply distinguishes the two cases where there is either no point in a cell (i.e., $N = 0$, this is an absence or nonoccurrence) or there is one or any number greater than one point in the cell (i.e., $N > 0$, this is a presence or occurrence). Thus, we can say that abundance is a first step of aggregating an underlying point pattern within some spatial discretization scheme, and occurrence is a second step in this aggregation over the spatial units. Alternatively, we can say that occurrence is a simple information-poor summary of abundance or "the poor man's abundance," where we only keep track of two abundance classes, one being zero (= "absence") and the other greater than zero (= "presence"). Thus, the relationships between a point pattern, abundance and occurrence are deterministic in one way only—if you know the full pattern and are given some spatial discretization scheme, you have full knowledge also about abundance; and if you know the spatial pattern of abundance, you also perfectly know the spatial pattern of occurrence. In contrast, things are not so straightforward the other way round, e.g., from knowing a presence/absence pattern you cannot perfectly infer the underlying abundance distribution, although you can make explicit statistical inferences about abundance from simple occupancy data (He and Gaston, 2000; Royle and Nichols, 2003; Royle et al., 2005; Ramsey et al., 2015).

We can describe the spatial abundance pattern that emerges from this underlying spatial point pattern by discretization of space and summarizing the mean and the variance of the individual values of N in each cell. The way that this simulation works (i.e., with a uniform intensity over the entire field), the resulting numbers N will follow a Poisson distribution with mean *lambda*, where *lambda* is estimated by the mean abundance (or density) over the 256 cells. In turn, the spatial presence/absence pattern will follow a Bernoulli distribution with a "success parameter" that we will later call "occupancy probability," and which corresponds to the expected proportion of occupied cells (that is, cells with nonzero abundance).

This is perhaps the simplest possible manner to explain by simulation the relationship between a point pattern, abundance, and distribution—we use a so-called homogenous Poisson process, which is one with a constant intensity. When modeling the point pattern aggregate of abundance, this is equivalent to adopting a Poisson generalized linear model with an intercept only for the cell values of abundance. In real life, homogenous intensity fields arguably never exist, instead intensity is patterned due to environmental heterogeneity, which can be described by spatially indexed covariates or spatially correlated random site effects. Much of ecological modeling in space, also in this book, is aimed at identifying the nature and strength of such covariate relationships. When modeling distribution or abundance from real data, we very often find that there are too many zeros. That is, a species is absent from more sites than what we would expect under our model. Some authors therefore make a clear distinction between "distribution," which is something like a potential distribution area where a species can occur in principle, and "abundance," which describes the number of individuals only at sites that belong to that distribution area. Such authors then typically adopt zero-inflated Poisson or related zero-inflated models to describe what they perceive of as two distinct processes, distribution and abundance.

This is very different from the way in which we look at the two concepts of distribution and abundance. As just explained, in our view, "distribution" naturally follows from any given spatial distribution of abundance. We think that it rarely ever makes sense to conceive of two distinct mechanisms underlying a realized abundance distribution in space. Instead, we think that in almost all cases where there are too many zeroes in a data set, this is simply due to a failure to include in our model all adequate covariates to model these zeroes through the Poisson (or negative binomial, etc.) mean. We think that it is not very interesting to try and attribute much biology to what in our view is merely a deficiency of our abundance model and which manifests itself by a too high frequency of zeros.

There may be rare exceptions, of course, where there are indeed two entirely distinct stochastic processes governing the abundance distribution in space. For instance, imagine the abundance of some terrestrial species in an archipelago. Clearly, any abundance greater than zero requires the colonization of an island beforehand and that is a stochastic process with binary outcome—either the island is colonized or it is not colonized. This may have nothing to do with the factors that determine abundance on that island once it is colonized, and, therefore in this example, it makes sense to imagine two separate mechanisms underlying the spatial variability of abundance as in a zero-inflated abundance model.

But in the vast majority of cases we think of such zero-inflated models simply as a modeling trick to make up for our lack of perfect knowledge of the covariates that really govern abundance. Therefore, we are happy to adopt zero-inflated models to account for the resulting lack of fit (see, e.g., Chapter 6), but we would not usually claim that there was much biology in the zero-inflation part of the model. Especially, we would not adopt complicated covariate models in the zero-inflation part and we would *never* use the same covariates in both the zero-inflation part and in the abundance part of the model (the resulting model is probably near-unidentifiable; see also Ghosh et al., 2012).

After this brief discussion of the meaning of zero-inflated models, let's now look further at the actual numbers in Figure 1.1. The intensity of the field underlying the point pattern is 0.5, hence we would expect to have a total of $M = 16^2 * 0.5 = 128$ individuals in the entire quadrat, which has an area of 256 units. However, due to the randomness in the number of points inherent in a point pattern model, we only have 114 individuals in this realization of the process. At the chosen grain size (i.e., with cell.size = 2), the abundance in the 256 cells varies from 0 to 6 individuals and the mean *realized* abundance is 1.78, while we would have expected $\lambda = 2^2 * 0.5 = 2$ (the difference is sampling variability). In addition, the variance of local abundance (N) is 1.86, while we would expect 2 under a Poisson distribution with expected value of 2. Finally, the *realized* proportion of occupied cells (occupancy) is 0.83, where under a Poisson process with constant intensity we would expect $\psi = 1 - \exp(-\lambda) \approx 0.86$ (i.e., 1 minus the expected proportion of zero abundance). As always with simulated data sets, we are neatly confronted with the difference between the *expected* value of the output from a stochastic process, i.e., the average over an infinite number of realizations, and the actual value in one particular realization of the process. The difference represents sampling variability.

In addition, you can use such simulation functions to learn something about the simulated process in a more general and fundamental way. For instance, sim.fn lets you study the relationships among the intensity of a field in a homogenous Poisson point process (intensity) and the grain (cell.size) of the measurement of distribution or abundance on one hand, and the resulting numerical values of abundance (N) and occurrence (z) on the other. The following sets of commands let you study some of the relationships in a more qualitative manner. To be able to average in your mind over the randomness of the process, you should execute every line multiple times.

```
# Effect of grain size of study on abundance and occupancy (intensity constant)
tmp <- sim.fn(quad.size = 10, cell.size = 1, intensity = 0.5)
tmp <- sim.fn(quad.size = 10, cell.size = 2, intensity = 0.5)
tmp <- sim.fn(quad.size = 10, cell.size = 5, intensity = 0.5)
tmp <- sim.fn(quad.size = 10, cell.size = 10, intensity = 0.5)
```

Although the underlying point pattern is identical on average, you see how both the mean abundance N (and the variance of abundance N) and the proportion of the occupied cells (ψ) increase with increasing grain size, provided that the quadrat size remains constant. When the cell size is equal to the quadrat size, we always observe 100% occupancy for a species that occurs at all.

```
# Effect of intensity of point pattern (intensity) on abundance and occupancy
tmp <- sim.fn(intensity = 0.1)   # chose default quad.size = 10, cell.size = 1
tmp <- sim.fn(intensity = 1)
tmp <- sim.fn(intensity = 5)
tmp <- sim.fn(intensity = 10)
```

Now, you will observe that when a species is very rare (intensity is low), the occurrence and the abundance patterns will be essentially identical, since rarely will a cell be inhabited by more than a single individual; see also Figure 1.2. However, the greater the intensity, the less informative will the spatial pattern of occurrence be about the spatial variation in population density.

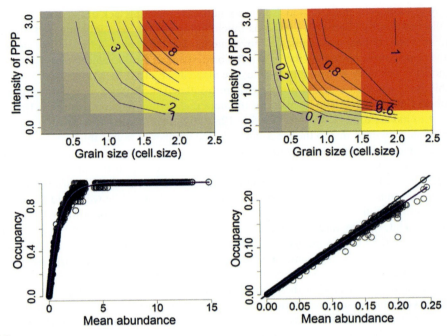

FIGURE 1.2

Relationships among intensity of the underlying Poisson point process and grain size and mean abundance per cell (top left), mean proportion of occupied cells (top right), and the relationship between mean occupancy and mean abundance for the full range of abundance created in the simulation (bottom left), and for a restricted range comprising only very small abundance values < 0.25 (bottom right); 1:1 line is added. Blue lines in bottom panels are smoothing splines with 4 d.f.

You can use a function such as this one for a formal simulation, to study the relationships among several quantities at a time. For instance, here is a little simulation to investigate the relationship between intensity and grain (cell.size) and the resulting mean density and occupancy proportion (ψ). We use the default quadrat size of 10 and vary both cell size and intensity in six steps each and record the mean abundance per cell and the realized proportion of occupied cells. We repeat this for a total of 100 data sets for each of the 36 combinations of the two factors grain and int(ensity). When you switch off the plotting in the function, you generate 36 * 100 data sets in barely four seconds!

```
simrep <- 100                          # Run 50 simulation reps
grain <- c(0.1,0.2,0.25,0.5,1,2)       # values will be fed into 'cell.size' argument
int <- seq(0.1, 3,,6)                  # values will be fed into 'lambda' argument
n.levels <- length(grain)              # number of factor levels in simulation
results <- array(NA, dim = c(n.levels, n.levels, 2, simrep)) # 4-D array !
for(i in 1:n.levels){                  # Loop over levels of factor grain
  for(j in 1:n.levels){                # Loop over levels of factor intensity
    for(k in 1:simrep){
```

```
      cat("\nDim 1:",i, ", Dim 2:", j, ", Simrep", k)
      tmp <- sim.fn(cell.size = grain[i], intensity = int[j], show.plot = F)
      results[i,j,1:2,k] <- c(mean(tmp$N), tmp$psi)
    }
  }
}
```

We visualize the results in two image plots that show the average abundance (over the 100 simulated data sets) as a function of the six levels of each simulation factor (Figure 1.2, left) and the same for the mean realized proportion of occupied quadrats (Figure 1.2, right; code not shown).

We learn three things from Figure 1.2. First, we see that both abundance and occurrence do contain some information about the intensity of the underlying point process. Second, both abundance and occupancy are scale dependent (Figure 1.2, top), and, hence, you don't need to be a genius to recognize that neither abundance nor occupancy make sense when you don't know the spatial scale (here, `cell.size`) at which it is expressed (Fithian and Hastie, 2013). And third, there is a strong positive relationship between the mean abundance in a grid and the proportion of occupied cells (occupancy). At very small mean abundance, occupancy is exactly identical to average abundance (there is a slope of 1), while with increasing density, the slope of the relationship becomes shallower and eventually even zero, when all cells are occupied. Then, occupancy is no longer informative at all about either the underlying abundance or about the fundamental point pattern.

We have said that we can use R code for data simulation to explain a model, but of course the reverse is true also—that any data simulation implies a specific statistical model. Clearly, this data simulation process represents one particular model with many specific assumptions; for instance, we assume a homogenous Poisson process, which involves three things: that the spatial variability in abundance follows a certain pattern (that of a Poisson distribution), that there is no spatial heterogeneity in the suitability of the habitat and finally, that individuals are occurring independently of each other. All three are idealizations that will strictly never be true in real life. For instance, individuals may occur more aggregated (with larger spatial variance) or more evenly (with smaller spatial variance) than stipulated under the Poisson, the environment will be heterogeneous and so will be the intensity of the process and there may be repulsion (from territoriality) or aggregation (e.g., from social attraction) among individuals, all of which will again be manifest in the variance of abundance N. Also, we simulated a certain geometry, a square grid with an integer number of nested and contiguous cells, and this may not be adequate for some things that you might perhaps want to learn from such a simulation. As always with models, you need to use abstraction wisely—leave out only the things that are unimportant and keep those that are important; the same applies for simulation models.

In summary, the important insights that we wanted to gain from this simple simulation exercise are therefore: (1) that at the base of all abundance and distribution data resides a spatial point pattern (and that species richness is a summary of a whole collection of such species-specific point patterns), (2) that to assign a value of "abundance" or "occurrence," one must have a spatial scale and this is *only* possible when you discretize space, and (3) there is a one-way deterministic relationship in the relationship among the three scales of aggregation {point pattern, abundance, occurrence}, where knowledge of the one on the left gives perfect knowledge about the quantity to the right, while in the other direction, there is some—sometimes considerable—loss of information and therefore no simple relationship. Hierarchical models are extremely suited for describing processes with multiple scales, including combinations of two or more scales in the triple: point pattern, abundance, occurrence (Begon et al., 1986).

1.2 META-POPULATION DESIGNS

Interestingly, without even knowing about the relationship between point processes and their areal summaries of abundance and occurrence, people have always liked to discretize their entire study area into smaller subunits, or, put in another way, to replicate their study areas in space. This gives rise to what we call a "meta-population design" (Royle, 2004a; Kéry and Royle, 2010). We are a tad shy about this term because we do not mean to imply that the animals living in such discrete spatial units necessarily behave according to a formal metapopulation (Hanski, 1998; Sutherland et al., 2012, 2014). Rather, we could not come up with a better and more concise term for the extremely common case where distribution or abundance is studied at a collection of spatially replicated sites or where a whole study area is subdivided into smaller subunits, which we typically call a "site." This is a "meta-population design" to us, and to avoid annoying metapopulation ecologists, we sometimes put the term in quotes and add a hyphen. Nevertheless, we emphasize that the general sampling situation does include the formal metapopulation situation, and any model we discuss in this book can apply to classical metapopulations. Especially the dynamic models for occurrence (in Chapters 16 and 22 in *AHM* volume 2) are exactly metapopulation models for colonization/extinction dynamics in a presence/absence pattern.

Such meta-population designs, or designs with spatially subdivided populations, are extremely common in ecology and all related sciences. In addition, they are adopted virtually everywhere in biological monitoring, where it is clear that you can't characterize the state of the environment from measurements taken only at a single site. Meta-population designs come in a large variety, and the number, size, and shape of cells (subunits) may all vary. Sometimes there is heterogeneity even within a single design, e.g., study sites in a collection differ in area and shape and also in their spatial configuration (e.g., intersite distance). Sites may be naturally defined by a habitat boundary and thus represent "habitat islands," such as ponds when you are studying fish or pond-breeding amphibians or mountaintops when you're interested in alpine plant life. This may then be the typical setting for formal metapopulations. Alternatively, sites may be defined arbitrarily, e.g., by laying a grid over a map and then calling a grid cell a site. Sites may come in two dimensions or they may be one-dimensional and follow linear structures such as rivers, coastlines, roads, or footpaths. Finally, one typically has some larger region that one wants to characterize in terms of the abundance or occurrence of some species, and the sampling fraction of a meta-population design may then differ between anything from almost zero to one, corresponding to the cases where only a small minority of the possible sites are surveyed on the one hand and the complete coverage of that region on the other.

Figure 1.3 shows just four examples among a myriad of possible "meta-population designs." In the top row we contrast coverage, with perfect regional coverage (all cells in the region of interest surveyed; left) and regional coverage of about 25% (right). In the bottom row, we contrast a systematic versus a random placement of the spatial replicates, with the actual spatial sample of 267 sites in the "meta-population design" of the Swiss breeding bird survey MHB (left; see Sections 6.9, 7.9, 10.9 and 11.3 for more information about that survey), while right is a hypothetical variant of that design where 267 sites are chosen randomly. In terms of the sampling fraction, the MHB has only about 0.64% coverage (267/42,000).

In addition, spatial subsampling is surprisingly common in meta-population designs, wherein each site (unit) is further subdivided into smaller spatial subunits, which may again cover the entire site or they may only cover part of the entire area of a site; see Sections 6.14 and 10.10, with Figure 10.13.

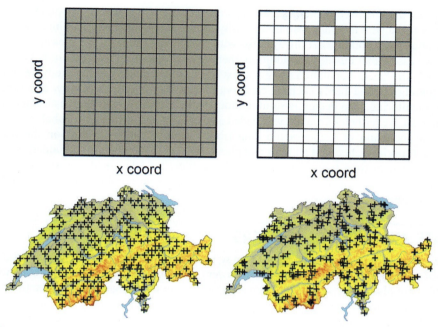

FIGURE 1.3

Four examples of meta-population designs: Top left: all cells (= sites) in a grid (= region) are surveyed; top right: only 25% of all sites in the region are surveyed; bottom left: the actual MHB meta-population design, with 267 1-km² quadrats laid out in an almost regular fashion that are surveyed from a region containing about 42,000 cells (= Switzerland); bottom right: a hypothetical variant of the MHB design where the same number of cells are placed randomly. In the bottom row increasing red means increasing elevation (the range is 200–4600 m).

And, finally, very often (especially in biodiversity monitoring) there is not only a spatial dimension in the study of distribution and abundance but also a temporal dimension, because there is an interest to study the dynamics of distribution or abundance over time. Thus, measurements of distribution and abundance are replicated temporally over longer time periods.

A final distinction may be made between the types of measurements that are taken at the collection of sites making up a whole "meta-population design." While an identical measurement protocol across units may perhaps be the most common approach, there are nevertheless many features of these measurements that may not be standardized, most of all the weather and other environmental conditions during which measurements are taken and commonly also the observer (i.e., it is rare that a single observer surveys all sites in a meta-population design). However, it is quite common also to measure several different things that are informative about the same underlying quantity. For instance, some sites may receive transect counts and others point counts (see Chapter 8). Or, there may be a combination of some sites where counts are conducted and others where only detection/nondetection observations (typically called "presence/absence data") are recorded. Of course, a very frequent source of such heterogeneity in sampling design is the need or the wish to combine the information from multiple schemes, where each scheme may be more homogenous by itself, but where there are systematic differences between schemes (Solymos et al., 2013). Thus, in meta-population designs we are commonly faced with not only spatial variability in the biological quantity of interest (e.g., spatial variation in abundance) but almost always also with spatial variability in the measurement protocol with which we assess that quantity. In addition, there may be temporal variation in both aspects, too.

In summary, the "meta-population design" must be one of the most common study designs in all of ecology including applied fields such as conservation biology, wildlife biology, and especially in biodiversity monitoring. The design is so ubiquitous that perhaps some may question the need for a special term at all. Nevertheless, we felt we needed a label for it and so this is our label. The *AHM* book deals with the modeling of demographic data from meta-population designs on the distribution, abundance, and species richness of animals and plants, and of the parameters that underlie the changes of these variables over time (see next section). Hierarchical models are ideally suited to describe a quantity such as distribution or abundance that is available from a collection of sites and possibly repeatedly over time.

1.3 STATE AND RATE PARAMETERS

Abundance, distribution, and species richness may perhaps be the single most widely studied quantities in all of ecology and related fields. However, very frequently there is an interest not only in these state variables but also in the parameters that govern the rate of change of these quantities, i.e., that drive population dynamics (rate variables). For instance, the single most important quantity in biodiversity monitoring seems to be the "trend," that is, some sustained rate of change over time of some quantity such as abundance or occurrence. And, a trend is only the simplest description of population dynamics; the most detailed description of the dynamics of an animal population is achieved by the four vital rates: birth rate, immigration rate, death rate, and emigration rate, and these may be stratified by age, sex, or possibly other classes in a study population. One vital rate that has received particular interest in animal ecology and other fields such as life-history evolution (Stearns, 1992) is survival probability. There is a very well worked out theory for estimating survival from temporally replicated samples of marked, wild animals, e.g., in the celebrated Cormack-Jolly-Seber model (Cormack, 1964; Jolly, 1965; Seber, 1965; Pollock et al., 1990), in its variant for ring-recovery data (Brownie et al., 1985) and in a multitude of generalizations including multistate (Arnason, 1972; Hestbeck et al., 1991; Brownie et al., 1993; Arnason and Schwarz, 1999) and related models (Barker, 1997; Pradel et al., 1997; Kendall et al., 2003; Pradel, 2005; Bonner and Schwarz, 2006).

In animal ecology there is fairly strong divide between models and methods (and interestingly also a little in the people applying these models) that target state variables (e.g., estimate abundance) or that aim at rate variables (e.g., estimate survival). This divide is mostly artificial and to a large part due to limitations of past and current models and the associated software to fit these models. Hence, up to very recently, there was a huge divide in ecological statistics between models for "closed populations" (which essentially meant abundance estimation) and models for "open populations" (which first and foremost meant survival estimation). (Indeed, you can still see this divide in the way in which we split up the two volumes of our book.) However, nowadays this distinction becomes increasingly blurred and especially in the context of hierarchical models you will see how easily we can bridge what may have been thought of as a deep divide perhaps only 10 years ago. In addition, the ease with which we can fit "hybrid open/closed" models to our data is largely due to the power of the Bayesian model fitting machinery and in practice, to BUGS software (Schofield et al., 2009). This is especially interesting also for population dynamics modeling such as population viability analysis (Beissinger, 2002) and related population modeling (Buckland et al., 2004b, 2007; Schofield and Barker, 2008; Tavecchia et al., 2009; Newman et al., 2006, 2014), including matrix projection models (Caswell, 2001; Link et al., 2003). Finally, another type of population model where the divide between open and closed populations is completely dropped is the fascinating area of *integrated population modeling* (IPMs; Baillie, 1991; Besbeas et al., 2002; Brooks et al., 2004; Schaub et al., 2007; Abadi et al., 2010a,b; Schaub and Abadi, 2011).

1.4 MEASUREMENT ERROR MODELS IN ECOLOGY

The error in a measurement is the difference between the measured value and the true value of some quantity. Probably by far the best-known type of measurement error in ecology is that associated with the measurement of continuous quantities such as body size, body mass, or the content of some pollutant in the air or the water. Their measurement is likely to be affected by a large number of small causes that act additively, giving rise to measurement errors that typically behave according to a normal distribution. An important consequence of this is that the measurements are unbiased, i.e., on average right on target—positive and negative errors simply cancel out in the average over repeated measurements. This type of measurement error seems to be the one that people have universally in mind when they think about this topic in ecology. For instance, such a type of measurement error is typically accommodated in the residual of a regression model.

However, things are very different for counts of discrete variables such as abundance, and this includes the binary variable "presence/absence," i.e., when you deal with the aggregation of data from an underlying point process. For them, you can undercount and overcount, and the mechanisms leading to the two types of errors are not the same, but can be very different. Thus, there is one set of mechanisms that lead to false-negative errors, when an individual is overlooked or a species is missed at a site where it occurs. This type of error cannot be reasonably described by a normal distribution but is typically described by a binomial or a Bernoulli distribution—given that there are N individuals out there and there is some probability p to detect any single one of them, the number of individuals detected (C) will be binomial:

$$C \sim Binomial(N, p) \quad \text{\# False-negative measurement error model for counts}$$

Here, p is the detection or encounter probability of an individual and it represents the complement of the false-negative error rate, i.e., the associated error rate is $1 - p$. Similarly, for the presence/absence state of a species at a site, z, where $z = 1$ denotes presence and $z = 0$ denotes absence, we can specify the following measurement error model for the presence/absence measurement or detection/nondetection datum y at an occupied site:

$$y \sim Bernoulli(p) \quad \text{\# False-negative measurement error model for det./nondet. obs.}$$

In either case, and unlike the normal model for measurement errors with continuous variables, the average of repeated measurements will *not* be unbiased with respect to the target quantities N (abundance) or z (presence/absence). Rather, the mean will be equal to Np for counts and equal to p for presence/absence measurements (detection/nondetection observations) at an occupied site. In contrast, the maximum among a series of n measurements will increasingly approach the true values N or z, when the number of replicate measurements n is increased. How quickly the maximum approaches N or z again depends on detection probability p (see Exercise 3).

False-negative detection error is "the" detection error that is addressed in the vast majority of capture-recapture and related methods that you have probably ever heard of (e.g., in Otis et al., 1978; Seber, 1982; Buckland et al., 2001, 2004a; Borchers et al., 2002; Williams et al., 2002; Amstrup et al., 2005; MacKenzie et al., 2006; Royle and Dorazio, 2008; King et al., 2009; Link and Barker, 2010; Kéry and Schaub, 2012; McCrea and Morgan, 2014; Royle et al., 2014). At the base of virtually all of these methods is the binomial or Bernoulli measurement error model from above. Arguably, false negative errors occur in almost all data sets of distribution and abundance, regardless whether they are for animals or for plants (Kéry and Gregg, 2003, 2004; Kéry, 2004; Kéry et al., 2005a, 2006; Chen et al., 2009, 2013).

In addition to false-negative errors, we may have false-positive errors, i.e., for abundance, we may overcount, most typically because we either count the same individual multiple times or because we mistake one species for another. For occurrence it means that we *think* we detected a species at a site where either it does not occur at all or we think we detected it at a site where it does occur, but what we saw was not the target species; see Chambert et al. (2015) for more on this subtle distinction. Methods that accommodate false-positive measurement errors in ecological models for abundance, occurrence and the the associated vital rates are still in their infancy and have essentially been developed in only two fields. For abundance, their development seems to have been restricted to genetic capture-recapture so far (e.g., Lukacs and Burnham, 2005; Wright et al., 2009; Yoshizaki et al., 2009; Link et al., 2010). For occurrence, the seminal paper by Royle and Link (2006) introduced occupancy models with both types of errors (i.e., false-negative and false-positive), but the main thrust of the development that led to more practically useful models came later with Miller et al. (2011, 2013b), Sutherland et al. (2013), and Chambert et al. (2015). However, at the time of writing the development of models with false-positive errors is both an active but also a difficult field, especially when (which is usually the case) we simultaneously have to confront both types of errors.

When we deal with areal-summary data from discretized "simple" point patterns, i.e., abundance or occurrence data, these are the two fundamental types of measurement errors. However, in addition, there are state classification errors when you classify individuals by age, size, or other states as in multistate models (Arnason, 1972; Hestbeck et al., 1991; Brownie et al., 1993) and when you distinguish different types of "occurrence" such as the occurrence of nonreproductive individuals versus that of reproductive individuals as in multistate occupancy models (Royle and Link, 2005; Nichols et al., 2007, see Chapter 18 in volume 2). (This means that you aggregate a *marked* point pattern, i.e., one where there are different types of points; Illian et al., 2008.) In these settings, the errors are simply a generalization of the two fundamental types of measurement error for discrete variables to multiple states.

When we deal directly with the underlying point pattern data, there is a third fundamental type of error: location error. That is, a difference between the true coordinates at which an individual is when you detect it and the coordinates that you record and feed into your model. Location error needs to be addressed in such spatially explicit models for abundance or density because otherwise biased estimators result. The simplest spatially explicit modeling framework for abundance, conventional distance sampling (see Chapter 8 in this book and Buckland et al., 2001), assumes away the problem by requiring zero location error as one of its main assumptions. Interestingly, once we "forget" the invididual's locations by aggregating a point pattern to become abundance or occurrence data, location error "vanishes" and is translated into either false-positive or false-negative error. If location error is such that an individual is erroneously recorded in a neighboring cell, then that record becomes a false

1.4 MEASUREMENT ERROR MODELS IN ECOLOGY

positive in that cell and will correspond to a false negative in the cell where the individual really is located. If location error does not lead to the recording of the individual in a different cell from the cell where the individual really is, it remains without a consequence in the modeling of abundance or occurrence.

And to finish our brief exegesis on measurement error in ecology, there is yet another type of measurement error: that in the covariables. This is quite different from the other measurement error types in this section, which are all associated with the response in a model, not with a covariate in the model. The issue of errors in covariables in models for distribution and abundance is exactly analogous to that in any other (regression) model; essentially, measurement errors in (continuous) covariables attenuates the slope estimate, that is, erroneously pulls the estimate towards zero. There is a pretty large body of research in statistics on this type of measurement error (Stefanski, 2000), and there are few if any novel considerations in the context of statistical ecology, and therefore we don't give it special attention.

Hence, it is typically *not* enough for an ecological model of distribution and abundance to simply describe the spatial variability of a process and possibly also the temporal dynamics in abundance or occurrence. Rather, to achieve unbiased inferences about the demographics of distribution and abundance it will be necessary to explicitly model the measurement error processes that underlie your data at hand. Studies employing meta-population designs typically face two sequential inferential steps (see Figure 1.4). The first is from the sample of surveyed sites to some larger, statistical "population" of sites in which we are interested (or the "region" we talked about in Section 1.2). We need a statistical model to describe the variability among these sites and the sampling of the surveyed sites to infer quantities in the entire region. And second, we need a second statistical model to describe the randomness in the measurement process, typically to estimate and therefore correct for false-negative and false-positive error rates. This two-step sampling procedure is ubiquitous in ecology and especially in biodiversity monitoring. It has been presented in a particularly lucid way in the seminal paper by Yoccoz et al. (2001), where they denote the two steps as "spatial variability and survey error" and "detection error."

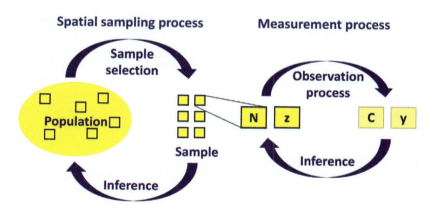

FIGURE 1.4

The two sampling processes in ecology that typically underlie the measurement of abundance or occurrence: first, spatial sampling and then the measurement of the desired quantity. N and z denote the typical quantities of interest (abundance and presence/absence) and C and y denote their measurements, a count or a detection/nondetection measurement, respectively (see also Yoccoz et al., 2001).

In summary, all of this calls for ecological models for distribution, abundance, and related demographic quantities that have at least two components: one submodel for the spatial and possibly also temporal variability in the focal quantity of interest and another submodel for the measurement process. Hierarchical models in this book achieve this aim in an admirable fashion.

1.5 HIERARCHICAL MODELS FOR DISTRIBUTION, ABUNDANCE, AND SPECIES RICHNESS

Hierarchical models are a sequence of probability models that are ordered by their conditional probability structure in the sense that they describe conditionally dependent random variables (see Chapter 2). In the context of models described and analyzed in this book, we use hierarchical models to describe both the true state of nature that is not observable (or only partly so) and also to describe the measurement error. Typically, our hierarchical models have one submodel for the true state of interest and another for measurement error although in some cases we might have more than one submodel for either. In the usual case where our model accommodates false-negative detection error only, the "bottom" level of the hierarchical model (where the data are) is a binomial (or Bernoulli) distribution where the "success probability" has the interpretation of the detection probability. In a sense, this turns most models in this book into some fancy sort of a logistic regression, but with possibly very complicated random effects structures. Other features of the spatiotemporal pattern of occurrence or abundance, or how we observe those, may be represented by additional levels in the model, especially groupings by site, species, etc.

Hierarchical models are our favorite framework for inference about distribution, abundance, species richness and related demographic quantities in populations, meta-populations, communities, and metacommunities. They are ideally suited to accommodate, in a single model, multiple data sets, multiple sources of variability such as spatial, temporal, and spatiotemporal variability, and multiple scales of measurement, while at the same time they rigorously propagate the combined uncertainty into every estimand from the model (Clark, 2007; Cressie et al., 2009; Cressie and Wikle, 2011; Hobbs and Hooten, 2015). Especially their Bayesian implementation with Markov chain Monte Carlo (MCMC) methods is almost limitless in its power to be applied to real data. In addition, hierarchical models represent a natural "compartmentalization" of a big, complex system into a sequence of smaller and usually far less complex subsystems. (In fact, 'sequential models' might be an equally fitting term for hierarchical models.) This is an ideal framework for describing jointly the true state and dynamics of an underlying system of interest, such as an animal or plant population or meta-population, and the potentially complex measurement processes with various and possibly heterogeneous types of error.

Understanding distribution and abundance and developing models that describe these things can be unified under a common framework of point process models (PPMs). Various models that we cover in this book either involve an explicit PPM (Chapters 8 and 9) or else we develop models for quantities that can only be sensibly understood as aggregations of an underlying point process (Chapters 6, 7, 10, and 11). Hence, we might call these two classes of models explicit and implicit PPMs.

1.6 SUMMARY AND OUTLOOK

In this chapter, we have clarified the meaning of three things: (1) of what distribution, abundance, and species richness are, (2) of what we call meta-population designs, and (3) of the types of measurement errors associated with studies of distribution, abundance, and related demographic quantities. First, we have seen that in spite of their foundational role for all of ecology, distribution, abundance and species richness are only derived quantities that can moreover only be defined if we discretize space. The first two are the result of aggregating an underlying point pattern over a study area or its subdivisions, and the last one is the result of aggregating not a single point pattern (for one species) but the point patterns of *all* occurring species at some site (or of some defined group of species, such as Red-list species.). We have derived this one-way deterministic relationship between a point pattern, abundance, and occurrence by way of a trivial little simulation in R. This simulation also emphasized that occurrence is nothing but an information-reduced summary of abundance, wherein instead of the full abundance distribution we simply keep track of two abundance states, $N = 0$ and $N > 0$. Finally, in this simulation we illustrated the relationship between occupancy probability and mean abundance, which is almost exactly linear at very low values of mean abundance and becomes increasingly shallow with higher mean abundance. Once mean abundance is so high that 100% of the spatial units are occupied, the slope is zero.

Second, the discretization of space required to even define abundance and occurrence based on an underlying point pattern is very often made at a collection of spatial replicates, leading to the study of what may be called one spatially subdivided population or of a collection of several spatially replicated populations. We called this design a "meta-population design," though it may or may not be inhabited by a true metapopulation in the technical sense of that term (e.g., Hanski, 1998; Sutherland et al., 2012, 2014).

Third, we have described the two fundamental types of errors for discrete model responses such as abundance and occurrence: false-negative and false-positive errors. We have discussed some of the typical mechanisms that give rise to these errors and have seen that location error in the underlying point pattern translates into one of these two types of errors for the aggregated data.

Finally, we have introduced hierarchical models, which are our modeling framework of choice to describe distribution, abundance, species richness, and other demographic quantities such as population trends or vital rates, including survival probability. Hierarchical models are perfectly suited to accommodate the spatial replication inherent in a meta-population design as well as all kinds of measurement error and idiosyncrasies of the measurement protocol. Thus, hierarchical models provide us with a tremendous power to describe and understand how populations, meta-populations, communities, or metacommunities vary over space and time. Add to this the ease with which you can specify these models in the BUGS language and you see golden times approach for ecological modelers.

Finally, the segmenting of a single big, complicated process into a sequence of linked, smaller, and simpler subprocesses, which is a hallmark of hierarchical modeling (or "sequential modeling"), has the potential to radically change the way in which you approach modeling and inference of ecological systems. It is our experience that hierarchical modeling not only invites you, but actually almost *forces* you, to adopt a much more mechanistic way of thinking about your study systems. Thus, hierarchical modeling may change, and improve, the very way in which you approach science. Hierarchical models are the subject of the next 10 chapters in this first volume of *Applied Hierarchical Modeling for Ecologists* and then of another 14 or so chapters in volume 2 of the book.

EXERCISES

1. Use that first R simulation function (`sim.fn`) to improve your intuition about the nature of distribution and abundance. NOTE: This function lacks one important feature that is always present in real-world abundance and abundance data: it does NOT model measurement error.

2. Function `sim.fn` is most appropriate for plants. Why? Because there is no movement. Can you think of ways in which the results from our simulations (e.g., Figure 1.2) would change if there was also movement? For volume 2, we envision developing a version of the function that also includes movement, so then you can try out whether your hunches were correct.
3. In Section 1.4, we said that the maximum among a series of n measurements will come increasingly close to the true value N or z with increasing number of measurements. Devise a little simulation in R to show the relationship between the maximum of a collection of counts as a function of both true abundance N and detection probability p.

CHAPTER 2

WHAT ARE HIERARCHICAL MODELS AND HOW DO WE ANALYZE THEM?

CHAPTER OUTLINE

- 2.1 Introduction 20
- 2.2 Random Variables, Probability Density Functions, Statistical Models, Probability, and Statistical Inference 21
  - 2.2.1 Statistical Models 23
  - 2.2.2 Joint, Marginal, and Conditional Distributions 24
  - 2.2.3 Statistical Inference 27
- 2.3 Hierarchical Models (HMs) 28
  - 2.3.1 Two Canonical HMs in Ecology 29
    - 2.3.1.1 The Occupancy Model for Species Distributions 29
    - 2.3.1.2 The N-Mixture Model for Abundance 30
  - 2.3.2 The Process of Hierarchical Modeling 31
  - 2.3.3 Inference for HMs 31
- 2.4 Classical Inference Based on Likelihood 31
  - 2.4.1 The Frequentist Interpretation 33
  - 2.4.2 Properties of MLEs 33
  - 2.4.3 The Delta Approximation 34
  - 2.4.4 Example: Classical Inference for Logistic Regression 36
  - 2.4.5 Bootstrapping 40
  - 2.4.6 Likelihood Analysis of HMs 42
    - 2.4.6.1 Discrete Random Effects 42
    - 2.4.6.2 A Continuous Latent Variable 44
  - 2.4.7 The R Package unmarked 46
- 2.5 Bayesian Inference 49
  - 2.5.1 Bayes' Rule 50
  - 2.5.2 Principles of Bayesian Inference 51
  - 2.5.3 Prior Distributions 52
  - 2.5.4 Computing Posterior Distributions 53
- 2.6 Basic Markov Chain Monte Carlo (MCMC) 54
  - 2.6.1 Metropolis–Hastings (MH) Algorithm 55
  - 2.6.2 Illustration: Using MH for a Binomial Model 56
  - 2.6.3 Metropolis Algorithm for Multiparameter Models 58
  - 2.6.4 Why We Need to Use MCMC for Logistic Regression 60

    2.6.5 Gibbs Sampling..62
    2.6.6 Convergence and Mixing of Markov Chains..63
        2.6.6.1 Slow Mixing and Thinning of Markov Chains 64
        2.6.6.2 Effective Sample Size and MC Error 64
    2.6.7 Bayesian Analysis of HMs 65
2.7 **Model Selection and Averaging**..**69**
    2.7.1 Model Selection by AIC...69
    2.7.2 Model Selection by Deviance Information Criterion...70
    2.7.3 Bayesian Model Averaging with Indicator Variables...71
2.8 **Assessment of Model Fit**..**72**
    2.8.1 Parametric Bootstrapping Example 73
    2.8.2 Bayesian *p*-value 75
2.9 **Summary and Outlook**...**76**
Exercises **77**

2.1 INTRODUCTION

In this chapter we provide a conceptual definition of the term *hierarchical model* (HM) as we use it throughout the book. We use the term to describe a coupled set of ordinary (or "flat") models that are conditionally related to each other. By conditional we mean in the sense of conditional probability: the probability distribution of one random variable depends on the other random variable, usually expressed as $f(y|z)$ or $[y|z]$, indicating that probabilities of outcomes of the random variable y depend on the outcome of the random variable z. Typically the models contain one component for the observations, or data, and then one or more additional components to describe latent variables or outcomes of some ecological process. In "integrated models" (Chapter 23), we have more than one observation model, one for each type of data. While our usage of the term HM is somewhat more specific than it is used in the statistical literature, we find that it serves to provide a conceptual unification of many different types of models, such as occupancy models (MacKenzie et al., 2002), N-mixture models (Royle, 2004b), distance sampling (Royle et al., 2004, Sillett et al., 2012), and many others (Royle and Dorazio, 2008; Kéry and Schaub, 2012).

    One of the overarching themes of this book is the analysis of HMs using both classical and Bayesian inference methods. We do this because we think that in order to be an effective modeler, you need to understand and be able to apply both, depending on the situation. There will be some problems that are most easily, or perhaps only, solvable using Bayesian analysis—by Markov chain Monte Carlo (MCMC) methods—and other problems that are best solved using classical likelihood methods. For instance, models that have spatially correlated random effects cannot easily be analyzed using classical likelihood methods, while many standard discrete mixture models such as the N-mixture are easily analyzed by likelihood methods, and therefore we can do things like Akaike's information criterion (AIC)-based model selection.

    What do we mean by "classical inference"? By that we mean, strictly speaking, direct analysis of the likelihood; e.g., estimation of parameters by maximum likelihood. We usually also mean the conceptual view of *frequentist* inference, in which the properties of estimators and procedures are evaluated by thinking about averaging over possible realizations of data. Therefore, in a straightforward practical sense, we mean likelihood estimation but we imagine that the frequentist evaluation of things is implied, especially as it relates to the consistency of maximum likelihood estimates (MLEs) or the interpretation of confidence intervals. Bayesian inference is becoming more and more familiar to practicing ecologists (McCarthy, 2007; Royle and Dorazio, 2008; Kéry, 2010; Kéry and Schaub, 2012; Hobbs and Hooten, 2015). The principal difference between Bayesian and classical inference has to do with how random variables are used to formulate models. In particular, classical inference regards parameters as fixed but unknown quantities, whereas Bayesians regard parameters as random variables endowed with a prior distribution, and as a result this enables the Bayesian to base inference on the probability distribution of the unknown quantity *given* the data; i.e., on the posterior distribution. Practical Bayesian analysis is now almost always carried out using MCMC methods that we introduce here, including some basic principles for constructing MCMC algorithms (in Chapter 5 we show how to do these things with BUGS software, and discuss some practicalities of using MCMC methods, such as convergence diagnostics and so forth). Finally, we discuss the important topics of model selection and model checking (or goodness-of-fit assessment) from both classical and Bayesian perspectives.

2.2 RANDOM VARIABLES, PROBABILITY DENSITY FUNCTIONS, STATISTICAL MODELS, PROBABILITY, AND STATISTICAL INFERENCE

Although not frequently taught to ecologists, random variables and probability density/mass functions are essential for statistical modeling. So what are they? A variable is a characteristic or feature that exhibits variability among units or elements that possess that characteristic or feature. For example, the height of an adult male, the number of peregrines counted during the spring peregrine survey, or the color of an M&M that I'm about to pull from a bag and eat. In other words, variables are quantities that can take on different values. *Random* variables are variables for which possible values are governed by probability distributions. The precise manner in which possible values are characterized is called the *probability density function* (pdf) if the variable is continuous (height of adult males) and it is called the *probability mass function* (pmf) if the variable is discrete (peregrine counts or colors of M&Ms). The pmf or pdf is a mathematical rule that governs possible outcomes of a random variable. Usually, the pmf (or pdf) depends on one or more quantities, called *parameters*, that affect its characteristic form. For example, one of our favorites is the binomial pmf for counts y with a natural upper limit N (representing, for instance, the number of samples or replicates or a population size):

$$f(y) = \frac{N!}{y!(N-y)!} p^y (1-p)^{(N-y)} \tag{2.1}$$

where y can take on integer values from $y = 0$ to $y = N$, and the single parameter p is known as the "success probability" which will typically be detection or encounter probability in many ecological sampling applications. This binomial pmf tells us the probabilities of each possible value of y, and so sometimes instead of $f(y)$ we will denote a pmf by writing $\Pr(y)$, $y \sim Binomial(p, N)$, or $y \sim Bin(p, N)$, where the $\sim$ is usually read "is distributed as." As an example, if the random variable is $y =$ "the number of times I observed at least one peregrine falcon" (Figure 2.1) during $N = 5$ visits to a nesting cliff, and I know that the value of the parameter $p = 0.2$, then the possible values of the random variable are $y = \{0,1,2,3,4,5\}$, and they occur with the probabilities given by Eq. (2.1), which in the R language can be computed as follows:

```
dbinom(0:5, size = 5, prob = 0.2)
[1] 0.32768 0.40960 0.20480 0.05120 0.00640 0.00032
```

FIGURE 2.1

The magnificent peregrine falcon (*Falco peregrinus*): adult female with fledged young in the Jura Mountains, Switzerland, 2008. *(Photo: Alain Georgy.)*

What this translates to, in words, is that the probability that I detect peregrines zero times is 0.32768, the probability that I detect peregrines one time is 0.40960, and so forth. We can obtain the probability of getting any number in a range of numbers, such as 2, 3, or 4, by summing the associated probabilities of each event in the set; e.g., $\Pr(y = \{2,3,4\}) = 0.20480 + 0.05120 + 0.00640 = 0.2624$.

The pdf is a similar rule but for continuous random variables. Continuous random variables have the peculiar feature that the probability of observing any *particular* value of the variable is 0 (and yet, in practice, we may observe some value multiple times but still use a continuous variable as a model). Thus, a sensible interpretation of a pdf is that it gives the probabilities of values that lie in prescribed intervals. One of the standard pdf's that we will use is the normal pdf that has this form:

$$f(y) = \frac{1}{\sqrt{2\pi}\sigma} \exp\left(-\frac{1}{2\sigma^2}(y - \mu)^2\right)$$

The normal probability distribution has two parameters, the mean μ and the standard deviation σ. If the height of adult males in a well-defined population is known to have mean $\mu = 190$ cm and standard deviation $\sigma = 10$ cm, then we can compute probabilities for specific ranges of y by integrating $f(y)$ over the desired range. For example, the probability that y is *less than* some value L is computed by integrating $f(y)$ from $-\infty$ to L, the probability that y lies between two values L and U is the integral

$\int_L^U f(y)dy$, and so forth. In R we can use the `pnorm` function to do these kinds of operations for us. To illustrate, the probability that $y < 200$ in this hypothetical population is:

```
pnorm(200, mean = 190, sd = 10)
[1] 0.8413447
```

So we expect that slightly more than 84% of the population is shorter than 200 cm. Or, the expected proportion of males with height between 180 and 200 (i.e., between -1 and $+1$ standard deviation (SD) of the mean) is:

```
pnorm(200, mean = 190, sd = 10) - pnorm(180, mean = 190, sd = 10)
[1] 0.6826895
```

You can compute this result directly by integrating the normal pdf using the R function `integrate` as follows:

```
f <- function(x, mu, sigma){
  (1 / sqrt(2*pi*sigma^2)) * exp( -((x-mu)^2)/(2*sigma^2))
}
integrate(f, lower = 180, upper = 200, mu = 190, sigma = 10)
0.6826895 with absolute error < 7.6e-15
```

The `integrate` function is useful in a lot of routine analyses of HMs that we do, although often its usage is hidden "under the hood" in R packages such as `unmarked` (Fiske and Chandler, 2011).

We will often use the shorthand "bracket notation" to represent probability distributions. For example, [y] is the marginal distribution of the random variable y, [y,x] is the joint distribution of y and x, and [y|x] is the conditional distribution of y given x. See also Section 2.2.2.

Some statistical distributions have a parameter that directly represents the mean expected response (e.g., the normal has parameter μ, and the Poisson has parameter λ), while some have a parameter that represents the variability, "spread," or dispersion of the response (e.g., σ in the normal). This allows us to model the mean tendency of a random variable or its spread by using covariates or other model structures such as random effects in a statistical model. However, for most distributions the mean and the dispersion are *not* represented by single parameters, but by functions of one or more parameters, such as Np for the binomial mean or $a/(a+b)$ for the mean of a beta distribution. The standard pdf can often be rewritten (or reparameterized) such that the mean (or the variance) of the response becomes a parameter that can then be modeled as a function of covariates (and adding of covariates is much of what we do when we do "statistical modeling"). This reparameterizing is called "moment matching" (Hobbs and Hooten, 2015), and can be very useful in modeling (see 5.12 and 7.6 for examples). A large number of statistical distributions, along with their pdf's (or pmf's) and the expressions for their means and variances, can be found on the Internet or in any number of statistics texts. These are the main building blocks of your statistical models (Bolker, 2008).

2.2.1 STATISTICAL MODELS

In most statistical models, a measured response is treated as a random variable endowed with a pdf that is assumed known. Hence, a statistical model for observed data is represented by a pdf. The pdf accommodates both the randomness (the unpredictable part) in the response and also the tendency for some values to occur more frequently than others—i.e., the part in the response that is predictable at least in an average sense. Almost always, we have some other variables that we assume cause some of the variability in the response, or at least are associated with the response in some stochastically predictable way. In the statistical model, the basic parameters of the pdf are then replaced by some linear or nonlinear function of these *covariates*.

Explicit specification of pdfs and pmfs may not be the typical description of a statistical model that you have seen previously in your training as an ecologist. In fact, it is quite likely that in your statistics classes at university you have *never* even encountered an explicit statistical model (i.e., one written as a pdf), but rather you had to memorize a certain range of study situations or data collection protocols that led to certain procedures that produced some numbers such as *p-values*. However, random variables and pdf's are the statisticians' way to think about and describe statistical models, and it is this way that ecologists also should think about statistical models. In part this is because viewing a model as a description of a single thing that is variable—i.e., a random variable described by a pdf—leads naturally to the adoption of HMs, where more than a single thing can vary simultaneously. In HMs, you will combine two or more random variables in sequence in a manner consistent with your scientific understanding of the modeled process, leading to what some have called a "scientific model" (Berliner, 1996). Hierarchical modeling represents a manner of "learning from data" that has tremendous power and considerable beauty, as we will see throughout this book.

2.2.2 JOINT, MARGINAL, AND CONDITIONAL DISTRIBUTIONS

In developing HMs, we are always interested in two or more random variables and how they are related to one another. Inferences about relationships among these random variables can be made using joint, marginal, or conditional probability distributions depending on the question being asked (note that much of the following material is modified from Chapter 2 in Royle et al., 2014). For our immediate purpose here of developing ideas related to these probability distributions, we will deviate slightly from our previously established notation by distinguishing between random variables and *realizations* of random variables. That is, let Y be a random variable, and let y indicate a particular realization of the random variable. Therefore we make statements such as $\Pr(Y = y)$ and so forth. This distinction is customarily done in classical statistical training, but in practice it is efficient to ignore it when the specific context is clear (which we do elsewhere in this book).

Conditional probability distributions provide an explicit description of the relationship between one random variable, say Y, and some other random variable, say X (or sometimes multiple other variables). Joint distributions describe the probabilities of simultaneously observing all unique combinations of both Y and X. This is easiest to conceptualize in the case of discrete random variables, where the joint distribution is the probability that X takes on the value x *and* that Y takes on the value y, which is written $[Y = y, X = x]$. The marginal distribution of a variable Y is its distribution "averaged over" all possible values of X. When X is a discrete random variable, we formally carry out this averaging, whereas if X is continuous, this operation is equivalent to integration over the range of X (see below).

To clarify these concepts of conditional, joint, and marginal distributions in more concrete terms, let us revisit our example where Y is the number of peregrine sightings after $N = 5$ survey visits to a nesting cliff, which we assumed to be a binomial random variable. Now, let us suppose that Y depends on the random variable X, which is the number of fledged young at a nesting cliff at the time of survey (assuming that all surveys take place after the young have fledged). Specifically, let us say that the probability of observing a peregrine, p, is related to a realized value of the random variable X (i.e., the value $X = x$), according to $\text{logit}(p) = -1.2 + 2x$. Furthermore, let us make the intuitive assumption

2.2 RANDOM VARIABLES, PROBABILITY DENSITY FUNCTIONS

that the number of fledged young around the cliff is a Poisson random variable with mean 0.4—i.e., $X \sim Poisson(0.4)$. Our model is now fully specified, and so we can answer the question: "What is the probability of observing peregrines y times *and* of there being x fledged young at the cliff?" This question may not be of interest by itself, but once we know how to answer this question we can answer much more useful questions as well, as we demonstrate below.

This *joint distribution* of Y and X is given by the product of the binomial pmf (with p determined by x) and the Poisson pmf with $\lambda = 0.4$. The following R commands create the joint distribution:

```
Y <- 0:5  # Possible values of Y (# surveys with peregrine sightings)
X <- 0:5  # Possible values of X (# fledged young)
p <- plogis(-1.2 + 2*X) # p as function of X
round(p, 2)

[1] 0.23 0.69 0.94 0.99 1.00 1.00

# Joint distribution [Y, X]
lambda <- 0.4
joint <- matrix(NA, length(Y), length(X))
rownames(joint) <- paste("y=", Y, sep="")
colnames(joint) <- paste("x=", X, sep="")
for(i in 1:length(Y)) {
   joint[,i] <- dbinom(Y, 5, p[i]) * dpois(X[i], lambda)
}

round(joint, 3)
```

	x=0	x=1	x=2	x=3	x=4	x=5
y=0	0.180	0.001	0.000	0.000	0.000	0
y=1	0.271	0.009	0.000	0.000	0.000	0
y=2	0.163	0.038	0.000	0.000	0.000	0
y=3	0.049	0.085	0.001	0.000	0.000	0
y=4	0.007	0.094	0.012	0.000	0.000	0
y=5	0.000	0.042	0.040	0.007	0.001	0

This matrix tells us the probability of all possible combinations of Y and X, and we see that the most likely outcome is $(y = 1, x = 0)$, which is to say that we are likely to detect peregrines only one time in our five visits to the nesting cliff, and the cliff is most likely to have zero fledged young (i.e., reproduction was not successful that year). Perhaps most peregrine watchers don't care about joint distributions, but a question that they might care about is whether they will have a successful peregrine watching trip, which we might phrase as the question "What is the probability of observing peregrines during all five days of our peregrine watching trip?" We know that this depends on the number of fledged young at a nesting cliff, but we don't know how many there are in advance, so this is a different question than "what are the most likely values of Y and X." This leads us to the marginal distribution, which is defined formally by

$$[Y] = \sum_X [Y, X] \qquad [X] = \sum_Y [Y, X]$$

for discrete random variables, and

$$[Y] = \int_{-\infty}^{\infty} [Y,X]dX \qquad [X] = \int_{-\infty}^{\infty} [Y,X]dY$$

for continuous random variables. The key idea here is that to get the marginal distribution of Y, we have to contemplate all possible values of X and average over them. Computing marginal distributions is a key step in maximizing likelihoods involving random effects, as we will discuss in Section 2.4.4. Here is some R code to compute the marginal distribution of X, the number of fledged young,

```
margX <- colSums(joint)
round(margX, 4)
   x=0    x=1    x=2    x=3    x=4    x=5
0.6703 0.2681 0.0536 0.0072 0.0007 0.0001
```

and for Y (i.e., the probability of seeing peregrines in $Y = y$ visits to the nesting cliff):

```
margY <- rowSums(joint)
round(margY, 4)
   y=0    y=1    y=2    y=3    y=4    y=5
0.1805 0.2792 0.2012 0.1352 0.1140 0.0899
```

The marginal pmf tells us that the most likely number of visits in which we will be successful at seeing peregrines is $y = 1$. We stand only a slim chance of seeing peregrines on all five visits to the cliff ($\Pr(Y = 5) = 0.0899$), and the probability that we see any peregrines at all is

$$\Pr(Y > 0) = 1 - \Pr(Y = 0) = \Pr(Y = 1) + \Pr(Y = 2) + \Pr(Y = 3) + \Pr(Y = 4) + \Pr(Y = 5) = 0.82.$$

We now turn to ideas of conditional probability, which we are interested in because in every application of hierarchical modeling, our model will involve an explicit specification of one or more *conditional* probability distributions. In a sense, conditional probability distributions are the elemental building blocks of HMs. In almost all applications we discuss in this book, we will find that certain natural conditional probability distributions lead to natural HMs for specific ecological problems. The conditional probability distribution is the distribution of one random variable, given (or conditional on) the realized value of the other. It is most easy to see how this is defined for the case of two discrete random variables. In that case, the conditional distribution may be written as $[Y = y|X = x]$—i.e., the probability of Y taking on the value y given the realized value of X being x. For simplicity, we will write this as $[Y|X]$, or most often for the rest of the book besides this chapter, we will ignore the upper- and lowercase convention and just write $[y|x]$. Conditional distributions are defined as follows:

$$[Y|X] = \frac{[Y,X]}{[X]} \qquad [X|Y] = \frac{[Y,X]}{[Y]}$$

That is, the conditional distribution of Y given X is the joint distribution divided by the marginal distribution of X.

```
YgivenX <- joint / matrix(margX, nrow(joint), ncol(joint), byrow=TRUE)
round(YgivenX, 2)
      x=0  x=1  x=2  x=3  x=4 x=5
y=0  0.27 0.00 0.00 0.00 0.00   0
y=1  0.40 0.03 0.00 0.00 0.00   0
y=2  0.24 0.14 0.00 0.00 0.00   0
y=3  0.07 0.32 0.03 0.00 0.00   0
y=4  0.01 0.35 0.23 0.04 0.01   0
y=5  0.00 0.16 0.74 0.96 0.99   1
```

Note that we have six probability distributions for Y, one for each possible value of X, and each pmf (column) sums to 1 as it should. In terms of its importance to bird watchers, we note that if there are >3 fledged young at this cliff, then you will expect to see peregrines nearly every day. The concepts of joint, marginal, and conditional distributions are explained in more detail in many other texts, Casella and Berger (2002) being one of our favorites.

The last point we wish to make in this section is that this made-up peregrine example *is* an HM, and we can put the pieces together using the following notation:

$$X \sim Poisson(0.4) \qquad \text{\# Submodel for no. young}$$
$$logit(p) = -1.2 + 2X \qquad \text{\# Relationship p and X}$$
$$Y|X \sim Binomial(5, p) \qquad \text{\# Submodel for observation}$$

From here on out, when you see such notation, you should immediately grasp the fact that X is a random variable independent of Y, but Y depends on X through the parameter p. Now you have the tools to make probability statements about the random variables in this system. The one caveat faced in reality is that we typically do not know the values of the parameters, and instead we have to estimate them, either by classical methods such as the MLE, or by Bayesian methods, both of which we address in this chapter.

2.2.3 STATISTICAL INFERENCE

An important use of probability, as introduced in the previous section, is for providing a description of natural systems using random variables and probability distributions. More specifically, in this book we are concerned with using random variables to describe outcomes of ecological processes (observed and unobserved) and probability distributions as models for the variability in these possible outcomes. Probability therefore forms our conceptual basis for *modeling* in ecology.

On the other hand, the field of *statistics* is concerned with the basic problem of learning about the parameters of probability distributions from observed outcomes (i.e., "data") of some variable. The two dominant paradigms of statistical inference, Bayesian and classical, share a common conceptual linkage as being distinct flavors of parametric inference. This means that both require that we make explicit probability model assumptions about the random variables that describe our system. Either paradigm then proceeds to carry out some inference task (estimation, prediction, model selection, testing) that *assumes those parametric assumptions are truth.*

In the following sections, we discuss basic ideas of both likelihood and Bayesian inference and then demonstrate their application to HMs.

2.3 HIERARCHICAL MODELS (HMs)

As a very general definition we can say that a hierarchical model, which we typically abbreviate as "HM" in this book, is a sequence of related models ordered by their conditional probability structure. What this means practically is that HMs have one or more "intermediate" models/levels/stages involving a latent variable (or random effect). By this definition, classical random effects models are clearly HMs. As an example, imagine a survey on a set of $i = 1, 2, ..., M$ spatial units, wherein we take $j = 1, 2, ..., J$ replicate observations y_{ij} of some ecological measurement. Naturally we might suppose that replicate measurements on the same unit are more similar than those among units, and this is the standard motivation for allowing for a random group effect α. We might therefore specify the model for the observations according to

$$y_{ij}|\alpha_i \sim Normal(\mu + \alpha_i, \sigma^2)$$

where the random effect α_i also has a distribution such as:

$$\alpha_i \sim Normal(0, \sigma_\alpha^2)$$

Note that we emphasized the conditional dependence of y on α in the expression of the model for y. In shorthand, we use the bracket notation $[y|\alpha]$ to represent the observation model, where the vertical bar | indicates explicit conditioning of one variable (y in this case) on another (α in this case), and $[\alpha]$ represents the model for the random effect. The HM is therefore the pair of submodels $[y|\alpha]$ and $[\alpha]$. We note that there is not necessarily a logical "top-down hierarchy" involved in a hierarchical model, such as when we model individuals nested within populations etc. Hence, in a sense, an equally or perhaps even better fitting term might be "sequential model" (SM) instead of hierarchical model. But of course, the latter term is so much common-place that we wouldn't want to change it.

In this book, we usually use the term HM in a more specific context, in which the observation model is specified conditional on a latent variable that represents an actual (i.e., real) biological process or the outcome thereof. For example, we might condition our observed data y on the true (but unobserved) population size N of a sample unit. We view population size N as the outcome of a real ecological process, or more realistically the aggregate outcome of many ecological processes such as survival and recruitment. In that sense, N is very distinct conceptually from the latent group effect α in the preceding example, in the sense that in some cases N might be observable given sufficient effort, resources, and proper design, whereas α is never observable because it doesn't represent a real state of nature, but rather is a purely hypothetical construct. We call the model with α an "implicit HM" to distinguish it from the explicit HM in which N results from explicit ecological processes and has an explicit ecological meaning such as "abundance." Thus, while both the classical random effects model and that which is conditional on an explicit ecological process are "HMs," we find the second case to be more interesting and directly relevant to ecology, if we have a choice between the two. Nevertheless, our "explicit HMs" will sometimes also contain random group effects to accommodate groups and other factors inducing correlations.

We use a bit of jargon in our discussions of HMs, including words like *process* or *state model*, *state variable*, *observation model*, and *observations* (same as *data*). As used here, the *state variable* is the random variable described by the *state model* (N in the preceding paragraph), and the *observations* (y) are described by the *observation model*. In the next section, we will elaborate on and clarify some of these aspects of HMs by discussing two typical examples that are extremely relevant in ecology, since they deal with distribution and abundance, which are defining concepts in ecology. Moreover, they lay the foundation for most models in this book.

2.3.1 TWO CANONICAL HMs IN ECOLOGY

We encounter a large number of distinct HMs in ecology, but two of the most prominent and widely used HMs are those for modeling species occurrence or distribution, often called "occupancy models" (MacKenzie et al., 2002; Tyre et al., 2003), and the N-mixture model for modeling species abundance (Royle, 2004b). We think of these as the foundational examples of "explicit" HMs because they are simple hierarchical extensions of ordinary (nonhierarchical, or "flat") models, logistic regression, and Poisson generalized linear models (GLMs), respectively, which are used throughout statistics to model ordinary binary and count data (see Chapter 3 for an applied introduction to GLMs).

2.3.1.1 The Occupancy Model for Species Distributions

We provide a brief introduction to the occupancy model (MacKenzie et al., 2002; Tyre et al., 2003), which is a natural HM for modeling species distributions (see also Chapter 10 and Kéry, 2011b). Imagine that a sample of M discrete sites is surveyed and the presence or absence of a species is noted. Let y_i denote the binary observations of presence/absence at site $i = 1, 2, ..., M$, where possible values are $y_i = 0$ ("absence") and $y_i = 1$ ("presence"). A key motivation for the development of occupancy models is that sometimes a species goes undetected even though it is present (in other words, the detection probability is <1). Thus, some of the absence *observations* should in practice correspond to instances of presence of the species. To specify the model for this situation, we introduce a new variable, denoted by z_i, which is the *true* presence/absence state of site i, where $z_i = 0$ is true absence, and $z_i = 1$ is true presence. This state variable is also binary.

The observation model links the observations to the state variable, and the standard observation model is a Bernoulli model with success probability $z_i p$:

$$y_i | z_i \sim Bernoulli(z_i p)$$

where p is the probability of detecting the species *given that it is present*. That is, if $z_i = 1$, then y_i is a Bernoulli trial with parameter p. Otherwise, y_i is a Bernoulli trial with probability 0, which is the same as saying that $y_i = 0$ with probability 1; i.e., if the species is truly absent, we must necessarily observe absence (we assume for now that the observers are well trained so that they do not misidentify the focal species, but see Chapter 19). The state model—i.e., the model for the ecological process of "true species presence or absence"—is similarly a simple Bernoulli model for a binary response:

$$z_i \sim Bernoulli(\psi_i)$$

where $\psi_i = \Pr(z_i = 1)$ is the probability of occurrence, or occupancy probability. Naturally, as this is itself an equivalent model to logistic regression (see Chapter 3), we model covariates that affect occurrence probability on the logit-linear scale (though see also 3.3.6). For example, if x_i is a measured habitat or landscape covariate for site i, then

$$\text{logit}(\psi_i) = \beta_0 + \beta_1 x_i.$$

As a practical matter, the parameter p and those of the model for z are not separately identifiable from a single observation of presence/absence at a site, unless we have some additional information. One way to obtain this information is through replicate surveys at each site, so that instead of a simple Bernoulli observation y_i, we have a sequence of repeated observations $y_{i1},...,y_{iJ}$, where there are J sample occasions. The standard assumption is that the sampling occurs over a sufficiently short period that a kind of "closure" with respect to occupancy status can be assured; i.e., occupancy status does not change over the J replicate surveys. In this case we can compute the total count, and for that we will abuse our notation and redefine y_i, so that now $y_i = \sum_{j=1}^{J} y_{ij}$, and the observation model is then a binomial model:

$$y_i | z_i \sim Binomial(J, z_i p)$$

This occupancy model has a transparent structure as a compound GLM—the model for the ecological state is a Bernoulli GLM and the model for the observations is a Bernoulli (or binomial) GLM *conditional* on the (in this case) partially observed state variable. This is arguably the simplest explicit HM imaginable because both the observations and the state variable are binary.

The occupancy model has become rather popular recently and is implemented in software MARK (White and Burnham, 1999), PRESENCE (Hines, 2006), and with the function `occu` in the R package `unmarked` (Fiske and Chandler, 2011). We discuss this model class in great detail in Chapters 10 and later.

2.3.1.2 The N-Mixture Model for Abundance

The *N*-mixture model (Royle, 2004b) is arguably the canonical HM for animal abundance. As for the occupancy model, the sampling design has an equivalent repeated measures structure where $i = 1, 2, ..., M$ sites are sampled for J occasions (e.g., point counts of birds on J mornings within the same season) and individuals of a species are counted. The observations are the counts y_{ij}, and the observation model is assumed to be a binomial distribution conditional on the true population size at site i:

$$y_{ij} | N_i \sim Binomial(N_i, p)$$

where p in this case is *individual*-level detection probability (as opposed to site-level detection in the case of the occupancy model). In general, we can model factors that influence detection using GLMs with the logit link function (see Chapter 3) but for now we neglect that generality. As with the occupancy model we just described, it is necessary to have repeated counts (= "measurements of abundance N") for at least some of the sites to ensure identifiability of model parameters.

The state variable here is "local population size," N_i, and the Poisson GLM is the natural framework for modeling variation in this state variable:

$$N_i \sim Poisson(\lambda_i).$$

In almost all studies there is interest in modeling the effect of measurable covariates, so we consider models for the log-transformed λ_i (i.e., the canonical link for Poisson GLMs; see Chapter 3):

$$\log(\lambda_i) = \beta_0 + \beta_1 x_i$$

where x_i is some site-level covariate. Of course, we may consider alternative models for local abundance such as the zero-inflated Poisson, Poisson lognormal, or negative binomial, all of which account for excess variation (or overdispersion) in local abundance relative to the Poisson model (Chapter 6 and Kéry et al., 2005b).

As with the occupancy model, the *N*-mixture model is clearly a type of compound GLM—both observation and process models are plain old GLMs, but they are linked together through the conditional dependence structure of the HM. The *N*-mixture models are implemented in program PRESENCE (Hines, 2006), program MARK (White and Burnham, 1999), and in `unmarked` with the function `pcount`, and they are the focus of a series of chapters starting with Chapter 6.

2.3.2 THE PROCESS OF HIERARCHICAL MODELING

HMs are the outcome of a certain process of building a model and a certain manner of thinking about the processes underlying observed data (Berliner, 1996). In ecological systems, we find that hierarchical modeling is a very appealing way to think about how to build models to idealized systems; i.e., simpler problems that we wish we had, and then linking these idealized models to observations. One way to think about constructing HMs is to ask this question: What unknown variable, if I had it, would simplify my problem? In all of the work we do, our attempt to answer this unifies the different HMs: (1) occupancy models—true occupancy state renders the model for observations as a simple logistic regression (2) N-mixture models—true population size renders the model for observations to be a simple logistic regression, and (3) spatial capture–recapture models (not covered here; see Royle et al., 2014)—the activity center renders the model for trap-level observations to be a simple logistic regression. Actually, the traditional binomial measurement error model underlying almost all capture–recapture and related models renders most of these models, at the level of the observed data, a type of logistic regression with a kind of complex random effects structure.

2.3.3 INFERENCE FOR HMs

HMs may be analyzed using both classical or Bayesian methods and either approach may be advantageous in some situations. Because HMs are specified conditional on one or more latent variables, Bayesian analysis proceeds directly with the use of Markov chain Monte Carlo (MCMC) algorithms that preserve the latent variables in the model. On the other hand, the classical method of likelihood inference for models with latent variables is based on marginal (or integrated) likelihood in which the latent variable is removed from the conditional likelihood (i.e., that containing the latent variable) by integration or, in the case of a discrete latent variable, by summation. In the following sections we address basic ideas of both classical and Bayesian inference.

2.4 CLASSICAL INFERENCE BASED ON LIKELIHOOD

Statistical inference for any system is based on the conceptual view that data we observe (or may potentially observe) are outcomes of random variables having a distribution with parameters that we would like to learn about from data. Most of the time we will denote the random variable as y, let us say the count of birds at some point count location. Absent some specific context, we will denote the basic probability model for y as follows:

$$y \sim f(y|\theta)$$

that we will read "the random variable y has probability distribution f," although normally the $\sim$ symbol is read "is distributed as" and we also usually do not say "the random variable." In this representation there is a parameter θ, or possibly a vector of parameters, that indexes the probability distribution f.

Once we go to a site and perform an actual point count of birds we obtain some realized value of this random variable, say $y = 5$ birds counted. The value $y = 5$ is our observation, data set, data, or a datum. To be completely clear, we must say this: When we use the $y \sim f(y|\theta)$ notation, we do not mean that the number 5 has the distribution f (which would be nonsense) but rather, and clearly, we mean that we are using a random variable having distribution f as a model for potential outcomes of the count, including the outcome(s) we have on hand ($y = 5$ in this case). In classical statistics, as we have noted before, this distinction is usually made explicitly with uppercase and lowercase letters that we are not generally adhering to, possibly at the expense of some clarity in certain instances.

We usually consider the possibility that we might obtain outcomes of multiple random variables having this distribution f, say a sample of size n, and we denote these multiple random variables by $y_1, \ldots, y_n$ or just the vector **y** (bold type here indicates a vector). In this case we assert, as our objective, the need to estimate the parameter θ from this sample of size n.

It is not possible to proceed without additional assumptions. In particular, while f is the distribution of a single random variable y, it does not provide a characterization of the distribution of multiple random variables. For that, we need to build their *joint distribution*. If we assume that each random variable is independent of each other, then the joint distribution of several random variables is the product of the distribution of each one. Therefore, the joint distribution of our sample of n random variables is:

$$f(y_1, y_2, \ldots, y_n) = \prod_{i=1}^{n} f(y_i|\theta)$$

Identifying the joint distribution of the sample of n random variables (those which may be observed) is the crucial step toward achieving formal inference procedures about θ from a data set. We cannot do anything statistical in terms of estimation or any other inference without being able to identify what this joint distribution is. This task is made simpler by assuming independence of the observable random variables, which implies directly the joint distribution from the simpler more elemental distribution of a single observation $f(y|\theta)$. But there will not generally be a direct correspondence between saying "the distribution of y is $f()$" and the joint distribution we need in order to develop inferences from a sample of n observations but we will often assume that random variables are independent so that the joint distribution is implied directly. Or more generally, we make explicit conditional independence assumptions, which is to say that the random variables are independent of one another conditional on some other variable. For example, in the N-mixture model in which N is the population size of individuals exposed to sampling and y_1 and y_2 are consecutive counts (e.g., on different days), we assume that y_1 and y_2 are independent of each other *conditional on N*. Then, we can make progress toward analyzing the N-mixture model by working with the joint distribution of y_1 and y_2, conditional on N, which is $[y_1, y_2|N] = [y_1|N][y_2|N]$; i.e., it is the product of two binomials. This still depends on the unknown quantity N, and additional consideration needs to be paid to this (see Section 2.4.4).

Having identified the joint distribution of our observable random variables we can proceed with obtaining an estimate of θ using the *method of maximum likelihood*, or the maximum likelihood estimator (MLE) (Edwards, 1992; Chapter 7 in Casella and Berger, 2002). As its name suggests, this method is based on the *likelihood*. Simply put, the likelihood is the joint distribution of the observable random variables, but regarded as a function of the parameter θ. We will denote this by $L()$ such as:

$$L(\theta; \mathbf{y}) \equiv f(y_1, y_2, \ldots, y_n|\theta) = \prod_{i=1}^{n} f(y_i|\theta).$$

Here we put a ";" between the arguments of L to distinguish that this is a function indexed by **y** and *not* conditional on **y** in the sense of conditional probability.

2.4 CLASSICAL INFERENCE BASED ON LIKELIHOOD

The value of θ that produces the highest likelihood for the observed data set is called the MLE and denoted by the parameter with a "hat" over it, $\hat{\theta}$. For practical (computational) reasons we often work in terms of the natural logarithm of the likelihood. We can do this because maximizing the likelihood or log-likelihood are equivalent operations, as is *minimizing* the *negative* of the log-likelihood. In the R software, the functions nlm or optim by default perform function minimization, and so you will see things multiplied by -1 in the definition of the likelihood function (see below).

The term "estimator" is an important one in statistics. An estimator is a rule for converting data to a guess for θ. This is different from an *estimate* that is a specific numerical value produced by the estimator. Estimators have statistical properties, so we say things about bias and variance when we talk about the rules (estimators) we use to produce estimates (specific guesses). We can think about an estimator in a more tangible way as an R function that takes data and returns some summary of the data. In that context then, we can evaluate the properties of the R function by simulating many data sets and executing the R function over and over again for each data set. The numerical result of the R function in some particular case is the estimate, while the function itself is the estimator. Note that strictly it is not correct to speak about an "unbiased estimate," because only the estimator can have the property of zero bias, not a single output from the estimator. However, we may sometimes see "unbiased estimate" as a short and somewhat colloquial way of saying "the estimate produced by an unbiased estimator."

2.4.1 THE FREQUENTIST INTERPRETATION

Classical inference, including inference based on likelihood, is justified by frequentist evaluation of statistical estimators. That is, the performance of an estimator (or other procedure) is evaluated by "averaging over" hypothetical realizations of the data y. For example, if $\hat{\theta}$ is an estimator of θ then the frequentist is interested in properties of $\hat{\theta}$ over replicate realizations of the data. The expected value, $E_y(\hat{\theta}|y)$, is used to characterize bias: If the expected value is equal to θ, then $\hat{\theta}$ is unbiased. The "sampling variance" of an estimator, $\text{Var}_y(\hat{\theta}|y)$, is what we use to characterize uncertainty, or precision, of MLEs. The variance is therefore taken with respect to hypothetical realizations of the data. The square root of the sampling variance is the standard error of an estimator (which may sometimes be a variance parameter, in which case we have a standard error of a variance).

In adopting a frequentist view of statistical inference (which is usually implicit by our adoption of MLE and asymptotic variances) we are not using probability directly to make inference about θ, and so we cannot make direct probability statements about θ. Instead, the probability statement is about the procedure itself. We see this in the standard explanation of a confidence interval which is "the probability that the interval contains the true value of the parameter is 95%"; i.e., it is a statement about the interval itself and not the parameter. So, if we were to repeat the experiment 100 times, then we expect to have our interval contain the true value 95 times.

2.4.2 PROPERTIES OF MLEs

Maximum likelihood estimation of parameters is a relatively easy proposition for many classes of models that ecologists are interested in. Using standard methods of maximizing the likelihood provides numerically precise parameter estimates, an easy model selection framework based on AIC (discussed in Section 2.6.1) and a number of other generally appealing theoretical asymptotic properties that we outline here. The term asymptotic means that they hold "as the sample size increases to infinity" which we usually interpret simply as "in large sample sizes." For example, to say that a certain formula gives the asymptotic variance of a parameter estimator means the variance is the correct variance for some hypothetical estimator as the sample size n "approaches infinity," which is the same as saying the error between this expression for variance and the actual variance of the MLE approaches 0 as our sample size increases. This is equivalent to saying that the result is not exact at all, for any specific n, just that we can make the error as small as we like, by increasing n.

One of the greatest things about MLEs, and something that is extremely important from a practical standpoint, is that the "uncertainty" of a MLE can be obtained easily: the variance of the MLE can be well approximated, again in large sample sizes, by the inverse of the *Fisher information* (matrix), which is the matrix of second derivatives of the log-likelihood with respect to the parameter(s) θ, evaluated at the MLE $\widehat{\theta}$. (This matrix of second derivatives is called the *Hessian* matrix). We refer to the inverse of the Fisher information matrix as the asymptotic variance–covariance matrix although typically the term "asymptotic" is omitted in practice and we understand our variances for MLEs as being reasonably valid under "large" sample sizes (without ever declaring or understanding what "large" is). From the asymptotic variance–covariance matrix we obtain the "Wald-type" confidence interval as

$$\widehat{\theta} \pm 1.96\sqrt{\text{Var}(\widehat{\theta})}.$$

where $\text{Var}(\widehat{\theta})$ is the asymptotic variance of $\widehat{\theta}$, taken from the diagonal of the asymptotic variance–covariance matrix. In addition to this simple method for obtaining the variance, MLEs possess the following desirable theoretical properties:

1. *Consistency*: The MLE is asymptotically unbiased. That is, if $\widehat{\theta}(\mathbf{y}_n)$ is an estimator depending on data $\mathbf{y}_n$, then $E_y\{\widehat{\theta}(\mathbf{y}_n)\} \to \theta$ as $n \to \infty$.
2. *Asymptotic normality*: The MLE is asymptotically normal with mean θ and variance–covariance matrix equal to the inverse of the Fisher information matrix.
3. *Efficiency*: The MLE achieves the famed Cramér-Rao lower bound as the sample size tends to infinity. What this means is that in a theoretical class of all consistent estimators, no other estimator has a lower asymptotic variance than the MLE.
4. MLEs are *invariant* to reparameterization. That is, if $\widehat{\theta}$ is the MLE of θ then the MLE of a function h of θ, $h(\theta)$, is $h(\widehat{\theta})$.

Basically what all of this means is that we expect to do alright using MLEs as long as n is "large," and furthermore MLEs possess a certain degree of analytic tractability in large sample sizes (we can characterize sampling distributions, compute variances, etc.).

2.4.3 THE DELTA APPROXIMATION

While MLEs are invariant to reparameterization, the variance of MLEs is *not* invariant to transformation. That is, if $\widehat{\theta}$ is the MLE of θ, then the MLE of $h(\theta)$ is $\widehat{h(\theta)} = h(\widehat{\theta})$ but $\text{Var}(\widehat{h(\theta)}) \neq h(\text{Var}(\widehat{\theta}))$. We often use the delta method of computing variances (Appendix 2 of Cooch and White (2014) is a good introduction along with technical background and context). Alternatively, we can also use bootstrapping, which we discuss in Section 2.4.3 below.

The delta method goes as follows: Suppose we wish to obtain the variance of a random variable y that is defined by a function of another variable x, denoted by $y = h(x)$ where h is some arbitrary (usually) nonlinear function. We derive a linear approximation to $h(x)$ using a Taylor series expansion of $h(x)$ around its mean $E(x) = \mu$. This produces

$$y = h(x) \approx h(\mu) + h'(\mu)(x - \mu)$$

where $h'(\mu)$ is the first derivative of h evaluated at μ. Therefore, because $h(\mu)$ is constant (no longer a function of the random variable x), we have

$$\text{Var}(y) \approx (h'(\mu))^2 \text{Var}(x - \mu)$$

So how do we apply this idea to an estimator of some parameter? We equate x to $\widehat{\theta}$ and $\mu = E(\widehat{\theta})$ and we assume $\widehat{\theta}$ is unbiased so that $\mu = E(\widehat{\theta}) = \theta$. The variable $y = h(\widehat{\theta})$ is what we seek the variance of in which case we see that

$$\text{Var}(y) \approx (h'(\theta))^2 \text{Var}(\widehat{\theta})$$

and because we don't know θ we plug in the estimate we have and find that

$$\text{Var}\left(h(\widehat{\theta})\right) \approx \left(h'(\widehat{\theta})\right)^2 \text{Var}(\widehat{\theta}).$$

In ecology we frequently encounter two specific cases: $h(x) = \exp(x)$ and $h(x) = \text{logit}^{-1}(x)$. The former arises in log-linear models such as Poisson regression and the latter of course is the inverse-logit transform for converting linear predictors in logistic regression back to probabilities. Because of their prevalence, we go through both of those cases here. In the case of $h(x) = \exp(x)$ we have $h'(x) = \exp(x)$ and so

$$\text{Var}(\exp(x)) \approx \exp(\mu)^2 \text{Var}(x)$$

and therefore

$$\text{Var}\left(\exp(\widehat{\theta})\right) \approx \exp(\widehat{\theta})^2 \text{Var}(\widehat{\theta})$$

The inverse logit transformation is $h(x) = \exp(x)/(1 + \exp(x))$ and we have to apply the quotient rule in order to differentiate this. We find that

$$h'(x) = \frac{\exp(x)}{(1 + \exp(x))^2}$$

so that

$$\text{Var}\left(\text{logit}^{-1}(\widehat{\theta})\right) \approx \frac{\text{Var}(\widehat{\theta}) \exp(\widehat{\theta})^2}{\left(1 + \exp(\widehat{\theta})\right)^4}$$

2.4.4 EXAMPLE: CLASSICAL INFERENCE FOR LOGISTIC REGRESSION

We now clarify some of the ideas related to the analysis of specific models using a simple logistic regression model. Let $y_1, y_2, \ldots, y_n$ denote our observations of presence/absence, which we suppose follow a basic logistic regression model, so that the observation model is:

$$y_i \sim Bernoulli(\psi_i)$$

with

$$\text{logit}(\psi_i) = \beta_0 + \beta_1 x_i$$

for sites $i = 1, 2, \ldots, M$ and where x_i is some habitat covariate. And further, we suppose that $y_1, \ldots, y_n$ are mutually *independent*, which as we tried to stress previously is an important part of the probability model specification for determining the joint probability distribution of the sample of n observations. Under independence, we can form the joint distribution as the product of the n constituent Bernoulli components:

$$f(y_1, y_2, \ldots, y_n; \beta_0, \beta_1) = \left\{ \prod_{i=1}^{n} Bernoulli(y_i; \boldsymbol{\beta}) \right\}$$

Here we emphasize the dependence on the parameters $\boldsymbol{\beta}$ by writing this as the argument of the Bernoulli distribution, keeping in mind that there is a logit-linear function relating the canonical parameter ψ and parameters $\boldsymbol{\beta}$. The likelihood is therefore:

$$L(\beta_0, \beta_1; y_1, y_2, \ldots, y_n) \equiv \left\{ \prod_{i=1}^{n} Bernoulli(y_i; \boldsymbol{\beta}) \right\}.$$

The main thing we usually do with the likelihood is maximize it over the parameter space (possible values of $\boldsymbol{\beta}$). The value of $\boldsymbol{\beta}$ that produces the maximum value is the maximum likelihood estimator, which we denote by $\hat{\boldsymbol{\beta}}$ (we will do this in R shortly). A key point to remember about the likelihood is that it is *not* a probability distribution for the parameters. Although it started out as a probability distribution for y, we have switched the arguments and now think about it as a function of $\boldsymbol{\beta}$.

The following set of R commands will allow us to simulate one data set with a covariate "vegetation height" (vegHt):

```
# Simulate a covariate called vegHt for 100 sites
set.seed(2014)   # Set seed so we all get the same values of vegHt
M <- 100         # Number of sites surveyed
vegHt <- runif(M, 1, 3)   # uniform from 1 to 3

# Suppose that occupancy probability increases with vegHt
# The relationship is described by an intercept of -3 and
#   a slope parameter of 2 on the logit scale
beta0 <- -3
beta1 <- 2
psi <- plogis(beta0 + beta1*vegHt) # apply inverse logit
# Now we go to 100 sites and observe presence or absence
z <- rbinom(M, 1, psi)
```

2.4 CLASSICAL INFERENCE BASED ON LIKELIHOOD

Now, to obtain the MLEs for the parameters β we need to maximize the likelihood for the set of observations that we have (or just simulated). The basic strategy for classical estimation based on likelihood that we adopt in this book is to express the likelihood as an R function and then use the standard functions `optim` or `nlm` to maximize it or, equivalently, minimize the negative of the log-likelihood. In the following bit of R code we write out the likelihood function (which we take the logarithm of and negate to return the negative of the log-likelihood) in a long-winded way, by writing the binomial pmf directly but we also show that this can be computed using the `dbinom` function (that line is commented out). Following our definition of the negative log-likelihood here we evaluate the function at different values of the parameter vector; e.g., $\beta = (0, 0)$ and $\beta = (-3, 2)$, and then we use the R function `optim` to obtain the MLEs. For illustration, we also show how the MLEs can be found by a brute-force grid search over the entire parameter space. Finally, we illustrate the use of the R function `glm` for the same, recognizing that the likelihood is that of a logistic GLM:

```
# Definition of negative log-likelihood.
negLogLike <- function(beta, y, x) {
   beta0 <- beta[1]
   beta1 <- beta[2]
   psi <- plogis(beta0 + beta1*x)
   likelihood <- psi^y * (1-psi)^(1-y) # same as next line:
#  likelihood <- dbinom(y, 1, psi)
   return(-sum(log(likelihood)))
}

# Look at (negative) log-likelihood for 2 parameter sets
negLogLike(c(0,0), y=z, x=vegHt)
negLogLike(c(-3,2), y=z, x=vegHt) # Lower is better!

# Let's minimize it formally by function minimisation
starting.values <- c(beta0=0, beta1=0)
opt.out <- optim(starting.values, negLogLike, y=z, x=vegHt, hessian=TRUE)
(mles <- opt.out$par)      # MLEs are pretty close to truth
    beta0      beta1
-2.539793   1.617025

# Alternative 1: Brute-force grid search for MLEs
mat <- as.matrix(expand.grid(seq(-10,10,0.1), seq(-10,10,0.1)))
                                   # above: Can vary resolution (e.g., from 0.1 to 0.01)
nll <- array(NA, dim = nrow(mat))
for (i in 1:nrow(mat)){
   nll[i] <- negLogLike(mat[i,], y = z, x = vegHt)
}
which(nll == min(nll))
mat[which(nll == min(nll)),]

# Produce a likelihood surface, shown in Fig. 2-2.
library(raster)
r <- rasterFromXYZ(data.frame(x = mat[,1], y = mat[,2], z = nll))
mapPalette <- colorRampPalette(rev(c("grey", "yellow", "red")))
```

```
plot(r, col = mapPalette(100), main = "Negative log-likelihood",
xlab = "Intercept (beta0)", ylab = "Slope (beta1)")
contour(r, add = TRUE, levels = seq(50, 2000, 100))

# Alternative 2: Use canned R function glm as a shortcut
(fm <- glm(z ~ vegHt, family = binomial)$coef)

# Add 3 sets of MLEs into plot
# 1. Add MLE from function minimisation
points(mles[1], mles[2], pch = 1, lwd = 2)
abline(mles[2],0)   # Put a line through the Slope value
lines(c(mles[1],mles[1]),c(-10,10))
# 2. Add MLE from grid search
points(mat[which(nll == min(nll)),1], mat[which(nll == min(nll)),2],
pch = 1, lwd = 2)
# 3. Add MLE from glm function
points(fm[1], fm[2], pch = 1, lwd = 2)
```

The likelihood surface is shown in Figure 2.2 along with the maximum value identified (actually, it is the negative log-likelihood). The three different numerical methods of obtaining the estimates yield essentially identical results (the points indicating each estimate are plotted on top of each other) as we expect. You can see how this process could easily be automated to do a simulation study for a particular situation, or even generalized to other types of models (e.g., Poisson regression instead of logistic regression), simply by changing the Bernoulli pmf to a Poisson pmf (i.e., using `dpois` instead of `dbinom`). Of course, in practice, the technical details of what we do are hidden in the guts of functions such as `glm` or `lm` and we hardly ever confront the likelihood directly; see Chapter 3 for more about these functions and the models they specify.

The `hessian=TRUE` option in the call to `optim` produces the Hessian matrix in the returned list `opt.out`, and so we can obtain the asymptotic standard errors (ASE) for the two parameters by doing this:

```
Vc <- solve(opt.out$hessian)    # Get variance-covariance matrix
ASE <- sqrt(diag(Vc))           # Extract asymptotic SEs
print(ASE)

    beta0      beta1
0.8687444  0.4436064

# Compare to SEs reported by glm() function (output thinned)
 summary(glm(z ~ vegHt, family = binomial))

Call:
glm(formula = z ~ vegHt, family = binomial)
Coefficients:
             Estimate Std. Error  z value  Pr(>|z|)
(Intercept)   -2.5397     0.8687   -2.923  0.003462 **
vegHt          1.6171     0.4436    3.645  0.000267 ***
```

2.4 CLASSICAL INFERENCE BASED ON LIKELIHOOD

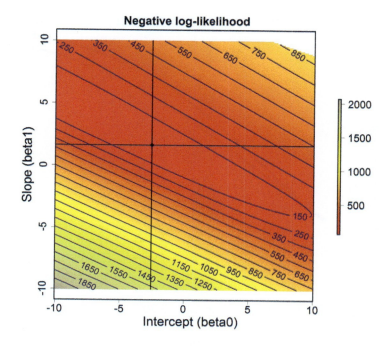

FIGURE 2.2

Likelihood surface for the logistic regression example. The MLE occurs at the point $(\hat{\beta}_0, \hat{\beta}_1) = (-2.54, 1.62)$, indicated by a point with intersecting lines through it. The estimates obtained by a grid search and using the `glm` function are also plotted with a point that falls right on top of the MLE.

We see consistency between our ASEs computed by inverting the Hessian directly and those produced by the `glm` function. Therefore, we have faith in our understanding of what `glm` is doing. We summarize some of our likelihood results in a table and produce the figure shown in Figure 2.3:

```
# Make a table with estimates, SEs, and 95% CI
mle.table <- data.frame(Est=mles,
                    ASE = sqrt(diag(solve(opt.out$hessian))))
mle.table$lower <- mle.table$Est - 1.96*mle.table$ASE
mle.table$upper <- mle.table$Est + 1.96*mle.table$ASE
mle.table
            Est       ASE       lower      upper
beta0  2.539793  0.8687444  -4.2425320  -0.8370538
beta1  1.617025  0.4436064   0.7475564   2.4864933

# Plot the actual and estimated response curves
plot(vegHt, z, xlab="Vegetation height", ylab="Occurrence probability")
plot(function(x) plogis(beta0 + beta1 * x), 1.1, 3, add=TRUE, lwd=2)
plot(function(x) plogis(mles[1] + mles[2] * x), 1.1, 3, add=TRUE,
     lwd=2, col="blue")
legend(1.1, 0.9, c("Actual", "Estimate"), col=c("black", "blue"), lty=1,
       lwd=2)
```

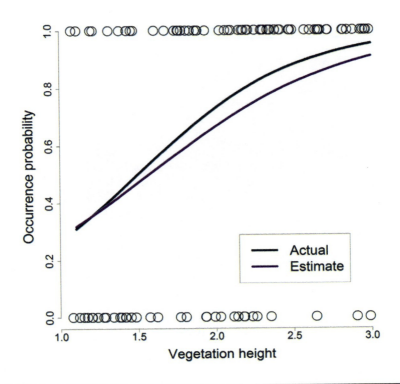

FIGURE 2.3
Actual (black) and fitted (blue) response curve, for expected occurrence $E(z) = \psi$ as a function of covariate "vegetation height" for a simulated data set. Circles show the observed presence (1) and absence (0) data.

2.4.5 BOOTSTRAPPING

For many classical analyses based on likelihood, it is easy enough to obtain variance estimates using the asymptotic results and Fisher Information. However, sometimes we may be worried about the validity of the asymptotic result or else have a really complicated function for which computing derivatives (required for the delta method approximation) is difficult. In such cases, we often compute standard errors using the method of *bootstrapping* (Efron, 1982; Dixon, 2006).

There are two flavors of bootstrapping: parametric and nonparametric. The idea of parametric bootstrapping is that we simulate data sets under the assumed model, but with parameters equal to the MLEs ($\hat{\theta}$) obtained from our observed data. For each simulated data set, we refit the model and obtain the (new) MLEs for the simulated data set. Each set of MLEs thus obtained is called a bootstrap sample, and together they form the bootstrap distribution that we can summarize to obtain quantities of interest. For example, the variance of the bootstrap samples is an estimate of the sampling variance (the squared standard error). The 0.025 and 0.975 percentiles of the bootstrap distribution form a 95% confidence interval. The difference between the mean of the bootstrap distribution and $\hat{\theta}$ is an estimate of the bias, and so forth. Nonparametric bootstrapping involves sampling randomly n values from the data set (also with n values) at hand *with replacement*, refitting the model in question, repeating that for a large number of times and summarizing the MLEs. In this latter case, the explicit parametric assumptions of the model being fitted are no longer relevant to the data generating process, hence the nonparametric label. The R package unmarked (Fiske and Chandler, 2011) contains generic parametric and nonparametric bootstrapping methods for certain HMs, including N-mixture, distance sampling, occupancy models, and so forth.

Bootstrapping has many purposes, including goodness-of-fit (GoF) evaluation (see Section 2.7), but for now we demonstrate its application to characterizing uncertainty of the MLEs of β_0 and β_1 from our logistic regression example. In addition, we define a parameter being "average occupancy" (psi.bar) for which we would like to construct a confidence interval. Otherwise, this quantity would require a delta approximation in order to obtain the ASE of the MLEs.

To carry out the parametric bootstrap for the logistic regression problem we do the following (note that the MLEs were previously computed and the objects mles and mle.table exist in your workspace):

```
nboot <- 1000    # Obtain 1000 bootstrap samples
boot.out <- matrix(NA, nrow=nboot, ncol=3)
dimnames(boot.out) <- list(NULL, c("beta0", "beta1", "psi.bar"))

for(i in 1:1000){
  # Simulate data
  psi <- plogis(mles[1] + mles[2] * vegHt)
  z <- rbinom(M, 1, psi)

  # Fit model
  tmp <- optim(mles, negLogLike, y=z, x=vegHt, hessian=TRUE)$par
  psi.mean <- plogis(tmp[1] + tmp[2] * mean(vegHt))
  boot.out[i,] <- c(tmp, psi.mean)
}

SE.boot <- sqrt(apply(boot.out, 2, var))   # Get bootstrap SE
names(SE.boot) <- c("beta0", "beta1", "psi.bar")

# 95% bootstrapped confidence intervals
apply(boot.out,2,quantile,c(0.025,0.975))
            beta0      beta1     psi.bar
2.5%     -4.490565  0.8379983  0.5728077
97.5%    -0.978901  2.5974839  0.7828377

# Boostrap SEs
SE.boot
     beta0       beta1      psi.bar
0.89156946  0.45943765  0.05428008

# Compare these with the ASEs for regression parameters
mle.table
            Est         ASE        lower       upper
beta0   -2.539793   0.8687444   -4.2425320   -0.8370538
beta1    1.617025   0.4436064    0.7475564    2.4864933
```

We see that the bootstrapped SEs are slightly larger than the ASEs. This is due to the fact that the bootstrap distribution is slightly skewed; i.e., not normal, as assumed in the calculation of the ASEs.

2.4.6 LIKELIHOOD ANALYSIS OF HMs
2.4.6.1 Discrete Random Effects

We seek now to extend the likelihood estimation framework to HMs, where we will see that a rote application of the basic idea leads us immediately to an insurmountable difficulty. For a case study we consider the occupancy model, our hierarchical extension of the logistic regression model. For this model, the observation model, which previously led directly to the likelihood, is now specified conditional on the random effects (or latent variables) z_i; i.e., it is $f(y_{ij}|z_i, p)$. Clearly it does not make any sense to think about maximizing this because (1) it contains $M + 1$ unknowns—all of the latent variables z_i as well as the parameter p and (2) the parameters of interest, β_0 and β_1 are nowhere to be seen. Therefore, building the likelihood directly from this observation model does not appear useful. Instead, classical analysis of models with random effects is based on the marginal (also called integrated) likelihood, in which we remove the random effects from the conditional likelihood by integration (for the case where the random effects are discrete, we replace our integration with a summation, but the principle is the same).

In general, if z is any random effect, the conditional likelihood is $f(y|z)$, then we would have to compute the marginal distribution of $\mathbf{y}_i$ *unconditional* on the random effect by doing this calculation:

$$f(\mathbf{y}_i|\boldsymbol{\beta},p) = \int \left\{ \prod_{j=1}^{J} f(y_{ij}|z_i, p) \right\} g(z_i|\boldsymbol{\beta}) dz_i \qquad (2.2)$$

where the result, the marginal distribution $f(\mathbf{y}_i|\boldsymbol{\beta}, p)$, conveniently, is no longer a function of z anymore but is a function of the parameters of interest, p and $\boldsymbol{\beta}$. We can now construct our likelihood from these independent marginal pieces $f(\mathbf{y}_i|\boldsymbol{\beta}, p)$ and do standard things to this likelihood, such as obtaining MLEs of parameters p or $\boldsymbol{\beta}$, their asymptotic variances, and so forth.

For the occupancy model, and indeed for many of the HMs that we will consider in this book, it is relatively easy to compute the marginal likelihood because the state variable is discrete, which means the marginal likelihood can be expressed as a summation over what is usually a moderate number of possible values (replacing the integral with a summation in the above expression). To compute the marginal likelihood, we apply the "law of total probability," because z is a discrete random variable (Royle and Dorazio, 2008, p. 34):

$$\Pr(y) = \Pr(y|z=1)\Pr(z=1) + \Pr(y|z=0)\Pr(z=0)$$

to see that the marginal probability for an observation y_i is:

$$[y_i|p, \psi] = Binomial(y_i; J, p)\psi + I(y_i = 0)(1 - \psi)$$

where we use the notation $I(y_i = 0)$ to be an indicator function that evaluates to 1 if the condition in parentheses holds, and 0 otherwise. The result in this case is the zero-inflated binomial pmf. It is the ordinary binomial, but with additional mass at the value $y = 0$ (Tyre et al., 2003). It is easy to express the marginal likelihood for this HM with binary latent effects using an R function that we provide in the following block of code:

```
set.seed(2014)
M <- 100                              # number of sites
vegHt <- runif(M, 1, 3)               # uniform from 1 to 3
psi <- plogis(beta0 + beta1 * vegHt)  # occupancy probability
```

2.4 CLASSICAL INFERENCE BASED ON LIKELIHOOD

```
z <- rbinom(M, 1, psi)          # realised presence/absence
p <- 0.6                        # detection probability
J <- 3                          # sample each site 3 times
y <- rbinom(M, J, p*z)          # observed detection frequency

# Define negative log-likelihood.
negLogLikeocc <- function(beta, y, x, J) {
    beta0 <- beta[1]
    beta1 <- beta[2]
    p <- plogis(beta[3])
    psi <- plogis(beta0 + beta1*x)
    marg.likelihood <- dbinom(y, J, p)*psi + ifelse(y==0,1,0)*(1-psi)
    return(-sum(log(marg.likelihood)))
}

starting.values <- c(beta0=0, beta1=0,logitp=0)
(opt.out <- optim(starting.values, negLogLikeocc, y=y, x=vegHt,J=J,
                  hessian=TRUE))

$par
     beta0      beta1     logitp
-2.4039229  1.4977667  0.5449382

$value
[1] 122.7039

$counts
function gradient
     170       NA

$convergence
[1] 0

$message
NULL

$hessian
            beta0     beta1     logitp
beta0   17.322109  32.45802   5.268095
beta1   32.458021  65.23173  11.393164
logitp   5.268095  11.39316  34.043000
```

The R function `optim` produces a list with a number of useful things, including the MLE, labeled `par`, the value of the negative log-likelihood (`value`), and the Hessian matrix (`hessian`) that we can invert to obtain the asymptotic variance–covariance matrix or standard errors, as follows:

```
sqrt(diag(solve(opt.out$hessian)))

    beta0     beta1    logitp
0.9258370 0.4799726 0.1770094
```

The function `occu` in the package `unmarked` will fit such occupancy models, which we demonstrate in some detail in Chapter 10.

2.4.6.2 A Continuous Latent Variable

When the latent variable or random effect is discrete, we saw that the marginal likelihood can be computed directly by summing up the likelihood for each possible value of the latent variable and multiplying that by the (prior) probability of each value of the latent variable. When the latent variable (or random effect) is continuous, we have to compute the marginal likelihood explicitly by integration (i.e., Eq. (2.2)). As an example, we consider analysis of a capture–recapture model known as "model M_h," in which individual detection probability contains a normal random effect. The data are the encounter frequencies of individuals over J replicate samples of a population, $y_i \sim Binomial(J, p_i)$ and $logit(p_i) \sim Normal(\mu, \sigma^2)$. The technical details of model M_h are tangential to the point here, but see Section 6.2 in Royle and Dorazio (2008). For our present purposes the key point is that, in order to compute the marginal likelihood, we have to calculate the integral of the binomial likelihood for each possible encounter frequency (from 0 to J) over the normal prior distribution. In R the commands for doing this integral for each value $j = 0, 1, ..., J$ look like this (the following code is not executable as is):

```
marg <- rep(NA, J+1)
for(j in 0:J){
marg[j+1] <- integrate(
    function(x){
        dbinom(j, J, plogis(x)) * dnorm(x, mu, sigma)},
        lower=-Inf,upper=Inf)$value
    }
}
```

The model M_h likelihood also includes a combinatorial term that is a function of unknown population size N ($N = n + n_0$, where n_0 is the number of unobserved individuals), and therefore the parameters of the model include N, μ, and σ^2. We demonstrate the application of this specialized type of HM using a data set on the flat-tailed horned lizard (Figure 2.4; Royle and Young, 2008; Section 6.2 in Royle and Dorazio, 2008) that includes detections of 68 individual lizards over $J = 14$ sample periods.

The R commands that create the data set, define the model M_h likelihood, and obtain the MLEs are as follows:

```
# nx = encounter frequencies, number inds. encountered 1, 2, ..., 14 times
nx <- c(34, 16, 10, 4, 2, 2, 0, 0, 0, 0, 0, 0, 0, 0)
nind <- sum(nx)      # Number of individuals observed
J <- 14              # Number of sample occasions

# Model Mh likelihood
Mhlik <- function(parms){
    mu <- parms[1]
    sigma <- exp(parms[2])
    # n0 = number of UNobserved individuals: N = nind + n0
    n0 <- exp(parms[3])

    # Compute the marginal probabilities for each possible value j=0,...,14
    marg <- rep(NA,J+1)
```

2.4 CLASSICAL INFERENCE BASED ON LIKELIHOOD

FIGURE 2.4

The flat-tailed horned *lizard* (Phrynosoma mcallii). *(Photo Kevin and April Young.)*

```
    for(j in 0:J){
       marg[j+1] <- integrate(
       function(x){dbinom(j, J, plogis(x)) * dnorm(x, mu, sigma)},
       lower=-Inf,upper=Inf)$value
    }

    # The negative log likelihood involves combinatorial terms computed
    # using lgamma()
    -1*(lgamma(n0 + nind + 1) - lgamma(n0 + 1) + sum(c(n0, nx) * log(marg)))
    }
    (tmp <- nlm(Mhlik, c(-1, 0, log(10)), hessian=TRUE))

    $minimum
    [1] -126.6951

    $estimate
    [1] -2.5648520 -0.2288488  3.6585131

    $gradient
    [1] -1.668833e-05 -1.929834e-05  2.742334e-06
```

```
$hessian
         [,1]      [,2]      [,3]
[1,] 64.58085 25.619056 29.506561
[2,] 25.61906 38.228472 -2.636507
[3,] 29.50656 -2.636507 24.290295

$code
[1] 1

$iterations
[1] 16
```

The likelihood is parameterized in terms of μ, $\log(\sigma)$ and $\log(n_0)$ and therefore we need to back-transform the MLEs to get them on the natural scale. In particular, we care mostly about the population size and so the MLE of n_0 is $\hat{n}_0 = \exp(3.6585) = 38.8$ and $\hat{N} = 68 + 38.8 = 106.8$ lizards. We use the delta approximation to obtain the standard error (SE) of $\hat{N}$ as follows:

```
(SE <- sqrt( (exp(tmp$estimate[3])^2)* diag(solve(tmp$hessian))[3] ) )
[1] 20.80025
```

Which we note is the SE of $\hat{n}_0$ but, since the number of captured individuals n is fixed (i.e., observed, known), the SE of $\hat{N}$ is precisely the same as the SE of $\hat{n}_0$.

2.4.7 THE R PACKAGE unmarked

The R package unmarked (Fiske and Chandler, 2011) provides a comprehensive platform for analysis of many of the HMs that we cover in this book. It implements likelihood estimation based on marginal likelihood and provides support functions for data organization, summary, and graphical analysis. The package was originally developed by Ian Fiske as a graduate student at North Carolina State University and more or less taken over by Richard Chandler (now at the University of Georgia) in about 2011. The unmarked package provides a unified framework for data manipulation, data exploration, model fitting (by maximum likelihood), model selection, model averaging, GoF evaluation, prediction, and it implements ideas of bootstrapping, prediction, empirical Bayes estimation, and other inference and analysis procedures. The core HMs available in unmarked are the following:

- Single-season ("static") site-occupancy model (MacKenzie et al., 2002; Tyre et al., 2003)
- Static Royle–Nichols model (Royle and Nichols, 2003)
- Static false-positive occupancy model (Royle and Link, 2006; Miller et al., 2011)
- Static penalized occupancy models (Hutchinson et al., 2015)
- Static binomial N-mixture model (Royle, 2004b)
- Static multinomial N-mixture model (Royle, 2004a; Dorazio et al., 2005; Langtimm et al., 2011)
- Static hierarchical distance sampling model (Royle et al., 2004)
- "Open-population" versions of many of the above: dynamic occupancy model (MacKenzie et al., 2003); multinomial mixture model with temporary emigration (Chandler et al., 2011); distance sampling with temporary emigration, dynamic N-mixture model (Dail and Madsen, 2011), dynamic distance sampling (Sollmann et al., 2015).

Budding hierarchical modelers often have difficulty in figuring out what type of model they should be concerned with for their specific problem. A simple dichotomous key can help guide the user to the appropriate unmarked functionality (Figure 2.5). A specific scientific question will often be stated

2.4 CLASSICAL INFERENCE BASED ON LIKELIHOOD

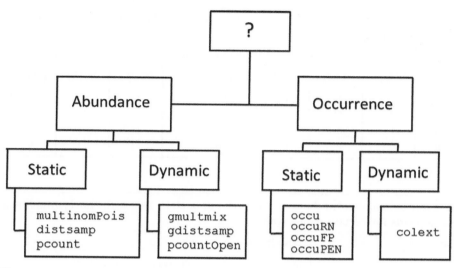

FIGURE 2.5

Decision tree for available models in R package unmarked. A scientific or management question (the question mark in the top square) first suggests a focus on either abundance or occurrence; second, the system is static or dynamic; and third, you use a particular sampling method; all of these determine the appropriate model for your study. *(Figure courtesy of R. Chandler.)*

explicitly in terms of abundance or occurrence (question 1), and whether the system is static or dynamic (question 2). Various factors interact to determine a sampling protocol/method (question 3), which then identifies one of the model fitting functions in unmarked as illustrated in Figure 2.5.

A unifying element of all HMs handled by unmarked is the underlying data structure. They all have "site × replicate" data characterized as repeated samples of sites. In addition, we have site-specific covariates such as "habitat," which only vary by site, or observation-specific covariates (such as "date" or weather conditions), which vary by site and sample occasion (dynamic models may have more complex covariates, such as "annual site-covariates," which we cover in later chapters). The typical data structure is illustrated in Table 2.1.

To use any of the unmarked model fitting functions, the data must be put in an object called an unmarkedFrame that has a typical implementation as follows:

```
umf <- unmarkedFrame(y = detectionData,
                     siteCovs = siteData,
                     obsCovs = list(wind = windData,
                                    date = dateData))
```

This special class of data frame is necessary because, in the types of HMs we are concerned with, the data structure is not usually amenable to simple matrices, vectors or lists. And, often important metadata such as distance-interval cut-points or measurement units need to be bundled with the data and the unmarkedFrame structure facilitates this. Finally, the unmarkedFrame structure makes it easy to detect errors in the data organization and allows for custom tools to summarize and visualize data. There is a specific unmarkedFrame constructor function for each fitting function, and most of them have the components shown in Table 2.2. Certain unmarkedFrames have additional components (Table 2.2).

CHAPTER 2 WHAT ARE HIERARCHICAL MODELS

Table 2.1 Typical data structure for the classes of HMs implemented in the `unmarked` package.

	Detection Data			Site Covariate	Observation Covariate		
	Visit1	Visit2	Visit3	Habitat	Date1	Date2	Date3
Site 1	1	1	1	Good	3	6	10
Site 2	0	0	0	Good	1	7	11
Site 3	1	0	0	Bad	2	9	12
Site 4	0	0	1	Bad	5	6	10

Table 2.2 Basic components of the `unmarkedFrame` data class. Below the horizontal line are optional arguments for certain model classes.

Component	Description
y	nSite x nObs matrix of observations (e.g., counts)
siteCovs	nSite x nCovs `data.frame` of site-specific covariates
obsCovs	A list with nCovs components. Each component is a nSite x nObs `data.frame` of covariate values
numPrimary	The number of primary periods (seasons, years, etc.)
yearlySiteCovs	A list with nCovs components. Each component is a nSite x nYear matrix of covariate values.
dist.breaks	A vector defining the distance intervals for a distance sampling analysis.
unitsIn	Measurement units for distance data

A summary of the different model classes, the name of the model fitting function, and the name of the `unmarkedFrame` constructor function is shown in Table 2.3.

The design and structure of `unmarked` accommodates a generic work flow that involves essentially three basic steps:

1. Import data, format it, create an `unmarkedFrame`
2. Fit some models
3. Conduct a summary analysis of the results: Assess model fit, do model selection, make predictions, map abundance, etc..

A conceptual example of this process (not with real data files, but see later chapters) goes like this:
STEP I (import data and create `unmarkedFrame`)

```
pointCountData <- read.csv("myPointCountData.csv")
siteCovariates <- read.csv("mySiteCovariate.csv")
pointCountUMF <- unmarkedFramePCount(y=pointCountData, siteCovs = siteCovariates)
```

STEP II (fit two models)

```
fm1 <- pcount(~1 ~1, data=pointCountUMF)
fm2 <- pcount(~vegHt ~vegHt, data=pointCountUMF)
```

Table 2.3 Model classes in the `unmarked` package and associated functions for fitting the model and creating the `unmarked` data frame.

Model	Fitting Function	unmarkedFrame
Closed populations		
Site-occupancy	`occu`	`unmarkedFrameOccu`
Royle–Nichols	`occuRN`	`unmarkedFrameOccu`
False-positive occupancy	`occuFP`	`unmarkedFrameOccuFP`
Penalized occupancy	`occuPEN`	`unmarkedFrameOccu`
N-mixture	`pcount`	`unmarkedFramePCount`
Removal (multinomial N-mix)	`multinomPois`	`unmarkedFrameMPois`
Capture-recapture (multinomial N-mix)	`multinomPois`	`unmarkedFrameMpois`
Hierarchical distance sampling	`distsamp`	`unmarkedFrameDS`
Open populations		
Dynamic site-occupancy	`colext`	`unmarkedMultFrame`
Dynamic N-mixture	`pcountOpen`	`unmarkedFramePCO`
Multi-scale (temporary-emigration) N-mixture	`gpcount`	`unmarkedFrameGPC`
Temporary-emigration removal	`gmultmix`	`unmarkedFrameGMM`
Temporary-emigration capture-recapture	`gmultmix`	`unmarkedFrameGMM`
Temporary-emigration, hierarchical distance sampling	`gdist`	`unmarkedFrameGDS`
Dynamic, hierarchical distance sampling	`gdistOpen`	`unmarkedFrameGDSO`

STEP III (model selection, model fit, prediction, mapping)

```
fms <- fitList(fm1, fm2)
modSel(fms)   # model selection
parboot(fm2)  # goodness of fit
predictions <- predict(fm2, type="state", newdata=mapdata)
levelplot(Predicted ~ x.coord+y.coord, data=predictions)
```

We don't dwell further on the capabilities of `unmarked` right now but instead introduce specific aspects of the package as we describe and analyze specific types of HMs in later chapters.

2.5 BAYESIAN INFERENCE

We turn now to a brief introduction to Bayesian inference. As with classical inference based on likelihood, the Bayesian inference process begins with some probability model for the observations that we use to construct the joint distribution of the data. Indeed, there is no difference at all between the joint distribution of the data analyzed by the non-Bayesian who is practicing likelihood inference and that which is analyzed by the Bayesian. The basic probability assumptions that go into creating a model for the observations are exactly the same no matter the inference paradigm being adopted.

Bayesian inference is characterized by two important features: (1) posterior inference and (2) regarding all parameters as realizations of random variables. Posterior inference—which is to say, inference based on the probability distribution of parameters conditional on the data, arises naturally by viewing parameters as being realizations of a random variable. When this is done, we can apply Bayes' rule (see next section) to combine the joint distribution of our observations with the distribution of the parameters (called the prior distribution, see Section 2.5.3), to directly compute the posterior distribution of the parameters in the model. The main conceptual distinction between Bayesian and classical inference, then, is that in Bayesian inference, we use probability directly to characterize uncertainty about parameters whereas in classical inference probability is used to characterize operating properties of statistical procedures. The Bayesian would argue (and so would we, the authors) that, just as it is natural to characterize uncertain outcomes of stochastic processes using probability, it seems natural also to characterize limited information about unknown parameters using probability.

Once this conceptual view of characterizing uncertainty about model parameters using probability is adopted, then inference about *anything* unknown reduces to a simple rote application of a very basic rule of probability, Bayes' rule, which we discuss next.

2.5.1 BAYES' RULE

As its name suggests, Bayesian analysis makes use of Bayes' rule in order to make direct probability statements about model parameters. Given two random variables z and y, Bayes' rule relates the two conditional probability distributions $[z|y]$ and $[y|z]$ by the relationship:

$$[z|y] = \frac{[y|z][z]}{[y]}. \tag{2.3}$$

Generally speaking, these component distributions are characterized as follows: $[y|z]$ is the conditional probability distribution of y given z, $[z]$ is the marginal distribution of z and $[y]$ is the marginal distribution of y. In the context of Bayesian inference we usually associate specific meanings in which $[y|z]$, the observation model, is thought of as "the likelihood," and $[z]$ as the "prior distribution." As such, we can think of this heuristically as a way of updating our prior knowledge about an unknown quantity (expressed by $[z]$) with information obtained from the data ($[y|z]$).

More generally though, Bayes' rule itself is a mathematical fact and there is no debate in the statistical community as to its validity and relevance to many problems. As an example of a simple application of Bayes' rule, we can use it to compute the probability that a site is occupied by a species, even if we have failed to observe it there, based on our canonical HM for species distributions, the occupancy model. Here z denotes species presence ($z = 1$) or absence ($z = 0$), and $\Pr(z = 1) = \psi$ is occupancy or occurrence probability. Let y be the *observed* presence ($y = 1$) or absence ($y = 0$) (or, strictly speaking, detection and nondetection), and let p be the probability that a species is detected in a single survey at a site given that it is present. Thus, $\Pr(y = 1|z = 1) = p$. The interpretation of this is that, if the species is present, we will only observe it with probability p. In addition, we assume here that $\Pr(y = 1|z = 0) = 0$. That is, the species cannot be detected if it is not present, which is a conventional view adopted in most biological sampling problems. If we survey a site J times yielding observations $y_1, \ldots, y_J$ but never detect the species, then this clearly does not imply that the species is absent ($z = 0$). Rather, our degree of belief in $z = 0$ should be made with a probabilistic statement, namely the conditional probability $\Pr(z = 1|y_1 = 0, \ldots, y_J = 0)$. If the J surveys are independent

Bernoulli trails, then the total number of detections is binomial with probability p and sample size J, and we can use Bayes' rule to compute the probability that the species is present given that it is not detected in J samples; i.e., $\Pr(z=1|y_1=0, ..., y_J=0)$. In words, the expression we seek is:

$$\Pr(\text{present}|\text{not detected}) = \frac{\Pr(\text{not detected}|\text{present})\Pr(\text{present})}{\Pr(\text{not detected})}$$

Mathematically, this is

$$\Pr(z=1|y=0) = \frac{\Pr(y=0|z=1)\Pr(z=1)}{\Pr(y=0)} = \frac{(1-p)^J \psi}{(1-p)^J \psi + (1-\psi)}.$$

The denominator here, the probability of not detecting the species, is composed of the probabilities of two distinct events: (1) not observing the species given that it is present (this occurs with probability $(1-p)^J \psi$) and (2) the species is not present (this occurs with probability $1-\psi$). To apply this result, suppose that $J=2$ surveys are done at a wetland for a species of frog, and the species is not detected there. Suppose further that $\psi = 0.8$ and $p = 0.5$ are obtained from a prior study. Then the probability that the species is present at this site, even though it was not detected, is $(1-0.5)^2 \times 0.8/((1-0.5)^2 \times 0.8 + (1-0.8)) = 0.5$.

That is, there is a 50/50 chance that the site is occupied despite the fact that the species wasn't observed there. We will see examples of this calculation in Chapter 10 for the occupancy model, and in Chapter 6 in a very similar context to estimate local abundance. In both cases, Bayes' rule forms the basis for estimating the values of the random effects, z and N.

In summary, Bayes' rule provides a simple linkage between the conditional probabilities $[y|z]$ and $[z|y]$, which is useful whenever we need to deduce one from the other.

2.5.2 PRINCIPLES OF BAYESIAN INFERENCE

Bayes' rule as a basic fact of probability is not disputed. What is controversial to some is the scope and manner in which Bayes' rule is applied by Bayesian analysts. Bayesian analysts assert that Bayes' rule is relevant, in general, to all statistical problems by regarding all unknown quantities of a model as realizations of random variables—this includes data (before they are collected), latent variables, missing values that are estimated, predictions, and of course also parameters. Classical (non-Bayesian) analysts sometimes object to regarding parameters as outcomes of random variables, preferring to think of them usually as "fixed but unknown" constants (using the terminology of classical statistics).

By regarding parameters as realizations of some random variable, Bayesians can use Bayes' rule to make direct probability statements about the unknown parameter values. To see the general relevance of Bayes' rule in this context, let y denote observations—i.e., data—and let $[y|\theta]$ be the observation model (often colloquially referred to as the "likelihood"). Suppose θ is a parameter of interest having (prior) probability distribution $[\theta]$. These are combined to obtain the posterior distribution using Bayes' rule:

$$[\theta|y] = \frac{[y|\theta][\theta]}{[y]}$$

The posterior distribution is a direct result of using Bayes' rule to combine the likelihood and the prior distribution. Once we admit the conceptual view of regarding parameters as random variables, this leads directly to the posterior distribution, a very natural quantity upon which to base inference about things we do not know—including parameters of statistical models. In particular, $[\theta|y]$ is a probability distribution for θ and we can therefore make direct probability statements to characterize uncertainty about θ.

The denominator of Bayes' rule, $[y]$, is the marginal distribution of the data y, which is the numerator with the unknowns eliminated by integration. We note without further explanation right now that, in many practical problems, this integration can be an enormous pain to compute. The main reason that the Bayesian paradigm has become so popular in the last 20 years or so is because methods have been developed for characterizing the posterior distribution that do *not* require that we possess a mathematical understanding of $[y]$, that is, methods that avoid this integration. This means that we never have to compute it or know what it looks like. Thus, while we can understand the conceptual basis of Bayesian inference merely by understanding Bayes' rule—that is really all there is to it—almost the entire rest of Bayesian analysis has to do with numerical methods of characterizing the posterior distribution in order to avoid having to deal with $[y]$. These methods are known collectively as MCMC methods, which we discuss in Section 2.6. But first we turn to a discussion of issues related to prior distributions.

2.5.3 PRIOR DISTRIBUTIONS

The prior distribution $[\theta]$ is an important feature of Bayesian inference. As a conceptual matter, the prior distribution characterizes "prior beliefs" or "prior (or external) information" about a parameter. Indeed, there is considerable disagreement over the generality and function of prior distributions. While it is true that prior distributions may be chosen to reflect subjective "beliefs," this does not often do us much good in practice and we probably want to avoid doing that as much as possible so that we might expect to obtain the same basic results as some other investigator who has the same data. More often, or rather in the vast majority of situations, the prior is chosen to express a lack of specific information, even if previous studies have been done and even if the investigator does in fact know quite a bit about a parameter, in order to "let the data speak" from the current study. Hence, a hardcore Bayesian might complain that most practicing Bayesians are mere "cryptofrequentists": they use the powerful Bayesian computing machinery (see below) to obtain the same old MLEs.

On the other hand, in some circumstances it may be useful to construct prior distributions from previous analyses of data sets conducted from comparable studies (Gopalaswamy et al., 2012). When the different studies are ordered sequentially in time (e.g., across years), the idea of using last year's posterior distribution as this year's prior distribution is eminently sensible. Indeed, in such a sequential analysis, using all data at once in a single analysis, with a flat prior distribution, will in theory lead to exactly the same posterior as when we sequentially use the posterior from the analysis of one part of the data as the prior for the next, until we have used all the data. This may allow one to break apart the analysis of a huge data set, which may perhaps not be possible using BUGS, into smaller subproblems that may be tractable.

A specific case where it might make sense to use an informative prior that is more based on subjective knowledge than previous data is when we have a parameter that is technically not identifiable (or nearly so). We may have a model that makes a good deal of scientific sense, but insufficient data to inform parameters at all levels of the model. In such cases the use of an informative prior is intermediate between estimating a parameter (the ideal case) and fixing it at some specific value, something that is quite often done in complex models, for instance Leslie-type projection matrix models. Fixing a parameter can be viewed as an informative prior with point mass at a specific value whereas using an informative prior loosens this up a little bit and formally allows for additional uncertainty in the inferences. In sense then, the use of an informative prior is like the use of a fixed value followed by a sensitivity analysis, but all at once.

Despite limited circumstances where using explicit information could be useful, we generally recommend the use of priors that are meant to reflect a lack of information. For example, for probability parameters such as p or ψ from a basic occupancy model, a natural uninformative prior is *Uniform*(0, 1) because it places equal probability on all possible values of p (or all intervals of equal width around any value). For regression coefficients and other parameters that do not have bounded support, we will sometimes use a *Uniform*($-\infty$, ∞) prior (also called a "flat" or "improper" prior) when we develop our own MCMC algorithms, but we generally do not use such priors in BUGS. Alternatively, we use what is usually called a diffuse or vague prior, which is some prior (e.g., normal) with a very large variance, or a uniform with suitably wide bounds. Technically, these contain some information, but (ideally) not enough to exert meaningful influence on the posterior.

While we tend to favor priors that express a lack of information, we note that you cannot possibly initiate a study in ecology without having pretty strong ideas of what you are going to observe based on familiarity with the species, related species, and general ecology. Thus it seems like we are not making the most efficient use of the Bayesian paradigm because we are not using this prior information. On the other hand, as noted above, incorporating subjective prior information into a formal analysis is a can of worms we think is better left unopened and that, as a matter of fact, is rarely done in ecology.

At times the situation arises where even what is thought of as being an uninformative prior can inadvertently impose substantial effect on the posterior of a parameter, and that is not desirable. In general, we need to be careful because prior distributions are not invariant to transformation of the parameter, and therefore neither are posterior distributions (see Link and Barker, 2010). Thus a prior that is meant to express a lack of information on one scale, may be very informative on another scale. For example, if we have a flat prior on logit(p) for some probability parameter p, this is very different from having a *Uniform*(0, 1) prior on p. Nonetheless, it is always possible to assess the influence of prior choice by doing a prior sensitivity analysis or by analyzing a model without any data and inspecting the induced priors for estimands other than the basic structural parameters (see Section 5.5.2). It is often the case that the influence of priors is negligible with sufficient data and in a structurally identifiable model.

2.5.4 COMPUTING POSTERIOR DISTRIBUTIONS

Once we have identified the model of interest and specified prior distributions, we need to compute the posterior distribution. In very limited cases, we can identify the posterior distribution analytically. For example, if $y = $ "number of times we detected some species at a sample site," and we assume $y \sim Binomial(J, p)$ with prior distribution $p \sim beta(a, b)$, then it can be shown that the posterior distribution of p given y is also a beta distribution:

$$[p|y] = beta(a + y, b + J - y).$$

The posterior mean is therefore $(a + y)/(a + b + J)$ (you can look this up on Wikipedia). We should recall from our basic statistics class that the MLE of a binomial success parameter is $\hat{p} = y/J$. Thus, the MLE and the posterior mean are equal if $a = b = 0$, which is in a sense a beta distribution with effective sample size 0. This is an improper prior distribution (it integrates to ∞), but despite this, the posterior distribution in this case is not unreasonable. It is worth noting also that the beta distribution with parameters $a + y$ and $b + J - y$ has mode $(a + y - 1)/(a + b + J - 2)$. If we use a uniform prior for p (i.e., with parameters $a = b = 1$) then the posterior mode is equivalent to the MLE. This is a general result: if you use a uniform prior then there is a correspondence between the posterior mode and the MLE. We will demonstrate this many times in the book, especially in Chapter 5.

In practice, we cannot usually obtain the posterior distribution analytically. This is because it is not feasible to compute the marginal probability distribution [y], the denominator resulting from application of Bayes' rule. For a long time this impeded the adoption of Bayesian methods by practitioners. The advent of Markov chain Monte Carlo (MCMC) methods has made it easier to characterize posterior distributions of parameters for any model. The power of MCMC is that it allows us to approximate the posterior using simulation without evaluating high dimensional integrals, and to directly sample from the posterior, even when the posterior distribution is not precisely known. The price can be that MCMC may be computationally expensive and most Bayesian analyses take longer to conduct than their corresponding likelihood analyses, where it is possible to do both. Although MCMC first appeared in the scientific literature in 1949 (Metropolis et al., 1953), it was not until the late 1980s that this technique gained widespread use thanks to advances in computational power and speed and important papers including Tanner and Wong (1987), Gelfand and Smith (1990) and others.

2.6 BASIC MARKOV CHAIN MONTE CARLO (MCMC)

Broadly speaking, MCMC (or Markov chain simulation) is a class of methods for drawing random samples (i.e., simulating) from the target posterior distribution (or any distribution for that matter). Thus, even though we might not recognize the posterior as a named distribution or be able to analyze its features analytically (to devise mathematical expressions for the mean and variance such as we did in the beta-binomial problem of the previous section) we can use these MCMC methods to obtain a large sample from the posterior, and then use that sample to characterize features of the posterior using Monte Carlo averages of the simulated values. For example, if we wish to obtain the posterior mean of the parameter p, and we are able to simulate a sample of T values $p^{(1)}, p^{(2)}, \ldots, p^{(T)}$ from the posterior distribution, then the Monte Carlo estimate of the posterior mean of p is the sample mean of the collection of simulated p's:

$$\tilde{E}(p|\mathbf{y}) = \frac{1}{T} \sum_t p^{(t)}.$$

In characterizing posterior quantities in this manner, we induce *Monte Carlo error* (MC error). That is, we expect a difference between the Monte Carlo average and its expected value (the feature it is meant to estimate). This is not the same as *estimation error* in the normal statistical sense because Monte Carlo error is controllable by increasing T and hence, the MC error decreases to 0 as T increases to infinity. The key useful feature of MCMC algorithms is that as T increases to infinity, the distribution of the samples at iteration t converges to the desired posterior distribution.

As a technical matter, MCMC usually produces *dependent* samples from the posterior distribution. The sequence of T values $\theta^{(1)}, \ldots, \theta^{(T)}$ is said to be a Markov chain, which is to say that $\theta^{(t)}$ depends on the preceding value in the chain, so the samples are autocorrelated. This does not have much practical effect on how we go about inference, and we still use Monte Carlo averages and summary statistics, but our effective sample size is less than the length of the MCMC run (T), which is what the effective sample size would be if we could obtain independent samples from the posterior.

The topic of MCMC is far too vast to cover in detail here, although we introduce some general-purpose methods in the following sections and then discuss, in Chapter 5, how to get the various BUGS engines to do your MCMC for you. We think BUGS will be satisfactory for almost all the problems you will need to deal with but, for the other few percent, it helps to be able to write your own MCMC algorithm or to know somebody who can do this for you (and then you have to understand that code). Also, you should always be on the lookout for new MCMC engines for Bayesian model fitting that are more efficient than BUGS.

2.6.1 METROPOLIS–HASTINGS (MH) ALGORITHM

The Metropolis–Hastings (MH) algorithm is a completely generic method for sampling from any distribution, say $[\theta]$. In our applications, $[\theta]$ will typically be the posterior distribution or full conditional distribution (defined below) of θ. The MH algorithm generates candidate values for the parameter(s) of interest from some proposal or "candidate-generating" distribution that may be conditional on the current value of the parameter. This candidate-generating distribution is denoted by $h(\theta^*|\theta^{t-1})$. Here, θ^* is the candidate or proposed value and θ^{t-1} is the value of θ at the previous time step; i.e., at iteration $t-1$ of the MCMC algorithm. The proposed value is accepted with probability

$$r = min\left(1, \frac{[\theta^*]h(\theta^{t-1}|\theta^*)}{[\theta^{t-1}]h(\theta^*|\theta^{t-1})}\right) \quad (2.4)$$

which is called the MH acceptance probability. The ratio in this expression can sometimes be >1, which is the reason for the $min(1,\cdot)$ construction. This means that if the product of likelihood and prior is greater for the proposed value of θ than for the current value, the former is accepted, while if it is smaller than the proposed value it may still be accepted, but is done so only with a probability equal to the ratio r.

It is useful to note that $h()$ can be any probability distribution, although there are good and bad choices and this candidate generator may or may not depend on the current value of the parameter θ^{t-1}. In practice, we often use a *random walk proposal* and set

$$h(\theta^*|\theta^{t-1}) = Normal(\theta^{t-1}, \delta^2)$$

for some tuning parameter δ. This just perturbs the current value of the parameter by some random noise. This particular proposal distribution is symmetric in the sense that $h(a|b) = h(b|a)$. Strictly speaking, when the proposal distribution is symmetric, we speak of a *Metropolis algorithm*, while for asymmetric proposal distributions, we get the MH algorithm (Hastings, 1970). We could as well use something like $Uniform(-B, B)$ for suitably large B as a proposal distribution; this doesn't depend on θ^{t-1}. This is allowable, although we would expect it to produce poor candidate values and therefore low acceptance probabilities and inefficient algorithms.

In the context of using the M/MH algorithms to do MCMC (in which case the target distribution is a posterior distribution), an important fact is, no matter the choice of $h()$, we can compute the MH acceptance probability directly because the marginal distribution of y cancels from both the numerator and denominator of r:

$$r = \frac{[\mathbf{y}|\theta^*]h(\theta^{t-1}|\theta^*)/[\mathbf{y}]}{[\mathbf{y}|\theta^{t-1}]h(\theta^*|\theta^{t-1})/[\mathbf{y}]} = \frac{[\mathbf{y}|\theta^*]h(\theta^{t-1}|\theta^*)}{[\mathbf{y}|\theta^{t-1}]h(\theta^*|\theta^{t-1})}$$

This is the magic of the MH algorithm: it obviates the need to evaluate the integrations required to compute the denominator of Bayes' rule. Thus, to use the MH algorithm we only ever have to evaluate known distributions, those being the likelihood and prior distributions. In addition, we have to be able to simulate from some proposal distribution which can be essentially anything, but in practice is often a simple normal.

In practice, any particular MCMC algorithm needs to run for some number of iterations before the Markov chain "converges" to the target posterior distribution. Samples from this early, transitional period, called the "burn-in" or "warm-up," are discarded and not used in characterizing posterior summaries. This issue is discussed further in Section 2.6.6.

2.6.2 ILLUSTRATION: USING MH FOR A BINOMIAL MODEL

We illustrate the application of the MH algorithm using a simple problem in which we have two observations from a binomial distribution with constant parameter p, and we wish to compute the posterior distribution of p. We will assume a *beta(a, b)* prior for p. For comparison, we will also compute the MLE of p and produce a plot of the likelihood and compare the two. To get things started we simulate some data and define a couple functions that will be needed for these two analyses. In particular, we need the joint distribution of the two observations and we need a function that defines the posterior distribution to within a constant of proportionality (i.e., *not* including the marginal distribution of the data). All of this goes as follows:

```
# Simulate data
set.seed(2016)
y <- rbinom(2, size=10, p = 0.5)

# Define the joint distribution (= likelihood) which we will maximize
jointdis <- function(data, J, p){
    prod(dbinom(data, size=J, p=p))
}

# Posterior is proportional to likelihood times prior
posterior <- function(p, data, J, a, b){
    prod(dbinom(data, size=J, p=p)) * dbeta(p, a, b)
}
```

We require several things in order to set up our MCMC algorithm. First we need a candidate generator or proposal distribution $h()$. For this we decide to use a completely naive and inefficient choice of choosing candidates according to a *Uniform*(0, 1) distribution (we will also consider an alternative shortly). Secondly, we need to pick a starting value for the parameter p. We can pick any value in the interval [0,1] for this and the algorithm will work effectively. For the analysis below we set $p = 0.2$. Then we repeatedly simulate from the candidate generator, compute the MH acceptance probability, and then decide whether to keep or reject the proposed value. Operationally, this final decision is done by flipping a coin with a probability that, in R, is the same as generating a random uniform deviate, say u, and then accepting the value if $u < r$. We execute these various steps for 100,000 iterations and save the current value of p at each iteration:

```
# Do 100,000 MCMC iterations using Metropolis algorithm
# Assume vague prior which is beta(1,1) = Unif(0,1)
mcmc.iters <- 100000
out <- rep(NA, mcmc.iters)
```

2.6 BASIC MARKOV CHAIN MONTE CARLO (MCMC)

```
# Starting value
p <- 0.2

# Begin the MCMC loop
for(i in 1:mcmc.iters){

    # Use a uniform candidate generator (not efficient)
    p.cand <- runif(1, 0, 1)

    # Alternative: random walk proposal
    # p.cand <- rnorm(1, p, 0.05)  # Need to reject if > 1 or < 0
    # if(p.cand < 0 | p.cand > 1 )  next

    r <- posterior(p=p.cand, y, J=10, a=1, b=1) / posterior(p=p, y, J=10, a=1, b=1)

    # Generate a uniform r.v. and compare with "r", this imposes the
        correct probability of acceptance
    if(runif(1) < r)
      p <- p.cand

    # Save the current value of p
    out[i] <- p
}
```

The vector `out` contains 100,000 posterior samples of the parameter p. We can use these samples to characterize features of the posterior distribution of p, such as the mean, standard deviation and any percentiles we care to by summarizing the Monte Carlo samples:

```
mean(out)
[1] 0.363855

sd(out)
[1] 0.1004638

quantile(out, c(0.025, 0.975))
    2.5%      97.5%
0.1817478 0.5708719
```

We see that the posterior mean of p is about 0.36, the standard deviation is about 0.10 and 95% of the posterior samples lie between 0.18 and 0.57.

We next evaluate the likelihood at a grid of 200 values of p and plot the likelihood and produce a density plot of the posterior distribution on the same graphics panel (Figure 2.6).

```
# Evaluate likelihood for a grid of values of p
p.grid <- seq(0.1, 0.9, , 200)
likelihood <- rep(NA, 200)

for(i in 1:200){
    likelihood[i] <- jointdis(y, J=10, p=p.grid[i])
}
```

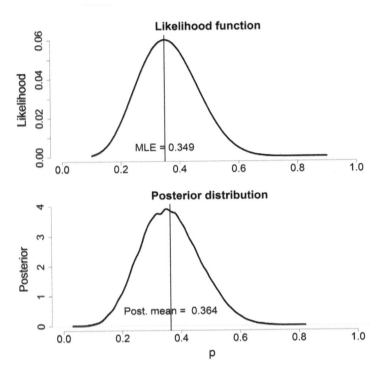

FIGURE 2.6

Likelihood function and posterior distribution for parameter *p* estimated from a sample of two independent binomial random numbers when vague priors are used in the Bayesian analysis.

```
par(mfrow=c(2,1), mar = c(5,5,3,2))
plot(p.grid, likelihood, xlab="", ylab="Likelihood", xlim=c(0,1), ty = "l", main =
"Likelihood function")
p.hat <- p.grid[likelihood == max(likelihood)]
abline(v = p.hat)
text(p.hat, 0.005, paste("MLE = ", round(p.hat, 3), sep= ""))

plot(density(out), xlim=c(0,1), main = "Posterior distribution", xlab = "p",
ylab = "Posterior")
p.mean <- mean(out)
abline(v = p.mean)
text(p.mean, 0.5, paste("Post. mean = ", round(p.mean, 3),sep=" "))
```

The MLE is $\hat{p} = 0.349$, which is slightly less than the posterior mean; this is typically the case when the posterior distribution is slightly right-skewed. In this particular instance we expect the mode of the posterior distribution to be precisely equal to the MLE except to within MC error.

2.6.3 METROPOLIS ALGORITHM FOR MULTIPARAMETER MODELS

Normally we have more than one parameter in our model and so we need to devise an MCMC algorithm for handling multiple parameters. How do we do this? In a nutshell, we repeat the single

2.6 BASIC MARKOV CHAIN MONTE CARLO (MCMC)

parameter Metropolis algorithm sequentially for each parameter, but holding each of the other parameters constant. We will demonstrate this approach here using our logistic regression example from Section 2.4.2.

Following our two observation binomial example, we need to make a function that is the unnormalized posterior; i.e., the *numerator* in Bayes rule (likelihood times prior). This time we will define this on the log scale, which will often be necessary for models with many parameters in order to avoid numerical over- or underflow problems. For the logistic regression model we make use of our negLogLike function that we used previously for MLE, and we assume *Normal*$(0, \sigma = 10)$ priors for both parameters. The log-posterior function is, to within an additive constant (that being the marginal distribution of the data):

```
log.posterior <- function(beta0, beta1, z, vegHt){
# Note: "z" and "vegHt" must be input
    loglike <- -1 * negLogLike(c(beta0, beta1), z, vegHt)
    logprior <- dnorm(c(beta0, beta1), 0, 10, log=TRUE)
    return(loglike + logprior[1] + logprior[2])
}
```

Note that, in addition to negLogLike from our previous analyses, we also simulated the data z and vegHt that should already exist in your R workspace (Section 2.4.2).

To get the MCMC algorithm running we need some initial values, which we will generate from a normal distribution, and we need candidate generators ($h()$) that, for each parameter, we will take to be normal distributions with mean being the current value and standard deviation $\delta = 0.3$; i.e.,

$$\beta_0^* \sim Normal(\beta_0, \delta)$$

Note that this normal distribution is *not* part of the model, but rather is a part of the MCMC algorithm which we may modify to improve algorithm performance.

Now we are ready to implement the algorithm, which we do in the following block of code executed for 50,000 iterations:

```
niter <- 50000
out <- matrix(NA, niter, 2, dimnames = list(NULL, c("beta0", "beta1")))

# Initialize parameters
beta0 <- rnorm(1)
beta1 <- rnorm(1)

# Current value of the log(posterior)
logpost.curr <- log.posterior(beta0, beta1, z, vegHt)

# Run MCMC algorithm
for(i in 1:niter){
    if(i %% 1000 == 0)                          # report progress
        cat("iter", i, "\n")
        # Update intercept (beta0)
        # Propose candidate values of beta
```

```
# If the proposal was not symmetric, would be Metrop-*Hastings*
beta0.cand <- rnorm(1, beta0, 0.3)   # 0.3 is tuning parameter
# Evaluate the log(posterior)
logpost.cand <- log.posterior(beta0.cand, beta1, z, vegHt)
# Compute Metropolis acceptance probability, r
r <- exp(logpost.cand - logpost.curr)
# Keep candidate if it meets criterion (u < r)
if(runif(1) < r){
   beta0 <- beta0.cand
   logpost.curr <- logpost.cand
}

# Update slope (beta1)
beta1.cand <- rnorm(1, beta1, 0.3)   # 0.3 is tuning parameter
# Evaluate the log(posterior)
logpost.cand <- log.posterior(beta0, beta1.cand, z, vegHt)

# Compute Metropolis acceptance probability
r <- exp(logpost.cand - logpost.curr)
# Keep candidate if it meets criterion (u < r)
if(runif(1) < r){
   beta1 <- beta1.cand
   logpost.curr <- logpost.cand
     }
   out[i,] <- c(beta0, beta1) # Save samples for iteration
 }

# Plot
layout(rbind(c(1,1),
          c(2,2),
          c(3,4)), respect = T) # <- play with these settings

par(oma=c(0,0,0,0),mar=c(5,4,1,1))
plot(out[,1], type="l", xlab="Iteration", ylab="beta0")
plot(out[,2], type="l", xlab="Iteration", ylab="beta1")
plot(density(out[,1]), xlab="beta0", main="")
plot(density(out[,2]), xlab="beta1", main="")
```

The trace plot, or MCMC history, for 50,000 iterations is shown in Figure 2.7 (top two panels). We can produce density plots from these 50,000 posterior samples using the density function, and those are shown in Figure 2.7 (bottom panel).

2.6.4 WHY WE NEED TO USE MCMC FOR LOGISTIC REGRESSION

We will elaborate a little bit on why we do the MCMC analysis of the logistic regression model shown in the previous section instead of just cooking up a formula for the posterior distribution. The logistic regression problem seems so basic and simple that one might think we should easily be able to

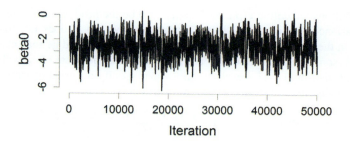

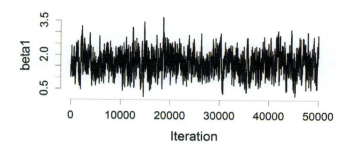

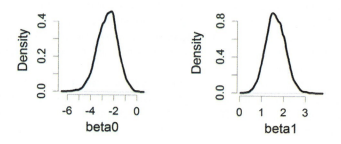

FIGURE 2.7

Output from running an MH algorithm for 50,000 iterations for a logistic regression model: trace plots for the intercept, β_0 (top panel), slope, β_1 (middle panel), and marginal posterior densities (bottom two panels).

compute the posterior distribution directly for this simple two-parameter model! Right? Well, let us see if we can do it!

To carry out a Bayesian analysis, we have to identify the elements needed for Bayes' rule—the "likelihood," the prior distribution for the parameters, and the marginal distribution of the data (the denominator of Bayes' rule). The likelihood is the joint distribution of the observations:

$$f(y_1, y_2, \ldots, y_n | \beta_0, \beta_1) = \left\{ \prod_{i=1}^{n} Bernoulli(\mathbf{y}_i | \boldsymbol{\beta}) \right\}$$

For prior distributions we define $g(\beta_0, \beta_1) \equiv [\beta_0, \beta_1]$ to be the product of two normal distributions each with some variance (we used 100 in the example above but it does not matter here). If our covariate x (vegHt in the example) is standardized then this prior should be reasonably flat where the likelihood has mass (sometimes called locally uniform prior) and therefore effectively noninformative. Identifying the joint distribution of the observations and specifying prior distributions allows us to identify the posterior distribution, using Bayes' rule, which is:

$$\pi(\beta_0, \beta_1 | y_1, y_2, \ldots, y_n) = \frac{f(y_1, y_2, \ldots, y_n | \beta_0, \beta_1) g(\beta_0, \beta_1)}{f(y_1, y_2, \ldots, y_n)}. \tag{2.5}$$

While we can easily obtain this conceptual representation of the posterior distribution using Bayes' rule, it will usually *not* be possible to recognize it as a standard named distribution. Therefore, computing posterior summaries such as the mean, or percentiles for a confidence interval, seems to be an elusive problem. The reason for this analytic intractability stems essentially from difficulty in computing the denominator of the Bayes' rule application. In the case of the posterior distribution expressed by Eq. (2.5), the denominator requires that we integrate the numerator over the bivariate prior distribution for β_0 and β_1 which is to say, we have to compute this two-dimensional integral:

$$[\mathbf{y}] = f(y_1, y_2, \ldots, y_n) = \int_{\beta_0} \int_{\beta_1} f(y_1, y_2, \ldots, y_n | \beta_0, \beta_1) g(\beta_0, \beta_1) d\beta_0 d\beta_1.$$

It may be that this can be done analytically (more likely we could do it numerically), and that some further analysis of Eq. (2.5) could be done by some mathematician to produce a simple result that allows us to obtain a formula for the posterior mean, variance, or other summaries. However, for the average analyst, including for the authors, we simply do not know enough math to be able to solve this problem and, importantly, we do not really want to make a big investment in figuring it out because, in the grand scheme of things, we have other more interesting things to do, including about 20 other analyses to finish just in this week alone.

While we cannot do this math (and maybe it cannot even be done, we do not know), we *can* do the MCMC that avoids having to compute this marginal distribution, which we showed in the previous section. In a sense then, MCMC makes Bayesian analysis accessible to the masses who may not be able or interested in solving fun math problems all day.

2.6.5 GIBBS SAMPLING

When we applied MH to multiparameter models we did this by sampling each parameter in succession but holding all other parameters constant, at their current value of the simulation. In our bracket notation, and using the logistic regression as an example, we applied the MH algorithm to sample from the distributions $[\beta_0|\beta_1,\mathbf{y}]$ and $[\beta_1|\beta_0,\mathbf{y}]$. These distributions are called the conditional posterior or "full conditional" distributions because they are the distributions of each parameter conditional on all other variables of the model. Such iterative sampling from full conditional distributions is often called Gibbs sampling (Gelfand et al., 1990). More typically, what we might call "pure" Gibbs sampling is when we can identify the form of each of these full conditional distributions as named distributions and sample from them directly, not using the MH algorithm that we used above. In the cases where we do use the MH algorithm then the resulting algorithm is usually called a

2.6 BASIC MARKOV CHAIN MONTE CARLO (MCMC)

Metropolis-within-Gibbs sampling algorithm. In general, any posterior distribution is completely specified by the set of full conditional distributions, one for *each* unknown quantity of the model (parameters and latent variables).

In most HMs we often have a case where one or more of the full conditionals can be identified as a named distribution, but others cannot. This is because the conditional structure of the model may allow some simplifications due to conditional independencies that are a feature of the model structure. We illustrate this using the occupancy model that we introduced in Section 2.3.1.1. The simple model in which p does not depend on time has the following two elements:

$$y_i \sim Binomial(J, pz_i); \quad i = 1, 2, \ldots, M$$
$$z_i \sim Bernoulli(\psi); \quad i = 1, 2, \ldots, M$$

and the model assumes each y is independent of each other and independent of z_j for $j \neq i$. (Note: We never really say these last two things, but unless a dependency is specified explicitly then we assume independence of variables in a model. This is the only sensible way to interpret written expressions of models.) The Gibbs sampler for this model will have $M + 2$ full conditional distributions: one for each variable z_i and one for each parameter p and ψ. We do not provide all of these here but we make the point that sometimes the full conditional distribution of a parameter has a simplified form by noting that basic probability arguments can be used to show that the full conditional of ψ is the following beta distribution:

$$[\psi|\mathbf{z}] = beta\left(1 + \sum z, 1 + M - \sum z\right)$$

Therefore we can draw samples of ψ directly from the full conditional distribution instead of using a Metropolis algorithm that, while it is very general and avoids having to specify the full conditional distributions, rejects some proposed values of the parameter and hence leads to some inefficiency in characterizing the posterior distribution. The various BUGS engines use some combination of Gibbs sampling from full conditionals, the M/MH algorithm and other more sophisticated MCMC methods.

2.6.6 CONVERGENCE AND MIXING OF MARKOV CHAINS

Once we have carried out an analysis by MCMC, there are many other practical issues that we have to confront. One characteristic of MCMC sampling is that Markov chains take some time to converge to their stationary distribution—in our case the posterior distribution for some parameter given data, $[\theta|y]$. Only when the Markov chain has reached its stationary distribution can the generated samples be used to characterize the posterior distribution. Thus, an important issue we need to address is "have the chains converged?"

Since we do not know what the stationary posterior distribution of our Markov chain should look like, we effectively have no means to assess whether or not it has truly converged to this desired distribution. Most MCMC algorithms only guarantee that, eventually, the samples being generated will be from the target posterior distribution, but no one can tell us how long this will take. There are several things we can do to increase the degree of confidence we have about the convergence of our Markov chains. Some problems are easily detected using simple plots, such as a time-series plot, where parameter values of each MCMC iteration are plotted against the number of iterations. We showed such a trace plot in Figure 2.7.

Typically a period of transience is observed in the early part of the MCMC algorithm, and this is usually discarded as the "burn-in" period (sometimes also called "warm-up"). A quick visual diagnostic to whether convergence has been achieved is that your Markov chains look "grassy" (see Figure 2.7). Another way to check convergence is to update the parameters some more and see if the posterior changes. If the chains have converged to the posterior, the posterior mean, confidence intervals, and other summaries should be relatively static as we continue to run the algorithm. Yet another option is to run several Markov chains and to start them off at different initial values that are overdispersed relative to the posterior distribution. In BUGS we do this with the n.chains option. Such multiple chains help to explore different areas of the parameter space simultaneously; if, after a while, all chains oscillate around the same average value, chances are good that they indeed converged to the posterior distribution. Gelman and Rubin (1992) came up with the so-called "R-hat" statistic ($\hat{R}$) or Brooks–Gelman–Rubin statistic that essentially compares within-chain and between-chain variance to check for convergence of multiple chains (Gelman et al., 2014). The R-hat statistic should be close to 1 if the Markov chains have converged and sufficient posterior samples have been obtained. We demonstrate these ideas with BUGS in later chapters of the book (in particular, see Chapter 5).

2.6.6.1 Slow Mixing and Thinning of Markov Chains

Some models exhibit "poor mixing" of Markov chains, in which case the samples might well be from the posterior (i.e., the Markov chains have converged to the proper stationary distribution) but simply mix or move around the posterior parameter space slowly. Poor mixing can happen for many reasons—when parameters are highly correlated (even confounded) or barely identified from the data, algorithms are inefficient, and probably for other reasons as well.

Slow mixing equates to high autocorrelation in the Markov chain—the successive draws are highly correlated, and thus we need to run the MCMC algorithm much longer to get an effective sample size that is sufficient for estimation, or to reduce the MC error (see below) to a tolerable level. A strategy often used to reduce autocorrelation is "thinning," where only every m^{th} value of the Markov chain output is retained for purposes of summarizing features of the posterior distribution. However, thinning is necessarily inefficient from the stand point of inference—you can always get more precise posterior estimates by using all of the MCMC output regardless of the level of autocorrelation (MacEachern and Berliner, 1994; Link and Eaton, 2012). Practical considerations might necessitate thinning, even though it is statistically inefficient. For example, in models with many parameters or other unknowns, the output files might be enormous and unwieldy to work with. In such cases, thinning is perfectly reasonable. In many cases, how well the Markov chains mix is strongly influenced by parameterization, standardization of covariates, and the prior distributions being used. Some things work better than others, and in Bayesian analyses as in life, the investigator should experiment with different settings and remain calm when things do not work out perfectly on the first attempt.

2.6.6.2 Effective Sample Size and MC Error

The subsequent samples generated from a Markov chain are not independent samples from the posterior distribution, due to the correlation among samples introduced by the Markov process. Thus, sample size has to be adjusted to account for the autocorrelation in subsequent samples (Chapter 8 in Robert and Casella, 2010). This adjusted sample size is referred to as the effective sample size.

Checking the degree of autocorrelation in your Markov chains and estimating the effective sample size your chain has generated should be part of evaluating your model output. This is part of the default output from running the `bugs` function (see Chapter 5) and can be done using the function `effectiveSize` in the `coda` package. If you find that your supposedly long Markov chain has only generated a very short effective sample, you should consider a longer run. The effective sample size determines the accuracy with which you can estimate features of the posterior (such as a 95% credible interval) from your sample.

Another diagnostic useful for assessing the effectiveness of your MCMC analysis is the time-series or MC error—the "noise" introduced into your samples by the stochastic MCMC process. The MC error is printed by default in summaries produced in the WinBUGS GUI. You want the MC error to be smallish relative to the magnitude of the parameter and what smallish means will depend on the purpose of the analysis. For a preliminary analysis you might settle for a few percent whereas for a final analysis then certainly less than 1% is called for. You can run your MCMC algorithm as long as it takes to achieve that (when you have time to wait). A consequence of the MC error is that even for the exact same model, results will usually be slightly different. Thus, as a good rule of thumb, you should avoid reporting MCMC results to more than two or three significant digits!

2.6.7 BAYESIAN ANALYSIS OF HMs

As we discussed in Section 2.4.4, likelihood analysis of HMs is based on the integrated or marginal likelihood, in which we remove the random effects by integration (or summation) from the likelihood that is specified conditional on the random effects. To do a Bayesian analysis of HMs, we apply MCMC directly to the model specified conditional on the random effects. This has the benefit that estimates of the random effects (usually called predictions) are produced directly as part of the MCMC estimation scheme.

For many HMs that we deal with we find that we can sample directly from full conditional distributions for some of the parameters or random effects but perhaps not others. Thus, normally, the MCMC algorithm we devise is a type of hybrid Metropolis-within-Gibbs algorithm. We demonstrate here using the basic occupancy model that we have been simulating from and analyzing in various ways.

The occupancy model: We apply MCMC to the problem of sampling from the joint distribution of our random variables, which now includes a set of random effects, z and therefore we have to include one additional level of full conditional distributions in the MCMC analysis. To sample from the posterior distribution we need to identify the joint distribution of the observations, latent variables and parameters, which has a slightly more complex form than before:

$$f(\mathbf{y}, \mathbf{z}, p, \psi) = \left\{ \prod_{i=1}^{n} \prod_{j=1}^{J} [y_{ij}|z_i, p] [z_i|\psi] \right\} [\psi]$$

(Note the latent variables z in here.) However, we still analyze the model by applying MCMC methods in order to obtain a Monte Carlo characterization of posterior summaries. Because of the additional model component for the latent variables, we have an extra full conditional distribution:

$$[z_i|\mathbf{y}, \mathbf{z}_{-i}, p, \psi]$$

for each i. Thus, in total, there are exactly $M+2$ unknown quantities: the parameters p, and ψ, and M latent variables z_i. Therefore, we need to come up with the $M+2$ full conditional distributions.

The conditional distribution of z_i given all other unknown parameters in the model, and the data, can be deduced by figuring out $Pr(z_i = 1|y_i, p, \psi)$. Two things need to be said: First, to the right of the vertical line the only data is y_i because the conditional independence assumptions will cause other terms to disappear. Second, we know that if $y_i = 1$ then it must be that $z_i = 1$ and so we only need to figure out this probability for the case $y_i = 0$. In fact, we can compute this probability by a direct application of Bayes' rule:

$$Pr(z_i = 1|y_i = 0, p, \psi) = \frac{Pr(y_i = 0|z_i = 1, p, \psi)Pr(z_i = 1)}{Pr(y_i = 0|z_i = 1)Pr(z_i = 1) + Pr(y_i = 0|z_i = 0)Pr(z_i = 0)}$$

We play around with this for a while and deduce that $z_i|y_i = 0$ is Bernoulli with probability

$$\frac{(1-p)^J \psi}{(1-p)^J \psi + (1-\psi)}.$$

This defines the full conditional for each of the M values of z_i. It remains to define the full conditional for ψ and p. Symbolically, the full conditional for the parameter p has this form:

$$[p|\mathbf{y}, \mathbf{z}, \psi] = \left\{ \prod_i \prod_j [y_{ij}|z_i, p] \right\} [p]$$

This is conditional on each value of z and logically we know that if $z = 0$ then $y = 0$ with probability 1 and so there is no information about p in such cases. Therefore, the full conditional of p is only informed by values of y for which $z = 1$. In fact under a beta prior distribution for p we can show with a little math that

$$[p|\mathbf{y}, \mathbf{z}, \psi] = beta\left(1 + \sum y, 1 + \left(\sum z\right) \times J - \sum y\right).$$

If we did not have covariates in our model that influence occupancy probability then, as we noted in Section 2.5.5, basic arguments can be made that the full conditional of ψ is the following beta distribution:

$$[\psi|\mathbf{z}] = beta\left(1 + \sum z, 1 + M - \sum z\right)$$

However, in keeping with our example in which covariate vegHt affects occupancy probability (on the logit scale) we have to update parameters β_0 and β_1. The update of these parameters is done as in Section 2.5.3 for the logistic regression parameters but the "data" for the MH update in this case are the current values of z_i.

Thus, the posterior distribution for this occupancy model is fully characterized by this set of conditional distributions that we iteratively sample from using MCMC algorithms such as the Metropolis-within-Gibbs algorithm. The slightly expanded MCMC algorithm reads as follows (with the output summarized in Figure 2.8:

```
# Simulate the data set
set.seed(2014)
M <- 100                    # number of sites
vegHt <- runif(M, 1, 3)     # uniform from 1 to 3
```

2.6 BASIC MARKOV CHAIN MONTE CARLO (MCMC)

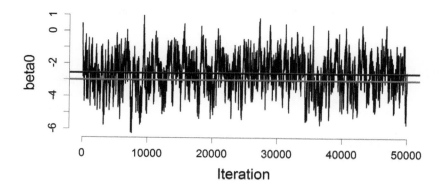

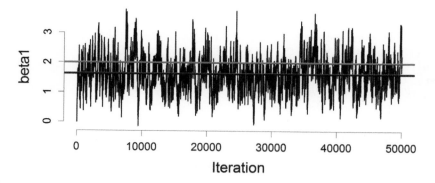

FIGURE 2.8

Output from the MCMC algorithm for site occupancy model using simulated data. Data-generating values (red) are $\beta_0 = -3$ and $\beta_1 = 2$ and posterior means are shown as a blue line. Note the "grassy look" (from the perspective of a vole in a field) indicating suitable mixing of the Markov chains.

```
psi <- plogis(-3 + 2*vegHt)    # occupancy probability
z <- rbinom(M, 1, psi)         # realised presence/absence
p <- 0.6                       # detection probability
J <- 3                         # sample each site 3 times
y <-rbinom(M, J, p*z)          # observed detection frequency

# Number of MCMC iterations to to
niter <- 50000

# Matrix to hold the simulated values
out <- matrix(NA, niter, 3, dimnames = list(NULL, c("beta0", "beta1", "p")))

# Initialize parameters, likelihood, and priors
starting.values <- c(beta0=0, beta1=0)
beta0 <- starting.values[1]
beta1 <- starting.values[2]
z <-ifelse(y>0, 1, 0)
p <- 0.2
```

```r
# NOTE: using logistic reg. likelihood function here (defined previously)
loglike <- -1*negLogLike(c(beta0, beta1), z, vegHt)
logprior <- dnorm(c(beta0, beta1), 0, 10, log=TRUE)

# Run MCMC algorithm
for(i in 1:niter) {
   if(i %% 1000 ==0 )   # report progress
   cat("iter", i, "\n")

   # PART 1 of algorithm -- same as before
   # Update intercept (beta0)
   # propose candidate values of beta
   beta0.cand <- rnorm(1, beta0, 0.3) # 0.3 is tuning parameter
   # evaluate likelihood and priors for candidates
   loglike.cand <- -1*negLogLike(c(beta0.cand, beta1), z, vegHt)
   logprior.cand <- dnorm(beta0.cand, 0, 10, log=TRUE)
   # Compute Metropolis acceptance probability (r)
   r <- exp((loglike.cand+logprior.cand) - (loglike + logprior[1]))
   # Keep candidate if it meets the criterion
   if(runif(1) < r){
      beta0 <- beta0.cand
      loglike <- loglike.cand
      logprior[1] <- logprior.cand
   }
   # Update slope (beta1)
   beta1.cand <- rnorm(1, beta1, 0.3) # 0.3 is tuning parameter
   # evaluate likelihood and priors for candidates
   loglike.cand <- -1*negLogLike(c(beta0,beta1.cand), z, vegHt)
   logprior.cand <- dnorm(beta1.cand, 0, 10, log=TRUE)
   # Compute Metropolis acceptance probability r
   r <- exp((loglike.cand+logprior.cand) - (loglike + logprior[2]))
   # Keep the candidates if they meet the criterion
   if(runif(1) < r) {
      beta1 <- beta1.cand
      loglike <- loglike.cand
      logprior[2] <- logprior.cand
   }

   # Part 2 of the algorithm
   # update z. Note we only need to update z if y=0.
   # The full conditional has known form

   psi <- plogis(beta0 + beta1 * vegHt)
   psi.cond <- dbinom(0,J,p) * psi /(dbinom(0, J, p) * psi + (1-psi))
   z[y==0] <- rbinom(sum(y==0), 1, psi.cond[y==0])
   loglike <- -1 * negLogLike(c(beta0, beta1), z, vegHt)
```

```
  # Part 3: update p
  ## The commented code will update p using Metropolis
  ## loglike.p <- sum(log(dbinom(y[z==1],J,p)))
  ## p.cand <- runif(1, 0, 1)
  ## loglike.p.cand <- sum(log(dbinom(y[z==1], J, p.cand)))
  ## if(runif(1) < exp(loglike.p.cand-loglike.p))
  ##    p <- p.cand
  ## This bit draws p directly from its full conditional
    p <- rbeta(1, 1+ sum(y), sum(z)*J +1 - sum(y) )

  # Save MCMC samples
    out[i,] <- c(beta0,beta1,p)
}

# Plot bivariate representation of joint posterior
pairs(out)

# Trace/history plots for each parameter (Fig. 2.8)
op <- par(mfrow=c(2,1))
plot(out[,1], type="l", xlab="Iteration", ylab="beta0")
abline(h = mean(out[,1]), col = "blue", lwd = 2)
abline(h = -3, col = "red", lwd = 2)
plot(out[,2], type="l", xlab="Iteration", ylab="beta1")
abline(h = mean(out[,2]), col = "blue", lwd = 2)
abline(h = 2, col = "red", lwd = 2)
```

2.7 MODEL SELECTION AND AVERAGING

Much of what we talk about in this book has to do with fitting models; i.e., obtaining estimates of parameters for specific models and sampling situations. However, we hardly ever have only a single model that we care to estimate the parameters of. Therefore an important problem that needs to be addressed after we have chosen a class of models and figured out how to fit them is that of choosing among several or many different models within the class, or combining the estimates from several different models ("model averaging"). The idea of model averaging is to average estimates of parameters or predictions over all models under consideration instead of relying on any particular model for making inferences. In this section we briefly address the topics of model selection and averaging under both Bayesian and classical inference paradigms.

2.7.1 MODEL SELECTION BY AIC

Using classical analysis based on likelihood, model selection is easily accomplished using AIC (Burnham and Anderson, 2002). The AIC maximizes "short-term predictive success"; i.e., it can be expected to select models that would best predict a similar data set as the one at hand, similar to leave-one-out cross-validation (Stone, 1977). The AIC of a model is computed as twice the negative log-likelihood evaluated at the MLE, penalized by the number of parameters (np) in the model:

$$\text{AIC} = -2\log L\left(\widehat{\boldsymbol{\beta}}|\mathbf{y}\right) + 2np$$

Models with small values of AIC are preferred. It is common to use AIC with an adjustment for small sample sizes, referred to as AIC_c, which is

$$\text{AIC}_c = -2\log L\left(\widehat{\boldsymbol{\beta}}|\mathbf{y}\right) + \frac{2np(np+1)}{n-np-1}$$

where n is the sample size. In applications that involve model selection based on AIC, it is typical to order models by AIC and produce AIC weights, which are exponentiated *differences* between the AIC value (or AIC_c value) for a model and that of the best model. That is, if AIC_m is the AIC value for model m and AIC_{min} is that of the best model, then define $\Delta_m = AIC_m - AIC_{min}$, and the model weight is

$$w_m = \frac{\exp(-(1/2)\Delta_m)}{\sum_m \exp(-(1/2)\Delta_m)}$$

(Burnham and Anderson, 2002, p. 124). The AIC weights are interpreted as model probabilities and are used to produce model-averaged values of parameters. The R package `AICmodavg` (Mazerolle, 2015) does model-averaging of parameters or predictions, for many of the HMs in the R package `unmarked` and other models based on AIC. As we use `unmarked` throughout the book, we will demonstrate application of AIC many times; e.g., see Section 6.9, 7.5.2 and elsewhere.

A big advantage of AIC for model selection is that it is automatic whenever we can compute the marginal likelihood, and it produces weights that can be used directly for model-averaging predictions or parameters that have a consistent interpretation across models. However, one potential problem for applying AIC to HMs, even in some cases when the marginal likelihood can be computed, is that it is not always clear what the effective sample size n should be in the calculation of AIC_c: i.e., is it the number of sites (most relevant for selecting structure for the occupancy part of the model) or the total number of samples (sites × replicates; most relevant for the detection part of the model).

2.7.2 MODEL SELECTION BY DEVIANCE INFORMATION CRITERION

Spiegelhalter et al. (2002) devised a Bayesian metric for model selection, analogous to AIC (in its application), called the deviance information criterion (DIC), which is a function of model deviance and a measure of effective number of parameters. Model deviance is defined as negative twice the log-likelihood; i.e., for a given model with parameters θ: $\text{Dev}(\theta) = -2\log L(\theta|\mathbf{y})$. The DIC is defined as the posterior mean of the deviance, $\overline{\text{Dev}}(\theta)$, plus a measure of model complexity, p_D:

$$\text{DIC} = \overline{\text{Dev}}(\theta) + p_D$$

The standard definition of p_D is

$$p_D = \overline{\text{Dev}}(\theta) - \text{Dev}(\overline{\theta})$$

where the second term is the deviance evaluated at the posterior mean of the model parameter(s), $\bar{\theta}$. The p_D here is interpreted as the effective number of parameters in the model. Gelman et al. (2003) suggest a different version of p_D based on one-half the posterior variance of the deviance:

$$P_V = \text{Var}(\text{Dev}(\theta)|\mathbf{y})/2.$$

This is what is produced from WinBUGS and JAGS if they are run using the R packages `R2WinBUGS` or `R2jags` or `jagsUI`, respectively (see Chapter 5). DIC summaries can also be obtained from package `rjags`. The DIC is widely used although its use has been called into question by several authors. Millar (2009) noted that, for HMs, DIC based on the conditional likelihood was invalid (using a case study of zero-inflated count models with overdispersion). Lunn et al. (2013) also noted problems with DIC related to the manner in which p_D is calculated, highlighting two specific issues:

> "1. p_D is not invariant to reparameterization, in the sense that if the model is rewritten into an equivalent form but in terms of a function $g(\theta)$, then a different p_D may arise since $p(y|\bar{\theta})$ will generally not be equal to $p(y|\overline{g(\theta)})$. This can lead to misleading values in some circumstances and even to negative values of p_D.
>
> 2. An inability to calculate p_D when θ contains a categorical parameter, since the posterior mean is not then meaningful. This renders the measure inapplicable to mixture models [....]."

The latent variables z in an occupancy model, or N in an N-mixture model, are exactly such categorical parameters and thus Royle et al. (2014) were perhaps a bit optimistic in their assessment of DIC when they stated:

> We think DIC is probably reasonable for certain classes of models that contain only fixed effects, or for which the latent variable structure is the same across models so that only the fixed effects are varied. However, it would be useful to see some calibration of DIC for some standardized model selection problems.

This does not seem generally true. For example, in capture–recapture models for estimating population size based on data augmentation, different models can suggest different population sizes, and we have observed that this has a strong influence on the model deviance and, in some cases, clearly superior models have much higher posterior deviance. A standard situation is that in which we compare an ordinary spatial capture–recapture model *without* a behavioral response to a model *with* a behavioral response. Even though the behavioral response is clearly important (is large in magnitude), the simpler model will often be favored by the posterior deviance and DIC because it has much lower N and therefore fewer encounter history contributions to the deviance. The use of data augmentation in capture–recapture models is a type of categorical covariate and so this seems consistent with Lunn et al.'s point 2 above.

2.7.3 BAYESIAN MODEL AVERAGING WITH INDICATOR VARIABLES

A convenient way to deal with model selection and averaging problems in Bayesian analysis by MCMC is to use the method of indicator variables (Kuo and Mallick, 1998; see Section 3.4.3 in Royle and Dorazio, 2008; Ntzoufras, 2009; O'Hara and Sillanpää, 2009; Link and Barker, 2010). Using this approach, we expand the model to include a set of prescribed models as specific reductions of a larger model. To implement the Kuo and Mallick approach, we expand the model to include the latent indicator variables, say w_m, for variable m in the model, such that $w_m = 1$ if the linear predictor contains covariate m, and $w_m = 0$ otherwise. We assume that the indicator variables w_m are mutually independent with

$$w_m \sim \text{Bernoulli}(0.5)$$

for each variable $m = 1, 2, \ldots$, in the model. For example, for our simulated example with vegetation height, which only has one covariate, the expanded model for occupancy probability takes this form:

$$\text{logit}(\psi_i) = \beta_0 + \beta_1 w_1 x_i.$$

The posterior probability of the event $w_1 = 1$ is a gauge of the importance of the variable x; i.e., high values of $\Pr(w_1 = 1)$ indicate stronger evidence to support that "x is in the model," whereas values of $\Pr(w_1 = 1)$ close to 0 suggest that x is less important.

In general, using this indicator variable formulation of the model selection problem we can characterize unique models by the sequence of w variables. For just one covariate as above, there are only two models, defined by $w_1 = 1$ or $w_1 = 0$. Considering a more general case; e.g., with three covariates, then each unique sequence (w_1, w_2, w_3) represents a model, and we can tabulate the posterior frequencies of each model by postprocessing the MCMC histories of (w_1, w_2, w_3). This method then produces posterior probabilities for each of the 2^3 models.

This method of indicator variables to do model selection is especially useful for producing model-averaged predictions of latent variables because the MCMC output for a variable represents a posterior sample from all possible models in the set defined by combinations of the w variables. On the other hand, when computing posterior summaries of parameters for specific models, one has to be sure that the MCMC output is subsetted according to the values of the indicator variables w, otherwise the output is a mixture of $w = 1$ and $w = 0$ values, which most of the time is nonsense. See Section 7.6 for detailed applications of the indicator variable approach.

One broader, technical consideration is that posterior model probabilities are well known to be sensitive to priors on parameters (Aitkin, 1991; Link and Barker, 2006). What might normally be viewed as vague priors are not usually innocuous or uninformative when evaluating posterior model probabilities. One solution is to compute posterior model probabilities under a model in which the prior for parameters is fixed at the posterior distribution under the full model (Aitkin, 1991). At a minimum, one should evaluate the sensitivity of posterior model probabilities to different prior specifications (Tenan et al., 2014a).

2.8 ASSESSMENT OF MODEL FIT

Once we decide on a model (or models) and come up with a method of fitting it, it is important to ask the question: Does my model fit the observed data? We define "fit" as follows: Do our data resemble, in some precisely defined manner, realizations from the model? We call the activity of investigating this question the assessment of model fit, or model checking, but a more conventional term is "goodness-of-fit" testing.

Conceptually, we can think of evaluation of model fit as follows: if we simulate data under the model in question, are simulated realizations consistent with ("similar to") the data set that we actually have? Thus in order to formalize the problem of assessing model fit, we need to articulate the manner in which we judge "consistency" in this sense. For either Bayesian or classical inference, the basic strategy to assessing model fit is to come up with a fit statistic that depends on the parameters and the data set, which we denote by $T(\mathbf{y}, \theta)$, and then we compute this for the observed data set, and compare its value with that computed for "perfect" data sets simulated from the correct model.

In the case of classical inference, we will almost always use parametric bootstrapping, in which we simulate data sets using the MLE of the model parameters and then fit the model to each simulated data set. For each simulated data set and fit we compute the value of a fit statistic and compare the value of the fit statistic for observed data with the distribution of that computed from the simulated data sets (this is called the "bootstrap distribution"). The R package unmarked (Fiske and Chandler, 2011) contains generic bootstrapping methods. In Bayesian analysis, we adopt the Bayesian p-value approach that has a similar feel to the bootstrap, in the sense that we compare the values of a fit statistic for simulated data sets with that computed for the data set at hand.

In either case (classical or Bayesian) we have to come up with one or more fit statistics to use as the basis for the fit assessment. Choice of fit statistic is model and problem-specific and may address only specific features of the model's performance (Gelman et al., 1996). More importantly, any particular fit statistic is unlikely to have good power under all reasonable alternatives and so, in practice, calibration studies may be important, and are needed, but have hardly ever been done for the types of HMs we discuss in this book. At the present time, fit assessment is as much art or intuition as science and we believe it is important and should be more actively investigated.

Perhaps more important than the question of whether a model "fits" is that of *where* does it fit and where does it not. For this, we need to look at patterns in some form of *residual*; i.e., the difference between the observed and the expected data. For HMs, whether in a Bayesian or a non-Bayesian analysis, residuals are not defined in an unambiguous way, but we feel that more can and should be done in this respect. We give some examples in later chapters.

2.8.1 PARAMETRIC BOOTSTRAPPING EXAMPLE

We provide an illustration of using a parametric bootstrapping fit assessment to our occupancy model. We will do this the hard way, by writing our own R code for the problem, although it can also be done easily using the parboot function in unmarked, which we will demonstrate in subsequent chapters as we introduce unmarked capabilities.

The first step in developing the fit assessment is to create a function that simulates data for a given set of parameters. The way this function is set up it always generates the same covariate values but generates a different stochastic response (y):

```
sim.data <- function(beta0 = -3, beta1 = 2, p = 0.6, x=NULL){
# Function allows input of covariate "x", or simulates new

M <- 100
if(is.null(x))
   vegHt <- runif(M, 1, 3) # uniform from 1 to 3
```

```
# Suppose that occupancy probability increases with vegHt
# The relationship is described (default) by an intercept of -3 and
#    a slope parameter of 2 on the logit scale
# plogis is the inverse-logit (constrains us back to the [0-1] scale)
psi <- plogis(beta0 + beta1*vegHt)

# Now we simulated true presence/absence for 100 sites
z <- rbinom(M, 1, psi)

# Now generate observations
J <- 3  # sample each site 3 times
y <- rbinom(M,J,p*z)

list(y=y, J=J, vegHt=vegHt)
}
```

Our function `negLogLikeocc` was defined earlier, but we reproduce it here to maintain the work flow:

```
# This is the negative log-likelihood based on the marginal distribution
# of y. It is the pmf of a zero-inflated binomial random variable.
#
negLogLikeocc <- function(beta, y, x, J) {
    beta0 <- beta[1]
    beta1 <- beta[2]
    p <- plogis(beta[3])
    psi <- plogis(beta0 + beta1*x)
    marg.likelihood <- dbinom(y, J, p) * psi + ifelse(y==0, 1, 0) * (1-psi)
    return(-sum(log(marg.likelihood)))
}
```

The previous analysis of the simulated data is as follows:

```
data <- sim.data()          # Generate a data set

# Let's minimize the negative log-likelihood
starting.values <- c(beta0=0, beta1=0, logitp=0)
opt.out <- optim(starting.values, negLogLikeocc, y=data$y, x=data$vegHt,J=data$J,
hessian=TRUE)
(mles <- opt.out$par)

# Make a table with estimates, SEs, and 95% CI
mle.table <- data.frame(Est=mles,
                        SE = sqrt(diag(solve(opt.out$hessian))))
mle.table$lower <- mle.table$Est - 1.96*mle.table$SE
mle.table$upper <- mle.table$Est + 1.96*mle.table$SE
mle.table
```

To implement the parametric bootstrap GoF analysis, we define a fit statistic as an R function and evaluate that for our data and MLEs. Then, we use the MLEs just obtained to simulate data 100 times and, for each simulated data set, compute the same fit statistic.

2.8 ASSESSMENT OF MODEL FIT

```
# Define a fit statistic
fitstat <- function(y, Ey){
   sum((sqrt(y) - sqrt(Ey)))
}
# Compute it for the observed data
T.obs <- fitstat(y, J*plogis(mles[1] + mles[2]*vegHt)*plogis(mles[3]))

# Get bootstrap distribution of fit statistic
T.boot <- rep(NA, 100)
for(i in 1:100){
   # Simulate a new data set and extract the elements. Note we use
   # the previously simulated "vegHt" covariate
   data <- sim.data(beta0=mles[1],beta1=mles[2],p=plogis(mles[3]),x=vegHt)
   # Next we fit the model
   starting.values <- c(0,0,0)
   opt.out <- optim(starting.values, negLogLikeocc, y=data$y, x= data$vegHt, J=data$J,
hessian=TRUE)
   (parms <- opt.out$par)
   # Obtain the fit statistic
   T.boot[i]<- fitstat(y, J*plogis(parms[1] + parms[2]*vegHt)*plogis(parms[3]) )
}
```

What does it all mean? Let us look at the observed value of the statistic and then summary statistics of the bootstrap distribution:

```
(T.obs)
[1] -22.67474

summary(T.boot)
   Min. 1st Qu. Median    Mean 3rd Qu.    Max.
 -28.15  -23.27  -21.91  -22.12  -20.85  -17.61
```

We see that our observed statistic T is somewhere between the mean and the first quartile of the bootstrap distribution and thus the observed data set seems consistent with data sets that were simulated under the model. We conclude that the model provides an adequate fit to the data.

2.8.2 BAYESIAN P-VALUE

To evaluate GoF in Bayesian analyses, we will typically use the Bayesian *p*-value as a summary of a posterior predictive check (Gelman et al., 1996). The basic idea is to define a fit statistic (or "discrepancy measure") and compare the posterior distribution of that statistic with the posterior distribution of that statistic for hypothetical perfect data sets for which the model is known to be correct. For example, with count frequency data, a standard measure of fit is based on the *Pearson residuals*:

$$D(y_i, \theta) = \frac{(y_i - \mathrm{E}(y_i))}{\sqrt{\mathrm{Var}(y_i)}}.$$

The fit statistic based on the squared residuals computed from the observations is

$$T(\mathbf{y}, \theta) = \sum_i D(y_i, \theta)^2$$

which can be computed at each iteration of a MCMC algorithm given the current values of parameters that determine the response distribution. At the same time (i.e., at each MCMC iteration), the equivalent statistic is computed for a "new" data set, say $\mathbf{y}^{new}$, simulated using the current parameter values. From the new data set, we compute the fit statistic:

$$T(\mathbf{y}^{new}, \theta) = \sum_i D(y_i^{new}, \theta)^2$$

and the Bayesian p-value is simply the posterior probability $\Pr(T(\mathbf{y}^{new}) > T(\mathbf{y}))$, which should be close to 0.5 for a good model—one that fits in the sense that the observed data set is consistent with realizations simulated under the model being fitted to the observed data. In practice we judge "close to 0.5" as being "not too close to 0 or 1" which admittedly is somewhat subjective. Another useful fit statistic is the Freeman–Tukey statistic, in which

$$D(\mathbf{y}, \theta) = \sum_i \left(\sqrt{y_i} - \sqrt{E(y_i)}\right)^2$$

(Brooks et al., 2000), where y_i is the observed value i and $E(y_i)$ its expected value. In contrast to a Chi-square discrepancy, the Freeman–Tukey statistic obviates the need to pool cells with small expected values and moreover is insensitive to unstable results due to small expected cell frequencies.

In summary, you can see that the Bayesian p-value is easy to compute, and it is widely used as a result. We will later provide several applications of assessing the fit of HMs using Bayesian p-values and parametric bootstrapping, including in Sections 6.7, 7.6 and elsewhere. As with the assessment of model fit using parametric bootstrapping, issues of sensitivity and power of the specific fit statistic and the need for calibration should be considered.

2.9 SUMMARY AND OUTLOOK

HMs represent a series of submodels linked by their conditional dependence structure. There are at least two main reasons for why hierarchical modeling is a sensible thing to do: first, the description of a complex stochastic system is often made much easier by representing it as a series of linked subsystems. This also naturally accommodates variability and uncertainty at all levels of the full system and therefore allows an honest assessment of uncertainty. However, perhaps an even more important reason for adopting HMs than simple practicality is a conceptual, philosophical one: we are convinced that thinking about a system in terms of linked subsystems naturally leads one to a more mechanistic way of statistical model building. HMs naturally enforce a focus on plausible mechanisms underlying the observed data. In this sense, HMs foster the adoption of a "scientific" approach to statistical modeling (Berliner, 1996) exemplified in this book by the clear segregation in the HMs into components describing the latent system and its dynamics and another part describing observation of the system.

There is a sense in which HMs are just glorified GLMs—or rather, a compounding of one GLM with another, or perhaps more than two separate GLMs. We saw that clearly with our two "canonical HMs"—the occupancy model for species distribution, and the N-mixture model for species abundance. The occupancy model is the compounding of a logistic regression for observed occupancy state with another logistic regression for true occupancy state, and the N-mixture model is a compounding of a logistic regression for observed population size with a Poisson regression for true population size. We will look at many more examples throughout this book including some HMs with three or four levels (our canonical HMs had only two levels each). Owing to the great importance of GLMs for hierarchical modeling, we will spend the entire next chapter on an applied summary of GLMs.

We find that both classical and Bayesian inference paradigms are useful for analysis of HMs. We think you should understand and be able to apply both paradigms because sometimes one has benefits compared with the other. For example, classical inference based on integrated likelihood is sometimes not practically feasible (e.g., for the multispecies occupancy model, Chapter 11). On the other hand, Bayesian analysis using MCMC is completely general and always works, at least in theory. Although, in such cases, MCMC may be very time consuming and lead to Markov chains that mix slowly. On the other hand, analysis of HMs by integrated likelihood provides a convenient and widely accepted framework for model selection based on AIC.

For many chapters of the book we will introduce a class of models (occupancy models, distance sampling, etc..) and then discuss and illustrate likelihood analysis of models using the `unmarked` package whenever we can, stressing consistent work flow and ease of doing standard things like prediction and model selection. We will then do a Bayesian analysis of the class of models using the BUGS language, and then an illustration of a type of model that cannot be done (easily, or in `unmarked`) using likelihood methods. We hope this gives you a sense of the importance of having a pluralistic outlook concerning inference paradigms.

EXERCISES

1. You should be able to apply Bayes' rule to the peregrine falcon example earlier in this chapter to compute the distribution of $X|Y$. That is: how many fledged young are there, given that we have detected birds on $0, 1, 2, \ldots, J$ visits?
2. For the bootstrap GoF analysis done on the occupancy model, we found that the model appears to fit the data well. Try fitting the wrong model; i.e., without the vegHt covariate, and see if the model *fails* the GoF test.
3. Where we talked about prior distributions and sensitivity we said "if we have a flat prior on logit(p) for some probability parameter p, this is very different from having a *Uniform*(0,1) prior on p." Evaluate this by simulating data for different priors on logit(p) and then back-transforming the simulated values to see what the *implied* prior is for p. Find a normal prior for logit(p) that is approximately uniform on [0,1] for p.
4. Use a Metropolis or MH algorithm to simulate *Normal*(0,1) random variables if all you have access to is a uniform random number generator. Of course you also know the mathematical form for the normal pdf but you do not have a normal random number generator at hand. Verify that your simulated data have the required normal distribution by making a histogram and computing summary statistics.

5. In Section 2.5.2 for a single parameter problem we did the MCMC to obtain the posterior of p and we compared it with the MLE. They appeared slightly different but, because the prior for p is uniform, the posterior mode should be exactly equal (in a mathematical sense) to the MLE. In practice, MC error will make this not so. Repeat the MCMC analysis for many more iterations and see how close the empirical mode of the posterior gets to the MLE. Note that computing the mode is not so easy because there is not a `mode` function in R. So do this in one of two ways: round the output to two or three decimal places and use the `table` function to compute the frequencies of each unique (rounded) value, and the mode is then the most frequent value. Alternatively, use the `density` function and take the value of x that has the highest value of y in the output.

CHAPTER 3

LINEAR MODELS, GENERALIZED LINEAR MODELS (GLMs), AND RANDOM EFFECTS MODELS: THE COMPONENTS OF HIERARCHICAL MODELS

CHAPTER OUTLINE

- 3.1 Introduction 79
- 3.2 Linear Models 83
  - 3.2.1 Linear Models with Main Effects of One Factor and One Continuous Covariate 84
  - 3.2.2 Linear Models with Interaction between One Factor and One Continuous Covariate 92
  - 3.2.3 Linear Models with Two Factors 96
  - 3.2.4 Linear Models with Two Continuous Covariates and Including Polynomials 99
- 3.3 Generalized Linear Models (GLMs) 102
  - 3.3.1 Poisson GLM for Unbounded Counts 104
  - 3.3.2 Offsets in the Poisson GLM 105
  - 3.3.3 Overdispersion and Underdispersion 107
  - 3.3.4 Zero-Inflation 108
  - 3.3.5 Bernoulli GLM: Logistic Regression for a Binary Response 109
  - 3.3.6 Modeling a Poisson Process from Presence/Absence Data Using a Bernoulli GLM with cloglog Link 110
  - 3.3.7 Binomial GLM: Logistic Regression for Bounded Counts 112
  - 3.3.8 The GLM as the Quintessential Statistical Model 113
- 3.4 Random Effects (Mixed) Models 114
  - 3.4.1 Random Effects for a Normal Data Distribution: Normal-Normal GLMM 117
  - 3.4.2 Random Effects for a Poisson Data Distribution: Poisson-Normal GLMM 120
- 3.5 Summary and Outlook 121
- Exercises 122

3.1 INTRODUCTION

In Chapter 2, we introduced hierarchical models (HMs) as consisting of a linked sequence of probability models for observed and unobserved random variables. We saw that the former are observed data, and the latter latent or partially latent variables or random effects such as the occurrence or abundance state of a local population, but that a random effect may also be some abstract "group effect." Typically, we model the structure in each random variable using link functions and linear functions of covariates—i.e., as simple *weighted sums of covariate values* (where coefficients represent the weights), exactly as in generalized linear models (GLMs; McCullagh and Nelder, 1989). Therefore, an HM can be described as a compound GLM—a sequence of two or more related GLMs. In this book, we use the R package unmarked (Fiske and Chandler, 2011) and BUGS software (Lunn et al., 2000; Plummer, 2003) for almost all fitting of HMs. Much of the power of HMs derives from the flexible ways in which we specify linear models at each level in the HM—e.g., for occupancy and for detection in an occupancy model. Thus, it is essential that you understand very well the specification of a wide range of linear models, in order to be an effective hierarchical modeler.

The main focus of this chapter is the linear model. We show how different linear models can be described in algebra and in the model definition language in R—e.g., using the functions lm and glm, which we assume you are familiar with. When fitting HMs with unmarked, you will specify linear models in exactly the same way regardless of the type of model. We also show how to specify the same linear models in the BUGS programming language. In so doing, we get a little ahead of ourselves because we don't formally introduce BUGS until Chapter 5. However, seeing the algebraic model description and the R and BUGS model descriptions alongside is illuminating, although you may perhaps fully understand the BUGS model specification only after you have read Chapter 5. Ecologists can find an excellent introduction to linear models in Chapter 6 of Evan Cooch's gentle introduction to program MARK (www.phidot.org/software/mark/docs/book).

We also provide a brief and applied introduction of two further crucial concepts in applied statistics that are almost always found in HMs: GLMs and random effects. Strictly speaking, random effects are nothing new in HMs: they are simply the outcome of an *unobserved* random variable, as opposed to the data that are the outcome of an *observed* random variable. The presence of an unobserved random variable is one defining feature of an HM or a mixed model. However, in our experience, ecologists often find random-effects or mixed models confusing, and hence we attempt to shed some light on this class of models here. We focus on GLMs and random effects mostly as they are relevant for your understanding of the types of HMs in this book. Our coverage of both topics is more summary than, for instance, in Chapters 3 and 4 of Kéry and Schaub (2012) or in textbooks such as McCullagh and Nelder (1989), Dobson and Barnett (2008) or Ntzoufras (2009) for GLMs, or in books like Pinheiro and Bates (2000), McCulloch and Searle (2001), Lee et al. (2006), Gelman and Hill (2007) or Littell et al. (2008), and in Bolker et al. (2009) for random-effects/mixed models.

Most ecologists learn statistical modeling best by example; hence, we use for illustration a toy data set and imagine that we had measured wingspan and body length of nine blue-eyed hooktail dragonflies (*Onychogomphus uncatus*; Figure 3.1) in three populations in the Spanish Pyrenees (Navarra, Aragon, and Catalonia). For each individual we also assessed sex, color intensity (proportion of body that is yellow as opposed to black), ectoparasite load (number of mites counted), and whether each of the four wings (two hind, two front) was damaged. Part of this toy data set and analysis is taken from Chapters 3 and 4 in Kéry and Schaub (2012).

In Section 3.2, we illustrate linear models by investigating the relationships between wingspan (a numerical metric response) and four explanatory variables: population, sex, body length, and color intensity. The first two are factors with three (population: 1 = Navarra, 2 = Aragon, 3 = Catalonia) and two (sex: 1 = male; 2 = female) levels, respectively. Factors are categorical explanatory variables in which numbers have no quantitative meaning; they are merely labels or group names. In contrast,

FIGURE 3.1

Male blue-eyed hooktail (*Onychogomphus uncatus*), Aragon/Spain, 2013. *(Photo: M. Kéry.)*

body length and color intensity are continuous explanatory variables, or covariates, where numbers do have a quantitative meaning. For instance, we can compute a difference in body length between two dragonflies, but we cannot compute a difference between their sexes. In this chapter, we will distinguish continuous and categorical explanatory variables by use of the terms "*covariates*" and "*factors*." We show how their pairwise effects are combined in linear models in an additive or interactive fashion. Of course, in your modeling, you will typically have more than just two explanatory variables, but the principles are best explained in the simplest possible setting.

In Section 3.3, we will use parasite load and wing damage as response variables to illustrate GLMs. Finally, in Section 3.4, we revisit some models from Sections 3.2 and 3.3 and turn them into random-effects, hierarchical, or generalized linear mixed models (GLMMs).

```
# Define data
pop <- factor(c(rep("Navarra", 3), rep("Aragon", 3), rep("Catalonia", 3)), levels
  = c("Navarra", "Aragon", "Catalonia"))                    # Population
wing <- c(10.5, 10.6, 11.0, 12.1, 11.7, 13.5, 11.4, 13.0, 12.9) # Wingspan
```

CHAPTER 3 LINEAR MODELS, GENERALIZED LINEAR MODELS (GLMs)

```
body <- c(6.8, 8.3, 9.2, 6.9, 7.7, 8.9, 6.9, 8.2, 9.2)    # Body length
sex <- factor(c("M","F","M","F","M","F","M","F","M"), levels = c("M", "F"))
mites <- c(0, 3, 2, 1, 0, 7, 0, 9, 6)                    # Number of ectoparasites
color <- c(0.45, 0.47, 0.54, 0.42, 0.54, 0.46, 0.49, 0.42, 0.57) # Color intensity
damage <- c(0,2,0,0,4,2,1,0,1)                           # Number of wings damaged

cbind(pop, sex, wing, body, mites, color, damage)         # Print out data set
     pop sex wing body mites color damage
[1,]  1   1   1  10.5  6.8    0   0.45    0
[2,]  1   2   2  10.6  8.3    3   0.47    2
[3,]  1   1   1  11.0  9.2    2   0.54    0
[4,]  2   2   2  12.1  6.9    1   0.42    0
[5,]  2   1   1  11.7  7.7    0   0.54    4
[6,]  2   2   2  13.5  8.9    7   0.46    2
[7,]  3   1   1  11.4  6.9    0   0.49    1
[8,]  3   2   2  13.0  8.2    9   0.42    0
[9,]  3   1   1  12.9  9.2    6   0.57    1
```

Internally in R, factor levels for pop and sex are expressed as integers ranging from 1 to the number of levels, but we can clearly see that R interprets these numbers as factor levels (i.e., mere names):

```
str(pop)
  Factor w/ 3 levels "Navarra","Aragon",...: 1 1 1 2 2 2 3 3 3
```

We plot the data for wingspan, parasite load, and damage (Figure 3.2).

```
par(mfrow = c(1, 3), cex = 1.2)
colorM <- c("red", "red", "blue", "green", "green")   # Pop color code males
colorF <- c("red", "blue", "blue", "green", "green")  # Pop color code females
```

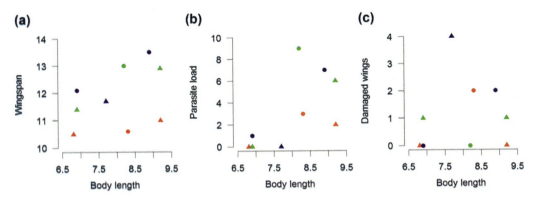

FIGURE 3.2

Relationships between wingspan (a), parasite load (b), and number of damaged wings (out of four; c), respectively, and body length, sex, and population in the dragonfly toy data set. Colors code for the three populations (Navarra—red, Aragon—green, and Catalonia—blue); circles denote females and triangles males.

```
plot(body[sex == "M"], wing[sex == "M"], col =colorM, xlim = c(6.5, 9.5), ylim = c(10, 14),
lwd = 2, frame.plot = FALSE, las = 1, pch = 17, xlab = "Body length", ylab = "Wingspan")
points(body[sex == "F"], wing[sex == "F"], col =colorF, pch = 16)
text(6.8, 13.8, "A", cex = 1.5)
plot(body[sex == "M"], mites[sex == "M"], col = colorM, xlim = c(6.5, 9.5), ylim = c(0, 10),
lwd = 2, frame.plot = FALSE, las = 1, pch = 17, xlab = "Body length", ylab = "Parasite load")
points(body[sex == "F"], mites[sex == "F"], col = colorF, pch = 16)
text(6.8, 9.7, "B", cex = 1.5)
plot(body[sex == "M"], damage[sex == "M"], col = colorM, xlim = c(6.5, 9.5), ylim = c(0, 4),
lwd = 2, frame.plot = FALSE, las = 1, pch = 17, xlab = "Body length", ylab = "Damaged wings")
points(body[sex == "F"], damage[sex == "F"], col = colorF, pch = 16)
text(6.8, 3.9, "C", cex = 1.5)
```

3.2 LINEAR MODELS

In a linear model, the effects of covariates or factors act on a response in a purely additive fashion—i.e., they are expressed as simple weighted sums of the values of the covariates, with the weights being the coefficients or parameters of the linear model. The model is said to be linear in the parameters, but it does not necessarily represent a straight line when shown as a graph. For instance, models with polynomial terms, such as quadratic or cubic, do not translate into a straight-line graph, but they are linear models. A linear model can be described in a variety of ways:

1. in words; e.g., when we say something like "population and body length act additively on wingspan,"
2. using specific names or labels for least-squares techniques that *imply* a certain linear model; e.g., when we say "t-test," "linear regression," "ANOVA," or "ANCOVA" (Mead, 1988),
3. in graphs; e.g., using line and bar plots,
4. in algebra; e.g., $y_i \sim Normal(\alpha_{j(i)} + \beta * x_i, \sigma^2)$,
5. using matrix algebra; e.g., $\mathbf{y} = \mathbf{X}\beta + \varepsilon$,
6. as a system of equations,
7. in the R model definition language; e.g., `lm(wing ~ pop + length)`, and finally
8. using the BUGS model definition language;

    e.g., `y[i] ~ dnorm(mu[i], tau)`

    `mu[i] <- alpha[pop[i]] + beta * length[i]`.

In statistical modeling, most of us have become accustomed to defining certain types of linear models using point-and-click techniques available in many statistical packages, or a slick model definition language such as the one in the R functions `lm` and `glm`, or in all functions in the package `unmarked`. Model definition languages make your life easier, because they allow you to define a large range of linear models quickly and error-free. The big drawback is that it is easy to fit linear models without actually understanding them or knowing what their parameters mean. In addition, you may not be able to fit certain nonstandard linear models, or you may be able to do so but only in a very complicated way. In contrast, when fitting models in BUGS, no such shortcuts to defining linear models are available. Hence, you must know exactly what kind of linear model you want to fit and in what parameterization (see below). This is one of the main stumbling blocks for beginners in BUGS. On the other hand, it is a great advantage, because it forces upon you a much clearer understanding of what these linear models mean.

To illustrate, we will look at how to describe a certain linear model in all the different ways listed above, and then extend that example to summarize some typical linear models that can be fitted for covariates and factors. We will also see how to combine factors and continuous covariates in additive and interactive ways (see also Chapter 6 in Kéry, 2010). We focus on the algebraic and R language descriptions of a linear model, because you are likely to know and use R already, and this will be the way in which you specify linear models within HMs in unmarked. In addition, we show the syntax for fitting the same linear models in the BUGS language.

3.2.1 LINEAR MODELS WITH MAIN EFFECTS OF ONE FACTOR AND ONE CONTINUOUS COVARIATE

We use our dragonfly example to illustrate the description and fitting of a linear model where pop and length act in an additive fashion on wingspan, namely according to the linear model underlying a least-squares technique known as a "main-effects analysis of covariance (ANCOVA)." In Section 3.2.2, we extend the main-effects model to an interaction-effects ANCOVA model. We can fit the main-effects model in R by issuing the following commands:

```
summary(fm1 <- lm(wing ~ pop + body))

Coefficients:
             Estimate Std. Error t value Pr(>|t|)
(Intercept)    6.5296     1.6437   3.973  0.01061 *
popAragon      1.8706     0.4521   4.138  0.00902 **
popCatalonia   1.7333     0.4489   3.861  0.01187 *
body           0.5149     0.1991   2.586  0.04908 *
[ ... ]
Residual standard error: 0.5498 on 5 degrees of freedom
Multiple R-squared: 0.8416,  Adjusted R-squared: 0.7465
F-statistic: 8.854 on 3 and 5 DF,  p-value: 0.01916
```

How many parameters does our model have? Where are the estimates of these parameters presented in the R output? What do these parameters mean? To understand this, let's look at the description of this model in algebra. We can write this model as follows:

$$wing_i = \mu + \alpha_j + \beta * body_i + \varepsilon_i$$

$$\varepsilon_i \sim Normal(0, \sigma^2)$$

Here, $wing_i$ is the response of unit i (= wingspan of dragonfly i, or a single data point), and $body_i$ is the value of covariate body length for dragonfly i. The intercept μ is a constant shared by the response of all nine dragonflies. Vector α_j has three elements corresponding to the number of levels, or populations, in the factor pop, and contains the *effects* of each population. (We could write $\alpha_{j(i)}$ to emphasize that population membership j is a function of the individual i.) For population 1 (Navarra), we have α_1; for population 2 (Aragon), α_2; and for population 3 (Catalonia), α_3. Parameter β is the expected change in wingspan for a one-unit change in body length. The unexplained part in the wingspan of dragonfly i, which is not accounted for by population membership or body length of dragonfly i, is the residual ε_i. The residuals for all nine dragonflies are assumed to be independent draws from the same normal distribution with constant variance. The residual variance (σ^2) is also an estimated model parameter. In the R output (under `Residual standard error`), we see the square root of that variance estimate (σ).

Hence, our model has four parameters (α_j and β) to describe the expected response and one parameter (σ) for the variability of the response around that mean. Importantly, when fitting linear models, statistical packages like R internally do not differentiate between the two types of explanatory variables, covariates and factors, although factors must be declared as such. All linear models are internally represented as a multiple linear regression, where the effects of a factor are broken apart into those of two or more *indicator or dummy variables* that contain 0s and 1s only. In our case for `pop`, we have three dummy variables, one for each level (population) of the `pop` factor, to code for the presence or absence of an effect of that population in the expected response of each dragonfly.

However, the model above is overparameterized or parameter-redundant (Mead, 1988; Ntzoufras, 2009): we try to estimate one parameter too many. Specifically, we cannot estimate both the intercept μ and a full parameter vector $\boldsymbol{\alpha} = \{\alpha_1, \alpha_2, \alpha_3\}$ of length 3. One of the population effects α_j has to be set to zero to make the model identifiable. Customarily in R, the parameter corresponding to the first factor level is set to zero—i.e., the constraint $\alpha_1 = 0$ is imposed. As a consequence, the intercept μ becomes the intercept of the regression line of dragonflies in population 1; this is the expected wingspan of a (hypothetical) dragonfly of length 0 in Navarra. The remaining two elements of the parameter vector $\boldsymbol{\alpha}$, which are not fixed but estimated, then become the differences between the intercepts of the regression lines in populations 2 and 3 relative to the intercept in population 1 (this is why seemingly only coefficient estimates for Aragon and Catalonia are shown in the R output above). In this model parameterization, population 1 (Navarra) serves as a baseline or reference, and the two parameters estimated for the population factor represent comparisons of the other populations with that one. This parameterization of the effects of the factor `pop` may therefore be called the *treatment contrast* or *effects parameterization*.

How do we write this model in the BUGS language? We will see in much more detail later (starting in Chapter 5) that BUGS is not a vectorized language: we cannot specify a linear model simply as y ~ x as in R. Rather, we have to specify the stochastic relationship between y and x for each element/observation in the two vectors, by looping over them from the first until the last element. Elements of vectors are indexed by square brackets, and hence our BUGS code will be this (this is *not* executable as is; and the text after the hash mark # is an explanatory comment, not part of the model specification):

```
for(i in 1:9){                                  # Loop over each individual
   wing[i] ~ dnorm(mean[i], tau)                # Stochastic relationship
   mean[i] <- mu + alpha[pop[i]] + beta * body[i]  # Deterministic relationship
}
alpha[1] <- 0                                   # Fix effect of 1st level at 0
```

This says that the wingspan of dragonfly i is a draw from a normal distribution with an expected value (or mean) called mean[i] and precision tau. In BUGS, the dispersion parameter of the normal distribution is not the variance or the standard deviation, but the precision (the reciprocal of the variance). The expected wingspan of dragonfly i, mean[i], is the sum of an intercept mu, a population-specific effect alpha[pop[i]], and an effect beta of body length. Thus, inside of the loop we write exactly what we might write in algebra to describe that model (see also below for different algebraic ways of writing the same model, especially variant 2).

We make the following three remarks:

1. In a statistical model, the observed data are assumed to be the outcome of a stochastic process (see Chapter 2). In line with this, observed data in BUGS must always be stochastic quantities—i.e., stand on the left side of the twiddle or tilde symbol, indicating a stochastic relationship. Residuals are only defined implicitly, and although correct in algebra, it would *not* be correct in BUGS to write the following:

```
for(i in 1:9){
   wing[i] <- mu + alpha[pop[i]] + beta * body[i] + eps[i]
   eps[i] ~ dnorm(0, tau)
   # In BUGS this is WRONG !
}
```

The reason is that you can't simply calculate an observed quantity such as wing[i]; it has to be stochastic—i.e. a draw from some distribution (though see the function dsum in the JAGS dialect of the BUGS language).

2. To specify effects of factor levels, or more generally parameters that are grouped in a vector or higher-dimensional array, we use *nested indexing* in the BUGS language: alpha[pop[i]], where pop is the indexing variable. Nested indexing can be best understood when read from the inside out: for instance, dragonfly $i = 1$ has (internally in R) a value of pop[1] = 1, hence its expected (mean) wingspan will get a contribution of alpha[1]. As another example, dragonflies 6 and 7 have (internal) values of the pop factor of 2 and 3, respectively. Hence, their expected wingspans will get contributions of alpha[2] and alpha[3], respectively. Be careful with the indices when writing a model in algebra, and similarly in the BUGS language; indexing of arrays is one of the more complicated things for a beginner to grasp. As said above, we might have written $\alpha_{j(i)}$ rather than simply α_j to clarify the algebraic description of the model. This would emphasize that the information given by index j (population membership of a dragonfly) is equivalent to that given by the factor pop in the analysis.

3. Factor levels in BUGS cannot be letters or words such as in our example here, but they must be integers ranging from 1 to the number of distinct levels. Thus, in a BUGS analysis, we would have to convert our factors pop and sex into numerical values and only (1,2,3) and (1,2), respectively, will be correct, while for instance, (1,2,4) for pop or (0,1) for sex will result in errors when using nested indexing.

There are several ways to write what is essentially the same linear model, and these are called *parameterizations* of a model; for instance, here is such a variant of the ANCOVA linear model:

$$wing_i = \alpha_j + \beta * body_i + \varepsilon_i$$

$$\varepsilon_i \sim Normal(0, \sigma^2)$$

In this parameterization of the model, everything is the same as before, except that the model no longer has the parameter μ, and the meaning of the parameters α_j representing the effects of the levels of factor pop has changed. Now, all three elements of this parameter vector are estimable, since there is no longer a separate intercept (μ). They now represent directly the intercepts of the three regression lines—i.e., the expected wingspans of a dragonfly at body length 0 in the three populations. We call this parameterization of the factor pop the *means parameterization*, because each parameter in the factor corresponds directly to a 'group mean'. We can fit it in R by "subtracting the intercept" in the model formula of the function call:

```
summary(fm2 <- lm(wing ~ pop-1 + body))

Coefficients:
             Estimate Std. Error t value Pr(>|t|)
popNavarra    6.5296     1.6437   3.973  0.01061 *
popAragon     8.4003     1.5916   5.278  0.00325 **
popCatalonia  8.2630     1.6437   5.027  0.00401 **
body          0.5149     0.1991   2.586  0.04908 *
[ ... ]
Residual standard error: 0.5498 on 5 degrees of freedom
Multiple R-squared: 0.9988,   Adjusted R-squared: 0.9979
F-statistic: 1053 on 4 and 5 DF,  p-value: 1.694e-07
```

This is an exactly equivalent model to that from before: the parameter estimates are either the same (for popNavarra, body, and the residual standard error) or they can be obtained by adding two of the parameters in the former parameterization. Hence, the value of popAragon here is obtained by adding the value of the intercept and of popAragon in the previous parameterization (i.e., 8.4002 = 6.5296 + 1.8706), and similar for popCatalonia (i.e., 8.2629 = 6.5296 + 1.7333; differences are due to rounding). The advantage of this parameterization is perhaps ease of interpretation. However, it cannot be adopted for models with multiple factors, for which we must choose the effects parameterization or impose sum-to-zero constraints (see below and Ntzoufras, 2009).

In BUGS, the means parameterization of the model is specified simply like this:

```
for(i in 1:9){
   wing[i] ~ dnorm(mean[i], tau)
   mean[i] <- alpha[pop[i]] + beta * body[i]
}
```

Hence, the BUGS code inside of the loop is again very similar to how we write the model in algebra. Indeed, it is identical to version 2 of the following three algebraically synonymous descriptions of the ANCOVA linear model

1. $wing_i \sim Normal(\alpha_j + \beta * body_i, \sigma^2)$
2. $wing_i \sim Normal(\mu_i, \sigma^2)$, with $\mu_i = \alpha_j + \beta * body_i$
3. $wing_i = \alpha_j + \beta * body_i + \varepsilon_i$, with $\varepsilon_i \sim Normal(0, \sigma^2)$

It is very important that you know how to write your linear models in algebra, because this greatly helps you understand what the parameters mean and, because the BUGS language model description very much resembles the model's representation in algebra; so much so, indeed, that once you know how to write the model in algebra, you have almost written it in the BUGS language! Furthermore,

BUGS software can be sensitive to the parameterization chosen: sometimes one may work well and another not as well. Therefore, it is good to be able to switch between different model parameterizations and try which works (best).

Another parameterization of a linear model for factors imposes a sum-to-zero constraint—i.e., the effects of the factor levels are estimated such that their combined effect is zero. The effects can then be interpreted as deviations from the common mean; this again allows a separate intercept to be estimated, as for the effects parameterization. This parameterization may have some advantages for ease of interpretation in models with multiple factors and is easily coded in BUGS (see Ntzoufras, 2009).

The intercept for each population is the expected wingspan of a dragonfly of body length 0. This is of course nonsensical, and a more meaningful model is obtained when we regress wingspan not on body length, but on *centered* body length—i.e., on a transformation of original body length obtained by subtracting the sample mean of body length. The intercept is then the expected wingspan of a dragonfly at the average observed body length; a much more meaningful parameter. Moreover, centering covariates is often helpful to obtain convergence, both in maximum likelihood and in Bayesian Markov chain Monte Carlo (MCMC) analyses. Often to avoid numerical problems in `unmarked` or BUGS, we have to standardize covariates; i.e., center them and then divide the result by some amount, typically the covariate sample standard deviation. Note that while centering only affects the intercept, standardizing affects both the intercept and the slope of a covariate. When scaling a covariate with the standard deviation of the observed values of a covariate, the slope will be the expected change in the response variable for a one-unit change in the scaled covariate, which corresponds to a one-standard-deviation change of the original covariate.

After fitting a model, we may often want to make predictions (to draw figures or even understand what our model is telling us about the processes underlying our data); i.e., we may want to compute the expected response given the parameter estimates for certain values of the covariates. This can easily be done by hand, or you can use the R function `predict` to do this for you automatically, including an assessment of prediction uncertainty (standard errors). It is trivial as well in BUGS, but there you have to do this "by hand." We will see many examples of predictions with R and BUGS in this book (this is emphasized, for example, in Chapter 5), but here as a mini example we simply point out that the expected wingspan of a dragonfly of body length 8 in Aragon can be estimated using the output of `fm2` as $8.4003 + 0.5149 * 8 = 12.5195$.

In statistics books, you are likely to see linear models described in matrix algebra; for instance, as $\mathbf{y} = \mathbf{X}\boldsymbol{\beta} + \boldsymbol{\varepsilon}$. What does this mean? Matrices and vectors are simply a manner in which numbers can be stored in an orderly way in an array. Hence, for our dragonfly example, $\mathbf{y}$ is a vector of length nine containing the responses—i.e., the wingspan of each dragonfly—$\mathbf{X}$ is the *design matrix* of the model (also called *model matrix* or *X matrix*), $\boldsymbol{\beta}$ is the parameter vector, and $\boldsymbol{\varepsilon}$ is the vector of residuals. We call *linear predictor* the result of the matrix multiplication of the design matrix and the parameter vector, $\mathbf{X}\boldsymbol{\beta}$; this is the expected wingspan given population membership and body length of each dragonfly, in the absence of the random noise $\boldsymbol{\varepsilon}$.

To better understand this, it is useful to write it out with the numbers of our data set plugged into the various matrices and vectors. We do this first for the effects parameterization and then for the means parameterization of our ANCOVA linear model. We will see that the structure of the linear model is exactly defined by the design matrix **X**, and that this moreover determines the interpretation of the fitted parameters. For the effects parameterization, we have for $\mathbf{y} = \mathbf{X}\boldsymbol{\beta} + \boldsymbol{\varepsilon}$:

$$\begin{pmatrix} 10.5 \\ 10.6 \\ 11.0 \\ 12.1 \\ 11.7 \\ 13.5 \\ 11.4 \\ 13.0 \\ 12.9 \end{pmatrix} = \begin{pmatrix} 1 & 0 & 0 & 6.8 \\ 1 & 0 & 0 & 8.3 \\ 1 & 0 & 0 & 9.2 \\ 1 & 1 & 0 & 6.9 \\ 1 & 1 & 0 & 7.7 \\ 1 & 1 & 0 & 8.9 \\ 1 & 0 & 1 & 6.9 \\ 1 & 0 & 1 & 8.2 \\ 1 & 0 & 1 & 9.2 \end{pmatrix} \begin{pmatrix} \mu \\ \alpha_2 \\ \alpha_3 \\ \beta \end{pmatrix} + \begin{pmatrix} \varepsilon_1 \\ \varepsilon_2 \\ \varepsilon_3 \\ \varepsilon_4 \\ \varepsilon_5 \\ \varepsilon_6 \\ \varepsilon_7 \\ \varepsilon_8 \\ \varepsilon_9 \end{pmatrix}, \text{ with } \varepsilon_i \sim Normal(0, \sigma^2)$$

From left to right, we have response vector, design matrix, parameter vector, and residual vector. To be able to do matrix multiplication, the number of columns in the design matrix must always match the length of the parameter vector. The design matrix contains indicator variables (0s and 1s) for the effects of factors, and the measured covariate values for continuous or numerical explanatory variables. Here, the first column of the design matrix defines the intercept (which is equal to the expected wingspan of an individual with body length zero in population Navarra) by a vector of all 1s, while in the next two columns, the 1s indicate individuals in populations Aragon and Catalonia, respectively. In this description of the linear model, we recognize particularly clearly the structure of a multiple linear regression with as many single degree-of-freedom terms (scalar parameters) as columns in the design matrix and rows in the parameter vector.

The design matrix multiplied with the parameter vector produces another vector, η_i, called the *linear predictor*, which is the expected wingspan for each dragonfly. For dragonfly 1, it is given by $1 * \mu + 0 * \alpha_2 + 0 * \alpha_3 + 6.8 * \beta$, and hence its expected wingspan contains one contribution from μ and a contribution of 6.8 times β. Likewise, for dragonfly 9, the value of the linear predictor is $1 * \mu + 0 * \alpha_2 + 1 * \alpha_3 + 9.2 * \beta$. With a fitted model such as fm1, we can obtain the expected wingspan "by hand" (using the powerful model.matrix function, which we describe in more detail below) or by using the predict function after fitting the linear model using lm; note that %*% denotes matrix multiplication:

```
cbind(model.matrix(~pop+body) %*% fm1$coef, predict(fm1)) # Compare two solutions
    [,1]     [,2] # 1st is 'by hand', 2nd from function predict
1 10.03068 10.03068
2 10.80297 10.80297
3 11.26635 11.26635
4 11.95280 11.95280
5 12.36469 12.36469
6 12.98252 12.98252
7 11.81550 11.81550
8 12.48482 12.48482
9 12.99968 12.99968
```

Here is the means parameterization in the vector-matrix description:

$$\begin{pmatrix} 10.5 \\ 10.6 \\ 11.0 \\ 12.1 \\ 11.7 \\ 13.5 \\ 11.4 \\ 13.0 \\ 12.0 \end{pmatrix} = \begin{pmatrix} 1 & 0 & 0 & 6.8 \\ 1 & 0 & 0 & 8.3 \\ 1 & 0 & 0 & 9.2 \\ 0 & 1 & 0 & 6.9 \\ 0 & 1 & 0 & 7.7 \\ 0 & 1 & 0 & 8.9 \\ 0 & 0 & 1 & 6.9 \\ 0 & 0 & 1 & 8.2 \\ 0 & 0 & 1 & 9.2 \end{pmatrix} \begin{pmatrix} \alpha_1 \\ \alpha_2 \\ \alpha_3 \\ \beta \end{pmatrix} + \begin{pmatrix} \varepsilon_1 \\ \varepsilon_2 \\ \varepsilon_3 \\ \varepsilon_4 \\ \varepsilon_5 \\ \varepsilon_6 \\ \varepsilon_7 \\ \varepsilon_8 \\ \varepsilon_9 \end{pmatrix}, \text{ with } \varepsilon_i \sim Normal(0, \sigma^2)$$

The expected wingspan of dragonfly 1 is given by $1 * \alpha_1 + 0 * \alpha_2 + 0 * \alpha_3 + 6.8 * \beta$, which is the same as before except for the small semantic change that we now call the intercept in population Navarra α_1 instead of μ. In contrast, for dragonfly 9 we get an expected wingspan of $0 * \alpha_1 + 0 * \alpha_2 + 1 * \alpha_3 + 9.2 * \beta$, and so we see that the intercept of the regression line in population 3 is now given by α_3, rather than by $\mu + \alpha_3$ as in the effects parameterization of the model. When we fit a linear model, we implicitly solve a system of equations subject to some constraints on the noise terms ε_i. Here is that system of equations:

$$10.5 = 1 * \alpha_1 + 0 * \alpha_2 + 0 * \alpha_3 + 6.8 * \beta + \varepsilon_1$$
$$10.6 = 1 * \alpha_1 + 0 * \alpha_2 + 0 * \alpha_3 + 8.3 * \beta + \varepsilon_2$$
$$11.0 = 1 * \alpha_1 + 0 * \alpha_2 + 0 * \alpha_3 + 9.2 * \beta + \varepsilon_3$$
$$12.1 = 0 * \alpha_1 + 1 * \alpha_2 + 0 * \alpha_3 + 6.9 * \beta + \varepsilon_4$$
$$11.7 = 0 * \alpha_1 + 1 * \alpha_2 + 0 * \alpha_3 + 7.7 * \beta + \varepsilon_5$$
$$13.5 = 0 * \alpha_1 + 1 * \alpha_2 + 0 * \alpha_3 + 8.9 * \beta + \varepsilon_6$$
$$11.4 = 0 * \alpha_1 + 0 * \alpha_2 + 1 * \alpha_3 + 6.9 * \beta + \varepsilon_7$$
$$13.0 = 0 * \alpha_1 + 0 * \alpha_2 + 1 * \alpha_3 + 8.2 * \beta + \varepsilon_8$$
$$12.0 = 0 * \alpha_1 + 0 * \alpha_2 + 1 * \alpha_3 + 9.2 * \beta + \varepsilon_9$$

Indeed, the least-squares model-fitting criterion used by the R function lm chooses the parameter values such that the sum of the squared residuals is minimized (hence the name). This technique does not make any distributional assumptions about the residuals, but the parameter estimates and standard errors are identical to those obtained using the maximum likelihood method when we make the explicit assumption that the residuals have a normal distribution.

The design matrix lies at the heart of a linear model, and it is imperative that we obtain a clear understanding of it when fitting linear models. R has the powerful model.matrix function, which shows you the design matrix for any linear model specified in the model definition language of R. This can be very useful to understand a model, and can even be useful when fitting the model in BUGS, for instance by clarifying what a linear model looks like (see also Sections 6.11 and 10.6 where we directly fit a design matrix in BUGS). Let's look at the design matrices of the two parameterizations of the main-effects ANCOVA model.

```
model.matrix(~ pop + body) # Effects parameterization
  (Intercept) popAragon popCatalonia body
1           1         0            0  6.8
2           1         0            0  8.3
3           1         0            0  9.2
4           1         1            0  6.9
5           1         1            0  7.7
6           1         1            0  8.9
7           1         0            1  6.9
8           1         0            1  8.2
9           1         0            1  9.2

model.matrix(~ pop-1 + body) # Means parameterization
  popNavarra popAragon popCatalonia body
1          1         0            0  6.8
2          1         0            0  8.3
3          1         0            0  9.2
4          0         1            0  6.9
5          0         1            0  7.7
6          0         1            0  8.9
7          0         0            1  6.9
8          0         0            1  8.2
9          0         0            1  9.2
```

Finally, the geometrical representation of the main-effects (i.e., additive) ANCOVA model in this section is that of a bundle of parallel regression lines, one for each population (Figure 3.3(a)). They are parallel because the slope is shared among all three populations.

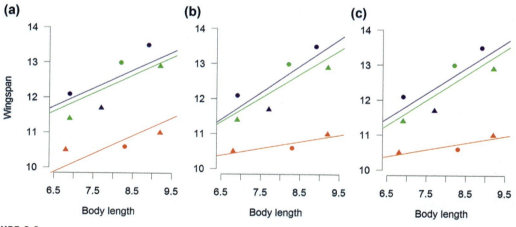

FIGURE 3.3

Geometric representation of the main-effects ANCOVA model (additive effects of population and body length; (a), the interaction-effects ANCOVA model (multiplicative effects of population and body length; b), and the partially interactive-effects ANCOVA model (c) in the *O. uncatus* toy data set. Color code and symbols as in Figure 3.2.

```
par(mfrow = c(1, 3), mar = c(5,4,2,2), cex = 1.2, cex.main = 1) # Figure 3.3 (a)
plot(body[sex == "M"], wing[sex == "M"], col = colorM, xlim = c(6.5, 9.5), ylim = c(10, 14),
lwd = 2, frame.plot = FALSE, las = 1, pch = 17, xlab = "Body length", ylab = "Wingspan")
points(body[sex == "F"], wing[sex == "F"], col = colorF, pch = 16)
abline(coef(fm2)[1], coef(fm2)[4], col = "red", lwd = 2)
abline(coef(fm2)[2], coef(fm2)[4], col = "blue", lwd = 2)
abline(coef(fm2)[3], coef(fm2)[4], col = "green", lwd = 2)
text(6.8, 14, "A", cex = 1.5)
```

The form of the design matrix determines the interpretation of the parameters. Table 3.1 gives an overview of a range of linear models that can be fitted for two explanatory variables when one is a factor and the other is a covariate. We also list the case of an interaction between pop and body, and briefly review the important concept of interaction, in the next section.

3.2.2 LINEAR MODELS WITH INTERACTION BETWEEN ONE FACTOR AND ONE CONTINUOUS COVARIATE

When two terms in a linear model interact, the effect of one depends on the value of the other and vice versa. We illustrate this in the context of our ANCOVA linear model with pop and body, which, when the two interact, geometrically represents a bundle of nonparallel regression lines (Figure 3.3(b)). Interaction also means that the effects of the terms are combined in a multiplicative rather than purely additive manner. The design matrix then contains additional columns that represent the products of the columns that stand for the main effects. In a sense, the polynomial terms in Section 3.2.4 represent such interactions of a covariate with itself. We first show the effects parameterization.

```
model.matrix(~ pop*body)  # Effects parameterization
  (Intercept) popAragon popCatalonia body popAragon:body popCatalonia:body
1           1         1            0  6.8            0.0               0.0
2           1         1            0  8.3            0.0               0.0
3           1         1            0  9.2            0.0               0.0
4           1         1            1  6.9            6.9               0.0
5           1         1            1  7.7            7.7               0.0
6           1         1            1  8.9            8.9               0.0
7           1         0            0  6.9            0.0               6.9
8           1         0            0  8.2            0.0               8.2
9           1         0            0  9.2            0.0               9.2
```

Here, the main effect of pop is represented by columns two and three, and that of body by column four, and the interaction terms represent the products of columns two and four and columns three and four, respectively. Next, the means parameterization.

```
model.matrix(~ pop*body-1-body)  # Means parameterization
# Output slightly trimmed
  popNavarra popAragon popCatalonia popNa:body popAr:body popCat:body
1          1         0            0        6.8        0.0         0.0
2          1         0            0        8.3        0.0         0.0
3          1         0            0        9.2        0.0         0.0
4          0         1            0        0.0        6.9         0.0
5          0         1            0        0.0        7.7         0.0
6          0         1            0        0.0        8.9         0.0
7          0         0            1        0.0        0.0         6.9
8          0         0            1        0.0        0.0         8.2
9          0         0            1        0.0        0.0         9.2
```

Table 3.1 Specification of different linear models in the R and BUGS model definition languages, when pop is a factor, i.e., a categorical explanatory variable with three levels indexed by j and body is a covariate, i.e., a continuous explanatory variable. The factor levels of pop can be numbers or letters (or names) in R, but they must be integers with first level coded as 1 in BUGS. Individuals are indexed by i in BUGS (i.e., $i = 1…9$). Models 6 and 7 can also be reparameterised similar to the move from model 2 to model 3, but this is not shown here. The full number of model parameters is one plus the number of coefficients, because it also includes the variance.

No.	Model in R	Model in Algebra	Model in BUGS	Traditional Name of Technique Based on That Linear Model	Number of Coefficients	Meaning
1	1	α	alpha	"Model of the mean"	1	Constant term (intercept) only.
2	pop	$\mu + \alpha_j$	mu + alpha[pop[i]] alpha[1] <- 0	One-way ANOVA	3	This is the default linear model parameterization of a factor in R, which has an intercept (= the value for the first population) and two constants that are the differences between the values of populations 2 and 1, and 3 and 1. In BUGS, the first level of the vector alpha must be manually set to zero to avoid overparameterization. In R, this is done automatically.
3	pop-1	α_j	alpha[pop[i]]	One-way ANOVA	3	This is a simple reparameterization of model 2. There are three constants, one for each population (called the level of the factor). Called a t-test if factor has only two levels. In R, this model is specified by "subtraction" of the intercept.
4	body	$\alpha + \beta * x_i$	alpha + beta * body[i]	Simple linear regression	2	An intercept plus a slope common to all three populations (i.e., no effect of pop).
5	body-1	$\beta * x_i$	beta * body[i]	Simple linear regression through the origin	1	"Subtract the intercept": because body is a continuous covariate, this is NOT a mere reparameterization of model 4. Regression through the origin; not usually a meaningful model.
6	pop+body	$\alpha_j + \beta * x_i$	alpha[pop[i]] + beta * body[i]	Main-effects ANCOVA	4	One separate intercept for each population and a common slope.
7	pop*body	$\alpha_j + \beta_j * x_i$	alpha[pop[i]] + beta[pop[i]] * body[i]	Interaction-effects ANCOVA	6	Three separate intercepts and three separate slopes. That is, fully separate regression of wing on body for each population.

Now, the main effect of pop is represented by columns one through three, and the interactive effect with body directly by columns four through six. Let's fit this latter parameterization and plot the lines of best fit (Figure 3.3(b)).

```
summary(fm3 <- lm(wing ~ pop*body-1-body))
Call:
lm(formula = wing ~ pop * body - 1 - body)
[ ... ]
Coefficients:
                 Estimate Std. Error t value Pr(>|t|)
popNavarra         9.1296     2.7626   3.305   0.0456 *
popAragon          6.4553     3.2116   2.010   0.1380
popCatalonia       6.9217     2.9023   2.385   0.0972 .
popNavarra:body    0.1939     0.3385   0.573   0.6070
popAragon:body     0.7632     0.4077   1.872   0.1580
popCatalonia:body  0.6805     0.3559   1.912   0.1518
[ ... ]
Residual standard error: 0.5805 on 3 degrees of freedom
Multiple R-squared: 0.9992,    Adjusted R-squared: 0.9976
F-statistic: 629.9 on 6 and 3 DF, p-value: 9.763e-05

# Plot (Figure 3.3 (b))
plot(body[sex == "M"], wing[sex == "M"], col = colorM, xlim = c(6.5, 9.5), ylim = c(10, 14),
  lwd = 2, frame.plot = FALSE, las = 1, pch = 17, xlab = "Body length", ylab = "")
points(body[sex == "F"], wing[sex == "F"], col = colorF, pch = 16)
abline(coef(fm3)[1], coef(fm3)[4], col = "red", lwd = 2)
abline(coef(fm3)[2], coef(fm3)[5], col = "blue", lwd = 2)
abline(coef(fm3)[3], coef(fm3)[6], col = "green", lwd = 2)
text(6.8, 14, "B", cex = 1.5)
```

In BUGS, the interactive model in the effects and in the means parameterizations just shown are specified as follows:

```
for(i in 1:9){    # Effects parameterization
   wing[i] ~ dnorm(mean[i], tau)
   mean[i] <- mu.alpha + alpha[pop[i]] + mu.beta * body[i] + beta[pop[i]] * body[i]
}
alpha[1] <- 0    # Impose 'corner constraints' for identifiability
beta[1] <- 0

for(i in 1:9){    # Means parameterization
   wing[i] ~ dnorm(mean[i], tau)
   mean[i] <- alpha[pop[i]] + beta[pop[i]] * body[i]
}
```

We can also fit what might be called partially interactive models. We illustrate this and show how we can directly fit a model matrix using R functions such as lm or glm. Note first that the fitted regression lines in Catalonia and Aragon are rather similar, while that in Navarra is different

(Figure 3.3(a)). So let's create a design matrix for a partially interactive model with identical slopes for Aragon and Catalonia and a separate slope for Navarra, while keeping the intercepts separate for all three populations. We can conduct a significance test to compare the two models (and find out that the partially interactive model is better) and then plot the best fit-lines under this model (Figure 3.3(c)). We take the last design matrix, copy parts of the column associated with the slope in Catalonia into the slope column for Aragon, and then delete the slope column for Catalonia.

```
# Create new design matrix
(DM0 <- model.matrix(~ pop*body-1-body))   # Original DM for means param
DM0[7:9,5] <- DM0[7:9,6]                   # Combine slopes for Ar and Cat
(DM1 <- DM0[, -6])                         # Delete former slope column for Cat
  popNavarra popAragon popCatalonia popNavarra:body popAragon:body
1          1         0            0             6.8            0.0
2          1         0            0             8.3            0.0
3          1         0            0             9.2            0.0
4          0         1            0             0.0            6.9
5          0         1            0             0.0            7.7
6          0         1            0             0.0            8.9
7          0         0            1             0.0            6.9
8          0         0            1             0.0            8.2
9          0         0            1             0.0            9.2

# Fit model with partial interaction
summary(fm4 <- lm(wing ~ DM1-1))

Call:
lm(formula = wing ~ DM1 - 1)
[ ... ]
Coefficients:
                   Estimate Std. Error t value Pr(>|t|)
DM1popNavarra        9.1296     2.4018   3.801   0.0191 *
DM1popAragon         6.8230     1.8491   3.690   0.0210 *
DM1popCatalonia      6.6320     1.9106   3.471   0.0256 *
DM1popNavarra:body   0.1939     0.2943   0.659   0.5461
DM1popAragon:body    0.7162     0.2331   3.072   0.0372 *
[ ... ]
Residual standard error: 0.5047 on 4 degrees of freedom
Multiple R-squared: 0.9992,  Adjusted R-squared: 0.9982
F-statistic: 1000 on 5 and 4 DF, p-value: 2.793e-06

# Do significance test
anova(fm3, fm4)              # F test between two models
Analysis of Variance Table

Model 1: wing ~ pop * body - 1 - body
Model 2: wing ~ DM1 - 1
  Res.Df    RSS Df  Sum of Sq      F Pr(>F)
1      3 1.0109
2      4 1.0187 -1 -0.0078683 0.0234 0.8882
```

```
# Plot (Figure 3.3.(c))
plot(body[sex == "M"], wing[sex == "M"], col = colorM, xlim = c(6.5, 9.5), ylim = c(10, 14),
  lwd = 2, frame.plot = FALSE, las = 1, pch = 17, xlab = "Body length", ylab = "")
points(body[sex == "F"], wing[sex == "F"], col = colorF, pch = 16)
abline(coef(fm4)[1], coef(fm4)[4], col = "red", lwd = 2)
abline(coef(fm4)[2], coef(fm4)[5], col = "blue", lwd = 2)
abline(coef(fm4)[3], coef(fm4)[5], col = "green", lwd = 2)
text(6.8, 14, "C", cex = 1.5)
```

In BUGS, we can use the function `inprod` to directly fit a design matrix, which must be provided to BUGS as data, see Sections 6.11.1 and 10.6. The design matrix and the parameter vector must be *compatible*, which means that the number of columns in the former must equal the length of the latter.

```
for(i in 1:9){
  wing[i] ~ dnorm(mean[i], tau)
  mean[i] <- inprod(DM1[i,], beta[]) # Requires ncol(DM1) = length(beta)
}
```

The ability to directly fit a design matrix in BUGS, whether directly generated by `model.matrix` in R, or after some modifications as in our example, can be quite useful (see Sections 6.11 and 10.6).

3.2.3 LINEAR MODELS WITH TWO FACTORS

Next, we briefly look at how the effects of two categorical explanatory variables (factors) can be combined in a linear model. For one factor and one (continuous) covariate, as just shown, this is simple: the columns in the design matrix are simply joined. However, when fitting two factors, we typically have to resolve some aliasing (parameter redundancy, nonidentifiability). Put simply, aliasing means that one parameter is identical to another one or can be expressed as a linear function of other model parameters. Hence, it cannot be estimated independently. For instance, in the effects parameterization of the main-effects ANCOVA model in Section 3.2.1, the intercept μ was aliased with the three population effects. To resolve this, we set the effect of the first population (α_1) to zero; μ then became the effect of population 1 (i.e., the intercept for Navarra).

When fitting linear models in R with `lm` or `glm`, or with functions in `unmarked`, aliasing is taken care of automatically. However, when fitting linear models in BUGS, we must know which columns of the full design matrix of a model correspond to redundant parameters and then not fit them (which is equivalent to setting them to zero). This can be difficult to understand at first. For models with two or more factors that you want to fit in BUGS, it may be helpful to inspect their design matrix in R, provided you know how to specify the linear model in R. You can then directly fit the modified design matrix in R using `lm`, and in BUGS using `inprod`, as we just saw, and will see again in Sections 6.11.1 and 10.6.

Let's now see how we can fit the effects of two factors in our analysis: `pop` with three levels and `sex` with two. We consider the models that underlie the least-squares techniques known as "two-way ANOVA with main effects" and "two-way interaction effects ANOVA" (Mead, 1988).

```
model.matrix(~ pop+sex)  # Design matrix of main-effects 2-way ANOVA
  (Intercept) popAragon popCatalonia sexF   # 1,3,5,7, and 9 are males
1      1           1          0        0
2      1           1          0        1
3      1           1          0        0
```

```
4          1         1                0    1
5          1         1                0    0
6          1         1                0    1
7          1         0                1    0
8          1         0                1    1
9          1         0                1    0
```

```
# Fit linear model with that design matrix
summary(fm5 <- lm(wing ~ pop + sex))
Coefficients:
              Estimate Std. Error t value Pr(>|t|)
(Intercept)    10.5000     0.4678  22.447 3.26e-06 ***
popAragon       1.5333     0.6375   2.405   0.0612 .
popCatalonia    1.7333     0.6125   2.830   0.0367 *
sexF            0.6000     0.5304   1.131   0.3093
---
Signif. codes:   0 '***' 0.001 '**' 0.01 '*' 0.05 '.' 0.1 ' ' 1

Residual standard error: 0.7501 on 5 degrees of freedom
Multiple R-squared:  0.7052,    Adjusted R-squared:  0.5283
F-statistic: 3.986 on 3 and 5 DF,  p-value: 0.08536
```

The parameters associated with the four columns in the design matrix are as follows (in this order): the intercept, which is the expected wingspan of a dragonfly with level 1 for both factors (i.e., males in Navarra); the difference between a male in Aragon and one in Navarra (that's the "effect" of Aragon); the difference between a male in Catalonia and in Navarra (the effect of Catalonia); and the difference between a female and a male (the effect of sex = "F"). We do not have a separate parameter for "male" in the factor sex, because it would be aliased with other model parameters. Hence, it is automatically dropped from the design matrix by R.

We write the following to fit this model in BUGS and set the effects of the first levels of each factor to zero:

```
for(i in 1:9){
   wing[i] ~ dnorm(mean[i], tau)
   mean[i] <- mu.alpha + alpha[pop[i]] + beta[sex[i]]
}
alpha[1] <- 0
beta[1] <- 0
```

We obtain the design matrix of the means parameterization of the same model in R as:

```
model.matrix(~ pop+sex-1) # Design matrix of the main-effects 2-way ANOVA
  popNavarra popAragon popCatalonia sexF
1          1         0            0    0
2          1         0            0    1
3          1         0            0    0
4          0         1            0    1
5          0         1            0    0
6          0         1            0    1
7          0         0            1    0
8          0         0            1    1
9          0         0            1    0
```

```
# Fit linear model with that design matrix
summary(fm6 <- lm(wing ~ pop + sex-1))
Coefficients:
              Estimate Std. Error t value Pr(>|t|)
popNavarra     10.5000     0.4678  22.447 3.26e-06 ***
popAragon      12.0333     0.5591  21.523 4.02e-06 ***
popCatalonia   12.2333     0.4678  26.152 1.53e-06 ***
sexF            0.6000     0.5304   1.131    0.309
---
Signif. codes:  0 '***' 0.001 '**' 0.01 '*' 0.05 '.' 0.1 ' ' 1

Residual standard error: 0.7501 on 5 degrees of freedom
Multiple R-squared: 0.9978,  Adjusted R-squared: 0.996
F-statistic:   565 on 4 and 5 DF,  p-value: 7.999e-07
```

Here, the parameters associated with the first three columns directly correspond to the expected wingspan of males in Navarra, Aragon, and Catalonia, respectively, and the last parameter in the mean model is the difference in wingspan between females and males. In BUGS, we can write the following (Note that `mu.alpha` from the previous parameterization of the model is dropped, which means that it is set to zero implicitly).

```
for(i in 1:9){
   wing[i] ~ dnorm(mean[i], tau)
   mean[i] <- alpha[pop[i]] + beta[sex[i]]
}
beta[1] <- 0
```

Finally, let's look at some possible design matrices of a model with interaction between two factors. The following three design matrices are all different parameterizations of such an interaction-effects two-way ANOVA model. They result in separate expected wingspan estimates for the six groups, corresponding to the pairwise combination of all levels of the two factors pop and sex. Perhaps the last parameterization is the easiest to understand in terms of the meaning of the parameters: the parameters associated with columns one through three denote the expected wingspan of males in Navarra, Aragon, and Catalonia, and columns four through six denote the same for females.

```
# Variant 1: Effects parameterization (R default)
model.matrix(~ pop*sex)
#model.matrix(~ pop + sex + pop:sex)          # Same
  (Intercept) popAragon popCatalonia sexF popAragon:sexF popCatalonia:sexF
1           1         0            0    0              0                 0
2           1         0            0    1              0                 0
3           1         0            0    0              0                 0
4           1         1            0    1              1                 0
5           1         1            0    0              0                 0
6           1         1            0    1              1                 0
7           1         0            1    0              0                 0
8           1         0            1    1              0                 1
9           1         0            1    0              0                 0
```

```
# Variant 2: Means param. for pop, effects param. for sex
model.matrix(~ pop*sex-1)
  popNavarra popAragon popCatalonia sexF popAragon:sexF popCatalonia:sexF
1          1         0            0    0              0                 0
2          1         0            0    1              0                 0
3          1         0            0    0              0                 0
4          0         1            0    1              1                 0
5          0         1            0    0              0                 0
6          0         1            0    1              1                 0
7          0         0            1    0              0                 0
8          0         0            1    1              0                 1
9          0         0            1    0              0                 0

# Variant 3 (output slightly trimmed): full means parameterization
model.matrix(~ pop:sex-1)
  popNav:sexM popAr:sexM popCat:sexM popNav:sexF popAr:sexF popCat:sexF
1           1          0           0           0          0           0
2           0          0           0           1          0           0
3           1          0           0           0          0           0
4           0          0           0           0          1           0
5           0          1           0           0          0           0
6           0          0           0           0          1           0
7           0          0           1           0          0           0
8           0          0           0           0          0           1
9           0          0           1           0          0           0
```

As an exercise, you may want to fit these linear models and compare the parameter estimates with the design matrix of the fitted model.

Variant 3 is perhaps also the easiest to code in BUGS. This shows that we can define parameters to come in multidimensional arrays: here, `alpha` is a matrix with three rows for population and two columns for sex. Thus, each combination of population and sex gets a separate parameter in the linear model.

```
for(i in 1:9){
  wing[i] ~ dnorm(mean[i], tau)
  mean[i] <- alpha[pop[i], sex[i]]   # note parameter may be a matrix
}
```

3.2.4 LINEAR MODELS WITH TWO CONTINUOUS COVARIATES AND INCLUDING POLYNOMIALS

Combinations of two or more continuous covariates are straightforward, and we illustrate them here mostly for the sake of completeness. In particular, the interaction between two covariates is simply their product. Thus, if we want to fit a model for wingspan with effects of body length and color intensity in an additive fashion, we have the following design matrix. The geometrical representation of this model is a plane in the three dimensions of wingspan, body length, and color intensity.

```
model.matrix(~ body + color) # main-effects of covariates
  (Intercept)  body  color
1           1   6.8   0.45
2           1   8.3   0.47
3           1   9.2   0.54
4           1   6.9   0.42
5           1   7.7   0.54
6           1   8.9   0.46
7           1   6.9   0.49
8           1   8.2   0.42
9           1   9.2   0.57

summary(fm7 <- lm(wing ~ body + color))  # Fit that model
Coefficients:
             Estimate  Std. Error  t value  Pr(>|t|)
(Intercept)   10.2184      3.7669    2.713    0.035 *
body           0.6579      0.4508    1.459    0.195
color         -7.5000      8.1414   -0.921    0.392
```

If we also want to fit the interaction between body and color, we get the following, the geometrical interpretation of which is a twisted surface in 3-D. The interaction term body:color determines the degree of twist of the surface (i.e., the deviation from a plane).

```
model.matrix(~ body*color) # Interaction between two covariates
  (Intercept) body color body:color
1           1  6.8  0.45      3.060
2           1  8.3  0.47      3.901
3           1  9.2  0.54      4.968
4           1  6.9  0.42      2.898
5           1  7.7  0.54      4.158
6           1  8.9  0.46      4.094
7           1  6.9  0.49      3.381
8           1  8.2  0.42      3.444
9           1  9.2  0.57      5.244

summary(fm8 <- lm(wing ~ body*color))  # Fit that model
Coefficients:
             Estimate  Std. Error  t value  Pr(>|t|)
(Intercept)     5.877      38.357    0.153     0.884
body            1.184       4.650    0.255     0.809
color           1.526      79.781    0.019     0.985
body:color     -1.088       9.555   -0.114     0.914
```

In BUGS, we can write these models as follows. For illustration, we group the regression parameters other than the intercept in the vector beta. We can compute the interaction between body and color as their product either inside of BUGS (variant 1) or outside in R, and then provide the product body.color as data (variant 2).

```
for(i in 1:9){    # Main effects
  wing[i] ~ dnorm(mean[i], tau)
  mean[i] <- alpha + beta[1] * body[i] + beta[2] * color[i]
}

for(i in 1:9){    # Interaction-effects (variant 1)
  wing[i] ~ dnorm(mean[i], tau)
  mean[i] <- alpha + beta[1] * body[i] + beta[2] * color[i] + beta[3] * body[i] * color[i]
}

for(i in 1:9){    # Interaction-effects (variant 2)
  wing[i] ~ dnorm(mean[i], tau)
  mean[i] <- alpha + beta[1] * body[i] + beta[2] * color[i] + beta[3] * body.color[i]
}
```

We often want to model covariate relationships that don't correspond to straight lines—e.g., when a response increases up to a certain covariate value and then declines again, or vice versa. Often in exploratory analyses, we may want to allow for some flexibility in the covariate effects, because we have no *a priori* knowledge about the true shape of the relationship. This is easily achieved by adding polynomial terms of a covariate (such as a quadratic or cubic) to account for deviations from a straight-line model for that covariate. We can then simply add, as explanatory variables, the square or the cube of that covariate. If we do this inside of R's lm function, we have to use the I() construct, whereas in BUGS, we use the function pow. The geometrical interpretation is a curvy line with one fewer inner extrema than the degree of the polynomial, with the convexity/concavity determined by the signs of the coefficients. For instance, a model with linear and quadratic terms is a parabola, which has a single extremum, while a model with cubic terms has two extrema. Note that a model with such higher-order terms should always have all the lower-order terms as well to obey the rules of marginality (McCullagh and Nelder, 1989).

```
# Cubic polynomial of body in R
body2 <- body^2         # Squared body length
body3 <- body^3         # Cubed body length
model.matrix(~ body + body2 + body3)
    (Intercept)  body   body2   body3
1             1   6.8   46.24  314.432
2             1   8.3   68.89  571.787
3             1   9.2   84.64  778.688
4             1   6.9   47.61  328.509
5             1   7.7   59.29  456.533
6             1   8.9   79.21  704.969
7             1   6.9   47.61  328.509
8             1   8.2   67.24  551.368
9             1   9.2   84.64  778.688
summary(fm9 <- lm(wing ~ body + body2 + body3))    # Fit that model
# summary(fm9 <- lm(wing ~ body + I(body^2) + I(body^3)))  # same
```

```
Coefficients:
            Estimate  Std. Error  t value  Pr(>|t|)
(Intercept)  165.4605   856.7892    0.193    0.854
body         -60.3906   323.7352   -0.187    0.859
body2          7.7949    40.5056    0.192    0.855
body3         -0.3306     1.6788   -0.197    0.852

Residual standard error: 1.252 on 5 degrees of freedom
Multiple R-squared: 0.1788,   Adjusted R-squared: -0.3139
F-statistic: 0.3629 on 3 and 5 DF,  p-value: 0.7833
```

In BUGS, we can similarly provide the values of the higher-order polynomials of wingspan as data, or we can compute them inside of BUGS using the power function. We show the latter here.

```
for(i in 1:9){ # Cubic polynomial of body in BUGS
    wing[i] ~ dnorm(mean[i], tau)
    mean[i] <- alpha + beta1 * body[i] + beta2 * pow(body[i],2) +  beta3 * pow(body[i],3)
}
```

For relatively simple covariate relationships, polynomials are perhaps the most widely used technique, though for potentially more complex ones, you may want more flexibility still and adopt an additive model (Hastie and Tibshirani, 1990; Wood, 2006). In BUGS, you may model such very "wiggly" relationships using a mixed-model representation of penalized splines (Crainiceanu et al., 2005); see Section 10.14 for an example in the context of a logistic-regression-type of model.

3.3 GENERALIZED LINEAR MODELS (GLMs)

GLMs extend the concept of a linear effect of covariates to response variables for which a statistical distribution other than a normal must be assumed, such as the Poisson, binomial/Bernoulli, gamma, or exponential distributions. Perhaps the key feature of GLMs is that the linear effect of covariates is expressed not for the expected response directly, but for a *transformation* of the expected response (McCullagh and Nelder, 1989). That transformation is called the *link function*. Generally, we can describe a GLM for response y_i in terms of three components:

1. *Random part of the response (or the error structure* of the model)—a statistical distribution f with parameter(s) θ:

$$y_i \sim f(\theta)$$

2. A *link function* g, which is applied to the expected response $E(y) = \mu_i$, with η_i known as the *linear predictor*:

$$g(E(y)) = g(\mu_i) = \eta_i$$

3. *Systematic part of the response* (or the *mean structure* of the model): the linear predictor (η_i), which contains a linear model, such as:

$$\eta_i = \alpha + \beta * x_i$$

We can combine elements 2 and 3 and define a GLM succinctly in just two lines:

$$y_i \sim f(\theta)$$
$$g(\mu_i) = \alpha + \beta * x_i$$

Hence, a response y follows a distribution f with parameter(s) θ, and a transformation g of the expected, or mean, response is modeled as a linear function of covariates; that's the GLM in a nutshell. And this is exactly the way in which a GLM is specified in the BUGS language, and the reason why BUGS is so great for helping us to *really* understand GLMs.

The GLM concept gives you considerable creative freedom in combining the three components of a GLM, but there are typically pairs of response distributions and link functions that go together particularly well. These latter are called the *canonical* link functions. They are the *identity link* for normal responses ($\eta_i = \mu_i$), the *log link* for Poisson responses ($\eta_i = log(\mu_i)$), and the *logit link* for binomial or Bernoulli responses ($\eta_i = log(\mu_i/(1 - \mu_i))$); "identity" simply means that the expected response is not transformed, but is directly modeled as a linear function of explanatory variables. Together, these three standard GLMs make up a vast number of statistical methods used in ecology and elsewhere; for overviews, see Dobson and Barnett (2008) or Ntzoufras (2009).

The vast scope of the GLM in applied statistical analysis is one reason for the great importance of the GLM to you. A second one is that GLMs unify a very large number of seemingly unrelated models and analysis techniques; hence, understanding GLMs helps you achieve a synthetic understanding of very many techniques and models. And finally, as we will see many times in this book, GLMs can be thought of as the main building blocks for all the HMs that we develop in later chapters, including occupancy and *N*-mixture models first introduced in Chapter 2. Many exciting ecological models for inference about populations or communities can be viewed simply as a sequence of coupled GLMs (Royle and Dorazio, 2008; Kéry and Schaub, 2012).

In Section 3.2, we described what is essentially a normal-response GLM, and here we illustrate three of the most typical GLMs with nonnormal responses: Poisson, Bernoulli, and binomial GLMs. All three are of fundamental importance for the HMs in this book, because the Poisson GLM is our typical model for spatiotemporal variation of abundance, and the Bernoulli GLM plays the analogous role for occurrence ("presence/absence" or distribution data). Furthermore, in models with population dynamics (see Chapters 9 and 12–14), the Poisson is a natural model for the recruitment process, and the binomial for the survival process. Finally, our canonical description of the observation process governed by false-negatives is the Bernoulli or the binomial distribution (Royle and Dorazio, 2008; Kéry and Schaub, 2012), although sometimes a Poisson, or a Bernoulli that *implies* an underlying Poisson process, may be more adequate; see Efford et al. (2013), Royle et al. (2014), and also Section 3.3.6 and Exercise 4.8 in Chapter 4. If you understand Poisson and Bernoulli GLMs, you are in good shape to understand all the HMs in this book.

Technically, GLMs are defined for all members of statistical distributions belonging to the so-called exponential family (McCullagh and Nelder, 1989; Dobson and Barnett, 2008; Ntzoufras, 2009), which includes the normal, Poisson, binomial/Bernoulli, multinomial, beta, gamma, lognormal, exponential, and Dirichlet distributions. Thus, the principles of linear modeling can be carried over to a vast number of "error models" other than the normal—i.e., to a very large number of stochastic processes that produce random outcomes that are predictable in some average sense only.

3.3.1 POISSON GLM FOR UNBOUNDED COUNTS

The Poisson distribution is the number of "things" from a Poisson process, which is a natural model for independent discrete "things" indexed by space or time (Gelman et al., 2014). "Things" is a placeholder for much of what scientists usually count. A Poisson GLM is a natural choice for unbounded counts—i.e., nonnegative integers that have no logical upper bound (unlike binomial counts, which cannot be greater than their sample size; see Section 3.3.7). We will use the parasite load in our toy dragonfly data set for illustration. For the number of mites counted on dragonfly i, C_i, we can write the main-effects ANCOVA linear model for factor pop and the body length covariate body within a Poisson GLM as follows:

1. Random part of the response (statistical distribution for the randomness in the response):
$$C_i \sim Poisson(\lambda_i)$$

2. Link of random and systematic part in response (link function):
$$\log(E(C)) = \log(\lambda_i) = \eta_i$$

3. Systematic part of the response (linear predictor η_i for link-transformed mean response):
$$\eta_i = \alpha_j + \beta * body_i$$

Here, λ_i is the expected (or mean) response for dragonfly i on the arithmetic scale ($E(C)$), η_i is the expected count on the log link scale, also called the linear predictor, $body_i$ is the body length of dragonfly i, and α and β are the two parameters (α being vector-valued) of the log-linear regression of the counts on factor pop and covariate body.

We fit this model in R and give the BUGS code for fitting it. Once again, we will see that the model descriptions in algebra and in the BUGS language are essentially identical. Here is the model in algebra in short; note that C corresponds to the variable mites in our toy data.

$$C_i \sim Poisson(\lambda_i)$$
$$\log(\lambda_i) = \alpha_j + \beta * body_i$$

```
summary(fm10 <- glm(mites ~ pop-1 + body, family = poisson))

Call:
glm(formula = mites ~ pop - 1 + body, family = poisson)
[ ... ]
Coefficients:
              Estimate  Std. Error  z value  Pr(>|z|)
popNavarra      -6.634       2.325   -2.854   0.00432 **
popAragon       -5.878       2.216   -2.653   0.00799 **
popCatalonia    -5.519       2.290   -2.410   0.01596 *
body             0.845       0.261    3.237   0.00121 **
[ ... ]
```

```
(Dispersion parameter for poisson family taken to be 1)

    Null deviance: 59.658  on 9  degrees of freedom
Residual deviance: 15.467  on 5  degrees of freedom
AIC: 42.591

Number of Fisher Scoring iterations: 6
```

We can specify the linear model for the factor `pop` in the means or effects parameterization, exactly as before. The methods for fitting and interpreting covariates, including two factors such as `pop` and `sex`, are exactly analogous for linear models and GLMs. To specify a Poisson GLM in the BUGS language using the means parameterization, we write:

```
for(i in 1:9){
  mites[i] ~ dpois(lambda[i])
  log(lambda[i]) <- alpha[pop[i]] + beta * body[i]
}
```

This is *exactly* how we just wrote the model in algebra, where `pop[i]` now takes the place of the index j for the intercept α; i.e., α_j in algebra is written as `alpha[pop[i]]` in the BUGS language.

3.3.2 OFFSETS IN THE POISSON GLM

The expected count C of a Poisson response is λ. Often we have a systematic component in the variation of C that is due to known variation in the size of the spatial or temporal unit, or effort, for which a count was obtained. For instance, the Poisson counts may have been obtained on sample plots of varying size, along transects of varying length or over temporal "observation windows" of varying duration. If we denote the size of the observation window as A, then all else equal it would be natural to expect a count of A times λ, where λ is the expected count per unit-sized observation window. For instance, if we expect 2 peregrine falcon pairs in a randomly placed 100 km^2 study plot in the Jura mountains, then we would probably expect 4 pairs for a study plot twice that size. When modeling variability in the expected count, we will then do this for $A * \lambda$ on the log scale. We illustrate the principle here for a simple loglinear regression on a covariate x:

$$C_i \sim Poisson(A_i * \lambda_i)$$

$$\log(A_i * \lambda_i) = \log(A_i) + \log(\lambda_i) = \log(A_i) + \alpha + \beta * x_i,$$

where the natural logarithm of A, $\log(A_i)$, is called an offset. An offset can be described as a covariate with coefficient fixed at 1, hence we can also describe the linear predictor as:

$$\log(A_i * \lambda_i) = \alpha + 1 * \log(A_i) + \beta * x_i$$

One consequence of adding an offset to a Poisson GLM is that now λ has the interpretation of a *density* relative to the units in which A is measured. The implied coefficient of 1 for the offset term in the linear model means that we expect density to be independent of the area. We could test this assumption by actually estimating a coefficient for $\log(A_i)$: if is greater than 1, density becomes greater in larger than in smaller areas, and vice versa.

Typically, the value of the offset variable A will be greater than 1, but sometimes it may be less than 1 (but greater than 0). For instance, we can specify an offset equal to log(p) to accommodate a known value of imperfect detection p in an *ad hoc* (two-step) HM for counts (Hedley and Buckland, 2004; Oedekoven et al., 2013; Solymos et al., 2013). With imperfect detection, we can write a count C as $C \sim Poisson(\lambda p)$; i.e., the expected count $E(C)$ is the product of the expected local abundance λ, and detection probability p. If we want to model variation in the expected abundance λ, while accounting for a known value of imperfect detection p (which will typically be an estimate coming from another analysis), we can express the log of the expected count as:

$$\log(E(C)) = \log(\lambda p) = \log(\lambda) + \log(p)$$

i.e., as the sum of the log(expected abundance) and log(detection probability). To specify models of the usual form for among-site variation in the expected abundance, we can then write (for the example of a loglinear regression of density λ at site i on a covariate x):

$$\log(E(C_i)) = \beta_0 + \beta_1 * x_i + \log(p_i)$$

Now, the coefficients β_0 and β_1 relate the covariate x to the expected abundance $E(N)$. This type of analysis assumes implicitly that density and detection probability are not affected by the same covariates. This is a drawback of a two-step analysis that is not shared by a joint model where both abundance and detection are formally estimated in a single hierarchical model, as in Royle et al. (2004) in the context of a distance sampling model (see also Chapters 8 and 9).

As an example for an offset in our dragonfly data set, let's standardize the mite counts by body size and regress it on another measure of size, wingspan. In this way, we are effectively modeling mite density per unit body length.

```
summary(fm10a <- glm(mites ~ pop-1 + wing, offset = log(body), family = poisson))
# summary(fm10a <- glm(mites ~ offset(log(body)) + pop-1 + wing, family = poisson))
# same

Call:
glm(formula = mites ~ pop - 1 + wing, family = poisson, offset = log(body))

Coefficients:
              Estimate  Std. Error  z value  Pr(>|z|)
popNavarra    -21.3736     6.7968   -3.145   0.00166 **
popAragon     -25.0056     8.3933   -2.979   0.00289 **
popCatalonia  -23.9713     8.1046   -2.958   0.00310 **
wing            1.8379     0.6271    2.931   0.00338 **
```

We see that mite density per unit body length increases significantly with wingspan. We also see that no coefficient is estimated for the offset term, since it is fixed at 1.

To specify in BUGS an offset for body length in a GLM of mite count on wingspan, we simply specify the linear model in one of the following ways:

```
for(i in 1:9){
   mites[i] ~ dpois(lambda[i])
   log(lambda[i]) <- log(body[i]) + alpha[pop[i]] + beta * wing[i]
#  log(lambda[i]) <- 1 * log(body[i]) + alpha[pop[i]] + beta * wing[i] # same
}
```

3.3.3 OVERDISPERSION AND UNDERDISPERSION

Neither the Poisson nor the binomial distributions have a dispersion parameter to govern the variability of the response. Rather, the magnitude of that variability is a fixed function of the mean: the variance is equal to the Poisson mean λ, equal to $Np(1-p)$ for the binomial and equal to $p(1-p)$ for the Bernoulli. Very often, we observe what is called overdispersion: that the variance is greater than expected from these expressions. In the Poisson, a simple index of overdispersion is the variance/mean ratio, which should be about 1. Note, however, that for the raw data this index is solely applicable for an intercepts-only model. The model's dispersion is what is "left over" after any structure in the mean response due to covariates has been accounted for. Hence, it would be wrong to observe a variance/mean ratio in the raw data that is greater than 1 and then conclude that a Poisson GLM was inadequate: it may well be that the "overdispersion" can be explained adequately with the available covariates. Overdispersion must be gauged *after* any known structure in the mean response has been accommodated in the model. Therefore, if we fit a Poisson model with covariates in R, we must inspect the ratio of the residual deviance to the residual degrees of freedom, and only then does a ratio substantially greater than 1 indicate overdispersion. The analogous applies to a binomial response (but note that overdispersion cannot be measured or modeled for a Bernoulli response; McCullagh and Nelder, 1989).

Overdispersion typically arises when some structure in the mean response is ignored. For instance, important covariates may have been left out of the model or hidden structure is ignored—e.g., due to clustering of individuals in groups, or due to spatial or temporal correlation. The result of overdispersion is usually that uncertainty assessments are too optimistic, because standard errors are too small and confidence intervals too short. It is important to accommodate this feature of the response. For instance, you could explicitly model such remaining structure in the expected response by adding these covariate effects or some random effects accommodating the missing grouping structure in your data. However, in practice you won't always have measured or even know them all, so various remedies are employed that simply measure the amount of overdispersion and adjust the uncertainty assessments accordingly. In frequentist analyses, quasi-likelihood is commonly employed, whereby it is assumed that the variance of the data differs from the, say, Poisson variance by a constant and estimable overdispersion factor ϕ (McCullagh and Nelder, 1989). This can also be done in more complex models; see for instance Johnson et al. (2010) for an application of overdispersion correction in a hierarchical distance sampling model (for which you may see Chapters 8 and 9). A more explicitly model-based alternative is to fit an additional zero-mean random effect that soaks up unexplained variation and in effect then acts as an additional variance parameter that is lacking in the nominal Poisson or binomial. When we fit a binomial mixture model with Poisson-lognormal distribution for abundance (see Section 6.11), we do just that. As an alternative to a Poisson, we could also adopt a negative binomial distribution that allows for extra-Poisson variability through its dispersion parameter. Indeed, a number of more flexible alternatives to the Poisson exist (e.g., see Wu et al., 2013, 2015).

Actually, "overdispersion" in the wider sense may sometimes be observed with almost any standard distribution including the normal, Poisson, and binomial, and all three have some classic alternatives that may be adopted in this case and allow for more variability (or more extreme values in the normal). For instance, when there are too many outliers in a normal model, a t-distribution is a more robust alternative (Jonsen et al., 2005). The classical overdispersed Poisson alternative is the negative binomial and an overdispersed alternative to the binomial is the beta-binomial (Gelman et al., 2014; Lynch et al., 2014). All of these can be fitted in BUGS.

The converse to overdispersion is underdispersion. This is much more rarely observed, and the usual advice is to simply ignore it. However, there may be an interest in accounting for the reduced variability in a response, because otherwise our uncertainty assessments will be too pessimistic (e.g., SEs become too wide). One example in population ecology is avian clutch size, which for most species varies somewhere between 1 and 10, and there is some upper limit that is very rarely exceeded. A Poisson with variance equal to the mean observed clutch size will frequently allow unrealistically high clutch sizes. One way of accounting for this reduced variability would be an adaptation to a Poisson (see Ridout and Besbeas, 2004; also Wu et al., 2013, 2015). A simple alternative in BUGS is to fit a binomial instead, where the trial size N is set equal to the highest clutch size ever observed. To model pattern in the mean clutch size, we can use *moment matching* (Hobbs and Hooten, 2015) and model $\log(Np)$ as a linear function of covariates. See Section 5.12 and also 7.6 for an example of moment matching in the modeling of a negative binomial response.

3.3.4 ZERO-INFLATION

One particular way in which an observed response can have a different dispersion structure than under the nominal distribution is called zero-inflation: there are more zeros than expected under the nominal distribution. This is very common for count data, but may also occur for continuous measurements. Zero-inflation can be thought to arise as a consequence of two linked processes, the first of which generates structural zeros. For instance, when modeling abundance with a Poisson distribution or species occurrence with a Bernoulli, there may be sites that are unsuitable in principle. These sites will necessarily produce zero abundance, beyond the sampling zeros arising from the random process represented by the Poisson or the Bernoulli distribution. Thus, a zero-inflated Poisson (ZIP) distribution for abundance N can be described as follows

$$s_i \sim Bernoulli(\theta) \quad \text{\# s is binary indicator of site suitability}$$

$$N_i \sim Poisson(s_i \lambda) \quad \text{\# model for realized abundance}$$

Only when a site is suitable (i.e., $s_i = 1$) can abundance be greater than zero, while when $s_i = 0$, we have $N_i \sim Poisson(0)$ and will only ever observe zero abundance. Under this model, zeros can arise from both processes: we naturally get $N_i = 0$ when a site is unsuitable (i.e., $s_i = 0$), but we may also obtain a zero by chance when rolling our "Poisson dice" for a site that is suitable in principle, where $s_i = 1$. Thus, it would be wrong to say that sites with zero abundance are unsuitable under this model. We find this a perfectly natural way of thinking about zero inflation arising due to a suitability process, but some don't like the fact that zero abundance can arise from both processes. An alternative is the use of hurdle models (e.g., Potts and Elith, 2006; Dorazio et al., 2013), which adopt a zero-truncated Poisson in the second (abundance) part of the HM. Hurdle models have zero abundance coming only from the suitability process and hence, sites with zero abundance can rightly be interpreted as unsuitable.

Similarly, when modeling occurrence in the presence of zero-inflation, we have

$$s_i \sim Bernoulli(\theta) \quad \text{\# s is binary indicator of site suitability}$$

$$y_i \sim Bernoulli(s_i p) \quad \text{\# model for observed presence/absence}$$

Such zero-inflated GLMs may readily be fitted in R using functions in package `pscl` (Jackman, 2012), and they are easy to implement in BUGS as an HM exactly analogous to the above expressions (Zuur et al., 2012). Indeed, the site-occupancy model (Chapter 10) is a special kind of zero-inflated binomial (ZIB) model, where sites at which a species is absent inflate the number of zeros in the observed detection/nondetection observations (Tyre et al., 2003).

Zero-inflated models are a powerful way to accommodate a typical feature of count data: that there often are too many zeros relative to a nominal distribution such as the Poisson or even the negative binomial. Moreover, it may sometimes make sense to imagine two linked processes that govern spatial variation in abundance: one process governs the suitability of a site and thus indirectly affects abundance by throwing into the abundance data structural zeros for N. It also seems sensible to assume that different environmental covariates affect the suitability and the abundance processes underlying the realized abundance. However, in a standard ZIP model with a single datum per site, we would strictly advise against using the same covariates in both parts of a zero-inflated model, since it seems this would induce some sort of nonidentifiability in the associated parameters. In contrast, for a ZIB model with multiple detection/nondetection observations available per site (i.e., the classical setting for a site-occupancy or an N-mixture model), this is perfectly fine since the two parts of the ZIB model really do represent two entirely distinct processes, and we have the information to differentiate between them. See also Section 1.1 for a general discussion of zero-inflated models and whether we can hope to find much biology in the zero-inflated part of such models.

3.3.5 BERNOULLI GLM: LOGISTIC REGRESSION FOR A BINARY RESPONSE

We use a Bernoulli distribution to describe patterns in binary responses—i.e., the outcomes of coin flip-like random processes. The Bernoulli distribution is also important because perhaps the more commonly encountered binomial distribution with trial size N is the sum of the outcomes of N independent Bernoulli trials. In this book, the Bernoulli (or alternatively, the binomial; see Section 3.3.7) is the canonical description of the measurement error process involved in all ecological studies of distribution or abundance, where the presence of a species or of an individual can be overlooked with probability equal to 1 minus the detection probability. That is, the probability of measurement error is the converse of detection probability. Unlike in most sciences where measurement error is typically modeled as a continuous random process (for instance using a normal distribution), for the measurement of distribution or abundance, measurement error is a binary process at the scale of each individual or the presence/absence state of species at a site. See Section 1.4 for an introduction to measurement error in population ecology.

For our dragonfly example data set, we can define a binary random variable, where the state "parasite presence" corresponds to a parasite load greater than zero. The event of a dragonfly with "parasite presence" is a so-called "success" (albeit less so for the dragonfly) and coded as 1, while the converse, "parasite absence," is traditionally called a "failure" and coded 0. We call y the indicator for a parasite load greater than zero; it contains a 1 if a dragonfly has at least one parasite and a 0 if it is parasite-free. Our Bernoulli GLM for the occurrence of at least one parasite on a dragonfly in relationship to population and body length is then:

$$y_i \sim Bernoulli(p_i)$$

$$\text{logit}(p_i) = \log\left(\frac{p_i}{1-p_i}\right) = \alpha_j + \beta * body_i$$

Here, p_i is the expected proportion of dragonflies with presence of parasites on the probability scale—i.e., with a parasite load of one or greater. This is the mean response at the level of each individual dragonfly. The linear model is applied to the logit transformation of that expected response. Thus, in a Bernoulli GLM we are modeling a proportion or probability.

```
presence <- ifelse(mites > 0, 1, 0)  # convert abundance to presence/absence
summary(fm11 <- glm(presence ~ pop-1 + body, family = binomial))
Call:
glm(formula = presence ~ pop - 1 + body, family = binomial)
[...]
Coefficients:
             Estimate  Std. Error  z value  Pr(>|z|)
popNavarra    -17.263      10.974   -1.573     0.116
popAragon     -16.710      10.557   -1.583     0.113
popCatalonia  -17.273      10.948   -1.578     0.115
body            2.297       1.416    1.622     0.105

(Dispersion parameter for binomial family taken to be 1)

    Null deviance: 12.4766 on 9 degrees of freedom
Residual deviance:  6.6209 on 5 degrees of freedom
AIC: 14.621

Number of Fisher Scoring iterations: 5
```

There is nothing new here in terms of the model specification in R. To specify this model in BUGS, we again write almost the identical thing as in algebra:

```
for(i in 1:9){
  presence[i] ~ dbern(p[i])
  logit(p[i]) <- alpha[pop[i]] + beta * body[i]
}
```

The logit link is the typical link employed in binomial or Bernoulli GLMs. Other link functions include the probit and the complementary log-log (cloglog), which are both available in R and BUGS. They give very similar results especially in the range of $0.1 < p < 0.9$, although the cloglog is asymmetric about 0.5. In custom-made MCMC algorithms, the probit can enjoy considerable efficiency benefits (Dorazio and Rodriguez, 2012; Johnson et al., 2013). The cloglog link leads to an astonishing interpretation of presence/absence data in terms of an underlying abundance; we describe this next.

3.3.6 MODELING A POISSON PROCESS FROM PRESENCE/ABSENCE DATA USING A BERNOULLI GLM WITH CLOGLOG LINK

The cloglog link function may sound obscure to most ecologists, but it has an intriguing feature: it effectively expresses the binomial success probability as a function of the intensity of an underlying

3.3 GENERALIZED LINEAR MODELS (GLMs)

Poisson process (Royle and Dorazio, 2008, p. 150; Baddeley et al., 2010). Thus, under suitable distributional assumptions of a Poisson point process, a cloglog Bernoulli GLM enables us to recover abundance estimates from mere presence/absence data! To see why, consider the fundamental relationship between occurrence (z) and abundance (N): we have an occurrence, or presence ($z = 1$), whenever abundance is greater than zero. Hence, for occupancy probability ψ we can write:

$$\psi = \text{Prob}(z = 1) = \text{Prob}(N > 0) = 1 - \text{Prob}(N = 0)$$

This expression is based on the fundamental relationship between occurrence and abundance, and does not depend on any statistical distribution assumed for N. If we assume a Poisson distribution with intensity λ, we have $\text{Prob}(N = 0) = \exp(-\lambda)$, and hence occurrence probability ψ is related to the expected abundance λ as follows:

$$\psi = 1 - \exp(-\lambda)$$

Adopting a cloglog link for occurrence probability ψ is equivalent to adopting a log link for λ, because we can rearrange the above expression to $\lambda = -\log(1 - \psi)$, and, taking logs, we get

$$\log(\lambda) = \log(-\log(1 - \psi)),$$

which is exactly the cloglog link for ψ. Hence, applying the cloglog link to the Bernoulli parameter ψ is equivalent to linear modeling of $\log(\lambda)$ for the intensity parameter of an underlying Poisson abundance model! We find this fascinating, since under suitable distributional assumptions, it allows you to estimate and model the abundance underlying an observed presence/absence process. We illustrate this for the occurrence models for mites in dragonflies by fitting the two equivalent models.

	Model for Mite Presence/Absence	**"Direct" Model for Mite Count**
Distribution	$presence_i \sim \text{Bernoulli}(\psi_i)$	$mites_i \sim \text{Poisson}(\lambda_i)$
Linear predictor	$\text{cloglog}(\psi_i) = \alpha_j + \beta * body_i$	$\log(\lambda_i) = \alpha_j + \beta * body_i$

Let's try this in practice and fit the "implicit Poisson model" to the presence/absence data via a Bernoulli GLM with cloglog link function.

```
summary(fm11a <- glm(presence ~ pop-1 + body, family = binomial(link = "cloglog")))
Coefficients:
              Estimate  Std. Error  z value  Pr(>|z|)
popNavarra    -12.447       8.545   -1.457     0.145
popAragon     -12.509       8.216   -1.523     0.128
popCatalonia  -12.410       8.459   -1.467     0.142
body            1.610       1.071    1.504     0.133
```

We compare with the Poisson model for the mite counts directly (this is `fm10`).

```
summary(fm10)
Coefficients:
              Estimate  Std. Error  z value  Pr(>|z|)
popNavarra     -6.634       2.325   -2.854   0.00432 **
popAragon      -5.878       2.216   -2.653   0.00799 **
popCatalonia   -5.519       2.290   -2.410   0.01596 *
body            0.845       0.261    3.237   0.00121 **
```

Admittedly, this is perhaps not too convincing: the coefficients of the two models don't agree very well. However, this should not be too surprising, since our sample is extremely small and we did not generate the mite counts using a Poisson distribution. In Exercise 4.8 in Chapter 4, you will see a much more convincing example of how the parameters of an underlying Poisson process can be estimated from simple presence/absence data using a Bernoulli GLM with a cloglog link.

In the BUGS language, a cloglog Bernoullli GLM is specified in either of the following:

```
for(i in 1:9){
  presence[i] ~ dbern(psi[i])
  psi[i] <- 1 - exp(-alpha[pop[i]] - beta * body[i])
}
```

```
for(i in 1:9){
  presence[i] ~ dbern(psi[i])
  cloglog(psi[i]) <- alpha[pop[i]] + beta * body[i]
}
```

Bernoulli GLMs with a cloglog link are important because they provide the natural link between occurrence and abundance, if the latter is approximately Poisson. Moreover, this model shows how occupancy probability at sites of different areas can be reconciled, namely by adoption of a complementary log-log Bernoulli regression with the logarithm of the area of the grid cells used as an offset (Baddeley et al., 2010; Dorazio, 2014; Fithian et al., 2014).

3.3.7 BINOMIAL GLM: LOGISTIC REGRESSION FOR BOUNDED COUNTS

The binomial distribution is the standard model for counts of independent and identically distributed discrete "things" that come in one of two types and which have a fixed total number (Gelman et al., 2014). As for the Poisson, "things" is again a very generic placeholder. Type refers to binary distinctions such as "presence/absence," "male/female," "dead/alive," "here/away," "reproductive/nonreproductive" or "detected/nondetected." The binomial distribution is adopted as a model for random variables that can be considered a sum of N independent Bernoulli trials, where N is called the sample or trial size. Typical examples may be the number of female nestlings in a nest with N nestlings, individuals parasitized in a group of N individuals scored for some parasite, individuals dying over a time period in a group of N individuals, or individuals detected among those present in a population of size N. A binomial count can never exceed the sample size, and hence it is naturally bounded. This is a major distinction from Poisson counts. Note also that the binomial is *not* an adequate model for every ratio of two numbers. For instance, it would not generally be adequate for a ratio of two areas—e.g., when your interest lies in the proportion of the leaf area chewed by some beetle. The binomial is a model for the ratio of two counts only.

We will see binomial GLMs many times throughout this book, since they or the elemental Bernoulli GLMs are our canonical descriptions of the observation process for distribution, abundance and species richness. At the level of the observed data, most models in this book have a binomial distribution; hence, there is a sense in which they can all be considered logistic regressions, albeit owing to latent variables such as presence or abundance, they typically have some fairly complicated random-effects structures.

For our toy dragonflies, we model as a binomial count the number of damaged wings in individual i from among the four wings (variable $damage_i$), in relation to population and body length:

$$damage_i \sim Binomial(N, p_i)$$

$$\text{logit}(p_i) = \log\left(\frac{p_i}{1 - p_i}\right) = \alpha_j + \beta * body_i$$

Here, $N = 4$ is the value of the binomial sample size, corresponding to the four wings. At the elemental level of a single wing, the expected response of this binomial (or Bernoulli) count is p_i; it is the expected proportion of damaged wings of an individual on the probability scale. Again, from observing counts, we model an underlying probability. The linear model is applied to the logit transformation of that expected response or probability. Note, however, that the expected value of $damage$ is $N * p$, or the sum of N Bernoulli trials each with probability p. In R, the response of a binomial GLM with sample size greater than one (i.e., not Bernoulli) is specified as the pair (number of successes, number of failures).

```
summary(fm12 <- glm(cbind(damage, 4-damage) ~ pop + body -1, family = binomial))
Call:
glm(formula = cbind(damage, 4 - damage) ~ pop + body - 1, family = binomial)
[...]
Coefficients:
              Estimate  Std. Error  z value  Pr(>|z|)
popNavarra     -4.7643     3.9351   -1.211    0.226
popAragon      -3.0047     3.6658   -0.820    0.412
popCatalonia   -4.7601     3.9255   -1.213    0.225
body            0.3837     0.4623    0.830    0.407

(Dispersion parameter for binomial family taken to be 1)

    Null deviance: 29.819  on 9  degrees of freedom
Residual deviance: 17.457  on 5  degrees of freedom
AIC: 32.833

Number of Fisher Scoring iterations: 4
```

In BUGS, a binomial GLM is specified exactly as in algebra above, except for the order of the arguments: success probability comes before the binomial sample size:

```
for(i in 1:9){
  damage[i] ~ dbin(p[i], 4)
  logit(p[i]) <- alpha[pop[i]] + beta * body[i]
}
```

3.3.8 THE GLM AS THE QUINTESSENTIAL STATISTICAL MODEL

There can be very few things in nature that are entirely predictable in space or time. Hence, essentially everything, in science and in life, has a chance element in it. As we have seen in Chapter 2, statisticians use probability to describe this element of chance and invoke the concept of a random variable. This is a hypothetical random process described by a probability density function. We imagine that the random variable has produced our observed data set, which we have an interest in explaining. The most common density functions include the normal, Poisson, and binomial distributions. A GLM simply represents the idea of adopting linear models for a link transformation of the expected response for certain types of random variables. Hence, the GLM is the quintessential statistical model—i.e., a formal description of a defined random process that produces some result that we are interested in explaining, describing, or predicting. The magnitude of the unpredictable part of the response is accounted for by the inherent variability of the random variable (this is the stochastic part of the response), while the average behavior of the response (its deterministic part) is accommodated by the linear model.

3.4 RANDOM EFFECTS (MIXED) MODELS

We previously defined an HM as a series of probability models for linked random variables where the outcome of one random variable is observed (this is the data), and the outcome of one or more additional random variables is not observed (or only partially so) and is thus latent. The outcomes of these entirely or at least partially latent random variables are called *random effects*. As we argued in Chapter 2, in HMs in this book, random effects arise naturally under a mechanistic view of the stochastic processes that underlie the observed data on occurrence and abundance in site-structured populations. A population at site *i* has a certain state (e.g., a species is present or absent, or we have abundance state N_i) that is observed imperfectly due to the peculiar, binary kind of measurement error induced by imperfect detection (and possibly also due to a false-positive error component). The latent occurrence or abundance state of a local population is viewed as the outcome of a stochastic process whose features we want to model; for instance, we want to express the occurrence probability or the expected abundance as a function of environmental covariates. Hence, it is natural to treat the population state as a random effect governed by some probability distribution. Notably, this type of random effect has a clear ecological meaning as one of the fundamental states in ecology: occurrence (or the element of a species distribution) and abundance.

In contrast, in the vast majority of examples of random-effects (or mixed) models in ecology, the random effects do *not* have a clear ecological interpretation. Rather, they are merely abstract constructs invoked to explain the fact that some measurements are more similar to each other than others are—i.e., to model correlations in the observed data. For instance, in a randomized-block ANOVA model, random "group effects" are introduced into the linear predictor for a response, to accommodate the fact that measurements taken in the same block or group all have a tendency to be higher or lower than the overall average (Mead, 1988; Littell et al., 2008). The shared group effect induces a correlation in the responses, and is treated as the outcome of a random process typically described by a normal distribution. Thus, random effects are effects, or parameters, that one may or may not want to estimate individually, and that are given a (prior) distribution (e.g., a normal) that itself has (hyper) parameters, such as the mean and the variance, that are estimated. We make the assumption that the group effects are *exchangeable*, which for practical purposes means independent and identically distributed.

The motivations to declare a set of effects as random—i.e., as the outcome of a stochastic process—are many and varied; see Kéry and Schaub, 2012, p. 77–82. We briefly summarize in the context of our dragonfly example from Sections 3.2.1 and 3.2.2, as far as possible:

- *Increased scope of inference and assessment of variability*: Treating the effects of population as random allows us to make an inference not only about the three particular populations sampled, but also about an entire "population of populations" from which the three populations are considered to be a sample. For instance, we can estimate the mean and the variability among the populations represented by our sample of three; these are the hyperparameters of the random effects prior.
- *Accounting for hidden structure in the data*: The classical example is the block effect in a randomized block ANOVA or in similar random-effects one-way ANOVA models (Mead, 1988; Chapter 9 in Kéry, 2010). Accounting for such hidden structure accommodates the correlations that exist and prevents pseudoreplication (Hurlbert, 1984).
- *Partitioning of variability*: Not only can we estimate the variability among populations, but we may also start to explain the differences among populations by measured covariates. Thus, whenever we want to *model* some parameters as a function of covariates, it is natural to assume that they are the outcome of a random process (i.e., random effects), the hyperparameters of which we can then model as being affected by those explanatory variables, by adopting a linear model at this higher level of the model; see Section 11.6.3 and Figure 11.12 for an example of this.
- *Modeling of correlations among parameters*: Not only can we model variability among a single set of parameters (e.g., population effects), but we can also model underlying correlations among two or more sets of parameters, such as survival and reproduction (Cam et al., 2002a; Schaub et al., 2013), juvenile and adult survival over time (Kéry and Schaub, 2012, p. 204–208) or pairs of values for occupancy and detection probability in a community occupancy model; see Section 11.6.2 for an example of how to code this in the BUGS language.
- *Modeling of spatial, temporal, or spatiotemporal correlation*: Spatial or temporal autocorrelation can be accommodated by adding correlated spatial or temporal random effects, with the correlation being a function of their distance in space or time (Banerjee et al., 2004; Cressie and Wikle, 2011; also see Chapter 21).
- *Partial pooling, borrowing strength, and "shrinkage"*: Treating population effects as random can be seen as a compromise between assuming all populations are equal (complete pooling of effects; corresponding to a model without pop effects at all) on the one hand, and assuming that they are totally unrelated (no pooling) on the other (Gelman, 2006; Gelman and Hill, 2007), corresponding to a model with fixed pop effects. By treating the effects as exchangeable—i.e., as similar but not identical—the estimate of each population is affected not only by the dragonflies measured in it, but also by the other six dragonflies measured in the other two populations. In this way, the estimate for each population "borrows strength" from the ensemble of populations. This means that random-effects estimates of pop will *not* be the same as fixed-effects estimates; rather, they will be pulled in towards their overall mean, and this is called "shrinkage" (Carlin and Louis, 2009; see Figure. 3.4).
- *Combining information*: Treating a set of parameters as random is a way of combining their information. For instance, parameters may be estimates from different studies. Treating them as random allows one to estimate their mean hyperparameter, which is a single measure of effect size that combines the information from all the studies (McCarthy and Masters, 2005; Schaub and Kéry, 2012). This is also called a meta-analysis in certain settings.

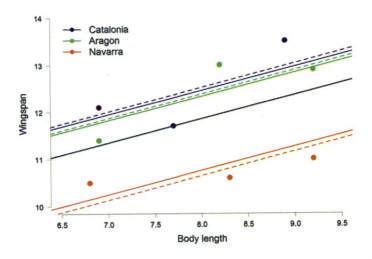

FIGURE 3.4

Comparison of lines of best fit under a pop+body linear model with fixed pop effects (dashed lines) and random pop effects (solid lines), respectively. The fixed pop effects (α_j) are estimated as three completely unrelated parameters, while when assumed random, they are estimated subject to the additional assumption $\alpha_j \sim Normal(\mu_\alpha, \sigma_\alpha^2)$—i.e., as related quantities. The random-effects lines are pulled in ("shrunk") towards the grand mean, μ_α, which is represented by the solid black line.

Treating sets of parameters as random has many advantages, but should perhaps not be done by default (though see Gelman, 2005): if the assumption of exchangeability does not hold, we will mix apples and oranges and may obtain meaningless estimates of hyperparameters, and hence random-effects estimates that are shrunk towards a nonsensical overall mean. Further, if interest lies in measuring the variation among random effects, a certain number is required; for instance, it does not make much sense to estimate a variance among populations in our dragonfly example with only three populations; we simply do this for illustrative purposes here. To obtain an adequate estimate of the among-population heterogeneity—that is, the variance parameter—at least 5–10 populations might be required. And finally, random-effects models are typically computationally (much) more expensive, both in frequentist and in Bayesian modes of analysis. Hence, there may be good reasons to not always treat all factors as random.

In the remainder of this chapter, we illustrate traditional random- or mixed-effects models, first for a normal and then for a Poisson response. These are mixed-effects models because they contain both random and fixed effects, and we call them traditional because they represent the typical motivation for random effects, to account for some hidden structure in the data: by treating the population effects as random, we account for correlated measurements among dragonflies in the same population. These population effects are not something with a clear ecological interpretation, but rather some elusive tendency for some populations to be above and others to be below the population average. Since these random population effects are continuous quantities, we assume a normal distribution for them. This is what is done in the vast majority of mixed-effects models in ecology and is also the default and only option in major standard software to fit such traditional mixed models, such as function lmer in the R package lme4 (Bates et al., 2014). However, as we have seen in the site-occupancy and *N*-mixture models in Chapter 2, random effects may be discrete and have distributions other than a normal; for instance, a Bernoulli or Poisson, respectively.

3.4.1 RANDOM EFFECTS FOR A NORMAL DATA DISTRIBUTION: NORMAL-NORMAL GLMM

The distribution assumed for the observed random variable—i.e., the data—is often called the data distribution or observation model. We will use our toy dragonfly example to illustrate a random-effects version of the linear regression model, and especially shrinkage; compared with their fixed-effects counterparts, random-effects estimates are pulled in (or "shrunk") towards the mean of their prior distribution. Here, we revisit the ANCOVA example from Section 3.2.1 and recreate the figure from before, but now without keeping track of the sex of the nine dragonflies. This is a normal-normal mixed model or an HM, because the distributions of the data and the random population effects are both normal.

```
# Plot data without distinguishing sex
plot(body, wing, col = rep(c("red", "blue", "green"), each = 3), xlim = c(6.5, 9.5), ylim =
c(10, 14), cex = 1.5, lwd = 2, frame.plot = FALSE, las = 1, pch = 16, xlab = "Body length", ylab =
"Wingspan")
```

The model adopted in Section 3.2 assumed a linear relationship between wingspan and length, a different baseline in each population, and residuals ε_i coming from a zero-mean normal distribution with variance σ^2.

$$wing_i = \alpha_j + \beta * length_i + \varepsilon_i$$

$$\varepsilon_i \sim Normal(0, \sigma^2)$$

As this model is written so far, the population effects α_j are estimated as completely unrelated numbers. On the other hand, if we assume that they are exchangeable, we can make the assumption that they are draws from a common distribution with shared (hyper)parameters that are estimated as part of the model fitting. That is, we add to the model the following assumption:

$$\alpha_j \sim Normal(\mu_\alpha, \sigma_\alpha^2)$$

This last assumption is the only difference between a fixed-effects version of the "ANCOVA linear model" and a random-effects version of it! We now use R to fit both models, and plot the resulting regression lines (Figure 3.4). For the random-effects model, we use the REML method, a variant of maximum likelihood that is better suited for mixed models (McCulloch and Searle, 2001; Littell et al., 2008), and fit the model with the function `lmer` in the R package `lme4`, which we load first. Note that in `lmer` (and function `glmer`; see below), terms appearing after the tilde (~) sign are assumed to be fixed effects except when they are between parentheses. For instance, a random intercept term for the levels of the factor `pop` is specified as (1 | pop), and a model with random intercepts *and* random slopes for each population is specified as (body | pop). Although this does not at all become clear from the output, by default in `lmer`, the latter defines a model where the covariance between intercepts and slopes is also an estimated parameter—i.e., where pairs of intercepts and slopes for the populations are treated as draws from a bivariate normal distribution (see p. 161−165 in Kéry, 2010). See below for how to specify the simpler model with two separate univariate normal distributions, one for the intercepts and one for the slopes. Let's now fit the two models first with α_j assumed fixed.

```
summary(lm <- lm(wing ~ pop-1 + body))    # Same as fm2
[ ... ]
Coefficients:
             Estimate  Std. Error  t value  Pr(>|t|)
popNavarra    6.5296    1.6437     3.973    0.01061 *
popAragon     8.4003    1.5916     5.278    0.00325 **
popCatalonia  8.2630    1.6437     5.027    0.00401 **
body          0.5149    0.1991     2.586    0.04908 *
[ ... ]
```

Next, we fit the model with population effects assumed random. With the function lmer, this model is fit in the following parameterization:

$$wing_i = \mu_\alpha + \gamma_j + \beta * length_i + \varepsilon_i$$

$$\varepsilon_i \sim Normal(0, \sigma^2)$$

$$\gamma_j \sim Normal(0, \sigma_\gamma^2)$$

Here, the mean μ_α is "pulled out" from random effects α_j, and instead random effects γ_j are expressed as zero-mean deviations from the intercept μ_α; hence, $\alpha_j = \mu_\alpha + \gamma_j$.

```
library(lme4)
summary(lmm1 <- lmer(wing ~ (1|pop) + body))   # Fit the model
ranef(lmm1)                                     # Print random effects

Linear mixed model fit by REML ['lmerMod']
Formula: wing ~ (1 | pop) + body
[ ... ]
Random effects:
 Groups   Name         Variance  Std.Dev.
 pop      (Intercept)  0.9861    0.9930    # This is sigma_alpha,
 Residual              0.3023    0.5498    # ... sigma,
Number of obs: 9, groups: pop, 3

Fixed effects:
            Estimate  Std. Error  t value
(Intercept)  7.7830    1.7034     4.569    # ... mu,
body         0.5084    0.1989     2.556    # ... beta,

ranef(lmm1)                                # ... and these are the gamma_j.
$pop
          (Intercept)
Navarra    -1.0894238
Aragon      0.6062096
Catalonia   0.4832142
```

We recover $\alpha_j = \mu + \gamma_j$, and compare fixed and random-effects estimates of the population-specific intercepts.

```
alpha_j <- fixef(lmm1)[1]+ranef(lmm1)$pop[,1]
cbind(fixed = coef(lm)[1:3], random = alpha_j)
                fixed   random
popNavarra   6.529633 6.693583
popAragon    8.400262 8.389216
popCatalonia 8.262966 8.266221
```

To better compare the fixed and random intercept (population) estimates, we plot the lines of best fit under the two versions of the model.

```
par(lwd = 3)
abline(lm$coef[1], lm$coef[4], col = "red", lty = 2)
abline(lm$coef[2], lm$coef[4], col = "blue", lty = 2)
abline(lm$coef[3], lm$coef[4], col = "green", lty = 2)
abline(alpha_j[1], fixef(lmm1)[2], col = "red")
abline(alpha_j[2], fixef(lmm1)[2], col = "blue")
abline(alpha_j[3], fixef(lmm1)[2], col = "green")
abline(fixef(lmm1), col = "black")
legend(6.5, 14, c("Catalonia", "Aragon", "Navarra"), col=c("blue", "green", "red"),
lty = 1, pch = 16, bty = "n", cex = 1.5)
```

In Figure 3.4, you can see how the random-effects regression lines are less extreme than the fixed-effects regression lines, and are pulled in towards their grand mean (i.e., the mean hyperparameter of the assumed distribution for the random effect, represented by the black line); this is shrinkage in action. In a normal-normal random-effects model, the degree of shrinkage, or pulling in towards the grand mean, depends on the ratio between the population variance σ_α^2 and the residual variance σ^2. If the population variance σ_α^2 is relatively large, we don't see much shrinkage, while if the residual variance σ^2 is relatively large, there is much shrinkage. Fixed-effects estimates can also be described as random-effects estimates with infinite variance hyperparameter σ_α^2.

To fit the random-intercepts model in BUGS, we write the following:

```
for(i in 1:9){          # Data model, loop over the individuals
  wing[i] ~ dnorm(mean[i], tau)
  mean[i] <- alpha[pop[i]] + beta * body[i]
}
for(j in 1:3){          # Parameter (random effects) model, loop over populations
  alpha[j] ~ dnorm(mu.alpha, tau.alpha) # Mean and precision = 1/variance
}
```

We could also treat the slopes as realizations of a random variable. This would result in the following model (note that now the slope β is indexed by j—i.e., it can vary by population).

$$wing_i = \alpha_j + \beta_j * length_i + \varepsilon_i$$
$$\alpha_j \sim Normal(\mu_\alpha, \sigma_\alpha^2) \qquad \text{\# Intercepts as random effects}$$
$$\beta_j \sim Normal(\mu_\beta, \sigma_\beta^2) \qquad \text{\# Slopes as random effects}$$
$$\varepsilon_i \sim Normal(0, \sigma^2) \qquad \text{\# Same old residual ``random effects''}$$

Another way of describing this model is that we fit three separate normal distributions, of which one has its mean fixed at zero. We can fit this model in R as follows:

```
summary(lmm2 <- lmer(wing ~ body + (1|pop) + (0+body|pop)))
# summary(lmm2 <- lmer(wing ~ body + (body||pop)))   # synonym
```

This notation specifies a model with random intercepts (`(1|pop)`) and random slopes (`(0+body|pop)`), but without a correlation between the α_j and the β_j. For our minute data set, this complex model does not make sense, and we don't fit it here (but see Exercise 3).

To fit the random-intercepts model in BUGS, we write the following:

```
for(i in 1:9){     # Data model, loop over the individuals
  wing[i] ~ dnorm(mean[i], tau)
  mean[i] <- alpha[pop[i]] + beta[pop[i]] * body[i]
}
for(j in 1:3){     # Parameter (random effects) model, loop over populations
  alpha[j] ~ dnorm(mu.alpha, tau.alpha) # Model for intercepts
  beta[j] ~ dnorm(mu.beta, tau.beta)    # Model for slopes
}
```

3.4.2 RANDOM EFFECTS FOR A POISSON DATA DISTRIBUTION: POISSON-NORMAL GLMM

Finally, we illustrate a traditional Poisson random-effects, or mixed, model by fitting to the mite counts a Poisson GLMM with random intercepts. This is a Poisson-normal mixed model or an HM, because the distribution of the data is Poisson and the random population effects are assumed to be draws from a normal:

$$mites_i \sim Poisson(\lambda_i)$$

$$\log(\lambda_i) = \alpha_j + \beta * length_i$$

$$\alpha_j \sim Normal(\mu_\alpha, \sigma_\alpha^2)$$

```
summary(glmm <- glmer(mites ~ body + (1|pop), family = poisson))
Generalized linear mixed model fit by maximum likelihood ['glmerMod']
  Family: poisson ( log )
Formula: mites ~ body + (1 | pop)
[ ... ]
Random effects:
 Groups Name        Variance Std.Dev.
 pop    (Intercept) 0.08535  0.2922
Number of obs: 9, groups: pop, 3

Fixed effects:
            Estimate Std. Error z value Pr(>|z|)
(Intercept)  -5.8718     2.2206  -2.644  0.00819 **
body          0.8351     0.2558   3.265  0.00110 **
```

The code in BUGS looks virtually identical to the algebraic description of the model:

```
for(i in 1:9){     # Data model, loop over the individuals
  mites[i] ~ dpois(lambda[i])
  log(lambda[i]) <- alpha[pop[i]] + beta * body[i]
}
for(j in 1:3){     # Parameter (random effects) model, loop over populations
  alpha[j] ~ dnorm(mu.alpha, tau.alpha) # Mean and precision = 1/variance
}
```

3.5 SUMMARY AND OUTLOOK

This completes our introduction to the main "ingredients" of HMs: linear models, GLMs, and random-effects models. Random effects are simply a consequence of having two or more linked probability models (i.e., an HM) for two or more outcomes from random variables, where the latent (unobserved) outcomes are the random effects. To model structure in the expected outcome of the random effects, we usually specify linear models, typically after applying a link function, exactly as for GLMs. Known structure in the mean of GLMs at each level of an HM can be accommodated with covariates, offsets, random effects, or possibly with more complex modeling of the variance (e.g., distance-related variance-covariance matrices; see Chapters 21 and 22). The ability to fit nonstandard linear models for factors and continuous covariates within the framework of the HMs in this book represents about half of the challenge for you to start to "walk" as a BUGS modeler. Moreover, if you have a good understanding of linear models, GLMs, and random effects, then you are in really good shape for most of the applied statistical analyses that are required of a modern ecologist.

Of course, you may also want to fit nonlinear models. These may be more mechanistic and directly suggested by subject matter knowledge (Pinheiro and Bates, 2000), and they may yield more reliable extrapolations beyond the observed range of the data. Though we do not show this here, nonlinear models may be fit naturally in BUGS (see, e.g., Chapter 19 in Gelman et al., 2014). In addition, generalized additive models (Hastie and Tibshirani, 1990; Wood, 2006) may be fit in BUGS using penalized splines, a sort of mixed model (Crainiceanu et al., 2005; Gimenez et al., 2006a,b; Strebel et al., 2014; see also Section 10.14).

In this chapter, we looked at some of the simplest traditional HMs, or GLMMs: a normal-normal mixed model and a Poisson-normal mixed model. In far more than 90% of applied statistical analyses in ecology, the distribution for the unobserved random variable—i.e., the random-effects distribution—is assumed to be normal. In sharp contrast, in most HMs in this book except for some models in Chapter 12, we will not have such traditional GLMMs, but rather will encounter nonstandard GLMMs with random effects that are assumed to be draws from a Bernoulli or Poisson distribution. This is perhaps surprising at first, and may strike you as strange; but nevertheless, the principle of hierarchical modeling, that a complex model is built as a sequence of simpler models, in the manner of elemental GLMs, is clearly retained for these hierarchical (or "sequential") models.

In the very simple HMs at the end of this chapter, specification of one or more group effects and a residual (which is implicit in the Poisson case) represents a simple form of modeling of the variance of the response by partitioning variance among hierarchical levels in the model. However, variances may also be modeled more directly—e.g., by known covariates in a linear model for the log-transformed variance. See Sections 6.11.2.2 and 11.6.3 for examples of such covariate modeling of a variance parameter in the context of an N-mixture and a community occupancy model, respectively. We may even add random effects into the linear predictor for variances of a model; see Lee et al. (2006) for examples of such doubly hierarchical GLMMs. Parametric modeling of spatial autocorrelation (see Chapters 21 and 22) could be described as a case where the variance of a response is modeled as a function of a distance covariate.

In Chapter 5, we will fit in BUGS most of the model types introduced in this chapter and compare their estimates with classical maximum likelihood estimates. Thus, Chapter 3 has provided you, in a very applied way, an illustration of the statistical theory for the GLMs and simple mixed models that we will fit with BUGS in Chapter 5. But first, in Chapter 4, we introduce data simulation, which is an extremely important topic for applied statistical analysis (Gelman and Hill, 2007), and assemble a data set to practice BUGS model fitting in Chapter 5.

EXERCISES

1. In Sections 3.2.1 and 3.2.2, fit ANCOVA models with `sex` instead of `pop`, and `color` instead of `body`.
2. Fit some or all of the linear models in Sections 3.2.1—3.2.4 to a Poisson, Bernoulli, and binomial response—i.e., to the mite counts, the indicator for heavy parasitization, and the damage variable (as per Sections 3.3.1—3.3.3).
3. See whether you can fit a model with random intercepts and random slopes (as in Section 3.4.1) to a clone of the toy data set that is 10 times larger. Here is that data set:

```
# Define and plot data (10 times larger data set than the toy data set)
clone.size <- 10       # clone size
pop <- factor(rep(c(rep("Navarra", 3), rep("Aragon", 3), rep("Catalonia", 3)),
   levels = c("Navarra", "Aragon", "Catalonia"), clone.size))
wing <- rep(c(10.5, 10.6, 11.0, 12.1, 11.7, 13.5, 11.4, 13.0, 12.9), clone.size)
body <- rep(c(6.8, 8.3, 9.2, 6.9, 7.7, 8.9, 6.9, 8.2, 9.2), clone.size)
sex <- rep(factor(c("M","F","M","F","M","F","M","F","M"), levels = c("M", "F")),
   clone.size)
mites <- rep(c(0, 3, 2, 1, 0, 7, 0, 9, 6), clone.size)
color <- rep(c(0.45, 0.47, 0.54, 0.42, 0.54, 0.46, 0.49, 0.42, 0.57), clone.size)
damage <- rep(c(0,2,0,0,4,2,1,0,1), clone.size)
```

4. Fit the analogous Poisson GLMM to the mite counts for the larger clone of the data set.
5. Fit a random-intercepts and a random-intercepts-and-slopes model (as in Section 3.4.2) for a binomial response (for `damage`).

CHAPTER

INTRODUCTION TO DATA SIMULATION

4

CHAPTER OUTLINE

4.1 What Do We Mean by Data Simulation, and Why Is It So Tremendously Useful? 123
4.2 Generation of a Typical Point Count Data Set 125
    4.2.1 Initial Steps: Sample Size and Covariate Values 127
    4.2.2 Simulating the Ecological Process and Its Outcome: Great Tit Abundance 128
    4.2.3 Simulating the Observation Process and Its Outcome: Point Counts of Great Tits 131
4.3 Packaging Everything in a Function 135
4.4 Summary and Outlook 140
Exercises 142

4.1 WHAT DO WE MEAN BY DATA SIMULATION, AND WHY IS IT SO TREMENDOUSLY USEFUL?

By data simulation, we simply mean the generation of random numbers from a stochastic process that is described by a series of distributional statements, such as $\alpha_i \sim Normal(\mu, \sigma_\alpha^2)$ and $y_{ij} \sim Normal(\alpha_i, \sigma^2)$, for a normal-normal mixed model; see Section 3.4.1. Data simulation is so exceedingly useful for your work as a quantitative ecologist, and moreover is done so frequently in this book, that we dedicate a whole chapter to it. Here is why:

1. *Truth is known*: We know the values of simulated parameters, and we can therefore compare the model estimates with those. Fitting a model and obtaining parameter estimates that resemble the input values is a good check that your coding in BUGS (or any other language) is right, and that the algorithmic MCMC black box in BUGS is doing the right thing.
2. *Calibrate derived model parameters*: Sometimes the effect of one or more parameters on the desired output of a model is unclear, such as the values of apparent survival and recruitment on the extinction probability of a population. Being able to tune the parameters such that meaningful data sets arise can be useful. Related to that, simulations can be regarded as controlled experiments, as simplified versions of a real system, in order to test how varying certain parameters affects estimates of other parameters—e.g., for a sensitivity analysis (T. L. Crewe, *pers. comm.*). Controlled experiments are impossible in many ecological studies, and simulation is one way to approximate them.
3. *Sampling error*: This can be described as the natural variability of the data and of the statistics computed from them, such as sample means or other parameter estimates (we find "error" a slightly

misleading term and think that "natural variability" or even noise would be better terms). This variability arises because we have not measured every member in a heterogeneous statistical population of interest. The magnitude of sampling error determines our measures of statistical precision (e.g., standard errors and confidence intervals), which can be computed based on theory even from a single sample (i.e., from a single data set). In our experience, most ecologists find it very challenging to grasp the concept of sampling error—i.e., the natural variability among a hypothetical large collection of replicate data sets—when all they have is usually just a single replicate. Repeatedly turning the handle on the data-generating stochastic process represented by your data simulation code, and observing the resulting variability among simulated data sets or things computed from them such as parameter estimates, is a fantastic learning experience because it allows you to actually *observe* sampling error directly.

4. *Check the frequentist operating characteristics of estimators (e.g., bias, precision)*: It can be very useful to see how good your estimates are expected to be for given sample sizes and parameter values. To assess the estimator bias (or "is my estimate on target on average?") or precision (or "how repeatable, or variable around their averages, are the estimates?"), it is most straightforward to analyze a large number of simulated data sets for different choices of sample size or parameter values, and compute the difference between the mean of the estimates and truth (bias), and the variance of the estimates (precision).

5. *Power analyses*: Power is the probability to detect an effect in the data when it is really there. Analyzing a large number of simulated data sets is the most flexible way to estimate the power of a sampling design and associated analysis method. A closely related problem is the determination of the necessary sample size to detect an effect of a certain magnitude with a chosen probability (power).

6. *Check the identifiability/estimability of model parameters*: The mere fact that we can fit a certain model to a certain data set and get some numbers out does not guarantee that we actually have the "right" kind of data to inform every parameter. A parameter may be intrinsically or extrinsically unidentifiable, the latter refers to the case where there is no information in our data set to obtain an estimate of it. Intrinsic unidentifiability refers to the case where the structure of a model does not allow us in principle to estimate a certain parameter from a certain type of data set. A famous example in ecological statistics is in the Cormack-Jolly-Seber model with full time-dependence, where the last survival and recapture probabilities are confounded, and only their product can be estimated (Kéry and Schaub, 2012, p. 217–220). With extrinsic unidentifiability, we cannot estimate a certain parameter because of the vagaries of a particular data set. For example, in an interaction-effects ANOVA model, we will not be able to estimate all interaction terms if some combinations of the factor levels are not observed. Identifiability is always a worry with complex models (Cressie et al., 2009; Lele, 2010), and it is compounded in Bayesian analyses where we will always obtain an estimate: when there is no information in the data set to inform the parameter estimate, the estimate will simply be determined by the prior (see Section 5.5.2 for a striking example). Hence, in a strictly technical sense, all parameters are always identifiable in a Bayesian analysis when proper prior distributions are used. Practically, however, we have exactly the same problem as we would have in a frequentist analysis, because a parameter estimate entirely determined by the prior will likely be unsatisfactory to most. To make things more complicated, there are intermediates between estimability and nonestimability (Catchpole et al., 2001). For simple models or for certain classes of models, a lot is known about the parameters that are intrinsically identifiable (Cole, 2012), and for some, relatively well-worked-out methods exist to check parameter identifiability (Catchpole and Morgan, 1997; Catchpole et al., 2002; Choquet and Cole, 2012; Cole, 2012). This is not the case for the vast majority of hierarchical models, however, and extrinsic unidentifiability can never be known beforehand. Hence, perhaps the most

straightforward approach to check estimability in practice is to generate many replicate data sets under a model for various sets of parameter values, estimate the parameters, and see whether the estimates cluster around the data-generating values as they should if they are estimable.

7. *Check for the robustness of estimators and effects of assumption violations*: An assumption violation can be loosely defined as the presence of an "important" effect in the data-generating model and its absence in the data-analysis model. "Important" means that this absence has a noticeable effect on the quality of the desired inference. A straightforward way to gauge the robustness of our model to the violation of an assumption such as "Y does not vary among individuals" is to: generate data under the more general model where Y *does* vary among individuals, and then analyze the data with the restricted model where Y does *not* vary among individuals. Repeating this a large number of times (e.g., 100–1000) will allow us to say how influential the violation of this assumption is for our inference, e.g. by introducing bias in the other estimates. Often, as few as 5–10 simulation replicates may be enough to give you a pretty good idea of the broad patterns.

8. Finally, *data simulation provides proof that you understand a model*: If you can simulate data under a certain model, then it is likely that you really understand that model. Simulating data sets is the opposite of analyzing a data set: you assemble a data set in the data simulation procedure and then break it down again using the analysis procedure. Analyzing data is like fixing a broken motorbike: to be able to fix the bike, you must really know all its parts and how they relate to each other. If you can take apart a motorbike and then reassemble the parts into a functioning bike, you prove to yourself that you understand how the bike works, and you will be better able to diagnose and fix problems when they arise. In a similar vein, if you can assemble a data set using its "ingredients" (parameters, covariates), you will certainly understand what the parameter estimates mean that come out of the analysis. In the same spirit, we often use data simulation in this book as a means of explaining a model or a process. We believe that this can be a very important type of explanation that is complementary to equations and may actually be much more easily understood by many ecologists than equations.

In the next section, we will use R to walk you through the generation of a data set in great detail. We build a typical point count data set, where we have one count from each of a number of sites for each of a number of surveys that are conducted over a short period such as a breeding season. After that, we package the essential parts of that R code into a function. Functions make it much easier to repeatedly execute the simulation code, and allow you to easily change key settings such as sample size or parameter values without altering the underlying code.

4.2 GENERATION OF A TYPICAL POINT COUNT DATA SET

The data set we assemble is perhaps the canonical example of a count data set in nature as we see it, and for which we employ hierarchical models to learn about the features of the two processes that have generated them: the ecological and observation models. Hence, in this chapter we also illustrate, using data simulation, the way in which observed counts of some imaginary species arise as a result of the actions of an ecological process governing spatiotemporal variation in abundance, and of an observation process embodied by two possible forms of binary measurement error. We also emphasize that occurrence is simply an information-reduced summary of abundance, something we have seen before (e.g., in Section 3.3.6). In essence, we generate a prototypical data set that embodies our view of how spatially replicated observations of the biological abundance state typically arise.

To be more concrete, we give a name to our imaginary species and call it a great tit (*Parus major*; Figure 4.1), a small passerine widespread in Eurasia. We generate a data set that contains J replicate

FIGURE 4.1

The famous great tit *(Parus major; J. Peltomäki)*.

counts at each of M sites under the "closure" assumption: that the counts at a site take place in such a short time that abundance N at the site does not change. See Section 6.9 for a typical breeding bird survey that produces data of this kind and for an analysis of a real data set on great tits.

You will recognize the same concepts in data simulation that we will later apply in model building for making inferences about distribution and abundance: we clearly distinguish the ecological process that generates a (partially) latent state of abundance, N, from the observation process that produces the observed data, the counts of great tits. We assume that the observation process is governed by imperfect detection only, and that false-positives are absent. To make the example a little more realistic, we further build into abundance the effects of elevation and forest cover, plus their interaction, by assuming that abundance decreases linearly with elevation and increases linearly with forest cover, and that the two covariates interact negatively. We introduce these abundance effects on the log scale, as is customary in Poisson GLMs (see Chapter 3).

In our simulation, we make explicit that not every great tit in the area sampled at each site will be detected; i.e., we are faced with the kind of binary measurement error described in Chapter 1. Some tits may not sing during our survey, or they may sing but we are distracted by another tit or another vocalizing species or … Indeed, there are a myriad of reasons for why we may fail to detect an individual bird in nature. Thus, we will only count an individual tit with a detection probability of p, which for the sake of illustration we will make dependent (on the logit scale) on one site covariate, elevation, and on one sampling covariate, wind speed. Birds sing less and are more difficult to hear in windier conditions. We assume that detection probability is positively related to elevation and negatively related to wind speed. Note that one covariate, elevation, affects both the ecological state (abundance) and its observation, or measurement (detection probability). We do this on purpose, because this is quite likely to happen in nature sometimes. The hierarchical models in this book are well able to tease apart such complex relationships (e.g., Kéry, 2008; and Chapters 12–13 in Kéry and Schaub, 2012). Finally, we could add an interaction effect between elevation and wind speed in detection, but we will not. In the R code, we allow you to build in such an interaction effect so that you can experiment with it if you like, but for now we drop this interaction effect by setting it to zero. In summary, we generate data under the following model, where sites are indexed i and replicate counts j, and α_3 is set to zero:

Ecological model:

$$N_i \sim Poisson(\lambda_i)$$

$$\log(\lambda_i) = \beta_0 + \beta_1 * elev_i + \beta_2 * forest_i + \beta_3 * elev_i * forest_i$$

Observation model:

$$C_{ij} \sim Binomial(N_i, p_{ij})$$

$$\operatorname{logit}(p_{ij}) = \alpha_0 + \alpha_1 * elev_i + \alpha_2 * wind_{ij} + \alpha_3 * elev_i * wind_{ij}$$

We will simulate the data from "the inside out" and from top to bottom. First, we choose the sample size and create values for the covariates. Second, we choose parameter values for the ecological model, assemble the expected abundance λ, and draw the Poisson random variables N. Third, we choose parameter values for the observation model, assemble detection probability p, and draw the binomial random variables C (i.e., the observed counts of great tits).

4.2.1 INITIAL STEPS: SAMPLE SIZE AND COVARIATE VALUES

We first choose the sample size; number of sites, and replicate counts at each site.

```
M <- 267     # Number of spatial replicates (sites)
J <- 3       # Number of temporal replicates (counts)
```

Next, we create values for the covariates. We have elevation and forest cover as *site covariates*: they differ by site only, not by survey. And we have wind speed, which is a *sampling* or *observational covariate*; it varies potentially by both site and replicate. We will simply fill arrays with zero-mean random numbers for these covariates, so that all three covariates are centered on zero and do not extend too far on either side of zero. In real data analyses, we typically center or scale our covariates to avoid numerical problems with finding the MLEs and getting convergence of the Markov chains. We here ignore one feature of real life, which is that the covariates are typically not independent of each other (e.g., forest cover may be related with elevation), but this is not relevant for us now.

To initialize the random number generator used to obtain our data, we can add the following line at the start of the simulation:

```
set.seed(24)     # Can choose seed of your choice
```

We then always obtain the same data set and therefore, up to Monte Carlo error in the MCMC analysis, also get the same estimates. For reasons explained here (http://www.mbr-pwrc.usgs.gov/pubanalysis/kerybook/), we often prefer not to set a seed for the random number generators when simulating a data set: basically, we feel that the actual data set at hand is fairly unimportant in most analyses. Your real interest usually lies in the stochastic process that has generated your data set simply as one *possible* realization, and this process is best studied by replicate data sets. By not singling out one particular data set as the "right" one, we emphasize the importance of the process that is behind the actual data set. Further, repeatedly running the code without a seed will give you a feel for sampling error. Still, for presentation purposes we use a seed here and generate values for elevation, forest cover, and wind speed that are all scaled to a range of $(-1,1)$.

```
elev <- runif(n = M, -1, 1)                          # Scaled elevation of a site
forest <- runif(n = M, -1, 1)                        # Scaled forest cover at each site
wind <- array(runif(n = M*J, -1, 1), dim = c(M, J))  # Scaled wind speed
```

We now have covariate vectors of the right dimensions, reflecting the chosen sample sizes.

4.2.2 SIMULATING THE ECOLOGICAL PROCESS AND ITS OUTCOME: GREAT TIT ABUNDANCE

To simulate the abundance of great tits at each site, we choose values for the parameters that govern the spatial variation in abundance, β_0 to β_3. The first parameter is the average expected abundance, on the log scale, of great tits when all covariates have a value of zero—i.e., the intercept of the abundance model. We usually prefer thinking about tits in terms of their abundance, rather than their log(abundance). Hence, we choose the average abundance first and then link-transform.

```
mean.lambda <- 2              # Mean expected abundance of great tits
beta0 <- log(mean.lambda)     # Same on log scale (= log-scale intercept)
beta1 <- -2                   # Effect (slope) of elevation
beta2 <- 2                    # Effect (slope) of forest cover
beta3 <- 1                    # Interaction effect (slope) of elev and forest
```

We apply the linear model and obtain the logarithm of the expected abundance of great tits, then use exponentiation to get the expected abundance of great tits, and plot everything (Figure 4.2).

```
log.lambda <- beta0 + beta1 * elev + beta2 * forest + beta3 * elev * forest
lambda <- exp(log.lambda)                # Inverse link transformation

par(mfrow = c(2, 2), mar = c(5,4,2,2), cex.main = 1)
curve(exp(beta0 + beta1*x), -1, 1, col = "red", frame.plot = FALSE, ylim = c(0, 18),
  xlab = "Elevation", ylab = "lambda", lwd = 2)
text(-0.9, 17, "A", cex = 1.5)
plot(elev, lambda, frame.plot = FALSE, ylim = c(0, 38), xlab = "Elevation", ylab = "")
text(-0.9, 36, "B", cex = 1.5)
curve(exp(beta0 + beta2*x), -1, 1, col = "red", frame.plot = FALSE, ylim = c(0, 18),
  xlab = "Forest cover", ylab = "lambda", lwd = 2)
text(-0.9, 17, "C", cex = 1.5)
plot(forest, lambda, frame.plot = FALSE, ylim = c(0, 38), xlab = "Forest cover", ylab = "")
text(-0.9, 36, "D", cex = 1.5)
```

To better show the joint relationships between the expected abundance, elevation, and forest cover, we can compute the expected abundance of great tits for a grid spanned by a range of observed values for both covariates, and then visualize the 3-D relationship (Figure 4.3(a)). Note that by doing

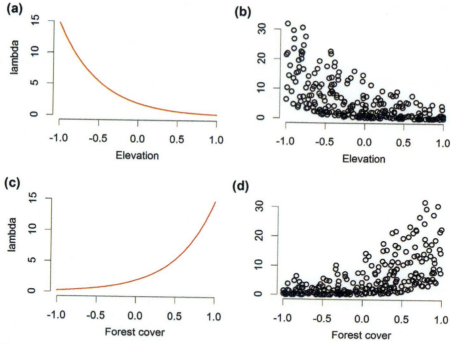

FIGURE 4.2

Two ways of showing the relationships between the expected abundance of great tits (lambda) and the two site covariates of elevation and forest cover. (a) Relationship of lambda–elevation for a constant value of forest cover (at the average of zero). (b) Relationship of lambda–elevation at the observed value of forest cover. (c) Relationship of lambda–forest cover for a constant value of elevation (at the average of zero). (d) Relationship of lambda–forest cover at the observed value of elevation.

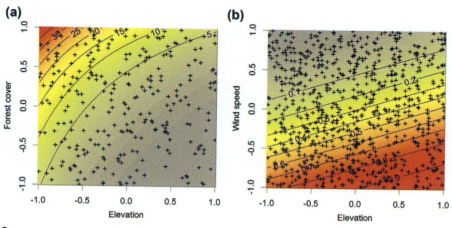

FIGURE 4.3

Relationships built into the simulated data between the expected abundance of great tits (lambda) and elevation and forest cover simultaneously (a), and between the expected detection probability of great tits (p) and elevation and wind speed simultaneously (b). Plus signs indicate the values of these covariates in the simulated data set. Note how the curvature of the contour lines in (a) indicates the presence of an interaction between the two covariates.

this, we are not changing anything with the simulated data; we simply execute this code to better visualize the relationships built into the simulated data between `lambda` (and N *and* C, below), `elev`, and `forest`.

```
# Compute expected abundance for a grid of elevation and forest cover
cov1 <- seq(-1,1,,100)                      # Values for elevation
cov2 <- seq(-1,1,,100)                      # Values for forest cover
lambda.matrix <- array(NA, dim=c(100, 100)) # Prediction matrix, for every combination
of values of elevation and forest cover
for(i in 1:100){
   for(j in 1:100){
      lambda.matrix[i,j] <- exp(beta0 + beta1 * cov1[i] + beta2 * cov2[j] + beta3 * cov1[i] * cov2[j])
   }
}

par(mfrow = c(1, 2), mar = c(5,4,3,2), cex.main = 1.6)
mapPalette <- colorRampPalette(c("grey", "yellow", "orange", "red"))
image(x = cov1, y = cov2, z = lambda.matrix, col = mapPalette(100), xlab = "Elevation",
   ylab = "Forest cover", cex.lab = 1.2)
contour(x = cov1, y = cov2, z = lambda.matrix, add = TRUE, lwd = 1)
matpoints(elev, forest, pch="+", cex=0.8)    # add observed cov values
```

Thus, we see that the slope of the positive effect of forest cover, and the negative effect of elevation on the expected abundance, are modified by the positive interaction (Figure 4.3 (a)): when both elevation and forest cover have either low or high values, the expected abundance is greater than what we expect from the main effects of these covariates. Without interaction, the contours would be straight lines (as in Figure 4.3 (b)).

So far, we have not built any stochasticity into the relationships between great tit abundance and the covariates: this comes only with the adoption of some statistical model, or a statistical distribution, to describe the random variability around the expected value `lambda`. As a typical choice, we draw the actual number of tits at each site i, N_i, from a Poisson distribution with the given expectation (λ_i).

```
N <- rpois(n = M, lambda = lambda)    # Realized abundance
sum(N)                                # Total population size at M sites
[1] 1507
table(N)                              # Frequency distribution of tit abundance
N
 0  1  2  3  4  5  6  7  8  9 10 11 12 13 14 15
61 41 27 19 14 16  7 14  5  6  5  5  5  5  3  4
16 17 18 19 20 21 22 23 24 25 26 27 28 32 38
 3  2  2  5  6  1  2  1  1  1  1  1  1  2  1
```

We have now created the result of the ecological process: site-specific abundance, N_i. We see that 61 sites are unoccupied, and the remaining 206 sites have a total of 1507 great tits, with 1 to 38 individuals each.

4.2.3 SIMULATING THE OBSERVATION PROCESS AND ITS OUTCOME: POINT COUNTS OF GREAT TITS

Abundance N is not what we usually get to see; rather, there is always a chance of overlooking or failing to hear an individual. Hence, there is a binary measurement error when "measuring" abundance (see Section 1.4). We assume here that we can make only one of the two possible observation or measurement errors: we can fail to see an individual great tit that is there, and thus detection probability is less than 1, and measurement error is affected by elevation and wind speed. We never record an individual that is not there, or count the same individual multiple times, so we assume there are no false-positives. To make explicit that we could build an interaction effect between the two covariates into our data, we next allow for an interaction effect in the code, but set it to zero, and so effectively drop it from the model used for data generation. We first choose the values for α_0 to α_3, where the first is the expected detection probability of an individual great tit, on the logit scale, when all detection covariates have a value of zero. We choose the intercept of the detection model and then link-transform. The result of this is not the same as the average detection probability, which will actually be higher in our simulation; see below.

```
mean.detection <- 0.3              # Mean expected detection probability
alpha0 <- qlogis(mean.detection)   # Same on logit scale (intercept)
alpha1 <- 1                        # Effect (slope) of elevation
alpha2 <- -3                       # Effect (slope) of wind speed
alpha3 <- 0                        # Interaction effect (slope) of elev and wind
```

Applying the linear model, we get the logit of the probability of detecting a great tit for each site and survey, and applying the inverse logit transformation, we get a matrix of dimension 267 by 3, with detection probability for each site i and survey j. Finally, we plot the relationships for detection probability in the data (Figure 4.4).

```
logit.p <- alpha0 + alpha1 * elev + alpha2 * wind + alpha3 * elev * wind
p <- plogis(logit.p)               # Inverse link transform
mean(p)                            # average per-individual p is 0.38

par(mfrow = c(2, 2), mar = c(5,4,2,2), cex.main = 1)
curve(plogis(alpha0 + alpha1*x), -1, 1, col = "red", frame.plot = FALSE, ylim = c(0, 1.1),
   xlab = "Elevation", ylab = "p", lwd = 2)
text(-0.9, 1.05, "A", cex = 1.5)
matplot(elev, p, pch = "*", frame.plot = FALSE, ylim = c(0, 1.1), xlab = "Elevation",
   ylab = "")
text(-0.9, 1.05, "B", cex = 1.5)
curve(plogis(alpha0 + alpha2*x), -1, 1, col = "red", frame.plot = FALSE, ylim = c(0, 1.1),
   xlab = "Wind speed", ylab = "p", lwd = 2)
text(-0.9, 1.05, "C", cex = 1.5)
matplot(wind, p, pch = "*", frame.plot = FALSE, ylim = c(0, 1.1), xlab = "Wind speed",
   ylab = "p")
text(-0.9, 1.05, "D", cex = 1.5)
```

We similarly produce a plot of the joint relationships between elevation, wind speed, and detection probability of an individual great tit (Figure 4.3(b)). Without an interaction, the relationship on the link scale is represented by a plane with constant slopes in the x and y directions, respectively.

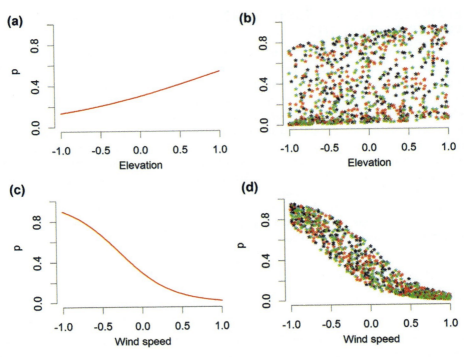

FIGURE 4.4

Two ways of depicting the relationships between the expected detection probability of an individual great tit (p) and the two covariates elevation and wind speed. (a) Relationship of p–elevation for a constant value of wind speed (at the average of zero). (b) Relationship of p–elevation at the observed value of wind speed. (c) Relationship of p–wind speed for a constant value of elevation (at the average of zero). (d) Relationship of p–wind speed at the observed value of elevation. In panels (b) and (d), colors represent different temporal replicates (surveys).

```
# Compute expected detection probability for a grid of elevation and wind speed
cov1 <- seq(-1, 1,,100)              # Values of elevation
cov2 <- seq(-1,1,,100)               # Values of wind speed
p.matrix <- array(NA, dim = c(100, 100))  # Prediction matrix which combines every value in
cov 1 with every other in cov2
for(i in 1:100){
   for(j in 1:100){
      p.matrix[i, j] <- plogis(alpha0 + alpha1 * cov1[i] + alpha2 * cov2[j] +
alpha3 * cov1[i] * cov2[j])
   }
}
image(x = cov1, y = cov2, z = p.matrix, col = mapPalette(100), xlab = "Elevation",
ylab = "Wind speed", cex.lab = 1.2)
contour(x = cov1, y = cov2, z = p.matrix, add = TRUE, lwd = 1)
matpoints(elev, wind, pch="+", cex=0.7, col = "black")   # covariate values in data set
```

4.2 GENERATION OF A TYPICAL POINT COUNT DATA SET

When "measuring" abundance, imperfect detection represents a binomial measurement error mechanism; i.e., each great tit is either detected with probability p or not detected with probability $1-p$. We apply this observation process now to produce replicate tit counts for each site.

```
C <- matrix(NA, nrow = M, ncol = J)     # Prepare array for counts
for (i in 1:J){                          # Generate counts
  C[,i] <- rbinom(n = M, size = N, prob = p[,i])
}
```

So here, at long last, are our simulated counts of great tits at 267 sites during three survey occasions! Let us look at them in tables and in a figure (Figure 4.5). Remember that sites are in the rows, and replicate surveys are in the columns. For comparison, we show the true abundance of tits in the first column.

```
head(cbind("True N" = N, "1st count" = C[,1], "2nd count" = C[,2], "3rd count" =
C[,3]), 10)              # First 10 rows (= sites)
      True N   1st count   2nd count   3rd count
 [1,]      9           8           0           1
 [2,]      9           1           2           7
 [3,]      7           0           7           5
 [4,]      0           0           0           0
 [5,]      0           0           0           0
 [6,]      0           0           0           0
 [7,]      8           0           5           1
 [8,]      5           1           3           4
 [9,]      0           0           0           0
[10,]      1           0           1           0
table(C)
C
  0   1   2   3   4   5   6   7   8   9  10  11  12  13
408 142  69  38  39  30  18  12  13   7   2   3   4   3
 14  15  16  17  21  22  23  30
  4   1   2   1   1   2   1   1

par(mfrow = c(2, 2), mar = c(5,4,2,2), cex.main = 1)
matplot(elev, C, pch = "*", frame.plot = FALSE, ylim = c(0, 38), xlab = "Elevation",
ylab = "Count (C)")
text(-0.9, 36, "A", cex = 1.5)
matplot(forest, C, pch = "*", frame.plot = FALSE, ylim = c(0, 38), xlab = "Forest cover", ylab
= "Count (C)")
text(-0.9, 36, "B", cex = 1.5)
matplot(wind, C, pch = "*", frame.plot = FALSE, ylim = c(0, 38), xlab = "Wind speed",
ylab = "Count (C)")
text(-0.9, 36, "C", cex = 1.5)
hist(C, breaks = 50, col = "grey", ylim = c(0, 460), main = "", xlab = "Count (C)")
text(3, 450, "D", cex = 1.5)
```

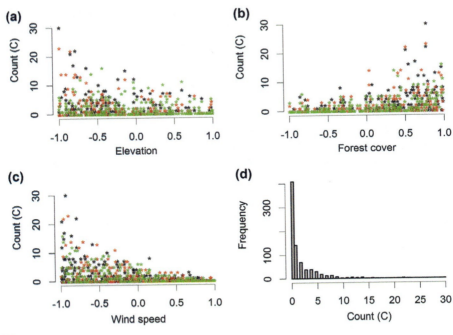

FIGURE 4.5

Relationships between the observed counts of great tits (C) and the three scaled covariates elevation (a), forest cover (b), and wind speed (c); and frequency distribution of the observed counts in the simulated data set for 267 sites with three surveys each (d). Different colors in a–c represent different temporal replicates (surveys).

Thus, we have created a data set where counts of great tits C are negatively related to elevation and wind speed and positively related to forest cover. The reason for these covariate relationships is fundamentally different. Site abundance, the logical target of ecological inference, is affected by forest cover and elevation, but *not* by wind speed, while detection probability, the parameter characterizing the measurement error process when taking measurements of abundance, is also affected by elevation, but by wind speed, too. Hence, we see that it may be challenging to disentangle the reasons for spatiotemporal variation in observed *counts*, since they can be affected by two entirely different processes: ecological and observational.

We can characterize the effects of imperfect detection in two ways: on perceived abundance and on perceived occurrence. The maximum count at each site is an observed measure for site-specific abundance of great tits, N. Hence, the sum of the maximum tit counts over sites is a readily available "detection-naïve" estimate of total abundance for all 267 sites combined.

```
sum(N)                   # True total abundance (all sites)
sum(apply(C, 1, max))    # 'Observed' total abundance (all sites)
[1] 1507
[1] 871
```

Thus, the combined estimation error for total great tit abundance of a procedure that does not model imperfect detection, but treats the counts as estimates of abundance, is (1507−871)/1507 = 0.42, representing an underestimation of great tit population size of 42%.

Occurrence or "presence," the "unit" of species distribution studies, denotes the case that $N_i > 0$ (i.e., that abundance at site i is greater than zero). Here are the corresponding figures for occurrence:

```
sum(N>0)                  # True number of occupied sites
sum(apply(C, 1, max)>0)   # 'Observed' number of occupied sites
[1] 206
[1] 190
```

Hence, in terms of great tit occurrence, the error in the total species distribution induced by imperfect detection in the sample of studied sites is only (206−190)/206, or about 8%, much less than the total error for the abundance measurement.

In later chapters, you will see hierarchical models that allow you to disentangle the parameters governing the ecological process from those of the observation process, especially in Chapter 6 where we cover the binomial N-mixture model (Royle, 2004b). This model is represented exactly by the two processes that we have assumed when simulating this data set. But this is not what we do first; for now, we generated this data set for three reasons: first, to illustrate how data sets can be simulated; second, to describe the two processes underlying all abundance measurements in nature; and third, to generate an example data set with which to introduce the Bayesian modeling software WinBUGS, OpenBUGS, and JAGS in Chapter 5.

4.3 PACKAGING EVERYTHING IN A FUNCTION

It can be very useful to package a simulation that you run repeatedly into a function. This will make your programming more concise and flexible, and it makes more transparent the settings used for data generation. Therefore, here we define a function to generate the same kind of data that we just created, assigning function arguments to any part of the simulation code that we might want to flexibly modify among simulated data sets, such as sample size, parameter values, presence/absence/magnitude of interaction terms, or detection error.

```
# Function definition with set of default values
data.fn <- function(M = 267, J = 3, mean.lambda = 2, beta1 = -2, beta2 = 2, beta3 = 1,
mean.detection = 0.3, alpha1 = 1, alpha2 = -3, alpha3 = 0, show.plot = TRUE){
#
# Function to simulate point counts replicated at M sites during J occasions.
# Population closure is assumed for each site.
# Expected abundance may be affected by elevation (elev),
# forest cover (forest) and their interaction.
# Expected detection probability may be affected by elevation,
# wind speed (wind) and their interaction.
# Function arguments:
#     M: Number of spatial replicates (sites)
```

```
#    J: Number of temporal replicates (occasions)
#    mean.lambda: Mean abundance at value 0 of abundance covariates
#    beta1: Main effect of elevation on abundance
#    beta2: Main effect of forest cover on abundance
#    beta3: Interaction effect on abundance of elevation and forest cover
#    mean.detection: Mean detection prob. at value 0 of detection covariates
#    alpha1: Main effect of elevation on detection probability
#    alpha2: Main effect of wind speed on detection probability
#    alpha3: Interaction effect on detection of elevation and wind speed
#    show.plot: if TRUE, plots of the data will be displayed;
#        set to FALSE if you are running simulations.

# Create covariates
elev <- runif(n = M, -1, 1)                         # Scaled elevation
forest <- runif(n = M, -1, 1)                       # Scaled forest cover
wind <- array(runif(n = M*J, -1, 1), dim = c(M, J)) # Scaled wind speed

# Model for abundance
beta0 <- log(mean.lambda)                  # Mean abundance on link scale
lambda <- exp(beta0 + beta1*elev + beta2*forest + beta3*elev*forest)
N <- rpois(n = M, lambda = lambda)         # Realised abundance
Ntotal <- sum(N)                           # Total abundance (all sites)
psi.true <- mean(N>0)                      # True occupancy in sample

# Plots
if(show.plot){
par(mfrow = c(2, 2), cex.main = 1)
devAskNewPage(ask = TRUE)
curve(exp(beta0 + beta1*x), -1, 1, col = "red", main = "Relationship lambda-elevation
\nat average forest cover", frame.plot = F, xlab = "Scaled elevation")
plot(elev, lambda, xlab = "Scaled elevation", main = "Relationship lambda-elevation
\nat observed forest cover", frame.plot = F)
curve(exp(beta0 + beta2*x), -1, 1, col = "red", main = "Relationship lambda-forest \ncover
at average elevation", xlab = "Scaled forest cover", frame.plot = F)
plot(forest, lambda, xlab = "Scaled forest cover", main = "Relationship lambda-forest
cover \nat observed elevation", frame.plot = F)
}

# Model for observations
alpha0 <- qlogis(mean.detection)           # Mean detection on link scale
p <- plogis(alpha0 + alpha1*elev + alpha2*wind + alpha3*elev*wind)
C <- matrix(NA, nrow = M, ncol = J)        # Prepare matrix for counts
for (i in 1:J){                            # Generate counts by survey
  C[,i] <- rbinom(n = M, size = N, prob = p[,i])
}
summaxC <- sum(apply(C,1,max))             # Sum of max counts (all sites)
psi.obs <- mean(apply(C,1,max)>0)          # Observed occupancy in sample
```

4.3 PACKAGING EVERYTHING IN A FUNCTION

```
# More plots
if(show.plot){
par(mfrow = c(2, 2))
curve(plogis(alpha0 + alpha1*x), -1, 1, col = "red", main = "Relationship p-elevation \nat average wind speed", xlab = "Scaled elevation", frame.plot = F)
matplot(elev, p, xlab = "Scaled elevation", main = "Relationship p-elevation\n at observed wind speed", pch = "*", frame.plot = F)
curve(plogis(alpha0 + alpha2*x), -1, 1, col = "red", main = "Relationship p-wind speed \n at average elevation", xlab = "Scaled wind speed", frame.plot = F)
matplot(wind, p, xlab = "Scaled wind speed", main = "Relationship p-wind speed \nat observed elevation", pch = "*", frame.plot = F)

matplot(elev, C, xlab = "Scaled elevation", main = "Relationship counts and elevation", pch = "*", frame.plot = F)
matplot(forest, C, xlab = "Scaled forest cover", main = "Relationship counts and forest cover", pch = "*", frame.plot = F)
matplot(wind, C, xlab = "Scaled wind speed", main = "Relationship counts and wind speed", pch = "*", frame.plot = F)
desc <- paste('Counts at', M, 'sites during', J, 'surveys')
hist(C, main = desc, breaks = 50, col = "grey")
}

# Output
return(list(M = M, J = J, mean.lambda = mean.lambda, beta0 = beta0, beta1 = beta1, beta2 = beta2, beta3 = beta3, mean.detection = mean.detection, alpha0 = alpha0, alpha1 = alpha1, alpha2 = alpha2, alpha3 = alpha3, elev = elev, forest = forest, wind = wind, lambda = lambda, N = N, p = p, C = C, Ntotal = Ntotal, psi.true = psi.true, summaxC = summaxC, psi.obs = psi.obs))
}
```

Once we have defined the function by executing the code of its definition in R, we can call the function repeatedly and send its results to the display, or more commonly assign them to an R object so that we can use the generated data set in an analysis.

```
data.fn()                                                    # Execute function with default arguments
data.fn(show.plot = FALSE)                                   # same, without plots
data.fn(M = 267, J = 3, mean.lambda = 2, beta1 = -2, beta2 = 2, beta3 = 1, mean.detection =
0.3, alpha1 = 1, alpha2 = -3, alpha3 = 0)                    # Explicit defaults
data <- data.fn()                                            # Assign results to an object called 'data'
```

Perhaps the simplest possible use of this function is to experience sampling error: the natural variability of repeated realizations (i.e., data sets) from our stochastic process, or of things calculated from these data sets. Let us simulate 10,000 data sets of great tit counts and see how they vary in terms of the true total population size of tits (Ntotal), and of the next-best observable thing: the sum of the max counts of tits (summaxC).

```
simrep <- 10000
NTOTAL <- SUMMAXC <- numeric(simrep)
```

```
for(i in 1:simrep){
    data <- data.fn(show.plot = FALSE)
    NTOTAL[i] <- data$Ntotal
    SUMMAXC[i] <- data$summaxC
}
plot(sort(NTOTAL), ylim = c(min(SUMMAXC), max(NTOTAL)), ylab = "", xlab = "Simulation",
col = "red", frame = FALSE)
points(SUMMAXC[order(NTOTAL)], col = "blue")
```

We see that different realizations from the identical stochastic process, as represented by the relations defined in the data-generating function, can yield dramatically different outcomes (Figure 4.6). Among the 10,000 simulation replicates, the total population size varied from about 1000 to 1850, and the sum of the maximum counts varied from 600 to 1100. On average, the ratio of the two was 58%, but varied from 50%–67%. This ratio is a measure of the detection probability over the three surveys combined; hence, summaxC clearly is not an unbiased estimator of Ntotal (for a finite number of surveys). We also see that for any given total population size N, the sum of the max counts varied by almost 200 among replicate data sets. In real life, we are typically faced with a single realization of a stochastic process such as this one, and we aim to estimate its key descriptors—i.e., the parameters in the statistical model. Seeing how much this can vary from one realization of a single process to the next, we understand that this can be a daunting task!

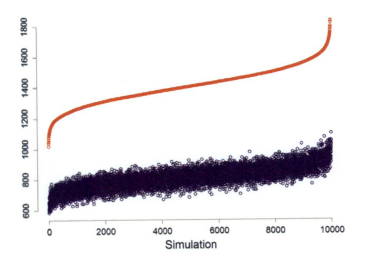

FIGURE 4.6

Natural variability (i.e., sampling error) of total population size (Ntotal; ordered by size, in red) and of the sum of the maximum counts over all sites (summaxC, in blue) in a simulated great tit population. The latter is a detection-naïve (observable) estimator of total population size. The figure shows the result of 10,000 calls to function data.fn.

4.3 PACKAGING EVERYTHING IN A FUNCTION

You can use this function to generate data sets under various sampling designs (e.g., number of sites or surveys) and for a variety of ecological and sampling situations (i.e., patterns in abundance and detection).

```
data.fn(J = 2)                    # Only 2 surveys
data.fn(J = 1)                    # No temporal replicate
data.fn(M = 1, J = 100)           # No spatial replicates, but 100 counts
data.fn(beta3 = 1)                # With interaction elev-wind on p
data.fn(M = 267, J = 3, mean.lambda = 2, beta1 = -2, beta2 = 2, beta3 = 1, mean.detection = 1)
                                  # No obs. process (i.e., p = 1, perfect detection)
data.fn(mean.lambda = 50)         # Really common species
data.fn(mean.lambda = 0.05)       # Really rare species
```

On every function execution, we get a different realization from the random process defined in the function. To get the same data set every time, we can set a random number seed.

```
set.seed(24)
data <- data.fn()                 # Default arguments
str(data)                         # Look at the object
List of 23
 $ M              : num 267
 $ J              : num 3
 $ mean.lambda    : num 2
 $ beta0          : num 0.693
 $ beta1          : num -2
 $ beta2          : num 2
 $ beta3          : num 1
 $ mean.detection : num 0.3
 $ alpha0         : num -0.847
 $ alpha1         : num 1
 $ alpha2         : num -3
 $ alpha3         : num 0
 $ elev           : num [1:267] -0.4149 -0.5502 0.4084 0.0378 0.3252 ...
 $ forest         : num [1:267] 0.432 0.446 0.97 -0.894 -0.278 ...
 $ wind           : num [1:267, 1:3] -0.8395 -0.0811 0.6636 -0.7518 -0.9838 ...
 $ lambda         : num [1:267] 9.093 11.474 9.129 0.3 0.547 ...
 $ N              : int [1:267] 9 9 7 0 0 0 8 5 0 1 ...
 $ p              : num [1:267, 1:3] 0.7784 0.2397 0.0809 0.8094 0.919 ...
 $ C              : int [1:267, 1:3] 8 1 0 0 0 0 0 1 0 0 ...
 $ Ntotal         : int 1507
 $ psi.true       : num 0.772
 $ summaxC        : int 871
 $ psi.obs        : num 0.712
```

To make the objects inside the list directly accessible to R, without having to address them as data$C for instance, you can attach data to the search path.

```
attach(data)      # Make objects inside of 'data' accessible directly
```

Remember to detach the data after use, and in particular before attaching a new data object, because more than one data set attached in the search path will cause confusion.

```
detach(data)     # Make clean up
```

4.4 SUMMARY AND OUTLOOK

The two main purposes served by this chapter were: to introduce data simulation and to reinforce, using R code, what we perceive as the two typical processes underlying *all* count data in ecology: an ecological process and an observation, or measurement, process. By data simulation, we understand the generation of a data set as the random outcome from a defined stochastic process. We described it by statistical distributions with effects of covariates built into the processes in the manner of GLMs. We have given an extended version of this code, where each step was explained and commented, and an abbreviated version where the key lines are packaged into a function with arguments that can be chosen upon calling the function. Using functions is often how we conduct data simulation, but we wanted to provide the extended version first to better illustrate the great importance of data simulation for applied statistics. We have pointed out some of the important benefits of data simulation.

- When we analyze a data set using a model similar to the one used for data simulation, we can compare the resulting inferences with the known truth. This can help avoid coding errors in the analysis, or allow us to diagnose problems with the MCMC algorithms. We will see this throughout the book.
- We have just seen how we can observe sampling error by repeatedly executing the data simulation code: unless using a seed, every single data set will differ, and so will things calculated from the observed data, such as mean counts or the significance of a parameter estimate. Sampling error can be observed by plotting or summarizing the distribution of those variables of interest across simulated data sets.
- As we will see in 6.6 for N-mixture models, and in 10.7 for site-occupancy models, we can check whether some estimator (for instance, the MLE or the posterior mean) is unbiased and gauge its precision for various sample sizes or magnitudes of measurement errors. In a similar vein, we have just seen that in the presence of imperfect detection, the sum of the maximum counts over a number of sites is not an unbiased estimator of the total abundance at these sites with three surveys. (However, it will become increasingly unbiased with increasing number of replicate surveys.)
- We can do power analyses by setting a parameter of interest at a particular value and creating and analyzing replicate data sets. Power is then estimated by the proportion of times that the parameter estimate is significant at a chosen significance level. Repeating this for different sample sizes such as number of sites or surveys can give information relevant for the design of a study.
- We can check the identifiability of a parameter.
- We can check estimator robustness and gauge the effects of assumption violations. For instance, in this chapter, we could generate data including an interaction between wind and elevation on detection, and then analyze the resulting data without the interaction to find out how robust estimates of the main effects are to omission of an existing interaction. In Chapter 6 we will see how the N-mixture model is affected by the presence of unmodeled effects, and in Chapter 16

(in volume 2), we will see how unmodeled detection heterogeneity affects the parameter estimates in a dynamic occupancy model.
- And finally, we repeat that if you *really* understand the data simulation under a specific model, you then understand that model. Hence, if you truly understand the simulation in this chapter, then you also understand the basic *N*-mixture model (Royle, 2004b) that is the subject of Chapter 6.

Just as we value the confirmation of field studies with lab experiments, we also believe that the benefits of simulating data are so great that almost any major data analysis project should be complemented with some simulation studies. Nevertheless, some may argue against the use of simulated data sets: First, ecologists often appear to immediately become less interested when they know a data set is simulated rather than "real," and second, simulated data may be "too simple" compared with real data sets. Of course, we are not saying you should *only* analyze simulated data sets; we see them as providing you with practice to be better able to manage the challenges of your real data sets. One way in which simulated data in this book are simpler than those of real data sets is that we usually do not create missing values. Rather, we generate balanced data sets; for instance, in this chapter we assumed the same number of replicate counts at every site. Of course, when you think that a certain complication such as missing data is needed in your analysis, you should simply build it into your data-generating procedure. For instance, when you worry that uneven numbers of counts among sites may affect the quality of your estimates, you could simply turn some of your data into missing values to mimic a pattern in which there are uneven numbers of counts per site, or even sites without counts. Then you can analyze these data sets to see whether there is an effect of a certain number or pattern of missing values on the estimates.

Simulating a data set has permitted us to illustrate the salient features of ecological count data: they are generated by an ecological process and an observation process. We like to call the latter a *measurement error process* because that is exactly what it is: it introduces error into our measurement C of abundance N. Although there are two possible kinds of measurement errors for counts (false-positives and false-negatives), we have only included the more common false-negative error for now, where the error rate is given by the complement of detection probability. Almost all models introduced in later chapters accommodate this type of measurement error, but some (see Chapter 19) will also accommodate false-positives. Incorporating false-positives is an active area of research in capture–recapture modeling.

Throughout this book, we will see many more examples of these topics, both data simulation, and the interplay of an ecological and a measurement error process underlying observed data on distribution and abundance. Some of these R functions will be considerably more complex and can be used to generate an even wider variety of data sets than the fairly simple function in this chapter. Data simulation is a vast field, and we have only shown a very simple example: you may want to use much more complicated ways of simulating data, either to analyze the resulting data sets or to study emerging properties of a system. One important field for the latter are so-called individual- or agent-based models (e.g., Grimm, 1999; Railsback and Grimm, 2012).

In Chapter 5, we use our function to generate a data set of counts of great tits, and analyze it to illustrate the fitting of several GLMs and simple random-effects models using WinBUGS, OpenBUGS, and JAGS, as well as using standard or restricted maximum likelihood procedures in R. Then, in Chapter 6, we will analyze exactly this type of data using the hierarchical *N*-mixture model, and we will encounter a much more complex version of a data simulation function that we will use in Chapters 6 and 7. If you understood the data simulation procedure in this chapter, then you fully understand the N-mixture model, which is a cornerstone of the hierarchical models in this book.

EXERCISES

1. *Sampling error*: Simulate count data with $p = 1$ and a single temporal replicate (and default function arguments otherwise). Fit the data-generating model using R function `glm`, and observe how variable the estimates are (perhaps run 1000 simulation replicates). This is sampling error and is quantified by the standard error of the estimates.
2. *Small-sample bias*: Repeat Exercise 1, but with a very small sample size (e.g., 5 or 10 sites), and inspect bias and precision of the maximum likelihood regression estimators.
3. *Power analysis*: With default arguments apart from $p = 1$ and a single survey per site, what is the power to detect the effects in abundance using a Poisson GLM fit with `glm`? Repeat with 20 sites and with 200 sites.
4. *Effects of ignoring the observation process*: Simulate data sets using the default arguments (but only one survey), and analyze them with a Poisson GLM with the data-generating structure in abundance. What is the effect of ignoring the measurement error (i.e., the observation process)?
5. *"Reliability" of count data (1)*: Run the data simulation function with a constant detection probability of 0.5 (i.e., set the coefficients for elevation and wind to zero). Does a constant measurement error of $(1-p) = 0.5$ standardize the counts? Will we always count the same number of tits at a particular site? Why/why not? What can be done to make the counts more reliable in the sense of "less variable"?
6. *"Reliability" of count data (2)*: Continuing, are replicated counts that are less variable more or less reliable as an index to the local population size N? Plot the SD of counts obtained with $p = 0, 0.1, \ldots, 0.9, 1$.
7. *When is the observed maximum count an unbiased estimator of abundance?* By trial and error, check how many surveys are required in the simulation at the end of Section 4.3 to make `summaxC` an unbiased estimator of `Ntotal`. Be prepared to wait for a while.
8. *Relationship of detection/nondetection data to counts and of cloglog-Bernoulli model to Poisson regression*: In 3.3.6, the Bernoulli model with complementary log-log link did not produce adequate estimates from the mere detection/nondetection data about the relationship between the actual mite counts and the covariates. We have claimed that this is probably due to the tiny sample size and serious departure from a Poisson. See whether the cloglog-Bernoulli model provides better estimates of an abundance relationship underlying detection/nondetection data for a large sample size, when the count data are in fact Poisson. Use the data simulation function in this chapter for 267 sites with a single replicate and perfect detection.
9. *Sensitivity of trends based on counts vs. trends based on detection/nondetection (presence/absence) data (see also Pollock, 2006)*: Simulate a population that declines at a chosen rate over a selected number of years. You can use parts of the simulation function and ignore the observation process (i.e., assume $p = 1$). Estimate the power to detect a population decline of your choice and for a selected number of years when you use the counts (you can use 1000 simulation reps). Then, do the same for detection/nondetection data by collapsing the counts to

0 and 1 and fitting a logistic regression. Fit both models for each simulated data set to enhance comparability (this is a more challenging exercise).

10. *Generation of overdispersion*: The variance of a Poisson random variable is equal to its mean. Use the data simulation function with the effects of all covariates set to 0, and mean detection to 1, in order to verify this. Then, add in first one, then two, and finally all three abundance covariates (keeping detection perfect all the time). Observe how the variance/mean ratio of the counts changes. What does this mean? Are these data no longer Poisson-distributed?

11. *Data simulation for hierarchical detection/nondetection data*: Modify the function such that you generate detection/nondetection (or "presence/absence") instead of abundance data, according to a site-occupancy model (see Chapter 10) with logit(psi) = covariate effects, and logit(detection) = covariate effects (see also Section 10.5).

CHAPTER 5

FITTING MODELS USING THE BAYESIAN MODELING SOFTWARE BUGS AND JAGS

CHAPTER OUTLINE

5.1 Introduction 145
5.2 Introduction to BUGS Software: WinBUGS, OpenBUGS, and JAGS... 146
5.3 Linear Model with Normal Response (Normal GLM): Multiple Linear Regression............................ 149
5.4 The R Package `rjags` 167
5.5 Missing values (NAs) in a Bayesian Analysis 169
    5.5.1 Some Responses Missing 170
    5.5.2 All Responses Missing 172
    5.5.3 Missing Values in a Covariate 174
5.6 Linear Model with Normal Response (Normal GLM): Analysis of Covariance (ANCOVA) 177
5.7 Proportion of Variance Explained (R^2) 183
5.8 Fitting a Model with Nonstandard Likelihood Using the Zeros or the Ones Tricks............................ 184
5.9 Poisson GLM 187
5.10 GoF Assessment: Posterior Predictive Checks and the Parametric Bootstrap 192
5.11 Binomial GLM (Logistic Regression) 198
5.12 Moment-Matching in a Binomial GLM to Accommodate Underdispersion 201
5.13 Random-Effects Poisson GLM (Poisson GLMM) 203
5.14 Random-Effects Binomial GLM (Binomial GLMM) 208
5.15 General Strategy of Model Building with BUGS 212
5.16 Summary and Outlook 213
Exercises 214

5.1 INTRODUCTION

In this chapter, we introduce BUGS software (WinBUGS, OpenBUGS, JAGS) by fitting some very basic models that typically serve as building blocks for more complex hierarchical models (HMs). We briefly outline the salient features of the BUGS language, point you to the use of WinBUGS and OpenBUGS as standalone software, but then run WinBUGS and JAGS exclusively from R to fit a series of basic linear and generalized linear models (GLMs). These include two examples with random effects, one with a Poisson data distribution (Poisson GLMM), and another with a binomial data distribution (binomial GLMM). This emphasizes the conceptual clarity enforced by the BUGS model definition language about what GLMs and random effects are. The two main groups of topics covered in this chapter are:

- *Bayesian MCMC engines and BUGS software*: WinBUGS, OpenBUGS and JAGS, and Bayesian MCMC-based inference; model specification, posterior inference, predictions, goodness-of-fit (GoF), and missing values (NAs), all shown through R; the great numerical resemblance between Bayesian estimates and maximum likelihood estimates (MLEs) when vague priors are used in a Bayesian analysis.
- *Applied statistics*: Linear models, GLMs, and traditional random-effects, or mixed, models (i.e., the same topics as in Chapter 3, but now illustrated with BUGS) and the things you can do with them—e.g., estimation, testing, predictions, and fit assessment using parametric bootstrap and posterior predictive distributions.

Clearly, the aim of this chapter is *not* to provide a detailed and comprehensive introduction to BUGS for GLMs and traditional mixed models: there are entire books dedicated to those topics (e.g., Gelman and Hill, 2007; McCarthy, 2007; Ntzoufras, 2009; Kéry, 2010). Even less do we strive for an overview of the vast field of linear models and GLMs, about which hundreds of books must have been written. Rather, we want to give a hands-on introduction to the use of BUGS software to fit those types of statistical models that are probably used most widely by ecologists. We also want to illustrate in practice the key topics dealt with in Chapters 2 and 3, such as how parameter estimation, testing, prediction, and model criticism are accomplished in the most widely used Bayesian software BUGS. We do expect you to have at least some applied knowledge of linear models, GLMs, and random effects, as we covered them in Chapter 3.

5.2 INTRODUCTION TO BUGS SOFTWARE: WinBUGS, OpenBUGS, AND JAGS

WinBUGS (Lunn et al., 2000, 2009, 2013), OpenBUGS (Thomas et al., 2006; Lunn et al., 2009, 2013), JAGS (Plummer, 2003) and NIMBLE (NIMBLE Development Team, 2015; de Valpine et al., in review) are generic Bayesian modeling software packages that allow you to describe almost any arbitrarily complex statistical model using the BUGS model definition language and to fit the model using MCMC techniques. The BUGS language is an ingeniously simple language, very similar to R, that lets you specify remarkably complex statistical models in a very concise and easy-to-understand way. Arguably, BUGS is currently the only software that enables ecologists without a formal training in statistics and computation to fit a very large range of fully custom statistical models with confidence. (Unfortunately, the powerful new STAN software (www.mc-stan.org; also see Gelman et al. (2014) and Korner-Nievergelt et al. (2015)) does *not* allow discrete random effects, such as we always have when modeling distribution and abundance using site-occupancy and *N*-mixture models. You can fit such models in STAN, but only when you explicitly specify the integrated likelihood. Arguably, doing this yourself is beyond the level of statistical understanding of most BUGS users.). Note that when we write "BUGS," we typically mean WinBUGS or OpenBUGS, but sometimes we mean JAGS as well and you may even imagine that we include NIMBLE, because this new software also uses a dialect of the BUGS language. We hope that the meaning of "BUGS" will become clear from the context. Functionality is very similar among the three BUGS sisters, with one major exception: WinBUGS and OpenBUGS have a module called `geoBUGS` for fitting spatial models to deal with spatial or temporal autocorrelation; see Chapters 21 and 22.

5.2 INTRODUCTION TO BUGS SOFTWARE: WinBUGS, OpenBUGS, and JAGS

BUGS and JAGS software is freely available and does four things for you:

1. Lets you describe almost any kind of statistical model in the simple and powerful BUGS **language**
2. Translates your BUGS language description of a statistical model into an **MCMC algorithm**
3. **Runs the algorithm** for as long as you wish (or have time to wait), and thereby accumulates samples of the desired joint posterior distribution for all unknown quantities in your model
4. Allows some processing of results, such as graphical or tabular **posterior summaries** and convergence assessment (not for JAGS)

WinBUGS and OpenBUGS exist as Windows standalone applications, and OpenBUGS is also available for Unix/Linux and Mac operating systems. They are available through each program's website (WinBUGS: www.mrc-bsu.cam.ac.uk/software/bugs/; OpenBUGS: www.openbugs.net/w/FrontPage). They come with a comprehensive hypertext manual and several volumes of worked and richly commented example analyses. Both manuals and example volumes are extremely rich in information, but it is fair to say that this wealth of information is not easy to navigate, most of all, because there is no index available. Development of WinBUGS ceased about 10 years ago and the developmental branch of the BUGS project has moved over to OpenBUGS. However, at the time of writing, OpenBUGS does not seem to have evolved very much beyond the capabilities of the original WinBUGS, in terms of the capabilities of the programming language or the speed of the MCMC algorithms. JAGS is developed by Martyn Plummer at the International Agency for Research on Cancer in Lyon, France. It does not have a Windows user interface, and its manual and other relevant things can be downloaded from sourceforge.net/projects/mcmc-jags/files/. NIMBLE is another project using the BUGS language see r-nimble.org/). For all three, there is a tremendous amount of documentation freely available on the web (try googling "BUGS and X," where X is your favorite model); also, there is a user group with a discussion list.

In this book, we use WinBUGS and JAGS, but note (and show in one example) that OpenBUGS can be run from R in exactly the same way as WinBUGS. We do not show how to run BUGS as a standalone program, because most people use it from R and because use of the standalone is explained in many books, including those by Woodworth (2004), McCarthy (2007), Ntzoufras (2009), Kéry (2010), and Woodward (2011). To run BUGS from R, you may use a number of packages including `R2WinBUGS` or `R2OpenBUGS` (Sturtz et al., 2005), `BRugs` (Thomas et al., 2006), `R2jags` (Su and Yajima, 2014), `rjags` (Plummer, 2015), `dclone` (Solymos, 2010), `jagsUI` (Kellner, 2015), or `runjags` (Denwood, 2015). We use `R2WinBUGS` for WinBUGS and `jagsUI` for JAGS, though we illustrate the use of `rjags` in Section 5.4. The new package `jagsUI` has similar capabilities as `R2jags`, its output is very similar to that of `R2WinBUGS`, and it avoids some problems in `R2jags` (at least as of mid-2014, the latter did not adequately separate the MCMC adaptation and burn-in phases). Package `runjags` has extensive facilities for parallel processing with JAGS, something that can be attractive given the sometimes extremely long run times of MCMC analyses. The packages `R2jags`, `jagsUI`, and `dclone` also have some facilities for running JAGS and/or WinBUGS in parallel.

Even though the three BUGS engines and also NIMBLE are totally separate programs and JAGS and NIMBLE are two entirely different enterprises, they use mere dialects of the BUGS language. So, what is the BUGS language? In a nutshell:

- The BUGS language serves to specify statistical models.
- It very much resembles R, but is not identical.
- There are "nodes" in statistical models; broadly, these are model quantities such as parameters, latent variables, predictions, and missing or observed data.

- Exactly two kinds of relations exist among nodes in a model: deterministic (represented by the arrow or assignment operator "<–"—*not* the equal sign) or stochastic (represented by the tilde " ~ "). All models that can be fitted in BUGS may be represented mathematically as so-called directed acyclic graphs, or DAGs (Gilks et al., 1994; Lunn et al., 2000, 2013).
- There is a moderate number of mathematical and statistical functions (see the manuals).
- There is a range of inbuilt statistical distributions (see the manuals); new distributions may be defined using the "zeros trick" or the "ones trick," provided you know how to write the likelihood explicitly (Lunn et al., 2013); see Section 5.8 below.
- The BUGS language is not really vectorized—we need loops to define the model for every element in vectors or multidimensional arrays. Indexing then becomes important—we use one or multiple indices to address elements of vectors or multidimensional arrays. (Note, though, that some vector or matrix operations are available, such as `inprod`).
- BUGS is a declarative language, and hence the order of statements does not matter, with the one main exception being what you put inside or outside of a loop.

This concludes our very brief first introduction to BUGS software and the BUGS language. We hope it suffices as a simple prelude to seeing BUGS "in action" for a series of basic and important statistical models in the remainder of this chapter (and in the rest of the book). We believe that learning by doing (or by watching others do), is the most powerful and quick way of learning a statistical programming language. Thus, in the remainder of this chapter we will fit a range of linear models, GLMs and random-effects (= hierarchical) models that are typical in the applied work of ecologists and indeed most empirical scientists. We illustrate, using BUGS, each type of model described in Chapter 3 using a data set of simulated counts of great tits (from Chapter 4), those being:

1. Multiple linear regression with normal errors (normal GLM): Sections 5.3–5.5, 5.8
2. Analysis of covariance (ANCOVA) model with normal errors (normal GLM): Sections 5.6 and 5.7
3. ANCOVA model with Poisson errors (Poisson GLM): Sections 5.9–5.10
4. ANCOVA model with binomial errors (logistic regression or binomial GLM): Sections 5.11–5.12
5. Poisson mixed model or Poisson generalized linear mixed model (Poisson GLMM): Section 5.13
6. Binomial mixed model or binomial generalized linear mixed model (binomial GLMM): Section 5.14

In this book, the normal distribution takes on a much less prominent role than in many statistics books, because we deal with models for counts. Counts are discrete and positive valued, and hence the normal is arguably an inadequate choice for describing such data, unlike the case for more conventional ecological data such as measurements on a continuous scale. Nevertheless, we will start by briefly illustrating the fitting of a model with normal response for the sake of illustration, and because ecologists are most familiar with a normal linear model. We do not necessarily endorse the application of every model in this chapter to the particular type of simulated data. However, we liked the idea of using a single data set to illustrate all models. Moreover, one can frequently see in the ecological literature all the kinds of analyses shown in this chapter—e.g., for better or worse the normal distribution is still quite often adopted for count data (see, e.g., state-space models for time series of population counts, Buckland et al., 2004b, 2007, and Chapter 5 in Kéry and Schaub, 2012).

The building blocks of much of what we do in hierarchical modeling of abundance, occurrence, and species richness are Poisson and binomial GLMs. We describe in the BUGS language some prototypical components of HMs; normal, Poisson, and binomial GLMs; and the specification of random effects. Almost all HMs in the remainder of this book are composed of these building blocks. We also illustrate the fitting of the same models using maximum likelihood (ML) to show the numerical similarities of Bayesian estimates and MLEs for reasonable sample sizes and vague priors adopted in Bayesian analyses. (Technically, we use least squares or iteratively reweighted least squares for linear models and GLMs, but the resulting estimates are equivalent to MLEs.)

5.3 LINEAR MODEL WITH NORMAL RESPONSE (NORMAL GLM): MULTIPLE LINEAR REGRESSION

The normal distribution does not take on a prominent role in many analyses of counts, since counts are neither continuous nor can they be negative (though the normal becomes a better approximation when the counts become large). With a data set created using the function in Chapter 4, we model the mean count of great tits assuming that they come from a normal distribution. This model is a special case of a GLM with a normal distribution and identity link, where we directly model the mean tit count as a linear function of elevation and forest cover and there is no link transformation of the expected response.

$$Cmean_i \sim Normal(\mu_i, \sigma^2)$$
$$\mu_i = \alpha_0 + \alpha_1 * elev_i + \alpha_2 * forest_i + \alpha_3 * elev_i * forest_i$$

```
# Generate data with data.fn from chapter 4
set.seed(24)
data <- data.fn()
str(data)
attach(data)

# Summarize data by taking mean at each site and plot
Cmean <- apply(C, 1, mean)
par(mfrow = c(1,3))
hist(Cmean, 50)                # Very skewed
plot(elev, Cmean)
plot(forest, Cmean)
```

To fit a model in WinBUGS or JAGS run from R and using packages `R2WinBUGS` (Sturtz et al., 2005) and `jagsUI` (Kellner, 2015), we must first prepare all the "ingredients" of the analysis. Then, we use the function `bugs` or `jags` to ship them over to BUGS and thereby instruct the BUGS engines on how they should run the analysis. After the desired number of MCMC draws has been produced by BUGS, a long list of samples from the joint posterior distribution is imported back into R and summarized in various convenient ways. This list of MCMC samples is also contained in the `coda.txt` files produced by WinBUGS. We strive for a consistent layout of all of these steps that lead to an analysis in BUGS. We do this here for the first time and hence give much more detail than in later examples.

We need to prepare the following objects: data set, model, a function for initial values for the Markov chains, a list of parameters that we want to estimate, and the MCMC settings (how many chains, for how long, etc.). The first thing we prepare is a list containing the data to be analyzed. Importantly, for WinBUGS we must include only data that are used in the model definition, while for JAGS the data package may include data not used in the model.

```r
# Package the data needed in a bundle
win.data <- list(Cmean = Cmean, M = length(Cmean), elev = elev, forest = forest)
str(win.data)                          # Check what's in win.data
```

Next, we use the `cat` function to write into the R working directory a named text file containing the model description in the BUGS language. The name of this text file is up to you (though ideally it should be shorter than the one we use here…).

```r
# Write text file with model description in BUGS language
cat(file = "multiple_linear_regression_model.txt",
"    # --- Code in BUGS language starts with this quotation mark ---
model {

# Priors
alpha0 ~ dnorm(0, 1.0E-06)          # Prior for intercept
alpha1 ~ dnorm(0, 1.0E-06)          # Prior for slope of elev
alpha2 ~ dnorm(0, 1.0E-06)          # Prior for slope of forest
alpha3 ~ dnorm(0, 1.0E-06)          # Prior for slope of interaction
tau <- pow(sd, -2)                  # Precision tau = 1/(sd^2)
sd ~ dunif(0, 1000)                 # Prior for dispersion on sd scale

# Likelihood
for (i in 1:M){
   Cmean[i] ~ dnorm(mu[i], tau)     # dispersion tau is precision (1/variance)
   mu[i] <- alpha0 + alpha1*elev[i] + alpha2*forest[i] + alpha3*elev[i]*forest[i]
}

# Derived quantities
for (i in 1:M){
   resi[i] <- Cmean[i] - mu[i]
}
}"# --- Code in BUGS language ends on this line ---
)
```

This is the first place we meet a full model written in the BUGS language; hence, a couple comments are in order. First, you must be absolutely clear about which code is R and which is BUGS. The way we write our code, everything between the quotes as indicated by the comments is in the BUGS language, and everything outside is R. The purpose of the R code is simply to write the BUGS model code to a text file. As far as R is concerned, the BUGS model file is simply a large character vector. Shortly, we will ask WinBUGS or JAGS to try to understand and do something useful with the character vector that is output by R.

In the BUGS model, you see the following quantities (nodes): `alpha0`, `alpha1`, `alpha2`, `alpha3`, `tau`, `sd`, and `mu`, as well as `Cmean`, `elev`, and `forest`. The relationship between the response `Cmean` (the observed mean count of great tits), and the mean and dispersion of the normal random variable, `mu` and `tau`, is a stochastic one; therefore, they are connected with a tilde ($\sim$)—that is, `Cmean` is drawn from a normal distribution with mean `mu` and dispersion `tau`. In contrast, the expected mean count (`mu`) is connected with covariates `elev` and `forest` in a deterministic manner, and therefore we use the arrow or assignment operator (<-), which is equivalent to an equal sign in the BUGS language.

We also see three functions: the power and the normal and the uniform distribution functions (a list of the functions available in BUGS and JAGS can be found in the manuals). Finally, we use a loop to describe the statistical model for the relationship between the covariates and the mean counts of great tits for each of the M sites in turn, with each data point indexed by `i`. We could have defined the priors after the loop over M (which is the likelihood), but not inside of that loop. Finally, in BUGS the normal distribution is defined in terms of the precision, which is the reciprocal of the variance. Hence, a prior with very small precision, such as one over one million, has a large standard deviation of 1000 and is pretty flat over a very large range of possible values of a parameter or where the likelihood has support (see also Section 5.5.2).

We try to keep the two main parts of the Bayesian analysis separate, namely the likelihood, which describes the relationship between the observed data and the unknown parameters, and the priors, which describe our knowledge about these parameters using a probability distribution. (Note that this distinction becomes blurred in HMs, as you will see later.) Priors must be specified for the primary unknown quantities in a Bayesian analysis, here, the four regression parameters, `alpha0` through `alpha3`, and the variance parameter, `tau` or `sd`. We typically use suitably wide uniform distributions or "flat" normal distributions with a small precision (corresponding to a large variance) to specify our wish to let only (or mainly) the data influence the parameter estimates. To be vague, a prior needs to be approximately flat only over the range of values where the likelihood has support—i.e., where the latter has values that are effectively nonzero. The question of whether a prior is vague for a given data set and model must be ascertained by trial and error for every new data set or model anew in a kind of informal sensitivity analysis—you make a guess about what is a vague prior, fit the model, and then make the priors more diffuse and refit the model, and if the posterior distributions do not change, then you have specified vague priors.

Next, we define a function that generates random initial values with some dispersion for at least some of the parameters. We do not need to give initial values for each parameter, since BUGS randomly initializes parameters for which no initial values are given based on the prior defined in the model. At the beginning, you may find the distinction between priors and initial values confusing; however, the two are fundamentally different. Priors are part of the model when you analyze it using Bayesian methods and they always affect the estimates (the posterior distributions), even if only very slightly. In contrast, initial values are *not* part of the model; they simply represent the starting points for the Markov chains. After chain convergence, initial values no longer have any effect on the properties of the posterior samples, and they, along with all chain values until convergence, are discarded as an MCMC "burn-in."

Even though initial values do not affect the posterior distribution, they can be of great practical importance especially for complex models. If we start the chains at too wildly improbable values for a parameter, the Markov chains may never find their way to the region in the parameter space where the likelihood does have some support—i.e., we may never get convergence. Or the MCMC algorithm may not start when unsuitable initial values are chosen. On the other hand, to gauge convergence we want the initial values of multiple chains to be "overdispersed"—i.e., to start at reasonably different places. If the chains move to the same region in the parameter space, we have much greater confidence in chain convergence. Therefore, we must strike a balance in the generation of initial values between not enough and too much "overdispersion."

Finally, initial values must not contradict the priors, the model, or the data. For instance, initializing a variance at a negative value would make BUGS crash. Similarly, in an occupancy model (see Chapter 10), if we initialize at zero (corresponding to an "absence") the latent presence/absence state z for a site where a species was detected, JAGS will crash (though not WinBUGS). This is a contradiction between model and data in the traditional occupancy model, which does not allow false-positives. Picking appropriate initial values is easy for simple models, but may become increasingly harder for more complex models (sometimes very much so). This is particularly so for JAGS, which does not forgive *any* contradictions. For complex models, there thus may be considerable trial and error involved in picking initial values. Here is a function that will create our initial values:

```
# Initial values
inits <- function() list(alpha0 = rnorm(1,0,10), alpha1 = rnorm(1,0,10), alpha2 =
rnorm(1,0,10), alpha3 = rnorm(1,0,10))
```

The next object is a list with the names of the parameters we want to estimate. While all unknown quantities described in the model are internally updated (i.e., estimated), the MCMC samples are only saved in the output for those in this list:

```
# Parameters monitored (i.e., for which estimates are saved)
params <- c("alpha0", "alpha1", "alpha2", "alpha3", "sd", "resi")
```

The final things to decide are the number of chains (nc), their length or number of iterations (ni), the length of the initial burn-in period (nb), and whether we want to discard any intermediate values ("thin") to produce a smaller but more information-dense sample from the posterior. For instance, setting nt = 10, we keep only every 10th value, which may be useful for saving disk space in very parameter-rich models. (However, there is no need to thin, since we always toss out information; MacEachern and Berliner, 1994, Link and Eaton, 2012.) The MCMC settings are also chosen by trial and error: when developing an analysis, you will often start with extremely short chains to ensure that everything is programmed fine. Then, you may run the chains longer to assess the necessary burn-in length, and finally run a production run that includes the required burn-in length and a sufficiently large post-burn-in sample:

```
# MCMC settings
ni <- 6000  ;  nt <- 1  ;   nb <- 1000  ;  nc <- 3
```

To observe convergence at the start of the chains and for testing the code, you could choose the following MCMC settings.

```
# ni <- 10  ;  nt <- 1  ;   nb <- 0  ;  nc <- 8 # not run
```

Now we have created all the objects that we need in order to use the bugs or jags functions to run an analysis. After loading the R2WinBUGS package, we run the analysis in WinBUGS first (check ?bugs if you need information about the last three function arguments in the function call below):

```
# Call WinBUGS from R (approximate run time (ART) <1 min)
library(R2WinBUGS)
bugs.dir <- "C:/WinBUGS14/"           # Place where your WinBUGS is installed
out1B <- bugs(win.data, inits, params, "multiple_linear_regression_model.txt",
n.chains = nc, n.thin = nt, n.iter = ni, n.burnin = nb, debug = TRUE,
bugs.directory = bugs.dir, working.directory = getwd())
```

The first thing you do should always be to visually inspect the trace plots to gauge convergence in WinBUGS (see Figure 5.1 for a sample). Then, we exit WinBUGS manually and look at the results

5.3 LINEAR MODEL WITH NORMAL RESPONSE (NORMAL GLM)

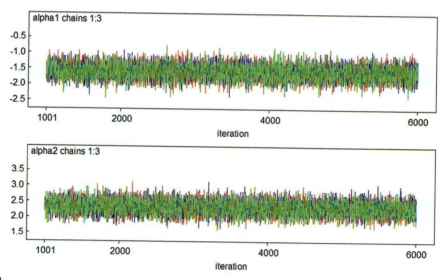

FIGURE 5.1

Trace plots produced by WinBUGS (when `debug = TRUE`) serve to visually check for convergence, here shown for two slope parameters in the multiple linear regression model. Oscillations around a horizontal level (or a "grassy" look) indicate convergence.

produced by `R2WinBUGS` in an overview. NOTE (this is important): with `debug = TRUE`, you have to manually exit WinBUGS before the BUGS results are imported into R. Before you do this, the R window is frozen.

If convergence has not been reached then we either repeat with longer chains and greater burn-in, or manually do an additional burn-in in R by tossing out additional draws at the start of the chains. You will only do the latter when a model takes very long to run; otherwise, it is much easier to simply rerun the model with changed MCMC settings.

The R object created is huge and contains many elements, of which the most frequently used are perhaps "sims.list," "summary," "mean," and "sd." Here are two overviews of the object.

```
# Overview of the object created by bugs
names(out1B)
 [1] "n.chains"         "n.iter"           "n.burnin"
 [4] "n.thin"           "n.keep"           "n.sims"
 [7] "sims.array"       "sims.list"        "sims.matrix"
[10] "summary"          "mean"             "sd"
[13] "median"           "root.short"       "long.short"
[16] "dimension.short"  "indexes.short"    "last.values"
[19] "isDIC"            "DICbyR"           "pD"
[22] "DIC"              "model.file"       "program"
```

```
str(out1B, 1)
List of 24
 $ n.chains       : num 3
 $ n.iter         : num 6000
 $ n.burnin       : num 1000
 $ n.thin         : num 1
 $ n.keep         : num 5000
 $ n.sims         : num 15000
 $ sims.array     : num [1:5000, 1:3, 1:273] 2.5 2.26 2.19 2.4 2.26 ...
  ..- attr(*, "dimnames")=List of 3
 $ sims.list      :List of 7
 $ sims.matrix    : num [1:15000, 1:273] 2.24 2.36 2.59 2.35 2.27 ...
  ..- attr(*, "dimnames")=List of 2
 $ summary        : num [1:273, 1:9] 2.36 -2.38 3.4 -1.98 2.11 ...
  ..- attr(*, "dimnames")=List of 2
 $ mean           :List of 7
 $ sd             :List of 7
 $ median         :List of 7
 $ root.short     : chr [1:7] "alpha0" "alpha1" "alpha2" "alpha3" ...
 $ long.short     :List of 7
 $ dimension.short: num [1:7] 0 0 0 0 0 1 0
 $ indexes.short  :List of 7
 $ last.values    :List of 3
 $ isDIC          : logi TRUE
 $ DICbyR         : logi FALSE
 $ pD             : num 4.98
 $ DIC            : num 1159
 $ model.file     : chr "multiple_linear_regression_model.txt"
 $ program        : chr "WinBUGS"
 - attr(*, "class")= chr "bugs"
```

Before we look any further into the results, we fit the same model in OpenBUGS using the bugs function in the R2OpenBUGS package, and finally in JAGS using the function jags in the package jagsUI (Kellner, 2015). For both, we can use exactly the same ingredients that we just prepared for the analysis in WinBUGS.

```
# Call OpenBUGS from R (ART <1 min)
library(R2OpenBUGS)
out1OB <- bugs(data=win.data, inits=inits, parameters.to.save = params, model.file =
"multiple_linear_regression_model.txt", n.chains = nc, n.thin = nt, n.iter = ni, n.burnin =
nb, debug = TRUE, working.directory = getwd())
detach("package:R2OpenBUGS", unload=T) # Otherwise R2WinBUGS is 'masked'

# Call JAGS from R (ART <1 min)
library(jagsUI)
?jags                   # Look at main function
out1J <- jags(win.data, inits, params, "multiple_linear_regression_model.txt",
n.chains = nc, n.thin = nt, n.iter = ni, n.burnin = nb)
```

5.3 LINEAR MODEL WITH NORMAL RESPONSE (NORMAL GLM)

We can easily run JAGS on multiple cores by setting the argument `parallel = TRUE`:

```
out1J <- jags(win.data, inits, params, parallel = TRUE,
"multiple_linear_regression_model.txt", n.chains = nc, n.thin = nt, n.iter = ni,
n.burnin = nb)
```

For models with long run times (several hours), this will speed up calculations by up to a factor of 3 when three cores are used (though usually the gain in speed is less). Shorter calculations may actually take longer, as the time to set up the parallel calculations outweighs the time saved by using multiple cores. The disadvantage of running JAGS in parallel is that the famous JAGS progress bar is no longer visible, thus diminishing our sense of accomplishment as we stare at our computer monitor.

For JAGS, we use the function `traceplot` to gauge convergence, either for a subset or for all parameters monitored.

```
par(mfrow = c(3,2))
traceplot(out1J, param = c('alpha1', 'alpha2', 'resi[c(1,3, 5:6)]'))  # Subset
# traceplot(out1J)                # All params
```

The object created by `jagsUI` is very similar to the one created when we run `R2WinBUGS`.

```
# Overview of object created by jags()
names(out1J)
 [1] "sims.list"   "means"         "sd"         "q2.5"        "q25"
 [6] "q50"         "q75"           "q97.5"      "overlap0"    "f"
[11] "Rhat"        "n.eff"         "pD"         "DIC"         "summary"
[16] "samples"     "modfile"       "model"      "parameters"  "mcmc.info"
[21] "run.date"    "random.seed"   "parallel"   "bugs.format"
```

Hence, using the R packages `R2WinBUGS`, `R2OpenBUGS`, and `jagsUI`, we can readily switch between WinBUGS, OpenBUGS, and JAGS for model fitting, and most of the time no change at all is required in the "ingredients" of the analysis. Next, we summarize the posterior distributions. Up to Monte Carlo (MC) error, estimates from BUGS and JAGS at convergence should be identical.

```
# Summarize posteriors from WinBUGS run
print(out1B, 2)
Inference for Bugs model at "multiple_linear_regression_model.txt", fit using WinBUGS,
 3 chains, each with 6000 iterations (first 1000 discarded)
 n.sims = 15000 iterations saved
          mean   sd   2.5%   25%    50%   75%   97.5% Rhat  n.eff
alpha0    1.66  0.12   1.44  1.58   1.66  1.74  1.89   1   15000
alpha1   -1.58  0.21  -1.98 -1.71  -1.58 -1.44 -1.17   1   15000
alpha2    2.35  0.20   1.95  2.21   2.35  2.48  2.73   1   15000
alpha3   -0.85  0.35  -1.55 -1.09  -0.86 -0.62 -0.15   1    5500
sd        1.86  0.08   1.71  1.80   1.86  1.91  2.03   1   15000
resi[1]  -0.48  0.17  -0.81 -0.59  -0.48 -0.37 -0.15   1   13000
resi[2]  -0.45  0.19  -0.82 -0.58  -0.45 -0.32 -0.08   1   11000
resi[3]   1.05  0.27   0.51  0.86   1.05  1.23  1.59   1   11000
```

```
[ output truncated ]
resi[265]   -0.61  0.13  -0.88  -0.70  -0.61  -0.52  -0.34   1  15000
resi[266]   -1.82  0.12  -2.06  -1.90  -1.82  -1.74  -1.58   1  15000
resi[267]    1.28  0.45   0.39   0.97   1.27   1.58   2.16   1  13000
deviance  1087.15  3.23 1083.00 1085.00 1086.00 1089.00 1095.00  1  15000
```

For each parameter, n.eff is a crude measure of effective sample size,
and Rhat is the potential scale reduction factor (at convergence, Rhat=1).

DIC info (using the rule, pD = Dbar-Dhat)
pD = 5.0 and DIC = 1092.2
DIC is an estimate of expected predictive error (lower deviance is better).

```
# Summarize posteriors from JAGS run
print(out1J, 2)
JAGS output for model 'multiple_linear_regression_model.txt', generated by jagsUI.
Estimates based on 3 chains of 6000 iterations,
burn-in = 1000 iterations and thin rate = 1,
yielding 15000 total samples from the joint posterior.
MCMC ran in parallel for 0.054 minutes at time 2014-11-11 14:58:27.

            mean     sd    2.5%     50%   97.5% overlap0     f  Rhat  n.eff
alpha0      1.66   0.11    1.44    1.66    1.88   FALSE   1.00    1  15000
alpha1     -1.58   0.20   -1.97   -1.58   -1.18   FALSE   1.00    1  15000
alpha2      2.34   0.20    1.96    2.34    2.73   FALSE   1.00    1  15000
alpha3     -0.85   0.36   -1.55   -0.85   -0.14   FALSE   0.99    1  10597
sd          1.86   0.08    1.71    1.86    2.03   FALSE   1.00    1   9993
resi[1]    -0.48   0.17   -0.80   -0.48   -0.15   FALSE   1.00    1  15000
resi[2]    -0.45   0.19   -0.82   -0.44   -0.08   FALSE   0.99    1  13961
resi[3]     1.05   0.28    0.50    1.05    1.60   FALSE   1.00    1  15000
[ output truncated ]
resi[265]  -0.61   0.13   -0.87   -0.61   -0.35   FALSE   1.00    1  15000
resi[266]  -1.82   0.12   -2.06   -1.82   -1.58   FALSE   1.00    1  15000
resi[267]   1.28   0.45    0.40    1.27    2.15   FALSE   1.00    1  15000
deviance 1087.09   3.18 1082.89 1086.45 1094.82   FALSE   1.00    1  15000
```

Successful convergence based on Rhat values (all < 1.1).
Rhat is the potential scale reduction factor (at convergence, Rhat=1).
For each parameter, n.eff is a crude measure of effective sample size.

overlap0 indicates if 0 falls within the 95% credible interval for the parameter.
f represents the proportion of a parameter's posterior distribution with the same
sign as the mean; i.e., our confidence that the parameter is positive or negative.

DIC info: (pD = var(deviance)/2)
pD = 5.1 and DIC = 1092.15
DIC is an estimate of expected predictive error (lower is better).

   The first thing to check in the work flow of our analysis is the penultimate column. It shows the value of the Gelman-Rubin convergence diagnostic Rhat, which is 1 at convergence (Gelman and Rubin, 1992; Brooks and Gelman, 1998); values below 1.1 or 1.2 are usually taken as indicating convergence, but not necessarily for producing posterior summaries with sufficiently low Monte Carlo error (for which, say, 1.01 or 1.001 might be better). The chains for different parameters converge at different rates, and it is entirely possible to obtain an `Rhat` value that indicates convergence for one parameter, but not for another. Similarly, `n.eff` typically varies strongly between different parameters and often even between replicate runs of an analysis. It is the estimated size of an equivalent, independent MCMC sample that contains the same amount of information as does your dependent sample. Chains with stronger autocorrelation are associated with higher values of `Rhat` and lower values of `n.eff`. You can visually check the autocorrelation for the first five parameters and up to lag n by typing this (though in this simple model, there is almost no autocorrelation):

```
n <- 10                    # maximum lag
par(mfrow = c(2, 3), mar = c(5,4,2,2), cex.main = 1)
for(k in 1:5){
   matplot(0:n, autocorr.diag(as.mcmc(out1B$sims.array[,,k]), lags = 0:n),
       type = "l", lty = 1, xlab = "lag (n)", ylab = "autocorrelation",
       main = colnames(out1B$sims.matrix)[k], lwd = 2)
   abline(h = 0)
}
```

Note also that the different BUGS engines are liable to perform very differently in terms of the MCMC efficiency (as measured by `n.eff`, or more accurately, `n.eff` per unit time). Thus, you should expect that sometimes one of the BUGS engines might not mix well, or maybe not even work for a certain class of model. For this reason, we think it is always a good idea to try more than one BUGS engine on a model, at least in the exploratory stage of an analysis.

To continue our discussion of the model output; the first two columns in the posterior summary contain the posterior means and standard deviations, which can be used for a Bayesian point estimate and an analogue to the standard error of the estimate in a frequentist analysis. Columns 3–7 show the percentiles of the posterior samples, of which the 2.5% and 97.5% form a customary 95% Bayesian confidence interval (often called a credible interval and abbreviated CRI). At the bottom of the tables we obtain the effective number of parameters (pD) and the value of the deviance information criterion (DIC; Spiegelhalter et al., 2002), which for non-HMs can be used for model selection in the same way as the Akaike's information criterion (Burnham and Anderson, 2002). Unfortunately, for HMs with strongly nonnormal posterior distributions of some parameters, (which includes those with discrete random effects such as occupancy and *N*-mixture models), and hence for essentially all models in this book, the DIC is not appropriate for model selection and so you won't see us using it.

We can easily get various graphical overviews for both analyses (Figure 5.2).

```
plot(out1B)                # For WinBUGS analysis from R2WinBUGS
plot(out1J)                # For JAGS analysis from jagsUI

par(mfrow = c(1, 2), mar = c(5,4,2,2), cex.main = 1)
whiskerplot(out1J, param = c('alpha0', 'alpha1', 'alpha2', 'alpha3', 'sd',
    'resi[c(1,3, 5:7)]'))  # For JAGS analysis from jagsUI
library(denstrip)          # Similar, but more beautiful, with package denstrip
plot(out1J$alpha0, xlim=c(-4, 4), ylim=c(1, 5), xlab="", ylab="", type="n", axes = F, main = "Density strip plots")
axis(1)
```

CHAPTER 5 FITTING MODELS USING THE BAYESIAN MODELING SOFTWARE

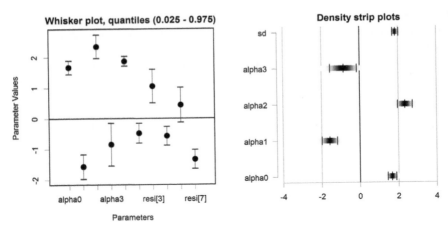

FIGURE 5.2

Two useful ways of graphically summarizing the posterior samples using the R functions whiskerplot (left) and denstrip (right).

```
axis(2, at = 1:5, labels = c('alpha0','alpha1','alpha2','alpha3','sd'), las = 1)
abline(v = c(-4,-2,2,4), col = "grey") ; abline(v = 0)
for(k in 1:5){
   denstrip(unlist(out1J$sims.list[k]), at = k, ticks = out1J$summary[k, c(3,5,7)])
}
```

For comparison, let us fit the same model using method of least-squares, which for normal models yields estimates that are identical to those obtained using the more general ML method.

```
(fm <- summary(lm(Cmean ~ elev*forest)))
Call:
lm(formula = Cmean ~ elev * forest)
[ ... ]
Coefficients:
             Estimate Std. Error t value Pr(>|t|)
(Intercept)   1.6603     0.1137   14.607  < 2e-16 ***
elev         -1.5765     0.2029   -7.771  1.76e-13 ***
forest        2.3440     0.1977   11.857  < 2e-16 ***
elev:forest  -0.8507     0.3540   -2.403  0.017 *
[ ... ]
Residual standard error: 1.849 on 263 degrees of freedom
Multiple R-squared: 0.447,   Adjusted R-squared: 0.4407
F-statistic: 70.86 on 3 and 263 DF, p-value: < 2.2e-16
```

We observe numerically very similar estimates (posterior means and sds from WinBUGS and from JAGS and MLEs with SEs from lm in R; note that no SE is given for the dispersion parameter using lm in R).

5.3 LINEAR MODEL WITH NORMAL RESPONSE (NORMAL GLM)

```
print(cbind(out1B$summary[1:5, 1:2], out1J$summary[1:5, 1:2], rbind(fm$coef[,1:2],
c(fm$sigma, NA))), 4) # WB cols 1-2, JAGS cols 3-4, lm cols 5-6
         mean      sd    mean       sd Estimate Std. Error
alpha0  1.6605 0.11523  1.660 0.11334   1.6603     0.1137
alpha1 -1.5765 0.20665 -1.575 0.20164  -1.5765     0.2029
alpha2  2.3450 0.19790  2.343 0.19852   2.3440     0.1977
alpha3 -0.8532 0.35456 -0.850 0.35716  -0.8507     0.3540
sd      1.8586 0.08223  1.858 0.08153   1.8495         NA
```

We see that when we use vague priors in a Bayesian analysis, we will typically obtain parameter estimates with great numerical resemblance to the MLEs. Moreover, the posterior standard deviations will be very similar numerically to the standard errors in the ML analysis. Thus, the fundamental thing is the model, which is the same whether analyzed in a non-Bayesian or in a Bayesian way. Whatever you may want to do with this model in a non-Bayesian analysis you can also do in a Bayesian analysis (such as, for instance, residual checks). In a frequentist analysis in R, you can do the following to do residual checks (and you will find out that there are too many large residuals for the assumed normal distribution):

```
plot(lm(Cmean ~ elev*forest))
```

In a Bayesian analysis, every unknown in a model is estimated and therefore has a posterior distribution. Residuals depend on the unknown parameters, and hence are unknown quantities themselves and have an entire posterior distribution. We can plot the residuals to check for normality visually, by seeing whether there is any evidence for lack of symmetry. We can also plot the residuals against their order in the data set and versus the predicted values for a visual check of variance homogeneity. These are three frequent residual diagnostic plots (Figure 5.3). For the predictions, we could simply have saved mu and then used its posterior mean, but since we saved the MCMC draws from its "ingredients" (the regression coefficients), we can compute the posterior mean of mu outside of BUGS in R.

```
mu <- out1B$mean$alpha0 + out1B$mean$alpha1 * elev + out1B$mean$alpha2 * forest +
out1B$mean$alpha3 * elev * forest          # Compute the posterior mean of mu

par(mfrow = c(2, 2), mar = c(5,4,2,2), cex.main = 1)
plot(1:M, out1B$summary[6:272, 1], xlab = "Order of values", ylab = "Residual",
frame.plot = F, ylim = c(-10, 15))
abline(h = 0, col = "red", lwd = 2)
segments(1:267, out1B$summary[6:272, 3], 1:267, out1B$summary[6:272, 7], col = "grey")
text(10, 14, "A", cex = 1.5)
hist(out1B$summary[6:272, 1], xlab = "Residual", main = "", breaks = 50, col = "grey",
xlim = c(-10, 15))
abline(v = 0, col = "red", lwd = 2)
text(-9, 48, "B", cex = 1.5)
qq <- qnorm(seq(0,0.9999,,data$M), mean = 0, sd = out1B$summary[5, 1])
plot(sort(qq), sort(out1B$summary[6:272, 1]), xlab = "Theoretical quantile", ylab =
"Residual", frame.plot = F, ylim = c(-10, 15)) # could also use qqnorm()
abline(0, 1, col = "red", lwd = 2)
text(-4.5, 14, "C", cex = 1.5)
```

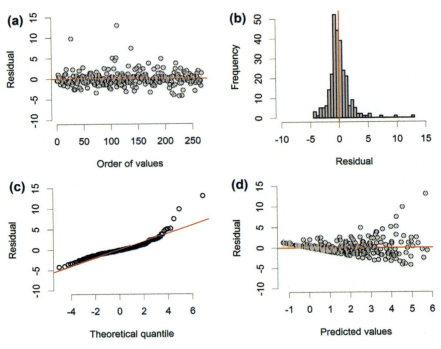

FIGURE 5.3

Residual diagnostic plots from the Bayesian analysis of the model: (a) residuals against their order, (b) histogram of residuals, (c) Q-Q plot, and (d) residuals against the fitted values. Red line shows zero, except in (c), where it is the 1:1 line.

```
plot(mu, out1B$summary[6:272, 1], xlab = "Predicted values", ylab = "Residual", frame.plot
 = F, ylim = c(-10, 15))
abline(h = 0, col = "red", lwd = 2)
segments(mu, out1B$summary[6:272, 3], mu, out1B$summary[6:272, 7], col = "grey")
text(-1, 14, "D", cex = 1.5)
```

Clearly, the homoscedasticity assumption underlying the normal model is violated somewhat, since the residuals do not form a patternless cloud around a value of zero and the Q-Q plot shows deviations from a straight line. A more appropriate analysis might use log(counts), or better a Poisson GLM, for the counts directly. However, since we show this normal model for the mean great tit count for illustration only, we are not overly concerned with this potential structural problem. The important message we want to relay is simply that any model diagnostic that you can do for a frequentist analysis, you can also do with a Bayesian analysis of the same model.

Hence, we ignore the lack of model fit and continue our illustration of a Bayesian analysis. Two things that we often want to do are (1) make a statement about how certain we are that a parameter has a particular value and (2) make predictions—i.e., compute (with uncertainty assessment) what

5.3 LINEAR MODEL WITH NORMAL RESPONSE (NORMAL GLM)

response we would expect for one value or for an entire range of values of the covariates. We illustrate this here and again compare with an ML analysis, where we can do a significance test for each of the regression coefficients.

```
fm
[...]
Coefficients:
              Estimate Std. Error t value  Pr(>|t|)
(Intercept)    1.6603     0.1137   14.607   < 2e-16 ***
elev          -1.5765     0.2029   -7.771  1.76e-13 ***
forest         2.3440     0.1977   11.857   < 2e-16 ***
elev:forest   -0.8507     0.3540   -2.403    0.017 *
```

We see that all three (and the intercept trivially so) are highly significant. We can also compute a confidence interval (CI), which has a somewhat contorted meaning in a non-Bayesian analysis: if we randomly sampled the same population of great tits many times, fitted our model, and computed a 95% CI each time, then 95% of these intervals would contain the true parameter value. Thus, the frequentist CI is an assessment of the reliability of a method, not a direct statement of uncertainty about an unknown quantity. In frequentist statistics, no probabilistic statements can be made about parameters, because parameters are not random variables. So, here are the CIs:

```
confint(lm(Cmean ~ elev*forest))
                  2.5 %      97.5 %
(Intercept)    1.436495    1.884119
elev          -1.975904   -1.177001
forest         1.954756    2.733260
elev:forest   -1.547748   -0.153619
```

In a Bayesian analysis, we use probability to express our degree of knowledge about uncertain quantities, such as the parameters in our regression model. This is the posterior distribution. Let us plot some posterior distributions and include the central range of the distributions containing 95% of their mass (Figure 5.4): this is the simplest way of computing a 95% CRI. Unlike a frequentist CI, a CRI is a direct probability statement about an unknown quantity; under the model, we can be 95% certain that the true parameter lies within this interval for the data set at hand.

```
par(mfrow = c(2, 2), mar = c(5,4,2,2), cex.main = 1)
hist(out1B$sims.list$alpha1, main = "", breaks = 100, col = "grey", freq=F)
abline(v = quantile(out1B$sims.list$alpha1, prob = c(0.025, 0.975)), col = "red", lwd = 2)
text(-2.4, 1.8, "A", cex = 1.5)
hist(out1B$sims.list$alpha2, main = "", breaks = 100, col = "grey", freq=F)
abline(v = quantile(out1B$sims.list$alpha2, prob = c(0.025, 0.975)), col = "red", lwd = 2)
text(1.7, 2, "B", cex = 1.5)
hist(out1B$sims.list$alpha3, main = "", breaks = 100, col = "grey", freq=F)
abline(v = quantile(out1B$sims.list$alpha3, prob = c(0.025, 0.975)), col = "red", lwd = 2)
text(-2.2, 1.2, "C", cex = 1.5)
hist(out1B$sims.list$sd, main = "", breaks = 100, col = "grey", freq=F)
abline(v = quantile(out1B$sims.list$sd, prob = c(0.025, 0.975)), col = "red", lwd = 2)
text(1.6, 4.9, "D", cex = 1.5)
```

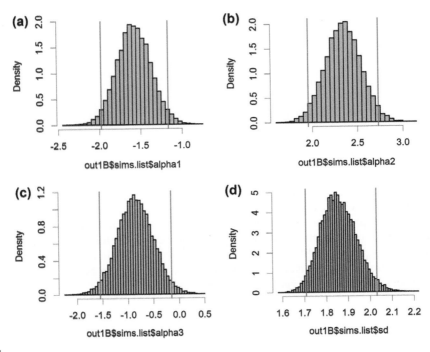

FIGURE 5.4

Marginal posterior distributions with percentile-based 95% CRIs (red) for four parameters in the normal model for mean tit counts. (a) Effect of elevation, (b) effect of forest cover, (c) interaction effect between elevation and forest cover, and (d) residual standard deviation.

Note that there are many different ways of constructing a 95% Bayesian CRI (Link and Barker, 2010), but the percentile method here is the easiest. One particular CRI is the highest-posterior density interval (HPDI), which is the shortest of such intervals, and can be computed using the `HPDinterval` function in the `coda` package.

```
HPDinterval(as.mcmc(out1B$sims.list$sd), prob = 0.95)    # HPDI
quantile(out1B$sims.list$sd, prob = c(0.025, 0.975))     # Percentile-based CRI
```

Although the fundamental meaning is quite different, CRIs in a Bayesian analysis can be used to do something analogous to a significance test in frequentist statistics: we see that zero is a fairly unlikely value for all four parameters shown and that zero is not included in the 95% CRI of any of them. Hence, we can say that we are quite certain that these parameters are different from zero. We would perhaps not want to say that the parameters are *significant*, because we feel that this term is associated too strongly with the frequentist technique of a significance test. Nevertheless, we often see Bayesians call this a significance test without even using quotes. This information is given in the posterior summary from `jags` in columns 6 and 7.

Although the posterior distribution is based on a degree-of-belief probability and does not have the same meaning as a probability ("long run frequency") in a frequentist analysis, Bayesian estimates are

5.3 LINEAR MODEL WITH NORMAL RESPONSE (NORMAL GLM)

often very well calibrated in a frequentist sense (Le Cam, 1990); i.e., their frequentist characteristics are often very good. That means, when replicated a great many times, a 95% Bayesian CRI will often contain the true value 95% of the time. Accordingly, frequentist CIs and Bayesian CRIs (when we use vague priors) are typically very similar numerically. Here they are side by side; the first two are the CI and the latter two form the CRI:

```
cbind(confint(lm(Cmean ~ elev*forest))[2:4,], out1B$summary[2:4, c(3,7)])
               2.5 %      97.5 %       2.5%       97.5%
elev        -1.975904  -1.177001  -1.981000  -1.167000
forest       1.954756   2.733260   1.954000   2.728025
elev:forest -1.547748  -0.153619  -1.551025  -0.150095
```

A particularly attractive feature when communicating the results of a Bayesian analysis is the ability to make direct probability statements about the magnitude of the parameters. For instance, from the posterior samples of `alpha1`, we can easily compute the probability that the slope of `elev` is more extreme than, say, −1.6, or that it lies between −1.8 and −1.6. Since the posterior is a probability distribution function and integrates to 1, the geometrical interpretation of these probabilities is simply the area under the posterior defined by these limits. e.g.,

```
mean(out1B$sims.list$alpha1 < -1.6)
[1] 0.4566
mean(out1B$sims.list$alpha1 < -1.6 & out1B$sims.list$alpha1 > -1.8)
[1] 0.3178667
```

Technically, what we obtain from the MCMC analysis is an estimate of a joint posterior distribution—i.e., the joint distribution of all estimated quantities. For n estimated quantities, you can imagine the joint posterior as a cloud in n-dimensional space. Let us look at this in an example in two dimensions only, for `alpha1` and `alpha2`. We can do "probability games" as before in two dimensions as well; e.g., we could test the hypothesis that the effect of `elev` lies between −1.9 and −1.6 and that of `forest` between 2.5 and 2.8. This is the proportion of samples that lie in the inner red square in Figure 5.5.

```
plot(out1B$sims.list$alpha1, out1B$sims.list$alpha2)
abline(h = c(2.5, 2.8), col = "red", lwd = 2)
abline(v = c(-1.9, -1.6), col = "red", lwd = 2)

mean(out1B$sims.list$alpha1 < -1.6 & out1B$sims.list$alpha1 > -1.9 &
out1B$sims.list$alpha2 > 2.5 & out1B$sims.list$alpha2 < 2.8)
0.08286667
```

Hence, that probability is about 8%. Thus, there is a lot of cool stuff that we can do with our posterior samples after running an MCMC algorithm that is not (easily) possible using a frequentist analysis. At many places in this book will we see examples of where we use posterior samples to compute derived quantities—e.g., functions of one or more parameters, by simply applying the function for each set of MCMC draws of the parameters. This can be done inside of the BUGS program, or outside, in R, if all the "ingredients" of the derived quantities have been sampled and saved.

Indeed, one of the neatest things in a Bayesian MCMC-based analysis is the ease with which such derived quantities can be obtained, along with a full assessment of their uncertainty. In non-Bayesian analyses, we have to use approximations like the delta rule (see Section 2.4.2) or else use bootstrapping

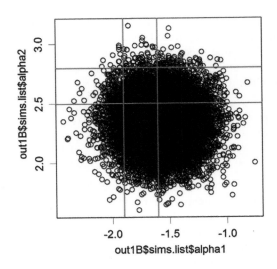

FIGURE 5.5
Joint posterior distribution of `alpha1` and `alpha2` with geometrical representation (red square) of the probability that the effect of `elev` lies between −1.9 and −1.6, and the effect of `forest` lies between 2.5 and 2.8.

(see Section 2.4.4), but in an MCMC-based analysis we can simply compute the function of interest for every step of the MCMC algorithm and then base the inference on the resulting posterior. To illustrate further, assume that we had some crazy theory that postulated the main effect of forest was more extreme than the main effect of elevation on the abundance of great tits. From the least-squares fit of the model, we can estimate the ratio between the two at 2.34/1.58 = 1.48; this is not difficult. But what about the uncertainty associated with this estimate? And, is the observed ratio of the absolute effects "significantly" different from 1 (i.e., is one effect clearly greater than the other)?

Both estimates have an associated estimation uncertainty represented by their standard error (SE). In a simple case, it would not be too hard to work out the SE of the ratio or to bootstrap it, but in a Bayesian analysis the uncertainty of the ratio is readily obtained for models or derived quantities of any complexity. Figure 5.6 gives a picture of the posterior distribution of the absolute ratio between the effects of forest cover and elevation along with a 95% CRI (red). Since the value 1 is outside of the 95% CRI, we can thus say that the data are in agreement with the crazy theory.

```
crazy.ratio <- out1B$sims.list$alpha2 / abs(out1B$sims.list$alpha1)
hist(crazy.ratio, main = "", breaks = 100, col = "grey", freq = F)
abline(v = quantile(crazy.ratio, prob = c(0.025, 0.975)), col = "red", lwd = 3)

mean(abs(out1B$sims.list$alpha2 / out1B$sims.list$alpha1) > 1)
[1] 0.9964667
```

Thus, we can be nearly 100% certain that the effect of forest cover is more extreme than that of elevation. Hence, very rich inferences are possible and indeed readily obtainable based on the MCMC sample of posterior distributions.

5.3 LINEAR MODEL WITH NORMAL RESPONSE (NORMAL GLM)

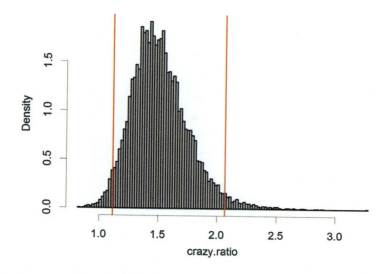

FIGURE 5.6

Posterior distribution of a crazy derived parameter, which is the absolute ratio of the effects of forest cover and elevation on the abundance of great tits, with 95% CRI shown with the red lines.

The final thing we illustrate here is the forming of predictions. Let us use our Bayesian parameter estimates to predict a response surface of the mean observed tit abundance as a function of elevation and forest cover and as a series of regression lines on elevation for selected values of forest cover (Figure 5.7(a) and (b)):

```
# Compute expected abundance for a grid of elevation and forest cover
elev.pred <- seq(-1, 1,,100)              # Values of elevation
forest.pred <- seq(-1,1,,100)             # Values of forest cover
pred.matrix <- array(NA, dim = c(100, 100))  # Prediction matrix
for(i in 1:100){
   for(j in 1:100){
      pred.matrix[i, j] <- out1J$mean$alpha0 + out1J$mean$alpha1 * elev.pred[i] +
out1J$mean$alpha2 * forest.pred[j] + out1J$mean$alpha3 * elev.pred[i] * forest.pred[j]
   }
}

par(mfrow = c(1, 3), mar = c(5,5,3,2), cex.main = 1.6, cex.axis = 1.5, cex.lab = 1.5)
mapPalette <- colorRampPalette(c("grey", "yellow", "orange", "red"))
image(x=elev.pred, y= forest.pred, z=pred.matrix, col = mapPalette(100), xlab =
"Elevation", ylab = "Forest cover")
contour(x=elev.pred, y=forest.pred, z=pred.matrix, add = TRUE, lwd = 1, cex = 1.5)
matpoints(elev, forest, pch="+", cex=1.5)
abline(h = c(-1, -0.5, 0, 0.5, 1))
```

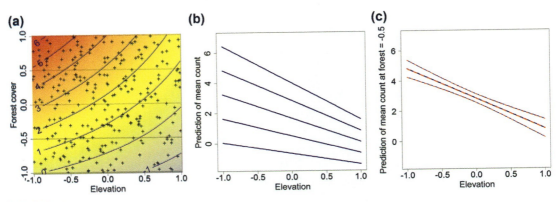

FIGURE 5.7

(a and b) Two ways of plotting predictions of the expected tit counts as a function of two continuous covariates: as a continuous response surface (a), and as a bundle of regression lines for a series of discrete values (here, for −1, −0.5, 0, 0.5, and 1 for the forest covariate) (b). Prediction with uncertainty (blue: Bayesian analysis; red: ML with 95% CRIs or CIs, respectively) is shown in (c). The curvature in (a) and nonparallelism in (b) indicate the presence of an interaction between elevation and forest cover. Note that the normal model for counts may predict impossible values such as negative tit counts in (a) and (b).

```
# Predictions for elev. at specific values of forest cover (-1,-0.5,0,0.5,1)
pred1 <- out1J$mean$alpha0 + out1J$mean$alpha1 * elev.pred + out1J$mean$alpha2 * (-1) +
out1J$mean$alpha3 * elev.pred * (-1)
pred2 <- out1J$mean$alpha0 + out1J$mean$alpha1 * elev.pred + out1J$mean$alpha2 * (-0.5) +
out1J$mean$alpha3 * elev.pred * (-0.5)
pred3 <- out1J$mean$alpha0 + out1J$mean$alpha1 * elev.pred + out1J$mean$alpha2 * 0 +
out1J$mean$alpha3 * elev.pred * 0
# pred3b <- out1J$mean$alpha0 + out1J$mean$alpha1 * elev.pred    # same
pred4 <- out1J$mean$alpha0 + out1J$mean$alpha1 * elev.pred + out1J$mean$alpha2 * 0.5 +
out1J$mean$alpha3 * elev.pred * 0.5
pred5 <- out1J$mean$alpha0 + out1J$mean$alpha1 * elev.pred + out1J$mean$alpha2 * 1 +
out1J$mean$alpha3 * elev.pred * 1
matplot(seq(-1, 1,,100), cbind(pred1, pred2, pred3, pred4, pred5), type = "l", lty=1, col =
"blue", ylab = "Prediction of mean count", xlab = "Elevation", ylim = c(-1.5, 7), lwd = 2)
```

These plots do not show the estimation uncertainty. This would be unwieldy in this type of plot, but it can easily be shown for simpler examples. We do this next, both for the Bayesian and the frequentist analysis, and plot the estimated relationship between the expected mean count of great tits and elevation at forest cover of 0.5 (Figure 5.7(c)). For this, we take all or a subset of the MCMC samples of the constituents of the predictions—i.e., the parameters `alpha0` through `alpha3`—and compute predictions using the values of the prediction covariate. We do this in a loop, but first prepare a matrix to contain the posterior samples for the predictions.

```
pred.mat <- array(dim = c(length(elev.pred), length(out1J$sims.list$alpha0)))
for(j in 1:length(out1J$sims.list$alpha0)){
    pred.mat[,j] <- out1J$sims.list$alpha0[j] + out1J$sims.list$alpha1[j] * elev.pred +
out1J$sims.list$alpha2[j] * 0.5 + out1J$sims.list$alpha3[j] * elev.pred * 0.5
}
```

To plot the 95% CRI of the Bayesian prediction we can use the point-wise 95% CRI:

```
CL <- apply(pred.mat, 1, function(x){quantile(x, prob = c(0.025, 0.975))})
plot(seq(-1, 1,,100), pred4, type = "l", lty = 1, col = "blue", ylab = "Prediction of mean
count at forest = -0.5", xlab = "Elevation", las =1, ylim = c(-1.5, 7), lwd = 3)
matlines(seq(-1, 1,,100), t(CL), lty = 1, col = "blue", lwd = 2)
```

We add the frequentist prediction along with the 95% confidence limits.

```
pred <- predict(lm(Cmean ~ elev*forest), newdata = data.frame(elev = seq(-1, 1,,100),
forest = 0.5), se.fit = TRUE, interval = "confidence")
lines(seq(-1, 1,,100), pred$fit[,1], lty = 2, col = "red", lwd = 3)
matlines(seq(-1, 1,,100), pred$fit[,2:3], lty = 2, col = "red", lwd = 2)
```

In real-life data analysis, what many find confusing is how predictions are presented for a covariate that has been transformed (e.g., scaled) before analysis. Then, we first have to form the predictions for the transformed covariates and then back-transform again for graphing. We show this in later parts of the book—e.g., in Section 5.9.

5.4 THE R PACKAGE rjags

We briefly illustrate model fitting in JAGS using the rjags package, including the production of trace plots and density plots, which is a fairly widely used way of running JAGS from R.

```
library(rjags)
load.module("glm")       # Careful with that package, see JAGS discussion list
load.module("dic")

# Have to explicitly list the deviance if want samples
params <- c("alpha0", "alpha1", "alpha2", "alpha3", "sd", "deviance")

# Adaptative phase to maximize MCMC efficiency
model <- jags.model(file = "multiple_linear_regression_model.txt", data = win.data,
inits = inits, n.chains = nc, n.adapt = 1000)

# Burnin
update(model, nb)

# Generate posterior samples
samples <- coda.samples(model = model, variable.names = params, n.iter = ni - nb, thin = nt)
```

```
# Get the summary statistics for the posterior samples
summfit <- summary(samples)
print(summfit, 2)
Iterations = 2001:7000
Thinning interval = 1
Number of chains = 3
Sample size per chain = 5000
```

1. Empirical mean and standard deviation for each variable, plus standard error of the mean:

	Mean	SD	Naive SE	Time-series SE
alpha0	1.66	0.114	0.00093	0.00093
alpha1	-1.58	0.204	0.00167	0.00170
alpha2	2.34	0.200	0.00163	0.00165
alpha3	-0.85	0.353	0.00288	0.00288
deviance	1087.14	3.237	0.02643	0.02923
sd	1.86	0.083	0.00068	0.00077

2. Quantiles for each variable:

	2.5%	25%	50%	75%	97.5%
alpha0	1.4	1.6	1.66	1.74	1.88
alpha1	-2.0	-1.7	-1.58	-1.44	-1.18
alpha2	2.0	2.2	2.34	2.47	2.73
alpha3	-1.5	-1.1	-0.85	-0.62	-0.14
deviance	1082.9	1084.8	1086.48	1088.79	1095.23
sd	1.7	1.8	1.86	1.91	2.03

```
# Traceplots and posterior densities
plot(samples[,1:4])

# Compute the Brooks-Gelman-Rubin statistic (R-hat)
gelman.diag(samples)
Potential scale reduction factors:
```

	Point est.	Upper C.I.
alpha0	1	1
alpha1	1	1
alpha2	1	1
alpha3	1	1
deviance	1	1
sd	1	1

Multivariate psrf

1

```
# Compute the effective sample size
effectiveSize(samples)
   alpha0    alpha1    alpha2    alpha3  deviance        sd
 15000.00  14541.42  14761.32  15000.00  12267.66  11601.70
```

```
# Secondary burnin can be applied (e.g. another 500 samples tossed out)
#samples <- window(samples, start = nb + 500 + 1, end = ni)

# More samples can be drawn (starting where the old chains stopped, not starting from 0)
newsamples <- coda.samples(model = model, variable.names = params, n.iter = 1500,
thin = nt)

# Combine the new samples with the old ones (ugly but works)
mc1 <- as.mcmc(rbind(samples[[1]], newsamples[[1]]))
mc2 <- as.mcmc(rbind(samples[[2]], newsamples[[2]]))
mc3 <- as.mcmc(rbind(samples[[3]], newsamples[[3]]))
allsamples <- as.mcmc.list(list(mc1, mc2, mc3))

# Mean deviance
Dbar <- summfit$statistics["deviance","Mean"]

# Variance of the deviance
varD <- summfit$statistics["deviance","SD"]^2

# Compute pD and DIC (according to A. Gelman, implemented in R2jags)
pD <- varD/2
DIC <- Dbar + pD

# Another DIC computation (according to M. Plummer). DIC = Penalized deviance
(dic.pD <- dic.samples(model, 2000, "pD"))

Mean deviance: 1088
penalty 3.993
Penalized deviance: 1092
```

Hence, we can readily use `rjags` directly for model fitting with JAGS or use a wrapper function like `jags` in `jagsUI`.

5.5 MISSING VALUES (NAs) IN A BAYESIAN ANALYSIS

How to deal with missing values (NAs) is the subject of a big field in statistics (Little and Rubin, 2002). An important distinction is between data that are missing at random (MAR) and those that are missing not at random (MNAR). With MAR, the probability for a datum to be missing does *not* depend on the value that it would have taken had it not been missing, and these data are therefore called ignorable missing values. The converse holds for MNAR, which are also called nonignorable missing data; to avoid biased inference, the missing-value-generating mechanism needs to be modeled for the data that are MNAR. An example for MAR data would be a study of the body mass in a population of mice where detection probability does not depend on body mass of an individual. In contrast, if heavier individuals are more likely to get caught and weighed, the missing value-generating mechanism—i.e., the mass-dependence of detection probability—needs to be modeled to avoid overestimating the population mean weight (Royle, 2008). In the following illustration, we assume data are MAR. Then the main difference in a Bayesian analysis is whether we have NAs in the response or in the covariates—i.e., whether the NAs are on the left- or right-hand side of the twiddle ($\sim$) in a BUGS model. The brief formula is then: a missing response is fine, but missing covariates must be dealt with.

5.5.1 SOME RESPONSES MISSING

When NAs are on the left of a twiddle, we can fit the same model and BUGS will simply update (=estimate, or predict) the missing responses. For illustration, we directly use code from Section 5.4. and analyze a modified data set where the first 10 responses (mean counts of great tits) are turned into NAs.

```
# Copy mean counts and turn first 10 into NAs
Cm <- Cmean        # Copy Cmean into Cm
Cm[1:10] <- NA     # turn first 10 into missing

# Bundle data (inside BUGS use Cm for Cmean)
win.data <- list(Cmean = Cm, M = length(Cm), elev = elev, forest = forest)
```

When we specify the response variable as a quantity to estimate, BUGS is (surprisingly) clever enough to know that only the missing elements of the response must be estimated, while the non-missing response data are fixed. Strangely though, in this case we have to set the argument DIC = FALSE to avoid the dreaded "NIL dereference (read)" crash in WinBUGS.

```
# Parameters monitored (i.e., for which estimates are saved)
params <- c("alpha0", "alpha1", "alpha2", "alpha3", "sd", "Cmean", "mu")

# ... or this to get a subset of the parameters
params <- c("alpha0", "alpha1", "alpha2", "alpha3", "sd", "Cmean[1:10]", "mu[1:10]")

# Call WinBUGS or JAGS from R (ART <1 min) and summarize posteriors
out1.1 <- bugs(win.data, inits, params, "multiple_linear_regression_model.txt",
   n.chains = nc, n.thin = nt, n.iter = ni, n.burnin = nb, debug = TRUE, bugs.directory =
   bugs.dir, working.directory = getwd(), DIC = FALSE)

out1.1 <- jags(win.data, inits, params, "multiple_linear_regression_model.txt",
   n.chains = nc, n.thin = nt, n.iter = ni, n.burnin = nb)

print(out1.1, 2)
JAGS output for model 'multiple_linear_regression_model.txt', generated by jagsUI.
Estimates based on 3 chains of 6000 iterations,
burn-in = 1000 iterations and thin rate = 1,
yielding 15000 total samples from the joint posterior.
MCMC ran for 0.072 minutes at time 2014-09-20 14:53:55.
```

	mean	sd	2.5%	50%	97.5%	overlap0	f	Rhat	n.eff
alpha0	1.67	0.12	1.44	1.67	1.90	FALSE	1.00	1	15000
alpha1	-1.59	0.21	-2.00	-1.59	-1.17	FALSE	1.00	1	15000
alpha2	2.34	0.21	1.93	2.34	2.75	FALSE	1.00	1	4321
alpha3	-0.87	0.37	-1.60	-0.87	-0.15	FALSE	0.99	1	10679
sd	1.89	0.08	1.73	1.89	2.06	FALSE	1.00	1	4599
Cmean[1]	3.48	1.89	-0.26	3.47	7.20	TRUE	0.97	1	15000
Cmean[2]	3.79	1.89	0.12	3.77	7.57	FALSE	0.98	1	15000
Cmean[3]	2.95	1.92	-0.81	2.96	6.70	TRUE	0.94	1	15000
Cmean[4]	-0.44	1.91	-4.15	-0.42	3.30	TRUE	0.59	1	7555

```
Cmean[5]    0.56 1.90  -3.20   0.56   4.25   TRUE  0.61   1 15000
Cmean[6]   -0.41 1.90  -4.12  -0.40   3.36   TRUE  0.58   1 15000
Cmean[7]    3.35 1.91  -0.41   3.35   7.09   TRUE  0.96   1 15000
Cmean[8]    2.37 1.92  -1.37   2.38   6.08   TRUE  0.89   1 15000
Cmean[9]    0.87 1.90  -2.83   0.88   4.61   TRUE  0.68   1 15000
Cmean[10]   0.93 1.90  -2.84   0.93   4.61   TRUE  0.69   1 15000
mu[1]       3.49 0.17   3.15   3.49   3.83   FALSE 1.00   1 15000
mu[2]       3.80 0.20   3.41   3.80   4.18   FALSE 1.00   1 15000
mu[3]       2.94 0.29   2.37   2.95   3.50   FALSE 1.00   1  5161
mu[4]      -0.45 0.23  -0.90  -0.45  -0.01   FALSE 0.98   1  6941
mu[5]       0.58 0.16   0.26   0.58   0.90   FALSE 1.00   1 15000
mu[6]      -0.43 0.30  -1.02  -0.43   0.17   TRUE  0.92   1 15000
mu[7]       3.36 0.17   3.03   3.36   3.68   FALSE 1.00   1 15000
mu[8]       2.38 0.28   1.82   2.38   2.93   FALSE 1.00   1  5856
mu[9]       0.88 0.18   0.53   0.88   1.23   FALSE 1.00   1 15000
mu[10]      0.92 0.21   0.51   0.92   1.34   FALSE 1.00   1 15000
```

Where does the information come from to estimate the missing responses Cmean[1:10]? As an aside, what is the difference between Cmean[1:10] and mu[1:10]? Let us compare the true values of the first 10 responses with their estimates for Cmean[1:10] and expectation mu[1:10].

```
print(cbind(Truth=Cmean[1:10], out1.1$summary[6:15,c(1:3,7)], out1.1$summary[16:25,
  c(1:3,7)]),3)
           Truth   mean    sd   2.5% 97.5%    mean    sd    2.5%   97.5%
Cmean[1]   3.000  3.476  1.89 -0.257  7.20   3.494 0.172   3.154  3.8321
Cmean[2]   3.333  3.792  1.89  0.117  7.57   3.800 0.196   3.412  4.1835
Cmean[3]   4.000  2.952  1.92 -0.813  6.70   2.942 0.288   2.373  3.4976
Cmean[4]   0.000 -0.436  1.91 -4.154  3.30  -0.453 0.227  -0.903 -0.0107
Cmean[5]   0.000  0.559  1.90 -3.201  4.25   0.581 0.162   0.263  0.8968
Cmean[6]   0.000 -0.414  1.90 -4.120  3.36  -0.429 0.302  -1.016  0.1666
Cmean[7]   2.000  3.347  1.91 -0.405  7.09   3.355 0.167   3.026  3.6806
Cmean[8]   2.667  2.367  1.92 -1.374  6.08   2.381 0.284   1.822  2.9327
Cmean[9]   0.000  0.873  1.90 -2.834  4.61   0.880 0.179   0.531  1.2335
Cmean[10]  0.333  0.926  1.90 -2.842  4.61   0.923 0.215   0.506  1.3435
```

To understand the difference between Cmean and mu, let us look at the relationship between the realized response *Cmean*$_i$ and the expected response μ_i (see Section 5.3).

$$Cmean_i \sim Normal(\mu_i, \sigma^2)$$

The information to estimate a missing response *Cmean* comes from the estimated regression relationship between the counts and the covariates elevation and forest: $\mu_i = \alpha_0 + \alpha_1 * elev_i + \alpha_2 * forest_i + \alpha_3 * elev_i * forest_i$. Based on that and the known covariate values, we can obtain an estimate for the missing responses. The difference between mu (μ_i) and Cmean (*Cmean*$_i$) is that between the expected response and the realized response, respectively. Up to Monte Carlo error, the point estimates of these quantities should be identical. However, the uncertainty about μ_i stems only from having to estimate the regression parameters α_0, α_1, α_2, and α_3, whereas the estimate for a realized response $Cmean_i$ also contains a contribution of the sampling variability of the data represented by the unknown residual (and the variance parameter σ^2) and the estimation uncertainty in σ^2. Hence, estimates of $Cmean_i$ are more variable than are those of the mean response μ_i. Clearly, even if we knew the expected response (μ_i), we would still be uncertain about the actual realized response. The 95% uncertainty interval around Cmean is also called a prediction interval, while that around mu is called a confidence interval (or, in the Bayesian analysis here, a credible interval).

The ability in a Bayesian analysis to produce estimates of the response variable provides us with a way to produce hypothetical replicate data sets under our model, which is at the root of a very general Bayesian GoF assessment technique; see Section 5.10. Note that the parameter estimates would have been the same if we had tossed out the data for the 10 NA responses altogether, since these data units do not contain any information about the regression relationship. This is the same as when we make predictions.

5.5.2 ALL RESPONSES MISSING

Next, let us look at an extreme case and fit the model when *all* responses are turned into NAs. This extreme case reiterates that in a Bayesian analysis the "result", i.e., the posterior distribution, is affected by the information in the data *and* that in the prior distribution. If there is no data set, parameter estimates are simply informed by the priors. In complex HMs, this can be useful to gauge the induced priors for those parameters for which we do not specify priors directly. We illustrate by monitoring the expected response mu (μ) for the first two data points.

```
# Bundle data: simply drop the response from list
win.data <- list(M = length(Cm), elev = elev, forest = forest)

# Alternatively, add all-NA data vector
win.data$Cmean <- as.numeric(rep(NA, length(Cmean)))
str(win.data)   # Cmean is numeric

# Parameters monitored
params <- c("alpha0", "alpha1", "alpha2", "alpha3", "sd", "mu[1:2]")

# Call WinBUGS from R (ART <1 min) and summarize posteriors
out1.2 <- bugs(win.data, inits, params, "multiple_linear_regression_model.txt",
 n.chains = nc, n.thin = nt, n.iter = ni, n.burnin = nb, debug = TRUE, bugs.directory =
 bugs.dir, working.directory = getwd(), DIC = FALSE)

print(out1.2, 2)
          mean      sd    2.5%       25%       50%    75%    97.5% Rhat n.eff
alpha0  -15.12 1003.24 -1976.05  -703.97   -15.59 663.62 1936.00    1 15000
alpha1    2.46  990.95 -1946.00  -656.18     5.81 674.78 1957.05    1 15000
alpha2   -8.87 1004.61 -1957.00  -690.30    -5.06 662.00 1953.05    1 15000
alpha3  -14.24  984.28 -1948.03  -684.70    -4.48 649.35 1886.00    1  3900
sd      500.49  288.57    25.50   251.30   504.45 749.20  976.10    1 15000
mu[1]   -17.42 1183.05 -2350.03  -814.40   -24.67 784.02 2312.02    1 10000
mu[2]   -16.93 1252.07 -2475.00  -863.08   -19.27 826.00 2449.07    1  9500
```

5.5 MISSING VALUES (NAS) IN A BAYESIAN ANALYSIS

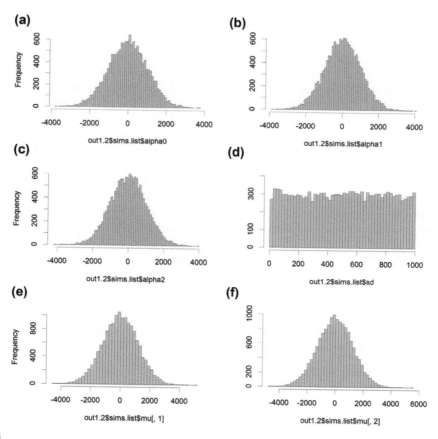

FIGURE 5.8

Induced priors can be studied by fitting a model with the response data unlinked from the analysis. For the first four parameters (a–d), priors are specified directly in the analysis, while for the last two (e–f), priors are induced.

The posterior distributions for the model parameters are directly the priors specified in the model, while those for mu are induced (Figure 5.8).

```
par(mfrow = c(3, 2), mar = c(5,5,3,2), cex.lab = 1.5, cex.axis = 1.5)
hist(out1.2$sims.list$alpha0, breaks = 100, col = "grey", main = "")
hist(out1.2$sims.list$alpha1, breaks = 100, col = "grey", main = "", ylab = "")
hist(out1.2$sims.list$alpha2, breaks = 100, col = "grey", main = "")
hist(out1.2$sims.list$sd, breaks = 100, col = "grey", main = "", ylim = c(0, 230),
  ylab = "")
hist(out1.2$sims.list$mu[,1], breaks = 100, col = "grey", main = "")
hist(out1.2$sims.list$mu[,2], breaks = 100, col = "grey", main = "", ylab = "")
```

5.5.3 MISSING VALUES IN A COVARIATE

When we have missing covariate values in our model, we cannot simply use the model to estimate them; there is no information available to estimate them and BUGS would crash with an "undefined node" error. Heuristically, in order to estimate missing covariates, we must add code so that somewhere in the model the missing covariate appears on the left-hand side of a twiddle—i.e., as a modeled quantity. Here, we only mention some of the simplest manners of dealing with missing covariates. The simplest of all (method 1) is often called mean imputation, where the NAs are replaced with the covariate mean. This method ignores the uncertainty stemming from having to estimate the missing covariates, and it does not use information about the association of the missing covariates with observed responses or observed values of the covariate. A slightly more refined method (2) is to estimate the missing covariate values as unrelated draws from a vague prior. This does incorporate the estimation uncertainty and the information stemming from the (possibly) observed response for a missing covariate. Finally, a third method is to specify a prior for all (observed and unobserved) covariate values and estimate hyperparameters of that prior. This method is the most refined and exploits information coming from both the observed values of the response and the observed covariates. Next, we show methods 2 and 3.

```
# Shoot 'holes' in the covariate data
ele <- elev        # copy of elevation covariate
ele[1:10] <- NA    # create some missing values in covariate elevation

# Bundle data: feed new 'ele' into 'elev' covariate inside of BUGS model
win.data <- list(Cmean = Cmean, M = length(Cmean), elev = ele, forest = forest)

# Specify model in BUGS language
cat(file = "missing_cov_imputation_model_1.txt","
model {

# Priors
alpha0 ~ dnorm(0, 1.0E-06)      # Prior for intercept
alpha1 ~ dnorm(0, 1.0E-06)      # Prior for slope of elev
alpha2 ~ dnorm(0, 1.0E-06)      # Prior for slope of forest
alpha3 ~ dnorm(0, 1.0E-06)      # Prior for slope of interaction
tau <- pow(sd, -2)
sd ~ dunif(0, 1000)             # Prior for dispersion on sd scale

# Likelihood
for (i in 1:M){
   Cmean[i] ~ dnorm(mu[i], tau)  # precision tau = 1 / variance
   mu[i] <- alpha0 + alpha1 * elev[i] + alpha2 * forest[i] + alpha3 * elev[i] * forest[i]
}

# Model for missing covariates
for (i in 1:M){
  elev[i] ~ dnorm(0, 0.001)
}
}")
```

5.5 MISSING VALUES (NAS) IN A BAYESIAN ANALYSIS

```
# Initial values
inits <- function() list(alpha0 = rnorm(1,,10), alpha1 = rnorm(1,,10), alpha2 =
rnorm(1,,10), alpha3 = rnorm(1,,10))

# Parameters monitored
params <- c("alpha0", "alpha1", "alpha2", "alpha3", "sd", "elev")

# MCMC settings
ni <- 6000  ;  nt <- 1  ;  nb <- 1000  ;  nc <- 3

# Call WinBUGS from R (ART <1 min)
out1.3 <- bugs(win.data, inits, params, "missing_cov_imputation_model_1.txt",
n.chains = nc, n.thin = nt, n.iter = ni, n.burnin = nb, debug = TRUE,
bugs.directory = bugs.dir, working.directory = getwd())
```

Method 2 estimates each NA of elevation as an unrelated quantity. In method 3, next, we additionally specify a model for the covariates, thereby the observed values of the covariate will contribute to estimation of the NAs of the covariate.

```
# Specify model in BUGS language
cat(file = "missing_cov_imputation_model_2.txt","
model {

# Priors
alpha0 ~ dnorm(0, 1.0E-06)          # Prior for intercept
alpha1 ~ dnorm(0, 1.0E-06)          # Prior for slope of elev
alpha2 ~ dnorm(0, 1.0E-06)          # Prior for slope of forest
alpha3 ~ dnorm(0, 1.0E-06)          # Prior for slope of interaction
tau <- pow(sd, -2)
sd ~ dunif(0, 1000)                 # Prior for dispersion on sd scale

# Likelihood
for (i in 1:M){
   Cmean[i] ~ dnorm(mu[i], tau)     # precision tau = 1 / variance
   mu[i] <- alpha0 + alpha1 * elev[i] + alpha2 * forest[i] + alpha3 * elev[i] * forest[i]
}

# Covariate mean as a model for missing covariates
for (i in 1:M){
   elev[i] ~ dnorm(mu.elev, tau.elev)   # Assume elevation normally distributed
}
mu.elev ~ dnorm(0, 0.0001)
tau.elev <- pow(sd.elev, -2)
sd.elev ~ dunif(0, 100)
}")

# Initial values
inits <- function() list(alpha0 = rnorm(1,,10), alpha1 = rnorm(1,,10), alpha2 =
rnorm(1,,10), alpha3 = rnorm(1,,10))
```

```
# Parameters monitored
params <- c("alpha0", "alpha1", "alpha2", "alpha3", "sd", "elev", "mu.elev", "sd.elev")

# MCMC settings
ni <- 6000  ;  nt <- 1  ;  nb <- 1000  ;  nc <- 3

# Call WinBUGS from R (ART <1 min)
out1.4 <- bugs(win.data, inits, params, "missing_cov_imputation_model_2.txt",
n.chains = nc, n.thin = nt, n.iter = ni, n.burnin = nb, debug = TRUE,
bugs.directory = bugs.dir, working.directory = getwd())
```

We compare the two approaches graphically (Figure 5.9). Both sets of estimates are about equally far away from the truth, but we see that when we can come up with a reasonable model for the covariates, we can obtain estimates with much reduced uncertainty.

```
par(cex = 1.5, lwd = 2)
plot(elev[1:10]-0.01, out1.3$summary[6:15,1], ylim = c(-10, 10), col = "red",
xlab = "True value of covariate", ylab = "Estimate (with 95% CRI)", frame.plot =F)
segments(elev[1:10]-0.01, out1.3$summary[6:15,3], elev[1:10]-0.01,
out1.3$summary[6:15,7], col = "red")
points(elev[1:10]+0.01, out1.4$summary[6:15,1], ylim = c(-3, 3), col = "blue")
segments(elev[1:10]+0.01, out1.4$summary[6:15,3], elev[1:10]+0.01,
out1.4$summary[6:15,7], col = "blue")
abline(0,1)
```

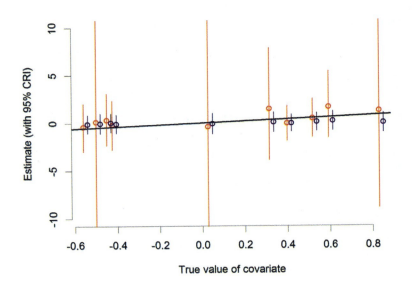

FIGURE 5.9

Estimates of missing covariates (with 95% CRI) when no model is specified for the covariate apart from a vague prior (red), and when missing covariate values are assumed exchangeable with the observed covariate values by specifying a normal prior with mean and variance estimated from the available covariate values (blue).

5.6 LINEAR MODEL WITH NORMAL RESPONSE (NORMAL GLM): ANALYSIS OF COVARIANCE (ANCOVA)

Returning to the illustration of common linear models, we next use BUGS to fit a linear model that underlies a technique called analysis of covariance (ANCOVA). Specifically, within a GLM with normal response, we fit to the mean tit counts the linear model underlying a fixed-effects ANCOVA with interaction effects. For this, we somewhat artificially first construct a factor that classifies the continuous covariate forest cover into four levels or groups, with level 1 for values between −1 and −0.5, level 2 corresponding to −0.49 and 0, etc.; see Figure 5.10(a) for the raw relationship between mean tit count and levels of the forest factor (facFor). Factors in BUGS must be labeled with integer numbers and not, for instance, with letters or words, and the numbering must start at 1 and end at the number of levels—i.e., have no jumps (e.g., 1, 2, 4, and 5 would cause a crash). We fit the following model in the effects and the means parameterization (see Chapter 3), where j indexes the four levels of the forest factor:

$$Cmean_i \sim Normal(\mu_i, \sigma^2)$$

$$\mu_i = \alpha_{0,j} + \alpha_{1,j} * elev_i$$

```
# Generate factor and plot raw data in boxplot as function of factor A
facFor <- as.numeric(forest < -0.5)      # Factor level 1
facFor[forest < 0 & forest > -0.5] <- 2  # Factor level 2
facFor[forest < 0.5 & forest > 0] <- 3   # Factor level 3
facFor[forest > 0.5] <- 4                # Factor level 4
table(facFor)                            # every site assigned a level OK
```

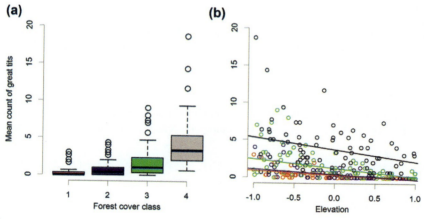

FIGURE 5.10

(a) Relationship between the mean count of great tits and the levels of the forest cover factor (facFor). Raw data are shown for each level of facFor. (b) Raw data and predicted relationship with elevation under the ANCOVA model with a least-squares fit. Colors denote the four levels of facFor.

```
par(mfrow = c(1, 2), mar = c(5,5,3,2), cex.lab = 1.5, cex.axis = 1.5)
plot(Cmean ~ factor(facFor), col = c("red", "blue", "green", "grey"), xlab = "Forest cover
class", ylab = "Mean count of great tits", frame.plot = F, ylim = c(0,20))

# Bundle data
win.data <- list(Cmean = Cmean, M = length(Cmean), elev = elev, facFor = facFor)
```

We can define the model in the effects or the means parameterization, and we show both. In either case, we define vector-valued parameters using the handy *nested indexing* in the BUGS language. We will fit the model in WinBUGS, JAGS, and compare with the MLEs obtained by using the least-squares method by way of the function lm in R.

```
# Specify model in BUGS language in effects parameterization
cat(file = "ANCOVA1.txt"."
model {

# Priors
alpha ~ dnorm(0, 1.0E-06)         # Prior for intercept = effect of level 1 of forest factor
beta2 ~ dnorm(0, 1.0E-06)         # Prior for slope = effect of elevation for level 1 of forest
                                   factor
beta1[1] <- 0                      # Set to zero effect of first level of facFor
beta3[1] <- 0                      # Set to zero effect of first level of facFor of elevation
for(k in 2:4){
  beta1[k] ~ dnorm(0, 1.0E-06)    # Prior for effects of factor facFor
  beta3[k] ~ dnorm(0, 1.0E-06)    # Prior for effects of factor facFor
}
tau <- pow(sd, -2)
sd ~ dunif(0, 1000)               # Prior for dispersion on sd scale

# Likelihood
for (i in 1:M){
  Cmean[i] ~ dnorm(mu[i], tau)    # precision tau = 1 / variance
  mu[i] <- alpha + beta1[facFor[i]] + beta2 * elev[i] + beta3[facFor[i]] * elev[i]
}
}
")
```

We must not give any initial values for fixed quantities (here, beta1[1] and beta3[1]); note that in place of the initial for the first element of the parameter vectors beta1 and beta3, we have an "NA."

```
# Initial values
inits <- function() list(alpha = rnorm(1,,10), beta1 = c(NA, rnorm(3,,10)), beta2 =
rnorm(1,,10), beta3 = c(NA, rnorm(3,,10)))

# Parameters monitored
params <- c("alpha", "beta1", "beta2", "beta3", "sd")

# MCMC settings
ni <- 6000 ; nt <- 1 ; nb <- 1000 ; nc <- 3
```

5.6 LINEAR MODEL WITH NORMAL RESPONSE (NORMAL GLM)

```
# Call WinBUGS or JAGS from R (ART <1 min)
out3 <- bugs(win.data, inits, params, "ANCOVA1.txt", n.chains = nc, n.thin = nt, n.iter =
ni, n.burnin = nb, debug = TRUE, bugs.directory = bugs.dir, working.directory = getwd())

out3J <- jags(win.data, inits, params, "ANCOVA1.txt", n.chains = nc, n.thin = nt,
n.iter = ni, n.burnin = nb)
# traceplot(out3J)

# Fit model using least-squares (yields equivalent estimates as MLE)
(fm <- summary(lm(Cmean ~ as.factor(facFor)*elev)))
```

Coefficients:

	Estimate	Std. Error	t value	Pr(>|t|)	
(Intercept)	0.3353	0.2301	1.457	0.14633	
as.factor(facFor)2	0.4244	0.3231	1.313	0.19028	
as.factor(facFor)3	1.2690	0.3083	4.115	5.2e-05	***
as.factor(facFor)4	3.7205	0.3162	11.766	< 2e-16	***
elev	-0.6013	0.4203	-1.431	0.15377	
as.factor(facFor)2:elev	-0.6866	0.5999	-1.145	0.25345	
as.factor(facFor)3:elev	-1.2116	0.5427	-2.232	0.02644	*
as.factor(facFor)4:elev	-1.6164	0.5708	-2.832	0.00499	**

Signif. codes: 0 '***' 0.001 '**' 0.01 '*' 0.05 '.' 0.1 ' ' 1

Residual standard error: 1.783 on 259 degrees of freedom
Multiple R-squared: 0.4941,   Adjusted R-squared: 0.4804
F-statistic: 36.13 on 7 and 259 DF, p-value: < 2.2e-16

```
# Summarize posteriors
print(out3, 3)
```

	mean	sd	2.5%	25%	50%	75%	97.5%	Rhat	n.eff
alpha	0.337	0.231	-0.114	0.183	0.337	0.491	0.799	1.001	15000
beta1[2]	0.422	0.324	-0.211	0.205	0.422	0.641	1.051	1.001	15000
beta1[3]	1.268	0.310	0.664	1.062	1.267	1.473	1.887	1.001	15000
beta1[4]	3.721	0.318	3.093	3.509	3.721	3.931	4.350	1.001	15000
beta2	-0.602	0.421	-1.442	-0.885	-0.600	-0.319	0.222	1.001	15000
beta3[2]	-0.687	0.605	-1.859	-1.101	-0.692	-0.277	0.503	1.001	15000
beta3[3]	-1.215	0.544	-2.290	-1.581	-1.218	-0.847	-0.158	1.001	15000
beta3[4]	-1.611	0.578	-2.744	-1.999	-1.610	-1.223	-0.456	1.001	6100
sd	1.791	0.078	1.648	1.737	1.788	1.842	1.953	1.001	15000
deviance	1067.483	4.339	1061.000	1064.000	1067.000	1070.000	1078.000	1.001	15000

DIC info (using the rule, pD = Dbar-Dhat)
pD = 9.0 and DIC = 1076.5
DIC is an estimate of expected predictive error (lower deviance is better).

We see the usual close numerical agreement between the Bayesian estimates and the MLEs obtained with function lm in R. Next, we fit the model using the means parameterization, where we fit directly the effect of each level of factor facFor (note the changed parameter naming in the output). We do not need any change in the data bundle. In addition, we also illustrate how we can estimate custom contrasts as derived quantities—i.e., differences or other functions of parameters. We estimate all pair-wise differences between the group means beta[1:4]. Of course, we could also easily compute these derived quantities in R using posterior samples of the vector beta produced by BUGS.

```
# Specify model in BUGS language
cat(file = "ANCOVA2.txt","
model {

# Priors
for(k in 1:4){
  alpha[k] ~ dnorm(0, 1.0E-06)    # Priors for intercepts
  beta[k] ~ dnorm(0, 1.0E-06)     # Priors for slopes
}
tau <- pow(sd, -2)
sd ~ dunif(0, 1000)               # Prior for dispersion on sd scale

# Likelihood
for (i in 1:M){
  Cmean[i] ~ dnorm(mu[i], tau)    # precision tau = 1 / variance
  mu[i] <- alpha[facFor[i]] + beta[facFor[i]] * elev[i]
}

# Derived quantities: comparison of slopes (now you can forget the delta rule !)
for(k in 1:4){
  diff.vs1[k] <- beta[k] - beta[1]   # Differences relative to beta[1]
  diff.vs2[k] <- beta[k] - beta[2]   # ... relative to beta[2]
  diff.vs3[k] <- beta[k] - beta[3]   # ... relative to beta[3]
  diff.vs4[k] <- beta[k] - beta[4]   # ... relative to beta[4]
}
}
")

# Initial values
inits <- function() list(alpha = rnorm(4,,10), beta = rnorm(4,,10))

# Parameters monitored
params <- c("alpha", "beta", "sd", "diff.vs1", "diff.vs2", "diff.vs3", "diff.vs4")

# MCMC settings
ni <- 6000  ;  nt <- 1  ;  nb <- 1000  ;  nc <- 3

# Call WinBUGS or JAGS from R (ART <1 min) and summarize posteriors
out4 <- bugs(win.data, inits, params, "ANCOVA2.txt", n.chains = nc, n.thin = nt,
  n.iter = ni, n.burnin = nb, debug = TRUE, bugs.directory = bugs.dir,
  working.directory = getwd())

system.time(out4J <- jags(win.data, inits, params, "ANCOVA2.txt", n.chains = nc,
  n.thin = nt, n.iter = ni, n.burnin = nb))
traceplot(out4J)

print(out4, 2)
Inference for Bugs model at "ANCOVA2.txt", fit using WinBUGS,
```

5.6 LINEAR MODEL WITH NORMAL RESPONSE (NORMAL GLM)

```
Current: 3 chains, each with 6000 iterations (first 1000 discarded)
Cumulative: n.sims = 15000 iterations saved
             mean   sd   2.5%    25%    50%    75%   97.5% Rhat  n.eff
alpha[1]     0.33  0.23  -0.13   0.18   0.33   0.49   0.79    1  15000
alpha[2]     0.76  0.23   0.31   0.61   0.76   0.91   1.21    1  15000
alpha[3]     1.60  0.21   1.19   1.46   1.60   1.74   2.00    1  15000
alpha[4]     4.06  0.22   3.64   3.91   4.05   4.20   4.49    1  15000
beta[1]     -0.60  0.42  -1.41  -0.88  -0.60  -0.31   0.23    1  15000
beta[2]     -1.29  0.43  -2.14  -1.58  -1.29  -0.99  -0.43    1  15000
beta[3]     -1.82  0.34  -2.49  -2.05  -1.82  -1.59  -1.13    1   6700
beta[4]     -2.22  0.39  -2.97  -2.48  -2.22  -1.96  -1.45    1  15000
sd           1.79  0.08   1.65   1.74   1.79   1.84   1.96    1   9900
diff.vs1[1]  0.00  0.00   0.00   0.00   0.00   0.00   0.00    1      1
diff.vs1[2] -0.69  0.60  -1.88  -1.10  -0.69  -0.28   0.48    1  15000
diff.vs1[3] -1.22  0.55  -2.30  -1.58  -1.22  -0.86  -0.15    1  15000
diff.vs1[4] -1.62  0.57  -2.73  -2.01  -1.63  -1.23  -0.49    1  14000
diff.vs2[1]  0.69  0.60  -0.48   0.28   0.69   1.10   1.88    1  15000
diff.vs2[2]  0.00  0.00   0.00   0.00   0.00   0.00   0.00    1      1
diff.vs2[3] -0.53  0.56  -1.63  -0.90  -0.53  -0.16   0.57    1  15000
diff.vs2[4] -0.93  0.58  -2.07  -1.32  -0.93  -0.54   0.21    1  15000
diff.vs3[1]  1.22  0.55   0.15   0.86   1.22   1.58   2.30    1  14000
diff.vs3[2]  0.53  0.56  -0.57   0.16   0.53   0.90   1.63    1  15000
diff.vs3[3]  0.00  0.00   0.00   0.00   0.00   0.00   0.00    1      1
diff.vs3[4] -0.40  0.52  -1.42  -0.75  -0.40  -0.05   0.62    1   9700
diff.vs4[1]  1.62  0.57   0.49   1.23   1.63   2.01   2.73    1  15000
diff.vs4[2]  0.93  0.58  -0.21   0.54   0.93   1.32   2.07    1  15000
diff.vs4[3]  0.40  0.52  -0.62   0.05   0.40   0.75   1.42    1   9700
diff.vs4[4]  0.00  0.00   0.00   0.00   0.00   0.00   0.00    1      1
```

```
# Fit model using maximum likelihood
(fm <- summary(lm(Cmean ~ as.factor(facFor)*elev-1-elev)))

Coefficients:
                         Estimate Std. Error t value Pr(>|t|)
as.factor(facFor)1         0.3353     0.2301   1.457 0.146328
as.factor(facFor)2         0.7596     0.2269   3.348 0.000935 ***
as.factor(facFor)3         1.6042     0.2052   7.816 1.37e-13 ***
as.factor(facFor)4         4.0558     0.2169  18.700  < 2e-16 ***
as.factor(facFor)1:elev   -0.6013     0.4203  -1.431 0.153772
as.factor(facFor)2:elev   -1.2880     0.4280  -3.009 0.002880 **
as.factor(facFor)3:elev   -1.8129     0.3433  -5.281 2.73e-07 ***
as.factor(facFor)4:elev   -2.2177     0.3862  -5.743 2.60e-08 ***
---
Signif. codes:  0 '***' 0.001 '**' 0.01 '*' 0.05 '.' 0.1 ' ' 1

Residual standard error: 1.783 on 259 degrees of freedom
Multiple R-squared: 0.6689,   Adjusted R-squared: 0.6587
F-statistic: 65.42 on 8 and 259 DF,  p-value: < 2.2e-16
```

We will often see the linear model that underlies an ANOVA (analysis of variance) or an ANCOVA using nested indexing in the BUGS language. Let us plot the predicted response as a function of the explanatory variables `facFor` and `elev` (Figure 5.10(b)). We use the parameter estimates from the least-squares fit (= MLEs), but clearly could also use the Bayesian posterior means.

```
plot(elev[facFor==1], Cmean[facFor==1], col = "red", ylim = c(0, 20), xlab = "Elevation",
  ylab = "",     frame.plot = F)
points(elev[facFor==2], Cmean[facFor==2], col = "blue")
points(elev[facFor==3], Cmean[facFor==3], col = "green")
points(elev[facFor==4], Cmean[facFor==4], col = "black")
abline(fm$coef[1,1], fm$coef[5,1], col = "red")
abline(fm$coef[2,1], fm$coef[6,1], col = "blue")
abline(fm$coef[3,1], fm$coef[7,1], col = "green")
abline(fm$coef[4,1], fm$coef[8,1], col = "black")
```

To further illustrate how simple it is to test custom hypotheses in an MCMC-based analysis, let us compute the probability that the difference in the slopes between level 3 of `facFor` and the other levels of that factor is greater than 1. We plot the histograms of these contrasts (Figure 5.11) and then compute the proportion of the area under the curve that lies to the right of 1.

```
attach.bugs(out4)    # Allows to directly address the sims.list
str(diff.vs3)
par(mfrow = c(1, 3), mar = c(5,5,3,2), cex.lab = 1.5, cex.axis = 1.5)
hist(diff.vs3[,1], col = "grey", breaks = 100, main = "", freq=F, ylim = c(0, 0.8))
abline(v = 1, lwd = 3, col = "red")
hist(diff.vs3[,2], col = "grey", breaks = 100, main = "", freq=F, ylim = c(0, 0.8))
abline(v = 1, lwd = 3, col = "red")
hist(diff.vs3[,4], col = "grey", breaks = 100, main = "", freq=F, ylim = c(0, 0.8))
abline(v = 1, lwd = 3, col = "red")
```

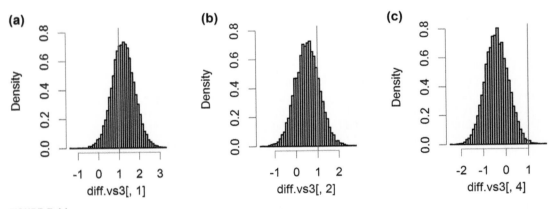

FIGURE 5.11

Posterior distributions of the difference between the slope of the regression of the mean tit counts on elevation in `facFor` level 3 versus levels 1 (a), 2 (b), and 4 (c). The probability that this difference is greater than 1 is represented by the area under the curve to the right of the red line for each posterior distribution.

```
# Prob. difference greater than 1
mean(diff.vs3[,1] > 1)
mean(diff.vs3[,2] > 1)
mean(diff.vs3[,4] > 1)
[1] 0.6554667
[1] 0.1981333
[1] 0.003733333
```

Hence, there is a 66% probability that the difference between the slopes in groups 1 and 3 of facFor is greater than 1, and we find corresponding probabilities of 20% and of essentially 0% for the analogous slope differences between group 3 and groups 2 and 4, respectively.

5.7 PROPORTION OF VARIANCE EXPLAINED (R^2)

Continuing in the ANCOVA example, we show how R^2 can be computed: the proportion of variance explained by a given model relative to a null model with an intercept only. For the least-squares analysis from the lm fit of the model using the effects parameterization, we get $R^2 = 0.49$ (the value of 0.67 from the means parameterization is not appropriate, since there is no intercept in that model). We can obtain a similar value in a Bayesian model by fitting a null model with only an intercept and then expressing as an R^2 value the proportional reduction in the residual variance (not the standard deviation) achieved by fitting a more complex model. We first fit the null model with an intercept only.

```
cat(file = "Model0.txt","
model {

# Priors
mu ~ dnorm(0, 1.0E-06)
tau <- pow(sd, -2)
sd ~ dunif(0, 1000)
# Likelihood
for (i in 1:M){
  Cmean[i] ~ dnorm(mu, tau)
}
}
")
inits <- function() list(mu = rnorm(1))
params <- c("mu", "sd")
ni <- 6000 ; nt <- 1 ; nb <- 1000 ; nc <- 3
out0 <- jags(win.data, inits, params, "Model0.txt", n.chains = nc, n.thin = nt, n.iter = ni,
n.burnin = nb)
print(out0)
                   mean      sd    2.5%      50%    97.5%  overlap0  f   Rhat   n.eff
mu              1.792   0.152   1.493    1.792    2.087     FALSE   1  1.000   15000
sd              2.485   0.108   2.282    2.481    2.705     FALSE   1  1.000   15000
deviance     1242.216   1.996 1240.258 1241.596 1247.474    FALSE   1  1.001    6256
```

We express the total unexplained variance around the mean in the null model as the residual variance. Then, we compute as R^2 the difference between the unexplained variance in the null model and that in the model with elevation and forest cover as a proportion of the total unexplained variance in the null model.

```
# Compute R2 from BUGS analysis
(total.var <- mean(out0$sims.list$sd^2))        # Total variance around the mean
[1] 6.185405
(unexplained.var <- mean(out3$sims.list$sd^2))  # Not explained by the ANCOVA
[1] 3.214631
(prop.explained <- (total.var - unexplained.var)/total.var)
[1] 0.4802878
```

Thus, a model with interaction effects of forest cover and elevation explains about half of the total variance in the mean counts of great tits. The idea of comparing the magnitude of a variance under a model with and without some covariate(s) is very general and can be used to express the explanatory power of covariates also in more complex models, such as HMs. For instance, in a survival analysis, we may express the effect of some climate variable on annual survival (phi) from a Cormack–Jolly–Seber model by the proportional reduction in the temporal variance of survival between a model that does and another that does not contain that covariate; i.e., compare model phi(random time) with model phi(covariate + random time)—see Grosbois et al. (2008) and Kéry and Schaub (2012, p. 189).

5.8 FITTING A MODEL WITH NONSTANDARD LIKELIHOOD USING THE ZEROS OR THE ONES TRICKS

Using the standard distribution functions in BUGS, you can fit a vast number of models. Moreover, by combining two or more of them, e.g., in an HM, you can extend the range of models considerably still. However, sometimes you may encounter a distribution that you cannot specify, or you may want to fit the integrated likelihood (see Chapter 2) of an HM directly. When you know how to write the likelihood of your model, you can fit it in BUGS using what is known as the "zeros trick" or the "ones trick" (Lunn et al., 2013, pp. 204–206).

For the zeros trick, imagine that you want to fit a model to a data set where observation i contributes a likelihood term L[i]. If we invent a dummy data set of all zeros and assume a Poisson(ϕ) distribution for it, then every observation has a contribution to the likelihood equal to $\exp(-\phi)$. If we then specify ϕ to be equal to the negative log-likelihood of observation i under our original model, we obtain the correct likelihood. In practice, we will have to add an arbitrary constant C to ensure nonnegativity of the Poisson mean. For the ones trick, we start with dummy data consisting solely of ones and specify a Bernoulli distribution with success parameter p_i, which is defined to be proportional to the desired likelihood term L[i] under the original model. Again, an arbitrary scaling constant C is usually required to ensure that $p_i \leq 1$. Here we illustrate both approaches for the trivial example of the normal response multiple linear regression from Section 5.3. We compare both solutions with the solution obtained using the standard distribution function for the normal.

5.8 FITTING A MODEL WITH NONSTANDARD LIKELIHOOD USING THE ZEROS

Remember the likelihood of an individual normal observation (see Section 2.2),

$$L(\mu, \sigma^2 | \mathbf{y}) = \sqrt{\frac{1}{2\pi\sigma^2}} \exp\left(-\frac{(y_i - \mu)^2}{2\sigma^2}\right),$$

and hence the negative log-likelihood is

$$NLL(\mu, \sigma^2 | \mathbf{y}) = -\log\left(\sqrt{\frac{1}{2\pi\sigma^2}}\right) + \frac{(y_i - \mu)^2}{2\sigma^2}.$$

Below we will use both equations in the likelihood specification, the former in the ones trick and the latter in the zeros trick.

```
# Package the data needed in a bundle
win.data <- list(Cmean1 = Cmean, Cmean2 = Cmean, zeros = rep(0, M), ones = rep(1, M),
M = length(Cmean), elev = elev, forest = forest) # note 2 copies of response

# Write text file with model description in BUGS language
cat(file = "multiple_linear_regression_model.txt",
"model {

# Priors
for(k in 1:3){ # Loop over three ways to specify likelihood
  alpha0[k] ~ dnorm(0, 1.0E-06)    # Prior for intercept
  alpha1[k] ~ dnorm(0, 1.0E-06)    # Prior for slope of elev
  alpha2[k] ~ dnorm(0, 1.0E-06)    # Prior for slope of forest
  alpha3[k] ~ dnorm(0, 1.0E-06)    # Prior for slope of interaction
  sd[k] ~ dunif(0, 1000)           # Prior for dispersion on sd scale
}
var1 <- pow(sd[1], 2)              # Variance in zeros trick
var2 <- pow(sd[2], 2)              # Variance in ones trick
tau <- pow(sd[3], -2)              # Precision tau = 1/(sd^2)

C1 <- 10000 # zeros trick: make large enough to ensure lam >= 0
C2 <- 10000 # ones trick: make large enough to ensure p <= 1
pi <- 3.1415926

# Three variants of specification of the likelihood
for (i in 1:M){
# 'Zeros trick' for normal likelihood
  zeros[i] ~ dpois(phi[i]) # likelihood contribution is exp(-phi)
#  negLL[i] <- log(sd[1]) + 0.5 * pow((Cmean1[i] - mu1[i]) / sd[1],2 )
  negLL[i] <- -log(sqrt(1/(2*pi*var1))) + pow(Cmean1[i]-mu1[i],2)/(2*var1)
  phi[i] <- negLL[i] + C1
  mu1[i] <- alpha0[1] + alpha1[1]*elev[i] + alpha2[1]*forest[i] + alpha3[1]*elev[i]*
  forest[i]

# 'Ones trick' for normal likelihood
  ones[i] ~ dbern(p[i])  # likelihood contribution is p directly
  L[i] <- sqrt(1/(2*pi*var2)) * exp(-pow(Cmean1[i]-mu2[i],2)/(2*var2))
```

```
   p[i] <- L[i] / C2
   mu2[i] <- alpha0[2] + alpha1[2]*elev[i] + alpha2[2]*forest[i] + alpha3[2]*elev[i]*forest
[i]

# Standard distribution function for the normal
  Cmean2[i] ~ dnorm(mu3[i], tau)
  mu3[i] <- alpha0[3] + alpha1[3]*elev[i] + alpha2[3]*forest[i] + alpha3[3]*elev[i]*forest
[i]
}
}"
)

# Initial values
inits <- function() list(alpha0 = rnorm(3, 0, 10), alpha1 = rnorm(3,0,10), alpha2 =
rnorm(3,0,10), alpha3 = rnorm(3,0,10))

# Parameters monitored (i.e., for which estimates are saved)
params <- c("alpha0", "alpha1", "alpha2", "alpha3", "sd")

# MCMC settings
ni <- 1200 ;  nt <- 1 ;  nb <- 200 ;  nc <- 3   # For JAGS
```

Much longer chains are required for BUGS to converge, so we fit the model in JAGS.

```
# Call JAGS
library(jagsUI)
outX <- jags(win.data, inits, params, "multiple_linear_regression_model.txt",
n.chains = nc, n.thin = nt, n.iter = ni, n.burnin = nb)
print(outX)
```

	mean	sd	2.5%	50%	97.5%	overlap0	f	Rhat	n.eff
alpha0[1]	1.662	0.115	1.441	1.663	1.884	FALSE	1.000	1.002	1417
alpha0[2]	1.660	0.114	1.433	1.660	1.874	FALSE	1.000	1.000	3000
alpha0[3]	1.657	0.115	1.436	1.660	1.878	FALSE	1.000	1.000	3000
alpha1[1]	-1.567	0.202	-1.957	-1.570	-1.176	FALSE	1.000	1.001	3000
alpha1[2]	-1.577	0.206	-1.980	-1.570	-1.187	FALSE	1.000	1.001	2180
alpha1[3]	-1.585	0.209	-1.989	-1.587	-1.170	FALSE	1.000	1.000	3000
alpha2[1]	2.346	0.191	1.966	2.347	2.719	FALSE	1.000	1.000	3000
alpha2[2]	2.349	0.202	1.963	2.345	2.753	FALSE	1.000	1.000	3000
alpha2[3]	2.346	0.199	1.957	2.342	2.720	FALSE	1.000	1.001	1435
alpha3[1]	-0.848	0.362	-1.574	-0.849	-0.153	FALSE	0.990	1.001	1226
alpha3[2]	-0.842	0.355	-1.524	-0.846	-0.128	FALSE	0.991	1.000	3000
alpha3[3]	-0.837	0.359	-1.548	-0.845	-0.119	FALSE	0.991	1.000	3000
sd[1]	1.861	0.083	1.706	1.858	2.033	FALSE	1.000	1.000	3000
sd[2]	1.861	0.083	1.702	1.858	2.032	FALSE	1.000	1.002	1333
sd[3]	1.860	0.084	1.704	1.856	2.030	FALSE	1.000	1.000	3000
deviance	5348179.889	5.641	5348171.008	5348179.076	5348192.794	FALSE	1.000	1.000	3000

Up to Monte Carlo error, we get identical answers for all three specifications of the normal likelihood. Finally, let us amuse ourselves by doing the analogous thing with ML—i.e., maximize the likelihood for the explicit description of the normal negative log-likelihood. We again find estimates that are numerically virtually identical to the posterior means.

```
# Define negative log-likelihood function
negLogLike <- function(param) {
 alpha0 = param[1]
 alpha1 = param[2]
 alpha2 = param[3]
 alpha3 = param[4]
 sigma = exp(param[5])    # Estimate sigma on log-scale
 mu = alpha0 + alpha1*elev + alpha2*forest + alpha3*elev*forest
 # -sum(dnorm(Cmean, mean=mu, sd=sigma, log=TRUE))  # cheap quick way
 sum(-log(sqrt(1/(2*3.1415926*sigma^2))) + (Cmean-mu)^2/(2*sigma^2))
}

# Find parameter values that minimize function value
(fit <- optim(par = rep(0, 5), fn = negLogLike, method = "BFGS"))
[1]   1.6603052 -1.5764599  2.3440036 -0.8506793  0.6073662

exp(fit$par[5])           # Backtransform to get sigma
[1] 1.83559
```

The zeros and ones tricks allow you to fit very general models, and you will sometimes see people fit HMs using these methods (e.g., Morales et al., 2004, for a hierarchical model for animal movement; Garrard et al. 2008; 2013 for time-to-detection occupancy models, and; Chelgren et al. 2011b for a distance sampling model).

5.9 POISSON GLM

We continue with a non-normal GLM and adopt a Poisson distribution with ANCOVA linear model for the counts of great tits. Since we have not one count per site, but three, we will follow a common approach and simply analyze the maximum count at each site, knowing that this must be the best non-model-based approximation, in the sense of being closest, to the true abundance of great tits at a site. We do *not* encourage this practice in general, but show this analysis here simply to illustrate a Poisson GLM. Let j index the four levels of facFor, and then we fit the following model:

$$Cmax_i \sim Poisson(\lambda_i)$$
$$\log(\lambda_i) = \alpha_{0,j} + \alpha_{1,j} * elev_i$$

where $Cmax_i$ is the maximum count for unit i and λ_i is the expected maximum count. We also compute Pearson residuals, $(Cmax_i - \lambda_i)/\sqrt{\lambda_i}$, which have the form of a raw residual divided by the standard deviation of unit i. To avoid numerical problems when the expected value becomes equal to zero, resulting in a division by zero in the Pearson residual, we add a small number e to the denominator. To emphasize the relatedness among different GLMs, we fit the same ANCOVA linear model as in the previous section, namely main and interaction effects of the forest factor (facFor) and elevation.

```
# Summarize data by taking max at each site
Cmax <- apply(C, 1, max)
table(Cmax)
Cmax
```

```
 0  1  2  3  4  5  6  7  8  9 10 11 12 13 14 15 16 17 22 30
77 55 30 12 17 23 11 10  6  6  2  2  2  3  4  1  2  1  2  1
```

```r
# Bundle data
win.data <- list(Cmax = Cmax, M = length(Cmax), elev = elev, facFor = facFor, e = 0.0001)

# Specify model in BUGS language
cat(file = "Poisson_GLM.txt","
model {

# Priors
for(k in 1:4){
  alpha[k] ~ dnorm(0, 1.0E-06)    # Prior for intercepts
  beta[k] ~ dnorm(0, 1.0E-06)     # Prior for slopes
}

# Likelihood
for (i in 1:M){
  Cmax[i] ~ dpois(lambda[i])      # note no variance parameter
  log(lambda[i]) <- alpha[facFor[i]] + beta[facFor[i]] * elev[i]
  resi[i] <- (Cmax[i]-lambda[i]) / (sqrt(lambda[i])+e)   # Pearson residual
}
}
")
```

This nonnormal GLM is specified in the first two code lines inside the loop under the heading of the likelihood. The first defines the data distribution and the second line specifies a log link and the linear predictor, with the latter corresponding simply to an ANCOVA linear model as before.

```r
# Initial values
inits <- function() list(alpha = rnorm(4,,3), beta = rnorm(4,,3))

# Parameters monitored
params <- c("alpha", "beta", "lambda", "resi")

# MCMC settings
ni <- 6000 ; nt <- 1 ; nb <- 1000 ; nc <- 3

# Call WinBUGS or JAGS from R and summarize posteriors
out5 <- bugs(win.data, inits, params, "Poisson_GLM.txt", n.chains = nc, n.thin = nt, n.iter
= ni, n.burnin = nb, debug = TRUE, bugs.directory = bugs.dir, working.directory = getwd())

out5J <- jags(win.data, inits, params, "Poisson_GLM.txt", n.chains = nc, n.thin = nt,
n.iter = ni, n.burnin = nb)
par(mfrow = c(4,2)) ;  traceplot(out5J, c("alpha[1:4]", "beta[1:4]"))
print(out5J, 3)
            mean     sd    2.5%     50%   97.5% overlap0      f   Rhat  n.eff
alpha[1]  -1.028  0.248  -1.526  -1.019  -0.569    FALSE  1.000  1.030     57
alpha[2]  -0.170  0.151  -0.484  -0.166   0.115     TRUE  0.874  1.012  10409
alpha[3]   0.754  0.091   0.569   0.755   0.927    FALSE  1.000  1.001  12407
```

```
alpha[4]   1.909  0.049   1.810  1.910   2.002    FALSE 1.000 1.001  2094
beta[1]   -2.294  0.398  -3.068 -2.287  -1.529    FALSE 1.000 1.023    97
beta[2]   -1.926  0.245  -2.418 -1.922  -1.461    FALSE 1.000 1.011  2797
beta[3]   -1.315  0.139  -1.595 -1.313  -1.051    FALSE 1.000 1.002  1818
beta[4]   -0.715  0.082  -0.875 -0.716  -0.555    FALSE 1.000 1.001  5423
```

We produce three residual plots as we did for the normal GLM (Figure 5.12). They do not look quite perfect; there are more large than small residuals. Since we fit this model for illustration only, we ignore this moderate lack of fit here.

```
par(mfrow = c(1, 3), mar = c(5,5,3,2), cex = 1.3, cex.lab = 1.5, cex.axis = 1.5)
hist(out5$summary[276:542, 1], xlab = "Pearson residuals", col = "grey", breaks = 50,
main = "", freq = F, xlim = c(-5, 5), ylim = c(0, 0.57))
abline(v = 0, col = "red", lwd = 2)

plot(1:267, out5$summary[276:542, 1], main = "", xlab = "Order of data", ylab = "Pearson
residual", frame.plot = F)
abline(h = 0, col = "red", lwd = 2)

plot(out5$summary[9:275, 1],out5$summary[276:542, 1], main = "", xlab = "Predicted
values", ylab = "Pearson residual", frame.plot = F, xlim = c(-1, 14))
abline(h = 0, col = "red", lwd = 2)
```

Next, we fit the model using iterative reweighted least-squares (the resulting solutions are equivalent to MLEs for a GLM) and find the usual comforting numerical similarity with the Bayesian estimates. Note how we specify the same means parameterization of the linear model that we chose in the BUGS fit.

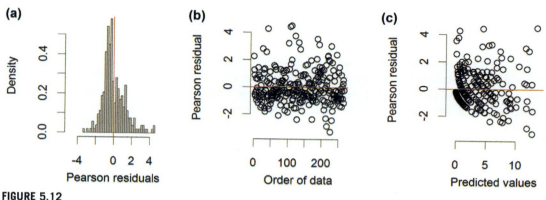

FIGURE 5.12

Three diagnostic plots for the Pearson residuals in a Poisson GLM: (a) histogram of the frequency distribution of the residuals, (b) residuals versus the order of the data, and (c) residuals versus predicted values. We could also plot the residuals against the covariates included in the analysis or against those left out of the model.

```
summary(glm(Cmax ~ factor(facFor)*elev-1-elev, family = poisson))
Coefficients:
                     Estimate  Std. Error  z value   Pr(>|z|)
factor(facFor)1      -0.99784  0.26366     -3.785    0.000154 ***
factor(facFor)2      -0.13993  0.15705     -0.891    0.372948
factor(facFor)3       0.76246  0.08999      8.472    < 2e-16 ***
factor(facFor)4       1.90985  0.04914     38.867    < 2e-16 ***
factor(facFor)1:elev -2.27293  0.41834     -5.433    5.54e-08 ***
factor(facFor)2:elev -1.90978  0.25114     -7.605    2.86e-14 ***
factor(facFor)3:elev -1.31210  0.13730     -9.556    < 2e-16 ***
factor(facFor)4:elev -0.69995  0.08343     -8.389    < 2e-16 ***
```

Derived quantities (= functions of parameters) may be computed inside of BUGS or outside in R, using results from a BUGS model fit. Sometimes it can be advantageous to do such calculations in R after fitting the model in BUGS. We illustrate this using the expected maximum count of great tits—i.e., lambda. The Poisson expectation is a deterministic function of the regression parameters alpha[1:4] and beta[1:4], for which we have 15,000 posterior samples each. We can use these to get samples of the posterior distribution of lambda for each unit in the data set. We create a data structure wherein we store the posterior samples for the newly calculated lambda, let us call it lambda2. We do this by applying the linear regression with each pair of MCMC samples from the two parameters (facFor and elev) and backtransforming using the inverse of the log link function. As an aside, note how we can use nested indexing in R exactly as in BUGS.

```
lambda2 <- array(dim = c(15000, 267))
for(j in 1:267){                       # Loop over sites
  lambda2[,j] <- exp(out5$sims.list$alpha[,facFor[j]] +
  out5$sims.list$beta[,facFor[j]] * elev[j])    # linear regression/backtransform
}
plot(out5$sims.list$lambda ~ lambda2, pch = ".")          # Check the two are identical
lm(c(out5$sims.list$lambda) ~ c(lambda2))
```

Finally, let us assume that we want to produce predictions of the expected maximum count versus elevation separately for each level of facFor (but we illustrate this for the first level only). We produce three plots (Figure 5.13) that illustrate different ways of graphing the point estimate of a prediction along with its uncertainty. Note that we have to sort pairs of elevation and predicted response according to the order of the values of elevation for easy plotting. In the third plot, we show the uncertainty around the prediction by plotting a random sample of 50 from their posterior predictive distribution.

```
sorted.ele1 <- sort(elev[facFor == 1])
sorted.y1 <- out5$summary[9:275,][facFor == 1,][order(elev[facFor == 1]),]

# Plot A
par(mfrow = c(1, 3), mar = c(5,5,3,2), cex.lab = 1.5, cex.axis = 1.5)
plot(elev[facFor == 1], jitter(Cmax[facFor ==1]), ylab = "Maximum count",
xlab = "Elevation (scaled)", frame.plot=F, ylim = c(0, 6))
```

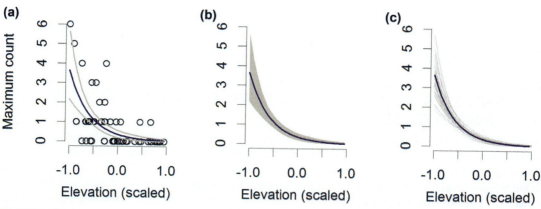

FIGURE 5.13

Three graphical summaries of the Poisson ANCOVA model fitted as the regression of the maximum count of great tits at each site on the scaled elevation and forest cover factor (shown for level 1 of facFor only; analogous plots could be produced for levels 2–4 of facFor). Posterior means of the predicted regression line of the mean tit count on scaled elevation are shown in blue, and the uncertainty around the prediction is depicted in grey ((a and b) 95% CRI; (c) a random sample of 50 posterior samples is shown). Note that the credible interval is for the expected count, not for observed counts; hence unsurprisingly, in the left plot most of the observed data lie outside of the interval (moreover, they are slightly jittered).

```
lines(sorted.ele1, sorted.y1[,1], col = "blue", lwd = 2) # Post. mean
lines(sorted.ele1, sorted.y1[,3], col = "grey", lwd = 2) # Lower 95% CL
lines(sorted.ele1, sorted.y1[,7], col = "grey", lwd = 2) # Upper 95% CL

# Plot B
plot(sorted.ele1, sorted.y1[,1], type='n', xlab = "Elevation (scaled)", ylab = "",
frame.plot = F, ylim = c(0, 6))
polygon(c(sorted.ele1, rev(sorted.ele1)), c(sorted.y1[,3], rev(sorted.y1[,7])),
col='grey', border=NA)
lines(sorted.ele1, sorted.y1[,1], col = "blue", lwd = 2)

# Plot C
elev.pred <- seq(-1,1, length.out = 200)   # Cov. for which to predict lambda
n.pred <- 50                               # Number of prediction profiles
pred.matrix <- array(NA, dim = c(length(elev.pred), n.pred))
for(j in 1:n.pred){
  sel <- sample(1:length(out5$sims.list$alpha[,1]),1) # Choose one post. draw
  pred.matrix[,j] <- exp(out5$sims.list$alpha[sel,1] + out5$sims.list$beta[sel,1] *
elev.pred)
}
```

```
plot(sorted.ele1, sorted.y1[,1], type='n', xlab = "Elevation (scaled)", ylab = "",
  frame.plot = F, ylim = c(0, 6))
matlines(elev.pred, pred.matrix, col = "grey", lty = 1, lwd = 1)
lines(sorted.ele1, sorted.y1[,1], col = "blue", lwd = 2)
```

5.10 GoF ASSESSMENT: POSTERIOR PREDICTIVE CHECKS AND THE PARAMETRIC BOOTSTRAP

We use the Poisson GLM to illustrate the assessment of the GoF of a model Bayesianly, using posterior predictive distributions (Rubin, 1984; Gelman et al., 1996). We also compare this approach with a related (purely) frequentist technique, the parametric bootstrap (Efron and Tibshirani, 1993; Dixon, 2006). We say "purely frequentist" because the Bayesian method has a decidedly frequentist flavor—it is based on hypothetical replicate data sets and hence could be called "semifrequentist" (or a "Bayesianly justifiable frequentist calculation," Rubin, 1984).

What is a posterior *predictive* distribution? This is simply the posterior distribution of anything that can be observed, here, of the data. We have seen earlier that using Bayesian analysis we can readily generate replicate data under our model when a response is missing. That is, using our model fit to the observed data and using the parameter values estimated at every iteration in the MCMC algorithm, we can create a "replicate" data set under the very same model with the same parameter values. As usual, we have an entire (posterior predictive) distribution for each replicate datum.

What does it mean to say "a model fits the data"? One reasonable answer is that the model is likely to produce data sets that are in some way "similar" to the data set we have at hand. Using MCMC, we can readily obtain the posterior predictive distribution of the data or of any function of the data. A posterior predictive check compares the observed data (or some function thereof) with their posterior predictive distribution (or some function thereof). And how do we measure "similarity," or conversely, dissimilarity? There is great flexibility in defining discrepancy measures (see, for instance, Gelman et al., 1996). For instance, we might choose as our discrepancy measure some statistic of the data that does not involve the estimated parameters, such as the mean, median, standard deviation, mean, maximum or range of the data. We might even use a visual impression of the data—i.e., graphs that are informative in some way about patterns in the data (e.g., pp. 154–155 in Gelman et al., 2014). However, the most common approach is to choose a test statistic that depends on the data and on the parameter estimates, and to compute an omnibus test that quantifies the "distance" between the observed data and the expected data under the model. Typically, such discrepancy measures are Chi-square, sums of squares, or the Freeman–Tukey statistic (see also Chapters 6 and 7).

We will see many applications of bootstrapping and posterior predictive distributions in later chapters and when using custom functions (such as `parboot` in `unmarked`), so it is good that we know how they work under the hood. Here, for the Poisson GLM from Section 5.9, we will compute a Chi-square discrepancy [i.e., the sum of (observed-expected)^2 / expected] as a posterior predictive check, and similarly in a parametric bootstrap. We will add to the expected values a very small number to avoid division by zero due to rounding errors. In addition, we will use the range of the observed values to check whether the spread of the data is sufficiently well represented by our model. We cannot

5.10 GoF ASSESSMENT

directly compute the range in WinBUGS, since the `min` and the `max` functions do not work for vectors like they do in R, but we can do so in OpenBUGS, and also in JAGS, when loading the module 'bugs.'

```
# Bundle data
win.data <- list(Cmax = Cmax, M = length(Cmax), elev = elev, facFor = facFor, e = 0.0001)

# Specify model in BUGS language
cat(file = "Poisson_GLM.txt","
model {

# Priors
for(k in 1:4){
  alpha[k] ~ dnorm(0, 1.0E-06)
  beta[k] ~ dnorm(0, 1.0E-06)
}

# Likelihood and computations for posterior predictive check
for (i in 1:M){
  Cmax[i] ~ dpois(lambda[i])
  log(lambda[i]) <- alpha[facFor[i]] + beta[facFor[i]] * elev[i]

# Fit assessments: Chi-square test statistic and posterior predictive check
  chi2[i] <- pow((Cmax[i]-lambda[i]),2) / (sqrt(lambda[i])+e)        # obs.
  Cmax.new[i] ~ dpois(lambda[i])     # Replicate (new) data set
  chi2.new[i] <- pow((Cmax.new[i]-lambda[i]),2) / (sqrt(lambda[i])+e) # exp.
}

# Add up discrepancy measures for entire data set
fit <- sum(chi2[])              # Omnibus test statistic actual data
fit.new <- sum(chi2.new[])      # Omnibus test statistic replicate data

# range of data as a second discrepancy measure
obs.range <- max(Cmax[]) - min(Cmax[])
exp.range <- max(Cmax.new[]) - min(Cmax.new[])
}
")

# Initial values
inits <- function() list(alpha = rnorm(4,,3), beta = rnorm(4,,3))

# Parameters monitored
params <- c("chi2", "fit", "fit.new", "obs.range", "exp.range")
params <- c("Cmax.new", "chi2.new", "chi2", "fit", "fit.new", "obs.range", "exp.range")

# MCMC settings
ni <- 6000  ;  nt <- 1  ;  nb <- 1000  ;  nc <- 3
```

We fit the model in JAGS and load the module 'bugs.'

```
# Call JAGS from R and summarize posteriors
out5.1 <- jags(win.data, inits, params, "Poisson_GLM.txt", n.chains = nc,
n.thin = nt, n.iter = ni, n.burnin = nb, modules = 'bugs')
```

```
print(out5.1, 2)
JAGS output for model 'Poisson_GLM.txt', generated by jagsUI.
Estimates based on 3 chains of 6000 iterations,
burn-in = 1000 iterations and thin rate = 1,
yielding 15000 total samples from the joint posterior.
MCMC ran for 0.235 minutes at time 2014-11-14 20:33:41.
              mean    sd    2.5%    50%    97.5%  overlap0 f Rhat n.eff
chi2[1]       9.80   1.43   7.12    9.74   12.75   FALSE 1   1    7655
chi2[2]       3.29   0.85   1.77    3.23    5.08   FALSE 1   1    5262
chi2[3]       1.74   0.69   0.58    1.68    3.28   FALSE 1   1    4379
[ ... ]
chi2[265]     0.24   0.15   0.03    0.22    0.59   FALSE 1   1    1871
chi2[266]     1.47   0.27   0.99    1.45    2.05   FALSE 1   1    4567
chi2[267]     0.02   0.01   0.00    0.01    0.05   FALSE 1   1    2478
fit         842.40  17.32 815.61  840.14  882.95   FALSE 1   1   15000
fit.new     416.91  45.66 332.45  414.98  512.72   FALSE 1   1   15000
obs.range    30.00   0.00  30.00   30.00   30.00   FALSE 1  NaN      1
exp.range    18.60   2.45  15.00   18.00   24.00   FALSE 1   1   15000
deviance   1004.05   4.05 998.12 1003.37 1013.55   FALSE 1   1   15000
```

First of all, we see that the Chi-square test statistic computed for data sets simulated under the assumptions of our model, and using the posterior distributions of the parameters estimated from our model (fit.new), is substantially smaller than that for our actual data set (fit). We can express the degree of lack of fit as the ratio of fit/fit.new, which is about 2. We can summarize this analysis by plotting the posterior distributions of the two fit statistics against each other (Figure 5.14(a)).

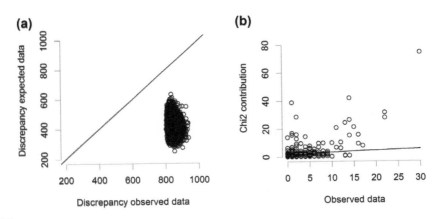

FIGURE 5.14

(a) Posterior predictive check for an omnibus fit statistic (Chi-square) for the Poisson model for maximum tit counts, showing the expected values of the statistic (computed from the replicate data sets) versus observed values (for the observed data set). (b) Contributions to the overall fit statistic of each individual datum, and expected values of the Chi-square contributions (this is the line).

```
par(mfrow = c(1, 2), mar = c(5,5,3,2), cex.lab = 1.5, cex.axis = 1.5)
plot(out5.1$sims.list$fit, out5.1$sims.list$fit.new, xlim = c(200, 1000), ylim =
c(200, 1000), main = "", xlab = "Discrepancy observed data", ylab = "Discrepancy
expected data", frame.plot = F, cex = 1.5)
abline(0,1, lwd = 2)
```

We can compute a Bayesian *p*-value, which is the probability to obtain, under the null hypothesis of a correctly specified model, a test statistic that is at least as extreme as the observed test statistic computed from the actual data. This is represented by the proportion of points above the diagonal equality line and here it is equal to zero.

```
(bpv <- mean(out5.1$sims.list$fit.new > out5.1$sims.list$fit))
[1] 0
```

A model that fits the data will have a Bayesian *p*-value around 0.5, and hence our *p*-value of zero suggests the model does not fit. The null distribution of Bayesian *p*-values is not uniform, so, we cannot say that a Bayesian *p*-value outside of the interval of, say, (0.025, 0.975) represents a significant lack of fit. Rather, we have to judge fit more qualitatively.

If this were a "real" analysis, then such a clear indication of lack of fit might be devastating to some. However, it would not be sensible to throw away the model and claim that the data set cannot be analyzed. First of all, we do not know whether the chosen fit statistic is really relevant for the objective of our modeling. And second, much more important than a black-and-white answer to the question "Does the model fit?" is an answer to "*Where* does the model fit and *where* does it not?" Hence, we are big fans of residual checks, where we can inspect patterns in the lack of fit. For instance, with count data it might be that the lack of fit is simple overdispersion—i.e., residuals that are too extreme and have no systematic pattern. We could then accommodate such overdispersion by adding a normal random effect in the linear predictor, thereby converting the model into a Poisson-lognormal model (see chapter 4 in Kéry and Schaub, 2012). We will not do so in this illustration of the technique, but want to point out again that model building is best seen as an iterative process in which we build a model, check whether it describes the salient features of our data well, fit an improved model and so forth.

Next, we illustrate such model diagnostics by looking at patterns in the point-wise contributions to the overall Chi-square GoF statistic `fit`. We plot these contributions versus the observed values (Figure 5.14(b)). For a fitting model, we would expect these contributions to be proportional to the square root of the data.

```
plot(Cmax, out5.1$mean$chi2, xlab = "Observed data", ylab = "Chi2 contribution",
frame.plot = F, ylim = c(0, 70), cex = 1.5)
lines(0:30, sqrt(0:30), lwd = 2)
```

There does not seem to be a clear pattern, but before we accept simple overdispersion in a "real" analysis, rather than a structural problem of the model, we should probably look a little harder for any patterns in the lack of fit.

We show this analysis to illustrate the mechanics of a posterior predictive check and how it can be summarized in a Bayesian *p*-value, but will not bother to track down a fitting model here. Rather, we next illustrate a parametric bootstrap to conclude this brief practical demonstration of two important methods for gauging GoF. In a parametric bootstrap, we repeat the following a large number of times (e.g., 1000–5000 times):

1. use the MLEs of the parameters from our data set to generate a replicate data set
2. refit the model to that new data set
3. compute a GoF discrepancy measure, such as Chi-square, using the simulated, new data set and associated MLEs

This yields the empirical distribution of the GoF statistic under the hypothesis of a fitting model. Exactly as for the posterior predictive check, we know that it does so because we generated this distribution for "perfect" data sets. They are perfect in the sense that they were generated under the exact assumptions of the model and using the point estimates from our data set. This is exactly like the replicated data sets in the posterior predictive checks earlier, except that the parametric bootstrap does not incorporate the parameter estimation uncertainty, while the Bayesian posterior predictive distributions do. If the test statistic obtained from our actual data set is not atypical in this reference distribution, then we conclude that the model fits, while if the test statistic is extreme (typically larger), then we conclude that the model does not fit. Let us do this now. Again, as for a posterior predictive check, we can be creative in our choice of test statistic. Here, we use Chi-square and the range of data as before.

```
# Fit model to actual data and compute two fit statistics
fm <- glm(Cmax ~ as.factor(facFor)*elev, family = poisson)   # Fit model
observed <- Cmax
expected <- predict(fm, type = "response")
plot(observed, expected)
abline(0,1)
chi2.obs <- sum((observed - expected)^2 / (expected + 0.0001))   # fit stat 1
range.obs <- diff(range(observed))                               # fit stat 2

# Generate reference distribution of fit stat under a fitting model
simrep <- 100000                             # Might want to try 1000 first
chi2vec <- rangevec <- numeric(simrep)       # vectors for Chi2 and maximum
for(i in 1:simrep){
  cat(paste(i, "\n"))
  Cmaxrep <- rpois(n = 267, lambda = expected)    # Generate replicate data set
  fmrep <- glm(Cmaxrep ~ as.factor(facFor)*elev, family = poisson) # Refit model
  expectednew <- predict(fmrep, type = "response")
  chi2vec[i] <- sum((Cmaxrep - expectednew)^2 / (expectednew + 0.0001))
  rangevec[i] <- diff(range(Cmaxrep))
}
```

Hence, the parametric bootstrap does something quite analogous to a posterior predictive check. It is instructive to look at the computed quantities side by side (Figure 5.15).

```
# Summarize bootstrap results and compare with posterior predictive dist.
par(mfrow = c(2, 2), mar = c(5,5,3,2), cex.lab = 1.5, cex.axis = 1.5)
hist(out5.1$sims.list$fit.new, col = "grey", main = "", breaks = 100, xlim = c(180, 900),
  freq = F, ylim = c(0, 0.01))
```

5.10 GoF ASSESSMENT

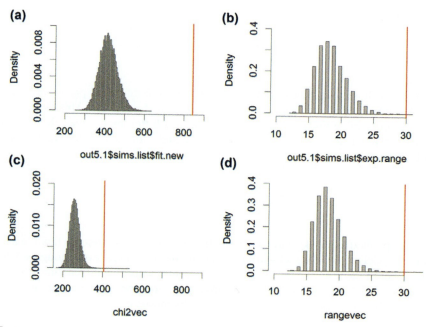

FIGURE 5.15

Posterior predictive checks (top) and parametric bootstrap (bottom) for a Chi-square test statistic (left) and for the range of the observed data (right). Histograms show the frequency distribution for replicate data sets simulated under the model. Red lines show the posterior means of the Chi-square test statistic for the actual data (a and c) or for the range of observed data (b and d).

```
abline(v = mean(out5.1$sims.list$fit), col = "red", lwd = 2)
hist(out5.1$sims.list$exp.range, col = "grey", main = "", breaks = 50, xlim = c(10, 32),
 freq = F, ylim = c(0, 0.40))
abline(v = mean(out5.1$sims.list$obs.range), col = "red", lwd = 2)
hist(chi2vec, col = "grey", main = "", breaks = 100, xlim = c(180, 900), freq = F,
 ylim = c(0, 0.02))
abline(v = chi2.obs, col = "red", lwd = 2)
hist(rangevec, col = "grey", main = "", breaks = 50, xlim = c(10, 32), freq = F)
abline(v = range.obs, col = "red", lwd = 2)
```

We see that the posterior predictive checks and the parametric bootstrap yield qualitatively very similar answers. However, the statistics are larger in magnitude for the former than for the latter, perhaps because the parametric bootstrap ignores some of the parameter estimation uncertainty, while in the posterior predictive check, the full parameter estimation uncertainty is propagated into the discrepancy measure. We can quantify the magnitude of the lack of fit by the ratio of the fit statistic for the observed data over that for the expected data.

```
# Lack of fit ratio in PPD and parboot
mean(out5.1$sims.list$fit/out5.1$sims.list$fit.new) # ppc
mean(chi2.obs/chi2vec)                              # parboot
[1] 2.045696
[1] 1.58963
```

For the parametric bootstrap, we can compute the *p*-value as 1 minus the percentile of the observed value in the reference distribution characterized by `chi2vec` and `rangevec`. This is the probability to obtain a more extreme value under the null hypothesis of a fitting model than the value of the test statistic for the observed data. This corresponds to the area under the curve to the right of the red line in Figure 5.15(d) and (e).

```
(pval1 <- 1-rank(c(chi2vec, chi2.obs))[simrep+1]/(simrep+1))
(pval2 <- 1-rank(c(rangevec, range.obs))[simrep+1]/(simrep+1))
[1] 0.0002199978
[1] 0.0001549985
```

Both posterior predictive checks and the parametric bootstrap are extremely flexible and powerful to test the fit of a model to your data set and should be more widely used, even though we do not know in general the power of the tests and, in the case of Bayesian *p*-value, we do not know their calibration either. These tests are perhaps best used in a qualitative way and, especially if an omnibus test suggests lack of fit, are best followed by more detailed residual diagnostics.

5.11 BINOMIAL GLM (LOGISTIC REGRESSION)

We next use BUGS to fit a logistic regression, also known as a binomial GLM. For this, we first quantize the great tit counts from one survey to obtain zeros and ones, indicating whether a count is zero or greater than zero. At several places in this book we emphasize that this is exactly how "presence/absence" or "detection/nondetection" data often arise in practice, namely as a simple summary of an abundance distribution (or here, of a distribution of counts). Once more, we fit the model using Bayesian and ML methods to emphasize their numerical similarity. We actually fit a Bernoulli GLM (i.e., a binomial GLM with trial size 1), but binomial GLMs with trial sizes >1 occur throughout the book, for instance in Chapter 6. We fit the following model to `y[,1]`, the detection/nondetection data for the first survey:

$$y_{i,1} \sim Bernoulli(\theta_i)$$

$$logit(\theta_i) = \alpha_{0,j} + \alpha_{1,j} * elev_i$$

As before, *j* indexes the four levels of `facFor`.

```
# Quantize counts from first survey and describe
y1 <- as.numeric(C[,1] > 0) # Gets 1 if first count greater than zero
table(y1)
```

5.11 BINOMIAL GLM (LOGISTIC REGRESSION)

```
y1
  0   1
137 130

mean(N > 0)        # True occupancy
mean(y1)           # Observed occupancy after first survey
```

Hence, true occupancy (proportion of occupied sites) is 77%, while the observed occupancy after the first survey is only 49%. We fit the ANCOVA linear model to the binary detection/nondetection response by specifying a Bernoulli GLM.

```
# Bundle data
win.data <- list(y1 = y1, M = length(y1), elev = elev, facFor = facFor)

# Specify model in BUGS language
cat(file = "Bernoulli_GLM.txt","
model {

# Priors
for(k in 1:4){
  alpha[k] <- logit(mean.psi[k])    # intercepts
  mean.psi[k] ~ dunif(0,1)
  beta[k] ~ dnorm(0, 1.0E-06)       # slopes
}

# Likelihood
for (i in 1:M){
  y1[i] ~ dbern(theta[i])
  logit(theta[i]) <- alpha[facFor[i]] + beta[facFor[i]] * elev[i]
}
}
")

# Initial values
inits <- function() list(mean.psi = runif(4), beta = rnorm(4,,3))  # Priors 2

# Parameters monitored
params <- c("mean.psi", "alpha", "beta", "theta")

# MCMC settings
ni <- 6000  ;  nt <- 1  ;  nb <- 1000  ;  nc <- 3

# Call WinBUGS or JAGS from R (ART <1 min)
out6 <- bugs(win.data, inits, params, "Bernoulli_GLM.txt", n.chains = nc,
n.thin = nt, n.iter = ni, n.burnin = nb, debug = TRUE, bugs.directory = bugs.dir,
working.directory = getwd())

out6J <- jags(win.data, inits, params, "Bernoulli_GLM.txt", n.chains = nc, n.thin = nt,
n.iter = ni, n.burnin = nb)
par(mfrow = c(4,2))   ;   traceplot(out6J, c("alpha[1:4]", "beta[1:4]"))
```

We can summarize the posterior distributions for the logit-linear regression parameters, though we can no longer directly compare them with the truth, since the data were generated under a different model than the one fitted.

```
print(out6, 2)
              mean   sd   2.5%   25%    50%    75%   97.5%  Rhat  n.eff
mean.psi[1]   0.17  0.06  0.08   0.13   0.17   0.21   0.29    1   3900
mean.psi[2]   0.25  0.06  0.14   0.21   0.25   0.30   0.39    1   4000
mean.psi[3]   0.54  0.06  0.43   0.50   0.54   0.58   0.66    1  15000
mean.psi[4]   0.82  0.05  0.73   0.80   0.83   0.86   0.91    1  15000
alpha[1]     -1.62  0.41 -2.50  -1.87  -1.59  -1.33  -0.88    1   4100
alpha[2]     -1.11  0.35 -1.82  -1.33  -1.10  -0.87  -0.47    1   4000
alpha[3]      0.17  0.24 -0.29   0.01   0.17   0.33   0.65    1  15000
alpha[4]      1.58  0.33  0.99   1.36   1.57   1.79   2.26    1  15000
beta[1]      -1.98  0.78 -3.62  -2.47  -1.93  -1.43  -0.57    1   2700
beta[2]      -2.56  0.74 -4.12  -3.05  -2.53  -2.05  -1.21    1   8500
beta[3]      -0.95  0.42 -1.77  -1.22  -0.94  -0.66  -0.13    1  11000
beta[4]       0.49  0.59 -0.63   0.10   0.49   0.88   1.68    1  15000
theta[1]      0.64  0.07  0.50   0.59   0.64   0.68   0.76    1  15000
theta[2]      0.66  0.07  0.52   0.62   0.67   0.71   0.79    1  15000
theta[3]      0.85  0.06  0.73   0.81   0.85   0.89   0.94    1  15000
[ ... ]

# Compare with MLEs obtained by iteratively-reweighted least-squares method
summary(glm(y1 ~ factor(facFor)*elev-1-elev, family = binomial))

[ .... ]
Coefficients:
                     Estimate Std. Error z value Pr(>|z|)
factor(facFor)1       -1.6194    0.4172  -3.881  0.000104 ***
factor(facFor)2       -1.1122    0.3481  -3.195  0.001398 **
factor(facFor)3        0.1774    0.2392   0.742  0.458303
factor(facFor)4        1.5873    0.3307   4.801  1.58e-06 ***
factor(facFor)1:elev  -1.9277    0.7860  -2.453  0.014180 *
factor(facFor)2:elev  -2.4480    0.7260  -3.372  0.000747 ***
factor(facFor)3:elev  -0.9154    0.4125  -2.219  0.026477 *
factor(facFor)4:elev   0.4924    0.5821   0.846  0.397600
[ .... ]
```

We overlay the estimated regression equations onto plots of the observed data (jittered for enhanced readability); see Figure 5.16. Note the slight contortions to get the ordering of the x and y variables right.

```
# Plot of observed response vs. two covariates
par(mfrow = c(1, 2), mar = c(5,5,3,2), cex.lab = 1.5, cex.axis = 1.5)
F1 <- facFor == 1 ; F2 <- facFor == 2 ; F3 <- facFor == 3 ; F4 <- facFor == 4
plot(jitter(y1,,0.05) ~ facFor, xlab = "Forest factor", ylab = "Observed occupancy
probability", frame.plot = F, ylim = c(0, 1.15))
```

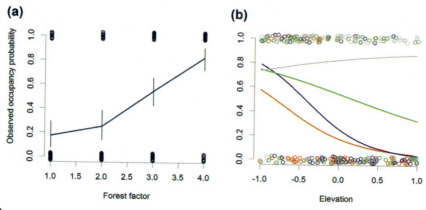

FIGURE 5.16

Predictions of the observed occupancy probability (theta) of simulated great tit data under the Bernoulli model; circles are observed data (jittered). (a) Observed detection/nondetection data versus `facFor`; line shows estimates of theta (with 95% CRI) at average scaled elevation. (b) Observed and estimated relationship of theta with elevation for all four levels of `facFor` (1—red, 2—blue, 3—green, 4—grey).

```
lines(1:4, out6$summary[1:4,1], lwd = 2)
segments(1:4, out6$summary[1:4,3], 1:4, out6$summary[1:4,7])

plot(elev[F1], jitter(y1,,0.1)[F1], xlab = "Elevation", ylab = "", col = "red",
frame.plot = F)
points(elev[F2], jitter(y1,,0.05)[F2], col = "blue")
points(elev[F3], jitter(y1,,0.05)[F3], col = "green")
points(elev[F4], jitter(y1,,0.05)[F4], col = "grey")
lines(sort(elev[F1]), out6$mean$theta[F1][order(elev[F1])], col="red", lwd=2)
lines(sort(elev[F2]), out6$mean$theta[F2][order(elev[F2])], col="blue", lwd=2)
lines(sort(elev[F3]), out6$mean$theta[F3][order(elev[F3])], col="green", lwd=2)
lines(sort(elev[F4]), out6$mean$theta[F4][order(elev[F4])], col="grey", lwd=2)
```

5.12 MOMENT-MATCHING IN A BINOMIAL GLM TO ACCOMMODATE UNDERDISPERSION

Assume for the sake of illustration that we want to model our observed counts with a distribution *less* variable than a Poisson. A good example in avian population ecology is provided by counts of chicks in the nest. One solution would be to directly use a variant of a Poisson that allows underdispersion (Ridout and Besbeas, 2004; Wu et al., 2013; Lynch et al., 2014). Another solution is to adopt a binomial instead and apply a very useful technique called *moment matching* (see Section 3.3.3 in this book and Section 3.4.4 in Hobbs and Hooten, 2015). In our example data, the maximum observed count is 30, so let us assume that for some reason we *knew* that values beyond 32 are essentially impossible. In this situation, a Poisson might not be adequate, because it might easily predict responses greater than 32. If we adopt a binomial with trial size 32 instead, no count greater than 32 is possible.

However, we then face a problem when we want to model variation in the mean response, since there is no parameter in the binomial that corresponds to the mean. Rather, the mean is a function of parameter p and trial size N: Np. Interestingly, we can simply apply a linear model to a link transformation of that mean! Since we want to avoid negative predicted counts in a binomial, we would typically adopt a log link, exactly as in the Poisson GLM. We illustrate moment matching by modeling the expected count by site elevation and want the response to be bounded by 0 and 32.

For count $y_{i,j}$ at site i during survey j we fit the following model:

$$y_{i,j} \sim Binomial(p_i, N).$$

Here, N is *not* population size but the upper limit we wish to impose on the counts, while p_i is the expected count at site i *as a proportion of that limit*. To model structure in the mean of that "squeezed" (between 0 and N) response, note first that the binomial mean is:

$$\mu_i = Np_i$$

We can then apply a linear model to a suitable transformation of the expected response as usual.

$$\log(\mu_i) = \alpha + \beta * elev_i$$

We rearrange this to express p_i a function of N and the parameters of the linear model:

$$p_i = \frac{1}{N} \exp(\alpha + \beta * elev_i)$$

Note that this formulation does not constrain p to be $<= 1$, because whenever the exponential becomes greater than N (here, 32) inadmissible values for p will result. If this is a problem with BUGS, we can likely diagnosed it by a crash. A simple workaround will then be to simply scale the covariates more, e.g., divide them by 10, 100 or 1000, as needed. In the BUGS code, to obtain predictions of the response for selected values of elevation, we also add a line that computes the expected binomial response Np_i under the model.

```
# Bundle data
win.data <- list(y = data$C, M = nrow(data$C), J = ncol(data$C), elev = elev, N = 32)

# Specify model in BUGS language
cat(file = "squeezed_count_GLM.txt","
model {

# Priors
alpha ~ dnorm(0, 1.0E-06)
beta ~ dnorm(0, 1.0E-06)

# Likelihood
for (i in 1:M){
   p[i] <- 1/N * exp(alpha + beta * elev[i])   # linear model for expected response
   (relative to bound N)
   mu[i] <- N * p[i]   # Expected response = binomial mean = first moment
   for(j in 1:J){
     y[i,j] ~ dbin(p[i], N)
   }
  }
 }
")
```

5.13 RANDOM-EFFECTS POISSON GLM (POISSON GLMM)

```
# Initial values
inits <- function() list(alpha = runif(1), beta = rnorm(1))
# Parameters monitored
params <- c("alpha", "beta", "mu")

# MCMC settings
ni <- 300 ; nt <- 1 ; nb <- 100 ; nc <- 3

# Call JAGS from R (ART <1 min)
library(jagsUI)
out7 <- jags(win.data, inits, params, "squeezed_count_GLM.txt", n.chains = nc,
n.thin = nt, n.iter = ni, n.burnin = nb)
par(mfrow = c(4,2)) ; traceplot(out7)
print(out7, 3)
```

As emphasized by Hobbs and Hooten (2015), moment matching is a very useful technique that gives you greatly flexibility in your modeling. For another example, see Section 7.6, where we use moment matching to reparameterize the BUGS implementation of the negative binomial distribution such that we have a mean and a variance parameter and can apply a linear model to the former. Another useful example Hobbs and Hooten (2015) give is for a beta distribution, with parameters α and β. The beta can be used to model ratios of numbers that are *not* counts but continuous measurements, say, to model coverage data in plant surveys. The mean (μ) of a beta distribution is $\alpha/(\alpha + \beta)$; hence, $\alpha = \mu\beta/(1 + \mu)$, and we can specify linear models for μ with an appropriate link function such as the logit. See also Cribari-Neto and Zeileis (2010) for an introduction to beta regression.

5.13 RANDOM-EFFECTS POISSON GLM (POISSON GLMM)

In the concluding two examples of the BUGS language, we illustrate the specification of random effects. Random effects are a defining feature of HMs, and the BUGS language makes particularly transparent what they are: realizations from an unobserved or partially observed random variable. In other words, random effects are parameters or latent variables that are given a distribution with (hyper) parameters that are estimated from the data.

We illustrate in the context of an HM that is somewhat related to the N-mixture model (see Chapter 6): a Poisson regression, or GLM, with random site effects fitted to all three simulated counts of great tits per site (Dennis et al., 2015a, also unpublished manuscript by W. A. Link). We include a random site effect, perhaps to avoid pseudoreplication and to account for the correlation of replicated counts made at the same site. In contrast to the *N*-mixture model, which is a binomial-Poisson mixture, this model is a Poisson–normal mixture model. Because we no longer aggregate the repeated measures, we can now directly fit both site covariates (elevation and forest cover) and sampling covariates (wind speed). We fit the following GLMM to the counts C_{ij} of great tits at site i during replicate survey j:

$$C_{ij} \sim Poisson(\lambda_{ij})$$

$$\log(\lambda_{ij}) = \alpha_{0,i} + \alpha_1 * elev_i + \alpha_2 * forest_i + \alpha_3 * elev_i * forest_i + \alpha_4 * wind_{ij}$$

$$\alpha_{0,i} \sim Normal(\mu_\alpha, \sigma_\alpha^2)$$

It is the last line that defines the regression intercepts as random effects: the alpha0 parameters are defined to be draws from a normal distribution with mean and variance as hyperparameters that are estimated.

We write the code in a slightly less verbose format now. We will also compute, as a derived quantity in the BUGS code, a zero-centered version of the random site effects alpha0—i.e., subtract the value of their hypermean—for direct comparison with the frequentist analyses below. Note that much longer chains are required to achieve convergence in random-effects models.

```
# Bundle data
win.data <- list(C = C, M = nrow(C), J = ncol(C), elev = elev, forest = forest, elev.forest =
elev * forest, wind = wind)

# Specify model in BUGS language
cat(file = "RE.Poisson.txt","
model {

# Priors
mu.alpha ~ dnorm(0, 0.001)      # Mean hyperparam
tau.alpha <- pow(sd.alpha, -2)
sd.alpha ~ dunif(0, 10)         # sd hyperparam
for(k in 1:4){
  alpha[k] ~ dunif(-10, 10)     # Regression params
}

# Likelihood
for (i in 1:M){
  alpha0[i] ~ dnorm(mu.alpha, tau.alpha)   # Random effects and hyperparams
  re0[i] <- alpha0[i] - mu.alpha           # zero-centered random effects
  for(j in 1:J){
    C[i,j] ~ dpois(lambda[i,j])
    log(lambda[i,j]) <- alpha0[i] + alpha[1] * elev[i] + alpha[2] * forest[i] +
alpha[3] * elev.forest[i] + alpha[4] * wind[i,j]
  }
}
}")

# Other model run preparations
inits <- function() list(alpha0 = rnorm(M), alpha = rnorm(4))   # Inits
params <- c("mu.alpha", "sd.alpha", "alpha0", "alpha", "re0")   # Params
ni <- 30000 ; nt <- 25 ; nb <- 5000 ; nc <- 3                   # MCMC settings

# Call WinBUGS or JAGS from R (ART 6-7 min) and summarize posteriors
out8 <- bugs(win.data, inits, params, "RE.Poisson.txt", n.chains = nc,
n.thin = nt, n.iter = ni, n.burnin = nb, debug = TRUE, bugs.directory = bugs.dir,
working.directory = getwd())

out8 <- jags(win.data, inits, params, "RE.Poisson.txt", n.chains = nc, n.thin = nt,
n.iter = ni, n.burnin = nb)
par(mfrow = c(3,2)) ; traceplot(out8, c("mu.alpha", "sd.alpha", "alpha[1:3]"))
```

5.13 RANDOM-EFFECTS POISSON GLM (POISSON GLMM)

```
print(out8, 3)
              mean     sd    2.5%    50%   97.5%  overlap0      f  Rhat  n.eff
mu.alpha    -0.839  0.072  -0.980 -0.840  -0.699     FALSE  1.000 1.001   2756
sd.alpha     0.267  0.049   0.169  0.268   0.362     FALSE  1.000 1.003   3000
alpha0[1]   -0.856  0.218  -1.306 -0.847  -0.444     FALSE  1.000 1.001    796
alpha0[2]   -0.910  0.208  -1.330 -0.903  -0.522     FALSE  1.000 1.004    650
alpha0[3]   -1.216  0.215  -1.657 -1.210  -0.809     FALSE  1.000 1.002   3000
[ ... ]
alpha0[265] -0.845  0.269  -1.409 -0.839  -0.340     FALSE  0.999 1.003    950
alpha0[266] -0.909  0.273  -1.466 -0.894  -0.387     FALSE  0.999 1.002   1298
alpha0[267] -0.836  0.276  -1.386 -0.831  -0.301     FALSE  0.999 1.001   3000
alpha[1]    -1.694  0.108  -1.900 -1.695  -1.486     FALSE  1.000 1.002   2944
alpha[2]     2.128  0.096   1.943  2.130   2.315     FALSE  1.000 1.003   2938
alpha[3]     1.390  0.161   1.076  1.391   1.702     FALSE  1.000 1.000   2721
alpha[4]    -1.711  0.063  -1.837 -1.711  -1.590     FALSE  1.000 1.001   3000
re0[1]      -0.017  0.206  -0.437 -0.014   0.386      TRUE  0.529 1.001    997
re0[2]      -0.071  0.195  -0.475 -0.064   0.298      TRUE  0.637 1.002    821
re0[3]      -0.377  0.195  -0.783 -0.368  -0.013     FALSE  0.981 1.001   3000
[ ... ]
re0[265]    -0.006  0.262  -0.534  0.003   0.491      TRUE  0.495 1.003   1243
re0[266]    -0.069  0.262  -0.598 -0.061   0.445      TRUE  0.596 1.002   1776
re0[267]     0.003  0.266  -0.544  0.003   0.524      TRUE  0.506 1.001   3000
```

If your computer runs out of memory, try running the model without the zero-centered random effects in the params list, and compute the posterior samples for re0 in R. You might also want to thin more to save fewer samples per parameter.

At last, we fit this model non-Bayesianly using ML with package lme4 (Bates et al. 2014). Function glmer requires the data to be input in a vector format, rather than as an array. Therefore, we first reformat our data set. Note how the frequentist model fit is much faster than using MCMC (though, admittedly, our MCMC settings are a little overkill).

```
Cvec <- as.vector(C)                    # Vector of M*J counts
elev.vec <- rep(elev, J)                # Vectorized elevation covariate
forest.vec <- rep(forest, J)            # Vectorized forest covariate
wind.vec <- as.vector(wind)             # Vectorized wind covariate
fac.site <- factor(rep(1:M, J))         # Site indicator (factor)
cbind(Cvec, fac.site, elev.vec, forest.vec, wind.vec) # Look at data

# Fit same model using maximum likelihood (NOTE: glmer uses ML instead of REML)
library(lme4)
summary(fm <- glmer(Cvec ~ elev.vec*forest.vec + wind.vec + (1| fac.site), family =
poisson))                               # Fit model
ranef(fm)                               # Print zero-centered random effects

Generalized linear mixed model fit by maximum likelihood (Laplace Approximation)
['glmerMod']
 Family: poisson ( log )
Formula: Cvec ~ elev.vec * forest.vec + wind.vec + (1 | fac.site)
[ ... ]
```

```
Random effects:
 Groups    Name         Variance Std.Dev.
 fac.site (Intercept) 0.06461  0.2542
Number of obs: 801, groups: fac.site, 267

Fixed effects:
                   Estimate Std. Error z value Pr(>|z|)
(Intercept)        -0.83243    0.07252  -11.48  <2e-16 ***
elev.vec           -1.69303    0.10480  -16.16  <2e-16 ***
forest.vec          2.12447    0.09546   22.25  <2e-16 ***
wind.vec           -1.70929    0.06326  -27.02  <2e-16 ***
elev.vec:forest.vec 1.39146    0.15582    8.93  <2e-16 ***
[ ... ]
```

We compare the fixed-effects estimates in a table below, and also the random-effects estimates from the Bayesian and the non-Bayesian analyses in a table and a graph (Figure 5.17(a)).

```
# Compare fixed-effects estimates (in spite of the confusing naming in glmer output),
Bayesian post. means and sd left, frequentist MLEs and SEs right
print(cbind(out8$summary[c(1:2, 270:273), 1:2], rbind(summary(fm)$coef[1,1:2],
c(sqrt(summary(fm)$varcor$fac.site), NA), summary(fm)$coef[c(2,3,5,4),1:2])), 3)
           mean      sd Estimate Std. Error
mu.alpha  -0.837 0.0758   -0.832    0.0725
sd.alpha   0.264 0.0507    0.254        NA
alpha[1]  -1.693 0.1079   -1.693    0.1048
alpha[2]   2.126 0.0976    2.124    0.0955
alpha[3]   1.389 0.1602    1.391    0.1558
alpha[4]  -1.712 0.0650   -1.709    0.0633
```

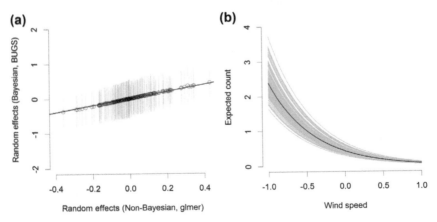

FIGURE 5.17

(a) Estimates of random site intercepts from a Bayesian (with 95% CRI in grey) and non-Bayesian analysis of the Poisson GLMM. (b) Bayesian estimates of site-specific relationships between the expected count of great tits and wind speed (black line is the population average).

5.13 RANDOM-EFFECTS POISSON GLM (POISSON GLMM)

```
# Compare graphically non-Bayesian and Bayesian random effects estimates
Freq.re <- ranef(fm)$fac.site[,1]      # Non-Bayesian estimates (MLEs)
Bayes.re <- out8$summary[274:540,]     # Bayesian estimates

par(mfrow = c(1, 2), mar = c(5,5,3,2), cex.lab = 1.5, cex.axis = 1.5) # Fig. 5.17 (a)
plot(Freq.re, Bayes.re[,1], xlab = "Non-Bayesian (glmer)", ylab = "Bayesian (BUGS)",
xlim = c(-0.4, 0.42), ylim = c(-2, 2), frame.plot = F, type = "n")
segments(Freq.re, Bayes.re[,3], Freq.re, Bayes.re[,7], col = "grey", lwd = 0.5)
abline(0, 1, lwd = 2)
points(Freq.re, Bayes.re[,1])
```

Once again, we note what many consider to be a comforting similarity between the non-Bayesian and the Bayesian parameter estimates, even with respect to their uncertainty (SEs and posterior standard deviations, respectively). In the Bayesian analysis, exact (i.e., not asymptotic) uncertainty estimates of the random effects are obtained easily (the grey error bars in Figure 5.17(a)), while for the non-Bayesian estimate, these would be harder to get (presumably by bootstrapping).

Finally, we plot predictions of the relationship between the expected counts and wind speed for each site separately and at a determined value of the other covariates (elevation and forest cover; Figure 5.17(b)). We could use the posterior samples of all parameters to produce uncertainty intervals as well, but this would be graphically unwieldy; hence, we plot only the posterior means. For clarity, we write the entire linear predictor, including the effects of the three other covariates set at zero (i.e., at their mean value).

```
wind.pred <- seq(-1, 1, , 1000)    # Covariate values for prediction
pred <- array(NA, dim = c(1000, 267))
for(i in 1:267){
  pred[,i]<- exp(out8$mean$alpha0[i] + out8$mean$alpha[1] * 0 + out8$mean$alpha[2] * 0 +
out8$mean$alpha[3] * 0 + out8$mean$alpha[4] * wind.pred)   # Predictions for each site
}

matplot(wind.pred, pred, type = "l", lty = 1, col = "grey", xlab = "Wind speed",
ylab = "Expected count", frame.plot = F, ylim = c(0, 4)) # Fig. 5.17 (b)
lines(wind.pred, exp(out8$mean$mu.alpha + out8$mean$alpha[4] * wind.pred),
col = "black", lwd = 3)
```

Note that for HMs, it is not straightforward to compute a quantity like the proportion of explained variation (R^2). The reason is that one may define such quantities for every level in the hierarchy of the model. For an attempt at defining R^2-like quantities in HMs, see Gelman and Pardoe (2006).

Often, fitting a model is only the first step of an analysis and much can be learnt by summarizing the results of that fit. This is true especially of a Bayesian analysis, where rich insights are possible based on the full posterior distributions and functions thereof, as we have already seen. The ability to make direct probability statements about parameters and the ease with which derived quantities such as predictions of the response for selected values of covariates can be computed, with a full assessment of the uncertainty, are great and underused benefits of a Bayesian analysis using MCMC.

5.14 RANDOM-EFFECTS BINOMIAL GLM (BINOMIAL GLMM)

We conclude our overview of "standard" GLMs and GLMMs by fitting a random-effects model with binomial response. We convert the counts for each survey at a site into "presence/absence," or more accurately "detection/nondetection," observations—i.e., "squash" the counts into zeros and ones, depending on whether they are equal to zero or greater.

```
# Get detection/nondetection data
y <- C
y[y > 0] <- 1
```

We then fit a kind of "naive" site-occupancy model, where we describe the detection/nondetection observations by a logistic regression with a continuous site-specific random effect assumed to be drawn from a normal distribution. This means that we assume that the effect of wind is specific to each site, but that the slopes of wind among all sites cluster around some common mean with a variance that is estimated. In contrast to the site-occupancy model (see Chapter 10), which is a Bernoulli–Bernoulli mixture, this model is a Bernoulli–normal mixture. We fit the following GLMM to the binary detection-nondetection observations y_{ij} at site i during replicate survey j:

$$y_{ij} \sim Bernoulli(\theta_{ij})$$

$$logit(\theta_{ij}) = \alpha_{0,i} + \alpha_1 * elev_i + \alpha_2 * forest_i + \alpha_3 * elev_i * forest_i + \alpha_{4,i} * wind_{ij}$$

$$\alpha_{0,i} \sim Normal(\mu_{\alpha 0}, \sigma^2_{\alpha 0})$$

$$\alpha_{4,i} \sim Normal(\mu_{\alpha 4}, \sigma^2_{\alpha 4})$$

For illustration, we now fit a slightly more complicated random-effects model than in the previous section, since in addition to the site-specific intercepts we now also specify the coefficients of wind to be random, by defining them to be draws from another (normal) prior distribution, with hyperparameters that we estimate. Other than that, we keep the linear model of this Bernoulli GLMM identical to that of the Poisson GLMM in the previous section, to emphasize that the linear model can be chosen entirely separately from the distribution chosen to describe the randomness in the response.

```
# Bundle data
win.data <- list(y = y, M = nrow(y), J = ncol(y), elev = elev, forest = forest,
elev.forest = elev * forest, wind = wind)
str(win.data)
```

In the BUGS code, we have to take the coefficient `alpha4` out of the vector `alpha`, since it is no longer a scalar, but a vector itself.

```
# Specify model in BUGS language
cat(file = "RE.Bernoulli.txt","
model {

# Priors
mu.alpha0 <- logit(mean.theta)     # Random intercepts
mean.theta ~ dunif(0,1)
```

5.14 RANDOM-EFFECTS BINOMIAL GLM (BINOMIAL GLMM)

```
  tau.alpha0 <- pow(sd.alpha0, -2)
  sd.alpha0 ~ dunif(0, 10)
  mu.alpha4 ~ dnorm(0, 0.001)              # Random slope on wind
  tau.alpha4 <- pow(sd.alpha4, -2)
  sd.alpha4 ~ dunif(0, 10)
  for(k in 1:3){
    alpha[k] ~ dnorm(0, 0.001)             # Slopes
  }

  # Likelihood
  for (i in 1:M){
    alpha0[i] ~ dnorm(mu.alpha0, tau.alpha0)   # Intercept random effects
    re00[i] <- alpha0[i] - mu.alpha0           # same zero-centered
    alpha4[i] ~ dnorm(mu.alpha4, tau.alpha4)   # Slope random effects
    re04[i] <- alpha4[i] - mu.alpha4           # same zero-centered
    for(j in 1:J){
      y[i,j] ~ dbern(theta[i,j])
      logit(theta[i,j]) <- alpha0[i] + alpha[1] * elev[i] + alpha[2] * forest[i] +
  alpha[3] * elev.forest[i] + alpha4[i] * wind[i,j]
    }
  }
}")

# Other model run preparations
inits <- function() list(alpha0 = rnorm(M), alpha4 = rnorm(M))# Inits
params <- c("mu.alpha0", "sd.alpha0", "alpha0", "alpha", "mu.alpha4", "sd.alpha4",
"alpha4", "re00", "re04")                    # Params
ni <- 30000 ; nt <- 25 ; nb <- 5000 ; nc <- 3    # MCMC settings

# Call WinBUGS from R .... and crash !
out9 <- bugs(win.data, inits, params, "RE.Bernoulli.txt", n.chains = nc,
n.thin = nt, n.iter = ni, n.burnin = nb, debug = TRUE, bugs.directory = bugs.dir,
working.directory = getwd())
```

While WinBUGS gets an "undefined real result" crash, JAGS works fine; this is an example that shows that JAGS is sometimes numerically more robust than WinBUGS. (We could get WinBUGS to run by adding into the model some numerical "stabilization"—see trick 15 in Appendix 1 of Kéry and Schaub, 2012.)

```
# Call JAGS from R (ART 2.5 min)
out9 <- jags(win.data, inits, params, "RE.Bernoulli.txt", n.chains = nc, n.thin = nt,
n.iter = ni, n.burnin = nb)
par(mfrow = c(2,2))
traceplot(out9, c("mu.alpha0", "sd.alpha0", "alpha[1:3]", "mu.alpha4", "sd.alpha4"))
print(out9, 3)
```

Again, if you run into memory problems, do not save the posterior samples for the zero-centered random effects, but compute them in R instead. Finally, we fit this model non-Bayesianly using maximum likelihood (ML) in the package lme4 after vectorizing the data.

```
yvec <- as.vector(y)              # Vector of M*J counts
elev.vec <- rep(elev, J)          # Vectorized elevation covariate
forest.vec <- rep(forest, J)      # Vectorized forest covariate
wind.vec <- as.vector(wind)       # Vectorized wind covariate
fac.site <- factor(rep(1:M, J))   # Site indicator (factor)
cbind(yvec, fac.site, elev.vec, forest.vec, wind.vec)   # Look at data

# Fit same model using maximum likelihood
library(lme4)        # Load package
summary(frem <- glmer(yvec ~ elev.vec*forest.vec + wind.vec + (wind.vec || fac.site),
  family = binomial))       # Fit model

Random effects:
 Groups     Name        Variance  Std.Dev.
 fac.site   (Intercept) 4.255     2.063
 fac.site.1 wind.vec    8.779     2.963
Number of obs: 801, groups:    fac.site, 267

Fixed effects:
                   Estimate Std. Error z value Pr(>|z|)
(Intercept)         -0.6381     0.2504  -2.548 0.010837 *
elev.vec            -3.8703     0.7855  -4.927 8.35e-07 ***
forest.vec           5.7574     1.0614   5.424 5.82e-08 ***
wind.vec            -5.9007     1.1714  -5.037 4.72e-07 ***
elev.vec:forest.vec  3.8181     1.0009   3.815 0.000136 ***

# Compare Bayesian and non-Bayesian estimates
print(out9$summary[c(1:2, 270:274),c(1:3,7:9)], 4)
              mean      sd    2.5%    97.5%   Rhat  n.eff
mu.alpha0  -0.6091  0.2414 -1.1101  -0.1457 0.9996  3000
sd.alpha0   2.0662  0.4821  1.2639   3.2134 1.0014  3000
alpha[1]   -3.8639  0.6980 -5.5057  -2.7587 1.0032  1633
alpha[2]    5.7575  0.9339  4.3289   7.9514 1.0057  1322
alpha[3]    3.7679  0.9383  2.1356   5.8646 1.0001  3000
mu.alpha4  -5.8058  1.0006 -8.1132  -4.3193 1.0067  1443
sd.alpha4   2.4177  1.0510  0.5137   4.6309 1.0034  1086
```

We find slightly less agreement between the two analyses now than in all the previous examples. This is perhaps not surprising given that binary data contain much less information about all the parameters of this fairly complex model. Next, we look at the estimates of the random effects.

```
(re <- ranef(frem))                # Print zero-centered random effects
$fac.site
       (Intercept)       wind.vec
1      0.01538103261  -0.5521873386140
2      0.13475066011  -0.0394227279709
3     -1.07204621369  -1.4696483198851
[...]
```

5.14 RANDOM-EFFECTS BINOMIAL GLM (BINOMIAL GLMM)

```
265   0.63961927035    2.0279735503384
266  -1.07504683084   -0.1964205985223
267  -0.00131161668    0.0019269593948
```

The random-effects estimates of the site-specific relationships between the observation of at least one great tit (i.e., our response) and wind speed can be plotted (Figure 5.18(a)).

```
par(mfrow = c(1, 3), mar = c(5,5,3,2), cex.lab = 1.5, cex.axis = 1.5)
pop.mean.int <- summary(frem)$coef[1,1]
pop.mean.slope <- summary(frem)$coef[4,1]
plot(sort(wind.vec), plogis(pop.mean.int + sort(wind.vec) * pop.mean.slope), type = "l",
xlab = "Wind speed", ylab = "Prob. to count >0 great tits", lwd = 3, frame.plot = F)
for(i in 1:267){
   lines(sort(wind.vec), plogis(pop.mean.int + re$fac.site[i,1] + sort(wind.vec) *
(pop.mean.slope + re$fac.site[i,2])), lwd = 1, col = i)
}
```

To conclude, we plot a response surface of the expected detection/nondetection probability for a grid of elevation by forest cover along with its uncertainty (Figure 5.18(b) and (c)).

```
# Compute expected detection/nondetection probability for a grid of elevation and forest
cover, at wind-speed = 0 (covariate average) and for hypermean of intercepts alpha0
n.sims <- length(out9$sims.list$mu.alpha0)
elev.pred <- seq(-1, 1, ,100)        # Values of elevation
forest.pred <- seq(-1,1, ,100)       # Values of forest cover
pred.array <- array(NA, dim = c(100, 100, n.sims))  # Prediction array
```

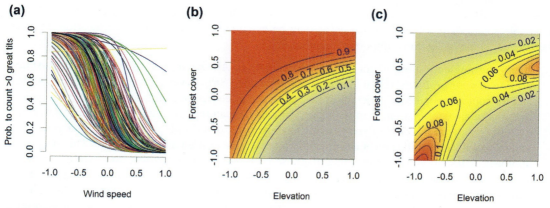

FIGURE 5.18

Site-specific estimates (ML) of the relationship between wind speed and the probability to count at least one great tit (a), and Bayesian predictions of detection/nondetection probability in a grid of elevation and forest cover ((b) posterior mean, (c) posterior standard deviation). Compare this with Figure 4.3.

```
for(i in 1:100){
  for(j in 1:100){
     pred.array[i,j,] <- plogis(out9$sims.list$mu.alpha0 + out9$sims.list$alpha[,1] *
elev.pred[i] + out9$sims.list$alpha[,2] * forest.pred[j] + out9$sims.list$alpha[,3] *
elev.pred[i] * forest.pred[j])
  }
}
pm.pred.array <- apply(pred.array, c(1,2), mean)   # Get posterior mean
psd.pred.array <- apply(pred.array, c(1,2), sd)    # Get posterior sd

mapPalette <- colorRampPalette(c("grey", "yellow", "orange", "red"))
image(x=elev.pred, y= forest.pred, z=pm.pred.array, col =mapPalette(100),
xlab = "Elevation", ylab = "Forest cover")
contour(x=elev.pred, y=forest.pred, z= pm.pred.array, add =TRUE, lwd = 1)

image(x=elev.pred, y= forest.pred, z=psd.pred.array, col =mapPalette(100),
xlab = "Elevation", ylab = "Forest cover")
contour(x=elev.pred, y=forest.pred, z= psd.pred.array, add =TRUE, lwd = 1)
```

5.15 GENERAL STRATEGY OF MODEL BUILDING WITH BUGS

We have introduced statistical modeling with BUGS for some of the simplest possible models: GLMs with just a few covariates and with zero to two random effects. When you start modeling your real data, there are a *lot* of things that can and frequently will go wrong. Of course, many problems can be avoided with a little experience. So, can we give a few tips to avoid the many possible pitfalls in BUGS modeling?

Unfortunately, this is not possible in general. We have given a list of BUGS survival tips elsewhere (see Appendix 1 in Kéry and Schaub, 2012). No doubt, there will be additional anecdotal wisdom available *somewhere*, which in some cases may be essential to success. Here we mention what we believe are some of the most important, general tips.

1. *By far* the single most important BUGS modeling tip is: ***Always start from the simplest possible version of your problem!*** (Or at least a very simple one.) Strip down your model to its bare bones and toss out any nonessential details at the start, such as covariates, time dependence, and spatial or temporal autocorrelation. Start with a very simple caricature of the model you want to get at; in a sense, *start with a model of your model*. Once you get this simple model to run and you understand it, add details one at a time until you reach the desired model structure. This stepwise approach will help you recognize and diagnose problems when something goes wrong—e.g., when some parameter estimates change dramatically when going from one model to a slight variant. Similarly, when something goes wrong in your modeling, go back to the last (simpler) model variant that did work and then try to identify what caused the problem in a new step-up approach. Arguably, this advice is essential for any kind of even moderately advanced statistical modeling and we acknowledge Gelman et al. (2003), who stress this in their book. There is also a heuristic advantage of such a step-up modeling strategy that carries over to all of science: there can hardly ever be a point in trying to understand a big version of a problem unless you first understand the simplest version of the problem. Thus, this modeling strategy is just a variant of a powerful model for the process of learning in general.

2. *Start with a template that runs.* Only very rarely will you write code for a new analysis from scratch; it is much too easy to get bogged down in myriad errors that are easy to commit. Rather, it is usually far more efficient to take code from a version of your problem, that you developed earlier for a related problem, that a colleague has given you, or that you found in a book or on the Internet, and then adapt that code toward the model you want to fit.
3. *Play with simulated data first.* If you get the right answers and understand your model, then you will be much better prepared to fit the model to your real data and understand it.
4. *Do not forget to google cryptic error messages* (B. Schmidt, pers. comm.).
5. *Write tidy code* (inspired by B. Schmidt).

5.16 SUMMARY AND OUTLOOK

In this chapter, we have covered much ground. In an applied way, we have wrapped up much of the contents of Chapters 2–4. Most importantly, we have presented a crash course in BUGS, which we use throughout the book to fit HMs. All three BUGS incarnations (WinBUGS, OpenBUGS, JAGS) as well as the new NIMBLE software (NIMBLE Development Team, 2015; de Valpine et al., in review) use virtually the same model definition language, the BUGS language, and hence from R you can readily send a model to any of them. Indeed, it often pays to try a model in more than one BUGS engine; sometimes an analysis does not work in one but will work in another, and typically one engine is faster. We have also introduced our typical work flow for a Bayesian analysis using BUGS run from R. We have illustrated many times the richness of posterior inferences in a Bayesian analysis. We fitted all models with maximum likelihood (ML) as well, thereby emphasizing that with reasonable sample size, Bayesian estimates with vague priors agree numerically very well with the corresponding MLEs. We have covered several types of linear models, GLMs, and two simple random-effects, or mixed, models (GLMMs).

To emphasize the value of generating and analyzing simulated data sets, we have worked exclusively with a simulated data set in this chapter, generated using the code from Chapter 4, and imagined these were counts of a small Eurasian passerine bird. We have encountered various ways in which such counts can be summarized or aggregated—e.g., to obtain presence/absence or species distribution data from the underlying counts. Finally, we have also covered quite an eclectic set of other topics that we think are important in applied Bayesian modeling, including prediction, GoF and residual diagnostics, the Bayesian treatment of missing values, proportion of variance explained, specification of nonstandard likelihoods using the zeros and the ones tricks, and alternative parameterizations of a model using moment matching. One potentially important topic that we have not touched upon in this chapter is model selection (see Burnham and Anderson, 2002; Kadane and Lazaar, 2004; O'Hara and Sillanpää, 2009; Tenan et al., 2014a; Hobbs and Hooten, 2015), but we introduced it in Chapter 2 and illustrate aspects of it throughout the book. We do not claim to cover any of these topics exhaustively. If Bayesian analysis, linear models, GLMs, or simple mixed models, and BUGS software is completely new to you, then you should complement your reading of this book with that of other books on these topics. At the very least, you should read the preceding Chapters 2–4 a couple of times.

We do not claim that all of the models in this chapter were the best possible analysis for our data set. For instance, the Poisson GLMM in Section 5.13 would probably not be our first choice; rather, we might adopt a binomial N-mixture model (see Chapter 6) for inference about abundance from such data. Similarly, the binomial GLMM in Section 5.14 is not the most natural model for such data.

Instead, we would naturally adopt a site-occupancy model (see Chapter 10) for inference about species occurrence from such data.

We wanted to introduce BUGS with the simplest models that you are likely to know well and use a lot: linear models, GLMs, and simple mixed models. These form the backbone of most of applied statistics in ecology, and yet, there are many aspects that ecologists may find confusing—e.g., the parameterization of factors, link functions, and random effects. The BUGS language can greatly enhance your understanding of these fundamental statistical models, because of the way these models are described in the BUGS language. BUGS may be an algorithmic black box, but it is the opposite of a modeling black box—in the BUGS language, the model specified is utterly transparent.

Nowadays, mixed models of a certain class in particular have become the workhorse models of many ecologists, and are widely fitted using functions such as (g)lmer in R or PROC MIXED in SAS. You can think of BUGS, with its simple but comprehensive model definition language, as a superpowerful lmer function for fitting virtually *any* kind of mixed (= hierarchical) model. The power of BUGS extends way beyond the "simple" mixed models with normal random effects, which are the only ones you can fit with lmer. It includes models with many nested levels of random effects, random effects that are nested or crossed, continuous or discrete-valued, and that can come from many distributions other than normal. Thus, BUGS allows you to easily specify a very large class of hierarchical (= mixed) models.

In almost all of the rest of the book, we will combine GLMs like those in this chapter to build HMs that exactly reflect the way in which we imagine the sequence of processes (e.g., ecological and observational) that produces the observed data. Hence, if you understand linear models, GLMs, and the concept of random effects, and know how to specify these modeling figures using the BUGS language, then you are very well prepared for building a vast number of HMs in a powerful and creative manner. Moreover, if you learn how to "walk" in BUGS, then you are likely to experience a completely exhilarating modeling freedom that you would never have dreamt of as an ecologist.

In other words, almost all of the rest of the book presents variants of HMs consisting of combinations of two to a couple of submodels, each of which can be described as a simple GLM in a sense. Their specific combination of GLM represents the basic structure of the particular model. Everything else then boils down to clever specification of a linear predictor to address the specific question and the specifics of your data collection protocol. In summary, a very good practical understanding of linear modeling and of GLMs is vital for your existence as a hierarchical modeler.

EXERCISES

1. In Section 5.3, fit the linear model that you get when typing in R lm(Cmean ~ (elev + I(elev^2)) * (forest + I(forest^2))); i.e., add quadratic effects of elevation and forest cover including all pair-wise interaction effects.
2. In Section 5.6, the residuals are seen to be heterogeneous in the different groups (levels of facFor); see Figure 5.10. Why is this? What can be done about this? Plot the residuals for each group for the model in that section. Then, in BUGS fit a normal model with heterogeneous variances for each level of facFor. Alternatively, specify a linear model with an effect of the continuous covariate forest in the variance to see whether in this way the heterogeneous variance can be accommodated.

3. Fit a Poisson GLM to one of the count vectors (e.g., from the first survey; Section 5.6), and check whether the residual plots look better than those for the maximum count.
4. In Section 5.9, the residuals in the Poisson regression of the maximum count have too many large values. Try to add an "extra residual," assumed to be normally distributed, and convert the Poisson GLM into a Poisson-lognormal GLM to soak up that variability, and inspect whether you get new residuals that are more nearly symmetrical around zero. Check how the posterior standard deviations of the coefficient estimates change and whether the Bayesian p-value now indicates a fitting model.
5. In the Poisson GLM (Section 5.9), fit a quadratic and a cubic term for elevation as well. Are these additional polynomial terms "significant"?
6. In Section 5.11, fit a binomial GLM to the detection frequency of the great tit—i.e., to the number of surveys (out of three) that a great tit was detected at a site.
7. In 5.13, fit a normal–normal mixed model to the logarithm of the replicated counts.
8. In 5.13, turn the Poisson GLMM into a Poisson GLM with fixed site effects.
9. Fit a binomial GLMM in Section 5.14 to the detection frequency data (as in Exercise 6).
10. In the Bernoulli GLMM in Section 5.14, model the site-specific effect of wind speed by some site-specific covariates, such as elevation. That is, fit a model with the last line as this $\alpha_{4,i} \sim Normal(\mu_{\alpha 4} + \beta_{\alpha 4} * elev_i, \sigma^2_{\alpha 4})$. What proportion of the variance of the slope of wind speed is explained by elevation?
11. In Section 5.14, check whether increasing the number of replicates per site (for instance, to 10 or 20) improves the agreement between the Bayesian and the frequentist estimates.
12. To find out which estimates under the Bernoulli GLMM are more trustworthy with small samples (as those in Chapter 5.14), you can analyze simulated occupancy data where you have a known truth to compare with (see Exercise 4.11 in Chapter 4 and Section 10.5).

PART 2

MODELS FOR STATIC SYSTEMS

CHAPTER

MODELING ABUNDANCE WITH COUNTS OF UNMARKED INDIVIDUALS IN CLOSED POPULATIONS: BINOMIAL N-MIXTURE MODELS

6

CHAPTER OUTLINE

6.1 Introduction to the Modeling of Abundance	220
6.2 An Exercise in Hierarchical Modeling: Derivation of Binomial N-mixture Models from First Principles	222
6.3 Simulation and Analysis of the Simplest Possible N-mixture Model	225
6.4 A Slightly More Complex N-mixture Model with Covariates	229
6.5 A Very General Data Simulation Function for N-mixture Models: `simNmix`	241
6.6 Study Design, Bias, and Precision of the Binomial N-mixture Model Estimator	245
6.7 Study of Some Assumption Violations Using Function `simNmix`	248
6.8 Goodness-of-Fit (GoF)	250
6.9 Abundance Mapping of Swiss Great Tits with `unmarked`	254
6.9.1 Setup of the Analysis	257
6.9.2 Model Fitting	258
6.9.3 Model Criticism and GoF	261
6.9.4 Analysis of Results	264
6.9.5 Conclusions on the Analysis with `unmarked`	277
6.10 The Issue of Space, or: What Is Your Effective Sample Area?	279
6.11 Bayesian Modeling of Swiss Great Tits with BUGS	282
6.11.1 Bayesian Fitting of the Basic Zero-Inflated Poisson (ZIP) N-mixture Model	283
6.11.2 Adding Random Effects in BUGS	290
6.11.2.1 Accounting for Overdispersion (OD) at Multiple Scales	290
6.11.2.2 Linear Modeling of a Variance in the N-mixture Model	294
6.12 Time-for-Space Substitution	298
6.13 The Royle-Nichols Model and Other Nonstandard N-mixture Models	299
6.13.1 The Royle-Nichols or Bernoulli/Poisson N-mixture Model	300
6.13.2 The Poisson/Poisson N-mixture Model	305
6.14 Multiscale N-mixture Models	306
6.15 Summary and Outlook	307
Exercises	311

CHAPTER 6 BINOMIAL N-MIXTURE MODELS

6.1 INTRODUCTION TO THE MODELING OF ABUNDANCE

This chapter, and indeed much of this book, is about the modeling of abundance. Abundance, or local population size, refers to the number of individuals at some place and time. We typically denote abundance by N_{it}, where i indexes space (the place) and t is an index for time. Surely, abundance must be the single most important state variable in all of ecology. This is nicely illustrated by the fact that several influential ecology textbooks have "abundance" in their title (e.g., Andrewartha and Birch, 1954; Krebs, 2009). This focus on N in ecology and its applications is certainly not unnatural. If you think of a population as analogous to your bank account, then a natural first question about the latter is "How much money do I have?" and similarly, the first question in ecology is likely to be "How many are there?"

We saw in Chapter 1 that fundamentally both abundance and distribution, the other major state variable in ecology (Krebs, 2009), are simply areal summaries of an underlying point pattern. Animals and plants are small at the scale of essentially any study area, hence they can be treated approximately as points. Then, abundance is the number of such points in the pattern, while occurrence z is an indicator for a greater than zero abundance (i.e., $z = I(N > 0)$; see Chapter 10). At a fundamental level, we could therefore model the number and location of every individual within a study area. Models for abundance at this most disaggregated level are called spatial point pattern (or process) models (Illian et al., 2008; Wiegand and Moloney, 2014), and there is much recent interest in this powerful class of models in ecology (Aarts et al., 2012; Dorazio, 2012, 2014; Fithian and Hastie, 2013; Renner and Warton, 2013; Renner et al., 2015). However, these models require more expensive data (i.e., location data), and are substantially more complex both conceptually and practically (i.e., to fit), than simpler models for areal aggregates of point patterns. Furthermore, it is not straightforward to describe a measurement error process for false-negative errors, nor to model temporal dynamics (see also Section 10.1). Hence, we see the more complicated point pattern models as complementary to the abundance models covered in our book, and indeed the two can be integrated in a single model (Dorazio, 2014). In addition, the new class of spatial capture-recapture (SCR) models (Efford, 2004; Borchers and Efford, 2008; Royle and Young, 2008; Royle et al., 2014) represents point process models; see Royle et al. (2011) for an example of such a model in a meta-population context. We present such models in Chapter 24 in volume 2.

When we count animals or plants, we *measure abundance*: a count C is a measurement of abundance. When measuring abundance, there are exactly two possible types of *measurement error*: (1) we can overlook or otherwise fail to detect an individual, or (2) we can count an individual multiple times or erroneously add some other species in the count (species misidentification). The first represents a false-negative (or detection) error and leads to a negative bias in the count relative to abundance, while the last two represent false-positive errors and lead to a positive bias in the count relative to abundance. These are the most fundamental types of errors with any measurement of abundance. There are other sources of errors; e.g., you may want to study the size of a population of territory-holding adults of some species, but may be unable to distinguish juveniles from adults or territorial from transient individuals. However, if you really cannot distinguish them, then in a sense this is not an error, but you are simply asking for too much. You need to redefine your N to encompass juveniles, adults, locals, and transients alike. Thus, this is a problem of how you define N. Perhaps you can come up with some method or data collection protocol that enables you to separate out those demographic classes. Regardless, we do not think of this kind of error as comparable to the two fundamental types of abundance measurement errors, which always affect your measurements once you have defined what you mean by N.

Over literally centuries, a large number of models, methods, and associated sampling designs have been developed that deal with one of the two kinds of abundance measurement errors: the one represented by imperfect detection—i.e., false-negative errors. Examples include the treatises by Otis et al. (1978), Seber (1982), Williams et al. (2002), Borchers et al. (2002), Royle and Dorazio (2008), King et al. (2009), McCrea and Morgan (2014), and Royle et al. (2014). Conceptually, almost all methods of estimating abundance in a closed (i.e., static) population that account for false-negative errors consist of obtaining an estimate of the fraction of individuals that are detected (p). Intuitively, it is clear that with a count C and an estimate of that detection probability, $\hat{p}$, we can estimate abundance N as the ratio of C over $\hat{p}$ (i.e., by the canonical abundance estimator, Williams et al., 2002). Most approaches to estimating p in the context of abundance require capture-recapture data—i.e., repeated observations of naturally or artificially marked individuals. Natural marks may be distinctive feather, fur or general color patterns, or genetic identity. Artificial marks include rings (called "bands" west of the Atlantic), PIT transponders, or many other marking methods that allow one to unambiguously determine the identity of an individual that is physically captured or that can be observed otherwise in such a way as to being able to record the mark. Typically, capture-recapture data are "expensive" data, because they may require physical capture of the animals and sometimes also the application of complex and expensive lab analysis methods such as microsatellite analysis. In addition, the handling of the animals may be risky in the case of rare or endangered species.

In the vast majority of studies of abundance, we do not have a single site, but multiple sites, and the interest is then in modeling variability in abundance among these sites and in explaining that variability by measured covariates or perhaps by hidden structure via random effects. Of course, the challenges of measuring N are compounded in the study of a "meta-population"—that is, when we have a collection of multiple sites that we want to compare in terms of their N. (In this book, we use the term *meta-population* rather loosely to refer to all individuals that live at a collection of sites, and to denote this we prefer to write the hyphen. Sometimes we use the term *site-structured population* interchangeably.)

This chapter deals with an appealing modeling framework and data collection protocol known as the N-mixture model or binomial-Poisson mixture model (Royle, 2004b); see Dénes et al. (2015) for a recent overview. In this model class, all that is required to estimate detection probability and abundance are counts of unmarked individuals that are replicated in two dimensions: there must be multiple sites and replicate observations (i.e., counts) for at least some of these sites. No individual recognition is required, but we must avoid counting the same individual twice within a single observation period. Such data are usually "cheap" to obtain compared with typical capture-recapture (though not necessarily when compared with distance sampling data; see Chapters 8–9). Therefore, this modeling framework and its associated data collection protocol may enable you to conduct inference about abundance at larger spatial or temporal scales than what you might be able to do with a more typical capture-recapture or distance sampling protocols.

As already shown in Chapters 1, 2, and 4, the N-mixture model is an HM that combines one submodel for the latent state, abundance N, with another submodel for the actual observations C that is conditional on N. The observation component of the model is usually the same old binomial distribution that forms the backbone of practically all of capture-recapture modeling, while the model for the latent abundance state is typically a Poisson, which is the workhorse model in statistics for abundance.

Often, we follow a meta-population over time, thereby adding a temporal dimension (t) to abundance; i.e., we have an interest in explaining N_{it}—that is, explaining spatiotemporal patterns in N. For the most part and with two small exceptions in Sections 6.12, 6.14 and 9.5–9.7, we defer the modeling of the temporal dimension of abundance to Chapters 12–14 in volume 2 and focus on static models for a meta-population observed at a single point in time. Note, however, that the temporal dimension in abundance can be accommodated by fitting a static model that is stratified by time (or, equivalently, simply fitting the model separately for each time point; Dodd and Dorazio, 2004). For instance, in a single model we could analyze data from a breeding bird survey conducted over multiple years (= breeding periods) and simply fit separate parameters for every year. In addition, you may then constrain the year-specific parameters to estimate a linear trend (Kéry et al., 2009; Zellweger-Fischer et al., 2010; Fujisaki et al., 2011); also see Chapter 12.

6.2 AN EXERCISE IN HIERARCHICAL MODELING: DERIVATION OF BINOMIAL *N*-MIXTURE MODELS FROM FIRST PRINCIPLES

One of the neatest things about HMs is that they *invite* you to build your own statistical models, exactly according to how *you* perceive the stochastic process *you* are interested in, for *your* data and *your* questions. You can build statistical models from first principles, by thinking about the processes that underlie the observed data set. Let us look at how replicated count data (C) typically arise as a combination of two processes, one ecological and one observational. In this section and the next, we show a variant of what we have already done in more detail in Chapter 4. But here we do this specifically to introduce the binomial *N*-mixture model in its simplest incarnation, to provide the gentlest possible introduction to this model. To motivate the *N*-mixture model, ask yourself the following questions:

1. Suppose you knew the value of abundance (N) for some species at each of 100 sites—e.g., 2, 5, 0, 2, 0,...—how would you describe this spatial variation? Or, in other words: what is a customary statistical distribution to describe the mean and the spatial variability of N?
2. Assume that we wanted to acknowledge the presence of measurement error of the nondetection kind—i.e., permit the observed counts C to usually be less than abundance N due to imperfect detection. For instance, at some hypothetical site 2 with $N = 5$, we might obtain the following *repeated measurements* of abundance N: 4, 2, 4, 3, ... What would be a useful starting point to model the temporal variability of these *measurements* of N?

First, most people, when confronted with the kind of data described in (1), would start with a Poisson distribution to model the spatial variation in abundance N. The Poisson distribution is the basic building block for modeling abundance (Royle and Dorazio, 2008). By adding covariate effects, zero inflation, or other latent structure (random effects), a Poisson model confers considerable flexibility to capture a wide variety of patterns in N (e.g., see Cressie et al., 2009; Zuur et al., 2012, and also Chapter 12 in volume 2). Second, knowing that there are N individuals at a site exposed to detection and that the detection probability is potentially less than 1 would lead most people to choose a binomial distribution to describe the temporal variation in the counts at a specific site. The binomial distribution is the canonical description of an observation process that contains false-negative detection error only. Again, GLM-type modeling (possibly including latent effects) of the binomial success parameter (detection probability) yields great flexibility in this model.

Excitingly, we have just reinvented the basic binomial *N*-mixture model from first principles! First, we have chosen the Poisson distribution as a description of the ecological (or state) process that

6.2 AN EXERCISE IN HIERARCHICAL MODELING

distributes individuals in space and thus creates the latent abundance state N. Second, we have chosen the binomial distribution as a description of the observation process, where we typically have a special kind of *measurement error*, one that does not cancel out but instead leads to observations (counts C) that are biased with respect to the measured quantity (abundance N, unless we do a large number of repeated measurements; see Exercise 2). Here is the simplest binomial N-mixture model written in algebra:

1. State process: $N_i \sim Poisson(\lambda)$
2. Observation process: $C_{ij}|N_i \sim Binomial(N_i, p)$

Here, N_i is the latent abundance state at site i ($i = 1, \ldots, M$) and λ is the expected abundance—i.e., the mean abundance over all sites (sampled or not). C_{ij} is the count at site i during survey j ($j = 1, \ldots, J$), and p is the per-individual detection probability.

The N-mixture model is a natural, hierarchical extension of the Poisson GLM. We use the Poisson GLM as the base model for N, but we regard N as a latent variable—i.e., as unobserved or only partially observed. We augment the Poisson GLM with a conditional model that describes how the observations C_{ij} are related to the latent variable N_i. The name, binomial mixture model, stems from the fact that the likelihood is a mixture of binomials each with a different sample size N. The resulting model is also called a binomial/Poisson mixture model. It is an HM consisting of two linked GLMs; a Poisson regression for the spatial variation in abundance and a binomial regression (a.k.a. logistic regression) for the variation of the observed counts at specific sites. Recognizing these two GLMs, we can immediately start doing all the things that we customarily do with GLMs, especially modeling structure in the two parameters (which then need to be indexed, λ_i, p_{ij}) as linear or other functions of covariates via a log and a logit link function, respectively. Other possibilities include the introduction of random effects to account for latent structure or correlations.

To ensure estimable parameters in the N-mixture model, we need data that are replicated both in space (i.e., $M > 1$) and in time (i.e., $J > 1$) (but see Section 6.12 on time-for-space substitution).

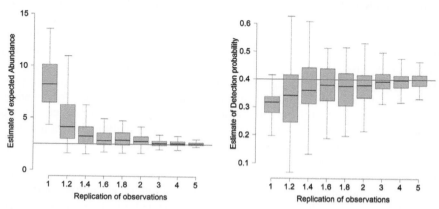

FIGURE 6.1

Effects of temporal replication in the N-mixture model on the estimators of expected abundance (λ) and detection probability (p) in a simulated example with 150 sites, $\lambda = 2.5$ and $p = 0.4$ (red lines), and no covariates. Replication is expressed as the average number of replicate surveys per site; e.g., 1.2 means that 20% of sites have two replicates and 80% only one. Each boxplot summarizes 100 simulated data sets.

Unreplicated data result in large biases (Figure 6.1), but even a small degree of replication improves the quality of estimators considerably (see also Knape and Korner-Nievergelt, 2015). You can easily repeat such a simulation for *your* sample sizes and the expected abundance and detection of *your* favorite species (see Section 6.6 for a template). It is hard to give general advice on how much replication is required to obtain reasonable estimates, but typically the number of sites (M) should be greater than 20, and you can very easily ascertain how good estimates get with simulation (see Section 6.6). Importantly, the number of temporal replicates (J) need only be two or more for *some*, but not necessarily for all sites. Typical ecological data sets are unbalanced in many ways, and most data sets to which we have applied N-mixture models did not have the same number of replicate surveys at all sites and times. The information about detection probability in the N-mixture model comes from the variability of repeated counts at a site, with more variability suggesting smaller p. In fact, the correlation coefficient among repeated measurements is a moment estimator of p when parameters are constant (Royle, 2004a). Thus, when our data set contains sites with replicated counts and without, then information about p will come from the sites with replication, and it will be propagated to the sites without replication. This means that the sites without replicated counts should constitute a random sample among the total number of sites. If this is not the case, the estimate of p may be based on a biased sample of sites with respect to all sites about which we would like to make inferences. This may lead to biased parameter estimates.

Some key assumptions of the N-mixture model, typical violations, and ways to mitigate them are:

1. *Closure assumption*: We assume that all within-site variation in counts C is attributable to detection probability p. When all counts at a site are constant, detection probability will be estimated at 1 (unless all counts are 0). So if N varies during a period when closure is assumed and the closure assumption is violated, the variability attributed to the binomial part of the model will be overstated. This will lead to underestimated p and overestimated N (see Section 6.7). Some lack of closure may not be disastrous, and one may then simply have to change the interpretation of the abundance parameter (λ) to have a superpopulation meaning: rather than the number of individuals that *permanently* reside within a site, λ will describe the total number of individuals ever associated with a site during a study period. This is analogous to site-occupancy models where random temporary emigration leads to a redefinition of the occupancy parameter from the probability of permanent occupation to the probability of use (see Section 10.2). Also, if you are sampling at two nested temporal scales, with closure satisfied among the secondary samples (i.e., when you have a robust design), then you can use multiscale N-mixture models to model the superpopulation size and temporary emigration jointly (Section 6.14). Some information about temporary emigration induced by within-territory movement may also be obtained from distance information (Johnson et al., 2014).

2. *No false-positive errors*: Implicit in the adoption of the binomial distribution as a description of the measurement error for abundance is the assumption that there are no false-positives. That is, we must not count another species as our target species, and we must not count the same individual multiple times during a single occasion. This assumption is required for the binomial to be an adequate description of the observation process. If the biology of a study species and/or the sampling protocol suggests substantial numbers of false-positive errors, the traditional N-mixture model is *not* an appropriate framework for inference about abundance. One example of this would be species with large flocks, where it would be hard to avoid double counts when measuring the size of flocks (plus independence of detection, mentioned next, would be violated). One possible way forward, however, might be a Poisson/Poisson mixture model (see Section 6.13.2).

3. *Independence of detection*: Similarly, implicit in the binomial distribution as a description of the observation process is that individuals are detected independently from each other. If they travel in pairs, groups, or flocks, detection of one individual will make it more likely that other group members are also detected, and the standard *N*-mixture model will no longer be an adequate representation of the observation process underlying the counts. A possible way forward might be the choice of a beta-binomial distribution for the observation process; see Martin et al. (2011) and Dorazio et al. (2013). Another way that this assumption may be violated is by density-dependent detection; for instance, if individual birds sing more frequently at higher density (Warren et al., 2013). See also our comments in Section 6.15.
4. *Homogeneity of detection among N_i individuals*: The *N*-mixture model uses as input not individual detection data, but simple counts of unique individuals. That is, the data distinguish sites *i* and sampling occasions *j*, but we do *not* keep track of the individuals in the counts C_{ij}. Hence, with two replicate counts at a site, such as 5 and 3, we have no way knowing whether the total number of individuals detected was 5, 6, 7, or 8. The individual ID is lost in the replicated count data, hence there is an implicit assumption that all N_i individuals present at *i* have an identical detection probability p_{ij} at survey *j*. Clearly, this can never be exactly true. For instance, this assumption probably offends all distance samplers (Chapters 8 and 9), whose fundamental assumption is that individuals closer to an observer are more easily detectable than individuals farther away (Efford and Dawson, 2009). The *N*-mixture model averages over such distance-related or indeed any other detection heterogeneity among the N_i individuals during occasion *j*. The effect of this can likely be deduced from the "second law of capture-recapture," which says that unmodeled detection heterogeneity will always lead to a negative bias in *N* (the first "law" is that things get harder as *p* gets smaller; see Section 6.6). If distance-related detection heterogeneity and its effects on *N* are a big concern, we may have to use distance sampling (Chapters 8 and 9) or spatial capture-recapture (SCR; Royle et al., 2014) for abundance estimation. We note, though, that in our small simulation study in Section 6.7 individual heterogeneity in detection probability did not seem to bias the abundance estimator from the *N*-mixture model, so perhaps more detailed investigations in this respect would be interesting.
5. *Parametric modeling assumptions*: The model makes specific distributional assumptions and if they deviate substantially from the true processes that generated the data set, biased estimators may result. These assumptions can and should be tested using conventional goodness-of-fit (GoF) techniques such as parametric bootstrap or posterior predictive checks (see Sections 2.4.3, 2.8.2, and 6.8). We note, though, that these techniques are much less well understood for HMs than for flat or nonhierarchical models.

6.3 SIMULATION AND ANALYSIS OF THE SIMPLEST POSSIBLE *N*-MIXTURE MODEL

We first illustrate the model in the simplest possible setting, with constant expected abundance and detection ($\lambda = 2.5$, $p = 0.4$), for a sample size of $M = 150$ sites and $J = 2$ repeated measurements of abundance. You will recognize that we have already shown exactly the same data-generation process in Chapter 4 for a more complex binomial *N*-mixture model, but without explicitly focusing on this model. Here, we simulate data in R and fit the model, using ML with `unmarked` and Bayesian inference in BUGS. As so often, we use data simulation to help explaining a model in the first part of this section.

```
# Choose sample sizes and prepare observed data array C
set.seed(24)            # So we all get same data set
M <- 150                # Number of sites
J <- 2                  # Number of abu. measurements per site (rep. counts)
C <- matrix(NA, nrow = M, ncol = J)  # to contain the obs. data

# Parameter values
lambda <- 2.5           # Expected abundance
p <- 0.4                # Probability of detection (per individual)

# Generate local abundance data (the truth)
N <- rpois(n = M, lambda = lambda)

# Conduct repeated measurements (generate replicated counts)
for(j in 1:J){
    C[,j] <- rbinom(n = M, size = N, prob = p)
}

# Look at data
# The truth ....
table(N)                # True abundance distribution
N
 0  1  2  3  4  5  6  7  8  9
11 37 39 37 10 10  3  1  1  1
sum(N)                  # True total population size at M sites
[1] 358
sum(N>0)                # True number of occupied sites
[1] 139
mean(N)                 # True mean abundance (estimate of lambda)
[1] 2.386667

# ... and the observations
table(apply(C, 1, max)) # Observed abundance distribution (max count)
 0  1  2  3  4
30 57 45 15  3
sum(apply(C, 1, max))   # Observed total population size at M sites
[1] 204
sum(apply(C, 1, max)>0) # Observed number of occupied sites
[1] 120
mean(apply(C, 1, max))  # Observed mean "relative abundance"
[1] 1.36

head(cbind(N=N, count1=C[,1], count2=C[,2])) # First 6 sites
     N count1 count2
[1,] 2      0      1
[2,] 1      1      0
[3,] 3      0      0
[4,] 2      0      1
[5,] 3      1      2
[6,] 5      3      3
```

Thus, in our simulation, the species occurs at 139 sites with a total of 358 individuals, but is detected at only 120 sites, and the maximum counts at each site add up to 204. Thus, the observed

6.3 ANALYSIS OF THE SIMPLEST POSSIBLE N-MIXTURE MODEL

"relative mean abundance" is 1.36, rather than 2.39, the true mean abundance in the M sampled sites, or 2.5 in the statistical population of sites from which we sampled 150. The overall measurement error at an occupancy level is therefore $(139 - 120)/139 \approx -14\%$ and at the abundance level, $(358 - 204)/358 \approx -43\%$. Among the first six sites, all are occupied, and under the assumption of no false-positives, we can only obtain counts $\leq N$. We undercount (i.e., $\max(C_i) < N_i$) at all sites except for site 2; indeed, at site 3 we fail to detect any of the three individuals present. Since our model has constant parameters, the correlation coefficient among repeated measurements should be approximately equal to p (Royle, 2004a); let us check this.

```
cor(C)[1,2]
[1] 0.4023582                          # Note excellent agreement with value of p
```

We use unmarked function pcount to fit the N-mixture model for replicated counts.

```
library(unmarked)                      # Load package
umf <- unmarkedFramePCount(y = C)      # Create um data frame
summary(umf)                           # Summarize
(fm1 <- pcount(~1 ~1, data = umf))     # Fit model: get estimates on link scale
backTransform(fm1, "state")            # Get estimates on natural scale
backTransform(fm1, "det")

Call:
pcount(formula = ~1 ~ 1, data = umf)

Abundance:
 Estimate    SE     z    P(>|z|)
    0.981  0.191  5.14  2.74e-07

Detection:
 Estimate    SE      z    P(>|z|)
   -0.565  0.298  -1.89   0.0582

AIC: 754.4465

Backtransformed linear combination(s) of Abundance estimate(s)
 Estimate    SE    LinComb  (Intercept)
     2.67  0.509    0.981        1

Backtransformed linear combination(s) of Detection estimate(s)
 Estimate    SE     LinComb  (Intercept)
    0.363  0.0689   -0.565        1
```

Next, a Bayesian analysis of the model with JAGS.

```
# Bundle and summarize data set
win.data <- list(C = C, M = nrow(C), J = ncol(C))
str(win.data)                          # Look at data

# Specify model in BUGS language
sink("model1.txt")
cat("
model {
```

```
# Priors
   lambda ~ dgamma(0.001, 0.001)
   p ~ dunif(0, 1)
# Likelihood
   for (i in 1:M) {
      N[i] ~ dpois(lambda)        # State model
      for (j in 1:J) {
         C[i,j] ~ dbin(p, N[i])   # Observation model
      }
   }
}
",fill = TRUE)
sink()

# Initial values
Nst <- apply(C, 1, max)       # Avoid data/model/inits conflict
inits <- function(){list(N = Nst)}

# Parameters monitored
params <- c("lambda", "p")

# MCMC settings
ni <- 25000 ; nt <- 20 ; nb <- 5000 ; nc <- 3

# Call JAGS (ART 1 min) and summarize posteriors
library(jagsUI)
fm2 <- jags(win.data, inits, params, "model1.txt", n.chains = nc, n.thin = nt, n.iter = ni,
   n.burnin = nb)
print(fm2, dig = 3)
          mean    sd    2.5%   50%    97.5%  overlap0  f   Rhat   n.eff
lambda    2.829  0.659  1.969  2.692  4.600  FALSE     1   1.010   1996
p         0.356  0.068  0.210  0.358  0.478  FALSE     1   1.002   3000
```

We get similar estimates from likelihood and Bayesian inference. To finish this section, note the striking similarity of an HM when written in algebra, in R when simulating the data, and in the BUGS language when fitting the model (Table 6.1). Hence, when you know how to write a model in algebra, you are almost there at fitting it in BUGS, and algebra, data simulation in R or possibly other computer languages, and the BUGS language are similarly precise and useful ways of describing and understanding an HM.

Table 6.1 Model descriptions in terms of algebra, data simulation in R, and in the BUGS language. See Table 10.1 for the analogous scheme in the case of the occupancy model.

	Algebraic Description	R Data Simulation	BUGS Model Statement
State model	$N_i \sim Poisson(\lambda)$	N <- rpois(M, lambda)	N[i] ~ dpois(lambda)
Obs. model	$y_{ij}\|N_i \sim Binomial(N_i, p)$	y[,j] <- rbinom(M,N,p)	y[i,j] ~ dbin(p, N[i])

6.4 A SLIGHTLY MORE COMPLEX N-MIXTURE MODEL WITH COVARIATES

The model in the previous section is rarely directly useful and we went through it mainly for conceptual reasons: to explain the basics of this HM. In practice, we will virtually always have covariates whose effects on abundance and/or detection we want to model. We illustrate this by simulating another data set, analyzing it, making predictions, and showing bootstrap error estimation. We illustrate one model with BUGS, but work mostly with unmarked. We simulate data under the following model:

$$N_i \sim Poisson(\lambda_i), \text{with } \log(\lambda_i) = \beta_0 + \beta_1 * \text{vegHt}_i$$

$$C_{ij}|N_i \sim Binomial(N_i, p_{ij}), \text{with } logit(p_{ij}) = \alpha_0 + \alpha_1 * \text{wind}_{ij}$$

Hence, expected abundance (λ) is affected by a site covariate (vegetation height), and detection probability is affected by a sampling or observational covariate (wind speed). You will note that much of this section is very similar to Section 10.3 for a similar HM for modeling presence/absence instead of abundance, the site-occupancy model. We do this on purpose to emphasize the strong structural similarity of the two.

```
# Choose sample sizes and prepare observed data array y
set.seed(1)              # So we all get same data set
M <- 100                 # Number of sites
J <- 3                   # Number of repeated abundance measurements
C <- matrix(NA, nrow = M, ncol = J)   # to contain the observed data

# Create a covariate called vegHt
vegHt <- sort(runif(M, -1, 1)) # sort for graphical convenience

# Choose parameter values for abundance model and compute lambda
beta0 <- 0               # Log-scale intercept
beta1 <- 2               # Log-scale slope for vegHt
lambda <- exp(beta0 + beta1 * vegHt)     # Expected abundance
plot(vegHt, lambda, type = "l", lwd = 3) # Expected abundance

# Draw local abundance and look at data so far
N <- rpois(M, lambda)
points(vegHt, N)         # Add realized abundance to plot
table(N)
N
 0  1  2  3  4  5  6  8  9
35 24 12  7  9  5  4  3  1

# Plot the true system state (Figure 6.2, left)
par(mfrow = c(1, 3), mar = c(5,5,2,2), cex.axis = 1.5, cex.lab = 1.5)
plot(vegHt, N, xlab="Vegetation height", ylab="True abundance (N)", frame = F, cex = 1.5)
lines(seq(-1,1,,100), exp(beta0 + beta1* seq(-1,1,,100)), lwd=3, col = "red")
```

Figure 6.2 (left) shows how the relationship between expected abundance and vegHt translates into a realized abundance pattern. In real life, both expected abundance (λ) and realized abundance (N) will usually be latent; i.e., neither will be directly observable, the former because it is an abstract construct

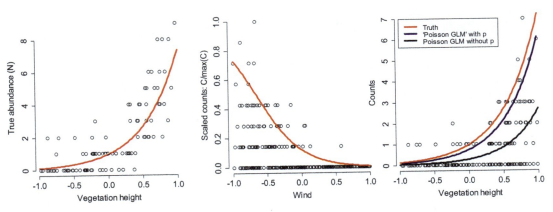

FIGURE 6.2

Left: The relationship between abundance and vegetation height (red line: expected abundance λ; circles: realized abundance N). Middle: The relationship between detection probability and wind speed (red line); circles are the scaled observed counts (by division with the maximum count). Right: Comparison between the true abundance–covariate relationship (red), and its estimate under the N-mixture model (blue) and a simple Poisson regression (black); circles represent the observed counts of each site.

(a parameter), and the latter because of measurement error. Although observable in principle, N is thus typically a latent variable, too. We simulate this next and assume that detection probability p depends on wind speed in a logit-linear relationship described by intercept -2 and slope -3, and that we make $J = 3$ repeated abundance measurements at each site (Figure 6.2, middle). We scale the counts in Figure 6.2 (middle) by dividing each by the maximum count to make the abundance measurements more directly comparable with detection probability.

```
# Create a covariate called wind
wind <- array(runif(M * J, -1, 1), dim = c(M, J))

# Choose parameter values for measurement error model and compute detectability
alpha0 <- -2                            # Logit-scale intercept
alpha1 <- -3                            # Logit-scale slope for wind
p <- plogis(alpha0 + alpha1 * wind)     # Detection probability
# plot(p ~ wind, ylim = c(0,1))         # Look at relationship

# Take J = 3 abundance measurements at each site
for(j in 1:J) {
   C[,j] <- rbinom(M, N, p[,j])
}

# Plot observed data and effect of wind on det. probability (Figure 6.2, middle)
plot(wind, C/max(C), xlab="Wind", ylab="Scaled counts: C/max(C)", frame = F, cex = 1.5)
lines(seq(-1,1,,100), plogis(alpha0 + alpha1*seq(-1,1,,100)), lwd=3, col="red")
```

It is instructive to compare the expected and realized abundance and the replicated abundance measurements; make sure you really understand what they all mean.

6.4 A SLIGHTLY MORE COMPLEX N-MIXTURE MODEL WITH COVARIATES

```
# Expected (lambda) and realized abundance (N) and measurements (C)
cbind(lambda=round(lambda,2), N=N, C1=C[,1], C2=C[,2], C3=C[,3])
     lambda N C1 C2 C3
[1,]   0.14 0  0  0  0
[2,]   0.15 0  0  0  0
[3,]   0.17 0  0  0  0
[4,]   0.17 2  1  0  0
[5,]   0.18 0  0  0  0
 [ output truncated]
[91,]  4.49 6  2  2  4
[92,]  4.50 4  3  3  0
[93,]  4.80 2  0  0  0
[94,]  4.92 8  3  0  7
[95,]  5.12 4  1  3  2
 [ output truncated]
```

Next, we use the *N*-mixture model to analyze these data using unmarked and BUGS. We start with unmarked. We also illustrate the fitting of two factors that are unrelated to the data (because the response was not generated with their effects "built in"): time will index the first through the third survey, while hab will contrast three imaginary habitat types.

```
# Create factors
time <- matrix(rep(as.character(1:J), M), ncol = J, byrow = TRUE)
hab <- c(rep("A", 33), rep("B", 33), rep("C", 34))    # assumes M = 100
```

We first package the data into an unmarked frame, noting the difference between site covariates (indexed by site only) and sampling or observational covariates (indexed by site and survey).

```
# Load unmarked, format data in unmarked data frame and summarize
library(unmarked)
umf <- unmarkedFramePCount(
    y = C,                                                  # Counts matrix
    siteCovs = data.frame(vegHt = vegHt, hab = hab),        # Site covariates
    obsCovs = list(time = time, wind = wind))               # Observation covs
summary(umf)
unmarkedFrame Object

100 sites
Maximum number of observations per site: 3
Mean number of observations per site: 3
Sites with at least one detection: 47

Tabulation of y observations:
  0   1   2   3   4   5   6   7 <NA>
214  43  19  17   2   3   1   1    0
```

```
Site-level covariates:
    vegHt            hab
 Min.   :-0.97322   A:33
 1st Qu.:-0.35384   B:33
 Median :-0.02438   C:34
 Mean   : 0.03569
 3rd Qu.: 0.53439
 Max.   : 0.98381

Observation-level covariates:
     time          wind
 1:100    Min.   :-0.99633
 2:100    1st Qu.:-0.54111
 3:100    Median :-0.10469
          Mean   :-0.03803
          3rd Qu.: 0.44824
          Max.   : 0.99215

# Fit model and extract estimates
# linear model for p follows first tilde, then comes linear model for lambda
summary(fm.Nmix1 <- pcount(~wind ~vegHt, data=umf, control=list(trace=T, REPORT=1)))

Abundance (log-scale):
            Estimate    SE       z     P(>|z|)
(Intercept)   -0.294   0.191   -1.53   1.25e-01
vegHt          2.141   0.271    7.91   2.53e-15

Detection (logit-scale):
            Estimate    SE       z     P(>|z|)
(Intercept)   -1.73    0.254   -6.83   8.27e-12
wind          -4.22    0.529   -7.97   1.61e-15

AIC: 333.3263
Number of sites: 100
optim convergence code: 0
optim iterations: 37
Bootstrap iterations: 0

Warning message:
In pcount(~wind ~ vegHt, data = umf, control = list(trace = TRUE,  :
  K was not specified and was set to 107.
```

Here K is the upper summation limit for the summation over the random effects in the integrated likelihood (Royle, 2004b, p. 110). In unmarked, the default choice of K is the maximum observed count plus 100. This should normally be enough, but some sensitivity analysis may be useful (e.g., by setting K as 200 plus the maximum count). For small data sets, there may be identifiability problems that can be diagnosed by estimates of λ that keep increasing when K is made larger (Couturier et al., 2013; Dennis et al., 2015a).

In unmarked, we may choose three alternative abundance models: the Poisson (default), negative binomial (NB) (Royle, 2004b), and zero-inflated Poisson (ZIP) (see Section 3.3.4; Wenger and

6.4 A SLIGHTLY MORE COMPLEX N-MIXTURE MODEL WITH COVARIATES

Freeman, 2008). We can compare them by AIC or a likelihood ratio test (LRT). Not surprisingly, given our way of data simulation, the Poisson comes out best.

```
fm.Nmix2 <- pcount(~wind ~vegHt, data=umf, mixture="NB",
control=list(trace=TRUE, REPORT=5))
fm.Nmix3 <- pcount(~wind ~vegHt, data=umf, mixture="ZIP",
control=list(trace=TRUE, REPORT=5))
cbind(AIC.P=fm.Nmix1@AIC, AIC.NB=fm.Nmix2@AIC, AIC.ZIP=fm.Nmix3@AIC)
       AIC.P    AIC.NB   AIC.ZIP
[1,] 333.3263 335.3284 335.327
```

The parameters of the linear model are defined on the log scale for abundance and on the logit scale for detection. Neither is a scale most of us like to think in, so to make better sense of what the parameter estimates mean, we can make predictions of λ and p for specified values of the covariates. The function predict uses the delta rule to compute SEs and 95% CIs (see Section 2.4.3).

```
# Predictions of lambda for specified values of vegHt, say 0.2 and 2.1
newdat <- data.frame(vegHt=c(0.2, 1.1))
predict(fm.Nmix1, type="state", newdata=newdat, append = T)
   Predicted       SE       lower      upper   vegHt
1    1.14387  0.1747408  0.8479004   1.543152   0.2
2    7.85632  1.4894445  5.4180585  11.391861   1.1

# ... or of p for values of wind of -1 to 1
newdat <- data.frame(wind=seq(-1, 1, , 5))
predict(fm.Nmix1, type="det", newdata=newdat, append = T)
   Predicted         SE         lower        upper    wind
1 0.922954766  0.033547623  0.8261414764  0.96794900  -1.0
2 0.592735702  0.065419999  0.4611115369  0.71227136  -0.5
3 0.150253394  0.032371147  0.0971288099  0.22518741   0.0
4 0.021030714  0.009093726  0.0089579531  0.04857659   0.5
5 0.002603173  0.001772162  0.0006844676  0.00984741   1.0
```

We can compute the expected values of the expected abundance λ and of detection probability p for the actual data set (i.e., for the specific, observed values of the vegHt and wind covariates). For the detection model, predictions are produced for each of (here) 300 observations. Predictions for the three surveys at site 1 are in rows 1–3 (i.e., p.hat[1:3,1] below, and *not* in rows 1, 101, and 201 as one might perhaps think.

```
# Predict lambda and detection for actual data set
(lambda.hat <- predict(fm.Nmix1, type="state"))   # lambda at every site
(p.hat <- predict(fm.Nmix1, type="det"))          # p during every survey
```

To visualize the covariate relationships in general, it is best to predict for a new data frame with a suitable range of covariate values.

```
# Predict lambda and detection as function of covs
newdat <- data.frame(vegHt=seq(-1, 1, 0.01))
pred.lam <- predict(fm.Nmix1, type="state", newdata=newdat)
newdat <- data.frame(wind=seq(-1, 1, 0.1))
pred.det <- predict(fm.Nmix1, type="det", newdata=newdat)
```

We may summarize the analysis by plotting the observed data, the true data-generating values, and the estimated relationship between expected abundance (i.e., lambda) and vegHt under the N-mixture model (a hierarchical Poisson regression that does account for imperfect detection p) and under a nonhierarchical Poisson regression that does not account for p and therefore only models an index of abundance (Johnson, 2008) (Figure 6.2, right). We see that ignoring imperfect detection leads to (1) underestimation of abundance and (2) a bias toward zero of the regression coefficient of vegHt.

```
# Fit detection-naive GLM to counts and plot comparison (Figure 6.2, right)
summary(fm.glm <- glm(c(C) ~ rep(vegHt, 3), family=poisson))  # p-naive model
matplot(vegHt, C, xlab="Vegetation height", ylab="Counts", frame = F, cex = 1.5, pch = 1,
  col = "black")
lines(seq(-1,1,,100), exp(beta0 + beta1* seq(-1,1,,100)), lwd=3, col = "red")
curve(exp(coef(fm.glm)[1]+coef(fm.glm)[2]*x), -1, 1, type ="l", lwd=3, add=TRUE)
lines(vegHt, predict(fm.Nmix1, type="state")[,1], col = "blue", lwd = 3)
legend(-1, 7, c("Truth", "'Poisson GLM' with p", "Poisson GLM without p"), col=c("red",
  "blue", "black"), lty = 1, lwd=3, cex = 1.2)
```

Predictions of λ represent an estimate of the expected abundance in a comparable site that has the chosen values of covariates. But we can also use the model to provide us with an estimate of abundance at the studied sites—i.e., for the realized abundance N_i. In the frequentist analysis, site-specific abundance, the random effect N_i, is lost by the summation over all possible states up to K (see above) of N_i in forming the integrated likelihood. We can obtain estimates of these random effects by application of the function ranef. This applies Bayes' rule for a conditional estimate of abundance N_i, given the model parameters and the observed data $\mathbf{y}_i$. That is, the function obtains the *best unbiased prediction* (BUP) of the random effects N_i as $\widehat{N}_i = E(N_i|\mathbf{y}_i, \lambda)$, with $\widehat{\lambda}$ used in place of λ (Royle, 2004a), based on the posterior distribution of N_i:

$$p(N_i|\mathbf{y}_i, \lambda_i, p_i) = \frac{f(\mathbf{y}_i|N_i, p_i)g(N_i|\lambda_i)}{q(\mathbf{y}_i|\lambda_i, p_i)}$$

Here, $f(\mathbf{y}_i|N_i, p_i)$ is the likelihood of the data y_{ij}, $g(N_i|\lambda_i)$ the prior for N_i, and $q(\mathbf{y}_i|\lambda_i, p_i)$ denotes the probability of the observed data, given the values of λ_i and p_i. The function ranef yields the results of this for each site.

```
ranef(fm.Nmix1)
         Mean    Mode 2.5% 97.5%
[1,]  0.037770455   0    0    1
[2,]  0.001489541   0    0    0
[3,]  0.081963672   0    0    1
[4,]  1.040688154   1    1    2
[5,]  0.021773600   0    0    0
[ ... ]
[91,] 4.075543688   4    4    5
[92,] 3.539540712   3    3    5
[93,] 1.969791630   1    0    5
[94,] 7.765449990   7    7   10
[95,] 4.280652414   4    3    7
[ ... ]
```

6.4 A SLIGHTLY MORE COMPLEX N-MIXTURE MODEL WITH COVARIATES

To clarify the above expression, we here do the same "by hand."

```
# calculate lambda.hat: exp(a0 + a1*vegHt)
lambda.hat <- predict(fm.Nmix1, type="state")[,1]

# calculate p.hat: plogis(b0 + b1*wind)
p.hat <- matrix(predict(fm.Nmix1, type="det")[,1], ncol=ncol(C), byrow=TRUE)

Ngrid <- 0:(100+max(umf@y, na.rm = TRUE))
posterior <- matrix(NA, nrow=nrow(C), ncol=length(Ngrid))
bup2 <- array(NA, dim = M)

for(i in 1:nrow(C)){      # Loop over sites
  # Compute prior using MLE
  gN <- dpois(Ngrid, lambda.hat[i])
  gN <- gN/sum(gN)

  # Compute likelihood for each possible value of N
  fy <- rep(NA, length(Ngrid))
  for(j in 1:length(Ngrid)){
     fy[j] <- prod(dbinom(C[i,], Ngrid[j], p.hat[i,]))
  }

  # Compute marginal of y. for denominator of Bayes rule
  qy <- sum(fy * gN)

  # Posterior
  posterior[i,] <- fy * gN / qy

  # N can't be less than max(C)
  if(max(C[i,] > 0))
     posterior[i,0:max(C[i,])]<- 0

  # Compute posterior mean (BUP)
  bup2[i] <- sum(posterior[i,] * Ngrid)
}

# Compare BUPS with true N and counts for first and last 5 sites
(bup1 <- bup(ranef(fm.Nmix1)))
cbind(N=N,count1=C[,1],count2=C[,2],count3=C[,3],BUP1=bup1,BUP2=bup2)[c(1:5,
91:95),]
        N count1 count2 count3       BUP1         BUP2
 [1,]   0      0      0      0  0.037770455  0.037770455
 [2,]   0      0      0      0  0.001489541  0.001489541
 [3,]   0      0      0      0  0.081963672  0.081963672
 [4,]   2      1      0      0  1.040688154  1.040688154
 [5,]   0      0      0      0  0.021773600  0.021773600

 [6,]   6      2      2      4  4.075543688  4.075543688
 [7,]   4      3      3      0  3.539540712  3.539540712
 [8,]   2      0      0      0  1.969791630  1.969791630
 [9,]   8      3      0      7  7.765449990  7.765449990
[10,]   4      1      3      2  4.280652414  4.280652414
```

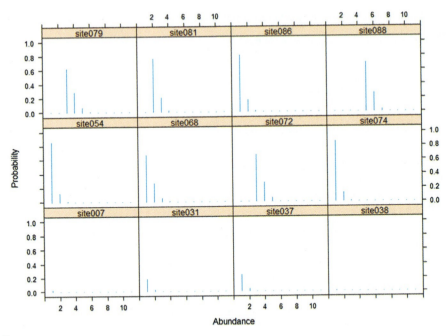

FIGURE 6.3

Estimated conditional posterior distributions of local abundance at a random selection of sites. The *best unbiased prediction* (BUP) is the mean of these distributions.

We can plot the empirical posterior distributions of N, shown here for a random sample of 12 sites (Figure 6.3).

```
plot(ranef(fm.Nmix1), xlim=c(0,12))[sort(sample(1:100, 12))]
```

These estimates are approximations, because their uncertainty does not incorporate the uncertainty stemming from the fact that the parameters had to be estimated. In contrast, the Bayesian estimates of N using BUGS (see later) are exact except for Monte Carlo error inherent in the MCMC analysis. We could also compute functions of the BUPs, for example, to estimate the combined population size in groups of sites (for instance, in habitats of type A, B and C) or over all of them (Lyons et al., 2012). The uncertainty around such derived quantities could be bootstrapped; see Section 10.4 for the analogous problem of estimating the number of occupied sites under an occupancy model.

We will next quickly illustrate the fitting of linear models involving factors inside an N-mixture model. We fit models in the "means parameterizations," where the parameters for the levels of the factors have the direct meaning of the means for each factor level (see Section 3.2.1). The linear models underlying these two models are called "main-effects analysis of covariance (ANCOVA)" and "interaction-effects ANCOVA" in other fields (Kéry, 2010; Section 3.2.2). For a factor A and a continuous covariate x, these linear models could be denoted "A + x" and "A * x," respectively.

```
# Main-effects ANCOVA: additive effects of factor and covariate
summary(fm.Nmix2 <- pcount(~ wind+time-1 ~ vegHt+hab-1, data=umf))
```

6.4 A SLIGHTLY MORE COMPLEX N-MIXTURE MODEL WITH COVARIATES

```
# Interaction-effects ANCOVA: multiplicative effects of factor and covariate
summary(fm.Nmix3 <- pcount(~ wind*time-1-wind ~ vegHt*hab-1-vegHt, data=umf))

# Get predictions for factor levels at average values of covariates
newdat <- data.frame(vegHt=0, hab = c("A", "B", "C"))
predict(fm.Nmix2, type="state", newdata=newdat, appendData = T)    # for abundance
  Predicted        SE       lower      upper    vegHt  hab
1 0.5527982 0.3188844  0.1784624  1.712326        0    A
2 0.7692908 0.1908027  0.4731208  1.250861        0    B
3 0.9158200 0.3858810  0.4010108  2.091530        0    C

newdat <- data.frame(time = c("1", "2", "3"), wind = 0)
predict(fm.Nmix3, type="det", newdata=newdat, appendData = T)    # for detection
  Predicted         SE        lower      upper    time  wind
1 0.2066560 0.06272938  0.10957280  0.3554218      1     0
2 0.1355945 0.04862346  0.06504282  0.2612868      2     0
3 0.1184702 0.04544869  0.05416939  0.2397505      3     0
```

A little confusingly, to get predictions only for selected levels of a factor such as habitat, you need to add *all* the levels (in the correct order!) to your factor in `newdata`:

```
newdat <- data.frame(vegHt=seq(0, 2, by=0.1), hab = factor("A", levels = c("A", "B", "C")))
predict(fm.Nmix2, type="state", newdata=newdat, appendData = T)
```

Model selection can be done using the AIC (see above) or by a likelihood ratio test.

```
LRT(fm.Nmix3, fm.Nmix1)
     Chisq DF Pr(>Chisq)
1 6.235671  8  0.6208518
```

In summary, `unmarked` gives us considerable flexibility in fitting a wide range of N-mixture models quickly and reliably, and makes available to us all the convenience of likelihood inference (e.g., tests and AIC). Despite the name of the model-fitting function (`pcount`), and even some published analyses that call this model "*the point count model*," applications of the binomial mixture model are of course *not* restricted to point count data. Instead, any kind of replicated counts for any reasonably well-defined spatial sampling units may be adequate for this model; e.g., counts along linear transects (Kéry et al., 2009), or areal counts with or without defined transects (Kéry et al., 2005b; Lyons et al., 2012); see also Figure 6.8.

We conclude this section with an illustration of BUGS fitted to the last one of these models, the one with an interaction-effects ANCOVA linear model structure. We do not need to add time into the data bundle, because we can use the column dimension of the two-dimensional data array containing the counts to specify models with seasonal variation in detection probability. We add four types of derived quantities: the number of occupied sites in the sample of 100 study sites (`Nocc`), the total population size (`Ntotal`), the habitat-specific total population size (`Nhab`), and predictions of expected abundance and of detection for a range of values of the covariates `vegHt` and `wind`, respectively, for which we provide two sets of evenly spaced covariate values (`XvegHt`, `Xwind`). We defer until later the fitting of alternative abundance models; e.g., the Poisson-lognormal (PLN), negative-binomial (NB), and zero-inflated Poisson (ZIP) (see Section 6.10).

One of the greatest things about a Bayesian analysis is the ease with which we can do calculations on latent variables, and in general make inferences about derived quantities or functions of parameters, with full propagation of all uncertainty involved in its computation via the posterior distribution (see also Sauer and Link, 2002, for a neat example). In our example we emphasize inference about functions of site-specific abundance N: the number of occupied sites (Nocc), the total abundance across all M sites (Ntotal), total abundance in the three habitat types (Nhab), as well as predictions of λ and p for a range of values of the vegHt and wind covariates, for each level of the two factors in the model.

As another, fairly crazy example of such derived quantities, suppose that a famous theoretical ecologist has just come up with a sophisticated theory about the extinction risk of metapopulations. His theory predicts that our metapopulation would go extinct if 75% or more of all patches had only two or fewer individuals. Is it possible to estimate the extinction probability of our metapopulation? It turns out that in a Bayesian analysis this is really simple: in the BUGS model, we simply define a statistic that codes for the desired condition, add it up over the M sites and then obtain posterior samples for this latter quantity. So as one solution to this difficult problem, we could add the following lines in the model:

```
critical[i] <- step(2-N[i])      # yields 1 whenever N is less or equal to 2
N.critical <- sum(critical[])    # Number of sites with critical size
```

Thus, whenever N_i is less than or equal to 2, the indicator critical evaluates to 1, and N.critical tallies the number of populations for which this is the case.

```
# Bundle data
win.data <- list(C = C, M = nrow(C), J = ncol(C), wind = wind, vegHt = vegHt,
    hab = as.numeric(factor(hab)), XvegHt = seq(-1, 1,, 100), Xwind = seq(-1, 1,,100) )
str(win.data)

# Specify model in BUGS language
cat(file = "model2.txt", "
model {
# Priors
for(k in 1:3){
    alpha0[k] ~ dunif(-10, 10)    # Detection intercepts
    alpha1[k] ~ dunif(-10, 10)    # Detection slopes
    beta0[k] ~ dunif(-10, 10)     # Abundance intercepts
    beta1[k] ~ dunif(-10, 10)     # Abundance slopes
}

# Likelihood
# Ecological model for true abundance
for (i in 1:M){
    N[i] ~ dpois(lambda[i])
    log(lambda[i]) <- beta0[hab[i]] + beta1[hab[i]] * vegHt[i]
    # Some intermediate derived quantities
    critical[i] <- step(2-N[i])   # yields 1 whenever N is 2 or less
    z[i] <- step(N[i]-0.5)        # Indicator for occupied site
    # Observation model for replicated counts
    for (j in 1:J){
        C[i,j] ~ dbin(p[i,j], N[i])
        logit(p[i,j]) <- alpha0[j] + alpha1[j] * wind[i,j]
    }
}
```

6.4 A SLIGHTLY MORE COMPLEX N-MIXTURE MODEL WITH COVARIATES

```
# Derived quantities: functions of latent variables and predictions
Nocc <- sum(z[])              # Number of occupied sites among sample of M
Ntotal <- sum(N[])            # Total population size at M sites combined
Nhab[1] <- sum(N[1:33])       # Total abundance for sites in hab A
Nhab[2] <- sum(N[34:66])      # Total abundance for sites in hab B
Nhab[3] <- sum(N[67:100])     # Total abundance for sites in hab C
for(k in 1:100){              # Predictions of lambda and p ...
    for(level in 1:3){        #    ... for each level of hab and time factors
        lam.pred[k, level] <- exp(beta0[level] + beta1[level] * XvegHt[k])
        logit(p.pred[k, level]) <- alpha0[level] + alpha1[level] * Xwind[k]
    }
}
N.critical <- sum(critical[])   # Number of populations with critical size
}")

# Initial values
Nst <- apply(C, 1, max)+1     # Important to give good inits for latent N
inits <- function() list(N = Nst, alpha0 = rnorm(3), alpha1 = rnorm(3), beta0 = rnorm(3),
beta1 = rnorm(3))

# Parameters monitored
params <- c("alpha0", "alpha1", "beta0", "beta1", "Nocc", "Ntotal", "Nhab",
"N.critical", "lam.pred", "p.pred") # could also estimate N, bayesian
counterpart to BUPs before: simply add "N" to the list

# MCMC settings
nc <- 3 ; ni <- 22000 ; nb <- 2000 ; nt <- 10

# Call JAGS, time run (ART 1 min) and summarize posteriors
library(jagsUI)
system.time(out <- jags(win.data, inits, params, "model2.txt", n.chains = nc,
n.thin = nt, n.iter = ni, n.burnin = nb, parallel = TRUE))
traceplot(out, param = c('alpha0', 'alpha1', 'beta0', 'beta1', 'Nocc', 'Ntotal',
'Nhab', 'N.critical'))
print(out, 2)
```

	mean	sd	2.5%	50%	97.5%	overlap0	f	Rhat	n.eff
alpha0[1]	-1.41	0.40	-2.23	-1.39	-0.67	FALSE	1.00	1.00	6000
alpha0[2]	-1.92	0.43	-2.82	-1.89	-1.15	FALSE	1.00	1.00	6000
alpha0[3]	-2.08	0.45	-3.01	-2.07	-1.26	FALSE	1.00	1.00	6000
alpha1[1]	-3.65	0.83	-5.43	-3.59	-2.21	FALSE	1.00	1.00	6000
alpha1[2]	-4.76	0.93	-6.76	-4.71	-3.13	FALSE	1.00	1.00	6000
alpha1[3]	-5.01	0.98	-7.09	-4.94	-3.29	FALSE	1.00	1.00	2789
beta0[1]	-3.75	1.87	-7.91	-3.55	-0.56	FALSE	0.99	1.01	361
beta0[2]	-0.35	0.28	-0.93	-0.34	0.16	TRUE	0.90	1.00	6000
beta0[3]	-0.36	0.52	-1.41	-0.36	0.64	TRUE	0.75	1.00	975
beta1[1]	-2.89	2.52	-8.10	-2.82	1.83	TRUE	0.88	1.01	457
beta1[2]	2.04	1.24	-0.33	2.03	4.50	TRUE	0.95	1.00	6000
beta1[3]	2.24	0.69	0.88	2.24	3.61	FALSE	1.00	1.00	825
Nocc	57.53	2.98	52.00	57.00	64.00	FALSE	1.00	1.00	6000

```
Ntotal         141.76  12.90  122.00  140.00  173.00    FALSE  1.00  1.00  4873
Nhab[1]          6.45   2.06    4.00    6.00   12.00    FALSE  1.00  1.00  1965
Nhab[2]         28.17   4.22   22.00   28.00   38.00    FALSE  1.00  1.00  6000
Nhab[3]        107.13  10.06   92.00  106.00  131.00    FALSE  1.00  1.00  6000
N.critical      76.43   2.45   71.00   77.00   81.00    FALSE  1.00  1.00  6000
[ output truncated ]
```

Hence, the number of occupied sites (Nocc) is estimated at 57.5 (CRI 52–64), the total abundance across all *M* sites (Ntotal) at 141.8 (CRI 122–173), and the habitat-specific total abundance at 6.45 (CRI 4–12) in habitat A, 28.2 (CRI 22–38) in habitat B, and 107.1 (CRI 92–131) in habitat C. Finally, the number of sites with critical population size (i.e., two or fewer individuals) is estimated at 76.4 (CRI 71–81); see Figure 6.4.

```
plot(table(out$sims.list$N.critical), xlab="Number of populations with critical size",
ylab="Frequency", frame = F)                 # Produces plot 6.4
abline(v = 74.5, col = "red", lwd = 3)
```

Our estimate of the metapopulation extinction risk is simply the proportion of MCMC samples for which N.critical ≥ 75.

```
(metapop.extinction.risk <- mean(out$sims.list$N.critical > 74))
[1] 0.8008333
```

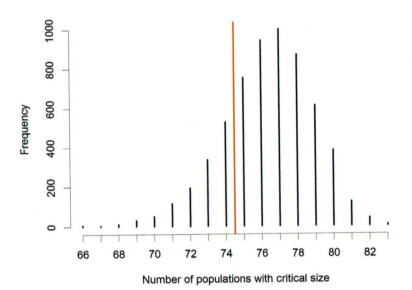

FIGURE 6.4

Example of a crazy derived variable in a Bayesian analysis: the posterior distribution of the number of populations that meet some hypothetical extinction threshold (here, two or fewer individuals), and the imaginary critical number (red line, 75) of such populations at which the metapopulation is going extinct. The metapopulation extinction risk is the relative mass to the right of the red line and amounts to about 80%.

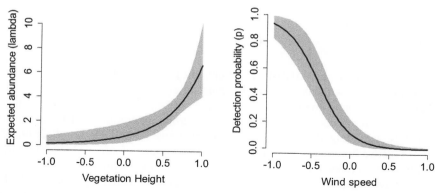

FIGURE 6.5

Predictions of expected abundance (λ) in habitat type C (left), and detection probability (p) for occasion 3 (right) (posterior means and 95% CRIs are shown).

Hence, the future is not bright for our metapopulation… let us produce some plots to illustrate the covariate relationships for one of the levels of the two factors (Figure 6.5).

```
par(mfrow = c(1,2), mar = c(5,5,3,2), cex.axis = 1.5, cex.lab = 1.5)
X <- seq(-1, 1,, 100)
plot(X, out$summary[219:318,1], xlab = "Vegetation Height", ylab = "Expected abundance
 (lambda)", ylim = c(0, 11), frame = F, type = "l")
polygon(c(X, rev(X)), c(out$summary[219:318,3], rev(out$summary[219:318,7])),
 col = "gray", border = F)
lines(X, out$summary[219:318,1], lty = 1, lwd = 3, col = "blue")
plot(X, out$summary[519:618,1], xlab = "Wind speed", ylab = "Detection probability (p)",
 ylim = c(0, 1), frame = F, type = "l")
polygon(c(X, rev(X)), c(out$summary[519:618,3], rev(out$summary[519:618,7])),
 col = "gray", border = F)
lines(X, out$summary[519:618,1], lty = 1, lwd = 3, col = "blue")
```

Thus, when fitting the model with the BUGS language, you have all the power and flexibility of the Bayesian approach to inference and the BUGS engines at your fingertips (see Sections 6.9 and 10.6 for easy implementation, in BUGS, of complex linear models using design matrices).

6.5 A VERY GENERAL DATA SIMULATION FUNCTION FOR *N*-MIXTURE MODELS: simNmix

In Section 4.3, we met a function that generates data under a binomial mixture model with covariates in both abundance and detection. Here, we introduce a much more general function that we have written to simulate data under both binomial and multinomial *N*-mixture models (Chapter 7), simNmix, which is in the applied hierarchical modeling (AHM) package. The default arguments are the following:

```
simNmix(nsite = 267, nvisit = 3, mean.theta = 1, mean.lam = 2, mean.p = 0.6,
  area = FALSE, beta1.theta = 0, beta2.theta = 0, beta3.theta = 0, beta2.lam = 0,
```

```
beta3.lam = 0, beta4.lam = 0, beta3.p = 0, beta5.p = 0, beta6.p = 0,
beta.p.survey = 0, beta.p.N = 0, sigma.lam = 0, dispersion = 10, sigma.p.site =
0, sigma.p.visit = 0, sigma.p.survey = 0, sigma.p.ind = 0, Neg.Bin = FALSE,
open.N = FALSE, show.plot = TRUE)
```

The default settings create an "MHB-type" of data set (267 sites, three replicate visits) under the simplest N-mixture model (see Section 6.3) with constant expected abundance ($\lambda = 2$) and detection ($p = 0.6$). Switching on some "toggles" such as area, Neg.Bin, and open.N, and choosing covariate coefficients and random effects SDs other than zero, permit generation of data sets under a large number of models, including different mixture distributions for abundance (Poisson, ZIP, PLN, ZIP-lognormal, negative binomial, zero-inflated NB), varying area of sites (which might then lead to an analysis with offset; see Section 3.3.2), violation of the closure assumption, and a large number of possible types of "overdispersion" (OD). One of the most general model types under which data sets can be generated is a zero-inflated PLN N-mixture model with covariates in suitability probability, expected abundance, and detection probability, and with OD in logit(detection) due to sites, surveys, visits, and individuals:

Suitability (or zero-inflation) model

$s_i \sim Bernoulli(\theta_i)$
$\quad logit(\theta_i) = \alpha_{0,\theta} + beta1.theta * Xsite1_i + beta2.theta * Xsite2_i + beta3.theta * Xsite3_i$

Abundance model (given suitability)

$N_i|s_i \sim Poisson(s_i * \lambda_i)$
$\quad log(\lambda_i) = log(A_i) + \alpha_{0,\lambda} + beta2.lam * Xsite2_i + beta3.lam * Xsite3_i + beta4.lam * Xsite4_i + eta.lam.site_i$

Detection model (given abundance)

$C_{ij}|N_i \sim Binomial(N_i, p_{ij})$
$logit(p_{ij}) = \alpha_{0,p} + beta3.p * Xsite3_i + beta5.p * Xsite5_i + beta6.p * Xsite6_i + beta.p.survey * Xsurvey_{ij}$
$\quad + beta.p.N * N.centered_i + eta.p.site_i + eta.p.visit_j + eta.p.survey_{ij} + eta.p.ind_n$

"Overdispersion" models

$eta.lam.site_i \sim Normal(0, sigma.lam^2)$	Abundance site random effects in $log(\lambda)$
$eta.p.site_i \sim Normal(0, sigma.p.site^2)$	Detection site random effects in $logit(p)$
$eta.p.visit_j \sim Normal(0, sigma.p.visit^2)$	Detection visit random effects in $logit(p)$
$eta.p.survey_{ij} \sim Normal(0, sigma.p.survey^2)$	Detection site/survey random effects in $logit(p)$
$eta.p.ind_n \sim Normal(0, sigma.p.ind^2)$	Detection individual random effects in $logit(p)$

This model describes three major, linked random variables, two latent and one observed: s_i denotes whether site i is suitable ($s_i = 1$) or not ($s_i = 0$; this is the zero-inflation part), $N_i|s_i$ is the abundance at site i given that site i is suitable, $C_{ij}|N_i$ denotes the observed counts at site i during replicate j given local abundance N_i, Xsite1 through Xsite6 are site-level covariates, and Xsurvey is an observational covariate. Coefficients of these covariates have names that start with beta and include as a suffix theta, lam, or p to indicate whether they are in the suitability (theta), abundance (lambda), or detection (p) model. Two covariates may affect more than one level in the model: Xsite2 can affect both suitability and abundance via the coefficients beta2.theta, beta2.lam, and Xsite3 can affect all three via coefficients beta3.theta, beta3.lam, and beta3.p. Covariate N.centered allows for a logarithmic relationship between local abundance and detection probability—i.e., for density-dependent detection probability. A_i is the area of a site; hence, unless $A = 1$ (the default when area = FALSE), the log of the supplied area will be used as an offset in the linear predictor of the abundance model; see below for an example. All eta terms are random

6.5 DATA SIMULATION FUNCTION FOR N-MIXTURE MODELS

effects that vary by what their name and their subscript says—i.e., by site, visit (= occasion), or site/visit (= survey). They are draws from separate zero-mean normal distributions with SDs that are selected as a function argument.

The last set of random effects, $eta.p.ind_n$, is specific to the n individuals ($n = 1, \ldots,$ Ntotal) occurring anywhere at the studied sites; i.e., it lets us simulate data under what is known as model M_h in the capture-recapture literature (Otis et al., 1978, Williams et al., 2002; Royle and Dorazio, 2008; King et al., 2009; see also Section 7.8.3). We introduce such individual effects by simulating the data *at the level of each individual.* To obtain the simulated count C_{ij}, we sum up the observed individuals for each site and visit. Therefore, function simNmix yields the individual-level detection histories for all individuals, including those that were never detected. These individual detection histories can be aggregated for the observed individuals and analyzed using multinomial mixture models (see Chapter 7), or aggregated to counts and analyzed using binomial mixture models.

Setting Neg.Bin = TRUE will generate data with an NB distribution for abundance, with dispersion equal to the dispersion argument. As another overdispersed alternative to a Poisson, we can choose a PLN distribution by setting sigma.lam to a nonzero value.

With open.N = TRUE, data are simulated under one specific form of an open-population model, where N in the first occasion is drawn from the specified distribution, and for all further occasions j we have $N_{ij} \sim Poisson(N_{i(j-1)})$. Data sets generated with this setting can be used to study the effects of a (fairly serious) violation of the closure assumption. With open.N = TRUE, we must have sigma.p.ind = 0 and nvisit >1.

Argument show.plot determines whether to show plots that visualize the data-generating processes including the final data. These should help you understand the effects of different settings for the function arguments. You must browse through the plots by pressing the Enter key; hence when running simulations, you should set show.plot = FALSE.

```
# Execute function and inspect results
data <- simNmix()                              # Default arguments
data <- simNmix(show.plot = FALSE)             # Default args, no plots
set.seed(24)
str(data <- simNmix(nsite = 267, nvisit = 3, mean.theta = 1, mean.lam = 2,
mean.p = 0.6, area = FALSE, beta1.theta = 0, beta2.theta = 0, beta3.theta = 0,
beta2.lam = 0, beta3.lam = 0, beta4.lam = 0, beta3.p = 0, beta5.p = 0, beta6.p =
0, beta.p.survey = 0, beta.p.N = 0, sigma.lam = 0, dispersion = 10, sigma.p.site
= 0, sigma.p.visit = 0, sigma.p.survey = 0, sigma.p.ind = 0, Neg.Bin = FALSE,
open.N = FALSE, show.plot = TRUE))             # All default args explicit
```

The object to which the function output is assigned first repeats the values of all function arguments, followed by most things created along the way, to generate the main output, the individual detection history matrix (DH) and the replicated counts (C).

```
List of 44
 $ nsite       : num 267
[ Output truncated: values of all function arguments ]
 $ site.cov    : num [1:267, 1:6] -0.8297 -1.1004 0.8169 0.0756 0.6505 ...
  ..- attr(*, "dimnames")=List of 2
  .. ..$ : NULL
  .. ..$ : chr [1:6] "cov1" "cov2" "cov3" "cov4" ...
 $ survey.cov  : num [1:267, 1:3] -1.209 1.111 -1.359 0.993 -1.304 ...
 $ log.lam     : num [1:267] 0.693 0.693 0.693 0.693 0.693 ...
```

```
$ s           : int [1:267] 1 1 1 1 1 1 1 1 1 1 ...
$ N           : int [1:267] 4 2 4 4 1 2 0 3 2 0 ...
$ p           : num [1:267, 1:3, 1:7] 0.6 0.6 0.6 0.6 0.6 0.6 NA ...
$ DH          : int [1:267, 1:3, 1:7] 1 1 0 1 1 1 NA 0 1 NA ...
$ N.open      : logi NA
$ C           : int [1:267, 1:3] 4 2 2 1 1 2 0 0 2 0 ...
[ more truncated output ]
$ Ntotal      : int 507
$ summax      : int 422
```

The function defaults to the simplest possible N-mixture model with constant lambda and p. By changing some default arguments, you can generate data under an extremely wide variety of models, as we will now show (we will omit default arguments and only set the arguments required for the stated kind of data-generating model). For the most complex models and large sample sizes, function execution may easily take 10–20 s, and you can ignore the warnings. We start our illustration with a few simple variants of the default Null model.

```
str(data <- simNmix())                                          # Null data-generating model
str(data <- simNmix(mean.theta = 0.60))                         # ZIP with 40% structural zeroes
str(data <- simNmix(sigma.lam = 1))                             # Poisson-lognormal (PLN) mixture
str(data <- simNmix(Neg.Bin = TRUE))                            # Negative-binomial mixture
str(data <- simNmix(mean.theta = 0.6, sigma.lam = 1))           # Zero-inflated PLN
str(data <- simNmix(mean.theta = 0.6, Neg.Bin = TRUE))          # Zero-infl. NegBin
str(data <- simNmix(mean.p = 1))                                # Perfect detection (p = 1)
str(data <- simNmix(mean.theta = 0.6, mean.p = 1))              # ZIP with p = 1
str(data <- simNmix(sigma.lam = 1, mean.p = 1))                 # PLN with p = 1
```

Other simple variants of the default model include the following.

```
areas <- runif(267, 1, 2)                                       # Generate vector with site area
str(data <- simNmix(nsite = 267, area = areas))                 # Sites with variable area
str(data <- simNmix(nvisit = 1))                                # Only one visit
str(data <- simNmix(sigma.p.site = 1))                          # Random site effects in p
str(data <- simNmix(sigma.p.visit = 1))                         # Random visit (= time) effects in p
str(data <- simNmix(sigma.p.survey = 1))                        # Random site-by-visit effects in p
str(data <- simNmix(sigma.p.ind = 1))                           # Random individual effects in p
str(data <- simNmix(mean.theta = 0.5, beta1.theta = 1))         # Site cov 1 in suitability
str(data <- simNmix(beta2.lam = 1))                             # Site covariate 2 in abundance process
str(data <- simNmix(beta3.p = 1))                               # Site covariate 3 in detection process
str(data <- simNmix(beta.p.N = 1))                              # Positive density-dep. in p
str(data <- simNmix(beta.p.N = -1))                             # Negative density-dep. in p
# Same covariate in suitability and abundance (see Phillips & Elith, Ecology, 2014)
str(data <- simNmix(mean.theta = 0.5, beta2.theta = 1, beta2.lam = -1))
# Same covariate in abundance and detection (see Kéry, Auk, 2008)
str(data <- simNmix(beta3.lam = 1, beta3.p = -1))
# Same covariate in all three levels of model (ouch!)
str(data <- simNmix(mean.theta = 0.5, beta3.theta = 1, beta3.lam = 1, beta3.p = -1))
```

Site covariates 2 and 3 may be chosen to affect both the suitability and the abundance processes, and sometimes one sees people fit such models to data. We suspect that identifiability in this case is due exclusively to the strict assumption about the link transformation (logit and log, respectively) and thus, that this estimation problem is analogous to that of estimating *N*-mixture or site-occupancy models from unreplicated data (Knape and Korner-Nievergelt, 2015).

```
# Use unmarked to fit some models to these data sets
cov <- data$site.cov
summary(umf <- unmarkedFramePCount(
    y=data$C, siteCovs= data.frame(cov1=cov[,1], cov2=cov[,2], cov3=cov[,3],
        cov4=cov[,4], cov5=cov[,5], cov6=cov[,6], area = data$area),
        obsCovs = list(survey.cov = data$survey.cov)))
summary(fm <- pcount(~1 ~1, umf))
summary(fm <- pcount(~1 ~1, umf, mixture = "ZIP"))
summary(fm <- pcount(~1 ~1, umf, mixture = "NB"))
summary(fm <- pcount(~cov1+cov2+cov3 ~ cov1+cov2+cov3, umf))
```

It would be straightforward to extend this function—e.g., to add other mixture distributions, more covariate effects including polynomial terms and their interactions, factors (categorical explanatory variables), or an individual covariate such as mass that affects the detection probability of each individual. Similarly, increased ecological realism could be included by shooting missing values in the resulting observed data—for instance, to emulate surveys that were not conducted.

6.6 STUDY DESIGN, BIAS, AND PRECISION OF THE BINOMIAL *N*-MIXTURE MODEL ESTIMATOR

As for many HMs, there is substantial scope for investigations about optimal sampling design. Some work has been done for occupancy models (e.g., MacKenzie and Royle, 2005; Bailey et al., 2007; Guillera-Arroita et al., 2010, 2014; Sanderlin et al., 2014; Ellis et al., 2015), but hardly anything has been published so far for *N*-mixture models (McIntyre et al., 2012; Yamaura, 2013; Yamaura et al., in press). A basic design question with practical importance is how many sites have to be visited or how many visits have to be conducted in order to obtain estimates of some desired quality. A related question is whether it is better to invest more in replicate sites or replicate visits. Both are best answered with simulation.

Further, while in principle, with decent data, the parameters of the *N*-mixture model may be estimated well, the quality of the estimates naturally depends on the sample size (number of sites, number of replicates) and on the magnitude of the parameters λ and p. For small sample sizes, the widely proclaimed unbiasedness of MLEs is lost (Le Cam, 1990), and it is not clear when this happens. In addition, there may be problems with the identifiability of the parameters for small samples or certain parameter values (Couturier et al., 2013; Dennis et al., 2015a). Simulation can give you extremely valuable insight, for exactly *your* sample sizes and presumed parameter values; the latter may come from previous studies, similar species or be based on expert opinion.

Here is a small simulation study that shows how the quality of estimates for particular scenarios (sample sizes, parameter values) can very easily be ascertained by simulation. In our example, we do this for the simplest *N*-mixture model, but such a simulation can easily be extended to more complex models (for instance, see the next section for a slightly more complex model). Also, see Section 10.7 for the analogous study with the site-occupancy model. Specifically, we want to know how bias and precision of the abundance estimator is affected by three factors: number of sites, number of visits, and detection probability (p). In our simulation, we treat the former two as factors (with three levels) and the latter as akin to a continuous explanatory variable by varying p continuously between 0.01 and 0.99.

```r
# Define simulation settings and arrays for sim results
simreps <- 1000                       # Simulate and analyse 1000 data sets
nsites <- c(20, 120, 250)             # Levels for nsites factor
nreps <- c(2, 5, 10)                  # Levels of nrep factor
estimates <- array(NA, dim = c(2, simreps, 3, 3))

# Fill p with random numbers between 0.01 and 0.99
p <- array(runif(n=simreps*3*3, 0.01, 0.99), dim = c(simreps, 3, 3))

# Launch simulation (takes about 6.3 hours)
for(s in 1:3){                        # Loop over levels of nsites factor
   for(r in 1:3){                     # Loop over levels of nreps factor
      for(i in 1:simreps){            # Simulate and analyse 1000 data sets
         cat("*** Simrep number", i, "***\n")
         data <- simNmix(nsite=nsites[s], nvisit=nreps[r], mean.lam = 5,
            mean.p=p[i,s,r], show.plot = F)   # Generate data set
         umf <- unmarkedFramePCount(y = data$C) # Bundle data for unmarked
         fm <- pcount(~1 ~1, umf)              # Fit model
         estimates[,i,s,r] <- coef(fm)          # Save estimates
      }
   }
}

# Visualization
par(mfrow = c(3,3), mar = c(4.5,4.5,2,2), cex.lab = 1.5, cex.axis = 1.3)
for(s in 1:3){              # Loop over nsites
   for(r in 1:3){           # Loop over nreps
      plot(p[,s,r], exp(estimates[1,,s,r]), xlab = "Detection probability",
         ylab = "lambda_hat", main = "", ylim = c(0, 75), frame = F)
      text(0.75, 60, paste("M = ", nsites[s], ", J = ", nreps[r], sep = ""),
         cex = 1.5)
      abline(h = 5, col = "red", lwd = 2)
      lines(smooth.spline(exp(estimates[1,,s,r])~p[,s,r]), col="blue", lwd=2)
   }
}
```

Figure 6.6 shows that there is considerable variation in the quality of the N-mixture model estimates: higher numbers of sites (M), visits (J), and values of detection probability are all beneficial for the quality of the estimates, and the effects of these factors interact. We can clearly recognize the "first law of capture-recapture": that things become more difficult when p gets small. The range of p over which we can expect unbiased estimates of λ (where the red and blue lines coincide) varies from 0.7–1 in the least information-rich scenario (20 sites surveyed twice) to essentially 0.1–1 for the most information-rich scenario (250 sites with 10 surveys each). We also note that for small values of M, J, and p, we get a higher incidence of "freak" estimates, where λ is estimated at some very large value.

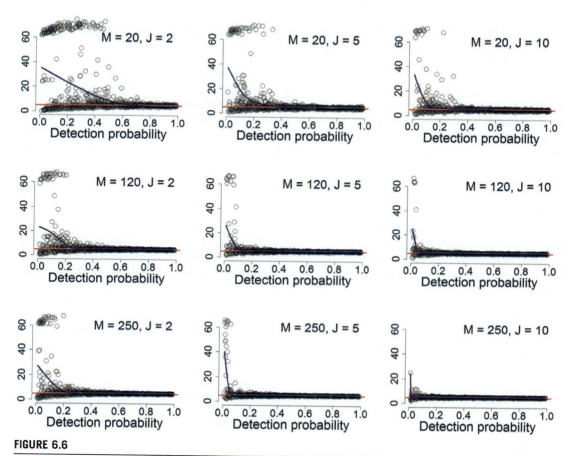

FIGURE 6.6

Yes, it is *that* easy to determine the required sample size for your study: simply simulate data sets where you vary the factors you are interested in, fit the model, and compare the estimates with the known truth. Then, choose the sample size that gives the desired quality of the estimates of the important parameters. Here are results from simulating/analyzing a total of 9000 data sets where we simultaneously varied two factors with three levels each (number of sites M and number of replicate surveys J) and detection probability p (continuously from 0.01 to 0.99). The red line shows truth ($\lambda = 5$), and the blue line is a smoothing spline fit to the estimates.

In the range of p where there are no such outliers, the N-mixture estimator appears unbiased, since the mean of the estimates (blue line) agrees well with the truth (red line).

These results are qualitatively identical to what we find for the occupancy model in Section 10.7. Computation time is much greater for the N-mixture models, but we never obtained crashes associated with a singular Hessian matrix. We could run this simulation with BUGS (and wait for a much *longer* time still) and would presumably observe a "regularizing" effect of the use of the posterior mean as a point estimator as we do in Section 10.7—i.e., observe fewer or no high, freak estimates.

6.7 STUDY OF SOME ASSUMPTION VIOLATIONS USING FUNCTION simNmix

Using simulation, it is also straightforward to study effects of assumption violations: we simply simulate data that include some extra feature such as random effects at some level or lack of closure, and then we analyze the data using a model that assumes the absence of these effects. The presence of an important effect in the data generation and its absence in the analysis model represents an assumption violation in the model. We illustrate this next. Two assumptions of the basic N-mixture models are population closure and the absence of unmodeled heterogeneity in λ and p. To test for the effects of a violation of these assumptions, we use function simNmix to generate six data sets with the following basic settings:

```
mean.lam = exp(1), mean.p = 0.5, beta2.lam = 1, beta3.p = 1, beta.p.survey = 1
```

That is, mean expected abundance is 2.71, mean detection is 0.5, and the slopes of the expected abundance on covariate 2, and of detection probability on covariate 3 and on the survey covariate are all equal to 1. In addition, data sets 1–5 contain one type of heterogeneity that can be created by setting the following arguments to a value of 1: sigma.lam (creating site random effects in abundance) and, for random effects in detection, sigma.p.site (site random effects), sigma.p.visit (occasion random effects), sigma.p.survey (random effects of site-by-occasion), or sigma.p.ind (random effects for each individual). For data set 6, we set open.N = TRUE and do not add any heterogeneity. We then use unmarked to fit a model that is correct in terms of the covariate structure but ignores the extra heterogeneity or openness. We repeat this for 1000 data sets. Code for case 6 (lack of closure) is shown here. This section builds partly upon Tanadini (2010).

```
simreps <- 1000                       # Number of data sets created/analysed
MLE <- array(dim = c(5, simreps))     # Array to hold MLEs

for(i in 1:simreps){                  # Create and analyse 1000 data sets
   cat("*** Simrep number", i, "***\n")
   # Create data set with some extra (here: open populations)
   data <- simNmix(mean.lam=exp(1), mean.p=0.5, beta2.lam=1,
      beta3.p=1, beta.p.survey=1, open.N=TRUE, show.plot=F)
   # Analyze data set with standard model (here: assuming closure)
   umf <- unmarkedFramePCount(y=data$C, siteCovs =
      data.frame(cov2=data$site.cov[,2], cov3=data$site.cov[,3]),
      obsCovs = list(survey.cov = data$survey.cov))
   fm <- pcount(~cov3+survey.cov ~cov2, umf, se = F)
   # Save MLEs
   MLE[,i] <- coef(fm)
}
```

Each type of assumption violation induced bias in some of the five estimated parameters, but not every parameter estimate was affected equally, or indeed at all (Table 6.2). With unmodeled site heterogeneity in abundance, we overestimated mean abundance by 50%, and detection and the slope of the observational covariate by 14%, but the slope estimates of the two site covariates on abundance and detection remained unbiased. Interestingly, Dorazio et al. (2008) found that when a model with lognormal Poisson abundance is assumed, but the site-specific heterogeneity in abundance has a more complex form, then abundance is also overestimated.

6.7 STUDY OF SOME ASSUMPTION VIOLATIONS USING FUNCTION `simNmix`

Table 6.2 Effects of six types of assumption violations on N-mixture model estimators. The first five effects are associated with unmodeled heterogeneity in the form of normal random effects in either log(λ) or logit(p), while the sixth case is lack of closure without any heterogeneity. For detection (p), slope 1 is for a site covariate and slope 2 for a sampling covariate. Mean and 2.5–97.5% percentiles are given for 1000 data sets that were simulated in every case.

Parameter	λ (Int.)	λ (Slope)	p (Int.)	p (Slope 1)	p (Slope 2)
True Value	2.71	1	0.5	1	1
Type of Assumption Violation					
(1) lambda site heterogeneity	4.04 (3.27–5.02)	0.99 (0.80–1.22)	0.57 (0.49–0.64)	1.02 (0.79–1.26)	1.14 (0.99–1.30)
(2) p site heterogeneity	2.60 (2.26–2.93)	1.00 (0.91–1.10)	0.52 (0.47–0.58)	0.84 (0.66–1.00)	0.89 (0.77–1.03)
(3) p occasion heterogeneity	3.76 (2.53–12.02)	1.01 (0.94–1.09)	0.40 (0.09–0.69)	0.81 (0.36–1.06)	0.77 (0.34–1.04)
(4) p survey heterogeneity	3.15 (2.77–3.64)	1.02 (0.94–1.10)	0.41 (0.35–0.46)	0.76 (0.63–0.90)	0.71 (0.60–0.85)
(5) p individual heterogeneity	2.62 (2.33–2.93)	1.00 (0.92–1.08)	0.52 (0.48–0.57)	0.86 (0.75–0.98)	0.89 (0.79–1.00)
(6) Open N (no closure)	9.03 (3.34–16.80)	1.01 (0.91–1.12)	0.20 (0.03–0.49)	0.00 (−0.08–0.08)	0.00 (−0.08–0.08)

Unmodeled site heterogeneity in detection leads to a slight underestimation of abundance, and attenuation (a pull toward zero) of both detection covariate slope estimates. Occasion-specific heterogeneity in detection induced a 39% overestimation of abundance, and bias in detection and attenuation in the estimate of both detection covariates. With survey-specific heterogeneity in detection, abundance was overestimated by 16%, detection underestimated, and both detection covariate estimates attenuated. Unmodeled detection heterogeneity at the level of the individual induced only a negligible bias in abundance and attenuation in the slopes of the detection covariate estimates. This is surprising in view of the "second law of capture-recapture", which says that unmodelled detection heterogeneity at the individual level induces a negative bias in the abundance estimator. Not surprisingly, the strong violation of the closure assumption represented by setting argument `open.N` to `TRUE` leads to strong positive bias in abundance, strong negative bias in detection, and maximal attenuation of the detection covariate estimates. Surprisingly, the slope of the single abundance covariate remained effectively unbiased over all scenarios tested, while estimates of the slopes of detection covariates were typically attenuated.

Obviously, this is not an exhaustive simulation for the study of the effects of assumption violations on the estimators of the binomial N-mixture model. However, it does yield some important insights. Moreover, it shows how this topic can be addressed in a very straightforward way using simulation.

Thus, there is an interest in getting right the main structures in the data and minimizing unexplained heterogeneity by use of covariates or by accommodating heterogeneity by additional random effects in the model. With the exception of individual heterogeneity (case 5), all forms of heterogeneity studied in this section may easily be specified in BUGS. For instance, occasion- and survey-specific detection heterogeneity can be coded by adding into the linear predictor for detection extra terms like `eta[j]` or `eta[i,j]` and then specifying a zero-mean normal prior for `eta` with an SD that is estimated; see Kéry et al. (2009), Zellweger-Fischer et al. (2011) and Sections 6.11, 7.6.3, and 7.8.4 for examples.

Openness can be accommodated by moving to a dynamic *N*-mixture model (see Chapters 12–14 in volume 2). The risk of bias for an ill-specified model emphasizes the importance of goodness-of-fit (GoF), which we cover next.

6.8 GOODNESS-OF-FIT (GoF)

The essence of GoF testing lies in comparing the observed data, or some function thereof, with the analogous quantity that we would expect to see under the model using some discrepancy measure. Ideally, this latter should be chosen such that it indicates a particular breakdown of the model that is relevant to *you*, because it affects *your* intended use of the model. For instance, it could be the observed and expected maximum response if you are interested in predicting extremes (Gelman et al., 1996). In addition, residual diagnostics should be used more often, since they can point out more specifically *where* a model fits and where it does not, and this may suggest ways to improve our model. However, most of the time, we simply calculate a single number that gives a combined measure of the discrepancy between the data and what the model says the data should look like. Typical examples of such statistics are Chi-square, sum of squares, or Freeman–Tukey, and these may be bootstrapped in a non-Bayesian analysis, or their posterior predictive distributions may be used in a Bayesian analysis (Sections 2.8.2 and 5.10).

We illustrate such a GoF test with *N*-mixture models, using again the function `simNmix`. We could directly use function `parboot` in `unmarked` to conduct a parametric bootstrap for a fit statistic of our choice, but instead use the wrapper function `Nmix.gof.test` in the R package `AICcmodavg`, which bootstraps a Chi-square statistic (Mazerolle, 2015). We illustrate with simulated data that represent different kinds of assumption violations with respect to the analyzing model. We start with the "true model" and repeat the exercise nine times to give you an idea of the sampling error in *p*-values from a GoF test. For illustration, we run only 100 bootstrap replicates, but for a "real" analysis, you should perhaps run at least 1000.

```
# Case 1: Test GoF of correct model
library(AICcmodavg)
par(mfrow = c(3,3))
for(i in 1:9){
    data <- simNmix(show.plot = F)              # Create data set
    fm <- pcount(~1 ~1, unmarkedFramePCount(y = data$C)) # Fit model
    pb.gof <- Nmix.gof.test(fm, nsim = 100)     # 100 bootstrap reps
}
```

Next, we look at the following assumption violations: in abundance, zero inflation, lognormal extra-Poisson dispersion, and an unmodeled site covariate, and in detection probability, logit-normal extra-binomial dispersion, and an unmodeled site or observational covariate. We run nine cases for each, with decreasing magnitudes of the assumption violation. When you execute the following code, you can observe how the GoF test statistic will come down *on average*, but that there is substantial sampling error.

```
# Case 2: Simulate data with zero inflation and analyse without
val.range <- seq(0.1, 1,,9)        # Much to no zero-inflation
for(i in 1:9){
    data <- simNmix(mean.theta = val.range[i], show.plot = F)
    fm <- pcount(~1 ~1, unmarkedFramePCount(y = data$C))   # Fit model
    pb.gof <- Nmix.gof.test(fm, nsim = 100)
}
```

```
# Case 3: Extra-Poisson dispersion in lambda
val.range <- seq(1, 0, ,9)          # Some to no extra-Poisson dispersion
for(i in 1:9){
   data <- simNmix(sigma.lam = val.range[i], show.plot = F)
   fm <- pcount(~1 ~1, unmarkedFramePCount(y = data$C))    # Fit model
   pb.gof <- Nmix.gof.test(fm, nsim = 100)
}

# Case 4: Site covariate in lambda
val.range <- seq(3, 0, ,9)          # Strong to no effect of covariate
for(i in 1:9){
   data <- simNmix(beta3.lam = val.range[i], show.plot = F)
   fm <- pcount(~1 ~1, unmarkedFramePCount(y = data$C))    # Fit model
   pb.gof <- Nmix.gof.test(fm, nsim = 100)
}

# Case 5: Extra-binomial dispersion in p (survey random effect)
val.range <- seq(1, 0, ,9)          # Strong to no effect extra-dispersion
for(i in 1:9){
   data <- simNmix(sigma.p.survey = val.range[i], show.plot = F)
   fm <- pcount(~1 ~1, unmarkedFramePCount(y = data$C))    # Fit model
   pb.gof <- Nmix.gof.test(fm, nsim = 100)
}

# Case 6: Site covariate in p
val.range <- seq(3, 0, ,9)          # Strong to no covariate effect
for(i in 1:9){
   data <- simNmix(beta3.p = val.range[i], show.plot = F)
   fm <- pcount(~1 ~1, unmarkedFramePCount(y = data$C))    # Fit model
   pb.gof <- Nmix.gof.test(fm, nsim = 100)
}

# Case 7: Observational covariate in p
val.range <- seq(3, 0, ,9)          # Strong to no covariate effect
for(i in 1:9){
   data <- simNmix(beta.p.survey = val.range[i], show.plot = F)
   fm <- pcount(~1 ~1, unmarkedFramePCount(y = data$C))    # Fit model
   pb.gof <- Nmix.gof.test(fm, nsim = 100)
}
```

The above simulation code may be valuable for you to gain an intuition into the working of such a GoF test and seems to suggest a decent sensitivity of our bootstrapped Chi-square statistic to most kinds of assumption violations. You could build on it to do a simulation study for gauging test sensitivity to certain kinds of assumption violations that are important to you.

With your real data sets, in addition to running a GoF test, we emphasize the usefulness of doing residual checks to highlight places *where* the model does not fit. For an unmarked fitted model object fm you can extract residuals by typing residuals(fm). Plotting residuals against predicted values or covariates can help determine why lack of fit is occurring and whether additional structure (e.g., additional covariates) should be added to help remedy lack of fit (ver Hoef and Boveng, 2015). Finally, if despite your best efforts you do not succeed in building a model that passes your GoF test, you have at least three options: (1) you can discard the data set and claim it is unanalyzable, (2) you can look for an entirely different type of model (and then test GoF again), or (3) you can stick to the nonfitting model in the hope that the lack of fit is inconsequential for your intended use of the model, and because it may still be the best (or only) model you can think of.

But perhaps you may assume that the lack of fit detected is simply unstructured noise (i.e., from over-dispersion (OD)), such that the mean structure of the model is correct but the variance is too large, something extremely common in count data (Lee and Nelder, 2000; ver Hoef and Boveng, 2007). One approach to accounting for OD is to estimate its magnitude—e.g., by dividing the observed Chi-square statistic by the mean of simulated values as an OD parameter (c-hat), and then multiplying the variance–covariance matrix by c-hat (MacKenzie and Bailey, 2004; McKenny et al., 2006; Johnson et al., 2010). Such an accommodation for up to moderate amounts of lack of fit (e.g., for c-hat up to 4, Mazerolle, 2015) is similar to what is commonly done for capture-mark-recapture models in program MARK, and widely used in Cormack-Jolly-Seber survival models (White and Burnham, 1999). See Sections 6.9 and 7.9.3 for illustrations in the context of HMs.

We go on to illustrate the use of posterior predictive checking for GoF when you fit an N-mixture model in a Bayesian analysis. Remember that at each iteration of the MCMC algorithm we randomly draw a new data set using the very model used to analyze the data set at hand and with the particular draws from the posterior distribution of all its parameters. For both the actual and the replicated ("perfect") data sets, we then compare some discrepancy measure between data and model. Then, we compare the posterior distributions of the two discrepancy statistics in a plot or by summarizing with a Bayesian p-value (Sections 2.8.2 and 5.10). We write the code for the simplest N-mixture model such that you can feed into it all types of data sets as created above and see whether a posterior predictive check based on a Chi-square discrepancy can pick up the known assumption violations. We illustrate with a data set that contains lognormal extra-Poisson dispersion in λ with `sigma.lam = 0.5` (case 3 above). We add a small constant (e) to the denominator of the Chi-square discrepancy in the BUGS code to avoid any division by zero.

```
# Bundle and summarize data set
str( win.data <- list(C = data$C, M = nrow(data$C), J = ncol(data$C), e = 0.001))

# Specify model in BUGS language
sink("model.txt")
cat("
model {
# Priors
  lambda ~ dgamma(0.001, 0.001)
  p ~ dunif(0, 1)
# Likelihood
  for (i in 1:M) {
    N[i] ~ dpois(lambda)       # State model
    for (j in 1:J) {
      C[i,j] ~ dbin(p, N[i])   # Observation model
    }
  }
```

6.8 GOODNESS-OF-FIT (GoF)

```
# Posterior predictive distributions of chi2 discrepancy
  for (i in 1:M) {
    for (j in 1:J) {
      C.sim[i,j] ~ dbin(p, N[i])    # Create new data set under model
      e.count[i,j] <- N[i] * p      # Expected datum
      # Chi-square discrepancy for the actual data
      chi2.actual[i,j] <- pow((C[i,j]-e.count[i,j]),2) / (e.count[i,j]+e)
      # Chi-square discrepancy for the simulated ('perfect') data
      chi2.sim[i,j] <- pow((C.sim[i,j]-e.count[i,j]),2) / (e.count[i,j]+e)
      # Add small value e to denominator to avoid division by zero
    }
  }
# Add up individual chi2 values for overall fit statistic
  fit.actual <- sum(chi2.actual[,])  # Fit statistic for actual data set
  fit.sim <- sum(chi2.sim[,])        # Fit statistic for a fitting model
  c.hat <- fit.actual / fit.sim      # c-hat estimate
  bpv <- step(fit.sim-fit.actual)    # Bayesian p-value
}
",fill = TRUE)
sink()

# Do other preps and run model with JAGS
inits <- function(){list(N = apply(data$C, 1, max)+1)}
params <- c("lambda", "p", "fit.actual", "fit.sim", "c.hat", "bpv")
ni <- 2500  ;  nt <- 2  ;  nb <- 500  ;  nc <- 3
fm <- jags(win.data, inits, params, "model.txt", n.chains = nc, n.thin = nt, n.iter = ni,
    n.burnin = nb)
print(fm, dig = 3)
```

	mean	sd	2.5%	50%	97.5%	overlap0	f	Rhat	n.eff
lambda	2.027	0.100	1.833	2.026	2.222	FALSE	1	1.000	3000
p	0.682	0.019	0.643	0.682	0.719	FALSE	1	1.000	3000
fit.actual	228.371	9.673	211.787	227.563	248.952	FALSE	1	1.001	1395
fit.sim	210.795	16.266	180.604	210.388	243.751	FALSE	1	1.000	3000
c.hat	1.087	0.061	0.977	1.084	1.213	FALSE	1	1.000	3000
bpv	0.070	0.255	0.000	0.000	1.000	TRUE	1	1.001	3000

For this realization of the random process represented by our data simulation process, doing a posterior predictive check based on a Chi-square discrepancy does not quite pick up the lack of fit here (i.e., the mismatch between the data-generating and the data-analyzing models): c.hat is estimated at only a little more than 1, and the Bayesian p-value of 0.070 almost suggests a fitting model.

We use our function ppc.plot to produce four plots that summarize the results from the Bayesian GoF test using posterior predictive distributions of the Chi-square discrepancy in this section (Figure 6.7). We see that the "lack-of-fit ratio," or c-hat, is estimated at 1.09, and that there is a 23% chance of observing a value of the test statistic that is more extreme than the one for the observed data set. A fitting model has a Bayesian p-value around 0.5 and a model that does not fit has a p-value close to or equal to 0 or 1. However, there are no rules to decide on a threshold value that distinguishes a fitting from a nonfitting model.

```
ppc.plot(fm)         # Produces Figure 6.7
```

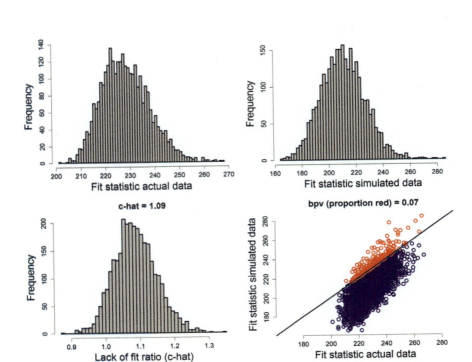

FIGURE 6.7

Results from a posterior predictive check of GoF (based on Chi-square discrepancy) for simulated data under an *N*-mixture model variant. Marginal posterior distributions are given for the test statistic for the actual and simulated data (top), and for their ratio (c-hat). Bottom right shows the joint posterior distribution of the fit statistics, with the proportion of the posterior mass above the 1:1 line (colored red) representing the value of the Bayesian *p*-value (bpv).

GoF relies on comparing expected and observed values. Link and Barker (2010, pp. 194–197) show a two-level posterior predictive check for a model that looks almost like a variation of an *N*-mixture model, except that both the Poisson and the binomial random variables are observed, so they can check for the fit of both model parts. This is not possible in the *N*-mixture model where the Poisson random variable is latent. See Section 7.9.3 for how we attempt to do this in a multinomial mixture model and Section 10.8 for GoF assessment in the related occupancy model. Despite its importance, GoF analysis is underdeveloped for HMs and some things we describe here should be understood as tentative. This is an area where more research is needed.

6.9 ABUNDANCE MAPPING OF SWISS GREAT TITS WITH unmarked

In this section, we use counts of great tits (Figure 4.1) collected in the Swiss breeding bird survey MHB from 2013. The Swiss common breeding bird survey MHB (Monitoring Häufige Brutvögel; Schmid et al., 2004) is based on a sample of 267 1-km^2 quadrats that are laid out as a grid over Switzerland, which has

6.9 ABUNDANCE MAPPING OF SWISS GREAT TITS WITH unmarked

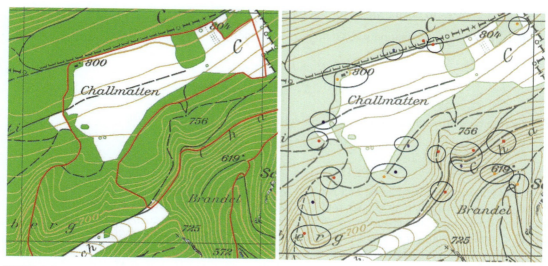

FIGURE 6.8

Example of a survey quadrat in the Swiss common breeding bird survey MHB with the surveyed transect route shown in red (left). Forest is shown in green and elevation in meters. Only the 4.0 km of transect and the birds inside the 1-km² quadrat are surveyed. Right: Combined survey data for the great tit *Parus major* in 2010. Dots represent detection locations, with colors indicating occasions 1–3, and ellipses are the putative territories.

an area a little over 41,000 km^2. Experienced volunteers survey a quadrat-specific, irregular transect route whose length varies from 1–9 km (mean 5.1); see Figure 6.8 (left) for an example. Each transect is surveyed three times during the breeding season (mid-April until early July) using the territory mapping method (Bibby et al., 2000). High-lying quadrats (above the tree line) are surveyed only twice. Surveys start at dawn and last for four hours on average (SD = 1 h). Observers record the location of each individual of each identified species on a map. Afterward, putative territories are determined based on the clustering of observations and for isolated records, based on the knowledge of typical territory sizes of each species (Figure 6.8, right). Here, we analyze quadrat counts of the number of great tit territories (y_{ij}) in quadrat *i* during survey *j* in 2013. We load a data set that contains MHB counts of all six Swiss tit species from 2004–2013. Virtually all quadrats are surveyed annually, but survey-specific counts are currently available for only a subset of quadrats that increases over the years. The data set includes site coordinates and covariates for elevation (m), forest cover (%), and route length (km), and as observational covariates the date (day 1 = April 1) and duration (min) of every survey.

The goals of our analysis are threefold:

1. Identify environmental factors that affect the abundance of Swiss great tits,
2. Produce a Swiss map of great tit abundance in 2013, and
3. Estimate the size of the Swiss population of great tits in 2013.

We use unmarked to obtain MLEs for the following HM:

1. Model for quadrat population size (state process model):

$$N_i \sim Poisson(\lambda_i)$$
$$\log(\lambda_i) = \beta_0 + \text{covariate effects}$$

2. Measurement error model (observation model):

$$y_{ij}|N_i \sim Binomial(N_i, p_{ij})$$
$$\text{logit}(p_{ij}) = \alpha_0 + \text{covariate effects}$$

Route length varies by quadrat and it is likely that quadrats with longer routes are covered more thoroughly than quadrats with shorter routes. To account for such variable coverage bias, we use the inverse of route length as a covariate on abundance, as did Royle and Dorazio (2006) and Royle et al. (2007b); see also Section 7.9. The idea is that the "exposed population," say, N_i, for a quadrat i with a given route location and length L is less than the actual population in the quadrat, say M_i, but as L goes to infinity, the quadrat becomes saturated with sampling effort and N_i goes to M_i.

Hence, the new quadrat population size model is

$$M_i \sim Poisson(\lambda_i)$$
$$\log(\lambda_i) = \beta_0 + \text{covariate effects}$$
$$N_i|M_i \sim Binomial(M_i, \phi(L_i))$$

where we model $\phi(L_i) = \exp(-\beta_5/L_i)$. Under this model the marginal distribution of N_i is

$$N_i \sim Poisson(\mu_i), \text{ with}$$
$$\mu_i = \phi(L_i)\lambda_i, \text{ and hence}$$
$$\log(\mu_i) = \beta_0 + \text{covariate effects} - \beta_5(1/L_i).$$

This means that we use inverse route length to account for $N_i \to M_i$ as L_i increases. For making predictions we set $1/L_i = 0$ (or possibly to any other route length). The prediction then applies to saturation sampling effort—i.e., to a 1-km² quadrat that is 100% covered (and quite possibly more than that; see Section 6.10).

We will consider two additional models for abundance, the NB and the ZIP distributions, both generalizations of a Poisson model. The NB model is a mixture of a Poisson and a gamma distribution (Lee and Nelder, 2000). Marginally, we have $N_i \sim NegBin(\mu_i, \alpha)$; here, α is the dispersion parameter, called "size" in R (we estimate $\log(\alpha)$ in unmarked). The variance of N is $\mu + \mu^2/\alpha$, so lower values of α imply higher variance and with $\alpha = \infty$ (or in practice with $\alpha \approx 100$) we are back to a Poisson. The ZIP model can be described by adding another hierarchical layer to our hierarchical model, which we can imagine describes the binary suitability of a site, with $w_i = 1$ denoting a suitable site and $w_i = 0$ one that is not suitable and that must therefore have zero abundance. The ZIP model in unmarked is this:

$$w_i \sim Bernoulli(1 - \phi) \quad \text{"Suitability" part of model}$$
$$N_i \sim Poisson(w_i\lambda_i) \quad \text{"Abundance" part of model}$$

6.9 ABUNDANCE MAPPING OF SWISS GREAT TITS WITH unmarked

Hence, the "zero-inflation parameter" ϕ here denotes the expected proportion of *unsuitable* sites that produce structural zero counts because they cannot be occupied by a species in principle. When $\phi = 0$, all sites are suitable in principle and we are back to a standard Poisson. The mean of a ZIP random variable is $(1 - \phi_i)\lambda_i$ (we will use this for prediction later). We could alternatively parameterize this model with $w_i \sim Bernoulli(\phi)$; now ϕ represents the expected proportion of *suitable* sites for which N has a standard Poisson distribution.

In unmarked, the zero-inflation or "suitability" part of the model cannot be modeled as a function of covariates, and sometimes there may be reasons for wanting to do so. However, there are several important, and we believe only little-known, caveats. First of all, fitting the same covariates in both the suitability and the abundance parts of the model results in a kind of borderline identifiability that is due exclusively to the critical parametric assumptions of the model (Ghosh et al., 2012). Arguably, this case is analogous to estimating abundance and detection from unreplicated count data (Knape and Korner-Nievergelt, 2015). Hence, we would try to avoid whenever possible using identical covariates in the suitability and the abundance parts of a ZIP model. The second reason for why we think you ought to be cautious with the use of ZIP models with covariates in the zero-inflation part is more philosophical. At many places (e.g., in Chapters 1 and 10), we stress the unity of abundance and distribution: species occurrence is simply an information-reduced summary of species abundance. Adopting elaborate covariate models for both the suitability and the abundance parts of a ZIP model seems to us to obscure this important conceptual point. It does not seem to be conducive to clear thinking either about the concepts of distribution and abundance, or about the factors governing spatial variation in the fundamental quantity, which is abundance. Notwithstanding, if you really, *really* feel you *must* put covariates also in the suitability part of the model, then you can do this fairly easily in BUGS and use code in Section 6.10 as a template.

In the ensuing analysis of great tits with unmarked, we will follow a four-step work flow: (1) setup of analysis, (2) model fitting, (3) GoF, model criticism, and model selection, and (4) analysis of results (especially the forming of predictions).

6.9.1 SETUP OF THE ANALYSIS

We read the file containing the counts of six Swiss tit species between 2004 and 2013, select the great tit data, and do some preliminaries.

```
tits <- read.table("SwissTits_mhb_2004_2013.csv", header = T, sep = ";")
show(tits)                    # Look at data
str(tits)                     # Overview of data file

table(tits$latname)           # Available species (Latin short)
PARATE   PARCAE   PARCRI   PARMAJ   PARMON   PARPAL
   267      267      267      267      267      267

table(tits$name)              # Same in English
Blue tit   Coal tit   Crested tit   Great tit   Marsh tit   Willow tit
     267        267           267         267         267         267

# Select Great tit and covariate data from 2013 and
# drop 4 sites not surveyed in 2013
tits <- tits[tits$latname == "PARMAJ",]                  # Get great tit data
NA.sites <- which(apply(is.na(tits[,41:43]), 1, sum) == 3)   # Unsurveyed sites
tits <- tits[-NA.sites,]                                 # Drop them
```

CHAPTER 6 BINOMIAL N-MIXTURE MODELS

```
y <- cbind(tits$count131,tits$count132, tits$count133)    # Counts
date <- cbind(tits$date131,tits$date132, tits$date133)    # Survey date
dur <- cbind(tits$dur131,tits$dur132, tits$dur133)        # Survey duration

# Plot observed data: counts vs survey date (Figure 6.9)
matplot(t(date), t(y), type = "l", lwd = 3, lty = 1, frame = F, xlab = "Survey date
(1 = April 1)", ylab = "Count of Great Tits")

# Load unmarked, create unmarked data frame and inspect result
library(unmarked)
time <- matrix(rep(as.character(1:3), nrow(y)), ncol = 3, byrow = TRUE)
umf <- unmarkedFramePCount(y = y, siteCovs=data.frame(elev=scale(tits[,"elev"]),
  forest=scale(tits[,"forest"]), iLength=1/tits[,"rlength"]),
  obsCovs=list(time = time, date = scale(date), dur = scale(dur)))
summary(umf)                             # Summarize unmarked data frame
summary(apply(y, 1, max, na.rm = TRUE))  # Summarize max counts
```

On executing this code, we are reminded that we have data from 263 sites, with up to 3 and on average 2.83 surveys per site. Great tits were detected at 191 sites, and counts ranged from 0–41 per survey. We also discover that the maximum count per site, a simplistic estimator of local abundance that ignores imperfect detection, ranges from 0–41, with an average of 8.4. We also see that counts seem to decline over the course of the season (Figure 6.9).

6.9.2 MODEL FITTING

We now fit our N-mixture model and explore different structures in the deterministic and the stochastic parts of the HM. Specifically, we investigate different covariate structures in abundance and detection,

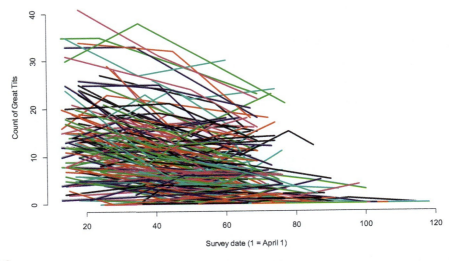

FIGURE 6.9

Time series of counts of great tits in 263 1-km^2 quadrats surveyed in 2013 as part of the Swiss breeding bird survey MHB.

and three different mixture distributions for abundance. To select a parsimonious covariate structure, we start with a highly parameterized model with a Poisson mixture and use backwards elimination guided by *p*-values and AIC scores. We start this process for the detection submodel and then proceed to the abundance submodel.

In the most complex abundance model, we fit the interactions of linear and quadratic effects of elevation and forest cover, as well as the main effect of the reciprocal of route length. In the detection model, we start with linear and quadratic effects and interactions of three continuous covariates (site elevation, and survey date and duration). We also fit main effects of the factor "time" (or occasion, representing the first, second, and third surveys), which accommodates temporal effects that are not captured by the continuous covariates survey date and duration.

```
fm1 <- pcount(~ (elev+I(elev^2)) * (date+I(date^2)) * (dur+I(dur^2)) + time-1
    ~ (elev+I(elev^2)) * (forest+I(forest^2)) + iLength,
    umf, control=list(trace=TRUE, REPORT=5))
summary(fm1)   ;   fm1@AIC
```

We first ensure that there are no identifiability issues and that the summation limit in the likelihood evaluation (K) is high enough at its default by refitting the model with K = 500 (Couturier et al., 2013; Dennis et al., 2015a). (Note the use of the formula slot in the fitted model object to simplify the definition of an identical covariate structure in two model fits.)

```
fm1.K500 <- pcount(fm1@formula, umf, control=list(trace=T, REPORT=5), K = 500)
summary(fm1.K500)   ;   fm1.K500@AIC
```

We find identical AIC scores of 3701.771, and hence the default K was sufficient. Looking at the estimates of model 1, we observe that none of the eight three-way interactions are significant (their *p*-values range from 0.28 to 0.96), so we drop them all and refit the model. We are using model selection rather informally here. Normally, we would drop only one term at a time, and use likelihood ratio rather than Wald tests as a formal significance test in a GLM or an HM. We use solutions from the earlier model as starting values for the next. This speeds up convergence and sometimes may even be necessary to find the MLEs at all. Of course, when doing this, you should take care to check the ordering of the coefficients so that they correspond to the terms in the current model. When your current model has, say, one more coefficient than the one whose solutions you used to provide starting values, then you may have to fill in an NA or some reasonable value at the right place.

```
fm2 <- pcount(~(elev+I(elev^2)) * (date+I(date^2)) * (dur+I(dur^2)) + time-1
    - elev:date:dur - elev:date:I(dur^2) - elev:I(date^2):dur
    - elev:I(date^2):I(dur^2) - I(elev^2):date:dur - I(elev^2):date:I(dur^2)
    - I(elev^2):I(date^2):dur - I(elev^2):I(date^2):I(dur^2)
    ~ (elev+I(elev^2)) * (forest+I(forest^2))
    + iLength, starts = coef(fm1)[1:31],
    umf, control=list(trace=TRUE, REPORT=5))
summary(fm2)                    # AIC = 3695.792
```

In fm2, the two-way interactions between quadratic terms (I(elev^2):I(date^2), I(elev^2):I(dur^2) and I(date^2):I(dur^2)) have *p*-values between 0.33 and 0.89, so we remove all three at once.

```
fm3 <- pcount(~(elev+I(elev^2)) * (date+I(date^2)) * (dur+I(dur^2)) + time-1
    - elev:date:dur - elev:date:I(dur^2) - elev:I(date^2):dur
    - elev:I(date^2):I(dur^2) - I(elev^2):date:dur - I(elev^2):date:I(dur^2)
    - I(elev^2):I(date^2):dur - I(elev^2):I(date^2):I(dur^2)
    - I(elev^2):I(date^2) - I(elev^2):I(dur^2) - I(date^2):I(dur^2)
    ~ (elev+I(elev^2)) * (forest+I(forest^2))
    + iLength, starts = coef(fm2)[-c(23, 27, 31)],
    umf, control=list(trace=TRUE, REPORT=5))
summary(fm3)                  # AIC = 3691.184
```

From `fm3`, we drop the interactions `elev:I(date^2)` and `I(date^2):dur`, with *p*-values of 0.92 and 0.54, respectively.

```
fm4 <- pcount(~(elev+I(elev^2)) * (date+I(date^2)) * (dur+I(dur^2)) + time-1
    - elev:date:dur - elev:date:I(dur^2) - elev:I(date^2):dur
    - elev:I(date^2):I(dur^2) - I(elev^2):date:dur - I(elev^2):date:I(dur^2)
    - I(elev^2):I(date^2):dur - I(elev^2):I(date^2):I(dur^2)
    - I(elev^2):I(date^2) - I(elev^2):I(dur^2) - I(date^2):I(dur^2)
    - elev:I(date^2) - I(date^2):dur
    ~ (elev+I(elev^2)) * (forest+I(forest^2))
    + iLength, starts = coef(fm3)[-c(21, 28)],
    umf, control=list(trace=TRUE, REPORT=5))
summary(fm4)         # AIC = 3687.565
```

At this point, respecting the rules of marginality (that main effects contained in a significant interaction must be kept in a model, etc.; McCullagh and Nelder, 1989; also see Section 3.2.4), we cannot simplify the detection model anymore, so we turn to the abundance model now, where we first drop the two-way interactions `elev2:forest2` and `elev2:forest` with $p = 0.30$ and $p = 0.42$.

```
fm5 <- pcount(~(elev+I(elev^2)) * (date+I(date^2)) * (dur+I(dur^2)) + time-1
    - elev:date:dur - elev:date:I(dur^2) - elev:I(date^2):dur
    - elev:I(date^2):I(dur^2) - I(elev^2):date:dur - I(elev^2):date:I(dur^2)
    - I(elev^2):I(date^2):dur - I(elev^2):I(date^2):I(dur^2)
    - I(elev^2):I(date^2) - I(elev^2):I(dur^2) - I(date^2):I(dur^2)
    - elev:I(date^2) - I(date^2):dur
    ~ (elev+I(elev^2)) * (forest+I(forest^2))+ iLength
    - I(elev^2):forest - I(elev^2):I(forest^2),
    starts = coef(fm4)[-c(9:10)],
    umf, control=list(trace=TRUE, REPORT=5))
summary(fm5)                 # AIC = 3686.094
```

The term `elev:I(forest^2)` has $p = 0.09$, so we try to drop it, but on doing so, the AIC goes up by 1 unit. Hence, we retain model 5 as our final model of the covariate structure. Next, we decide on an appropriate variance structure by comparing Poisson, NB, and ZIP (Wenger and Freeman, 2008) mixtures.

```
# Negative binomial (NB) mixture
fm5NB <- pcount(fm5@formula, starts = c(coef(fm5),0),
    umf, control=list(trace=TRUE, REPORT=5), mixture = "NB")
summary(fm5NB)               # AIC = 3181.046
```

6.9 ABUNDANCE MAPPING OF SWISS GREAT TITS WITH unmarked

```
# Zero-inflated Poisson (ZIP) mixture
fm5ZIP <- pcount(fm5@formula, starts = c(coef(fm5),0),
      umf, control=list(trace=TRUE, REPORT=5), mixture = "ZIP")
summary(fm5ZIP)                        # AIC = 3636.058
```

These models predict quite different average abundances and detection errors. In fact, under the NB mixture, mean abundance is estimated twice as high as under the two Poisson and ZIP models, and, consequently, the mean detection probability during occasions 1–3 is estimated to be much lower as well. Hence, our decision about which variance structure is most appropriate will really matter for our population assessment of the great tits in Switzerland.

```
cbind(rbind("Poisson" = exp(coef(fm5)[1]), "NegBin" = exp(coef(fm5NB)[1]), "ZIP"
= exp(coef(fm5ZIP)[1])), rbind(plogis(coef(fm5)[15:17]),
plogis(coef(fm5NB)[15:17]), plogis(coef(fm5ZIP)[15:17])))
         lam(Int)   p(time1)   p(time2)   p(time3)
Poisson  14.45832   0.6286573  0.5303886  0.4513780
NegBin   31.14023   0.3076560  0.2581500  0.2177134
ZIP      15.74359   0.5934831  0.4955699  0.4188714
```

In addition, the ZIP model estimates that `plogis(coef(fm5ZIP)[25])`, or about 8%, of all sites are not suitable for great tits in principle, while the NB model estimates the dispersion parameter α at `exp(coef(fm5NB)[25])`, or about 2. (As often, you have to execute all this code to see this.)

6.9.3 MODEL CRITICISM AND GoF

The NB is favored by AIC by a very large margin, followed by the ZIP ($\Delta AIC = 455$) and the Poisson ($\Delta AIC = 505$). We run a parametric bootstrap GoF test on a Chi-square discrepancy...

```
library(AICcmodavg)
system.time(gof.P <- Nmix.gof.test(fm5, nsim=100))       # 65 min
system.time(gof.NB <- Nmix.gof.test(fm5NB, nsim=100))    # 131 min
system.time(gof.ZIP <- Nmix.gof.test(fm5ZIP, nsim=100))  # 69 min
gof.P    ; gof.NB   ; gof.ZIP                            # print results
```

... and find that none of the models formally passes the GoF test, with the magnitude of overdispersion (OD, `c-hat`) ranging from 3.82 (Poisson), to 2.47 (ZIP) and to 1.79 (NB). This is disappointing, but before we discard the data set and declare it unanalyzable, we want to see whether there are any patterns in this lack of fit by investigating the residuals of each model. Perhaps we can treat this lack of fit simply as unstructured OD and then adjust for it by inflating the uncertainty around the estimates? We start by using our AHM function `plot.Nmix.resi` to graph the fitted values (i.e., the expected value of each datum) vs. the observed data and the residuals vs. the fitted values under each model (Figure 6.10). It can also be useful to directly compare data, fitted values, and residuals.

```
# Look at data, fitted values and residuals and produce plots
print(cbind(y, fitted(fm5), residuals(fm5)), 2)   # For Poisson model
plot.Nmix.resi(fm5, fm5NB, fm5ZIP)                # Produces Figure 6.10
```

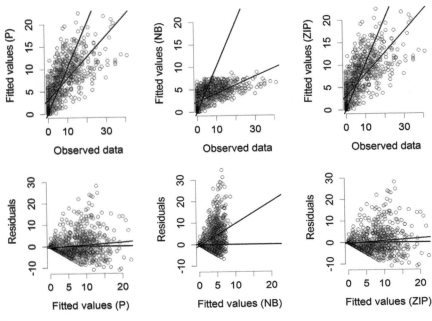

FIGURE 6.10

Residual diagnostics for the three N-mixture models fitted to the 2013 Swiss great tit counts. Top: Fitted values (= expected data) versus observed counts; the black line shows a 1:1 relationship and the blue line is the linear regression line of best fit. Bottom: Residuals versus fitted values (black line denotes a zero residual and the blue line is the linear regression line).

Surprisingly, we find a much better agreement between observed and expected data under the Poisson and ZIP models than under the NB. For the two former, fitted values cluster almost symmetrically around the 1:1 line of correspondence, while for the NB model, fitted values are much too small for a large range of observed data (Figure 6.10, top). When plotting residuals against fitted values (Figure 6.10, bottom), we find more or less symmetry of positive and negative residuals for the Poisson and ZIP models; this is shown by the blue regression line that essentially covers the black zero line. In contrast, there is a preponderance of positive residuals in the NB model; this is emphasized by the strong mismatch of the blue and the black lines. Accordingly, the root mean square error (RMSE) is about 40% larger for the NB than for the Poisson and ZIP mixtures.

```
# Compute RMSE for all three models
(RMSEP <- sqrt(mean((y - fitted(fm5))^2, na.rm = TRUE)))        # Poisson
(RMSENB <- sqrt(mean((y - fitted(fm5NB))^2, na.rm = TRUE)))     # NB
(RMSEZIP <- sqrt(mean((y - fitted(fm5ZIP))^2, na.rm = TRUE)))   # ZIP
[1] 5.18944
[1] 7.139032
[1] 5.228797
```

6.9 ABUNDANCE MAPPING OF SWISS GREAT TITS WITH unmarked

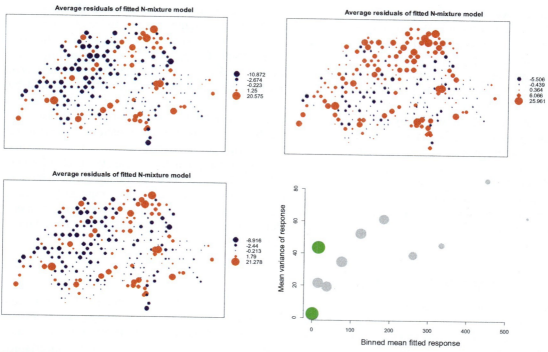

FIGURE 6.11

Maps of the residuals (averaged over replicate surveys) under the AIC-best Poisson, NB, and ZIP N-mixture models for Swiss great tits in 2013, and of the mean variance of the response vs. the mean fitted response (bottom right) under the Poisson (gray) and the NB (green) models.

To investigate any spatial structure in the lack of fit of the three models, we use our AHM function `map.Nmix.resi` to produce a map of the residuals (averaged over occasions; Figure 6.11). We see less spatial structure (local pockets of high or low values) in the residuals of the Poisson and the ZIP models, and more such pattern, and more consistently positive residuals, for the NB model.

```
map.Nmix.resi(fm5)        # Map of average residuals for Poisson model
map.Nmix.resi(fm5NB)      # Map of average residuals for NB model
map.Nmix.resi(fm5ZIP)     # Map of average residuals for ZIP model
```

As a final check on model adequacy, we plot the squared residuals against the fitted values, to investigate the variance/mean structure in the data (ver Hoef and Boveng, 2007; code not shown). Under the Poisson models, the variance should be equal to the mean, while the NB stipulates a quadratic variance/mean relationship. We produce plots that show the means for suitably chosen bins, to smooth out the ruggedness in the raw data (Figure 6.11, bottom right).

We see some scatter and only few points for the NB, but the variance/mean relationship seems much better described by a straight line than by a quadratic curve. This indicates a constant degree of OD rather than an NB error structure (Lee and Nelder, 2000; ver Hoef and Boveng, 2007).

Hence, we feel much more comfortable to base our inference about the abundance of Swiss great tits on one of the two Poisson models than on the NB. The NB predicts almost twice as many great tits as do Poisson or ZIP; and this discrepant behavior of the NB is consistent with earlier observations (Kéry et al., 2005b; ver Hoef and Boveng, 2007; Joseph et al., 2009). In addition, the residual diagnostics for the Poisson and ZIP models look much better than those of the NB, intriguingly despite the much better fit (measured by c-hat) and predictive ability (measured by AIC) of the NB.

We are a little puzzled by these somewhat conflicting findings in terms of AIC, GoF, RMSE, and the graphical fit assessments for these mixture distributions. Moreover, these patterns appear to be quite common and not restricted to the great tit data at all. In many of our own analyses, the NB clearly won in terms of AIC over the Poisson and ZIP mixtures, yet it predicted unrealistic levels of abundance. Thus, we often observe the "good fit/bad prediction dilemma" with binomial N-mixture models and suggest caution when using an NB, even when it is greatly favored by AIC, *especially* when it produces substantially higher abundance estimates than a Poisson or ZIP model. Overall, we repeat our call for more study of the topics of GoF and model selection in N-mixture and similar hierarchical models.

6.9.4 ANALYSIS OF RESULTS

Overdispersion (OD) due to unmodeled covariates or ignored dependency structures is very common in analyses of count data (ver Hoef and Boveng, 2007). We could not find strong patterns in the residuals of the Poisson and the ZIP models. Hence, we now assume that the lack of fit revealed by the GoF tests is simply additional noise, and we want to correct for it by adjusting upwards the uncertainty statements (SEs, CIs) in our inference (e.g., Johnson et al., 2010). This can be done by multiplying the variance–covariance matrix by the estimate of c-hat under the ZIP model (e.g., where it is 2.47), which we prefer over the simple Poisson for its better result in the GoF test. All three main goals of our analysis require us to predict abundance over a range of covariate values. We will thus use the ZIP model for making these predictions, but adjust the prediction uncertainty for an OD factor (c-hat) of 2.47.

A little confusingly, there are currently three functions in `unmarked` and `AICcmodavg` to form predictions for an N-mixture model: `predict`, `predictSE`, and `modavgPred`. We here give an overview of the working of these functions because prediction is extremely important in statistical practice and yet is often felt to be confusing. For illustration, we predict λ along an elevational gradient from 200–2250 m and p for survey date from April 1 to the end of June (days 1–90). The values of all other covariates are set at their average and for the time factor in p, we choose the middle occasion 2. We define two data frames for use in the predictions below.

```
# Two new data sets for prediction: for lambda and for p (200 data points each)
lamNewData <- data.frame(elev = (seq(200, 2250,,200) - mean(tits[,"elev"]))/
sd(tits[,"elev"]), forest = 0, iLength = 1/5.1)
pNewData <- data.frame(elev = 0, time = factor("2", levels = c("1", "2", "3")),
dur = 0, date = (seq(1,90,,200) - mean(date, na.rm = T))/sd(date, na.rm = T))
```

6.9 ABUNDANCE MAPPING OF SWISS GREAT TITS WITH unmarked

We have three options to produce predictions (although it is likely that future versions of unmarked and AICcmodavg will have more functionality in this respect):

1. Function predict in unmarked yields point predictions with SEs and 95% CIs for λ and p for Poisson and NB mixtures, but for ZIP mixtures currently only gives point estimates (though we have a beta version of the function in the AHM package that gives SEs and CIs for the ZIP as well). Function predict cannot inflate SEs and CIs to accommodate an OD correction factor.

   ```
   # Predictions for lambda and p, with SE and CIs, but no overdispersion
   predict(fm5, type="state", newdata=lamNewData)    # Poisson model
   predict(fm5, type="det", newdata=pNewData)
   predict(fm5NB, type="state", newdata=lamNewData)  # NegBin model
   predict(fm5NB, type="det", newdata=pNewData)
   predict(fm5ZIP, type="state", newdata=lamNewData) # ZIP
   predict(fm5ZIP, type="det", newdata=pNewData)
   ```

2. Function predictSE in package AICcmodavg produces point predictions and SEs for ZIP mixtures, but currently no CIs, and allows specification of an OD correction c-hat, which is used to inflate the SEs. Only predictions of λ are currently available and only on the response, but not the link scale.

   ```
   # Predictions for lambda only, incl. SE and CIs, with overdispersion
   predictSE(fm5ZIP, newdata=lamNewData, print.matrix = TRUE, type="response",
   parm.type = "lambda", c.hat = 2.47)
   ```

3. Function modavgPred in AICcmodavg can be tricked into producing point predictions with SEs and 95% CIs by providing a candidate list with a single model only. As of August 2015, no CIs are available, though for some cases 95% CIs can be obtained by adding/subtracting twice the SEs from the point estimates on the link scale and then applying the inverse link transformation.

   ```
   # Predictions for lambda and p, incl. SE and with overdispersion, but no CIs
   # Poisson model, for natural and for link scale of both lambda and p
   modavgPred(cand.set = list(fm5), newdata=lamNewData, parm.type = "lambda", type
   = "response", c.hat = 3.82)
   modavgPred(cand.set = list(fm5), newdata=lamNewData, parm.type = "lambda", type
   = "link", c.hat = 3.82)      # Could be used to get 95% CIs
   modavgPred(cand.set = list(fm5), newdata=pNewData, parm.type = "detect", type =
   "response", c.hat = 3.82)
   modavgPred(cand.set = list(fm5), newdata=pNewData, parm.type = "detect", type =
   "link", c.hat = 3.82)        # Could be used to get 95% CIs

   # NegBin model, for natural and for link scale of both lambda and p
   modavgPred(cand.set = list(fm5NB), newdata=lamNewData, parm.type = "lambda",
   type = "response", c.hat = 1.79)
   modavgPred(cand.set = list(fm5NB), newdata=lamNewData, parm.type = "lambda",
   type = "link", c.hat = 1.79)    # Could be used to get 95% CIs
   ```

```
modavgPred(cand.set = list(fm5NB), newdata=pNewData, parm.type = "detect",
type = "response", c.hat = 1.79)
modavgPred(cand.set = list(fm5NB), newdata=pNewData, parm.type = "detect",
type = "link", c.hat = 1.79)      # Could be used to get 95% CIs

# ZIP model, for natural and for link scale of both lambda and p
modavgPred(cand.set = list(fm5ZIP), newdata=lamNewData, parm.type = "lambda",
type = "response", c.hat = 2.47)
modavgPred(cand.set = list(fm5ZIP), newdata=lamNewData, parm.type = "lambda",
type = "link", c.hat = 2.47)      # Not yet implemented (August 2015)
modavgPred(cand.set = list(fm5ZIP), newdata=pNewData, parm.type = "detect", type =
"response", c.hat = 2.47)
modavgPred(cand.set = list(fm5ZIP), newdata=pNewData, parm.type = "detect", type =
"link", c.hat = 2.47)      # this works, so we could get 95% CIs
```

Our model enables us to gauge coverage bias in the MHB survey: i.e., of the proportion of the nominal sample area, represented by the 1-km^2 sample plot, that is actually covered by the survey route. We specify a relationship between the size of the exposed population within a nominal 1-km^2 sample plot and a saturation population that is informed by variation in route length and estimate a coefficient of the reciprocal of route length of −1.3412. We first form predictions of λ for this crucial relationship to see how it looks (Figure 6.12).

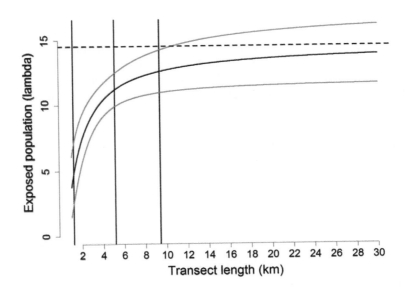

FIGURE 6.12

Estimated relationship (with 1-SE bounds) between the expected numbers of great tit territories (λ) that are exposed to detection by a 1-km^2 quadrat as a function of transect length. Vertical black lines show the minimum, mean, and maximum route lengths in the Swiss MHB (1.2, 5.1, and 9.4, respectively). The associated exposed population represents 33%, 77%, and 87% of the exposed population at saturation sampling (horizontal dashed line), when a hypothetical route of length infinity is squeezed into the 1-km^2 sampling quadrat.

6.9 ABUNDANCE MAPPING OF SWISS GREAT TITS WITH unmarked

```
rlength <- seq(1, 30, 0.01)          # Vary route length from 1 to 30 kms
newData <- data.frame(elev=0, forest=0, iLength=1/rlength)
pred <- predictSE(fm5ZIP, parm.type="lambda", newdata=newData, c.hat = 2.47)
par(mar = c(5,5,3,2), cex.lab = 1.5, cex.axis = 1.3)
plot(rlength, pred[[1]], type = "l", lwd = 3, col = "blue", frame = F, xlab =
"Transect length (km)", ylab = "Exposed population (lambda)", ylim = c(0, 16), axes = F)
axis(1, at = seq(2,30,2))             ;      axis(2)
abline(v = c(1.2, 5.135, 9.4), lwd = 2)
matlines(rlength, cbind(pred[[1]]-pred[[2]], pred[[1]]+pred[[2]]), type = "l",
lty = 1, lwd = 2, col = "gray")
```

By doing the prediction with the value for iLength set at 0, we compute the expected abundance of great tits at saturation sampling (i.e., with infinite route length). We do this and add this value to the plot.

```
sat.pred <- predictSE(fm5ZIP, parm.type="lambda", newdata=data.frame(elev=0, forest=0,
iLength=0), c.hat = 2.47)
abline(h = sat.pred$fit, lwd = 2, lty = 2)
```

```
# Inspect the numbers
print(cbind("Route length" = rlength, "Exp. pop" = pred[[1]], "Rel. exp. pop" = pred[[1]] /
sat.pred$fit), 3)
```

We estimate a relative expected population size of 77% for the average route length of 5.1 km. This percentage rises to 87% for the maximum route length of 9.4, and to 91%, 94%, and 95% at route lengths of 15, 20, and 25 km, respectively. By setting the value of iLength to zero, we estimate the expected size of the local great tit population at saturation sampling effort—i.e., λ under the assumption that we survey a route of length infinity within each 1-km^2 sample quadrat. For all other covariates, we predict at their average value—i.e., at zero—since they were standardized in the analysis. For the time factor, we choose the middle level 2; i.e., we predict for occasion 2. As said above, we cannot currently produce 95% CIs easily for ZIP models with OD, and hence we compute intervals of +/− 1 SE.

```
# Create covariate vectors for prediction and standardise as in analysis
ep.orig <- seq(200, 2250, length.out=100)      # Elevation between 200 and 2250 m
(elev.mean <- mean(tits[,"elev"]))
(elev.sd <- sd(tits[,"elev"]))
ep <- (ep.orig - elev.mean) / elev.sd          # Standardized for prediction
fp.orig <- seq(0, 100, length.out=100)         # Forest cover between 0 and 100%
(forest.mean <- mean(tits[,"forest"]))
(forest.sd <- sd(tits[,"forest"]))
fp <- (fp.orig - forest.mean) / forest.sd      # Standardised for prediction
date.orig <- seq(1, 90, length.out=100)        # Survey date from 1 April - 1 July
(date.mean <- mean(date, na.rm = TRUE))
(date.sd <- sd(date, na.rm = TRUE))
datep <- (date.orig - date.mean) / date.sd     # Standardised for prediction
dur.orig <- seq(90, 420, length.out=100)       # Survey duration from 90 - 420 min
(dur.mean <- mean(dur, na.rm = TRUE))
(dur.sd <- sd(dur, na.rm = TRUE))
durp <- (dur.orig - dur.mean) / dur.sd         # Standardised for prediction
```

```
# Do predictions along single covariate gradient
newData1 <- data.frame(elev=ep, forest=0, iLength=0, date=0, dur=0, time =
factor("2", levels = c("1", "2", "3")))
pred1 <- predictSE(fm5ZIP, newdata=newData1, c.hat = 2.47)
pred2 <- modavgPred(cand.set = list(fm5ZIP), newdata=newData1, parm.type =
"detect", type = "response", c.hat = 2.47)
newData3 <- data.frame(elev=0, forest=fp, iLength=0, date=0, dur=0, time =
factor("2", levels = c("1", "2", "3")))
pred3 <- predictSE(fm5ZIP, newdata=newData3, c.hat = 2.47)
newData4 <- data.frame(elev=0, forest=0, iLength=0, date=datep, dur=0, time =
factor("2", levels = c("1", "2", "3")))
pred4 <- modavgPred(cand.set = list(fm5ZIP), newdata=newData4, parm.type =
"detect", type = "response", c.hat = 2.47)
newData5 <- data.frame(elev=0, forest=0, iLength=0, date=0, dur=durp, time =
factor("2", levels = c("1", "2", "3")))
pred5 <- modavgPred(cand.set = list(fm5ZIP), newdata=newData5, parm.type =
"detect", type = "response", c.hat = 2.47)
newData6 <- data.frame(elev=0, forest=0, iLength=0, date=0, dur=0,
time = c("1", "2", "3"))
pred6 <- modavgPred(cand.set = list(fm5ZIP), newdata=newData6, parm.type =
"detect", type = "response", c.hat = 2.47)

# Plot these predictions along single covariate gradient
par(mfrow = c(3,2), mar = c(5,5,3,2), cex.lab = 1.3, cex.axis = 1.3)
plot(ep.orig, pred1[[1]], type = "l", lwd = 2, col = "blue", xlab = "Elevation (m)",
ylab = "Expected abundance", las = 1, ylim = c(0,50), frame = F)
matlines(ep.orig, cbind(pred1[[1]]-pred1[[2]], pred1[[1]]+pred1[[2]]), type = "l",
lty = 1, lwd = 1, col = "gray")
plot(fp.orig, pred3[[1]], type = "l", lwd = 2, col = "blue", xlab = "Forest cover (%)",
ylab = "Expected abundance", las = 1, ylim = c(0, 18), frame = F)
matlines(fp.orig, cbind(pred3[[1]]-pred3[[2]], pred3[[1]]+pred3[[2]]), type = "l",
lty = 1, lwd = 1, col = "gray")
plot(ep.orig, pred2[[1]], type = "l", lwd = 2, col = "blue", xlab = "Elevation (m)",
ylab = "Expected detection", las = 1, ylim = c(0,1), frame = F)
matlines(ep.orig, cbind(pred2[[1]]-pred2[[2]], pred2[[1]]+pred2[[2]]), type = "l",
lty = 1, lwd = 1, col = "gray")
plot(date.orig, pred4[[1]], type = "l", lwd = 2, col = "blue", xlab = "Survey date (1 =
April 1)", ylab = "Expected detection", las = 1, ylim = c(0,1), frame = F)
matlines(date.orig, cbind(pred4[[1]]-pred4[[2]], pred4[[1]]+pred4[[2]]), type = "l",
lty = 1, lwd = 1, col = "gray")
plot(dur.orig, pred5[[1]], type = "l", lwd = 2, col = "blue", xlab = "Survey duration
(min)", ylab = "Expected detection", las = 1, ylim = c(0,1), frame = F)
matlines(dur.orig, cbind(pred5[[1]]-pred5[[2]], pred5[[1]]+pred5[[2]]), type = "l",
lty = 1, lwd = 1, col = "gray")
barplot(pred6[[1]], names.arg = c("1", "2", "3"), ylim = c(0,1),
ylab = "Expected detection", xlab = "Survey Number")
segments(c(0.7,1.9, 3.1), pred6[[1]]-pred6[[2]], c(0.7,1.9, 3.1), pred6[[1]]+
pred6[[2]], lwd = 2)
```

6.9 ABUNDANCE MAPPING OF SWISS GREAT TITS WITH unmarked

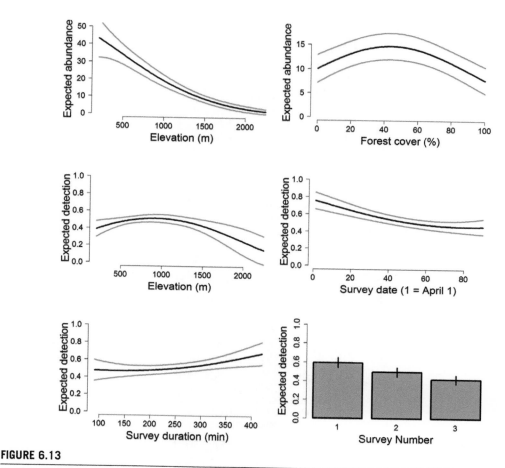

FIGURE 6.13

Predictions of expected abundance (λ) and detection (p) for Swiss great tits in 2013 under model `fm5ZIP` along single covariate gradients. Gray lines and black error bars show 1-SE bounds.

We see in Figure 6.13 that the expected abundance of great tits in Switzerland declines strongly with altitude. Interestingly, there is an intermediate optimum forest cover at a little over 40%. Detection probability has an optimum relationship with elevation, with a maximum around 1000 m, declines over the survey season from about 0.8 to 0.4, and increases only slightly with increasing survey duration. Irrespective of these effects, detection declines from the first through the third survey.

As we stress at many places, plotting such predictions is often virtually the only way that we can understand the meaning of linear models with polynomial terms, interactions, and/or nonlinear link functions. Sometimes it can be interesting to produce predictions when we vary more than one

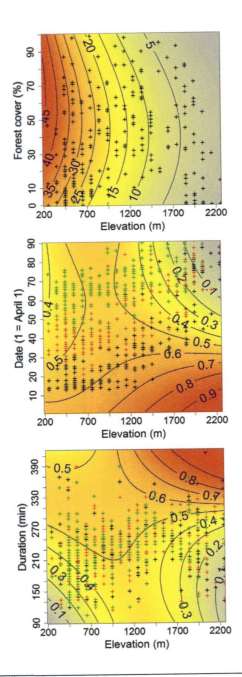

FIGURE 6.14

Bivariate predictions of expected abundance and detection when two covariates are varied simultaneously. Plus signs show values of covariates for the 263 quadrats or for every survey in 2013, with colors in the the middle and bottom panels denoting surveys 1 (black), 2 (red), and 3 (green).

6.9 ABUNDANCE MAPPING OF SWISS GREAT TITS WITH unmarked

covariate. We illustrate this now with the joint relationship of expected abundance, forest cover, and elevation, and of expected detection, elevation, and survey date (Figure 6.14). We will not bother with prediction uncertainty now and simply use the `predict` function in unmarked.

```
# Make predictions along two covariate gradients
# (1) Expected abundance (lambda) for forest and elevation
pred.matrix1 <- array(NA, dim = c(100, 100))
for(i in 1:100){
  for(j in 1:100){
    newData <- data.frame(x=0, y=0, elev=ep[i], forest=fp[j], iLength=0)
    pred.matrix1[i,j] <- predict(fm5ZIP,type="state", newdata=newData)[1,1]
  }
}

# (2) Expected detection (p) for elevation and survey date
pred.matrix2 <- array(NA, dim = c(100, 100))
for(i in 1:100){
  for(j in 1:100){
     newData <- data.frame(elev=ep[i], date=datep[j], dur=0,
     time = factor("2", levels = c("1", "2", "3")))
     pred.matrix2[i,j] <- predict(fm5ZIP, type="det", newdata=newData)[1,1]
  }
}

# (3) Expected detection (p) for elevation and survey duration
pred.matrix3 <- array(NA, dim = c(100, 100))
for(i in 1:100){
  for(j in 1:100){
    newData <- data.frame(elev=ep[i], date=0, dur=durp[j],
    time = factor("2", levels = c("1", "2", "3")))
    pred.matrix3[i,j] <- predict(fm5ZIP, type="det", newdata=newData)[1,1]
  }
}

# Plot these prediction matrices
par(mfrow = c(1, 3), mar = c(5,5,2,2), cex.lab = 1.5, cex.axis = 1.5)
mapPalette <- colorRampPalette(c("grey", "yellow", "orange", "red"))

image(x=ep.orig, y=fp.orig, z=pred.matrix1, col = mapPalette(100), axes = F,
xlab = "Elevation (m)", ylab = "Forest cover (%)")
contour(x=ep.orig, y=fp.orig, z=pred.matrix1, add = T, col = "blue", labcex = 1.5,
lwd = 1.5)
axis(1, at = seq(min(ep.orig), max(ep.orig), by = 250))
axis(2, at = seq(0, 100, by = 10))
box()
points(tits$elev, tits$forest, pch="+", cex=1.5)

image(x=ep.orig, y=date.orig, z=pred.matrix2, col = mapPalette(100), axes = F,
xlab = "Elevation (m)", ylab = "Date (1 = April 1)")
```

```
contour(x=ep.orig, y=date.orig, z=pred.matrix2, add = T, col = "blue", labcex = 1.5,
lwd = 1.5)
axis(1, at = seq(min(ep.orig), max(ep.orig), by = 250))
axis(2, at = seq(10, 120, by = 10))
box()
matpoints(tits$elev, date, pch="+", cex=1.5)

image(x=ep.orig, y=dur.orig, z=pred.matrix3, col = mapPalette(100), axes = F,
xlab = "Elevation (m)", ylab = "Duration (min)")
contour(x=ep.orig, y=dur.orig, z=pred.matrix3, add = T, col = "blue",
labcex = 1.5, lwd = 1.5)
axis(1, at = seq(min(ep.orig), max(ep.orig), by = 250))
axis(2, at = seq(90, 420, by = 20))
box()
matpoints(tits$elev, dur, pch="+", cex=1.5)
```

The panels for detection in Figure 6.14 are especially interesting: they show that the relatively mild effects of survey date and duration on detection probability at the *average* of the other covariates including elevation (at 1190 m a.s.l.), and depicted in Figure 6.13, are modified substantially by elevation: both effects become stronger with increasing elevation. Thus, again, forming predictions, in one or in multiple dimensions, is *the key* to understanding what complicated regression models are telling us about covariate relationships.

Let us now move to our second goal and produce abundance maps of great tits in Switzerland in 2013 (Figure 6.15). For this we will project the estimated covariate relationships for abundance onto a map of Switzerland, using the values of these covariates that are known for every 1-km^2 quadrat. We load the Swiss landscape data from unmarked.

```
data(Switzerland)          # Load Swiss landscape data in unmarked
CH <- Switzerland
```

We scale all four covariates using the same mean and sd that we used for the covariates in the analyzed data set and predict expected abundance for every Swiss km^2 quadrat.

```
# Predictions for lambda, with overdispersion
newData <- data.frame(elev = (CH$elev-elev.mean)/elev.sd, forest = (CH$forest-
forest.mean)/forest.sd, iLength = 0, date=0, dur=0, time = factor("2",
levels = c("1", "2", "3")))
predCH <- predictSE(fm5ZIP, newdata=newData, print.matrix = TRUE, type="response",
parm.type = "lambda", c.hat = 2.47)
```

We check for unrealistic abundance predictions as they can occur quite frequently with linear models with a log link. We are prepared to censor any prediction greater than about 60, because a great tit density greater than that is deemed rare in Switzerland (Schmid et al., 1998). Depending on the model, freak predictions may occur and are particularly common with greater elevation. However, in our case there are none.

```
max(predCH[,1])            # Look at the max prediction --- 43.8
sum(predCH[,1] > 60)       # How many are > 60 ?  --- none
```

6.9 ABUNDANCE MAPPING OF SWISS GREAT TITS WITH unmarked

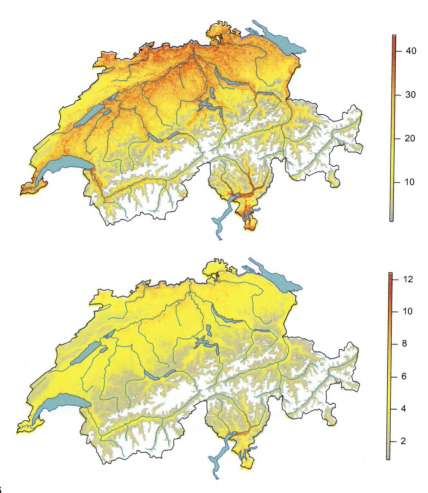

FIGURE 6.15

Species distribution maps for the great tit (*Parus major*; see Figure 4.1) in Switzerland in 2013 under the best *N*-mixture model (`fm5ZIP`) for data modeled at the 1-km² scale. Top: map of point estimates (expected number of territories (λ) per 1 km²); bottom: map of prediction uncertainty (SE).

```
plot(CH$elev, predCH[,1])           # Relationship with elevation
predCH[1][predCH[1] > 60] <- 60     # Censor freak predicions (not req'd here)

# Prepare Swiss coordinates and produce rudimentary
library(raster)

# Define a new dataframe with coordinates and outcome to be plotted
PARAM1 <- data.frame(x = CH$x, y = CH$y, z = predCH[,1])
```

```r
# Convert the DataFrame into a raster object
r1 <- rasterFromXYZ(PARAM1)

# Create mask for elevation (mask areas > 2250 m)
elev <- rasterFromXYZ(cbind(CH$x, CH$y,CH$elevation))
elev[elev > 2250] <- NA
r1 <- mask(r1, elev)

# Create custom color palette
mapPalette <- colorRampPalette(c("grey", "yellow", "orange", "red"))

# Map expected abundance of great tits in Switzerland in 2013
par(mfrow = c(1,2), mar = c(1,1,2,4))
plot(r1, col = mapPalette(100), axes = F, box = FALSE, main ="")
```

Although this is very rarely done in the species distribution modeling world, we should always plot the uncertainty associated with a species distribution model—e.g., as a map of the SEs or the width of 95% CIs of these predictions (see Webster et al., 2008; also see Figures 8.8 and 11.27). With a parametric statistical model, such an uncertainty assessment is easily obtained.

```r
# Prepare raster with prediction SEs
r2 <- rasterFromXYZ(data.frame(x = CH$x, y = CH$y, z = predCH[,2]))
elev <- rasterFromXYZ(cbind(CH$x, CH$y,CH$elevation))
elev[elev > 2250] <- NA
r2 <- mask(r2, elev)

# Map prediction SEs of expected abundance
plot(r2, col = mapPalette(100), axes = F, box = FALSE, main ="")
```

We could also produce a map of the predicted detection probability of great tits in Switzerland (Olea and Mateo-Tomas, 2011; Kéry et al., 2013). However, the only spatially indexed covariate in detection in our model is elevation, and hence a map of predictions of detection probability does not yield much more insight into spatial variation of p than does Figure 6.13. Hence, we omit a map of detection probability, but you could produce one as an exercise and use the following R code as a start.

```r
# Predictions for p with overdispersion
newData <- data.frame(elev = (CH$elev-elev.mean)/elev.sd, date=0, dur=0, time =
factor("2", levels = c("1", "2", "3")))
predCHp <- modavgPred(cand.set = list(fm5ZIP), newdata = newData, parm.type =
"detect", type = "response", c.hat = 2.47)
```

We move on to the third goal of our analysis: come up with a formal estimate of the national population size of great tits in Switzerland. Assuming independence of quadrats, this estimate is naturally obtained by adding up, for all 1-km^2 quadrats in Switzerland, the predictions under our model (see Section 6.10 for a discussion of this innocuous seeming assumption). We can do this easily (and exclude quadrats at >2250 m elevation—where we know there are no great tits).

```r
(N <- sum(predCH[-which(CH$elev > 2250),1]))     # National population size
[1] 655509.5
```

6.9 ABUNDANCE MAPPING OF SWISS GREAT TITS WITH unmarked

But wait, some quadrats lie mostly in lakes. So let us also exclude quadrats with >50% water.

```
out <- which(CH$water > 50 | CH$elev > 2250)
(N <- sum(predCH[-out,1]))      # 'Terrestrial' population size < 2250 m elevation
[1] 618840.3
```

Hence, we estimate about 0.62 million pairs of great tits in Switzerland in 2013, right in the middle of the estimate of 500,000–700,000 pairs put forward during the last Swiss breeding bird atlas project (Schmid et al., 1998). Our estimate is conditional on the explicit assumptions of our model, which also includes our specification of an infinite route length representing saturation sampling density within a quadrat of exactly 1 km^2. In Sections 6.9.5 and 6.10 we discuss this crucial assumption further, but for now want to obtain an estimate of the uncertainty in our Swiss population size estimate of great tits. We use the bootstrap for this. To speed up computation we do not use function predict, but do the required calculations "by hand."

```
# Remind ourselves of the relevant coefficients in the model
cbind(coef(fm5ZIP)[25])      # Zero-inflation model
cbind(coef(fm5ZIP)[1:8])     # Abundance model
```

We prepare covariates for the Swiss map of elevation and forest cover that are scaled identically to those in our analysis.

```
pelev <- (CH$elev - elev.mean)/elev.sd
pforest <- (CH$forest - forest.mean)/forest.sd
```

To run a bootstrap, we must define a function that produces a national population size estimate based on the output of our model. Function Nhat returns three such estimates: one estimate of the national population size at elevations below 2250 m a.s.l. when water bodies and potential nonsensical high estimates are ignored (N1), another when nonsensical high estimates are censored at a biologically realistic ceiling for great tits of 60 territories (N2), and a third when we also exclude quadrats with >50% water (N3). There is a second function argument that enables us to accommodate variation in effective sampling areas associated with differences in route lengths: the default is infinite route length, corresponding to saturation density. The third argument gives the effective sample area associated with a 1-km^2 quadrat (see below) and is set at 1 by default.

```
Nhat <- function(fm = fm5ZIP, iLength = 0, area = 1) {
   betavec <- coef(fm)[1:8]
   DM <- cbind(rep(1,length(pelev)), pelev, pelev^2, pforest,
      pforest^2, rep(iLength,length(pelev)), pelev*pforest, pelev*pforest^2)
   pred <- exp(DM %*% betavec) * (1-plogis(coef(fm)[25])) * (1/area)
   pred2 <- pred
   N1 <- sum(pred[-which(CH$elev > 2250),])   # Drop quads > 2250 m
   pred2[pred2 > 60] <- 60      # Censor freak-high estimates
   N2 <- sum(pred2[-which(CH$elev > 2250),])   # Estimate with freaks censored
   out <- which(CH$water > 50 | CH$elev > 2250)
   N3 <- sum(pred2[-out,])      # Estimate excluding water bodies
   return(c(N1 = N1, N2 = N2, N3 = N3))
}
```

We test the function for our actual data set and recover identical point estimates to those from predictSE, then we launch a bootstrap with 2500 replications.

```
(Nest <- Nhat(fm5ZIP))             # For default saturation density
(Nest <- Nhat(fm5ZIP, 1/10))       # For 10 km routes
      N1        N2        N3
 655509.5  655509.5  618840.3
      N1        N2        N3
 573235.2  573235.2  541168.4

# Launch the bootstrap (takes about 30h)
system.time(pb.N <- parboot(fm5ZIP, Nhat, nsim = 2500, report=5))
```

Once we are done, we focus on the bootstrapped values for N3, where any freak high estimates are censored at 60 and main Swiss water bodies and quadrats > 2250 m elevation are excluded, and summarize them, first by sample statistics and then in a plot (Figure 6.16).

```
bs <- pb.N@t.star[,3]             # Extract the bootstrapped vals of N3

# Sample statistics
summary(bs)
quantile(bs, prob = c(0.025, 0.975))   # Get 95% CI
> summary(bs)
   Min.  1st Qu.  Median    Mean  3rd Qu.    Max.
 412500   571300  617200  623300   668500  933100
```

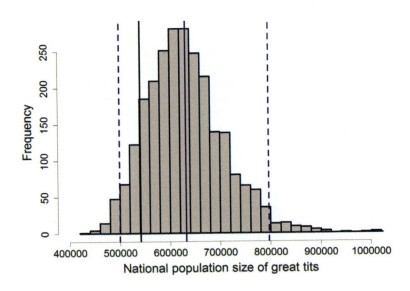

FIGURE 6.16

Bootstrapped sampling distribution of the number of pairs of great tits in Switzerland under model fm5ZIP. The black line shows the estimate for the actual data set and the solid blue line the mean of the 2500 bootstrap replicates. The dashed blue lines show the bootstrapped 95% CIs.

6.9 ABUNDANCE MAPPING OF SWISS GREAT TITS WITH unmarked

```
> quantile(bs, prob = c(0.025, 0.975))    # Get 95% CI
    2.5%    97.5%
 489623.2 784803.2

# Plot (produces Fig. 6.16)
hist(bs, breaks = 100, col = "grey", xlab = "National population size of great tits",
 cex = 1.5, main = "", xlim = c(400000, 1000000))
abline(v = Nest[3], lwd = 3)
abline(v = mean(bs), lwd = 3, col = "blue")
abline(v = quantile(bs, prob = c(0.025, 0.975)), lwd = 3, col = "blue", lty = 2)
```

This bootstrap does not account for OD (c-hat = 2.47). Therefore, we next attempt to propagate this additional uncertainty into the national population total by multiplying the bootstrap SD with the square-root of c-hat and computing approximate 95% CIs by adding/subtracting twice that product from the mean. As a check for how reasonable this approach might be, we first use such a normal approximation to compute 95% CIs without OD, so that we can compare them with the percentile-based CIs: we see that they are pretty similar.

```
(CI1 <- quantile(bs, prob = c(0.025, 0.975)))         # Percentile-based CI
(CI2 <- c(mean(bs-2*sd(bs)), mean(bs+2*sd(bs))))      # Normal approx.-based CI

    2.5%    97.5%
 489623.2 784803.2    # Percentile-based CI
 473448.8 773055.6    # Normal approximation-based CI
```

Presumably, this is a reasonable approximation when the bootstrap distribution is approximately normal. We now do something analogous with an OD correction to make this interval wider, to account for the extra-uncertainty represented by OD.

```
c.hat <- gof.ZIP$c.hat.est
(CI <- c(mean(bs-2*sqrt(c.hat)*sd(bs)), mean(bs+2*sqrt(c.hat)*sd(bs))))
[1] 387673.2 858831.2
```

Hence, based on our *ad hoc* OD-correction of the CIs, we conclude that were 388,000–859,000 pairs of great tits in Switzerland in 2013.

6.9.5 CONCLUSIONS ON THE ANALYSIS WITH unmarked

In this section, we have seen a couple of important things:

- We can use unmarked to flexibly fit a large range of *N*-mixture models for spatially and temporally replicated count data.
- We can use AIC or LRTs to tell us which model has the most support from the data; though we should never trust such numbers blindly (see comments below on the NB model).
- We can use parametric bootstrapping to compute measures of GoF, uncertainty in derived quantities such as predictions, or the total population size of Swiss great tits.
- We have looked into the calculation of predicted values of abundance or detection under the model. This is an extremely important topic, because the resulting plots are an informative way of presenting the results from such an analysis. Also, when your model contains

- polynomials or interactions of covariates, we cannot usually understand what the model is telling us from staring at the parameter estimates alone. Rather, the only way to understand what the model is telling us is usually by producing a picture of the response in relation to the covariate values.
- As always, whenever we have spatially indexed covariates in an analysis, we can compute the expected value of a response such as abundance for each pixel in some region and then project these predictions, producing a map (Kéry et al., 2013). Hence, the N-mixture model is naturally also a *species distribution model*, primarily for abundance. However, we could also compute the probability of occurrence (as 1 minus the probability that $N > 0$ under the abundance model) and plot that; see Royle et al. (2005). Since there is a logical equivalence of abundance and occurrence, we can use the N-mixture model for abundance to produce a traditional map of species occurrence probability (Dorazio, 2007).
- Estimates of national population sizes are naturally obtained by predicting abundance for each pixel in the area of interest and then summing them up. Of course, if you are interested in smaller-scale totals—e.g., for certain regions or for different habitat types or other categorizations of space—then you proceed in exactly the same manner: you simply add up the predictions over the pixels in each group to obtain the total population size in each. You will see an example of this in Section 6.11, where we estimate the population size of great tits per elevation band of 100 m. Parametric bootstrapping is a flexible frequentist method of obtaining uncertainty intervals for such derived quantities.
- We have seen that the Poisson mixture models did not formally fit the data based on the bootstrap GoF. This means that the observed responses were more variable than expected under the Poisson model. Allowing for extra-Poisson dispersion in the abundance part of the model by use of an NB distribution bought us a better fitting model that won the AIC contest by a very large margin, but also predicted much too high abundances, leading to the "good fit/bad prediction dilemma" fairly often observed in this model (Kéry et al., 2005b; Joseph et al., 2009). We suggest caution when using the NB distribution for abundance in binomial N-mixture models and think that using a Poisson or ZIP model with moderate lack of fit (e.g., values of c-hat less than about 4) followed by an "OD correction" may be a safer choice for inferences about abundance. At any rate, the NB seems to always predict greater or equal abundance, compared with Poisson or ZIP binomial N-mixture models (though perhaps not for multinomial mixture models; see Chapter 7). Hence, if you erroneously choose one of the latter over an NB model, you will err on the side of caution when predicting population sizes (e.g., when modeling a species of conservation concern).

This is only a relatively rough sketch of a thorough spatial analysis of the abundance of a species. For instance, in your study you may have many more covariates. Then, the model selection part may become heavier or you may want to model-average your predictions to propagate model selection uncertainty into your abundance maps and into estimates of regional or national totals. Or you may want to adopt more elaborate models—e.g., to model spatial autocorrelation (see Chapter 21) or "wiggly" covariate relationships using splines (see Section 10.14). These and many other extensions are conceptually fairly straightforward, but in some cases at least, they may require you to code up your model in BUGS.

6.10 THE ISSUE OF SPACE, OR: WHAT IS YOUR EFFECTIVE SAMPLE AREA?

Any calculation that adds site-specific estimates of local density up to a population total makes the implicit assumption that we know the precise area of each site. In the Swiss MHB a site is 1 km^2—or is it really? Is this a sensible assumption? There are two potential issues: first, "holes" of unsurveyed areas inside a quadrat and second, "edge effects" represented by home ranges that sit on the quadrat boundary. Look for instance at Figure 6.8: just from the location of the detected great tit territories you get a pretty good idea of where the transect route lies. Presumably, then, there were other territories away from the survey route that had a zero or at least a very low probability of being detected. This leads to "holes" in the quadrat coverage and will *reduce* the effective sample area relative to the nominal 1-km^2 area. In the preceding analysis, we have taken account of this effect by including into our model the relationship between the exposed population and route length.

On the other hand, clearly the sample quadrat is not an island, but simply an arbitrary square cut out from a more or less continuous landscape. Hence, it would be naive to believe that all home ranges lie either exclusively inside or exclusively outside of a quadrat; rather, some territories are likely to straddle the quadrat boundary. You can actually see one such home range near the top-right corner of the quadrat (Figure 6.8, right). The presence of such edge territories will tend to make the effective sample area bigger than the 1 km^2 associated with the nominal MHB sample area. The shape of the effectively sampled area is more likely to resemble that of a sunflower, with petals representing edge home ranges. Both the presence and the magnitude of "holes" and of "sunflower effects" depend on the home range size of a species. Species with a small home range, such as small passerines, will range much less beyond the quadrat edge and will more likely have holes inside the quadrat than do species with a large home range such as crows or raptors, for which the sunflower effect will be more severe, but there will be fewer and smaller holes. See Figure 6.17 for a schematic representation of these ideas.

As a consequence, strictly speaking we very rarely know the effective size and shape of a sampled area in continuous habitat, regardless of how authoritative we feel when we mark its boundaries on a map. And worse, both the size and the shape of the effective sampling area associated with a nominal study area will differ by species, habitat, and a multitude of potential factors that may affect the movement behavior of an organism within its home range and especially the home range size. These problems have long been known in closed population estimation (Dice, 1938; Efford and Dawson, 2012) and their knowledge has given rise to the development of the novel class of spatial capture-recapture (SCR) models (Efford, 2004; Borchers and Efford, 2008; Royle and Young, 2008; Royle et al., 2014). Only distance sampling (Chapters 8, 9, 14 and 24) and SCR models (Chapter 24) are immune to the general problems induced by the impossibility of defining the limits of a study area unequivocally. However, these methods usually require data with extra information that is more "expensive" to collect (though this cost is arguably lower for distance sampling).

In the absence of such extra information in the MHB great tit data set, we must either make assumptions about the effective sample area (for instance, in our case, that it is exactly 1 km^2) or, alternatively, accept that the effective sample area is smaller than 1 km^2 (because the area of 'holes' probably exceeds that of the 'sunflower petals'), and that our totals will be conservative; i.e., they will underestimate the true N. A third possibility is to try and accommodate space in a model, as we just showed when we adjusted for "holes," representing incomplete sample coverage, in our great tit

FIGURE 6.17

Schematic of "holes" (top row) and "sunflower effects" (bottom row) in biological transect surveys in continuous habitat as a function of the size of a territory or activity range (left: small territories, right: large territories). The survey route is shown in red (see Figure 6.8 for more detail of this sample quadrat). For species (or study sites) where territory size is small, there are many "holes" (represented by 15 yellow territories that are unavailable for detection; top left). For large territories, the identical transect route will be associated with fewer "holes" (only one yellow territory unavailable for detection; top right). For species with small territories (bottom left), the presence of edge territories (marked yellow) will not increase the effective sample area so much as for species with large territories (bottom right).

analysis. Modeling sampling coverage as a function of a measured covariate such as route length is a useful way to correct for incomplete quadrat coverage (Royle and Dorazio, 2006; Royle et al., 2007b). On the other hand, extrapolating density to saturation coverage in the MHB almost certainly makes the effective sampling area greater than the nominal 1 km². This may not be the case in the absence of adjustment for coverage bias, since the holes inside of the nominal sampling area may compensate for any extra sampled area outside—i.e., for the sunflower "petals." In our analysis, if the effective sample area in the Swiss MHB resembles a sunflower with greater than 1 km² area, then regional or national

6.10 THE ISSUE OF SPACE, OR: WHAT IS YOUR EFFECTIVE SAMPLE AREA?

population totals obtained by adding up of estimates from adjacent quadrats represent an overestimate. Next, we try a somewhat *ad hoc* method to correct for this.

Great tit breeding home ranges in Switzerland vary from 0.3–0.5 ha in the insect-richest habitats (lowland deciduous forests), to 2–5 ha in mixed forests at medium elevations (600–800 m), and to 5–10 ha at high elevations around 1800 m (B. Naef-Daenzer, pers. comm.). Assuming circular home ranges, this corresponds to home range radii of between 31 and 178 m. Can we translate this knowledge into an "effective sample area" that adjusts for the sunflower effect? Actually, we can follow the same reasoning that folks in closed population estimation did before the advent of SCR: they added a boundary strip to the nominal sampling area of a width equal to half the diameter of an average home range (e.g., Dice, 1938; Karanth and Nichols, 1998; Karanth et al., 2006; Trolle and Kéry, 2003). For instance, for a species with a 100-m home range radius, a nominal sample area of 1 km^2 would then become a square with a side of length 1.2 km and an area of 1.44 km^2. It is instructive to plot the relationship between the average home range size of a species and the effective sample area in the MHB. Here we do the calculations for a range of home range sizes between 0.1 ha (perhaps representing *Regulus* spec., the smallest European bird species) and about 100 km^2 (representing a moderate to large raptor; Figure 6.18, left).

```
AHR <- seq(0.001, 100, 0.1)          # Area home range (in km2): 0.1-100
RHR <- sqrt(AHR/pi)                  # Translate to home range radius (in km)
ESA <- (1 + 2*RHR)^2                 # Eff. sample area for 1km2 nominal area
par(mfrow = c(1,2), mar = c(5,5,3,2), cex.lab = 1.3, cex.axis = 1.3)
plot(AHR, ESA, xlab = "Home range size (km2)", ylab = "Effective sample area (km2)",
   type = "l", lwd = 3, frame = F)
abline(h = 1, col = "red", lwd = 3)  # Nominal sample area
abline(h = 0, col = "grey", lwd = 3)
```

Using this reasoning for our great tits, home range radii of 31–178 m will translate to effective sample areas of an MHB quadrat for the great tit of 1.062^2–1.356^2 or about 1.13–1.84 km^2. Hence, the quadrat population sizes associated with our saturation sampling effort can be assumed to refer to this area. Continuing this *ad hoc* argument, the expected density per 1 km^2 nominal sample area will correspond to a proportion between 0.54 and 0.88 of that and our quadrat-specific predictions would have to be reduced by 12–46% before they could be added up to provide a Swiss national population total. We can make that adjustment with our `Nhat` function using the argument `area`, which denotes the effective sample area.

Actually, multiplying our original estimate of 618,840 Swiss tits with `1/area` yields the same estimate, but the function argument can be useful if we want to bootstrap an uncertainty interval around it. We compute and plot the estimated national population size for a range of average home range sizes between 0.3 and 10 ha, corresponding to 1.13–1.84 km^2. We see a considerable reduction in our estimate of the Swiss national population size of great tits when we do this adjustment for the effective sampling area being greater than 1 km^2 (Figure 6.18, right). Hence, our estimate is quite sensitive to our assumptions about the size of the sampled area.

```
GT.HR <- seq(0.003, 0.1,, 1000)      # Great tit home ranges in km2
GT.rad <- sqrt(GT.HR/pi)             # Great tit home range radius (in km)
ESA.GT <- (1 + 2*GT.rad)^2           # Effective sample area for Great tit
NTOT <- numeric(length(ESA.GT))      # Adjusted national total population
```

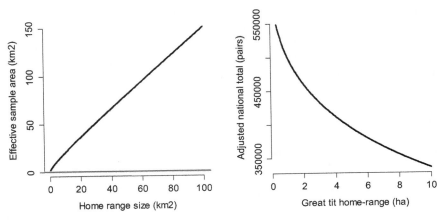

FIGURE 6.18

Left: Relationship between the average size of a species' home range (assumed to be circular) and the effective sample area of a 1-km^2 quadrat obtained by buffering the nominal area with a strip of width equal to half the home range diameter. Gray and red lines show value of 0 and of the nominal sample area of 1 km^2. Right: Relationship between the adjusted Swiss national population size of great tits in 2013 and the average home range size of great tits.

```
for(i in 1:length(NTOT)){
   NTOT[i] <- Nhat(fm5ZIP, 0, ESA.GT[i])[3]
}
plot(100*GT.HR, NTOT, xlab = "Great tit home-range (ha)", ylab = "Adjusted national
total (pairs)", type = "l", lwd = 3, frame = F)
```

Our inability to unambiguously define the effective sample area and its consequences are *not* a defect of our explicit estimation approach. Rather, any nonspatial method of inference about regional or national totals would suffer similarly from this challenge—e.g., even a simple method that might multiply the average number of tit territories detected per quadrat, 8.4, by the number of quadrats summed over. Only methods that explicitly deal with space, such as distance sampling and spatial capture-recapture, are immune to this problem. Finally, note that Chandler and Royle (2013) have developed a spatial *N*-mixture model that estimates the parameters of an underlying spatial point pattern, representing home range or activity centers, from the spatial correlation of replicated counts in adjacent sample areas. To fit this model you need a fairly large number of sampling sites that touch each other, and hence the model would not directly be applicable to spatial grid samples as in the Swiss MHB. It might be possible, however, to subdivide, say, each 1 km^2 main quadrat into 25 sub-quadrats with 200 m length. Combining such data for a large number of main quadrats may provide enough information to obtain parameter estimates with reasonable precision Ramsey et al. (2015) have extended that model to "presence/absence" data.

6.11 BAYESIAN MODELING OF SWISS GREAT TITS WITH BUGS

In Section 6.11.1, we start by fitting our great tit model of choice in BUGS to compare the estimates with those from the analysis in unmarked. After that, we illustrate the accommodation of OD with additional random effects in the linear predictors of abundance or detection in BUGS. You may use this analysis as a template for models with more complex random effects in BUGS generally. The coding of extra-Poisson and extra-binomial dispersion may be important in allowing for additional sources of heterogeneity and achieving a model that passes a formal GoF test. In addition, you may be interested in decomposing additional variability in expected abundance (λ) and detection probability (p). Finally, we also show how you can code "switches" in BUGS that turn some model components on or off. Finally, we give an example of the covariate modeling of a heterogeneous variance term.

6.11.1 BAYESIAN FITTING OF THE BASIC ZERO-INFLATED POISSON (ZIP) N-MIXTURE MODEL

We start with the Bayesian implementation of model `fm5ZIP` from Section 6.9. We will provide another example of how easy it is to compute derived quantities in an MCMC-based analysis by estimating the proportion of the Swiss great tit population that lives in each 100-m elevation band between 200 and 2250 m. Elevation gradients of occurrence or abundance have always fascinated ecologists and are topical in climate change research, when fairly big changes can be expected along this dimension of species distributions. We will compute the derived quantities in R, outside of BUGS, using MCMC samples of all involved quantities, because sometimes it is easier to compute posterior samples of derived quantities after the analysis in BUGS.

```
# Prepare data for BUGS data bundle
elev <- umf@siteCovs$elev    ; elev2 <- elev^2
forest <- umf@siteCovs$forest ; forest2 <- forest^2
date <- matrix(umf@obsCovs$date, ncol = 3, byrow = TRUE)
dur <- matrix(umf@obsCovs$dur, ncol = 3, byrow = TRUE)
date[is.na(date)] <- 0   ; date2 <- date^2
dur[is.na(dur)] <- 0     ; dur2 <- dur^2
iRoute <- umf@siteCovs$iLength

# Design matrix for abundance model (no intercept)
lamDM <- model.matrix(~ elev + elev2 + forest + forest2 + elev:forest +
elev:forest2 + iRoute)[,-1]
```

In the BUGS model, we add a couple of special features: (1) we add a posterior predictive GoF check, (2) we add "switches" to turn on/off specific components of the model, and (3) we simplify the fitting of the linear model for abundance by using a design matrix (see Section 10.6 for more on this useful trick). "Switches" in BUGS programming can be quite useful because they allow you to fit a much larger variety of statistical models using a single BUGS model. We build three switches, `hlam.on`, `hp.site.on`, and `hp.survey.on` (note that h stands for 'heterogeneity'), to add to the model normally distributed random effects in abundance (site-specific) and in detection (site-specific or site-by-occasion specific). We discuss these switches in the next section, where we actually turn some of them on; for now, they are all turned off—i.e., set to zero.

```
# Specify model in BUGS language
sink("ZIPNmix.txt")
cat("
model {
```

```
# Specify priors
# zero-inflation/suitability
phi ~ dunif(0,1)              # proportion of suitable sites
theta <- 1-phi                # zero-inflation (proportion unsuitable)
ltheta <- logit(theta)

# abundance
beta0 ~ dnorm(0, 0.1)         # log(lambda) intercept
for(k in 1:7){                # Regression params in lambda
   beta[k] ~ dnorm(0, 1)
}
tau.lam <- pow(sd.lam, -2)
sd.lam ~ dunif(0, 2)          # site heterogeneity in lambda

# detection
for(j in 1:3){
   alpha0[j] <- logit(mean.p[j])
   mean.p[j] ~ dunif(0, 1)    # p intercept for occasions 1-3
for(k in 1:13){               # Regression params in p
   alpha[k] ~ dnorm(0, 1)
}
tau.p.site <- pow(sd.p.site, -2)
sd.p.site ~ dunif(0, 2)       # site heterogeneity in p
tau.p.survey <- pow(sd.p.survey, -2)
sd.p.survey ~ dunif(0, 2)     # site-survey heterogeneity in p

# ZIP model for abundance
for (i in 1:nsite){
   a[i] ~ dbern(phi)
   eps.lam[i] ~ dnorm(0, tau.lam)    # Random site effects in log(abundance)
   loglam[i] <- beta0 + inprod(beta[], lamDM[i,]) + eps.lam[i] * hlam.on
   loglam.lim[i] <- min(250, max(-250, loglam[i]))   # 'Stabilize' log
   lam[i] <- exp(loglam.lim[i])
   mu.poisson[i] <- a[i] * lam[i]
   N[i] ~ dpois(mu.poisson[i])
}

# Measurement error model
for (i in 1:nsite){
  eps.p.site[i] ~ dnorm(0, tau.p.site) # Random site effects in logit(p)
  for (j in 1:nrep){
    y[i,j] ~ dbin(p[i,j], N[i])
    p[i,j] <- 1 / (1 + exp(-lp.lim[i,j]))
    lp.lim[i,j] <- min(250, max(-250, lp[i,j]))      # 'Stabilize' logit
    lp[i,j] <- alpha0[j] + alpha[1] * elev[i] + alpha[2] * elev2[i] +
       alpha[3] * date[i,j] + alpha[4] * date2[i,j] +
       alpha[5] * dur[i,j] + alpha[6] * dur2[i,j] +
       alpha[7] * elev[i] * date[i,j] + alpha[8] * elev2[i] * date[i,j] +
       alpha[9] * elev[i] * dur[i,j] + alpha[10] * elev[i] * dur2[i,j] +
```

6.11 BAYESIAN MODELING OF SWISS GREAT TITS WITH BUGS

```
          alpha[11] * elev2[i] * dur[i,j] + alpha[12] * date[i,j] * dur[i,j] +
          alpha[13] * date[i,j] * dur2[i,j] +
          eps.p.site[i] * hp.site.on + eps.p.survey[i,j] * hp.survey.on
          eps.p.survey[i,j] ~ dnorm(0, tau.p.survey) # Random site-survey effects
      }
}
# Posterior predictive distributions of Chi2 discrepancy
for (i in 1:nsite) {
   for (j in 1:nrep) {
      y.sim[i,j] ~ dbin(p[i,j], N[i])    # Create new data set under model
      e.count[i,j] <- N[i] * p[i,j]      # Expected datum
      # Chi-square discrepancy for the actual data
         chi2.actual[i,j] <- pow((y[i,j]-e.count[i,j]),2) / (e.count[i,j]+e)
      # Chi-square discrepancy for the simulated ('perfect') data
         chi2.sim[i,j] <- pow((y.sim[i,j]-e.count[i,j]),2) / (e.count[i,j]+e)
      # Add small value e to denominator to avoid division by zero
   }
}
# Add up individual Chi2 values for overall fit statistic
fit.actual <- sum(chi2.actual[,])    # Fit statistic for actual data set
fit.sim <- sum(chi2.sim[,])          # Fit statistic for a fitting model
bpv <- step(fit.sim-fit.actual)      # Bayesian p-value
c.hat <- fit.actual/fit.sim          # c-hat estimate

# Derived parameters: Total abundance at 263 sampled sites
Ntotal263 <- sum(N[])
}
",fill = TRUE)
sink()
```

Here are the remaining ingredients for the analysis.

```
# Initial values
Nst <- apply(y, 1, max, na.rm = T) + 1
Nst[is.na(Nst)] <- round(mean(y, na.rm = TRUE))
Nst[Nst == "-Inf"] <- round(mean(y, na.rm = TRUE))
inits <- function(){ list(N = Nst, beta0 = 0, mean.p = rep(0.5,3), beta = runif(7, 0,0), alpha
= runif(13, 0,0))}

# Parameters monitored
params <- c("theta", "ltheta", "phi", "beta0", "beta", "sd.lam", "alpha0", "mean.p",
"alpha", "sd.p.site", "sd.p.survey", "fit.actual", "fit.sim", "bpv", "c.hat",
"Ntotal263")
```

It is quite difficult to find initial values that satisfy JAGS' high standards, so we fit the model in good old WinBUGS. The final three elements in the data bundle select the type of ZIP model that we are fitting by turning on or off additional random effects. We start with exactly the ZIP N-mixture model that we fit with unmarked in the last Section 6.9—i.e., the base model without any OD in either λ or p—and call this model 1. It has all switches set to "off" ($= 0$).

```
# Bundle data and choose to fit simple ZIP model (model 1)
win.data1 <- list(y = y, nsite = nrow(y), nrep = ncol(y),
    lamDM = lamDM, elev = elev, date = date, dur = dur, elev2 = elev2,
    date2 = date2, dur2 = dur2, e = 1e-06, hlam.on = 0, hp.site.on = 0,
    hp.survey.on = 0)

# MCMC settings
ni <- 50000  ;  nt <- 4  ;  nb <- 10000  ;  nc <- 3

# Call WinBUGS from R (ART 93 min) and summarize posteriors
out1 <- bugs(win.data1, inits, params, "ZIPNmix.txt", n.chains = nc, n.thin = nt,
n.iter = ni, n.burnin = nb, debug = TRUE, bugs.directory = bugs.dir,
working.directory = getwd())
print(out1, dig = 3)
```

We compare the MLEs from unmarked with the Bayesian posterior inference in a table and in a graph (Figure 6.19). Essentially, with a large enough data set we always find that the posterior means from the Bayesian analysis with vague priors numerically agree very well with the MLEs. Perhaps this may allay the fears of some about the perceived dangers of Bayesian as opposed to non-Bayesian statistical analyses (e.g., Dennis, 1996; Lele and Dennis, 2009; Lele, 2015)?

```
# Compare MLEs and Bayesian posterior means (order first): table and graph
tmp <- summary(fm5ZIP)
ord.MLE <- rbind(tmp$psi[,1:2], tmp$state[,1:2], tmp$det[c(7:9, 1:6, 10:16),1:2])
ord.Bayes <- out1$summary[-c(1,3,12,16:18,32:39), 1:2]
cbind(ord.MLE, ord.Bayes)
par(mar = c(5,5,3,2), cex.lab = 1.5, cex.axis = 1.5)
```

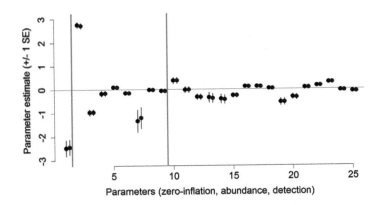

FIGURE 6.19

Comparison of MLEs (black; +/− 1 SE) and Bayesian posterior means (blue; +/− 1 posterior SD) of the parameters of the ZIP N-mixture model from Section 6.9 (model fm5ZIP). Gray vertical lines separate the parameters for the zero-inflation, abundance, and detection parts of the model (in this order).

6.11 BAYESIAN MODELING OF SWISS GREAT TITS WITH BUGS

```
plot(ord.MLE[,1], ylim = c(-3,3), pch = 16, col = "black", main = "", frame = F,
xlab = "Parameters (zero-inflation, abundance, detection)", ylab =
"Parameter estimate (+/- 1 SE)", cex = 1.5)
segments(1:25, ord.MLE[,1]-ord.MLE[,2], 1:25, ord.MLE[,1]+ord.MLE[,2], lwd = 2)
abline(h = 0)
abline(v = c(1.5, 9.5), col = "grey")
points((1:25)+0.3, ord.Bayes[,1], pch = 16, col = "blue", cex = 1.5)
segments(1:25+0.3, ord.Bayes[,1]-ord.Bayes[,2], 1:25+0.3, ord.Bayes[,1]+ord.
Bayes[,2], col = "blue", lwd = 2)
```

Next, we want to make predictions—i.e., plot distribution maps and elevation gradients of population size—under the Bayesian model fit, and also compute a national population size estimate to compare with the non-Bayesian estimate in Section 6.9. We could do the necessary computations for these derived quantities in BUGS directly, but this would be terribly slow. Here is an example of where it is more efficient to take the posterior MCMC samples from some parameters and then compute the derived quantities in R. To obtain the posteriors of the expected abundance (λ) that incorporate the zero-inflation part of the model, we have to do the following computation for every MCMC draw: (1-theta) * exp(beta0 + beta1 * covariate 1 + ...). We recycle some code from Section 6.9.4.

```
library(unmarked)
data(Switzerland)          # Load Swiss landscape data in unmarked
CH <- Switzerland
```

We collect the covariates for the predictions into a design matrix and standardize all covariates using the same mean and sd that we used for the covariates in the analyzed data set.

```
ELEV <- (CH$elev-elev.mean)/elev.sd
ELEV2 <- ELEV^2
FOREST <- (CH$forest-forest.mean)/forest.sd
FOREST2 <- FOREST^2
CHdata <- cbind(elev = ELEV, elev2 = ELEV2, forest = FOREST, forest2 = FOREST^2, iRoute =
rep(0, length(CH$elev)), elev.forest = ELEV * FOREST, elev.forest2 = ELEV * FOREST2)
str(CHdata)                # This is a design matrix
```

Next, for every step in the MCMC output we compute the expected abundance (accounting for zero inflation). Before we do this, we define an array to hold the results and we subsample the MCMC output to prevent a weakling computer from crashing. We write the code such that we can input BUGS model fits from other models below, by substituting out1 with out7, for instance.

```
MCMCout <- out1            # Choose results output from model 1
(nsamp <- length(MCMCout$sims.list$theta))    # how many MCMC samples do we have ?

# Subsample Markov chains
sub.sample.size <- 3000    # choose sample of 3000
selection <- sort(sample(1:nsamp, sub.sample.size))

# Array to hold predictions for every Swiss 1km2 quadrat
lamPred <- array(NA, dim =c(length(CH[,1]), sub.sample.size))
```

```
# Fill the array
for(i in 1:sub.sample.size){
   MCMCstep <- selection[i]
   lamPred[,i] <- (1-MCMCout$sims.list$theta[MCMCstep]) *
   exp(MCMCout$sims.list$beta0[MCMCstep] +
      CHdata %*% MCMCout$sims.list$beta[MCMCstep,1:7])
}
```

What have we done now? We have computed 3000 samples from the posterior distributions of the number of great tit territories in every single one of 42,275 1-km^2 quadrats in Switzerland. We can summarize these posterior distributions in suitable ways. For instance, we can map the posterior mean in every quadrat to obtain a map of abundance, exactly as we did in the last section (again using some of that code for Figure 6.15). We check for predictions of more than 100 pairs of great tits per 1 km^2 to censor them at 100, but we find that there are not any.

```
# Get posterior means for every quadrat and check if sensible
meanlam <- apply(lamPred, 1, mean)    # Get posterior mean
max(meanlam)                           # Check maximum
sum(meanlam > 100)                     # Are any predictions >100 ?
plot(CH$elev, meanlam, ylim = c(0, max(meanlam))) ; abline(v=2250, col="red")
plot(CH$elev, meanlam, ylim = c(0,100)) ; abline(v = 2250, col = "red", lwd = 2)
meanlam[meanlam > 100] <- 100          # Censor all predictions
lamPred[lamPred > 100] <- 100

# Produce map of posterior mean of lambda
library(raster)
r <- rasterFromXYZ(data.frame(x = CH$x, y = CH$y, z = meanlam))
elevation <- rasterFromXYZ(cbind(CH$x, CH$y, CH$elevation))
elevation[elevation > 2250] <- NA
r <- mask(r, elevation)
mapPalette <- colorRampPalette(c("grey", "yellow", "orange", "red"))
par(mfrow = c(1,2), mar = c(5,5,1,5))
plot(r, col = mapPalette(100), axes = F, box = F, main = "")
```

Instead of summarizing the abundance distribution in two dimensions, we can summarize it in the third dimension—i.e., along the elevation gradient. We do this here and produce a histogram of the estimated number of great tit territories per 100-m elevation band in Switzerland.

```
elev.class <- 100*(CH$elev %/% 100 + 1)   # Elevation class of each km2
tmp <- aggregate(lamPred, by = list(elev.class), FUN = sum)
N.elev <- as.matrix(tmp[,-1])              # Posterior sample of Ntotal per band
band <- tmp[,1]                            # Elevation band (in m)
meanN <- apply(N.elev, 1, mean)
barplot(meanN, col = "grey", horiz = T, xlab = "Number of Great tit territories",
   ylab = "Elevation band (100m)", xlim = c(0, 200000))
axis(2, at = 1:length(band), labels = band)
```

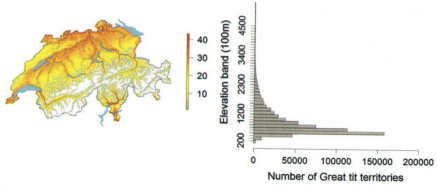

FIGURE 6.20

Summaries of the Bayesian posterior distributions of the expected abundance (λ, including zero-inflation) of great tits in Switzerland in 2013 in two dimensions, along the coordinate gradients (left), and in the third dimension along the elevation gradient (right, in number of pairs). Resolution of modeled data is 1 km². You can compare this map with its frequentist version in Figure 6.15.

Not surprisingly, Figure 6.20 (left) is essentially the same map as with the ML fit of the same model (Figure 6.15, left). It would be trivial to also produce a map of the posterior SD in the Bayesian analysis. In the elevation histogram (Figure 6.20, right), we see that although great tits live at highest densities in the lowlands, their largest numbers overall occur at medium elevations, because the area of these elevation bands in Switzerland is much greater than that of the lowest elevation bands.

Finally, we produce a national population size estimate again. To ensure comparability with the analyses in Section 6.9, we again exclude quadrats with more than 50% water and those at heights >2250 m a.s.l.

```
# Posterior distribution of total number of great tit territories in 2013
keep <- which((CH$water < 50) & (CH$elev < 2251))
Ntot <- apply(lamPred[keep,], 2, sum)
hist(Ntot, breaks = 100, col = "grey", main = "Posterior of national population size")

# Point estimate and 95% CRI
mean(Ntot)
quantile(Ntot, prob = c(0.025, 0.975))

> mean(Ntot)
[1]   601510.4
> quantile(Ntot, prob = c(0.025, 0.975))
    2.5%      97.5%
494574.8   729662.7
```

These estimates are pretty similar to those obtained in the non-Bayesian analysis in Section 6.9.

6.11.2 ADDING RANDOM EFFECTS IN BUGS

One of the great assets of Bayesian inference using MCMC is the ease with which you can add random effects to a model. There is a large number of reasons for why you may want to do that (Section 3.4), including the wish to accommodate overdispersion (OD), to investigate sources of variability (i.e., to partition variability in abundance or detection not explained by the covariates), or to accommodate grouping structures that induce dependencies in the data. We give two examples here. In the first, we allow for OD at multiple scales in our original model. In the second, we look at structure in the OD in abundance by fitting a linear model to a *variance* term of the model. This example underlines the frequently ignored fact that you may not only do covariate modeling for the mean structure of a response, but you may also explain patterns in the variance by adopting a linear or other covariate model for a variance term.

6.11.2.1 Accounting for Overdispersion (OD) at Multiple Scales

We start our illustration of random effects in BUGS by adding up to three sets of random effects into our N-mixture model 1 (corresponding to `fm5ZIP`): we add site random effects in abundance, and site- or site-by-survey random effects in detection. We wrote the BUGS code such that there are switches to turn on or off these additional random effects. These switches are set in the data bundle: when a switch is set to "on" ($= 1$), these sources of additional variability in λ and p are estimated from the data, whereas when it is "off" ($= 0$), the associated random effects are zeroed out and not part of the fitted model—i.e., not "connected to the data"—and therefore are not influencing the parameters that are connected to the data. Note that you will not get a zero estimate for the heterogeneity terms that are switched off, but will have to remember which nonzero estimate of the additional random effects is a data-free estimate based solely on the prior.

We will fit six variants of the basic model 1 that all accommodate some form of extra-Poisson or extrabinomial dispersion. With additional random effects, we typically need to run the chains for (much) longer to achieve convergence and it is not a typical to fail to obtain convergence at all. The following, "heavy" MCMC settings were sufficient to achieve convergence (Rhat < 1.1) for the main structural parameters of the model in all except three instances (when Rhat was up to 1.2). This may be felt to be worrisome and some may wonder whether these models are estimable at all with our MHB sample size of 267 sites and three replicates. However, our own unpublished simulations and results reported in Tanadini (2010) suggest estimability of these models even with two replicates. You could fairly easily run your own simulation study: simply use function `simNmix` repeatedly to simulate data with certain random effects and then see whether the distribution of the estimates obtained with BUGS is centered on the data-generating value.

```
# MCMC settings
ni <- 10^6  ;  nt <- 80  ;  nb <- 200000  ;  nc <- 3
```

As our second model, we fit a binomial mixture model with a zero-inflated Poisson-lognormal (PLN) mixture for abundance; i.e., we allow for random site effects in the abundance model and call this model 2.

```
# Bundle data and select model 2
win.data2 <- list(y = y, nsite = nrow(y), nrep = ncol(y), lamDM = lamDM, elev = elev,
    date = date, dur = dur, elev2 = elev2, date2 = date2, dur2 = dur2, e = 1e-06, hlam.on = 1,
    hp.site.on = 0, hp.survey.on = 0)
```

```
# Call WinBUGS from R (ART 4050 min) and summarize posteriors
out2 <- bugs(win.data2, inits, params, "ZIPNmix.txt", n.chains = nc, n.thin = nt,
   n.iter = ni, n.burnin = nb, debug = F, bugs.directory = bugs.dir)
print(out2, dig = 3)
```

Third, we fit a binomial mixture model with a ZIP mixture for abundance and random normal site effects in detection and call this model 3.

```
# Bundle data and select model 3
win.data3 <- list(y = y, nsite = nrow(y), nrep = ncol(y), lamDM = lamDM, elev = elev,
   date = date, dur = dur, elev2 = elev2, date2 = date2, dur2 = dur2, e = 1e-06, hlam.on = 0,
   hp.site.on = 1, hp.survey.on = 0)

# Call WinBUGS from R (ART 4200 min) and summarize posteriors
out3 <- bugs(win.data3, inits, params, "ZIPNmix.txt", n.chains = nc, n.thin = nt,
   n.iter = ni, n.burnin = nb, debug = F, bugs.directory = bugs.dir)
print(out3, dig = 3)
```

Fourth, the model with random normal site-by-survey effects in detection (this is model 4).

```
# Bundle data and select model 4
win.data4 <- list(y = y, nsite = nrow(y), nrep = ncol(y), lamDM = lamDM, elev = elev,
   date = date, dur = dur, elev2 = elev2, date2 = date2, dur2 = dur2, e = 1e-06,
   hlam.on = 0, hp.site.on = 0, hp.survey.on = 1)

# Call WinBUGS from R (ART 4020 min) and summarize posteriors
out4 <- bugs(win.data4, inits, params, "ZIPNmix.txt", n.chains = nc, n.thin = nt,
   n.iter = ni, n.burnin = nb, debug = F, bugs.directory = bugs.dir)
print(out4, dig = 3)
```

Fifth, a model with random normal site effects both in abundance and in detection (model 5).

```
# Bundle data and select model 5
win.data5 <- list(y = y, nsite = nrow(y), nrep = ncol(y), lamDM = lamDM, elev = elev,
   date = date, dur = dur, elev2 = elev2, date2 = date2, dur2 = dur2, e = 1e-06, hlam.on = 1,
   hp.site.on = 1, hp.survey.on = 0)

# Call WinBUGS from R (ART 4250 min) and summarize posteriors
out5 <- bugs(win.data5, inits, params, "ZIPNmix.txt", n.chains = nc, n.thin = nt,
   n.iter = ni, n.burnin = nb, debug = F, bugs.directory = bugs.dir)
print(out5, dig = 3)
```

Sixth, the model with site random effects in abundance and site-by-survey random effects in detection (model 6).

```
# Bundle data and select model 6
win.data6 <- list(y = y, nsite = nrow(y), nrep = ncol(y), lamDM = lamDM, elev = elev,
   date = date, dur = dur, elev2 = elev2, date2 = date2, dur2 = dur2, e = 1e-06, hlam.on = 1,
   hp.site.on = 0, hp.survey.on = 1)
```

```
# Call WinBUGS from R (ART 4230 min) and summarize posteriors
out6 <- bugs(win.data6, inits, params, "ZIPNmix.txt", n.chains = nc, n.thin = nt,
n.iter = ni, n.burnin = nb, debug = F, bugs.directory = bugs.dir)
print(out6, dig = 3)
```

And finally, we fit the model with all three types of over-dispersion (model 7).

```
# Bundle data and select model 7
win.data7 <- list(y = y, nsite = nrow(y), nrep = ncol(y), lamDM = lamDM, elev = elev,
    date = date, dur = dur, elev2 = elev2, date2 = date2, dur2 = dur2, e = 1e-06, hlam.on = 1,
    hp.site.on = 1, hp.survey.on = 1)

# Call WinBUGS from R (ART 4625 min) and summarize posteriors
out7 <- bugs(win.data7, inits, params, "ZIPNmix.txt", n.chains = nc, n.thin = nt,
n.iter = ni, n.burnin = nb, debug = F, bugs.directory = bugs.dir)
print(out7, dig = 3)

# Look at posteriors for random effects
MCMCout <- out7          # Choose which model you want to plot random effects
par(mfrow = c(1,3))
hist(MCMCout$sims.list$sd.lam, breaks = 60, col = "grey")
hist(MCMCout$sims.list$sd.p.site, breaks = 60, col = "grey")
hist(MCMCout$sims.list$sd.p.survey, breaks = 60, col = "grey")
```

We provide an overview of the fit and the inference about abundance and the random effects terms under all six models in Table 6.3(a) and (b). We also extrapolate the inferences to the rest of Switzerland as in the last section, to obtain an estimate of the national population size of great tits under each model.

Table 6.3(a) Comparison of seven ZIP models for the abundance of Swiss great tits in 2013. The term c-hat is the magnitude of OD, bpv the Bayesian *p*-value, theta the zero-inflation parameter (the expected proportion of sites with structural zero counts), and lambda (intercept) is the intercept of lambda (on the log scale). The final two columns give the estimates of the total number of Great tit territories in the 263 surveyed quadrats and in all of Switzerland (excluding quadrats with >50% water and those with elevation >2250 m).

Model (Type of Heterogeneity)	c-hat	bpv	theta	lambda (Intercept)	Pop. Size (263 quads)	National Pop. Size (in 000s with 95% CRI)
(1) Basic ZIP	1.54	0.02	0.08	2.73	3,549	591 (491–706)
(2) λ(*site*)	1.05	0.36	0.04	3.07	13,303	1824 (836–3713)
(3) *p*(*site*)	1.05	0.43	0.06	4.00	11,294	1663 (1080–2486)
(4) *p*(*site.survey*)	0.99	0.56	0.07	3.05	3,664	611 (478–791)
(5) λ(*site*), *p*(*site*)	1.06	0.40	0.04	3.75	10,403	1537 (863–2753)
(6) λ(*site*), *p*(*site.survey*)	1.04	0.41	0.04	2.98	10,102	1398 (536–3756)
(7) λ(*site*), *p*(*site*), *p*(*site.survey*)	0.99	0.59	0.03	3.27	5,651	829 (527–1368)

6.11 BAYESIAN MODELING OF SWISS GREAT TITS WITH BUGS

Table 6.3(b) Comparison of ZIP models in terms of their extra variance components (Posterior means and 95% CRI of SDs are given).

Model (Type of Heterogeneity)	sd.lam	sd.p.site	sd.p.survey
(1) Basic ZIP	–	–	–
(2) λ(site)	0.70 (0.60–0.82)	–	–
(3) p(site)	–	0.87 (0.75–1.02)	–
(4) p(site.survey)	–	–	0.95 (0.81–1.10)
(5) λ(site), p(site)	0.30 (0.00–0.71)	0.78 (0.13–1.05)	–
(6) λ(site), p(site.survey)	0.69 (0.59–0.81)	–	0.12 (0.01–0.31)
(7) λ(site), p(site), p(site.survey)	0.38 (0.18–0.55)	0.95 (0.70–1.19)	0.31 (0.17–0.46)

We see that the introduction of any of the three random effects brings down the OD coefficient to about 1 and results in a fitting model according to a Bayesian *p*-value. However, the inferences about abundance in terms of the intercept of the abundance model and the estimated population sizes for the 263 surveyed quadrats as well as that extrapolated to the whole of the country vary considerably (the latter two about fourfold). The random-effects SD estimates are also sensitive to model structure—i.e., to other random effects terms in the model. In some instances, some unexplained variance seems to be "soaked up" by one term when another was not in the model. Finally, there is some "competition" between zero-inflation and site random effects in lambda; with site random effects in lambda present, the estimate of the proportion of structural zero counts (theta) was clearly lower.

As is so often the case with Bayesian inference, we are thus confronted with a sort of embarrassment of riches: with BUGS it is usually easy to fit even very sophisticated models but then we typically lack easy and unequivocal ways of doing model selection. (Although the ability to use AIC for model selection in the analysis in `unmarked` did not help us much either, since the AIC pointed to the NB model with unrealistic predictions of lambda.) Although there is perhaps a danger of circularity, we think that few would dispute that common sense and subject matter knowledge must play important parts in model selection. The Swiss population is unlikely to number much more than about one million pairs of great tits, so from this perspective, only models 1, 4, 6, and 7 are potentially reasonable. However, they predict rather different elevation profiles for the expected abundance across Switzerland (Figure 6.21, code not shown). The models with additional random effects (4, 6, and 7) have a "hook" for very high elevations beyond the tree line around 2250 m where no great tits can live—we are not so much bothered by this, because it represents an extrapolation of the model with quadratic terms beyond observed covariate values. Models 4 and 7 in particular have more variable values of predicted λ at any given elevation: this makes sense given the additional random effects. But model 4 and especially model 7 make unrealistically high predictions around the tree line (red vertical line). Under model 7 λ is even predicted to *increase* again in the elevation range from about 1900–2250 m; this is not a realistic pattern. In our opinion, this rules out models 4 and 7 as a reasonable basis for inference, since they do not seem to perform well for one central goal of our analysis. The elevation profile of abundance predicted by model 6 looks broadly reasonable; however, the predicted density at low elevations is certainly too high. Perhaps the quadratic function of elevation is too rigid and we ought to try a more flexible semiparametric (spline) model; see Section 10.14.

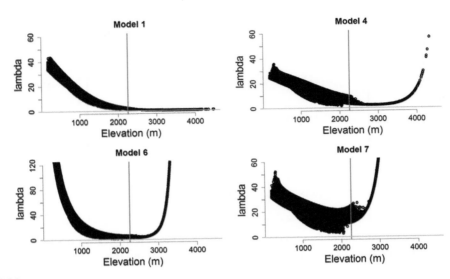

FIGURE 6.21

Comparison between the estimated elevation profiles of the expected abundance of great tits across Switzerland among models 1, 4, 6, and 7. The red vertical line represents the tree line at around 2250 m.

This leaves us with the base model 1, which does not formally fit but seems to produce sensible predictions of abundance overall. Hence, in this case adding random effects into the model to accommodate unexplained variability in abundance or detection did not help us so much. Nevertheless, in many cases this may be a fruitful avenue when looking for a fitting model. Remember, that all estimates of national total population sizes suffer from the issues of space discussed in Section 6.10 and that accounting for an effective sample area greater than 1 km² reduces these crude estimates. Assuming an average home range size of 2 ha for great tits, we get an effective sample area of 1.344 km² associated with each 1-km² MHB sample quadrat and the Swiss national population size estimate of great tits under model 6 in Table 6.3 (a) would have to be reduced by a factor of 1.344.

6.11.2.2 Linear Modeling of a Variance in the N-mixture Model

We end this section on random effects with an example where we take the best model from the previous section (model 6) and investigate possible patterns in the OD in abundance—i.e., in the parameter sd.lam. Specifically, we look at the elevation profile of this OD in abundance, by fitting a log-linear model to the extra-Poisson variance:

$$\log(\sigma^2_{\lambda,i}) = \alpha_{var.lambda} + \beta_{var.lambda} * elev_i$$

Thus, we adopt a linear model for the variance on a log-transformed scale, since a variance is positive-valued like a Poisson mean, in a model that accounts for heteroskedasticity of a variance component by modeling it as a function of a covariate. As always, the first step when we want to model a parameter (i.e., explain it by covariates) is that we have to allow it to vary in the model. This is done by indexing it in algebra, denoting the dimension along which we want to explain its variability. We want to model pattern among sites, and hence we must index the variance by i. We do this in a stripped-down

6.11 BAYESIAN MODELING OF SWISS GREAT TITS WITH BUGS

version of the original ZIP model: we drop code for the GoF, the switches, and also the zero-inflation part, since with random effects in lambda, the estimated proportion of structural zero sites (i.e., unsuitable sites) is very close to 1. The following code is the result of preliminary exploratory analyses in which we identified priors for the linear model parameters for the variance that were just wide enough to count as vague, but not so wide as to cause trouble for the model to converge.

```
# Bundle and summarize data set
str( win.data8 <- list(y = y, nsite = nrow(y), nrep = ncol(y), lamDM = lamDM, elev = elev,
    date = date, dur = dur, elev2 = elev2, date2 = date2, dur2 = dur2) )

# Specify model in BUGS language
sink("Nmix.special.txt")
cat("
model {

# Specify priors
# abundance
beta0 ~ dnorm(0, 0.1)      # log(lambda) intercept
for(k in 1:7){             # Regression params in lambda
    beta[k] ~ dnorm(0, 1)
}
# Model for unexplained variance in lambda among sites
for (i in 1:nsite){
    tau.lam[i] <- 1/var.lam[i]
    log(var.lam[i]) <- alpha.var.lam + beta.var.lam * elev[i]
}
# Priors for intercept and slope of linear model for variance
alpha.var.lam ~ dunif(-1, 1)
beta.var.lam ~ dunif(0, 3)

# detection
for(j in 1:3){
    alpha0[j] <- logit(mean.p[j])
    mean.p[j] ~ dunif(0, 1)    # p intercept for occasions 1-3
}
for(k in 1:13){                # Regression params in p
    alpha[k] ~ dnorm(0, 1)
}
tau.p.survey <- pow(sd.p.survey, -2)
sd.p.survey ~ dunif(0, 1)      # site-survey heterogeneity in p

# Poisson-lognormal model for abundance
for (i in 1:nsite){
    eps.lam[i] ~ dnorm(0, tau.lam[i])    # Random site effects in log(abundance)
    loglam[i] <- beta0 + inprod(beta[], lamDM[i,]) + eps.lam[i]
    loglam.lim[i] <- min(250, max(-250, loglam[i]))    # 'Stabilize' log
    mu.poisson[i] <- exp(loglam.lim[i])
    N[i] ~ dpois(mu.poisson[i])
}
```

```
# Binomial measurement error model with extra-binomial dispersion
for (i in 1:nsite){
  for (j in 1:nrep){
    y[i,j] ~ dbin(p[i,j], N[i])
    p[i,j] <- 1 / (1 + exp(-lp.lim[i,j]))
    lp.lim[i,j] <- min(250, max(-250, lp[i,j]))       # 'Stabilize' logit
    lp[i,j] <- alpha0[j] + alpha[1] * elev[i] + alpha[2] * elev2[i] +
      alpha[3] * date[i,j] + alpha[4] * date2[i,j] +
      alpha[5] * dur[i,j] + alpha[6] * dur2[i,j] +
      alpha[7] * elev[i] * date[i,j] + alpha[8] * elev2[i] * date[i,j] +
      alpha[9] * elev[i] * dur[i,j] + alpha[10] * elev[i] * dur2[i,j] +
      alpha[11] * elev2[i] * dur[i,j] + alpha[12] * date[i,j] * dur[i,j] +
      alpha[13] * date[i,j] * dur2[i,j] + eps.p.survey[i,j]
    eps.p.survey[i,j] ~ dnorm(0, tau.p.survey) # Random site-survey effects
  }
}
}
",fill = TRUE)
sink()
```

Here are the remaining ingredients for the analysis. For the parameters associated with the variance components, we use solutions from earlier fits of a similar model as initial values.

```
# Initial values
Nst <- apply(y, 1, max, na.rm = T) + 1
Nst[is.na(Nst)] <- round(mean(y, na.rm = TRUE))
Nst[Nst == "-Inf"] <- round(mean(y, na.rm = TRUE))
inits <- function(){ list(N = Nst, beta0 = 0, mean.p = rep(0.5,3), beta = runif(7, 0,0), alpha
  = runif(13, 0,0), alpha.var.lam = 0, beta.var.lam = 1.5, sd.p.survey = 0.3)}

# Parameters monitored
params <- c("beta0", "beta", "alpha.var.lam", "beta.var.lam", "alpha0", "mean.p",
"alpha", "sd.p.survey")

# MCMC settings
ni <- 180000   ;   nt <- 100   ;   nb <- 10000   ;   nc <- 3

# Call WinBUGS from R (ART 374 min) and summarize posteriors
out8 <- bugs(win.data8, inits, params, "Nmix.special.txt", n.chains = nc, n.thin = nt,
n.iter = ni, n.burnin = nb, debug = TRUE, bugs.directory = bugs.dir,
working.directory = getwd())
print(out8, dig = 3)
```

As an informal check of whether the model gives reasonable estimates we compare its main estimates with those under the related model 6 and do indeed find fairly similar estimates, with the exception of the coefficients for elev and elev2.

```
print(cbind(out6$summary[c(4:11,13:31,33),1:2],
out8$summary[c(1:8,11:30),1:2]),2)
```

6.11 BAYESIAN MODELING OF SWISS GREAT TITS WITH BUGS

The estimate of the slope of the site-specific extra-Poisson variance (or SD) on elevation is clearly positive (posterior mean 1.48, CRI 1.21–1.82). Hence, at higher elevations of Switzerland, we have more unexplained among-site variability in the expected abundance of great tits than we do at lower elevations. To better understand this interesting relationship, we form predictions of the lambda OD against elevation and plot them.

```
# Predict site variance as a function of elevation
# Get posterior distribution of predictions first
orig.elev.pred <- seq(200, 2250, 50)
elev.pred <- (orig.elev.pred - mean(tits$elev)) / sd(tits$elev)
(n.mcmc <- length(out8$sims.list$alpha.var.lam))   # how many MCMC samples ?
post.sd.lam <- array(NA, dim = c(length(elev.pred), n.mcmc))
for(i in 1:length(elev.pred)){
   post.sd.lam[i,] <- sqrt(exp(out8$sims.list$alpha.var.lam +
   out8$sims.list$beta.var.lam * elev.pred[i]))
}

# Plot posterior mean and a sample of 500 regression lines from posterior
show <- sample(1:n.mcmc, 500)
matplot(orig.elev.pred, post.sd.lam[,show], xlab = "Elevation (m)", ylab = " sd.lam",
   type = "l", lty = 1, lwd = 1, col = "grey", frame = F, ylim = c(0, 6))
lines(orig.elev.pred, apply(post.sd.lam, 1, mean), lwd = 3, col = "blue")
```

We see a considerable increase in the extra-Poisson dispersion in lambda with increasing elevation (Figure 6.22). At low elevations up to about 500 m, the distribution of N is fairly close to a Poisson because the amount of OD is small, but at higher elevations the Poisson model with the specified covariates becomes increasingly poor to explain all the site-specific variability in latent abundance N.

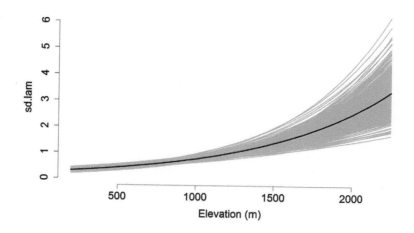

FIGURE 6.22

Relationship between unexplained site-specific heterogeneity in expected abundance (sd.lam) of Swiss great tits in 2013 and elevation. The blue line shows the posterior mean and the gray lines give a random sample of size 500 from the posterior distribution of the regression lines of the overdispersion standard deviation on (scaled) elevation.

Thus, the power of BUGS allowed us to uncover, in a rigorous parametric model, a quite fascinating pattern in the spatial variation of abundance of Swiss great tits.

This example concludes our illustration of the Bayesian treatment of the N-mixture model. A Bayesian treatment for these models becomes essential whenever we consider any of a number of extensions of the model.

6.12 TIME-FOR-SPACE SUBSTITUTION

Sometimes we have very sparse spatial replication or even none: all we have is a time series of counts from a single site. We can still jointly estimate abundance and detection if we have the necessary short-term replication within a period of closure. We simply treat the ("large-scale") temporal replication as a substitute for the usual spatial replication. Yamaura et al. (2011) did this in the context of multi-species N-mixture models (see Chapter 11). We illustrate with a small simulation in which we imagine that we had counted a snake species at a single site for 12 summers, with four counts conducted at each. Assuming closure for a single summer, we simply fit the standard model as if we had surveyed 12 sites in a single year with four temporal replicates. We use our AHM function simpleNmix to generate a data set with the following default arguments:

```
simpleNmix(nyear = 12, nrep = 4, beta0 = 2, beta1 = 0.1, alpha0 = 0.5, alpha1 = -0.1,
   alpha2 = 1)
```

This generates data for 12 years, with four replicate surveys in each, where there is a log-linear trend in abundance with $log(\lambda_i) = \beta_0 + \beta_1 * Time_i$, and where detection probability is logit-linearly affected by year and the temperature during a survey according to $log(p_{ij}) = \alpha_0 + \alpha_1 * Time_i + \alpha_2 * temp_{ij}$. We generate the data with opposing trends over time in abundance and detection to make the estimation problem harder.

We generate 2500 data sets and analyze them both with a simple Poisson GLM (with covariates time and temperature) and with the data-generating N-mixture model. These simulation settings represent averages of abundance, detection, and counts of around 15, 0.47, and 6.8.

```
library(unmarked)
simrep <- 2500                                  # Number of simreps
results <- array(NA, dim = c(simrep, 8))        # Array for results
for(i in 1:simrep){
  cat("Simrep", i, "\n")
  data <- simpleNmix(nyear = 12, nrep = 4)      # Simulate a data set
  umf <- unmarkedFramePCount(y = data$C,  siteCovs = data.frame(Time = data$Time),
      obsCov = list(temp = data$temp))
  fm1 <- pcount(~Time+temp ~Time, data = umf)
  fm2 <- glm(c(data$C)~rep(data$Time,data$nrep)+c(data$temp),family='poisson')
  results[i, 1:5] <- coef(fm1)
  results[i, 6:8] <- coef(fm2)
}
colnames(results) <- c(names(coef(fm1)), names(coef(fm2)))
```

We compare the intercept (abundance) and the slope on time (population trend) for expected abundance (λ) in Figure 6.23 (code not shown). We notice apparently unbiased estimates for both

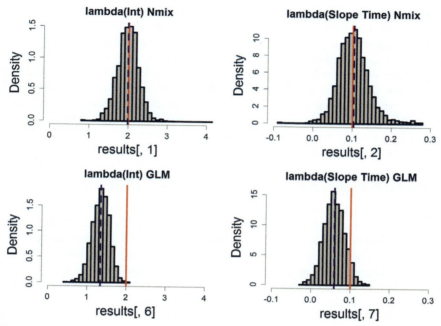

FIGURE 6.23

Summary of simulation exercise for a time-for-space substitution of the N-mixture model for the abundance intercept (left) and the population trend (right) estimated from count data with four replicates in each of 12 years. Top: N-mixture model; bottom: simple Poisson GLM. Both data generation and analysis also contained an effect of temperature on detection. Red line: truth in data generation; blue line: mean of the 2500 simulations.

intercept and population trend when using the *N*-mixture model, and substantial bias in a simple Poisson GLM. In the GLM we corrected for the effects of temperature on detection, but in this nonhierarchical analysis, the separate effects of time on abundance and detection cannot be teased apart, hence the bias in the trend estimate in the GLM.

The ability to fit the *N*-mixture models even in such extreme situations is interesting. However, in practice you should perhaps accompany the use of the model for real data with such borderline sample sizes with simulations to get a better feel for how good your estimates are likely to be.

6.13 THE ROYLE-NICHOLS MODEL AND OTHER NONSTANDARD N-MIXTURE MODELS

N-mixture models are HMs for inference about abundance that adjust explicitly for some kind of measurement error. Typically, they adopt a Poisson or similar distribution for spatial, temporal, or spatiotemporal variation in the latent abundance states *N*. However, depending on the protocol applied for measuring abundance, the observation model may vary quite a bit. In most of this chapter, we cover the protocol that accommodates false-negative detection error, leading to a conditional binomial distribution for the observation process. In Chapters 7–9, we cover various protocols leading to the adoption of a multinomial distribution for the observation process. Here we cover two further designs leading to two further variants of N-mixture models. In the first (Section 6.13.1), instead of observing counts, we only have information on whether the count was zero or greater than zero—i.e., detection/nondetection (or "presence/absence") data—leading to a Bernoulli/Poisson mixture model. In the second variant (Section 6.13.2), we may observe a member of N multiple times, effectively representing a kind of false-positive error in addition to the usual false-negatives. This leads to the adoption of a Poisson/Poisson mixture model.

6.13.1 THE ROYLE-NICHOLS OR BERNOULLI/POISSON N-MIXTURE MODEL

In Section 3.3.6 we saw that when suitable parametric assumptions are met, we can estimate abundance from mere detection/nondetection ("presence/absence") data. The Royle-Nichols model (or RN model for short) is another such example (Royle and Nichols, 2003). The basic N-mixture model in this chapter is a binomial/Poisson mixture for the observed counts, whereas the RN model is a Bernoulli/Poisson mixture for observed detection/nondetection data. It is an important HM in the family tree of models in this book, because it provides a conceptual link between the hierarchical N-mixture models for abundance and the HM for occurrence (i.e., the classical occupancy model in Chapter 10). The RN model may be useful to estimate abundance from replicated measurements of presence/absence, to accommodate detection heterogeneity when focusing on occupancy (Dorazio, 2007) or to link occupancy data with abundance data in an integrated model (Conroy et al., 2008) and looks like this:

1. Model for quadrat population size (state process model):

$$N_i \sim Poisson(\lambda_i)$$

$$\log(\lambda_i) = \beta_0 + \text{covariate effects}$$

2. Measurement error model (observation model):

$$y_{ij}|N_i \sim Bernoulli\left(P_{ij}^*\right)$$

$$P_{ij}^* = 1 - \left(1 - p_{ij}\right)^{N_i}$$

$$\text{logit}\left(p_{ij}\right) = \alpha_0 + \text{covariate effects}$$

The observations y_{ij} are binary, indicating that your species was detected or not during survey j at site i. The probability to detect the species (P_{ij}^*; this is the "occupancy" or "per-quadrat detection probability") is expressed as a function of the number of individuals at the site, N_i, and the per-individual detection probability p_{ij} (which in the literature is typically denoted r; Royle and Nichols, 2003; Dorazio, 2007). This relationship is based on a simple binomial argument that assumes independence and identical detection probability p_{ij} for all individuals N_i at site i during occasion j. The RN model is often described as an occupancy model because the modeled data are identical to those for an occupancy model, or as a heterogeneity occupancy model, because the relationship between abundance and per-individual detection probability adjusts for site-specific heterogeneity in per-quadrat detection probability P_{ij}^*. However, we prefer to describe the RN model as an N-mixture model with a modified observation process for detection/nondetection data.

We illustrate the RN model first with a simulation and then we fit it to Swiss tit data. We use function occuRN in unmarked, but the model can be fit in BUGS too (Yamaura et al., 2011). We use our AHM function playRN to illustrate the RN model in the experimental setting of a simulation. The function internally uses function simNmix to generate replicated count data under the standard N-mixture model of Royle (2004b) and then "degrades" these data to detection/nondetection and fits the RN model using unmarked. It also computes the BUPs for site-specific abundance N and compares them with the known true local abundances in a graph and with an informal linear regression (informal because it does not account for nonindependence of the N estimates). For a perfect RN model, and indeed for any HM, this regression of predicted vs. true latent variables should have a slope somewhat less than 1 because of shrinkage in the BUPs. As additional output you get the parameter estimates of the RN model. Use of playRN with varying values of sample size (number of sites M, number of repeated surveys J) and of mean.abundance and mean.detection will train your intuition about the RN model. (If you want to change the effects of three covariates then you have to change the definition of playRN.)

```
# Execute the function using various settings
playRN(M = 100, J = 3, mean.abundance = 0.1)    # Increasing abundance
playRN(M = 100, J = 3, mean.abundance = 1)
playRN(M = 100, J = 3, mean.abundance = 5)
playRN(M = 100, J = 3, mean.detection = 0.3)    # Increasing detection
playRN(M = 100, J = 3, mean.detection = 0.5)
playRN(M = 100, J = 3, mean.detection = 0.7)
playRN(M = 100, J = 20)                         # More visits
playRN(M = 1000, J = 3)                         # More sites
```

We generate 100 MHB-type data sets (with 267 sites and three visits) for a range of values of mean abundance between 0.1 and 10 and with the default mean detection of 0.3.

```
# Run simulation
lam <- c(0.1, 0.5, 1, 2.5, 5, 10)        # 6 levels of mean abundance
simrep <- 100
results <- array(NA, dim = c(length(lam), simrep, 6))
for(i in 1:6){
   for(j in 1:simrep){
      cat(paste("\n *** lambda level", lam[i], ", simrep", j, "***\n"))
      tmp <- playRN(mean.abundance = lam[i])
      results[i,j,1:5] <- tmp$coef        # Coefficients of RN model
      results[i,j,6] <- tmp$slope         # Slope of regression of Nest on Ntrue
   }
}
```

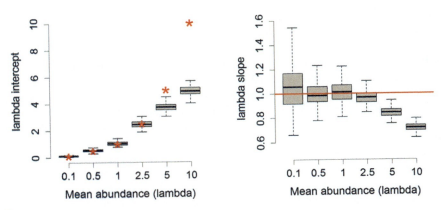

FIGURE 6.24

Summary of simulation for the RN model when mean abundance (λ) is varied (see text for details). Each boxplot summarizes estimates from 100 simulated data sets. Red asterisks and the line show the true values used in the data generation.

```
# Summary of results for abundance (Figure 6.24)
par(mfrow = c(1, 2), mar = c(5,5,3,2), cex.lab = 1.5, cex.axis = 1.5)
boxplot(t(exp(results[,,1])), names = as.character(lam), outline = F, frame = F, col =
"grey", xlab = "Mean abundance (lambda)", ylab = "lambda intercept", ylim = c(0,10))
points(1:6, lam, pch = "*", col = "red", cex = 3)
boxplot(t(results[,,2]), names = as.character(lam), outline = F, frame = F, col = "grey",
xlab = "Mean abundance (lambda)", ylab = "lambda slope", ylim = c(0.5, 1.6))
abline(h = 1, col = "red", lwd = 3)
```

We summarize the sampling distributions of the intercept and the slope in the abundance model (Figure 6.24) and find that the model recovers unbiased and reasonably precise estimates of the abundance parameters up to a certain mean abundance (λ). This makes sense since as λ increases there is less and less heterogeneity in P_{ij}^* and so we lose information about model parameters. It is interesting to note that λ is to some extent controllable *by design*—i.e., by changing the size of the sample unit. These simulation results suggest that there is an optimal size of a quadrat for any species—the size should produce around one individual per sample unit. This concept may appear quite strange but the direct connection between plot size and statistical efficiency is an interesting use of these models.

Next we take the 2013 counts of Swiss tits and fit both the binomial/Poisson model to the counts directly, and the Bernoulli/Poisson (= RN) model to the counts turned into detection/non-detection data. We fit the initial model from Section 6.9 with a Poisson mixture since occuRN does not support ZIP or NB mixtures. We first organize the tit counts into a 3-D array y3D, where slices 1–6 represent great, blue, coal, crested, marsh, and willow tits, in that order. Function occuRN does not allow sites with all observations missing, so we only keep the 263 sites surveyed in 2013.

6.13 THE ROYLE-NICHOLS MODEL

```
# Load data on Swiss tits
tits <- read.table("SwissTits_mhb_2004_2013.csv", header = T, sep = ";")
str(tits)                  # Overview of data file
table(tits$latname)        # Available spec (Latin short) (not in order !)
table(tits$name)           # Same in English

# Select 2013 counts (all species) and put into 3D array
y <- as.matrix(tits[,c("count131","count132", "count133")])
# Create same NA pattern in counts y as in survey date
y[is.na(cbind(tits$date131,tits$date132, tits$date133))] <- NA
# List of sites that were surveyed in 2013
keep <- which(apply(is.na(y[1:267,]), 1, sum) != 3)   # surveyed sites

# Get covariate data (site and observational) and drop unsurveyed sites
elev <- tits[keep,14]
route <- tits[keep,15]
forest <- tits[keep,17]
date <- cbind(tits$date131,tits$date132, tits$date133)[keep,] # Survey date
dur <- cbind(tits$dur131,tits$dur132, tits$dur133)[keep,]  # Survey duration

# Put counts (all species) into 3D array
y3D <- array(NA, dim = c(267, 3, 6))    # Define 3D array
for(i in 1:6){                          # Fill in slices
   y3D[,,i] <- y[((i-1)*267+1):(i*267),]
}
y3D <- y3D[keep,,]                      # only keep sites surveyed in 2013

# 'Degrade' counts to mere detection/nondetection data
y3DRN <- y3D              # Copy counts
y3DRN[y3DRN > 1] <- 1     # Overwrite any count >1 with 1 (for RN model)
```

Next, we write code to loop over all six species (you have to continue by pressing the Enter key after each species), plot the data, fit the standard *N*-mixture model (with function pcount) and the RN model (with occuRN), and compare the two sets of estimates.

```
library(unmarked)
spec.names <- c("Great tits", "Blue tits", "Coal tits", "Crested tits", "Marsh tits",
"Willow tits")

# Loop over 6 species of tits
par(mfrow = c(2,2), mar = c(5,4,3,1))
for(k in 1:6){
cat("\n*** Analysis for ", spec.names[k], "***\n")
# Plot observed data: counts vs survey date
matplot(t(date), t(y3D[,,k]), type = "l", lwd = 3, lty = 1, frame = F, xlab =
 "Survey date (1 = April 1)", ylab = "Observed counts", main = paste("Counts of",
spec.names[k], "as a function of survey date"))

# Fit standard Nmix model (Nmix1)
time <- matrix(rep(as.character(1:3), 263), ncol = 3, byrow = T)
```

```r
summary(umf1 <- unmarkedFramePCount(y = y3D[,,k],
siteCovs=data.frame(elev=scale(elev), forest=scale(forest), iLength=1/route),
obsCovs=list(time = time, date = scale(date), dur = scale(dur))) )
Nmix1 <- pcount(~(elev+I(elev^2)) * (date+I(date^2)) * (dur+I(dur^2)) + time-1
    ~ (elev+I(elev^2)) * (forest+I(forest^2))+ iLength,
    umf1, control=list(trace=TRUE, REPORT=5, maxit = 250))
(tmp1 <- summary(Nmix1))

# Fit RN model (Nmix2)
summary(umf2 <- unmarkedFrameOccu(y = y3DRN[,,k],
siteCovs=data.frame(elev=scale(elev), forest=scale(forest), iLength=1/route),
obsCovs=list(time = time, date = scale(date), dur = scale(dur))))
# Use solutions from Nmix1 as inits for Nmix2
Nmix2 <- occuRN(~(elev+I(elev^2)) * (date+I(date^2)) * (dur+I(dur^2)) + time-1
    ~ (elev+I(elev^2)) * (forest+I(forest^2))+ iLength,
    umf2, control=list(trace=TRUE, REPORT=5), starts = coef(Nmix1))
(tmp2 <- summary(Nmix2))

# Compare estimates under both models
# Table with MLEs and SEs
print(cbind(rbind(tmp1$state[,1:2], tmp1$det[,1:2]), rbind(tmp2$state[,1:2],
tmp2$det[,1:2])))

# Plot of all RN estimates versus all Nmix estimates
plot(coef(Nmix1), coef(Nmix2), xlab = "Coefficients Nmix", ylab = "Coefficients RN",
main = spec.names[k])
abline(0,1, lwd = 2)
abline(h = 0, lwd = 1, col = "grey")
abline(v = 0, lwd = 1, col = "grey")
abline(lm(coef(Nmix2) ~ coef(Nmix1)), lwd = 2, col = "blue")
browser()

# Overall discrepancy measure (for state model only): slope of regression and r2
print(slope <- coef(lm(coef(Nmix2)[1:10] ~ coef(Nmix1)[1:10]))[2])    # Slope
print(r <- cor(coef(Nmix2)[1:10], coef(Nmix1)[1:10]))      # Correlation
}
```

When you execute this code, you will see that the RN quite often produces reasonable estimates (judged by their similarity to those under the binomial/Poisson model). Table 6.4 shows that the slope of the RN abundance estimates on the N-mixture model-based abundance estimates is close to 1 in three out of six species and that their correlation is high in five out of six. Perhaps not surprisingly, the correspondence between the two sets of estimates was by far the worst for the most common species, the coal tit, and best for the rarest of them, the marsh tit. On the other hand, there was also a good agreement for the second most common species, the great tit, so rarity does not seem to be the only factor ensuring success of abundance estimation based on the RN model.

In practice, the RN model has not been extensively used to actually estimate abundance. Arguably, it is more sensitive to parametric assumptions and it will fail when abundance and/or detection are high. Moreover, adopting mixture distributions other than the Poisson, such as the NB, leads to unstable estimates and often also to biologically unreasonably high estimates for abundance

Table 6.4 Comparison of estimates for abundance models for six species of Swiss tits under binomial/Poisson (= traditional N-mixture) and Bernoulli/Poisson (= RN) models. Tits are ordered from most to least abundant, based on the exponentiated intercept of the abundance model under the binomial/Poisson model for counts. The second and third columns give the slope of the RN estimates of the abundance model on the corresponding estimates of the traditional N-mixture model, and the Pearson correlation coefficient of these two sets of 10 estimates each in the linear predictor for abundance.

Species	lambda (Intercept)	Slope RN Estimates on Nmix Estimates	Correlation Between RN Estimates and Nmix Estimates
Coal tit	19.5	0.24	0.35
Great tit	14.3	1.01	0.92
Willow tit	5.6	0.49	0.86
Blue tit	4.7	0.72	0.96
Crested tit	3.3	0.93	0.82
Marsh tit	2.6	0.99	0.96

(Royle and Nichols, 2003). Hence, the RN model may be most useful as an occupancy model that accounts for part of the site-specific heterogeneity in detection (Dorazio, 2007) or in integrated models to link occupancy data with data that are directly informative about abundance (Conroy et al., 2008). Note also that Yamaura et al. (2011) have used it as the basis for their community models (Chapter 11).

6.13.2 THE POISSON/POISSON N-MIXTURE MODEL

In this model for latent abundance based on counts of individuals, we allow not only false-negatives (we may fail to detect an individual), but also false-positives in the sense that an individual may be counted more than once during a single occasion. Typical examples for studies producing such data are camera traps without individual recognition, when an individual may produce multiple pictures, or snow track surveys where an individual may cross a transect multiple times. One natural model for count y_{ij} at site i at occasion j is this (Stanley and Royle, 2005; Guillera-Arroita et al., 2011):

1. Process model:

$$N_i \sim Poisson(\lambda_i)$$
$$\log(\lambda_i) = \beta_0 + \text{covariate effects}$$

2. Measurement error model:

$$y_{ij}|N_i \sim Poisson(N_i \phi_{ij})$$
$$\text{logit}(\phi_{ij}) = \alpha_0 + \text{covariate effects}$$

The first part of the model is the same as in any N-mixture model, but the second part expresses the counts as the product of the number of individuals N_i and a rate parameter ϕ_{ij}, which is the detection rate—i.e., the expected number of times that a member of N_i appears in count y_{ij}. We believe that this is a very sensible model conceptually, and it can be easily fit in BUGS. Moreover, its parameters can be estimated with a single visit to each site—i.e., no repeated measurements are required. However, there may be problems with serial correlation in the detection rate, and in several of our unpublished applications, the model has not always seemed to produce sensible estimates of abundance. We suggest that you apply it with caution only or try out more advanced variants of such Poisson/Poisson models that accommodate serial correlation (Guillera-Arroita et al., 2011, 2012).

6.14 MULTISCALE N-MIXTURE MODELS

As we will discuss in some more detail in Section 10.10 for the occupancy case, various forms of temporal or spatial subsampling are common protocols in studies of occurrence and abundance. This means that sampling in time or in space occurs at two or possibly even more scales. For instance, we may do two repeat point counts at the same site at each of multiple times within a breeding season (temporal subsampling) or we may count at multiple sites within each of a series of several larger areas (spatial subsampling; see Figure 10.13). In both cases we can imagine an intrinsically nested structure of subunits within units. There may also be a combination of the two, when we have both spatial and temporal subsamples. This raises two potential issues: first, there may be a dependency between subunits in the same unit. For the temporal case, we would expect both abundance and detection for multiple, repeated counts at the same site to be more similar than multiple repeated counts at different sites, because the population may simply not turn over so quickly. Similarly, for the spatial case we might expect both parameters to be more similar for two sites that lie within the same area than for two sites that are located in two different areas, because we would expect the habitat to be more similar within the same area. And second, the additional information provided by two-scale sampling may allow us to fit more complex models. In the temporal case, we may adopt a model that allows for temporary emigration (the complement of availability), a common form of violation of the closure assumption. Adoption of such a model naturally accommodates the increased similarity of subunits within the same unit.

Multiple scales in models for abundance have received even less attention so far than multiple scales in occupancy models. Chandler et al. (2011) have developed a three-level abundance model for estimation of temporary emigration based on a multinomial mixture model. This model can be fit in unmarked using function gmultmix (see Section 7.5.3). The same model structure has been extended to multinomial mixture models for distance sampling protocols (function gdistsamp; see Section 9.5.3) and for binomial mixture models for repeated counts with function gpcount. All three functions may be useful for temporal subsampling by adopting a simple model of random temporary emigration that assumes a superpopulation of size M_i that is available during some larger time period at site i. At any given point in time j, only a fraction ϕ_{ij} is available, leading to a population size of N_{ij}. This is then repeatedly sampled with detection probability p_{ijk} during a short period of closure. When $\phi_{ij} = 1$, there is no temporary emigration and we are back to the two-level model for a fully closed (meta)population. For more information on these models, see Chandler et al. (2011), the help text for functions gpcount, gmultmix and gdistsamp and Sections 7.5.3 and 9.5.3. These models would be easy to fit in BUGS. Hence, for the temporal case of subsampling, the adoption of a structurally fully different model, with an additional layer in the HM representing a somewhat meaningful biological process (temporary emigration), is fairly straightforward. Johnson et al. (2014) used the observed maximum distances at each of a series of point counts to model spatial variability in availability probability separately from 'real' detection probability (or 'perceptibility'; Marsh and Sinclair, 1989).

In contrast, for the spatial case of subsampling a more phenomenological approach inspired by nested ANOVA, with random unit effects, may appear more straightforward. Imagine point count stations (subunits) nested in each of a number of reserves or national parks, or other spatial "blocks" or "units." To account for the dependency among points within the same park and avoid "pseudoreplication" (Hurlbert, 1984) you can simply specify normal random effects for the "units" (i.e., reserve, park) in the linear predictors for abundance or detection (see Section 10.10). Adding such additional random effects into the N-mixture model cannot be done in `unmarked`, but it is trivially easy in BUGS and works fully analogous to the random effects in 6.11.2. For example, imagine that the MHB sample quadrats in Switzerland had been selected in one of 10 distinct regions of the country and we wanted to accommodate in our analysis this nested (or blocking) structure of the sample quadrats. We could then simply have provided a covariate `region`, which for each site codes for the region the site belongs to, and modified the linear model for abundance to include a region random effect `eps.region`.

```
loglam[i] <- beta0 + inprod(beta[], lamDM[i,]) + eps.lam[i] * hlam.on +
             eps.region[region[i]]
```

In the priors we would then add the following lines:

```
for(r in 1:10){
    eps.region[r] ~ dnorm(0, tau.region)
}
tau.region <- pow(sd.region, -2)
sd.region ~ dunif(0, U)
```

with the upper bound (U) of the uniform prior for the heterogeneity among regions chosen to be suitably large.

People have often wondered whether small-scale spatial replicate counts in each of a set of larger sites could provide information about detection probability, similar to the space-for-time substitution in the related occupancy model (see Section 10.11). We don't think that one could directly treat small-scale replicates akin to the usual temporal replicates in the traditional N-mixture model. However, it seems likely that we could separately estimate abundance and detection in a model that specifies spatial correlation in abundance (provided there is some at the scale of the study). This would seem like a similar problem to that of estimating abundance and detection over multiple years, without replicate counts within years, in the model of Dail and Madsen (2011): in their case, temporal correlation in abundance is accommodated in the Markovian dynamics model and allows both abundance and detection to be estimated even with a single count per site and year. One way of parameterizing spatial correlation is by explicitly modeling the movement process as in the latent spatial point process models of Chandler and Royle (2013) and Ramsey et al. (2015).

6.15 SUMMARY AND OUTLOOK

We have covered the binomial N-mixture model of Royle (2004b) for closed populations. This is simply a Poisson or related model for abundance, but with a binomial measurement error model attached. It can be fit to counts of individuals that are replicated in space and in time, where the time span is short so that the abundance at each site can be assumed constant. This is our canonical HM for inference about abundance from replicated counts. If all goes well—i.e., if the model assumptions are not violated too strongly—the binomial mixture model permits unbiased inferences about that central quantity in ecology and its applications: abundance. This sets this HM apart from a host of other HMs for abundance that only model some index of abundance or "relative abundance" (ver Hoef and Frost, 2003; ver Hoef and Jansen, 2007; Cressie et al., 2009; Zuur et al., 2012). In its most basic form, the model is a mixture between a Poisson distribution for abundance and a conditional binomial distribution for false-negative measurement error.

As always with HMs, we can readily switch one model component for another if this is dictated by the data or the goals of our analysis. For instance, to model excess variability in the latent abundance relative to a Poisson, we may change the Poisson for a negative-binomial (NB), zero-inflated Poisson (ZIP) (Wenger and Freeman, 2008), or a Poisson-lognormal (PLN) (Kéry et al., 2009) distribution and we have given examples for all of these. We have even shown an example of where some sites have little and others have substantial OD (Section 6.11.2.2). In that model, we expressed the site-specific extra-Poisson variance in lambda as a linear function of a covariate (elevation). Other possibilities for the abundance part of the model include a hurdle model (see Section 3.3.4; Ghosh et al., 2012; Dorazio et al., 2013), which unlike a ZIP permits only one type of zero abundance, that coming from the zero-inflation part, a Conway–Maxwell–Poisson distribution (Wu et al., 2013), which unlike NB and PLN allows both over- *and* underdispersion, or a Dirichlet process prior for abundance (Dorazio et al., 2008); the latter being a form of Bayesian nonparametrics for the abundance part of the model (Chapter 23 in Gelman et al., 2014).

Similarly, depending on the data collection protocol, you may choose any of a number of descriptions for the measurement error process. We have given examples with a binomial distribution (resulting in the classical binomial *N*-mixture model), a Bernoulli distribution (resulting in the RN model), and a Poisson distribution, which represents a measurement error model that combines false-negative and false-positive errors. In addition, individuals may not be detected independently, for instance, because they occur in groups or when there is temporal autocorrelation in individual detection probability in counts along a transect. You may then look up an *N*-mixture model with beta-binomial observation model (Martin et al., 2011; Dorazio et al., 2013) or else move to Markov-modulated point process models in the case of serial correlation (Guillera-Arroita et al., 2012). Finally, there is a host of protocols for the measurement of abundance that all lead to a multinomial distribution for the observation process. They include hierarchical capture–recapture and hierarchical distance sampling and are covered in great detail over the next three chapters.

The combination of two or more statistical distributions in the form of a mixture model determines the basic structure of our HM, and at each level of the model hierarchy we can recognize a sort of GLM (Chapter 3). This makes it clear that in principal we can apply all the usual GLM extensions to both the abundance and the detection parts of the model, such as inclusion of covariates, including the "wiggly" form of a GAM (see Section 10.14), or the addition of random effects to account for additional variability or hidden correlation structures, including spatial or temporal autocorrelation (Chapters 21 and 22).

We like to emphasize the modularity of these HMs. For instance, if you drop the measurement error part in an *N*-mixture model, then you are back to a traditional "flat" or (if your *N*-mixture model contains extra random effects) back to a conventional hierarchical Poisson or related GLM. Such variants of Poisson and related GLMs form the backbone of inference methods about unbounded count data in the vast majority of examples in ecology (e.g., Chapter 6 in McCullagh and Nelder, 1989; ver Hoef and Frost, 2003; ver Hoef and Jansen, 2007; Cressie et al., 2009; Zuur et al., 2012; and also see Chapter 12 in volume 2). Often you may lack the necessary replicate measurements or other forms of extra information such as distance measurements, that permit the joint modeling of abundance and measurement error process. Or perhaps you do not like our emphasis on an explicit description of the measurement error process underlying your count data? Perhaps you find the resulting models too difficult to understand or you are worried about their robustness… or you work in the tropics where some have recently claimed that things are all different anyway and that such models are too hard to apply (Banks-Leite et al., 2014). But even then you can still use all of our models for abundance by simply dropping the measurement error part—i.e., by eliminating the observation model. Of course, you ought to be honest then and qualify your analysis as dealing with an *index* of abundance or with "*relative abundance*" (Johnson, 2008) Any direct statement about *abundance* is then out of reach for you. Speaking about *abundance*, when in reality your inference is about the unknown product of abundance and measurement error (i.e., on relative abundance), would be dishonest and is scientifically indefensible.

Two further topics that we have emphasized in this chapter are the power of simulation, as we do throughout the book, and the sensitivity of model inferences to distributional assumptions, especially about the latent abundance state. First, we have given several examples that show how simulation can give you extremely rich insights into the model, indeed, we set out to explain the model by simulation code in the R language. Further, we used simulation to study the effects of assumption violations and for study design. We hope that the code we provide will serve as a template for your own investigations for your specific needs.

Second, in practical applications of the N-mixture model there is the frequent dilemma that a standard Poisson model does not formally fit our data according to an omnibus test such as Chi-square, but the NB does, or at least fits substantially better, and moreover is much better supported by AIC, but then predicts unrealistically high levels of abundance. This is not a new observation (Kéry et al., 2005b; ver Hoef and Boveng, 2007; Joseph et al., 2009). Looking at residual diagnostics and empirical variance/mean relationships, we found that the NB was not adequate for our great tit data, although both GoF and the AIC model selection criterion selected it way ahead of the other two mixture distributions available in `unmarked` (Poisson, ZIP). In situations where the NB predicts very high abundance or (if you do not have any *a priori* knowledge about likely abundance levels) if the NB predicts substantially higher than the Poisson or ZIP, we would tentatively prefer to use one of the latter with a somewhat *ad hoc* correction for OD, provided that the lack of fit subsumed into such "overdispersion" is not extreme. Since use of the NB virtually always results in predicted abundance at least as high as that under the standard Poisson model, you will therefore err on the side of caution. We gave examples of how to propagate this added uncertainty into predictions of abundance and functions thereof, such as national population size estimates. Part of this material is tentative; it makes sense to us, but is not based on formal theory, nor has it been rigorously tested. Therefore, we repeat our call for more research on model diagnostics, GoF, and robustification in HMs such as the N-mixture model.

We have also covered the frequently forgotten fact that we almost never know the exact size and configuration of our study areas. There may be unsurveyed "holes" inside our nominal study area where detection probability for an individual is essentially zero, leading to negative coverage bias (effective sample area is smaller than nominal sample area). Conversely, we may erroneously add in our count individuals that have activity ranges on the edge of our nominal study area (the 'sunflower effect'), leading to positive coverage bias. We presented a method where we use a covariate (the length of a transect route) to adjust for negative coverage bias, but thereby we almost certainly exacerbated the problem of positive coverage bias; i.e., we estimated the population size in an area that is *larger* (to an unknown extent) than the nominal sample area. We then presented another *ad hoc* method to account for this positive bias based on ideas that have been used in closed population estimation for a long time (Dice, 1938; Karanth and Nichols, 1998). Again, our procedure makes sense to us and seems to be a way forward, but we make no claims that it is optimal or that it will work in all circumstances. The spatial reference of counts is another area where more research is needed, perhaps along the lines of the seminal paper by Chandler and Royle (2013), in which they introduce a spatially explicit N-mixture model. Their current model is challenged by real-world sample sizes and moreover needs adjacent sample sites, but perhaps some sort of "integrated analysis" (where different types of data sets are combined in a model) or else spatial subsampling might be a way forward.

Two further areas where more research would be fruitful are in the modeling of false-positives and density-dependent detection. The Poisson/Poisson model (Section 6.13.2) allows for the combination of false-negative and false-positive error rates, because a member of N may be counted more than once (the detection model is Poisson, and moreover the rate ϕ may be greater than 1). However, this model has had mixed success in our attempts to fit it to real data. Perhaps it is possible to formulate false-positive abundance models similar to those for occupancy (Royle and Link, 2006; Miller et al., 2011; Sutherland et al., 2013)—i.e., with an observation process for a count C that separates out the two processes:

$$C = C^- + C^+,$$

where $C^- \sim Binomial(N, p)$ and $C^+ \sim Poisson(N\lambda)$.

Here, C^- is the usual false-negative detection error part of the observation process, while C^+ describes false-positives that are expressed as a rate relative to the true abundance N. We doubt whether such a model would be identifiable from the data to which the standard N-mixture model is fit, but perhaps with some extra information it might be: as in the occupancy model of Miller et al. (2011), it might be possible to identify subsets of the data that are devoid of one of the error types, and a combined analysis might then be able to estimate all parameters.

Density-dependent detection is quite likely on biological grounds: for instance, it makes sense to assume that birds sing—i.e., advertise and defend their territory—more when they live at higher densities (Warren et al., 2013) and this has been shown experimentally in the field (Penteriani, 2003). Although there seems to be some form of circularity involved, you can actually specify in BUGS a model where the latent abundance N_i at a site i, perhaps log-transformed, is treated as an explanatory variable in the linear predictor for detection. Surprisingly to us, when we experimented with such a model it seemed to be estimable with simulated data and the effect of ignoring density-dependent detection was always a negative bias in abundance. This makes sense when we realize that such abundance effects on detection will, when not modeled, cause unmodeled site-specific detection heterogeneity that will bias low the estimates of abundance (see Section 6.7). However, when we fit the model to the Swiss tits, we failed to obtain reasonable estimates. It would be interesting to follow up this line of research.

In the following three Chapters 7–9, we retain the basic structure of the model for abundance and simply vary the observation model to reflect a change in the data collection protocol—i.e., the manner in which we *measure* abundance. We will collect some extra information that enables us to specify a different model of the false-negative detection process. This extra information will be individual identity across occasions for capture–recapture designs (Chapter 7). (Note that we also require individual identity for the binomial N-mixture model, but only within any occasion, to rule out multiple counts of the same individual.) Capture-recapture designs do not help us with our problem of defining the effective sampling area. In contrast, for the powerful distance sampling protocol (Chapters 8–9) the extra information comes in the form of measured distances, and this manifestation of space enables us to define the effective sampling area unambiguously. In Chapter 10, you will encounter an analogous HM where the latent state is presence/absence instead of abundance (i.e., the site-occupancy model). Finally, in Chapter 11, we cover the community versions of both site-occupancy and N-mixture models. In Chapters 12–14 (in volume 2) we will generalize the closed-population N-mixture models to open populations by modeling the population dynamics at each site as a function of gains and losses (Dail and Madsen, 2011; Chandler and King, 2011; Zipkin et al., 2014a,b; Hostetler and Chandler, 2015; Sollmann et al., 2015; Bellier et al., in review) or with reduced-parameter versions of such explicit population dynamics models.

EXERCISES

1. Take a count data set of your choice (ideally real data) and use `unmarked` to fit some binomial N-mixture models.
2. Based on code given in 6.3, devise a simulation that investigates the relationship between detection probability (p), the number of sites (M), the number of replicate surveys (J) and the bias of the maximum count at a site i as an estimator of N_i. In other words, how many surveys are required, on average, for the maximum count to amount to at least 95% of N_i on average.
3. Based on 6.4, devise a simulation that investigates the relationship between M, J, p, and the likelihood of estimability issues in the N-mixture model. You can diagnose the latter by a sensitivity of the numerical MLEs to the choice of `K` (the upper summation limit in the likelihood evaluation in `unmarked`).
4. In the example at the end of Section 6.4, imagine that the theoretical ecologist's theory predicts that the meta-population goes extinct when the (spatial) variance in population size becomes greater than 3.84. What is the probability of extinction of that meta-population? (Use JAGS for this exercise.)
5. Play around with different settings of function `simNmix` to get a feel for how count data arise in the presence of imperfect detection. Perhaps it is a good idea to write your own R code to visualize and summarize the simulated data.
6. Repeat the simulation study (Section 6.6) in BUGS, using reasonably vague priors for the parameters, and compare the incidence of freak high estimates of lambda with that when using MLEs.
7. Along the lines of Section 6.6, conduct a simulation study of your own that investigates some design question about the N-mixture model that might be relevant to your own work.

8. Conduct another simulation to investigate the effects of density-dependent detection probability on the estimators of the N-mixture model (similar to Section 6.7); see also Warren et al. (2013). Is there a difference between positive and negative density dependence in the direction of bias? Can you account for this in a model fit in BUGS? (If you can it might be worth a paper!)
9. In Section 6.8, see whether you can detect diagnostic patterns in the residuals of the models for specific assumption violations. (Again, this might be the start of a useful paper.)
10. In the modeling of Swiss great tits, determine the best day of the season to go out and survey for the birds. Do this for different elevation classes.

CHAPTER 7

MODELING ABUNDANCE USING MULTINOMIAL N-MIXTURE MODELS

CHAPTER OUTLINE

- 7.1 Introduction 314
- 7.2 Multinomial N-Mixture Models in Ecology 316
  - 7.2.1 Covariate Models 317
  - 7.2.2 Types of Multinomial Models 318
- 7.3 Simulating Multinomial Observations in R 320
- 7.4 Likelihood Inference for Multinomial N-Mixture Models 323
- 7.5 Example 1: Bird Point Counts Based on Removal Sampling 325
  - 7.5.1 Setting Up the Data for Analysis 328
  - 7.5.2 Fitting Models Using Function `multinomPois` 329
  - 7.5.3 Fitting Models Using Function `gmultmix` 330
  - 7.5.4 Assessing Model Fit in `unmarked` 333
- 7.6 Bayesian Analysis in BUGS Using the Conditional Multinomial (Three-Part) Model 334
  - 7.6.1 Goodness-of-Fit Using Bayesian p-Values..337
  - 7.6.2 Model Selection in BUGS 338
  - 7.6.3 Poisson Formulation of the Multinomial Mixture Model 342
- 7.7 Building Custom Multinomial Models in `unmarked` 346
- 7.8 Spatially Stratified Capture-Recapture Models..349
  - 7.8.1 Example 2: Fitting Models M_0, M_t, and M_x to Chandler's Flycatcher Data 351
  - 7.8.2 Models with Behavioral Response (M_b) 355
  - 7.8.3 Models with Individual Heterogeneity (M_h) 356
  - 7.8.4 Bayesian Analysis Using Data Augmentation: Heterogeneity Models in BUGS 358
  - 7.8.5 Analysis of Data Generated with the `simNmix` Data Simulation Function 363
- 7.9 Example 3: Jays in the Swiss MHB 367
  - 7.9.1 Setting Up the Data and Preparing for Analysis 370
  - 7.9.2 Fitting Some Models 371
  - 7.9.3 Analysis of Model Fit 375
  - 7.9.4 Summary Analyses of Jay Models...379
  - 7.9.5 Spatial Prediction 383
  - 7.9.6 Population Size of Jays in Switzerland...386
  - 7.9.7 Bayesian Analysis of the Jay Data 387

CHAPTER 7 MODELING ABUNDANCE USING MULTINOMIAL N-MIXTURE MODELS

7.10 Summary and Outlook 390
Exercises 391

7.1 INTRODUCTION

In the previous chapter, we introduced a class of mixture models for meta-population studies in which the observation model was a simple binomial counting model, but with several such binomial counts being made in temporal succession at each site. These models did not require us to keep track of individuals other than to ensure that no individual is doubly counted during each occasion. In this chapter, we consider a similar class of models, but substituting a multinomial observation model for the binomial. These models require us to keep track of the individual identity across occasions. This extra information should lead to more precise estimates (Kéry and Royle, 2010, also see Exercise 3). As with the N-mixture model of the previous chapter, we assume that M sites are sampled by some protocol, and we imagine that the population size at site i, N_i, is the realization of a random variable such as a Poisson or negative binomial. The data collected at each site are assumed to be outcomes of a multinomial distribution with cell probabilities π determined by the specific protocol being used. In general, the sampling protocol defines the cell probabilities to be a function of some encounter probability parameters p_{ij} that depend on site, sample occasion, observer, or some other factor. Thus, for the Poisson case, the basic multinomial mixture models that we discuss in this chapter have the form:

$$\mathbf{y}_i | N_i \sim Multinomial(N_i, \pi(p))$$
$$N_i \sim Poisson(\lambda)$$

where λ is the expected number of individuals at site i (i.e., "local population size") and $\mathbf{y}_i$ is a vector containing the number of individuals at site i each having a unique observable encounter history. We consider various sampling protocols (corresponding to different functions $\pi(p)$), models for encounter probability p, and models for local abundance N.

Multinomial observation models have been used in wildlife sampling problems for decades, and they remain important in many ecological studies. For example, multiple observer sampling methods (Cook and Jacobson, 1979) are common in transect surveys of ungulates and marine mammals and have become widely used in bird surveys over the last decade (Nichols et al., 2000). All classical capture-recapture models (Otis et al., 1978) involve specific multinomial observation models. Further, a very special type of multinomial model arises under a distance sampling protocol. This is such an important method that we devote four entire chapters (the following two in volume 1 and two more in volume 2) to its study and its relationship to the classical multinomial mixture model. While multinomial observation models have been around for a long time, their application to meta-population sampling designs is fairly recent. A number of authors have applied multinomial models to data arising from spatial sampling (Nichols et al., 2000; Farnsworth et al., 2002) but these authors did not exploit that spatially structured design in their development of models or inference strategies. Instead, they relied on a two-stage procedure of pooling data from sites to estimate detection probabilities and then adjusting counts to produce an estimate of total population size among sites. Technically, this is a reasonable thing to do because combining multinomial observations yields a multinomial observation model (Royle, 2004a), but it is not very flexible for modeling covariates or considering general models of abundance. A formal modeling framework for spatially structured multinomial sampling based on marginal likelihood was described by Royle et al. (2004), who considered distance sampling models, and Dorazio et al. (2005), who developed the framework for removal counts (see also Etterson et al., 2009). Koneff et al. (2008) and Langtimm et al. (2011) considered double-observer multinomial mixture models. Some general context related to bird sampling can be found in Royle (2004a). Capture-recapture protocols were studied by Royle et al. (2007b), Webster et al. (2008) and Kéry and Royle (2010), which we consider in Section 7.8. Spatially stratified studies of species richness in a metacommunity context (see Chapter 11) are another common example of such a data type, where species take the place of individuals.

In this chapter, we start as usual by showing how you can simulate data from multinomial models in R so that you can carry out simulation studies that mimic your specific problem. As always, data simulation is important for you to develop an intuition for a data collection protocol and associated model. In addition, the multinomial distribution is pretty mysterious to many ecologists so seeing multinomial data being generated by simulation should be helpful for your understanding of this important statistical distribution. We will discuss classical (likelihood)-based inference using the unmarked package (Fiske and Chandler, 2011), and we will show how to carry out Bayesian inference in BUGS. We show some examples including double-observer, removal, and capture-recapture type models. As for any of your statistical modeling, BUGS provides extreme flexibility in the specification of multinomial N-mixture models. We illustrate this in Chapter 21 (and elsewhere) by accommodating random effects in the model for expected abundance (λ). For example, it is common for sample locations to be grouped or clustered in some meaningful way (often based on geographic strata), and it would be sensible to model this group structure with random effects (see Section 10.10).

Central to this chapter is the multinomial distribution, which is the multivariate extension of the binomial distribution. The binomial distribution is the canonical distribution for frequencies of a random variable with two possible outcomes (Bernoulli trials: success or failure, 0 or 1, dead or alive, married or not, etc.), and the multinomial distribution is the corresponding distribution for *aggregations* of multinomial trials, discrete variables having >2 possible outcomes. To illustrate, consider a variable having $H = 3$ distinct values such as (blue, green, red), (small, medium, large), (captured by observer 1, captured by observer 2, captured by observer 3). Whatever the possible outcomes are we index them by $h = 1, 2, 3$ and organize them in a binary vector $\mathbf{y}$ of length H having elements 0 but for a single 1 to indicate the realized value. Therefore, possible realizations of this vector are $\mathbf{y} = \{(0,0,1), (0,1,0), (1,0,0)\}$. These are called multinomial trials. If we aggregate N of these multinomial trials, where N is a quantity referred to as the *multinomial sample size parameter*, or *multinomial index*, the vector of aggregated frequencies, $\mathbf{y} = \sum_{i=1}^{N} \mathbf{y}_i = (y_1, y_2, \ldots, y_H)$, has a multinomial distribution with sample size N and cell probabilities that we denote by $\boldsymbol{\pi} = (\pi_1, \pi_2, \ldots, \pi_H)$ where $\pi_H = 1 - \pi_1 - \pi_2 - \ldots, \pi_{H-1}$. The probability mass function (pmf) of the vector $\mathbf{y}$ has the form:

$$[\mathbf{y}|N, \boldsymbol{\pi}] \propto \frac{N!}{\prod y_h!} \left\{ \prod_{h=1}^{H} \pi_h^{y_h} \right\}$$

where the cell probabilities are constrained to sum to 1, which we write as $\pi_H = 1 - \sum_{h=1}^{H-1} \pi_h$. Further, the multinomial frequencies must sum to N, and so $y_H = N - \sum_{h=1}^{H-1} y_h$. To denote that vector $\mathbf{y}$ has this multinomial distribution, we write the shorthand:

$$\mathbf{y} \sim Multinomial(N, \boldsymbol{\pi}).$$

We often make use of the *categorical distribution*, which has an equivalence to the multinomial distribution with sample size 1. If we consider a multinomial trial and identify the element or position (in the vector of length H) of the sole "success" (1), then this is a categorical random variable. In this transformation of the multinomial trial, we write $y \sim Cat(\boldsymbol{\pi})$. The various relationships and quantities underlying binomial and multinomial distributions are illustrated in Figure 7.1.

CHAPTER 7 MODELING ABUNDANCE USING MULTINOMIAL N-MIXTURE MODELS

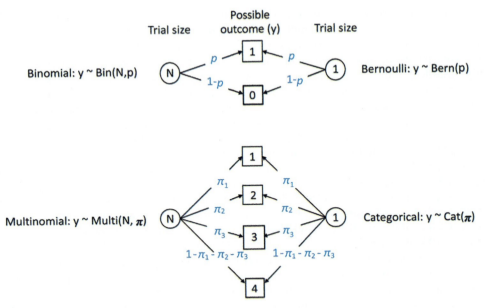

FIGURE 7.1
Relating the binomial, multinomial, and categorical distributions, and Bernoulli and multinomial trials. The response, y, on the left-hand side of the graph consists of the *aggregated frequency* of the elemental response on the right, either the Bernoulli (in the case of the binomial response) or categorical (in the case of the multinomial) outcome. Note that in the lower half we could as well have denoted π as p, but we write π for consistency with the main text, where the cell probabilities π are described as a function of the underlying detection probability parameters, p. (Figure courtesy of M. Schaub.)

7.2 MULTINOMIAL N-MIXTURE MODELS IN ECOLOGY

In the previous chapter, we formulated the N-mixture models, which we also referred to as binomial mixture models because they involve the mixture of a binomial observation model with respect to some model for local abundance, N. Now, we consider a parallel class of models with a multinomial observation model in place of the (product) binomial observation model. A basic form of the multinomial mixture model, in the case of a Poisson distribution for N, is

$$\mathbf{y}_i | N_i \sim Multinomial(N_i, \pi(p))$$
$$N_i \sim Poisson(\lambda)$$

where λ is the expected number of individuals at site i (i.e., "local population size") and $\mathbf{y}_i$ is a vector containing the observed multinomial frequencies at sample location or site $i = 1, 2,..., M$. In the context of animal sampling, the multinomial categories usually correspond to distinct observation states, or encounter histories, and the multinomial cell probabilities are the probabilities of those encounter histories occurring in the population of N individuals. The cell probabilities are functions of more fundamental parameters that relate directly to encounter probability or covariates as we discuss below. We made this explicit in our notation where the cell probabilities, $\pi(p)$, are functions of some other parameters, p. Each specific multinomial sampling protocol has its own *pi function*, relating the detection probability parameters to the cell probabilities. We illustrate this using a standard capture-recapture protocol based on $J = 2$ sample occasions. In this case the observable encounter histories can be denoted by "10" (= "individual detected on occasion 1, but not on occasion 2"), "01" (= "individual detected only on occasion 2"), and "11" (= "individual detected on both occasions"). The probabilities of observing each specific encounter history are $\pi_{10} = p_1(1 - p_2)$, $\pi_{01} = (1 - p_1)p_2$, and $\pi_{11} = p_1 p_2$. There is one remaining unobservable encounter history, which is "00" (= "individual never detected"), and the probability of this encounter history is one minus the sum of the others: $\pi_{00} = 1 - p_1(1 - p_2) - (1 - p_1)p_2 - p_1 p_2 = (1 - p_1)(1 - p_2)$. Thus, the *pi function* $\pi(p)$ converts the canonical parameters (p_1, p_2), which define the observation model, into the multinomial cell probabilities $(\pi_{10}, \pi_{01}, \pi_{11}, \pi_{00})$.

7.2.1 COVARIATE MODELS

The encounter probability parameters, p, may vary fundamentally with the basic replicate structure of the data collection protocol (observer or time period), which are covariates we refer to as *observation covariates*. In addition, in many cases there should also be covariates at the level of the sample unit, which affect encounter probability for that site. We call these "site covariates." A good example of an observation covariate for a bird survey would be an "observer effect" or date if the sampling occurs over multiple days during a breeding season. Covariates of effort are also common, e.g., if sampling or counting at sites was done for a variable amount of time. We will model such effects using a typical logistic regression formulation in which the probability of detection at site i and replicate (observer, occasion) j is p_{ij}, then

$$\text{logit}(p_{ij}) = \alpha_0 + \alpha_1 x_{1,i} + \alpha_2 x_{2,ij}$$

where $x_{1,i}$ is a measured site covariate (e.g., percent forest cover within 100 m of the point) and $x_{2,ij}$ is a measured observation covariate (e.g., day of sampling).

Following the formulation of the N-mixture model, we consider an additional model component that describes variation in the local abundance, N_i, across sites, say,

$$N_i \sim g(N; \theta).$$

This meta-population structure is, in essence, a prior distribution for the unobserved parameters $\{N_i; i = 1, 2, \ldots, M\}$ and the additional model structure provides the framework for combining a large number of spatially referenced count surveys. A purely statistical and equivalent solution to managing the larger number of abundance parameters is to view $\{N_i; i = 1, 2, \ldots, M\}$ as a collection of latent variables or random effects—in effect nuisance parameters to be removed from the likelihood by integration. Our default model for $g(N)$ is the Poisson distribution, which is the canonical distribution for elements of a population that are randomly distributed in space. The Poisson distribution is a somewhat inflexible model for ecological processes because its mean and variance are equal. As such, we also consider negative binomial models (Kéry et al., 2005b) to allow for overdispersion. Both Poisson and negative binomial models are presently implemented in unmarked, while with BUGS negative binomial, Poisson-lognormal and many other models for overdispersion are readily specified.

We are often interested in modeling the effect of explicit covariates on abundance (e.g., forest cover or other variables that describe landscape structure). To model such effects, we include them as linear effects on the natural logarithm of the expected abundance $\lambda_i = E(N_i)$:

$$\log(\lambda_i) = \mathbf{x}_i'\boldsymbol{\beta}$$

where $\mathbf{x}_i$ are measured values for a vector of L covariates for site i and $\boldsymbol{\beta}$ is the $(L \times 1)$ parameter vector of coefficients to be estimated.

7.2.2 TYPES OF MULTINOMIAL MODELS

We discuss several specific multinomial sampling protocols that are widely used in practice. One popular protocol is based on the use of **independent multiple observers** (Caughley, 1974; Magnusson et al., 1978) in which two or more observers make independent detections of individuals during point counts of a fixed time and then, after the counting, reconcile their lists. For example, each observer might record marks and notes on a map, and the observers will identify unique individuals by notes on their respective maps. This allows for the creation of individual encounter histories from the sampling, which in essence represent basic capture-recapture encounter histories.

For example, with two observers, there are $H = 4$ possible encounter histories, of which three are observable, and the observation vector has elements: y_1 = the number of birds seen by observer 1 (but not observer 2), y_2 = the number of birds seen by observer 2 (but not 1), y_3 = the number of birds seen by both observers, and the "all zero" encounter history is y_4. We denote the complete vector of frequencies by $\mathbf{y} = (y_1, y_2, y_3, y_4)$. In general, with J observers, there are $2^J - 1$ observable encounter histories. Sometimes, when it is helpful for clarity, we may index the specific quantities by the observable encounter history so that we might use y_{10} to indicate the encounter history "seen by observer 1 but not observer 2" and so forth.

To construct the cell probabilities we think about the process of sampling birds and we ask, "What is the probability of observing the encounter history 10?" With independent observers, the probability should be the product of the probabilities of observer 1 detecting the bird and observer 2 *not* detecting the bird. If we denote these basic probabilities by p_1 and p_2, then $\Pr(10) = p_1(1 - p_2)$. Similarly, $\Pr(01) = (1 - p_1)p_2$, and so on. The probabilities of all four distinct encounter histories are given in the following table:

Encounter History	Frequency	Cell Probability	Function of p
10	y_1 or y_{10}	π_1	$p_1(1 - p_2)$
01	y_2 or y_{01}	π_2	$(1 - p_1)p_2$
11	y_3 or y_{11}	π_3	$p_1 p_2$
00	y_4 or y_{00}	π_4	$(1 - p_1)(1 - p_2)$

A slightly different double observer protocol exists, the **dependent double observer** protocol (Cook and Jacobson, 1979; Nichols et al., 2000). Using this protocol, one observer serves as a "recorder," recording each detection by a primary observer and, in addition, recording birds missed by the primary observer. The two observers switch roles for some of the samples, so that all parameters of the model are identifiable. The key feature is that it results in two different multinomial samples (i.e., with different cell probabilities). When observer 1 is the recorder, there are two count frequencies produced, y_{1-} and y_{01}, being individuals counted by observer 1 and individuals missed by observer 1

7.2 MULTINOMIAL N-MIXTURE MODELS IN ECOLOGY

but counted by the recorder, respectively. The cell probabilities are p_1 and $(1 - p_1)p_2$. When the observer and recorder switch roles, two additional frequencies are produced, which we'll label y_{-1} and y_{10}, which have probabilities p_2 and $(1 - p_2)p_1$.

Another popular protocol is **removal sampling** (Hayne, 1949; Zippin, 1956). Under this protocol, a population is sampled successively and individuals are removed during each sample so that, on average, a depletion of encounter frequencies is observed in each successive removal. Historically, removal has corresponded to catch or harvest, but more recently the removal has been either temporary, in which individuals are put in a bucket until after a number of removal samples (Dorazio et al., 2005), or "mental" in the case of bird counting in which the time interval that individuals are first detected is noted (Farnsworth et al., 2002). In this situation, j indexes the time interval of first detection, i.e., y_{i1} is the number of birds first seen in interval 1, y_{i2} in interval 2, and y_{i3} in interval 3, etc. For a three-removal interval study, the cell probabilities have the following form:

Encounter History	Frequency	Cell Probability	Function of p
1--	y_1	π_1	p_1
01-	y_2	π_2	$(1-p_1)p_2$
001	y_3	π_3	$(1-p_1)(1-p_2)p_3$
000	y_4	π_4	$1-\pi_1-\pi_2-\pi_3$

We see that for the removal sampling protocol over $J = 3$ removal periods there are $H = 4$ possible "encounter" histories. We also see that the removal design an be considered as a special case of a capture-recapure design where all data after the first detection of an individual are discarded or simply ignored.

For **capture-recapture** studies that may also occur in a meta-population sampling context (Royle et al., 2007b; Webster et al., 2008; Kéry and Royle, 2010; Converse and Royle, 2012), the data structure is analogous to that obtained under multiple observer sampling except the capture histories are organized in time, i.e., j indexes sampling occasions instead of observers. For example, in a study with $J = 3$ sample occasions there are $H = 8$ possible encounter histories including the all-zero encounter history:

Encounter Frequency	$j=1$	$j=2$	$j=3$
y_1	1	0	0
y_2	0	1	0
y_3	0	0	1
y_4	1	1	0
y_5	1	0	1
y_6	0	1	1
y_7	1	1	1
y_8	0	0	0

Capture-recapture protocols are sufficiently important that we focus on them in Section 7.8 (and, in the context of CJS models for open populations, in Chapter 15). Finally, we note that **distance sampling** when distances are binned into distance categories also leads to a multinomial observation model. We discuss these models in some detail in Chapters 8 and 9 in volume 1 and in Chapters 14 and 24 in volume 2.

7.3 SIMULATING MULTINOMIAL OBSERVATIONS IN R

To let you experience the full joy of data simulation (see Chapter 4) also for multinomial observation protocols, we next cover the simulation of multinomial random variables. There are a number of ways to simulate multinomial samples in R, which we describe here. The first method uses the function rmultinom, which is called like this: rmultinom(n, size, prob) where, here, n is the sample size (number of multinomial samples to draw), size is the number of multinomial trials upon which the multinomial variable is based (i.e., N from above), and prob is an $H \times 1$ vector of the multinomial cell probabilities. If $n = 1$ then a single vector of length H is returned, but if $n > 1$ then an $H \times n$ matrix is returned where each column has one of the multinomial outcomes in it. As an example, we generate 10 samples from a multinomial distribution with $H = 3$ and probabilities $(0.1, 0.2, 0.7)$ for $h = 1, 2, 3$, and with each multinomial sample being based on a sample size of 5:

```
rmultinom(10, 5, c(0.1, 0.2, 0.7))
     [,1] [,2] [,3] [,4] [,5] [,6] [,7] [,8] [,9] [,10]
[1,]   0    2    0    2    0    1    1    1    0     1
[2,]   1    1    1    1    1    0    0    0    0     1
[3,]   4    2    4    2    4    4    4    4    5     3
```

Another easy way to do this is by using the R function sample for directly generating categorical outcomes, which is called like this: sample(x, size, replace, prob). The argument x specifies the possible values of the categorical random variable, size is as above, replace=TRUE indicates that sampling should be done with replacement so that the same element may appear multiple times, and also prob is as above. To illustrate using sample to generate a single multinomial observation, we do this:

```
sample(1:3, 5, replace=TRUE, prob = c(0.1, 0.2, 0.7))
[1] 3 3 3 2 2
```

The returned value is a vector of five categorical outcomes that we can aggregate after the fact (e.g., using the table function) to produce a multinomial random variable. In this case, our multinomial frequencies are (0, 2, 3) indicating that two outcomes of our categorical variable had the value $y = 2$ and three outcomes had the value $y = 3$.

The third method is based on formulating the multinomial as a product of conditionally dependent binomial distributions. This is really handy for implementing some models in BUGS, which we demonstrate in Section 7.6 below. Consider above where we simulated multinomial outcomes based on a sample size of $N = 5$, with $H = 3$ possible values having cell probabilities $\pi_1 = 0.1$, $\pi_2 = 0.2$, and $\pi_3 = 0.7$. We can express the multinomial for (y_1, y_2, y_3) by the sequence of binomial random variables that are conditionally related to one another:

$$y_1 \sim Binomial(5, \pi_1)$$
$$y_2|y_1 \sim Binomial(5 - y_1, \pi_2/(\pi_2 + \pi_3))$$
$$y_3|y_1, y_2 \sim Binomial(5 - y_1 - y_2, 1)$$

In general, the binomial probabilities for the successive counts after $h = 1$ and before $h = H$ have to be "proportional to what's left over." In the first case, for simulating y_1, there have been no previous removals from $N = 5$ and so we require that proportion π_1 be allocated to category $h = 1$, on average.

7.3 SIMULATING MULTINOMIAL OBSERVATIONS IN R

For y_2, however, we have to allocate the remaining $5 - y_1$ individuals to both classes $h = 2$ and $h = 3$. Naturally, the proportion going to $h = 2$ should be proportional to the remaining total $\pi_2 + \pi_3$. Finally, if there are any trials not allocated by the time we get to $h = 3$, then we allocate all of them there, hence the binomial probability is 1 for the last class.

We note that it is also possible to construct multinomial random variables from independent Poisson random variables, which can sometimes be useful. For example, if x_1, x_2 are independent Poisson random variables with means λ_1 and λ_2, then the joint distribution of x_1 and x_2 *conditional* on their total, $T = x_1 + x_2$, is a multinomial (binomial in this case of two random variables) random variable with probabilities $\pi_1 = \lambda_1/(\lambda_1 + \lambda_2)$ and $\pi_2 = \lambda_2/(\lambda_1 + \lambda_2)$. See Section 7.8.4 for an illustration.

Finally, we can simulate repeated binary detection events for each of a number of individuals and aggregate the resulting detection histories. This is the approach taken in the simNmix data simulation function introduced in Section 6.5; see below and Section 7.8.5 for an illustration with multinomial data.

In the remainder of this section, we apply our newly acquired wealth of knowledge about simulating multinomial random variables to simulate bird sampling data of the type that might arise in practice. Suppose we sample $M = 100$ sites using a double-observer protocol, and we suppose that local population sizes for sites $i = 1, 2, ..., 100$ are outcomes of a Poisson random variable with mean λ_i related to a habitat covariate x_i, which we will simulate as a normal random variable:

```
set.seed(2015)                          # Initialize RNG

# Simulate covariate values and local population size for each point
x <- rnorm(100)
N <- rpois(100, lambda=exp(-1 + 1*x))   # Intercept and slope equal to 1
table(N)                                # Summarize
N
 0  1  2  3  4  5  6
72 17  6  1  2  1  1
```

We see that 72 sites are unoccupied (i.e., have $N = 0$), one site each has a local abundance of $N = 5$ and $N = 6$, and so on. Now we subject each sample site to a double-observer sampling protocol and suppose that one of our observers is quite skilled and has $p_1 = 0.8$ whereas the other observer is less skilled and has $p_2 = 0.6$. We now simulate some data that are consistent with this scenario.

```
# Define detection probabilities (p) for both observers
p1 <- 0.8
p2 <- 0.6

# Construct the multinomial cell probabilities (pi)
cellprobs <- c(p1*p2, p1*(1-p2), (1-p1)*p2, (1-p1)*(1-p2))

# Create a matrix to hold the data
y <- matrix(NA, nrow=100, ncol=4)
dimnames(y) <- list(1:100, c("11", "10", "01", "00"))

# Loop over sites and generate data with function R rmultinom()
for(i in 1:100){
   y[i,] <- rmultinom(1, N[i], cellprobs)
}
```

```
# Remove 4th column ("not detected") and summarize results
y <- y[,-4]
apply(y, 2, sum)
11 10 01
23 17  6
```

We note that of the $H = 4$ possible outcomes only three of them are observable, the fourth outcome corresponds to the event "not detected," and so we discard those counts because of course those are the ones we cannot see in practice. We see here that our simulated data set has 23 birds that were observed by both observers, 17 birds observed by observer 1 but not observer 2, and 6 birds detected by observer 2 but not observer 1. The key part of this process that allows us to simulate data for any multinomial sampling situation is the construction of the multinomial cell probabilities. However we define those to be, the rmultinom function will produce a frequency vector of the proper dimension and according to the specified cell probabilities.

We can use our simNmix data simulation function to generate multinomial counts under several protocols (including capture-recapture, independent double-observer, and removal) for a very wide variety of covariate or heterogeneity models; see Section 6.5 for a general description of that function. The data are simulated at the level of an individual and, hence, individual detection histories are obtained, which can then easily be postprocessed to obtain the required table of counts. For instance, here we generate removal and capture-recapture data under a model where the Poisson mean is affected by one continuous covariate and there is time dependence of detection probability. Although we can't specify p for each occasion individually, we can get random temporal variation by setting the standard deviation for visit ($=$ occasion) random effects in p to a non-zero value; this will naturally lead to time variation.

```
# Generate specific pseudo-random data set
set.seed(2014)
data <- simNmix(mean.lam = exp(1), beta3.lam = 1, mean.p = plogis(0),
     sigma.p.visit = 1, show.plot=FALSE)
str(data$DH)

# View detection histories for site with max abundance (here, N = 30)
t(data$DH[min(which(data$N == max(dim(data$DH)[3]))),,])
```

The detection histories come in the format site $\times$ occasion $\times$ individual, and the length of the third dimension of the array is the maximum local population size N realized over all M sites. Hence, this regular array contains a lot of missing values, which have to be discarded before the analysis. In addition, the unobserved, all-zero detection histories also must be tossed out. The following code generates counts for a capture-recapture protocol for the default three visits, and it must be adapted "by hand" when the number of visits is different from three:

```
# Get detection history frequencies for each site (for exactly 3 surveys)
dhfreq <- array(NA, dim = c(data$nsite, 7),
    dimnames = list(NULL, c("100", "010", "001", "110", "101", "011", "111")))
for(i in 1:data$nsite){
   dhfreq[i,] <- table(factor(paste(data$DH[i,1,], data$DH[i,2,],
   data$DH[i,3,], sep = ""),
   levels = c("100", "010", "001", "110", "101", "011", "111")))
}
```

```
head(dhfreq)         # Data for first 6 sites
      100 010 001 110 101 011 111
[1,]    0   0   2   0   0   1   0
[2,]    0   1   1   0   0   4   0
[3,]    0   0   0   0   0   0   0
[4,]    0   0   0   0   0   1   0
[5,]    0   0   0   0   0   0   0
[6,]    0   0   4   0   0  11   2
```

To format the same detection histories into a table of removal counts, you simply do this (this code works for any number of occasions).

```
# Get occasions with first detection of each individual
f <- apply(data$DH, c(1,3), function(x) min(which(x != 0)))
head(f)  ;  str(f)  ;  table(f)    # Inspect result

# Produce removal counts
y <- array(NA, dim = c(data$nsite, data$nvisit), dimnames = list(NULL,
    as.factor(1:data$nvisit)))
for(i in 1:data$nsite){
   y[i,] <- table(factor(f[i,], levels = as.character(1:data$nvisit)))
}
head(y)      # Data for first 6 sites
     1  2  3
[1,] 0  1  2
[2,] 0  5  1
[3,] 0  0  0
[4,] 0  1  0
[5,] 0  0  0
[6,] 2 11  4
```

As shown in Section 6.5, you can use this function to generate binomial and multinomial counts under an extremely wide variety of models. Here is an example where abundance follows a negative binomial with mean affected by site covariates 2, 3, and 4, and with time variation in detection, in addition to an effect on detection of site covariate 3, which has the opposite sign to that of the covariate on abundance. You can postprocess the resulting detection histories to obtain multinomial counts under a removal or capture-recapture protocol in an exactly analogous way to the above, and input these data into unmarked later:

```
set.seed(24)
data <- simNmix(mean.lam = exp(0), beta2.lam = 1, beta3.lam =-1,
   beta4.lam = 0.2, dispersion = 1, mean.p = plogis(0),
   beta3.p = 1, sigma.p.visit = 1, Neg.Bin = TRUE)
```

7.4 LIKELIHOOD INFERENCE FOR MULTINOMIAL N-MIXTURE MODELS

With multinomial sampling protocols both detection probability parameters and N for each site are estimable from the site-specific data alone. Therefore, in principle, we could obtain maximum likelihood estimates of N separately for each site. However, in practice, it will not generally be feasible to estimate the collection of abundance parameters $\{N_i\}_{i=1}^{M}$ as distinct parameters primarily because, for many species, the site-specific abundance parameters N_i are small owing to small sampled area, and low densities of many species, especially territorial birds. Consequently, the sample size (# of individuals observed) at each site will also be small (often even 0). One solution is to pool the data and use all the data to estimate encounter probability parameters using conditional likelihood ideas (White, 2005; Royle, 2004a; Converse and Royle, 2012) and then obtain independent estimates of N for each site by "adjusting" the observed count of individuals. This approach has been frequently applied in the context of small-sample capture-recapture studies based on replicate trapping grids. An alternative, which we advocate here, is to view the N_i parameters as latent variables, for which we can specify explicit models, such as Poisson or negative binomial. As with all hierarchical models, we can adopt a classical inference framework based on marginal likelihood, in which the collection of latent abundance parameters, $\{N_i\}_{i=1}^M$, is removed from the conditional-on-N likelihood by summation. The marginal distribution of the data for each site, $\mathbf{y}_i$, can be computed by summing over all possible values of the latent variable N_i:

$$f(\mathbf{y}_i|\mathbf{p},\theta) = \sum_{N_i=0}^{\infty} Multinomial(\mathbf{y}_i|N_i, \boldsymbol{\pi}(\mathbf{p})) g(N_i|\theta).$$

The joint likelihood for the data from all $i = 1, 2, \ldots, M$ sites is then the product of M such pieces:

$$f(\mathbf{y}_1, \mathbf{y}_2, \ldots, \mathbf{y}_M|\mathbf{p},\theta) = \prod_{i=1}^{M} f(\mathbf{y}_i|\mathbf{p},\theta).$$

As a practical matter we may truncate the summation to some large number, say N_{max}, and calculate the marginal probability as follows:

$$f(\mathbf{y}_i|\mathbf{p},\theta) = \sum_{N_i=0}^{N_{max}} Multinomial(\mathbf{y}_i|N_i, \boldsymbol{\pi}(\mathbf{p})) g(N_i|\theta). \quad (7.1)$$

If N_{max} (called K in unmarked functions pcount and gmultmix) is chosen to be too small, the likelihood is truncated and the estimates will not be the MLEs under the specified model but, rather, correspond to MLEs under the corresponding model truncated at N_{max} (Section 6.4). While this may not be unreasonable, we prefer to avoid sensitivity to choice of N_{max}. In practice, when we might have doubt about whether N_{max} was chosen to be sufficiently large, we should fit the model for some value of N_{max} and then increment N_{max} a little bit to make sure that the estimates don't change.

For the special case where N_i is Poisson with mean λ_i, we can compute the marginal likelihood (Eq. (7.1)) analytically (Royle, 2004a; Dorazio et al., 2005) and avoid having to deal with truncating the summation. For the Poisson case, the likelihood reduces to that of a collection of independent Poisson random variables:

$$f(\mathbf{y}_1, \mathbf{y}_2, \ldots, \mathbf{y}_M|\mathbf{p},\theta) = \prod_i \prod_h Poisson(\lambda_i \pi_h)$$

Therefore, the multinomial/Poisson mixture models are basically just Poisson GLMs but with expected value that has a contribution both from the observation model (the π_h piece) and the abundance model (the λ_i piece), and π_h effectively serves as an offset (see also Section 3.3.2), but one that is a function of unknown model parameters.

It is easy to maximize this likelihood using standard optimization utilities in R such as `optim` or `nlm`. The R package `unmarked` has two basic functions for fitting multinomial mixture models (in addition, for those for distance sampling, see next chapter): `multinomPois` and `gmultmix`. The former is, as the name suggests, restricted to the multinomial-Poisson mixture, whereas the latter will also fit negative binomial mixtures and a certain type of open population model (see Chapter 13) allowing for temporary emigration or, equivalently, availability less than 1 (Chandler et al., 2011). The two distinct functions for fitting multinomial mixtures are completely redundant from a technical standpoint (there is nothing you can't do with `gmultmix` that you can with `multinomPois`) but the Poisson version of the function is an earlier implementation of this class of models that is extremely fast and therefore might be useful in specific cases with very large data sets. The formula structure is slightly different for the two functions with `gmultmix` formula elements being separated by commas but no commas for `multinomPois`. The order of the formulas for the `multinomPois` function has the detection probability model first, followed by the abundance model. For the `gmultmix` function, the order is: detection model, availability model, abundance model. Therefore, to fit a basic model with constant parameter values to the ovenbird data (see next section) we do this, after having created the appropriate `unmarked` data frames:

```
multinomPois( ~ 1 ~ 1, data = ovenFrame)
gmultmix( ~1, ~1, ~1, data = ovenFrame)
```

The argument `data` is an "unmarked data frame" (here named `ovenFrame`). We create these for multinomial mixture models using helper functions `unmarkedFrameGMM` and `unmarkedFrameMPois`, which help set up your data for use by either of the two model fitting functions. We show the details of this in the next section where we construct the object `ovenFrame`. Either of these helper functions to create the `unmarked` data frame requires that you specify the structure of the cell probabilities (the argument `type=`), either by setting `type=removal` or `type=double` (for double-observer sampling). These options automatically create the pi function to convert p to multinomial cell probabilities π. Other types of models, such as capture-recapture, can be fitted, but special attention has to be given to these; see Section 7.7 below where we discuss constructing custom pi functions. Typically, other arguments, such as covariate data, may be passed to the `unmarkedFrame` constructor functions, too, depending on the specific analysis at hand (we provide examples shortly).

7.5 EXAMPLE 1: BIRD POINT COUNTS BASED ON REMOVAL SAMPLING

Here we analyze bird point count data collected using a mental removal protocol (Farnsworth et al., 2002) by which the observer records first detections of individuals into a number of successive time intervals, e.g., the observer might do a five-minute point count and record individuals into one-minute subcounts. Implicit in the use of this protocol is that the observer, upon first seeing or hearing an individual, mentally blocks it from subsequent observation so that the individual only appears in one time interval. Naturally, the observer might also obtain a full encounter history for each individual (Webster et al., 2008), which might be more informative about model parameters (see Section 7.8). However, this is less commonly done, presumably because it is relatively easier to obtain removal interval data instead of full capture-history data. We will here provide analyses of a standard removal data set using both `unmarked` functions `multinomPois` and `gmultmix`.

This particular data set was collected in Maryland forest units comprising Catoctin Mountain National Park (NP) and Frederick Co. Wildlife Management Area (WMA) by our colleagues Deanna Dawson (USGS) and Scott Bates (NPS). The study area and the 70 point count locations (35 in each of the two land units, randomly selected for sampling from a grid of points, spaced by 250 m), are shown in Figure 7.2. The data were processed into four removal intervals of three minutes each, although a version of the data set based on distance sampling was analyzed by Royle et al. (2004).

Summarizing the data structure, we have removal frequencies $\mathbf{y}_i = (y_{i1}, y_{i2}, y_{i3}, y_{i4})$ (removed in intervals 1, 2, 3, and 4, respectively) for $i = 1, 2, \ldots, 70$ point count locations.

```
data(ovendata)
ovendata.list$data[11:20,]   # Look at a snippet of data set for sites 11-20
       [,1] [,2] [,3] [,4]
 [1,]    0    0    0    0
 [2,]    0    0    0    0
 [3,]    2    0    1    0
 [4,]    1    0    0    0
 [5,]    1    0    0    0
 [6,]    0    0    0    0
 [7,]    0    1    0    1
 [8,]    0    1    0    0
 [9,]    0    0    0    0
[10,]    0    0    0    0

apply(ovendata.list$data,2,sum)   # Removals in occasion 1-4
[1]  49  16   5   7
```

For the removal protocol, the observation model is multinomial and the cell probabilities for the four observable frequencies (removed on occasion 1–4, respectively), are:

$$\pi_1 = p$$
$$\pi_2 = (1-p)p$$
$$\pi_3 = (1-p)^2 p$$
$$\pi_4 = (1-p)^3 p$$

The focus of the study had to do with effects of deer management on vegetation characteristics and hence bird populations and diversity (Zipkin et al., 2010). Deer are not hunted in the Catoctin Mountain NP but they are in the Frederick City Watershed Cooperative WMA and where deer harvest is allowed the understory should be richer. So we have two habitat covariates measured in the vicinity of each point count location, which we think might affect local abundance: understory foliage cover (UFC) and basal area of large trees (TRBA). We look at data for the ovenbird (*Seiurus aurocapillus*; Figure 7.3), which is provided with the unmarked package and can be loaded simply by executing this command data(ovendata). We will consider models for local abundance of the form:

$$N_i \sim Poisson(\lambda_i)$$
$$\log(\lambda_i) = \beta_0 + \beta_1 \text{UFC}_i + \beta_2 \text{TRBA}_i$$

7.5 EXAMPLE 1: BIRD POINT COUNTS BASED ON REMOVAL SAMPLING

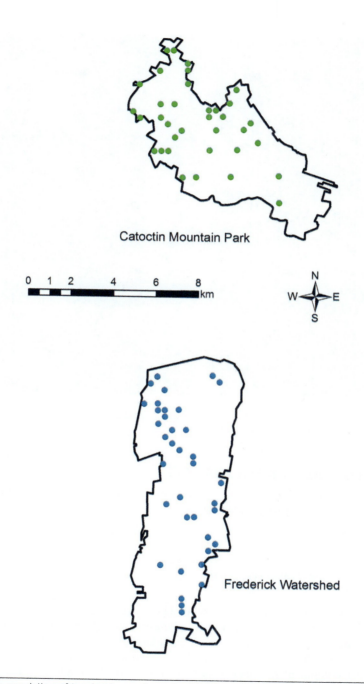

FIGURE 7.2

Catoctin study area consisting of two land parcels each with 35 point count locations: The Frederick City Watershed Cooperative Wildlife Management Area (south) and the Catoctin Mountain National Park (north).

FIGURE 7.3

The Ovenbird (*Seiurus aurocapillus*). *(Photo R. Royse.)*

7.5.1 SETTING UP THE DATA FOR ANALYSIS

To load the data and set things up for an analysis as an `unmarkedFrame` using the function `multinomPois` we issue the following commands:

```
library(unmarked)
data(ovendata)
ovenFrame <- unmarkedFrameMPois(y = ovendata.list$data,
    siteCovs = as.data.frame(scale(ovendata.list$covariates[,-1])),
    type = "removal")
```

The necessary arguments to the constructor function `unmarkedFrameMPois` are the multinomial encounter frequencies y, and the `type` of multinomial observations ("removal" or "double" are built-in options). The argument `siteCovs` is a matrix of covariates that vary by site, and for the ovenbird analysis we're inputting a 70 × 2 matrix having the standardized UFC and TRBA covariates (above, we remove the first column of the data, which are site IDs). Although not present here, another covariate argument is `obsCovs`, which is a list having 1 or more $M \times J$ matrices, each one containing covariates that vary by site and sample occasion (or observer, or removal interval).

7.5.2 FITTING MODELS USING FUNCTION multinomPois

We follow this up by fitting a few basic models involving our two covariates. We consider a set of main effects models and interaction models and, furthermore, we also consider the effect of UFC, package these into a fitList and then produce an AIC table.

```
# Fit models: multinomPois order of formulas: detection, abundance
fm0 <- multinomPois(~ 1 ~ 1, ovenFrame)
fm1 <- multinomPois(~ 1 ~ ufc, ovenFrame)
fm2 <- multinomPois(~ 1 ~ trba, ovenFrame)
fm3 <- multinomPois(~ 1 ~ ufc + trba, ovenFrame)
fm4 <- multinomPois(~ 1 ~ ufc + trba + ufc:trba, ovenFrame)
fm5 <- multinomPois(~ ufc ~ ufc + trba, ovenFrame)
fm6 <- multinomPois(~ ufc ~ ufc + trba + ufc:trba, ovenFrame)

# Rank models by AIC
ms <- fitList(
"lam(.)p(.)"                      = fm0,
"lam(ufc)p(.)"                    = fm1,
"lam(trba)p(.)"                   = fm2,
"lam(ufc+trba)p(.)"               = fm3,
"lam(ufc+trba+ufc:trba)p(.)"      = fm4,
"lam(ufc+trba)p(ufc)"             = fm5,
"lam(ufc+trba+ufc:trba)p(ufc)"    = fm6)

(ms1 <- modSel(ms))
                              nPars     AIC  delta AICwt cumltvWt
lam(trba)p(.)                     3  324.77   0.00 0.284     0.28
lam(ufc)p(.)                      3  325.73   0.96 0.176     0.46
lam(ufc+trba)p(.)                 4  326.14   1.37 0.143     0.60
lam(.)p(.)                        2  326.28   1.51 0.134     0.74
lam(ufc+trba)p(ufc)               5  326.63   1.86 0.112     0.85
lam(ufc+trba+ufc:trba)p(.)        5  327.17   2.40 0.086     0.93
lam(ufc+trba+ufc:trba)p(ufc)      6  327.72   2.95 0.065     1.00

# Table with everything you could possibly need
coef(ms1)[,1:4] # Only first 4 columns shown
                              lambda(Int) lambda(trba) lambda(ufc) lambda(ufc:trba)
lam(trba)p(.)                   0.1062626   -0.2202827          NA               NA
lam(ufc)p(.)                    0.1134562           NA   0.1789146               NA
lam(ufc+trba)p(.)               0.1023681   -0.1708609   0.1002939               NA
lam(.)p(.)                      0.1296349           NA          NA               NA
lam(ufc+trba)p(ufc)             0.1069856   -0.1704169   0.1333443               NA
lam(ufc+trba+ufc:trba)p(.)      0.1547003   -0.1499694   0.1563209        0.1298563
lam(ufc+trba+ufc:trba)p(ufc)    0.1568600   -0.1503608   0.1854883        0.1264758
```

You can obtain a table that has coefficients, standard errors, AIC, AIC weights, and other summary information as follows:

```
output <- as(ms1, "data.frame")
```

Looking at the basic model selection table above we see that the models containing the single variables TRBA and UFC are the top models and then the model containing both. We see a negative effect of TRBA and a positive effect of UFC, which makes sense for the ovenbird, an understory nesting and foraging species. The variables are highly (negatively) correlated, which may explain why the model with both variables only ranks third in terms of AIC.

7.5.3 FITTING MODELS USING FUNCTION gmultmix

The unmarked package allows for a more general class of multinomial-mixture models that can be fitted using the gmultmix function ("g" here for *generalized*). This function accommodates the following models:

- The negative binomial abundance model. The negative binomial parameterization used in gmultmix is by the mean, λ, and logarithm of the negative binomial "size" parameter, say $\log(\tau)$. The variance is $\lambda + \lambda^2/\tau$. Therefore, as $1/\tau \to 0$ (or $\tau \to \infty$), the negative binomial tends to the Poisson (i.e., no overdispersion is indicated).
- A type of *open* model (with temporary emigration) with an argument numPrimary, which is the number of primary sampling periods among which closure is not satisfied; see Chandler et al. (2011).
- Closed population models by specifying numPrimary=1 (these models can be fit with function *multinomPois*).

To set up the data for analysis using gmultmix, we use the constructor function unmarkedFrameGMM as follows:

```
ovenFrame <- unmarkedFrameGMM(ovendata.list$data,
    siteCovs=as.data.frame(scale(ovendata.list$covariates[,-1])),
    numPrimary=1,type = "removal")
```

Now we are free to fit various different models using gmultmix. However, note that there is an additional formula in this more general model, including not just a formula for λ and p but also a formula for a new parameter ϕ, which is the "availability" parameter (the complement of temporary emigration probability). Models in which $\phi = 1$ are ordinary closed population models whereas, in general, temporary emigration is allowed when $\phi < 1$. The more general models are described in Chapter 13. When the data are based on a single sample occasion (numPrimary = 1), then ϕ is internally fixed at $\phi = 1$. Using gmultmix we specify the model by a set of three formulae as follows, in the case of a model with constant parameters.

```
fm0 <- gmultmix(lambdaformula = ~1, phiformula = ~1, pformula = ~1,
    data=ovenFrame)
```

in which, as we noted above, we see here that the component model expressions are separated by commas unlike in the more specific function multinomPois.

We repeat some of the analyses using the Poisson abundance model as before but using the gmultmix function. In addition, we fit an equivalent set of negative binomial abundance models, and we'll package all of the results into a fitList and provide a model selection table based on AIC (note how it takes longer now to fit these models than with multinomPois, and even longer when the mixture is NB):

```
# Fit Poisson models
fm1 <- gmultmix(~ ufc, ~ 1, ~ 1, data = ovenFrame)
fm2 <- gmultmix(~ trba, ~ 1, ~ 1, data = ovenFrame)
```

7.5 EXAMPLE 1: BIRD POINT COUNTS BASED ON REMOVAL SAMPLING

```
fm3 <- gmultmix(~ ufc + trba, ~ 1, ~ 1, data = ovenFrame)
fm4 <- gmultmix(~ ufc + trba + ufc:trba, ~ 1, ~ 1, data = ovenFrame)
# Maybe p also depends on understory foliage?
fm5 <- gmultmix(~ ufc + trba, ~ 1, ~ ufc, data = ovenFrame)
fm6 <- gmultmix(~ ufc + trba + ufc:trba, ~ 1, ~ ufc, data = ovenFrame)

# Fit analogous NegBin models
fm0nb <- gmultmix(~ 1, ~ 1, ~ 1, mixture = "NB", data = ovenFrame)
fm1nb <- gmultmix(~ ufc, ~ 1, ~ 1, mixture = "NB", data = ovenFrame)
fm2nb <- gmultmix(~ trba, ~ 1, ~ 1, mixture = "NB", data = ovenFrame)
fm3nb <- gmultmix(~ ufc + trba , ~ 1, ~ 1, mixture = "NB", data = ovenFrame)
fm4nb <- gmultmix(~ ufc + trba + ufc:trba, ~ 1, ~ 1, mixture = "NB",
        data = ovenFrame)
# maybe p also depends on understory foliage?
fm5nb <- gmultmix(~ ufc + trba, ~ 1, ~ ufc, mixture = "NB",
        data = ovenFrame)
fm6nb <- gmultmix(~ ufc + trba + ufc:trba, ~ 1, ~ ufc, mixture = "NB",
        data = ovenFrame)

# Rank models by AIC
gms <- fitList(
"lam(.)p(.)"                            = fm0,
"lam(ufc)p(.)"                          = fm1,
"lam(trba)p(.)"                         = fm2,
"lam(ufc+trba)p(.)"                     = fm3,
"lam(ufc+trba+ufc:trba)p(.)"            = fm4,
"lam(ufc+trba)p(ufc)"                   = fm5,
"lam(ufc+trba+ufc:trba)p(ufc)"          = fm6,
"NB,lam(.)p(.)"                         = fm0nb,
"NB,lam(ufc)p(.)"                       = fm1nb,
"NB,lam(trba)p(.)"                      = fm2nb,
"NB,lam(ufc+trba)p(.)"                  = fm3nb,
"NB,lam(ufc+trba+ufc:trba)p(.)"         = fm4nb,
"NB,lam(ufc+trba)p(ufc)"                = fm5nb,
"NB,lam(ufc+trba+ufc:trba)p(ufc)"       = fm6nb)

(gms1 <- modSel(gms))

# Table with everything you could possibly need
output <- as(gms1, "data.frame")

# Summary results
gms1
                                nPars    AIC  delta AICwt cumltvWt
lam(trba)p(.)                       3 302.35  0.00  0.208    0.21
lam(ufp)p(.)                        3 303.32  0.96  0.128    0.34
lam(ufp+trba)p(.)                   4 303.72  1.37  0.105    0.44
lam(.)p(.)                          2 303.86  1.51  0.097    0.54
lam(ufp+trba)p(ufp)                 5 304.21  1.86  0.082    0.62
NB,lam(trba)p(.)                    4 304.35  2.00  0.076    0.70
lam(ufp+trba+ufp:trba)p(.)          5 304.75  2.40  0.063    0.76
lam(ufp+trba+ufp:trba)p(ufp)        6 305.30  2.95  0.047    0.81
```

```
NB,lam(ufp)p(.)                         4 305.31  2.95 0.047  0.85
NB,lam(ufp+trba)p(.)                    5 305.72  3.37 0.038  0.89
NB,lam(.)p(.)                           3 305.78  3.43 0.037  0.93
NB,lam(ufp+trba)p(ufp)                  6 306.21  3.86 0.030  0.96
NB,lam(ufp+trba+ufp:trba)p(.)           6 306.75  4.40 0.023  0.98
NB,lam(ufp+trba+ufp:trba)p(ufp)         7 307.32  4.97 0.017  1.00
```

We see that the Poisson models are favored by AIC, i.e., no overdispersion is indicated. This is supported by inspecting the summary results for the best negative binomial model fm2nb:

```
fm2nb

Call:
gmultmix(lambdaformula = ~trba, phiformula = ~1, pformula = ~1,
    data = ovenFrame, mixture = "NB")

Abundance:
            Estimate    SE       z      P(>|z|)
(Intercept)    0.106  0.119   0.894     0.371
trba          -0.221  0.121  -1.831     0.067

Detection:
 Estimate    SE      z    P(>|z|)
    0.288  0.233  1.24    0.216

Dispersion:
 Estimate   SE     z    P(>|z|)
     4.57 16.9  0.271   0.786

AIC: 304.3497
```

The negative binomial "size" parameter is exp(4.57) ≈ 97, indicating negligible overdispersion relative to the Poisson model. We note also that the AIC values produced by gmultmix are *not* comparable to those produced by multinomPois, even for the Poisson models. This is because the likelihood is constructed in a different manner in the two functions and therefore **AIC scores should not in general be compared across different** unmarked **fitting functions**.

It is often useful to examine the coefficients on the different covariates, and especially their signs, to inspect the sensitivity of our conclusions to model choice. We do this for the ovenbird analysis using the coef summary function (output slightly formatted to fit on the page):

```
> print(coef(gms1), digits = 2)
                     alpha(alpha) lambda(Int) lambda(trba) lambda(ufc) lambda(ufc:trba) p(Int) p(ufc)
lam(trba)p(.)                  NA        0.11        -0.22          NA               NA   0.29     NA
lam(ufc)p(.)                   NA        0.11           NA        0.18               NA   0.29     NA
lam(ufc+trba)p(.)              NA        0.10        -0.17        0.10               NA   0.29     NA
lam(.)p(.)                     NA        0.13           NA          NA               NA   0.29     NA
lam(ufc+trba)p(ufc)            NA        0.11        -0.17        0.13               NA   0.34  -0.34
NB,lam(trba)p(.)              4.6        0.11        -0.22          NA               NA   0.29     NA
[... rows deleted ...]
NB,lam(ufc)p(.)               4.1        0.11           NA        0.18               NA   0.29     NA
NB,lam(ufc+trba)p(.)          5.3        0.10        -0.17        0.10               NA   0.29     NA
NB,lam(.)p(.)                 2.9        0.13           NA          NA               NA   0.29     NA
[... rows deleted ...]
```

We find a positive understory foliage (UFC) effect on $E(N)$ (= expected abundance, lambda) and a corresponding negative effect of tree basal area (TRBA) on $E(N)$, which are sensible results given our understanding of the species. Further, there is a positive interaction (though not "significant") indicating that the effect of UFC becomes stronger (more positive) as TRBA increases (we could visualize these joint effects by producing two-dimensional prediction plots as in Section 6.9.4). There is some evidence of a negative effect of UFC on encounter probability, although this is not a strong effect. A useful feature of hierarchical models such as the multinomial mixture model considered here is that they permit modeling effects of the same covariate on both encounter probability *and* the state variable of interest (abundance in this case; Kéry, 2008). The potential for effects on encounter probability (a nuisance effect) to cancel out important biological effects is often given as a critique against using simple regression approaches or just raw "index" summaries, and it seems to us to be a clear advantage of hierarchical models. Finally, in comparing the Poisson and negative binomial models for the ovenbird data, we find no evidence of extra-Poisson variation.

7.5.4 ASSESSING MODEL FIT IN unmarked

The unmarked package has general functions for bootstrap goodness-of-fit (GoF) testing; see, e.g., Sections 2.8.1, 5.10, 6.8. The parboot function takes an R function defining the fit statistic(s), and an unmarked fit object, and carries out a parametric bootstrap goodness-of-fit evaluation. This involves using the parameter estimates provided by the fit object to simulate new data sets, fit the model to each new data set, and compute the fit statistic(s) for each of the resulting model fits. To use the parboot function we need to define a function that computes fit statistics from the fit object in question. We use the function fitstats (see the unmarked helpfile ?parboot), which computes fit statistics based on error sums-of-squares, the standard Chi-square and the Freeman-Tukey statistic, and then call parboot using our best fit object fm2, which is a Poisson model with covariate TRBA on expected population size.

```
set.seed(2015)
(gof <- parboot(fm2, fitstats, nsim = 1000, report = 1))
[... reporting output deleted ....]
Call: parboot(object = fm2, statistic = fitstats, nsim = 1000, report = 1)

Parametric Bootstrap Statistics:
             t0  mean(t0 - t_B)  StdDev(t0 - t_B)  Pr(t_B > t0)
SSE         72.4         -3.2696            12.05         0.595
Chisq      307.0         30.4393            28.95         0.126
freemanTukey 64.2         -0.0364             5.43         0.520

t_B quantiles:
             0%  2.5%  25%  50%  75%  97.5%  100%
SSE          46    55   67   75   83    102   124
Chisq       193   226  257  273  293    343   381
freemanTukey 48    54   60   64   68     74    79
```

These results indicate that the best model appears to fit the data reasonably well with the bootstrap *p*-value not being extreme (not so close to 0 or 1) for any of the three fit statistics. That being said, we don't know whether any of these fit statistics have reasonable power under any specific alternative, so some basic calibration studies might be useful (as in Section 6.8).

7.6 BAYESIAN ANALYSIS IN BUGS USING THE CONDITIONAL MULTINOMIAL (THREE-PART) MODEL

Bayesian analysis of multinomial observation models does not pose any novel technical difficulty. BUGS has a multinomial distribution and, therefore, in principle, we could think about specifying the multinomial/Poisson or other mixture directly like this:

```
# Set-up data with a missing value for element "not captured"
y[i,] <- c(y[i,],NA)

# Then, in BUGS, do this:
y[i,] ~ dmulti(probs[i,], N[i])
N[i] ~ dpois(lambda[i])
```

However, this construction *does not work* in BUGS because a random sample size argument is not allowed in the multinomial distribution (this is different in the binomial, as we saw in Chapter 6). There are a number of solutions to get around this deficiency, including the following three approaches:

1. Use the multinomial/Poisson analytic result and express the model for each count as a Poisson random variable with mean $\pi_{hi}\lambda_i$. We may even model overdispersion directly on λ_i in this way; see Section 7.6.3.
2. There is a data augmentation formulation of this from Converse and Royle (2012) that is based on individual-level encounter histories.
3. We can express the model in terms of the conditional multinomial observation model, i.e., condition on n_i = number of individuals captured at site i. This method breaks the multinomial observation model up into two pieces: a multinomial conditioned on a known sample size (being the number of encountered individuals) and a binomial with the unknown sample size (see also Section 8.5.3).

We first demonstrate the conditional multinomial approach (method 3) here using the ovenbird data. The model is formally expressed by the following three component models:

$$y_i | n_i \sim Multinom(n_i, \pi_i^c)$$
$$n_i \sim Binomial(N_i, 1 - \pi_0)$$
$$N_i \sim Poisson(\lambda_i)$$

where $\pi_h^c = \pi_h/(1 - \pi_0)$ and $1 - \pi_0$ is total capture probability (pcap in BUGS below). The BUGS model specification is given in the following bit of code where we have labeled each of the three parts of the model with comments and where we add two GLM-type regressions for detection probability and expected abundance:

```
# Harvest the data and bundle it up for sending to BUGS
y <- as.matrix(getY(ovenFrame))
ncap <- apply(y, 1, sum)    # number of individuals removed per point
data <- list(y = y, M = nrow(y), n = ncap, X=as.matrix(siteCovs(ovenFrame)))
str(data)                   # Good practice to always inspect your BUGS data
```

7.6 BAYESIAN ANALYSIS IN BUGS USING THE CONDITIONAL MULTINOMIAL

```
# Write BUGS model
cat("
model {

# Prior distributions
p0 ~ dunif(0,1)
alpha0 <- logit(p0)
alpha1 ~ dnorm(0, 0.01)
beta0 ~ dnorm(0, 0.01)
beta1 ~ dnorm(0, 0.01)
beta2 ~ dnorm(0, 0.01)
beta3 ~ dnorm(0, 0.01)

for(i in 1:M){ # Loop over sites
   # Conditional multinomial cell probabilities
   pi[i,1] <- p[i]
   pi[i,2] <- p[i]*(1-p[i])
   pi[i,3] <- p[i]*(1-p[i])*(1-p[i])
   pi[i,4] <- p[i]*(1-p[i])*(1-p[i])*(1-p[i])
   pi0[i] <- 1 - (pi[i,1] + pi[i,2] + pi[i,3] + pi[i,4])
   pcap[i] <- 1 - pi0[i]
   for(j in 1:4){
     pic[i,j] <- pi[i,j] / pcap[i]
   }

   # logit-linear model for detection: understory cover effect
   logit(p[i]) <- alpha0 + alpha1 * X[i,1]

   # Model specification, three parts:
   y[i,1:4] ~ dmulti(pic[i,1:4], n[i])   # component 1 uses the conditional
                                         #   cell probabilities
   n[i] ~ dbin(pcap[i], N[i])            # component 2 is a model for the
                                         #   observed sample size
   N[i] ~ dpois(lambda[i])               # component 3 is the process model

   # log-linear model for abundance: UFC + TRBA + UFC:TRBA
  log(lambda[i])<- beta0 + beta1*X[i,1] + beta2*X[i,2] + beta3*X[i,2]*X[i,1]
}
}
",fill=TRUE, file="model.txt")

# Initial values
inits <- function(){
  list (p0 = runif(1), alpha1 = runif(1), beta0 = runif(1), N = ncap+2)
}

# Parameters monitored
params <- c("p0", "alpha0", "alpha1", "beta0", "beta1", "beta2", "beta3")
```

```
# MCMC settings
nc <- 3 ; ni <- 6000 ; nb <- 1000 ; nt <- 1

# Experience the power of BUGS and print posterior summary
library(R2WinBUGS)
bd <- "c:/Program Files/WinBUGS14/"
out <- bugs(data, inits, params, "model.txt", n.thin = nt,
    n.chains = nc, n.burnin = nb, n.iter = ni, debug = TRUE,
    bugs.directory = bd)
print(out, 3)

Inference for Bugs model at "model.txt", fit using WinBUGS,
 3 chains, each with 6000 iterations (first 1000 discarded)
 n.sims = 15000 iterations saved
            mean      sd     2.5%     25%     50%     75%   97.5%  Rhat n.eff
p0         0.569   0.061    0.443   0.530   0.572   0.612   0.681 1.002  2900
alpha0     0.283   0.253   -0.230   0.119   0.291   0.457   0.759 1.001  3400
alpha1    -0.383   0.332   -1.128  -0.581  -0.352  -0.157   0.196 1.002  2200
beta0      0.155   0.137   -0.116   0.065   0.154   0.247   0.424 1.001  8200
beta1      0.214   0.154   -0.077   0.111   0.211   0.311   0.533 1.002  2100
beta2     -0.152   0.140   -0.424  -0.245  -0.152  -0.058   0.126 1.001  3400
beta3      0.134   0.138   -0.127   0.040   0.132   0.226   0.416 1.001 15000
deviance 156.838  16.040  132.697 145.000 154.200 166.000 194.602 1.002  2900

# To make trace plots using CODA functionality
plot(as.mcmc.list(out), ask = TRUE)
```

Next we fit this model in JAGS using the `jagsUI` package (Kellner, 2015).

```
library(jagsUI)
out <- jags(data, inits, parameters, "model.txt", n.thin=nt,
    n.chains = nc, n.burnin = nb, n.iter = ni)
print(out, 3)
            mean      sd     2.5%     50%   97.5% overlap0     f  Rhat n.eff
p0         0.569   0.061    0.437   0.571   0.681    FALSE 1.000 1.001  1981
alpha0     0.281   0.254   -0.254   0.287   0.760     TRUE 0.869 1.001  2047
alpha1    -0.379   0.314   -1.036  -0.368   0.209     TRUE 0.895 1.001  3222
beta0      0.150   0.133   -0.114   0.152   0.410     TRUE 0.867 1.002   847
beta1      0.210   0.156   -0.089   0.210   0.518     TRUE 0.914 1.003  7738
beta2     -0.146   0.137   -0.412  -0.146   0.125     TRUE 0.860 1.008   261
beta3      0.131   0.138   -0.132   0.127   0.406     TRUE 0.830 1.001 15000
deviance 156.745  15.656  132.856 154.082 194.066    FALSE 1.000 1.002  1355
```

It is relatively straightforward to fit more complex models that accommodate overdispersion, such as negative binomial, zero-inflated Poisson, or Poisson/lognormal. For the latter, we add a normal random effect to the log-linear model for $E(N)$ (or lambda), which requires only a couple of minor modifications to the previous BUGS model specification (see also Section 6.11). We have to define the normal random effects `eta` within the M loop, and then add it to the linear predictor:

```
eta[i] ~ dnorm(0, tau)
log(lambda[i]) <- beta0 + beta1*X[i,1] + ... + eta[i]
```

7.6 BAYESIAN ANALYSIS IN BUGS USING THE CONDITIONAL MULTINOMIAL

In addition, we require a prior distribution for the parameter tau, which could be a gamma prior distribution or a uniform prior on a transformation to the standard deviation scale. The negative binomial model could be fitted directly using the dnegbin distribution. However, the negative binomial implemented in WinBUGS or JAGS is not the "mean/variance" parameterization that is convenient for modeling covariate effects. Instead, we would have to reparameterize the model in terms of the mean and variance or standard deviation. This example of "moment-matching" (Hobbs and Hooten 2015) was borrowed from Jim Albert's website (http://bayes.bgsu.edu/multilevel/jags/nb$/ %$20regression/JAGSnbregression.R; Jim Albert cites the book by Jackman (2009) for the original example). A sketch of the implementation to do this is as follows (we use this for an analysis in Chapter 9) (for another example of moment-matching see Section 5.12):

```
model{
for(i in 1:n){
   mu[i] <- linear predictor for mean abundance on log scale
   lambda[i] <- exp(mu[i])
   Q[i] <- r / (r + lambda[i])
   y[i] ~ dnegbin(Q[i], r)
}
r ~ dunif(0, 50)
# Other priors here too.....
}
```

You could easily adapt this construction to the model specification given above for the ovenbird data (see Exercise 1). We have been told that fitting negative binomial models does not pose problems in WinBUGS, but JAGS requires the following setting (executed in R before running BUGS) in order to fit these models (R. Sollmann, personal communication; see also Section 9.7.2.1):

```
set.factory("bugs::Conjugate", FALSE, type="sampler")
```

7.6.1 GOODNESS-OF-FIT USING BAYESIAN p-VALUES

We implement a standard Bayesian p-value approach (Sections 2.8 and 6.8) to assess model fit for the multinomial mixture models. For the conditional formulation described in the previous section, in which we have distinct observation models for y conditional on n and for the total number of individuals encountered, n, it is natural to think of overall "fit" as having two components: the fit of the encounter model (i.e., how well does the model for y conditional on n fit) and the fit of the model for data n. This second bit is more of a test for the abundance part of the model since the variation in N shows up in the expected value of n. We use the Bayesian p-value framework for evaluating the two components of fit separately.

To implement the Bayesian p-value, we need to define a statistic that quantifies fit (or lack thereof), and for that purpose we use the Freeman–Tukey statistic:

$$FIT = \sum \left(\sqrt{y} - \sqrt{E[y]}\right)^2.$$

At each MCMC iteration (say, m) we compute this statistic for the observed data, call this $FIT_{data}^{(m)}$, and simultaneously we obtain a new data set simulated from the posterior distribution and compute the fit statistic for the new data set at iteration m, $FIT_{pred}^{(m)}$. The Bayesian p-value is the probability of getting a more extreme value of the test statistic under the Null hypothesis of a fitting model, i.e., the fraction of times $FIT^{(m)}_{pred} > FIT^{(m)}_{data}$. For implementation in BUGS, we require a few additional lines of BUGS model specification added to the model specification given previously (just copy this into the BUGS model file):

```
for(i in 1:M){
   n.pred[i] ~ dbin(pcap[i], N[i])
   y.pred[i, 1:4] ~ dmulti(pic[i,1:4], n[i]) #note this is cond'l on ncap[i]
   for(k in 1:4){
      e1[i,k] <- pic[i,k] * n[i]
      resid1[i,k] <- pow(pow(y[i,k], 0.5)-pow(e1[i,k], 0.5), 2)
      resid1.pred[i,k] <- pow(pow(y.pred[i,k], 0.5) - pow(e1[i,k], 0.5), 2)
   }
   e2[i] <- pcap[i] * lambda[i]
   resid2[i] <- pow(pow(n[i], 0.5) - pow(e2[i], 0.5), 2)
   resid2.pred[i] <- pow(pow(n.pred[i], 0.5) - pow(e2[i], 0.5), 2)
}
fit1.data <- sum(resid1[,])           # Fit statistic for observed data y
fit1.pred <- sum(resid1.pred[,])

fit2.data<- sum(resid2[])             # Fit statistic for new data n
fit2.pred<- sum(resid2.pred[])
```

Conducting the analysis proceeds as before except we must retain the fit statistics in the params argument and then summarize the posterior probability that the fit statistic for the posterior simulated "new" data is larger than that for the observed data. The additional commands to do all of this are:

```
params <- c("N", "p0", "beta0", "beta1", "beta2", "beta3",
    "fit1.data", "fit1.pred", "fit2.data", "fit2.pred")
out <- bugs (data, inits, params, "model.txt",n.thin=nt, n.chains=nc,
    n.burnin=nb, n.iter=ni, debug=TRUE, bugs.dir = bd)

mean(out$sims.list$fit1.pred > out$sims.list$fit1.data)
[1] 0.714

mean(out$sims.list$fit2.pred > out$sims.list$fit2.data)
[1] 0.4066667
```

We see that the Bayesian *p*-values for the observation part of the model and the abundance part of the model are 0.714 and 0.407, respectively. This suggests that the model in question provides an adequate fit to the data.

7.6.2 MODEL SELECTION IN BUGS

Posterior model probabilities can be easily obtained using the Kuo and Mallick (1998) indicator variable method (see Section 2.7.3). Using this approach we expand the model to include a set of binary indicator variables, say $w_x = 1$ if variable "*x*" is in the model. Model selection then is equivalent to estimating the posterior probabilities $Pr(w_x = 1)$. Examples of this in the context of occupancy models are in Royle and Dorazio (2008, 3.4.3), O'Hara and Sillanpää (2009), Tenan et al. (2014a), and

7.6 BAYESIAN ANALYSIS IN BUGS USING THE CONDITIONAL MULTINOMIAL

Hooten and Hobbs (2015). To illustrate using the ovenbird model that we've been developing, we can express the model as follows:

$$\log(\lambda_i) = \beta_0 + w_1 * \beta_1 * x_{1i} + w_2 * \beta_2 * x_{2i} + w_1 * w_2 * w_3 * \beta_3 * x_{1i} * x_{2i} \quad (7.2)$$

All possible models are characterized by the possible values of $(w_1, w_2, w_1 * w_2 * w_3)$. Note that to obey the rules of marginality (McCullagh and Nelder, 1989, p. 89) the interaction only appears if both w_1 and w_2 are equal to 1.

It remains to specify prior probabilities for the w_x variables. Normally we use $w_x \sim$ *Bernoulli*(0.5) so there is an equal probability of x being included in the model as not being included. The choice of prior distributions for each w_x induces nonuniform prior probabilities for interaction variables defined in the manner of Eq. (7.2). In particular, because the interaction is only included if all three indicator variables equal 1, the prior probability of including the interaction is ⅛. This may or may not be desirable. As an alternative, we could enumerate the specific models of interest and assign probabilities in model space, and then define the w_x variables as derived parameters depending on the model (e.g., see Royle et al., 2014, Section 8.2.4).

We illustrate the use of the indicator variables to carry out Bayesian model selection for the abundance part of the ovenbird model that we've been working with. The R commands for importing and processing the data are the same as in the previous analyses, but the BUGS model file changes slightly to accommodate the indicator variables. The new BUGS model file reads as follows:

```
# Write BUGS model
cat("
model {

# Prior distributions
p0 ~ dunif(0,1)
alpha0 <- log(p0/(1-p0))
alpha1 ~ dnorm(0, 0.1)
beta0 ~ dnorm(0, 0.1)
beta1 ~ dnorm(0, 0.1)
beta2 ~ dnorm(0, 0.1)
beta3 ~ dnorm(0, 0.1)
w1 ~ dbern(0.5)
w2 ~ dbern(0.5)
w3 ~ dbern(0.5)

for(i in 1:M){
   # Conditional multinomial cell probabilities
   pi[i,1] <- p[i]
   pi[i,2] <- p[i] * (1-p[i])
   pi[i,3] <- p[i] * (1-p[i]) * (1-p[i])
   pi[i,4] <- p[i] * (1-p[i]) * (1-p[i]) * (1-p[i])
   pi0[i] <- 1 - (pi[i,1] + pi[i,2] + pi[i,3]+ pi[i,4])
   pcap[i] <- 1 - pi0[i]
   for(j in 1:4){
      pic[i,j] <- pi[i,j] / pcap[i]
   }
```

```
# logit-linear model for detection: understory cover effect
    logit(p[i]) <- alpha0 + alpha1 * X[i,1]

# Model specification, 3 parts:
    y[i,1:4] ~ dmulti(pic[i,1:4], n[i])   # component 1 uses the conditional
                                          #    cell probabilities
    n[i] ~ dbin(pcap[i], N[i])            # component 2 is a model for the
                                          #    observed sample size
    N[i] ~ dpois(lambda[i])               # component 3 is the process model

# log-linear model for abundance: effects of UFC and TRBA, with weights
    log(lambda[i]) <- beta0 + w1*beta1*X[i,1] + w2*beta2*X[i,2] +
w1*w2*w3*beta3*X[i,2]*X[i,1]
}
}
",fill=TRUE,file="model.txt")
```

We fit the model as follows (only new parts shown):

```
# Parameters monitored
params <- c("p0", "alpha0", "alpha1", "beta0", "beta1", "beta2",
    "beta3", "w1", "w2", "w3")

# Initial values
set.seed(2015)
inits <- function(){
  list (p0 = runif(1), alpha1 = runif(1), beta0 = runif(1), N = ncap+2,
    w1=1, w2=1, w3=1)
}
# Call WinBUGS from R and summarize marginal posteriors
out <- bugs(data, inits, params, "model.txt", n.thin = nt,
    n.chains = nc, n.burnin = nb, n.iter = ni, debug = TRUE, bugs.dir=bd)
print(out, digits = 3)
```

In order to obtain the posterior model probabilities, we have to post-process the results to obtain model frequencies by combining the unique values of w_1, w_2, and w_3, of which there are a total of eight unique combinations of the three binary variables. i.e., (1,0,0), (0,1,0), (1,1,0), etc. The following R commands accomplish this summarization into model probabilities:

```
# Extract the indicator variables for model selection
w1 <- out$sims.list$w1
w2 <- out$sims.list$w2

# Create a new w3 variable which takes on the value 1 if the
#     interaction is in the model, also requires w1 = 1 AND w2 = 1
w3 <- out$sims.list$w3 * w1 * w2

# Combine into a model indicator string and tabulate posterior frequencies
mod <- paste(w1, w2, w3, sep = "")
table(mod)
```

7.6 BAYESIAN ANALYSIS IN BUGS USING THE CONDITIONAL MULTINOMIAL

```
mod
 000  010  100  110  111
9867 2505 2493  133    2

# Convert to probabilities
table(mod) / length(out$sims.list$w1)
mod
         000          010          100          110          111
0.6578000000 0.1670000000 0.1662000000 0.0088666667 0.0001333333
```

While your results will vary slightly from this, we see that the model without any effects has most of the posterior probability and the models with one or the other receive only 17% (UFC) and about 8% (TRBA). This is quite a different result compared to what we found using model selection based on AIC. A number of possible explanations for this exist. One is that our model set here doesn't include the covariate on encounter probability. Another possible explanation has to do with the prior distribution on the model set. Our implied prior probabilities for the five different models can be computed directly from the prior probabilities on each indicator variable. For the case where $\Pr(w = 1) = 0.5$, we have the following prior probabilities on the set of possible models:

Model	Prior Model Probability
No effects	$0.5 * 0.5 = 0.25$
UFC only	$0.5 * 0.5 = 0.25$
TRBA only	$0.5 * 0.5 = 0.25$
both UFC and TRBA	$0.5 * 0.5 * 0.5 = 0.125$
UFC + TRBA + UFC:TRBA	$0.5 * 0.5 * 0.5 = 0.125$

Conversely, AIC is known to correspond to a specific prior in which the probabilities decrease exponentially with the number of parameters (Kadane and Lazar, 2004; Link and Barker, 2006). Yet another explanation for the difference between AIC and posterior model probabilities could be the inherent sensitivity of posterior model probabilities to prior distributions on parameters. It is well known that posterior model probabilities are sensitive to the prior distribution for parameters (Link and Barker, 2010), and our results here are based on a normal prior with variance 10. To assess this sensitivity, we can see that if we consider alternative prior distributions, then we obtain different statements of posterior model probabilities. For example, we ran the above analysis, but using normal priors with variance 5 and 100; these produce the following different posterior model frequencies (note that specific numerical results will vary by chance):

```
Prior: beta ~ dnorm(0, 0.2)    # Variance 5

 000  010  100 110  111
9122 3112 2575 171   20

Prior: beta ~ dnorm(0, 0.01)   # Variance 100

  000  010  100 110
12779 1049 1166    6
```

From this we see a substantial effect of changing the prior for the coefficients on posterior model probabilities which is typical of Bayesian model selection (Link and Barker, 2006). As the precision decreases, the posterior model probability increases on the null model and gets pulled away from the other models in the model set. In most cases this would not be regarded as desirable, but it is a basic reality of Bayesian analysis.

A potential solution to this sensitivity-to-prior issue is to run the full model and then use estimates of the posterior as the prior distribution in a model-selection phase (Aitkin, 1991; see also Tenan et al., 2014a). Alternatively, we could use different prior distributions depending on whether the variable is part of the model or not. There is a good discussion and explanation of this in Hooten and Hobbs (2015), which we reproduce here with permission, using notation modified slightly to conform to ours:

> In implementing an indicator variable selection model, one would be tempted to use independent priors for z_j and θ_j; for example, we might specify
>
> $$z_j \sim Bernoulli(\phi)$$
> $$\theta_j \sim Normal(0, \tau^2)$$
>
> for all $j = 1, 2, \ldots, p$, assuming the covariates are standardized. However, an independent prior specification can cause computational problems if the prior for θ_j is too vague (i.e., the prior variance, τ^2, is large) because when $z_j = 0$ in an MCMC algorithm, θ_j will be sampled from its prior and the subsequent sampling of future $z_j = 1$ will rarely occur since the θ_j is likely to be far from the majority of posterior mass. Thus, to alleviate these computational problems, others (e.g., George and McCulloch, 1993; Carlin and Chib, 1995) have suggested joint priors for z_j and θ_j that include explicit dependence between the indicators and coefficients.
>
> In Gibbs variable selection, Carlin and Chib (1995) and Dellaportas et al. (2002) suggest decomposing the joint prior distribution $[z_j, \theta_j] = [\theta_j|z_j][z_j]$. In this joint prior specification, the Bernoulli prior for z_j is retained, but the prior for θ_j conditional on z_j is written as
>
> $$\theta_j|z_j \sim z_j \text{Norm}(0, \tau^2) + (1 - z_j)\text{Norm}(\mu_{tune}, \sigma^2_{tune})$$
>
> which has the form of a mixture distribution and is often referred to as a "slab and spike" prior (Miller, 2002). The Gibbs variable selection procedure then involves choosing the tuning parameters μ_{tune} and σ^2_{tune} such that $\text{Norm}(\mu_{tune}, \sigma^2_{tune})$ is near the posterior so that the MCMC algorithm exhibits better mixing. Surprisingly, the seemingly informative prior does not actually influence the posterior for θ_j, but rather only influences the behavior of the MCMC algorithm (Carlin and Chib, 1995).

7.6.3 POISSON FORMULATION OF THE MULTINOMIAL MIXTURE MODEL

In Section 7.4 we noted that, under the Poisson abundance model, the marginal likelihood can be computed analytically. In particular, if $N_i \sim Poisson(\lambda_i)$ and $\mathbf{y}_i|N_i \sim Multinom(N_i, \boldsymbol{\pi})$, then we can regard each element of the multinomial count vector as an *independent* Poisson count with expected value $\lambda_i \pi_h$. Using this Poisson formulation of the model, N_i no longer appears in the model (recall, in

7.6 BAYESIAN ANALYSIS IN BUGS USING THE CONDITIONAL MULTINOMIAL

the multinomial formulation above, we had $n_i \sim Bin(N_i, p_{cap})$, and so we do not directly specify a model for N_i, but we can still obtain predictions of it for every site. We can express the Poisson model directly in the BUGS language as we demonstrate here for the ovenbird model for which we gave the multinomial version in the previous section (using the BUGS data bundle from before):

```
# Specify model in BUGS language
cat("
model {

# Prior distributions
p0 ~ dunif(0,1)
alpha0 <- logit(p0)
alpha1 ~ dnorm(0, 0.01)
beta0 ~ dnorm(0, 0.01)
beta1 ~ dnorm(0, 0.01)
beta2 ~ dnorm(0, 0.01)
beta3 ~ dnorm(0, 0.01)

for(i in 1:M){
  # logit-linear model for detection: understory cover effect
  logit(p[i]) <- alpha0 + alpha1 * X[i,1]
  # log-linear model for abundance: UFC + TRBA + UFC:TRBA
  log(lambda[i]) <- beta0 + beta1*X[i,1] + beta2*X[i,2] + beta3*X[i,2]*X[i,1]

  # Poisson parameter = multinomial cellprobs x expected abundance
  pi[i,1] <- p[i] * lambda[i]
  pi[i,2] <- p[i] * (1-p[i]) * lambda[i]
  pi[i,3] <- p[i] * (1-p[i]) * (1-p[i]) * lambda[i]
  pi[i,4] <- p[i] * (1-p[i]) * (1-p[i]) * (1-p[i]) * lambda[i]

  for(j in 1:4){
      y[i,j] ~ dpois(pi[i,j])
  }
  # Generate predictions of N[i]
  N[i] ~ dpois(lambda[i])
}
}
",fill=TRUE,file="modelP.txt")

# Bundle up the data and inits
data <- list(y = y, M = nrow(y), X = as.matrix(siteCovs(ovenFrame)))
inits <- function(){
  list (p0 = runif(1), alpha1=runif(1), beta0=runif(1), beta1=runif(1),
beta2=runif(1), beta3=runif(1))
}

# Define parameters to save and MCMC settings
params <- c("p0", "alpha1", "beta0", "beta1", "beta2", "beta3", "N")
nc <- 3 ; ni <- 6000 ; nb <- 1000 ; nt <- 1
```

In the model specification given above, note that even though N_i no longer appears as an explicit parameter in the model, we can still generate predictions of each N by adding the line of BUGS code N[i] ~ dpois(lambda[i]). These are not conditional on the observed number of individuals but, rather, are predictions at sites as if they were not sampled (i.e., they are not the Bayesian analogue to frequentist BUPs; see Section 6.4).

```
# Call WinBUGS from R and summarize marginal posteriors
out <- bugs(data, inits, params, "modelP.txt", n.thin=nt,
    n.chains = nc, n.burnin = nb, n.iter = ni, debug = TRUE, bugs.dir=bd)
print(out, 3)

Inference for Bugs model at "modelP.txt", fit using WinBUGS,
 3 chains, each with 6000 iterations (first 1000 discarded)
 n.sims = 15000 iterations saved
            mean     sd    2.5%     25%     50%     75%   97.5%   Rhat  n.eff
p0         0.570  0.062   0.441   0.530   0.573   0.613   0.683  1.001  15000
alpha1    -0.399  0.317  -1.093  -0.591  -0.383  -0.183   0.192  1.002   1700
beta0      0.155  0.134  -0.114   0.069   0.155   0.242   0.414  1.001   4800
beta1      0.217  0.154  -0.083   0.114   0.212   0.317   0.528  1.001  14000
beta2     -0.153  0.140  -0.429  -0.248  -0.150  -0.059   0.119  1.001   4200
beta3      0.136  0.137  -0.122   0.042   0.132   0.227   0.415  1.001  15000
deviance 321.885  3.570 317.100 319.200 321.200 323.800 330.600  1.002   2000

For each parameter, n.eff is a crude measure of effective sample size,
and Rhat is the potential scale reduction factor (at convergence, Rhat=1).

DIC info (using the rule, pD = Dbar-Dhat)
pD = 6.0 and DIC = 327.9
DIC is an estimate of expected predictive error (lower deviance is better).
```

We conclude the following things: While there is a slight negative response of understory cover on detection probability, the posterior distribution overlaps 0 to a large extent. We see similar effects on abundance as before: a positive effect of understory cover and a negative effect of basal area of large trees.

One of the great things about BUGS is that you're not limited by fitting models that someone else has figured out the likelihood construction or MCMC algorithm for. All you have to be able to do is describe the model in the BUGS model specification language and you can build fairly complex models without having to know much about algorithms or computing. We illustrate this point here by considering the Poisson/lognormal model as a means of modeling overdispersion in local population size, and we fit the model to the ovenbird data. Note how the simple addition of an "extra residual" eta to the linear predictor of the abundance model (and specification of eta as a draw from a zero-mean normal distribution with an estimated variance) is all the change that is needed.

```
# Specify model in BUGS language
cat("
model {

# Prior distributions
p0 ~ dunif(0,1)
alpha0 <- logit(p0)
```

7.6 BAYESIAN ANALYSIS IN BUGS USING THE CONDITIONAL MULTINOMIAL

```
alpha1 ~ dnorm(0, 0.01)
beta0 ~ dnorm(0, 0.01)
beta1 ~ dnorm(0, 0.01)
beta2 ~ dnorm(0, 0.01)
beta3 ~ dnorm(0, 0.01)

tau ~ dgamma(0.1,0.1)  # Excess-Poisson variation (precision)
sigma <- sqrt(1 / tau)

for(i in 1:M){
    # logit-linear model for detection: understory cover effect
    logit(p[i]) <- alpha0 + alpha1 * X[i,1]
    # Normal random effects
    eta[i] ~ dnorm(0, tau)  # 'residuals' for extra-Poisson noise
    # log-linear model for abundance: UFC + TRBA + UFC:TRBA + eta
    log(lambda[i]) <- beta0 + beta1*X[i,1] + beta2*X[i,2] +
beta3*X[i,2]*X[i,1] + eta[i]       # note 'extra-residual' for overdispersion

    # Poisson parameter = multinomial cellprobs x expected abundance
    pi[i,1] <- p[i] * lambda[i]
    pi[i,2] <- p[i] * (1-p[i]) * lambda[i]
    pi[i,3] <- p[i] * (1-p[i]) * (1-p[i]) * lambda[i]
    pi[i,4] <- p[i] * (1-p[i]) * (1-p[i]) * (1-p[i]) * lambda[i]

    for(j in 1:4){
        y[i,j] ~ dpois(pi[i,j])
    }
    # Generate predictions of N[i]
    N[i] ~ dpois(lambda[i])
  }
}
",fill=TRUE,file="modelP.txt")

# Inits
inits <- function(){
  list(p0 = runif(1), alpha1=runif(1), beta0=runif(1), beta1=runif(1),
      beta2=runif(1), beta3=runif(1), tau = 1) }

# Define parameters to save and MCMC settings
params <- c("p0", "alpha1", "beta0", "beta1", "beta2", "beta3", "sigma")
nc <- 3 ; ni <- 32000 ; nb <- 2000 ; nt <- 1

# Call WinBUGS from R and summarize marginal posteriors
out <- bugs(data, inits, params, "modelP.txt", n.thin=nt,
    n.chains = nc, n.burnin = nb, n.iter = ni, debug = TRUE, bugs.dir = bd)
print(out, digits=3)

Inference for Bugs model at "modelP.txt", fit using WinBUGS,
 3 chains, each with 32000 iterations (first 2000 discarded)
 n.sims = 90000 iterations saved
```

```
          mean     sd    2.5%     25%     50%     75%   97.5%   Rhat  n.eff
p0       0.569  0.062   0.438   0.530   0.573   0.613   0.682  1.001  84000
alpha1  -0.376  0.313  -1.040  -0.571  -0.359  -0.164   0.200  1.001  90000
beta0    0.075  0.160  -0.252  -0.027   0.080   0.182   0.379  1.001  90000
beta1    0.214  0.170  -0.111   0.101   0.212   0.324   0.554  1.001  69000
beta2   -0.161  0.158  -0.478  -0.267  -0.159  -0.055   0.151  1.001  23000
beta3    0.142  0.157  -0.155   0.034   0.136   0.244   0.461  1.001  90000
sigma    0.406  0.149   0.181   0.293   0.385   0.497   0.744  1.002   2700
deviance 312.738 7.881 296.100 307.700 313.200 318.100 327.300 1.001  13000

[..output truncated..]
plot(density(out$sims.list$sigma), frame = F, main = "") # Figure 7.4
```

We don't find a lot of extra-Poisson dispersion (Figure 7.4). This is perhaps not surprising, given that our GoF assessment earlier let us conclude that the simpler Poisson (rather than the NB) model was adequate for the data.

7.7 BUILDING CUSTOM MULTINOMIAL MODELS IN unmarked

By default unmarked can accommodate two types of multinomial sampling models: double observer and removal sampling. These are specified by the type = removal or type = double options when using the constructor functions (e.g., unmarkedFrameGMM). These options automatically create the pi function that converts encounter probability parameters p to multinomial cell probabilities π. Internally, unmarked has a special function (called the piFun) that builds the multinomial cell probabilities depending on the specified type of sampling protocol. The piFun function converts a matrix of

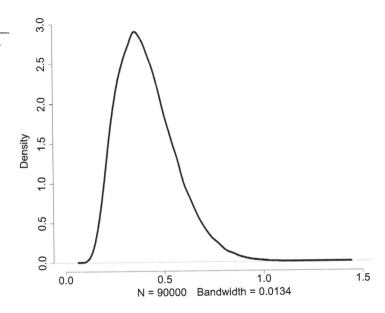

FIGURE 7.4

Posterior distribution of the excess-Poisson variance parameter sigma (σ) fitted to the ovenbird data.

detection probabilities having *J* columns (sample occasions) into a matrix of cell probabilities with $H = 2^J - 1$ columns (corresponding to the number of *observable* encounter histories). Each column corresponds to the probability of observing the encounter history *h*. That is, piFun maps "per sample" detection probabilities to multinomial cell probabilities. The pi function for the removal and double observer models are the defaults available in unmarked, and you can investigate what these two functions do with the following bit of code that is available in the help file ?piFuns.

```
# Removal model: capture probs for 5 sites, with 3 removal periods
(pRem <- matrix(0.5, nrow=5, ncol=3))
     [,1] [,2] [,3]
[1,]  0.5  0.5  0.5
[2,]  0.5  0.5  0.5
[3,]  0.5  0.5  0.5
[4,]  0.5  0.5  0.5
[5,]  0.5  0.5  0.5

removalPiFun(pRem)   # Multinomial cell probabilities for each site
     [,1] [,2]  [,3]
[1,]  0.5 0.25 0.125
[2,]  0.5 0.25 0.125
[3,]  0.5 0.25 0.125
[4,]  0.5 0.25 0.125
[5,]  0.5 0.25 0.125

# Double observer model: capture probs for 5 sites, with 2 observers
(pDouble <- matrix(0.5, 5, 2))
     [,1] [,2]
[1,]  0.5  0.5
[2,]  0.5  0.5
[3,]  0.5  0.5
[4,]  0.5  0.5
[5,]  0.5  0.5

doublePiFun(pDouble)  # Multinomial cell probabilities for each site
     [,1] [,2] [,3]
[1,] 0.25 0.25 0.25
[2,] 0.25 0.25 0.25
[3,] 0.25 0.25 0.25
[4,] 0.25 0.25 0.25
[5,] 0.25 0.25 0.25
```

The multinomial cell probability matrix created by removalPiFun gives, for each site (row), the probability of being removed in the first pass, second pass, and third pass, respectively. The matrix created by doublePiFun is organized according to: the probability observer 1 but not observer 2 detects the object (column 1); the probability that observer 2 but not observer 1 detects the object (column 2); and the probability that both detect it (column 3).

Instead of specifying one of the two default multinomial types, we can create custom multinomial cell probabilities by writing our own piFun to define the probabilities and then use the fitting functions for multinomial mixture models to fit a wide array of models that arise in practice. In the next section, we use this capability to create general pi functions for fitting capture-recapture models. For now, we provide an illustration (thanks to the esteemed Prof. Richard Chandler) of a custom pi function that is given in the documentation for `PiFuns`, which provides a user-defined pi function for removal sampling when the removal intervals differ in length. The specific example is for the situation where you have a 10-minute point count divided into three intervals of length 2, 3, and 5 minutes. The function expresses the detection probability for each time interval relative to a base per-minute detection probability and is defined as follows (see `?PiFuns`):

```
instRemPiFun <- function(p){
    M <- nrow(p)
    J <- ncol(p)
    pi <- matrix(NA, M, J)
    p[,1] <- pi[,1] <- 1 - (1 - p[,1])^2
    p[,2] <- 1 - (1 - p[,2])^3
    p[,3] <- 1 - (1 - p[,3])^5
    for(i in 2:J) {
        pi[,i] <- pi[, i - 1]/p[, i - 1] * (1 - p[, i - 1]) * p[, i]
    }
    return(pi)
}
```

When we execute the function using the previously created 5 × 3 matrix of probabilities, we find that the multinomial cell probabilities change substantially to reflect the variable length removal intervals:

```
instRemPiFun(pRem)
     [,1]   [,2]      [,3]
[1,] 0.75 0.21875 0.03027344
[2,] 0.75 0.21875 0.03027344
[3,] 0.75 0.21875 0.03027344
[4,] 0.75 0.21875 0.03027344
[5,] 0.75 0.21875 0.03027344
```

When providing a user-defined pi function, we also need to provide information about how to handle missing values. That is, if we have a missing value in a covariate, we need to tell unmarked which values of y should be excluded from the likelihood. The need for this will be made clear in a moment, with specific examples. In unmarked, control of how missing values are handled can be done by supplying a matrix called obsToY as an argument to the appropriate constructor function. The obsToY matrix needs to be a matrix of zeros and ones with the number of rows equal to the number of columns of obsCov, and the number of columns equal to the number of columns in the $M \times (H - 1)$ matrix of encounter frequencies **Y**. If obsToY[j,h] = 1, then a missing value in obsCov[i,j] removes y[i,h] from the likelihood calculation. For a standard double-observer data set with observer ID being an obsCov, all elements of obsToY should generally be 1, i.e., we define the obsToY matrix like this (two observers, three observable encounter histories):

```
o2y <- matrix(1, 2, 3)
```

The consequence of this is that if some covariate specified as an `obsCov` is missing in any sample occasion, no element of the encounter frequency vector for that site is contributing anything to the likelihood. This is because, in a double-observer type of protocol, *all* p_j parameters are required to compute any of the multinomial cell probabilities (see Section 7.2.2). Therefore, even though the encounter history frequencies are nonmissing, the fact that there is a missing covariate causes `unmarked` to discard that observation.

Now consider the handling of missing covariate values in removal sampling. Unlike in the case of the double-observer model just discussed, we can have missing covariate values that do *not* cause us to discard the likelihood contribution for the whole site (but we typically will not have observation covariates in a removal sampling situation). For a removal study with $J = 3$ removal periods, the standard `obsToY` matrix looks like this:

```
o2y
     [,1] [,2] [,3]
[1,]   1    1    1
[2,]   0    1    1
[3,]   0    0    1
```

The relationship between the rows and columns of this matrix determines the linkage between parameters p_j and the cell probabilities. Therefore, each row corresponds to a p_j and each column corresponds to an encounter history probability. We put labels on rows and columns here along with notes to the right about what each row means:

```
     |-encounter history-|
      1xx  01x  001
p1     1    1    1      p1 appears in all 3 encounter histories
p2     0    1    1      p2 only appears in the last 2 encounter histories
p3     0    0    1      p3 appears up in the 3rd encounter history only
```

This tells `unmarked`: Covariates in sample occasion 1, which affect p_1, influence the probabilities of all three encounter frequencies (1xx, 01x, 001). But covariates in occasion 2, which affect p_2, only influence the encounter frequencies 01x and 001, and covariates in occasion 3, which affect p_3, only influence encounter history 001. So think about this in the context of a real study involving a removal sampling study of salamanders along a segment of stream, where we do three passes of the stream and we have a covariate that is the amount of time spent searching for salamanders. The underpaid technician forgot to set his stopwatch on the third pass and therefore the covariate entry for this stream segment reads 45 32 NA. The effect of this is that `unmarked` will not count the third removal frequency for that segment. If you inadvertently used an `obsToY` matrix that was all 1's, then `unmarked` would delete the whole observation for that stream segment.

7.8 SPATIALLY STRATIFIED CAPTURE-RECAPTURE MODELS

Previously we noted that capture-recapture models have exactly the same multinomial structure as double-observer methods, but they occur in practice in somewhat more generality than double- or multiple-observer sampling. For example, small mammal trapping studies based on replicate trapping grids produce spatially indexed capture-recapture data, and many bird studies involve arrays of mist nets that produce data of a similar structure. When replicate observations are available, meta-community studies (see Chapter 11) also produce data that correspond exactly to the capture-recapture protocol and we simply have species taking the place of individuals. We can analyze capture-recapture data in some generality using the multinomial mixture models in unmarked (Chandler, 2015). In our development, here we reproduce aspects of the analysis from Chandler (2015), which is an unmarked vignette available at: http://cran.r-project.org/web/packages/unmarked/vignettes/cap-recap.pdf and (in the next section) we develop an analysis of data from the Swiss survey of common breeding birds (Monitoring Häufige Brutvögel, MHB; Schmid et al., 2004) and produce glorious color maps of population density.

To recapitulate the model, we have the multinomial N-mixture model of the form:

$$\mathbf{y}_i | N_i \sim Multinomial(N_i, \pi(p))$$
$$N_i \sim Poisson(\lambda)$$

where, in this equation, $\pi(p)$ is a function that converts the basic encounter probability parameters p to multinomial cell probabilities. For capture-recapture studies, the observable data are individual encounter histories $\mathbf{y} = \{y_j\}$, a vector of 0's and 1's indicating whether an individual was encountered in each of $j = 1, 2, ..., J$ sample occasions. The canonical parameters of the model are the occasion-specific encounter probabilities p_j, which can be converted to multinomial cell probabilities according to:

$$Pr(\mathbf{y}) = \prod_j p_j^{y_j} (1 - p_j)^{1-y_j}$$

for any observable encounter history $\mathbf{y}$. For example, if $J = 2$, there are only three observable encounter histories, (10, 01, 11), precisely the same as for double-observer sampling, and the multinomial cell probabilities are:

$$\pi(p) = \{p(1-p), (1-p)p, p^2\}.$$

In addition, the probability of not capturing an individual is $(1-p)^2$. When p is assumed to be constant, this is what is referred to as model M_0 in the capture-recapture literature. Alternatively, p may vary by sampling occasion (referred to as model M_t), may depend on whether an individual has been previously encountered (a "behavioral response," model M_b), there may be individual heterogeneity (model M_h), or there may be combinations of these different influences (Otis et al., 1978; Williams et al., 2002; Royle and Dorazio, 2008; Kéry and Schaub, 2012).

To make use of the capabilities of unmarked to fit spatially stratified capture-recapture models, we need to build a custom pi function that builds the multinomial cell probabilities from a matrix of occasion-specific p_j parameters. Shortly we will analyze data based on $J = 3$ sample occasions, and so we provide the piFun for that case:

```
crPiFun <- function(p) {
    p1 <- p[,1]
    p2 <- p[,2]
    p3 <- p[,3]
    cbind("001" = (1 - p1) * (1 - p2) *      p3,
          "010" = (1 - p1) *      p2  * (1 - p3),
          "011" = (1 - p1) *      p2  *      p3,
          "100" =      p1  * (1 - p2) * (1 - p3),
          "101" =      p1  * (1 - p2) *      p3,
```

```
        "110" =  p1 *  p2 * (1 - p3),
        "111" =  p1 *  p2 *      p3)
}
```

The input matrix p is supposed to have three columns, one for each sample occasion, and the guts of the function rearrange those columns in specific ways to compute the multinomial cell probabilities. The input matrix p is generated internally by unmarked depending on the formula given. This particular function will allow us to fit a number of specific capture-recapture models, which we illustrate shortly. To demonstrate how the basic function works, imagine that we surveyed two sites and detection probability was constant ($p = 0.4$) among sites and survey occasions. The function converts these capture probabilities to multinomial cell probabilities:

```
p <- matrix(0.4, 2, 3)
crPiFun(p)
        001    010    011   100    101    110   111
[1,] 0.144  0.144  0.096 0.144  0.096  0.096 0.064
[2,] 0.144  0.144  0.096 0.144  0.096  0.096 0.064

# To compute pi0, the probability of being uncaptured, we do this:
(pi0 <- 1 - rowSums(crPiFun(p)))
[1] 0.216 0.216
```

Hence, at both sites the probability of *not* detecting an individual at all is 0.216.

7.8.1 EXAMPLE 2: FITTING MODELS M_0, M_t, AND M_x TO CHANDLER'S FLYCATCHER DATA

We illustrate the fitting of some spatially stratified capture-recapture models using the example from Chandler (2015; see also Chandler et al., 2009a,b) based on point count data on the alder flycatcher (*Empidonax alnorum*). Capture-recapture data can be obtained using point count methods by keeping track of individual birds during a single point count divided into shorter intervals (e.g., a three-minute point count divided into one-minute intervals). Like "mental removal" this is a mental capture-recapture protocol, where the mark exists only in our head as the point count is being conducted (Alldredge et al., 2007; Webster et al., 2008). Chandler's alder flycatcher data were collected at 50 point count locations, using 15-minute point counts, divided into five-minute intervals. (Each point was also surveyed three times on different days during 2005, which allows certain kinds of open models to be fitted, see Chapter 13.) We import the capture histories for 98 individuals (over all three days).

```
alfl <- read.csv(system.file("csv", "alfl.csv", package="unmarked"))
head(alfl, 5)
        id survey interval1 interval2 interval3
1 crick1_05      1         1         1         1
2 crick1_05      3         1         0         1
3   his1_05      1         0         1         1
4   his1_05      1         1         1         1
5   his1_05      2         0         1         1
```

We note that the "id" column represents the point count ID and not the ID of the *individual* bird. The five rows of data displayed here represent encounter histories of five unique birds detected at two different point count locations over the three survey days. For example, bird 1, detected at location crick1_05, during survey day 1, was detected in all three of the five-minute intervals. Associated with the bird point count data are site- and visit-specific covariates for each of the 50 sites. Next, we import the covariate data into our R work session.

```
alfl.covs <- read.csv(system.file("csv", "alflCovs.csv",package="unmarked"),
       row.names=1)
head(alfl.covs)
          struct  woody  time.1  time.2  time.3  date.1  date.2  date.3
crick1_05   5.45   0.30   8.68    8.73    5.72      6      25      34
his1_05     4.75   0.05   9.43    7.40    7.58     20      32      54
hisw1_05   14.70   0.35   8.25    6.70    7.62     20      32      47
hisw2_05    5.05   0.30   7.77    6.23    7.17     20      32      47
kenc1_05    4.15   0.10   9.57    9.55    5.73      8      27      36
kenc2_05    9.75   0.40   9.10    9.12    9.12      8      27      36
```

Each row of the data matrix contains the covariate values for a given point count location, defined as follows:

- struct is a measure of vegetation structure
- woody is the percent cover of woody vegetation at each of the 50-m radius plots
- time.x is the time of day for each of the three sample occasions
- date.x is the day of each sample occasion.

To format the data for unmarked, we need to tabulate the frequency of each unique encounter history for each point count location or site. Operationally this is done (in R) by pasting together the binary encounter events from each interval. Then we'll stuff those into a new column called captureHistory and specify the potential levels of the factor using the levels function. At the same time, we will also set the levels for the "id" (point count location) column.

```
alfl$captureHistory <- paste(alfl$interval1, alfl$interval2, alfl$interval3,
    sep="")
alfl$captureHistory <- factor(alfl$captureHistory,
      levels=c("001", "010", "011", "100", "101", "110", "111"))
alfl$id <- factor(alfl$id, levels=rownames(alfl.covs))
```

Specifying the levels of captureHistory ensures that when we tabulate the encounter histories, we will include zeros for histories that were not observed. Similarly, setting the levels of alfl$id tells R that there were some sites where no ALFL were detected. This way, when we tabulate the data, the result will contain a frequency of 0 for sites where no detection occurred. In general, this "padding the data with zeros" where sampling produced a count of no birds is extremely important, and not doing this is probably the most severe problem that commonly occurs in collecting or processing such data sets for an analysis.

We extract data from the first primary period and tabulate the encounter histories:

```
alfl.v1 <- alfl[alfl$survey==1,]
alfl.H1 <- table(alfl.v1$id, alfl.v1$captureHistory)
head(alfl.H1, 5)
```

7.8 SPATIALLY STRATIFIED CAPTURE-RECAPTURE MODELS

```
           001 010 011 100 101 110 111
crick1_05    0   0   0   0   0   0   1
his1_05      0   0   1   0   0   0   1
hisw1_05     0   0   0   0   0   0   0
hisw2_05     0   0   0   0   0   0   1
kenc1_05     0   0   0   0   0   0   0
```

The object `alfl.H1` contains the tabulated capture history frequencies for each site in the format required by `unmarked`.

Now that we have organized the ALFL data, we can create an `unmarkedFrame` and fit some capture-recapture models. We will use the `multinomPois` function to fit models, and the associated constructor function `unmarkedFrameMPois` to which we will input our capture-recapture pi function defined previously.

To allow for the possibility that p varies by sample occasion ("model M_t"), we will construct a covariate that represents the time interval, which we call `intervalMat` below, and this will be specified as an observation covariate ("obsCov" for short). We also provide the two site-specific covariates `struct` and `woody`.

```
intervalMat <- matrix(c('1','2','3'), 50, 3, byrow=TRUE)
class(alfl.H1) <- "matrix"
o2y <- matrix(1, 3, 7)
umf.cr1 <- unmarkedFrameMPois(y=alfl.H1,
    siteCovs=alfl.covs[,c("woody", "struct", "time.1")],
    obsCovs=list(interval=intervalMat), obsToY=o2y, piFun="crPiFun")
summary(umf.cr1)
```

Here the `obsToY` matrix is a 3 × 7 matrix of 1's, which means that each p_j parameter appears in each observable encounter history, so that if a covariate is missing for any of the three sample periods, then the encounter history probability cannot be calculated and the whole record must be discarded.

Now that we have our data formatted we can fit some models. The following correspond to spatially structured versions of model M_0 and model M_t. We also fit a model with a continuous covariate effect on p (we call this model M_x). For this case, the covariate `time.1` is the time of sampling for the first day (the first "primary period").

```
M0 <- multinomPois(~ 1 ~ 1, umf.cr1)
Mt <- multinomPois(~ interval - 1 ~ 1, umf.cr1)
Mx <- multinomPois(~ time.1 ~ 1, umf.cr1)
```

We can now extend the basic capture-recapture models to accommodate variation in abundance and detection probability among sites. For example, we fit a model with "percent woody vegetation" as a covariate on site-specific abundance:

```
(M0.woody <- multinomPois(~ 1 ~ woody, umf.cr1))

Call:
multinomPois(formula = ~1 ~ woody, data = umf.cr1)
```

```
Abundance:
             Estimate     SE       z      P(>|z|)
(Intercept)  -0.962    0.325   -2.96    0.003059
woody         2.587    0.680    3.80    0.000143

Detection:
 Estimate    SE       z      P(>|z|)
   1.43    0.216    6.63    3.42e-11

AIC: 245.9301
```

Various other interesting models can be fitted easily in the same way. For now we focus on the four models that we have and consider model selection for this model set. To do that, we package all of these models up into a `fitList` and produce an AIC table using the `modSel` command.

```
fl <- modSel(fitList(M0, Mt, Mx, M0.woody))

         nPars     AIC   delta   AICwt   cumltvWt
M0.woody     3  245.93    0.00  0.99634      1.00
M0           2  258.24   12.31  0.00212      1.00
Mx           3  259.67   13.74  0.00104      1.00
Mt           4  261.10   15.17  0.00051      1.00
```

We see that the `M0.woody` model has nearly all of the AIC weight, and it indicates that alder flycatcher abundance increases with the percent cover of woody vegetation. We can plot this relationship (Figure 7.5) by predicting abundance at a sequence of woody vegetation values.

```
nd <- data.frame(woody=seq(0, 0.8, length=50))
E.abundance <- predict(M0.woody, type="state", newdata=nd, appendData=TRUE)
plot(Predicted ~ woody, E.abundance, type="l", ylim=c(0, 6),
```

FIGURE 7.5

Response of mean flycatcher abundance ($E(N)$) to woody vegetation. They love woody vegetation.

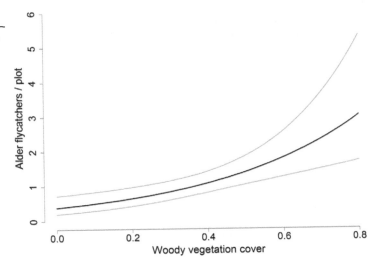

```
        ylab="Alder flycatchers / plot", xlab="Woody vegetation cover",
        frame = F)
   lines(lower ~ woody, E.abundance, col=gray(0.7))
   lines(upper ~ woody, E.abundance, col=gray(0.7))
```

Since there was no evidence of variation in p, we can produce a summary of detection probability simply by back-transforming the logit-scale estimate to obtain:

```
backTransform(M0.woody, type="det")

Backtransformed linear combination(s) of Detection estimate(s)

 Estimate     SE   LinComb  (Intercept)
    0.808  0.0336    1.43         1

Transformation: logistic
```

The corresponding multinomial cell probabilities can be computed by plugging this estimate of detection probability into our `piFun`. The `getP` function makes this easy to do:

```
round(getP(M0.woody), 2)[1,]
  001   010   011   100   101   110   111
 0.03  0.03  0.13  0.03  0.13  0.13  0.53
```

These are consistent with the high detection probability indicated by the observed data (where we see many "111" encounter histories): The encounter probability most likely to be observed was 111. The probability of not detecting an alder flycatcher in three replicate samples was essentially zero, $(1 - 0.808)^3 = 0.0071$.

7.8.2 MODELS WITH BEHAVIORAL RESPONSE (M_b)

A permanent behavioral response to capture is that in which capture probability changes as a result of initial capture (note that sometimes one may also want to model ephemeral, or transient, trap response or some trap response that is intermediate in duration, see Yang and Chao, 2005). Both trap avoidance (trap shyness) and trap attraction (trap happiness) are frequently observed in many studies based on physical capture or when traps are baited. A simple model of these two behaviors is known as model M_b (Otis et al., 1978). The model assumes that newly captured individuals are captured with probability p_{naive} and then are subsequently recaptured with probability p_{wise}. If $p_{wise} < p_{naive}$, then animals exhibit trap avoidance and, otherwise, they exhibit what is called trap happiness. To fit model M_b in `unmarked`, we need to create a new `piFun` and we need to provide an occasion-specific covariate (`obsCov`) that distinguishes the two capture probabilities, p_{naive} and p_{wise}. This is illustrated in the following bit of R code (thanks to R. Chandler):

```
crPiFun.Mb <- function(p) {
 pNaive <- p[,1]
 pWise <- p[,3]
 cbind("001" = (1 - pNaive) * (1 - pNaive) *     pNaive,
       "010" = (1 - pNaive) *      pNaive  * (1 - pWise),
       "011" = (1 - pNaive) *      pNaive  *      pWise,
       "100" =      pNaive  * (1 - pWise)  * (1 - pWise),
       "101" =      pNaive  * (1 - pWise)  *      pWise,
       "110" =      pNaive  *      pWise   * (1 - pWise),
       "111" =      pNaive  *      pWise   *      pWise)
}
```

An important thing to note about this function is that it takes the input matrix p and it only uses the first and third columns. This is because the behavioral response model builds the encounter histories for all seven observable histories from only the two basic parameters p_{naive} and p_{wise} and thus the pi function is a mapping of two fundamental parameters to the seven encounter history probabilities. We could also change this to use the first and second columns or the second and third also; it doesn't matter, the different formulations are handled the same way internally by unmarked.

This function crPiFun.Mb allows capture probability to be modeled as:

$$\text{logit}(p_{ij}) = \alpha_{naive} + \alpha_{wise} * \text{Behavior}_j$$

where Behavior_j is simply a dummy variable indicating previous capture or not. Thus, p_{ij} is either p_{naive} or p_{wise}. We construct a new unmarkedFrame and fit model M_b to the alder flycatcher data.

```
behavior <- matrix(c('Naive', 'Naive', 'Wise'), 50, 3, byrow=TRUE)
umf.cr1Mb <- unmarkedFrameMPois(y=alfl.H1,
    siteCovs=alfl.covs[,c("woody", "struct", "time.1")],
    obsCovs=list(behavior=behavior), obsToY=o2y, piFun="crPiFun.Mb")
M0 <- multinomPois(~1 ~1, umf.cr1Mb)
(Mb <- multinomPois(~behavior-1 ~1, umf.cr1Mb))

Call:
multinomPois(formula = ~behavior - 1 ~ 1, data = umf.cr1Mb)

Abundance:
 Estimate     SE       z    P(>|z|)
  0.00955  0.142  0.0674      0.946

Detection:
               Estimate     SE     z    P(>|z|)
behaviorNaive      1.31  0.356  3.69   2.25e-04
behaviorWise       1.50  0.276  5.44   5.27e-08

AIC: 260.0599
```

It is tempting to now fit a model with behavioral response and an observation covariate time.1 or other covariates on detection probability. This is easy enough to do and unmarked produces results, BUT THEY ARE WRONG! This is because the accommodation of temporal covariates in the behavioral response pi function is not possible as we are using the columns of the "p matrix" input into the pi function to hold parameters of behavioral response not of time specificity (see Section 3.3 "Caution, Warning, Danger" in Chandler, 2015). (Note: in BUGS we could fit this model to a data set formatted such that rows correspond to individual detection histories, as in Section 7.8.3 and then we could easily combine behavioral response effects with other covariate or time effects in the detection part of the model.)

7.8.3 MODELS WITH INDIVIDUAL HETEROGENEITY (M_h)

So far, we have considered a suite of basic capture-recapture models that allow for explicit covariates that affect p or N, including time effects or behavioral response and habitat effects. Another common class of models in capture-recapture are models that contain individual heterogeneity in detection probability, so called "model M_h" (e.g., Otis et al., 1978; Norris and Pollock, 1996; Coull and Agresti, 1999; Dorazio and Royle, 2003). One possible model, the logit-normal model (Coull and Agresti, 1999), assumes normally distributed logit-scale detection probabilities for individuals (completely

7.8 SPATIALLY STRATIFIED CAPTURE-RECAPTURE MODELS

analogous (except for the logit scale) to our specification of extra-Poisson, site-specific variability in abundance in the model in Section 7.6.3):

$$\text{logit}(p_i) \sim Normal\left(\mu_p, \sigma_p^2\right)$$

To compute the likelihood under this model, in which the p_i parameters are random effects, we must remove them from the likelihood by integration to obtain the marginal probability of the capture frequencies. This can be done using the `integrate` function in R (see Section 2.4.4), and we can create a pi function (function `MhPiFun` in the supplemental material) that obtains the marginal probabilities as a function of only two parameters. As before when modeling behavioral response, this is not a completely general pi function and it should *not* be used to try to model additional effects on detection probability beyond heterogeneity (Chandler, 2015). In particular, this pi function does not allow for temporal variation in capture probability (covariates or time effects) because it uses the second column of the input p matrix for the parameter σ, governing the variance of the random effects. Having defined our new pi function, we can fit models such as the following, which has woody vegetation as a covariate on $E(N)$ and individual heterogeneity in detection probability:

```
parID <- matrix(c('p','sig','sig'), 50, 3, byrow=TRUE)
umf.cr2 <- unmarkedFrameMPois(y=alfl.H1,
        siteCovs=alfl.covs[,c("woody", "struct", "time.1")],
        obsCovs=list(parID=parID), obsToY=o2y, piFun="MhPiFun")
multinomPois(~ parID-1 ~ woody, umf.cr2)

Call:
multinomPois(formula = ~parID - 1 ~ woody, data = umf.cr2)

Abundance:
            Estimate    SE      z       P(>|z|)
(Intercept) -0.84       0.363   -2.31   0.02078
woody       2.59        0.680   3.81    0.00014

Detection:
          Estimate  SE     z      P(>|z|)
parIDp    1.637     0.645  2.54   0.0112
parIDsig  0.841     0.622  1.35   0.1762

AIC: 242.3731
```

In the detection model, `parIDp` and `parIDsig` denote the estimates of μ_p and σ_p^2. This model is more than three AIC units better than model `M0.woody`, indicating significant heterogeneity in detection probability. Note that the σ parameter is estimated on the log scale and therefore the MLE is $\hat{\sigma} = \exp(0.841) = 2.32$. Shortly we will compare this estimate with a Bayesian estimate from BUGS.

One possible application of this model is for the estimation and modeling of local species richness in metacommunity studies (see Chapter 11) by treating species as analogous to "individuals" in this chapter. In such an analysis, we would allow for species-specific detection heterogeneity, which we argue in Kéry (2011a) is required when using capture-recapture models for species richness, and at the same time could directly model effects of the environment on the local community size (i.e., the abundance parameter in the model). The only drawback in such an analysis in `unmarked` would be that we would not be allowed to add any other effects into detection probability, beyond individual heterogeneity. But if this was necessary, we could simply move to BUGS for the fitting of such a metacommunity model. These approaches would allow you to model species richness directly as a parameter, rather than as a derived quantity as in Section 11.6.4.

7.8.4 BAYESIAN ANALYSIS USING DATA AUGMENTATION: HETEROGENEITY MODELS IN BUGS

Chandler's ALFL data exist in their raw form as individual encounter histories, which we aggregated into encounter history frequencies for each sample location in order to fit the multinomial mixture models. In the BUGS language, we can analyze models with the multinomial mixture structure (i.e., a model on site-level abundance) using a formulation of the model based on "data augmentation" (DA; Royle et al., 2007a; Converse and Royle, 2012; Royle and Dorazio, 2012; Royle and Converse, 2014). This formulation of the model is especially useful when there are individual-level effects, such as individual detection heterogeneity in model M_h, individual covariates (Royle, 2009), or location information such as in spatial capture-recapture models (Efford, 2004; Borchers and Efford, 2008; Royle and Young, 2008). A closely related class of models is distance sampling, which we address in the next chapter (and later) using the individual-level formulation of the multinomial mixture models based on data augmentation.

To explain the formulation of multinomial mixture models using data augmentation, *we have to modify our usual notation a little bit* because here we have individuals *and* sites and so we need a new subscript. We introduce the subscript s for site (or, in general, a stratum or grouping level of individuals), and there are a total of S sites, and for now we use i for individual. To implement DA for a multinomial mixture problem, we take the individual encounter histories from all S sites and pool them into one big data set where the total number of encounter histories is n, the number of individuals observed among all sites. Under DA this pooled data set is augmented with a large number of $M - n$ individuals that have all-zero encounter histories (i.e., are never detected). A set of latent data augmentation variables z_i are introduced such that $z_i = 1$ if individual i is a "real" individual that was exposed to sampling and $z_i = 0$ if the encounter history is a structural zero. In essence, the model with latent variables serves to partition the augmented all-zero encounter histories into fixed or "excess" zeros, and stochastic or sampling zeros due to imperfect detection. The model assumes that $z_i \sim Bernoulli(\psi)$ where ψ is the "data augmentation parameter." Under this formulation, M is supposed to be (much) larger than the extreme upper limit of the total population size among all sites. The parameter ψ is related to the total population size among all S sites, N, in the sense that $N \sim Binomial(M, \psi)$.

The key idea (Converse and Royle, 2012; Royle and Converse, 2014) is that under the assumption $N_s \sim Poisson(\lambda_s)$ then group structure of a pooled data set can be modeled simply by including an individual group (= site here) membership variable g_i which has a categorical distribution with probabilities related directly to the group means $\log(\lambda)_s = \beta_0 +$ covariates. The following two results hold:

1. Group membership has a categorical distribution $g_i \sim Categorical(\pi)$ with cell probabilities

$$\pi_s = \frac{\lambda_s}{\sum_s \lambda_s}$$

where λ_s is the mean of the Poisson abundance just as in an ordinary multinomial or binomial N-mixture model.

2. The data augmentation parameter ψ is derived from the λ_s parameters in the following way:

$$\psi = \frac{\sum_s \lambda_s}{M}$$

Thus, the parameter ψ is a derived parameter, being a function of β_0 and whatever other parameters influence the local population means.

7.8 SPATIALLY STRATIFIED CAPTURE-RECAPTURE MODELS

For additional discussion and details of this individual-level model based on data augmentation, see Chapters 11 and 14 in Royle et al. (2014).

We use this approach to fit a couple of models to the ALFL data. For this we use JAGS and the `jagsUI` package because it is somewhat more efficient than WinBUGS. First we fit a basic model with the site covariate `woody` as a covariate on N and the covariates `struct` and `time.1` as covariates on p. Next, we include an individual heterogeneity term on p similar to model M_h, which we demonstrated in the previous section.

Before fitting the model, we do a little bit of processing of the ALFL data, first extracting the raw encounter history data, but we do *not* summarize into frequencies as before. Second, we do the data augmentation. In this case we augment the data set up to $M = 400$ individuals by padding the encounter history matrix with a bunch of all-zero encounter histories. We also create a "site identity" index (an integer) for each encounter history, which takes on the value 1–50 depending on the site where the individual was observed. Because the sites are in alphabetical order in the data set, we can use the `as.numeric` function to convert the factor site name into an integer index that corresponds to a row index of the covariate matrix. (Note: if the site names were not in alphabetical order, then they would *not* correspond to row numbers of the covariate matrix, so be careful!). In addition, we have to augment the site identity vector, which we do by padding it with up to 400 missing values. Finally, we extract the matrix of three site covariates (`woody`, `struct`, `time.1`) and standardize them. We didn't do this when carrying out the analysis in `unmarked`, but the mixing of the MCMC is terrible if we don't standardize here. We also have to specify starting values for the data augmentation variables z_i such that they don't contradict the model and the data (i.e., that an observed individual is initialized at $z = 0$); we do so below by setting the first 100 equal to 1 and the remaining equal to 0.

```
# Extract data and do data augmentation up to M = 400
y <- as.matrix(alfl[,c("interval1","interval2","interval3")] )
nind <- nrow(y)
M <- 400
y <- rbind(y, matrix(0, nrow=(M-nind), ncol=3))

# Site ID
# Make site ID into an integer: This only works ok here because sites are in
#  alphabetical order in the data set!
site <- as.numeric(alfl$id)
site <- c(site, rep(NA, M-nind))

# Next we extract the covariates and standardize them
sitecovs <- scale(as.matrix(alfl.covs[,c("woody", "struct", "time.1")]))

# Bundle data for BUGS
data <- list(y = y, J = 3, M = M, nsites = 50, X = sitecovs, group=site)
str(data)

# Specify model in BUGS language
cat("
model {

# Prior distributions
p0 ~ dunif(0,1)
alpha0 <- log(p0 / (1-p0))   # same as logit(p0)
```

```
alpha1 ~ dnorm(0, 0.01)
alpha2 ~ dnorm(0,0.01)
beta0 ~ dnorm(0,0.01)
beta1 ~ dnorm(0,0.01)
psi <- sum(lambda[]) / M       # psi is a derived parameter

# log-linear model for abundance: lambda depends on WOODY
for(s in 1:nsites){
   log(lambda[s]) <- beta0 + beta1 * X[s,1]
   probs[s] <- lambda[s] / sum(lambda[])
}

# Model for individual encounter histories
for(i in 1:M){
   group[i] ~ dcat(probs[])    # Group == site membership
   z[i] ~ dbern(psi)           # Data augmentation variables

   # Observation model: p depends on 2 covariates: STRUCT + TIME
   for(j in 1:J){
      logit(p[i,j]) <- alpha0 + alpha1 * X[group[i],2] + alpha2*X[group[i],3]
      pz[i,j] <- p[i,j] * z[i]
      y[i,j] ~ dbern(pz[i,j])
   }
}
}
",fill=TRUE,file="model.txt")

# Parameters monitored
params <- c("p0", "alpha0", "alpha1", "alpha2", "beta0", "beta1", "psi")

# Initial values
inits <- function(){
   list (p0 = runif(1), alpha1 = runif(1), alpha2 = rnorm(1),beta0=runif(1),
     beta1=rnorm(1), z= c( rep(1,100), rep(0, 300))) }

# MCMC settings
ni <- 11000 ; nb <- 1000 ; nt <- 4 ; nc <- 3

# Call JAGS from R and summarize marginal posteriors
library("jagsUI")
out <- jags(data, inits, params, "model.txt", n.thin = nt,
   n.chains = nc, n.burnin = nb, n.iter = ni)
print(out, digits = 3)

JAGS output for model 'model.txt', generated by jagsUI.
Estimates based on 3 chains of 11000 iterations,
burn-in = 1000 iterations and thin rate = 4,
yielding 7500 total samples from the joint posterior.
MCMC ran for 3.166 minutes at time 2015-01-21 16:17:34.
```

	mean	sd	2.5%	50%	97.5%	overlap0	f	Rhat	n.eff
p0	0.712	0.030	0.651	0.713	0.768	FALSE	1.000	1.000	7276
alpha0	0.911	0.146	0.621	0.911	1.199	FALSE	1.000	1.000	7387
alpha1	0.119	0.137	-0.138	0.118	0.391	TRUE	0.809	1.001	1266
alpha2	0.096	0.148	-0.195	0.097	0.388	TRUE	0.742	1.001	3637
beta0	0.563	0.103	0.351	0.566	0.760	FALSE	1.000	1.000	7500
beta1	0.491	0.100	0.293	0.492	0.685	FALSE	1.000	1.000	5715
psi	0.251	0.022	0.208	0.250	0.297	FALSE	1.000	1.000	7500
deviance	1104.198	13.652	1082.091	1103.081	1134.686	FALSE	1.000	1.000	7500

[...output truncated...]

The upper 95% credible limit of the estimate of ψ is well below one. This suggests that we did too much data augmentation and we could really improve the speed of things by reducing M considerably, perhaps to $M = 200$ or even less. Otherwise, we see consistency with the results obtained previously. The effect of the covariate `woody` on local abundance is strongly positive. The effects of the covariates `struct` and `time.1` on detection probability are negligible.

Next, we use this data augmentation formulation of the model to fit an individual heterogeneity model and the `woody` covariate on $E(N_s)$. This would be the Bayesian equivalent of the model we fitted in `unmarked` using the `MhPiFun` pi function that we created. We have to modify the starting values and parameters to be monitored to include the heterogeneity parameter σ. Also, we keep the unused parameters α_1 and α_2 in the model instead of deleting them, which means these parameters will simply be drawn from their prior distributions (and posterior summaries of these parameters will be uninteresting). To dump the model out and set these things up, we do as follows (new bits in red):

```
# Specify model in BUGS language
cat("
model {

# Prior distributions
p0 ~ dunif(0,1)
alpha0 <- log(p0/(1-p0))
alpha1 ~ dnorm(0, 0.01)
alpha2 ~ dnorm(0, 0.01)
beta0 ~ dnorm(0, 0.01)
beta1 ~ dnorm(0, 0.01)
psi <- sum(lambda[])/M
tau ~ dgamma(0.1,0.1)   # New parameter, precision of ind. random effects
sigma <- 1/sqrt(tau)

# log-linear model for abundance: lambda depends on WOODY
for(s in 1:nsites){
  log(lambda[s])<- beta0 + beta1*X[s,1]
  probs[s]<- lambda[s]/sum(lambda[])
}
```

```
# Model for individual encounter histories
for(i in 1:M){
    eta[i] ~ dnorm(alpha0, tau)   # Individual random effect
    group[i] ~ dcat(probs[])      # Group == site membership
    z[i] ~ dbern(psi)             # Data augmentation variables
    # Observation model: p depends on STRUCT + TIME + ind. heterogeneity
    for(j in 1:J){
       logit(p[i,j]) <- alpha1 * X[group[i],2] + alpha2*X[group[i],3] + eta[i]
       pz[i,j] <- p[i,j] * z[i]
       y[i,j] ~ dbern(pz[i,j])
    }
  }
}
",fill=TRUE,file="model.txt")

# Parameters monitored: add sigma
params <- c("p0", "alpha0", "alpha1", "alpha2", "beta0", "beta1", "psi", "sigma")

# Initial values: add tau
inits <- function(){
list (p0 = runif(1), alpha1 = runif(1), alpha2 = rnorm(1),beta0=runif(1),
      beta1=rnorm(1),z= c( rep(1,100), rep(0, 300)), tau = 1) }

# MCMC settings (others as before)
ni <- 50000 ; nb <- 10000 ; nt <- 2 ; nt <- 10

# Call JAGS from R and summarize marginal posteriors
out <- jagsUI(data, inits, params, "model.txt", n.thin = nt,
    n.chains = nc, n.burnin = nb, n.iter = ni)
print(out, digits = 3)
             mean      sd     2.5%     50%    97.5%   overlap0    f     Rhat    n.eff
p0          0.387   0.186    0.048   0.391   0.710   FALSE     1.000   1.012    205
alpha0     -0.587   0.992   -2.991  -0.444   0.898   TRUE      0.695   1.020    161
alpha1      0.477   0.423   -0.343   0.465   1.344   TRUE      0.881   1.001   1371
alpha2      0.572   0.439   -0.253   0.550   1.487   TRUE      0.912   1.000   7804
beta0       1.046   0.226    0.642   1.035   1.539   FALSE     1.000   1.013    221
beta1       0.490   0.103    0.286   0.490   0.688   FALSE     1.000   1.001   2440
psi         0.415   0.097    0.275   0.399   0.664   FALSE     1.000   1.020    176
sigma       3.183   0.854    1.646   3.151   4.964   FALSE     1.000   1.013    230
deviance 1075.126  19.517 1039.701 1073.938 1115.869 FALSE     1.000   1.000   3539
[. . . output truncated . . .]
```

The posterior mean of sigma is 3.183, which is quite a bit larger than that produced by unmarked, although we have not fitted an exactly equivalent model so far. We note that the MhPiFun presently does not accommodate detection covariates, and it may be difficult to achieve that extension given current internal programming of unmarked. So, we see that BUGS enables you to easily go beyond the capabilities of a canned program.

7.8.5 ANALYSIS OF DATA GENERATED WITH THE simNmix DATA SIMULATION FUNCTION

In Section 7.3 we illustrated the use of the simNmix data simulation function to generate multinomial counts under the capture-recapture and the removal protocol, and now we briefly illustrate the analysis, using unmarked. Here is an analysis of the capture-recapture format of the first data set generated in Section 7.3.

```
# Fit model in unmarked
library(unmarked)
time <- matrix(as.character(1:3), data$nsite, 3, byrow = T)

# Define pifun for J=3 occasion capture-recapture protocol
crPiFun <- function(p) {
    p1 <- p[,1]   # Extract the columns of the p matrix, one for
    p2 <- p[,2]   # each of J = 3 sample occasions
    p3 <- p[,3]
    cbind(  # define multinomial cell probabilities:
    "100" = p1 * (1-p2) * (1-p3),
    "010" = (1-p1) * p2 * (1-p3),
    "001" = (1-p1) * (1-p2) * p3,
    "110" = p1 * p2 * (1-p3),
    "101" = p1 * (1-p2) * p3,
    "011" = (1-p1) * p2 * p3,
    "111" = p1 * p2 * p3)
}

# Define mapping function for missing values
o2y <- matrix(1, 3, 7)

# Create unmarked frame and fit couple of models
umf <- unmarkedFrameMPois(y = dhfreq, siteCovs = data.frame(cov3 = data$site.cov[,3]),
obsCovs = list(time = time), obsToY = o2y, piFun = "crPiFun")
fm1 <- multinomPois(~1 ~1, umf)    # detection model before abundance model
fm2 <- multinomPois(~time-1 ~1, umf)
fm3 <- multinomPois(~1 ~cov3, umf)
fm4 <- multinomPois(~time-1 ~cov3, umf)

# Assemble the models into a fitList and rank using AIC
ms <- fitList(
"lam(.)p(.)" = fm1,
"lam(.)p(time)" = fm2,
"lam(cov3)p(.)" = fm3,
"lam(cov3)p(time)" = fm4)

(AICtable <- modSel(ms))
                 nPars     AIC    delta AICwt cumltvWt
lam(cov3)p(time)     5 2556.64     0.00  1.00     1.00
lam(.)p(time)        4 4127.68  1571.04  0.00     1.00
lam(cov3)p(.)        3 4606.29  2049.65  0.00     1.00
lam(.)p(.)           2 6177.33  3620.69  0.00     1.00
```

Hence, the data-generating model has overwhelming AIC weight. We can compare the estimates with the "truth" (the values used to simulate the data set).

```
summary(fm4)
p.true <- qlogis(data$p[min(which(data$N>0)),,1])
tmp <- cbind(rbind(lam0 = log(data$mean.lam), beta3 = data$beta3.lam, logit.p1 =
p.true[1], logit.p2 = p.true[2], logit.p3 = p.true[3]), coef(fm4))
colnames(tmp) <- c("Truth", "MLEs")
tmp
                Truth        MLEs
lam0        1.0000000   0.8069142
beta3       1.0000000   1.1489853
logit.p1   -1.9829093  -1.9196300
logit.p2    0.8704367   0.9379833
logit.p3    2.0491658   2.1247917
```

Here is an analysis of a new data set when the removal protocol is assumed. We fit directly the data-generating model:

```
# No temporal variation in p and effect of site-covariate on lambda
set.seed(24)
data <- simNmix(mean.lam = exp(1), beta2.lam = 1, beta3.lam = 1, mean.p = plogis(0.2),
beta3.p = -1, beta.p.survey = -1)
str(data$DH)

# Get occasions with first detection of each individual
f <- apply(data$DH, c(1,3), function(x) min(which(x != 0)))
head(f)   ;   str(f)

# Produce removal counts (for any number of occasions)
y <- array(NA, dim = c(data$nsite, data$nvisit), dimnames = list(NULL,
as.factor(1:data$nvisit)))
for(i in 1:data$nsite){
   y[i,] <- table(factor(f[i,], levels = as.character(1:data$nvisit)))
}
y                 # Look at removal data set

# Create and look at um data frame
summary(umf <- unmarkedFrameMPois(y = y, siteCovs = data.frame(cov2 = data$site.cov[,2],
cov3 = data$site.cov[,3]), obsCovs = list(obs.cov = data$survey.cov), type = "removal"))
# Fit models in unmarked
fm1 <- multinomPois(~1 ~1, umf)   # Detection model before abundance model
(fm4 <- multinomPois(~cov3 + obs.cov ~cov2 + cov3, umf))

Call:
multinomPois(formula = ~cov3 + obs.cov ~ cov2 + cov3, data = umf)

Abundance:
             Estimate      SE      z     P(>|z|)
(Intercept)    0.955  0.0512   18.7    8.45e-78
cov2           0.989  0.0312   31.6    7.85e-220
cov3           1.056  0.0382   27.6    4.31e-168
```

7.8 SPATIALLY STRATIFIED CAPTURE-RECAPTURE MODELS

```
Detection:
             Estimate     SE       z      P(>|z|)
(Intercept)    0.300    0.0941    3.19    1.44e-03
cov3          -1.079    0.0834  -12.93    3.09e-38
obs.cov       -0.965    0.0417  -23.16    1.04e-118

AIC: 1696.602
```

These estimates can be compared, in this order, with the "truth" in the data simulation process.

```
print(c(log(data$mean.lam), data$beta2.lam, data$beta3.lam,
     logit(data$mean.p), data$beta3.p, data$beta.p.survey))

[1] 1.0 1.0 1.0 0.2 -1.0 -1.0
```

At the end of Section 7.3 we generated spatially stratified capture-recapture data under a negative binomial distribution for abundance, with covariates on mean abundance as well as on detection. To adopt the negative binomial in a multinomial mixture model in unmarked, we need to use the gmultmix function. We re-create that data set, organize the table containing the frequencies of each of the seven observable detection histories, and fit a couple of models. (In case you play around with the data-generating function settings, don't forget that R has trouble fitting negative binomial models with a dispersion parameter substantially smaller than 1.)

```
# Simulate detection frequency data as at the end of section 7.3
set.seed(24)
data <- simNmix(mean.lam = exp(0), beta2.lam = 1, beta3.lam = -1,
   beta4.lam = 0.2, dispersion = 1, mean.p = plogis(0),
   beta3.p = 1, sigma.p.visit = 1, Neg.Bin = TRUE)
dhfreq <- array(NA, dim = c(data$nsite, 7),
   dimnames = list(NULL, c("100", "010", "001", "110", "101", "011", "111")))

for(i in 1:data$nsite){
  dhfreq[i,] <- table(factor(paste(data$DH[i,1,], data$DH[i,2,],
  data$DH[i,3,], sep = ""),
  levels = c("100", "010", "001", "110", "101", "011", "111")))
}
dhfreq                   # Look at resulting data set

# Bundle data in unmarked frame
time <- matrix(as.character(1:3), data$nsite, 3, byrow = T)
summary(umf <- unmarkedFrameGMM(y = dhfreq, numPrimary = 1,
       siteCovs = data.frame(cov2 = data$site.cov[,2],
       cov3 = data$site.cov[,3], cov4 = data$site.cov[,4]),
       obsCovs = list(time = time), obsToY = o2y, piFun = "crPiFun"))

# Fit a couple of models, first for detection
fm1 <- gmultmix(lambdaformula = ~1, phiformula = ~1, pformula = ~1, mix =
"NB", data = umf)
fm2 <- gmultmix(~1, ~1, ~time-1, mix = "NB", data = umf)
fm3 <- gmultmix(~1, ~1, ~time-1+cov3, mix = "NB", data = umf)

# ... then for abundance,
fm4 <- gmultmix(~cov2, ~1, ~1, mix = "NB", data = umf)
fm5 <- gmultmix(~cov2+cov3, ~1, ~1, mix = "NB", data = umf)
fm6 <- gmultmix(~cov2+cov3+cov4, ~1, ~1, mix = "NB", data = umf)
```

```
# ... and the data-generating model
fm7 <- gmultmix(~cov2+cov3+cov4, ~1, ~time-1+cov3, mix = "NB", data = umf)

# Compare models with AIC
ms <- fitList(
"lam(.)p(.)" = fm1,
"lam(.)p(time)" = fm2,
"lam(.)p(time+cov3)" = fm3,
"lam(cov2)p(.)" = fm4,
"lam(cov2+cov3)p(.)" = fm5,
"lam(cov2+cov3+cov4)p(.)" = fm6,
"lam(cov2+cov3+cov4)p(time+cov3)" = fm7)

(AICtable <- modSel(ms))
```

	nPars	AIC	delta	AICwt	cumltvWt
lam(cov2+cov3+cov4)p(time+cov3)	9	74.96	0.00	1.0e+00	1.00
lam(.)p(time+cov3)	6	275.97	201.01	2.2e-44	1.00
lam(.)p(time)	5	340.90	265.94	1.8e-58	1.00
lam(cov2+cov3+cov4)p(.)	6	957.66	882.70	2.1e-192	1.00
lam(cov2+cov3)p(.)	5	964.59	889.63	6.6e-194	1.00
lam(cov2)p(.)	4	1029.72	954.75	4.8e-208	1.00
lam(.)p(.)	3	1129.75	1054.79	9.0e-230	1.00

Perhaps not surprisingly, the data-generating model again comes up on top of the list. We note that, for some of these model fits, warning messages are produced of this sort: "In log(cp[i, t, R+1]) : NaNs were produced", which simply means that the last cell probability is near 0 (i.e., you can ignore them).

```
# Compare data-generation truth with estimates
p.true <- qlogis(data$mean.p) + data$eta.p.visit
Truth <- rbind(lam0 = log(data$mean.lam), beta2.lam = data$beta2.lam, beta3.lam =
data$beta3.lam, beta4.lam = data$beta4.lam, logit.p1 = p.true[1], logit.p2 = p.true[2],
logit.p3 = p.true[3], beta3.p = data$beta3.p, log.dispersion = log(data$dispersion))
MLEs <- coef(fm7)
print(cbind(Truth, MLEs), 3)
                           MLEs
lam0             0.000   0.0253
beta2.lam        1.000   1.0755
beta3.lam       -1.000  -1.0250
beta4.lam        0.200   0.2302
logit.p1         1.318   1.1623
logit.p2        -0.775  -0.9937
logit.p3        -2.108  -2.1222
beta3.p          1.000   0.9281
log.dispersion   0.000  -0.0285
```

Thus, we recover estimates that resemble well the input in the data-generating function. In summary, we see that we can use function simNmix to enhance our intuition about the data-generation and the multinomial-mixture models under various designs. Furthermore, we can use it to experience the full joy of simulating and analyzing data sets as described in Chapter 4 and illustrated a fair bit in Chapter 6 (e.g., Sections 6.6–6.8).

7.9 EXAMPLE 3: JAYS IN THE SWISS MHB

We consider an analysis of data from the Swiss survey of common breeding birds (Monitoring Häufige Brutvogel, "MHB") that is based on Royle et al. (2007b); see Section 6.9 for an overview of this monitoring scheme. We cover genuine spatial models for the MHB data in chapter 21 (in volume 2). Here we analyze the data from a sample of 238 quadrats surveyed in 2002. The survey route length in this data set varied between about 3.5 km and 12 km. Observers walk the same route three times between mid-April and late June, although some high elevation sites are surveyed only twice, typically in late May and June. All bird observations are noted on a map of the quadrat, including sex and behavioral observations, if any. The survey uses a type of territory mapping method (Bibby et al., 2000) that allows for putative territory encounter histories to be constructed (Kéry et al., 2005a; Royle et al., 2007b; Kéry and Royle, 2010). In contrast to Section 6.9 (where we use replicated counts that ignore territory ID) and Section 10.9 (where we use replicated detection/nondetection, or "presence/absence," data that ignore abundance altogether), here we analyze the quadrat-specific territory encounter history frequencies for the Eurasian jay (*Garrulus glandarius*, Figure 7.6), using the capture-recapture multinomial mixture modeling capabilities of `unmarked`, and we focus on addressing the following specific objectives: (1) We determine what landscape features affect breeding abundance of this species; (2) We produce a breeding abundance map of the species using fitted models; (3) We assess model fit; (4) We produce an estimate of total breeding population size.

FIGURE 7.6

The Eurasian jay (*Garrulus glandarius*). (*Photo Markus Varesvuo.*)

We load the Eurasian jay data into the R workspace and inspect the encounter history data for the first few quadrats.

```
data(jay)              # Load data
str(jay)               # Inspect data list
dim(jay$caphist)       # Look at detection history data
[1] 238 10

jay$caphist[1:4,]
  100 010 001 110 101 011 111 10x 01x 11x
1   1   0   0   0   0   0   0  NA  NA  NA
2   0   0   1   0   0   0   0  NA  NA  NA
3   0   2   0   0   0   0   0  NA  NA  NA
4   1   0   0   0   0   0   0  NA  NA  NA
```

That some quadrats were only sampled twice is accommodated by defining 10 distinct encounter histories (instead of the expected seven based on $J = 3$ replicate samples). The encounter histories for the quadrats having only $J = 2$ replicates are denoted such as 10x, etc. and these entries contain missing values when the quadrat was sampled with three surveys and vice versa. We accommodate this special encounter history structure by defining a new pi function below. We imagine this sort of "mixed replication" should be pretty common in practice, and therefore, hopefully, the structure of the analysis we are about to do is relevant beyond this specific application. In particular, an analogous situation would arise whenever multiple multinomial protocols (e.g., capture-recapture and removal) are used simultaneously i.e., in a form of *integrated model* (see Chapter 23 in volume 2). In such cases, a pi function needs to be created that will allow for multiple distinct conversions of a p matrix into cell probabilities.

The MHB data set contains a number of covariates, a sample of which is shown here:

```
|--abun covs-----|  |---- sampling/observation  covariates ------------|
elev forest length day.1 day.2 day.3 dur.1 dur.2 dur.3 int.1 int.2 int.3
1300   32     6.1    35    58    75   260   270   290  42.6  44.3  47.5
1270   66     4.5    28    39    61   176   145   150  39.1  32.2  33.3
 380   45     6.2    21    40    86   180   160   165  29.0  25.8  26.6
 550   31     6.9    20    42    63   195   240   270  28.3  34.8  39.1
 390    8     4.6    35    50    75   150   130   140  32.6  28.3  30.4
1380   78     3.7    54    75    99   160   160   150  43.2  43.2  40.5
```

These variables are defined as follows:
- elev: average elevation of the quadrat in meters
- forest: forest cover (percent)
- length: length of the sample route through the quadrat in kilometers
- day.x: integer day of the sample (day 1 = April 1)
- dur.x: duration of the survey (minutes)
- int.x: survey intensity (duration divided by length)

We will use unmarked to fit a sequence of multinomial *N*-mixture models allowing for variation in territory abundance across sites. The hierarchical model consists of component models for the quadrat population size:

$$N_s \sim Poisson(\lambda_s)$$
$$\log(\lambda_s) = \beta_0 + covariates$$

with covariates including forest cover, elevation, and route length, and the multinomial observation model where $\mathbf{y}_s = (y_{001,s}, y_{010,s}, y_{100,s}, y_{011,s}, y_{101,s}, y_{110,s}, y_{111,s}, y_{10x,s}, y_{01x,s}, y_{11x,s})$ are the frequencies of each encounter history for quadrat s, but note that this vector does not have a single multinomial distribution. Rather, the first seven elements and the last three elements go together, say $\mathbf{y}_s = (\mathbf{y}_s^{(1)}, \mathbf{y}_s^{(2)})$ and

$$\mathbf{y}_s^{(1)} \sim Multinomial\left(N_s, \pi_s^{(1)}(\mathbf{p})\right)$$
$$\mathbf{y}_s^{(2)} \sim Multinomial\left(N_s, \pi_s^{(2)}(\mathbf{p})\right)$$

While the dimension of the multinomial is different in the two cases, they both depend on the same fundamental parameters, the probability of encounter for each of three sampling occasions, $\mathbf{p} = (p_{1,s}, p_{2,s}, p_{3,s})$.

We standardized the elevation and forest cover covariates. Duration of sampling should affect the probability of detecting a territory, but we scaled by total route length defining `intensity = duration/length`. Thus, we consider the `intensity` covariate in models for encounter probability. The parameterization of route length merits some discussion. As in Section 6.9, we use the *inverse* of route length as a covariate on abundance. The idea is that the "exposed population," say N_s, for a quadrat is less than the actual population of the quadrat, say M_s, but as $L \to \infty$ (the quadrat becomes saturated with sampling effort) then $N_s \to M_s$. So we imagine the quadrat population size model in terms of some superpopulation size M_i:

$$M_s \sim Poisson(\lambda_s)$$
$$log(\lambda_s) = \beta_0 + \beta_1 Elev_s + \beta_2 Forest_s$$
$$N_s | M_s \sim Binomial(M_s, \phi(L_s))$$

where we model

$$\phi(L_s) = \exp(-\beta_3/L_s)$$

Under this model the marginal distribution of the available population size, N_i, is

$$N_s \sim Poisson(\phi(L_s)\lambda_s)$$

And, in our implementation of the model we assume

$$log(E(N_s)) = \beta_0 + \beta_1 Elev_s + \beta_2 Forest_s - \beta_3(1/L_s)$$

Therefore, this argument that route length affects the superpopulation of individuals exposed to sampling suggests that we should use the inverse of route length to account for $N_s \to M_s$ as L_s increases. For making predictions for the entire, effective sampling area associated with the nominal 1 km² sampling area, we set $1/L_s = 0$, so that the prediction applies to saturation sampling effort. See the important discussion of the spatial aspect of this problem in section 6.10.

More generally, this is a three-level hierarchical model with abundance, availability, and detection error (Pollock et al., 2002; Johnson et al., 2014), very similar to models that can be fitted using `gmultmix` (Chandler et al., 2011). To fit hierarchical models with an intermediate (i.e., between state and detection) availability level, we usually need additional information that comes in the form of an additional level of replication, for instance, at each site, two nested temporal scales of sampling (as needed for `gmultmix`) or a distance sampling protocol at every occasion, as needed for `gdistsamp` (see Chapters 8 and 9). Thus, our model here is a special case of more general class of availability models where the information about the middle level of the hierarchy (availability) comes from the assumed deterministic relationship with a known covariate (route length). See also Section 10.10 about occupancy models with an availability level and Section 10.11.1 about 'magical covariates.'

7.9.1 SETTING UP THE DATA AND PREPARING FOR ANALYSIS

We first need to create a new pi function that accommodates that our capture-recapture survey with $J = 3$ sampling occasions leads to ten observable encounter history (because the third sampling is missing for some sites). The new pi function is defined as follows:

```
crPiFun <- function(p) {
   p1 <- p[,1]   # Extract the columns of the p matrix, one for
   p2 <- p[,2]   #   each of J = 3 sample occasions
   p3 <- p[,3]
   cbind(        # define multinomial cell probabilities:
      "100" = p1 * (1-p2) * (1-p3),
      "010" = (1-p1) * p2 * (1-p3),
      "001" = (1-p1) * (1-p2) * p3,
      "110" = p1 * p2 * (1-p3),
      "101" = p1 * (1-p2) * p3,
      "011" = (1-p1) * p2 * p3,
      "111" = p1 * p2 * p3,
      "10x" = p1*(1-p2),
      "01x" = (1-p1)*p2,
      "11x" = p1*p2)
}
```

By this definition we see that only p_1 and p_2 go into constructing the cell probabilities for the encounter histories based on $J = 2$ sample occasions. We define the obsToY matrix to be a 3×10 matrix of 1's:

```
o2y <- matrix(1, 3, 10)
```

The consequence of having this specific obsToY matrix is that we will toss out an encounter history whenever a covariate value is missing for *any* of the three sample occasions. Clearly this is not generally desirable when we are missing encounter histories based on mixed two and three samples because a $J = 2$ sample quadrat must necessarily have a missing covariate value for the third sample! To fix this and make everything work cleanly we can pad any missing third covariate values with any numerical value because, according to our crPiFun definition, p_3 is not used to construct the encounter history at any of the quadrats for which $J = 2$.

Now that we have defined the pi function that we require for the mixed two and three sample occasion study, we will harvest our data objects from the jay data, standardize covariates, and prepare things for analysis:

```
# Grab the data objects
covinfo <- jay$covinfo
gridinfo <- jay$gridinfo
sitecovs <- jay$sitecovs
caphist <- jay$caphist
```

```
# Get observation covariates to use in model
# Day of year, sample intensity and survey duration.
# Standardize them.
day <- scale(covinfo[,c("date1", "date2", "date3")])
dur <- as.matrix(covinfo[,c("dur1", "dur2", "dur3")])
dur[is.na(dur)] <- mean(dur, na.rm=TRUE) # Pad the 6 missing values
intensity <- dur / sitecovs[,"length"] # Sample rate = duration/length
dur <- scale(dur)
intensity <- scale(intensity)
```

Now we change the missing values (NAs) in the third replicate survey for those quadrats only surveyed twice to any arbitrary value. Here we use 0. This has no effect because as per our custom pi function above day[,3] never enters the likelihood for any of the two-sample encounter histories (see above). We also define a variable iLength, the inverse of route length (see above), and we use this new covariate as one of our site covariates.

```
reps <- apply(!is.na(day), 1, sum)
day[reps==2,3] <- 0
dur[reps==2,3] <- 0
intensity[reps==2, 3] <- 0

# Store the observation covariates in a list
obscovs <- list(intensity = intensity, dur = dur, day = day)

# Standardize site covariates
sitecovs[,"elev"] <- scale(sitecovs[,"elev"])
sitecovs[,"forest"] <- scale(sitecovs[,"forest"])
# NOTE: length is NOT standardized
sitecovs[,"iLength"] <- 1 / sitecovs[,"length"]

# Create unmarkedFrame (need crPiFun above and unmarked loaded)
caphist <- as.matrix(caphist)
mhb.umf <- unmarkedFrameMPois(y=caphist, siteCovs=as.data.frame(sitecovs),
    obsCovs=obscovs, obsToY=o2y, piFun="crPiFun")
```

7.9.2 FITTING SOME MODELS

Now that we have created the unmarkedFrame we can easily fit a suite of models using the multinomPois or gmultmix functions. In the following set of commands we fit a set of models having various effects on detection probability, and then we keep the best model from among that set and extend the model to include effects on abundance. At each step we apply AIC-based model selection by assembling the various fits into a fitList.

```
# Fit a series of models
fm1 <- multinomPois(~1 ~1, mhb.umf)
fm2 <- multinomPois(~day ~1, mhb.umf)
fm3 <- multinomPois(~day + I(day^2) ~1, mhb.umf)
fm4 <- multinomPois(~intensity ~1, mhb.umf)
fm5 <- multinomPois(~intensity + I(intensity^2) ~1, mhb.umf)
fm6 <- multinomPois(~day + intensity ~1, mhb.umf)
fm7 <- multinomPois(~day + I(day^2) + intensity + I(intensity^2) ~1, mhb.umf)
```

```
# Assemble the models into a fitList and rank them by AIC
mspart1 <- fitList(
  "lam(.)p(.)" = fm1,
  "lam(.)p(day)" = fm2,
  "lam(.)p(day+day^2)" = fm3,
  "lam(.)p(intensity)" = fm4,
  "lam(.)p(intensity + intensity^2)" = fm5,
  "lam(.)p(day + rate)" = fm6,
  "lam(.)p(data + day^2 + intensity + intensity^2)" = fm7 )

(mspart1 <- modSel(mspart1))
```

	nPars	AIC	delta	AICwt	cumltvWt
lam(.)p(data + day^2 + intensity + intensity^2)	6	2057.62	0.00	4.9e-01	0.49
lam(.)p(day+day^2)	4	2057.73	0.11	4.6e-01	0.95
lam(.)p(day)	3	2063.01	5.39	3.3e-02	0.99
lam(.)p(day + rate)	4	2064.78	7.16	1.4e-02	1.00
lam(.)p(intensity + intensity^2)	4	2147.18	89.56	1.7e-20	1.00
lam(.)p(intensity)	3	2155.25	97.63	3.1e-22	1.00
lam(.)p(.)	2	2157.57	99.95	9.7e-23	1.00

We see the most complex model, which includes a quadratic of both day and sampling intensity, has 49% of the weight, and the second-best model, which only has a quadratic of day, has another 46% of the AIC weight. We use the top model as the basis for expanding the model to include covariates on abundance.

```
# Fit a series of models with abundance covariates
fm7  <- multinomPois(~day + I(day^2) + intensity + I(intensity^2) ~1, mhb.umf)
fm8  <- multinomPois(~day + I(day^2) + intensity + I(intensity^2) ~elev, mhb.umf)
fm9  <- multinomPois(~day + I(day^2) + intensity + I(intensity^2) ~forest, mhb.umf)
fm10 <- multinomPois(~day + I(day^2) + intensity + I(intensity^2) ~iLength, mhb.umf)
fm11 <- multinomPois(~day + I(day^2) + intensity + I(intensity^2) ~forest + elev, mhb.umf)
fm12 <- multinomPois(~day + I(day^2) + intensity + I(intensity^2) ~forest + iLength,
   mhb.umf)
fm13 <- multinomPois(~day + I(day^2) + intensity + I(intensity^2) ~elev + iLength, mhb.umf)
fm14 <- multinomPois(~day + I(day^2) + intensity + I(intensity^2) ~forest + elev + iLength,
   mhb.umf)
fm15 <- multinomPois(~day + I(day^2) + intensity + I(intensity^2) ~elev + I(elev^2),
   mhb.umf)
fm16 <- multinomPois(~day + I(day^2) + intensity + I(intensity^2) ~forest + elev +
   I(elev^2), mhb.umf)
fm17 <- multinomPois(~day + I(day^2) + intensity + I(intensity^2) ~forest + elev +
   I(elev^2) + iLength, mhb.umf)

# Assemble the models into a fitList
mspart2 <- fitList(
  "lam(.)p(best)"          = fm7,
  "lam(elev)p(best)"       = fm8,
  "lam(forest)p(best)"     = fm9,
  "lam(length)p(best)"     = fm10,
```

```
"lam(forest + elev)p(best)"                      = fm11,
"lam(forest + iLength)p(best)"                   = fm12,
"lam(elev + iLength)p(best)"                     = fm13,
"lam(forest + elev + length)p(best)"             = fm14,
"lam(elev + elev^2)p(best)"                      = fm15,
"lam(forest + elev + elev^2)p(best)"             = fm16,
"lam(forest + elev + elev^2 + iLength)p(best)"   = fm17)

# Rank them by AIC
(mspart2 <- modSel(mspart2))
                                              nPars      AIC    delta    AICwt  cumltvWt
lam(forest + elev + elev^2 + iLength)p(best)     10  1927.79     0.00  1.0e+00      1.00
lam(forest + elev + elev^2)p(best)                9  1942.53    14.74  6.3e-04      1.00
lam(forest + elev + iLength)p(best)               9  1947.20    19.41  6.1e-05      1.00
lam(forest + elev)p(best)                         8  1956.81    29.02  5.0e-07      1.00
lam(elev + elev^2)p(best)                         8  1991.30    63.51  1.6e-14      1.00
lam(forest + iLength)p(best)                      8  1992.53    64.75  8.7e-15      1.00
lam(forest)p(best)                                7  2012.67    84.88  3.7e-19      1.00
lam(elev + iLength)p(best)                        8  2018.26    90.47  2.3e-20      1.00
lam(elev)p(best)                                  7  2029.44   101.65  8.4e-23      1.00
lam(iLength)p(best)                               7  2039.10   111.31  6.7e-25      1.00
lam(.)p(best)                                     6  2057.62   129.83  6.4e-29      1.00
```

We find that, on top of the detection model already selected, the most favored model includes forest cover, a quadratic elevation effect, and route length. This model contains 100% of the AIC weight. Let's look at the coefficients and see if they all make sense.

```
fm17

Call:
multinomPois(formula = ~day + I(day^2) + intensity + I(intensity^2) ~
    forest + elev + I(elev^2) + iLength, data = mhb.umf)

Abundance:
             Estimate      SE      z     P(>|z|)
(Intercept)     1.828  0.2505   7.30    2.92e-13
forest          0.370  0.0530   6.98    2.86e-12
elev           -0.674  0.0941  -7.16    7.89e-13
I(elev^2)      -0.480  0.1091  -4.40    1.09e-05
iLength        -4.392  1.1045  -3.98    6.98e-05

Detection:
                 Estimate      SE       z    P(>|z|)
(Intercept)     -0.495411  0.1065  -4.6502  3.32e-06
day             -0.287148  0.0889  -3.2318  1.23e-03
I(day^2)         0.000859  0.0658   0.0131  9.90e-01
intensity        0.254052  0.0979   2.5949  9.46e-03
I(intensity^2)  -0.047959  0.0798  -0.6009  5.48e-01

AIC: 1927.787
```

The interpretation of this is straightforward: The abundance of jay territories increases with forest cover and decreases as the inverse of route length increases (i.e., increases for longer route lengths). The effect of elevation is not immediately clear but see the next section where we plot the response curve. Territory detection probability is suggested to decrease over the survey season (having a very negative coefficient on day) and it increases with sample intensity (the amount of time spent per unit route length) although with a slightly quadratic response indicating a decline for higher levels of intensity. These are sensible results.

Next, we fit some negative binomial models to the Jay data using the unmarked function gmultmix. Note that for versions of unmarked prior to and including v. 0.10-6, the following analyses would not work with the redundant multinomial cell probabilities specified by our custom pi function (but would work with other pi functions). As of version 0.11-0 this has been fixed. Here we fit the Poisson and negative binomial models using gmultmix so as to compare the AICs directly. (Remember: you cannot compare AIC from multinomPois and gmultmix as the likelihood is constructed differently.)

```
mhb.umf2 <- unmarkedFrameGMM(y=caphist, numPrimary = 1,
      siteCovs=as.data.frame(sitecovs), obsCovs=obscovs, obsToY=o2y, piFun="crPiFun")

fm1NB <- gmultmix(~1, ~1, ~1, mix = "NB", data = mhb.umf2)

fm17P <- gmultmix(~forest + elev + I(elev^2) + iLength, ~1, ~day + I(day^2) + intensity +
   I(intensity^2), mix = "P",  data = mhb.umf2)

fm17NB <- gmultmix(~forest + elev + I(elev^2) + iLength, ~1, ~day + I(day^2) + intensity +
   I(intensity^2), mix = "NB",  data = mhb.umf2)

fm17P    # AIC-best Poisson mixture model
Call:
gmultmix(lambdaformula = ~forest + elev + I(elev^2) + iLength,
    phiformula = ~1, pformula = ~day + I(day^2) + intensity +
        I(intensity^2), data = mhb.umf2, mixture = "P")

Abundance:
              Estimate      SE      z    P(>|z|)
(Intercept)     1.828   0.2505   7.30   2.92e-13
forest          0.370   0.0530   6.98   2.86e-12
elev           -0.674   0.0941  -7.16   7.89e-13
I(elev^2)      -0.480   0.1091  -4.40   1.09e-05
iLength        -4.392   1.1045  -3.98   6.98e-05

Detection:
                Estimate      SE        z    P(>|z|)
(Intercept)    -0.495411  0.1065  -4.6502  3.32e-06
day            -0.287148  0.0889  -3.2318  1.23e-03
I(day^2)        0.000859  0.0658   0.0131  9.90e-01
intensity       0.254052  0.0979   2.5949  9.46e-03
I(intensity^2) -0.047959  0.0798  -0.6009  5.48e-01

AIC: 1757.454

fm17NB    # NegBin version of AIC-best Poisson mixture model
Call:
gmultmix(lambdaformula = ~forest + elev + I(elev^2) + iLength,
    phiformula = ~1, pformula = ~day + I(day^2) + intensity +
        I(intensity^2), data = mhb.umf2, mixture = "NB")
```

```
Abundance:
              Estimate     SE       z    P(>|z|)
(Intercept)      1.786  0.2970   6.01  1.82e-09
forest           0.393  0.0673   5.84  5.36e-09
elev            -0.663  0.1031  -6.43  1.30e-10
I(elev^2)       -0.486  0.1262  -3.86  1.16e-04
iLength         -4.203  1.2976  -3.24  1.20e-03

Detection:
                 Estimate     SE        z   P(>|z|)
(Intercept)       -0.4963 0.1080  -4.5969  4.29e-06
day               -0.2830 0.0905  -3.1250  1.78e-03
I(day^2)          -0.0038 0.0666  -0.0571  9.54e-01
intensity          0.2235 0.1002   2.2312  2.57e-02
I(intensity^2)    -0.0325 0.0823  -0.3943  6.93e-01

Dispersion:
  Estimate   SE    z   P(>|z|)
      1.71 0.381  4.5  6.88e-06

AIC: 1746.688
```

The results are surprisingly consistent between the two models in terms of the parameter estimates, although the NB model is superior by about 10 AIC units, and the estimated "size" parameter of the negative binomial model is $\exp(1.71) = 5.52$. This suggests a moderate amount of extra-Poisson variation (see Section 7.5.3 for the interpretation of this). The good agreement between the parameter estimates under a Poisson and a NB multinomial mixture model contrast with the much larger discrepancy that we often observe in a similar comparison in binomial mixture models (see Chapter 6). This is something that warrants further research.

7.9.3 ANALYSIS OF MODEL FIT

We investigate whether the best model provides a reasonable fit to the data using the `parboot` function to do a goodness-of-fit evaluation. We use the three fit statistics defined previously (sum-of-squared errors, Chi-square, Freeman–Tukey; see Sections 2.8.2 and 6.8), and in addition we construct an analogous set of fit statistics to evaluate the fit of the model for predicting n_s, the observed number of individuals at quadrat s. We look at fit of this component of the model because it is easier to visualize and diagnose issues with a quadrat-specific summary of the model. Note that $E(n_s) = (1 - \pi_0)\lambda_s$ and so, in a sense, the adequacy of the model for predicting n_s measures its adequacy for both the spatial model of local abundance (the λ_s part) and also "total" encounter probability (the π_0 part). We conduct the analysis by executing the following commands, with a modified version of the function `fitstats`:

```
# Define new fitstats function
fitstats2 <- function(fm) {
  observed <- getY(fm@data)
  expected <- fitted(fm)
  resids <- residuals(fm)
  n.obs <- apply(observed,1,sum,na.rm=TRUE)
```

```
n.pred <- apply(expected,1,sum,na.rm=TRUE)
sse <- sum(resids^2,na.rm=TRUE)
chisq <- sum((observed - expected)^2 / expected,na.rm=TRUE)
freeTuke <- sum((sqrt(observed) - sqrt(expected))^2,na.rm=TRUE)
freeTuke.n <- sum((sqrt(n.obs)-sqrt(n.pred))^2,na.rm=TRUE)
sse.n <- sum( (n.obs -n.pred)^2,na.rm=TRUE)
chisq.n <- sum((n.obs - n.pred)^2 / expected,na.rm=TRUE)

out <- c(SSE=sse, Chisq=chisq, freemanTukey=freeTuke,
    SSE.n = sse.n, Chisq.n = chisq.n, freemanTukey.n=freeTuke.n)
return(out)
}

(pb.mhb <- parboot(fm17, fitstats2, nsim=1000, report=1)) # Takes a while

Call: parboot(object = fm17, statistic = fitstats2, nsim = 1000, report = 1)

Parametric Bootstrap Statistics:
                  t0    mean(t0 - t_B)    StdDev(t0 - t_B)    Pr(t_B > t0)
SSE              528        9.23e+01           2.89e+01         0.00200
Chisq           1704        2.36e+02           1.71e+02         0.02398
freemanTukey     409        2.98e+01           1.25e+01         0.00899
SSE.n           1512        2.13e+02           1.26e+02         0.05994
Chisq.n        44163       -5.21e+06           1.17e+08         0.09990
freemanTukey.n   203        3.69e+01           1.12e+01         0.00000

t_B quantiles:
                  0%    2.5%    25%    50%    75%    97.5%    100%
SSE              340    383    416    435    454      499    5.4e+02
Chisq           1258   1310   1398   1450   1501     1692    4.2e+03
freemanTukey     334    355    370    379    388      402    4.2e+02
SSE.n            879   1060   1209   1300   1381     1552    1.7e+03
Chisq.n        30267  32429  35406  37385  39780   164257    3.6e+09
freemanTukey.n   136    145    158    165    174      188    2.0e+02

t0 = Original statistic computed from data
t_B = Vector of bootstrap samples
```

In fact, the results suggest that this model does not adequately fit the data, having bootstrap p-values near 0 for each of the three fit statistics based on the encounter history frequencies. The fit is marginally better for predicting the total number of observed territories, but still not acceptable. To investigate this further, we could do the following things:

1. See if we can improve fit by throwing everything in the model or modifying the model in some way by accommodating zero inflation or using a negative binomial. We take this approach here shortly.
2. We can look at the patterns in the residuals and see if we can identify a specific mode for lack of fit. For purposes of residual analysis of multinomial models, we recommend looking at the residuals in predicting n_s, which is what we do below.

3. If we can identify a few extreme residuals, then perhaps the model fit is improved by not counting these "outliers." If the model is adequate for the preponderance of observations then it should generally predict well and we may be satisfied with this result.
4. Finally, we can accept the poor fit as is (or even if discarding outliers) and adjust the model standard errors by an overdispersion ratio (see Section 6.9.3 and below), c-hat, which can be computed from the bootstrapped Chi-square values as the ratio of the observed to the mean of the bootstrapped Chi-square values, which here is about c-hat = 1704/1467.4 = 1.16 (= pb.mhb@t0 [2]/mean (pb.mhb@t.star[,2])). A fitting model has an expected c.hat of 1; hence, despite the clear result of the GoF significance test, the *magnitude* of the lack of fit seems to be fairly small.

We check the goodness of fit for the equivalent negative binomial abundance model:

```
(pb.mhbNB <- parboot(fm17NB, fitstats2, nsim=1000, report=1)) # takes even longer ...

Call: parboot(object = fm17NB, statistic = fitstats2, nsim = 1000, report = 1)

Parametric Bootstrap Statistics:
                 t0     mean(t0 - t_B)   StdDev(t0 - t_B)   Pr(t_B > t0)
SSE              529       5.85e+01         3.89e+01          0.0779
Chisq            1706      1.71e+02         9.86e+01          0.0360
freemanTukey     409       2.19e+01         1.50e+01          0.0679
SSE.n            1544      1.96e+01         1.97e+02          0.4236
Chisq.n          43863     -2.08e+06        5.74e+07          0.3776
freemanTukey.n   203       3.38e+00         1.51e+01          0.3976

t_B quantiles:
                 0%     2.5%    25%    50%    75%    97.5%     100%
SSE              354    402     442    469    497    552     5.9e+02
Chisq            1327   1383    1475   1526   1582   1739    2.5e+03
freemanTukey     334    359     377    387    397    418     4.4e+02
SSE.n            998    1173    1390   1506   1655   1920    2.3e+03
Chisq.n          33123  36032   39928  42456  45755  216363  1.8e+09
freemanTukey.n   160    171     189    198    209    230     2.5e+02

t0 = Original statistic compuated from data
t_B = Vector of bootstrap samples

# Compute c-hat
pb.mhbNB@t0[2] / mean(pb.mhbNB@t.star[,2])
   Chisq
1.111119
```

The result is encouraging, suggesting that the NB model improves the fit considerably and it passes all of our fit tests (except for one). We might therefore be satisfied with using the NB model as a basis for inference. In particular, we think the model is adequately describing the spatial variation in abundance given the *p*-values indicated for the three fit statistics based on quadrat-specific observed sample sizes.

In general, especially when models don't fit well, we might investigate the source of lack of fit by investigating patterns in the residuals using a few basic residual plots or plots of observed *vs.* predicted values. For example, here we provide a plot of the predicted n for each plot as a function of the observed n, under both the Poisson and NB models (see Figure 7.7):

```
n.obs <- apply(caphist, 1, sum, na.rm=TRUE)
n.predP <- apply(fitted(fm17P), 1, sum, na.rm=TRUE)
n.predNB <- apply(fitted(fm17NB), 1, sum, na.rm=TRUE)
plot(n.obs, n.predP, frame = F)
abline(0,1)
points(smooth.spline(n.predP ~ n.obs, df = 4), type = "l", lwd = 2, col = "blue")
points(smooth.spline(n.predNB ~ n.obs, df=4), type= "l", lwd = 2, col = "red")
```

In this case, the abundance predictions under a Poisson model and an overdispersed model (here, the negative binomial) were hardly different, although the latter model passed a GoF assessment, while the former did not. However, as we have argued in Chapter 6, in general, if a fitting overdispersed model predicts unrealistic abundance, then we may settle for the poor fitting model and some kind of a quasi-likelihood adjustment for lack of fit ("lack-of-fit ratio") to inflate the standard errors associated with parameter estimates (see Section 6.9; Chapter 5 in Cooch and White, 2014; Section 12.3 in Kéry and Schaub, 2012). This may be an adequate solution if we only care about assessing the significance of some covariate effects or account for the additional uncertainty induced by such "overdispersion" when making predictions (see below).

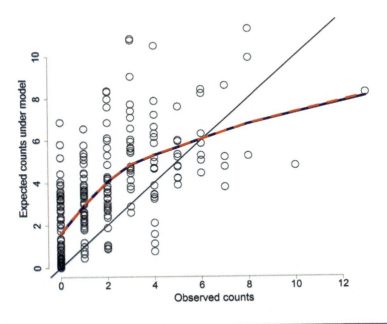

FIGURE 7.7

Fitted and observed number of jay territories in each MHB quadrat for both the Poisson (blue) and negative binomial models (red). The predicted values are very similar.

7.9.4 SUMMARY ANALYSES OF JAY MODELS

We will use our estimates from the top model for several interesting summary analyses, including making a response curve of the expected abundance, $E(N)$, as a function of elevation, and computing the optimal elevation for jays, and also producing an abundance map of the species for Switzerland. To make a response curve we need to generate a grid of predictor variables at which to evaluate the model. Recall that we had standardized the elevation and forest variables. We can find the range of the standardized elevation variable as follows:

```
range(siteCovs(mhb.umf)$elev)
[1] -1.440050 2.429829
```

Then we set forest cover to 0, so that predictions are at the mean level of forest (which was standardized for the analysis). With respect to inverse route length (covariate iLength), we could set it to 0, so that we would be predicting at infinite sample route length, conceptually as if we had saturated the quadrat with effort. However, in Sections 6.9 and 6.10 we have argued that we would then very likely predict to a greater effective sampling area than the nominal 1 km^2 sampling area. Therefore, here we compute two sets of predictions: first, at the unknown effective sample area that corresponds to the average route length of 5.1 km (with iLength = 1/5.1), and second, at the unknown effective sample area that corresponds to saturation sampling effort (with iLength = 0). We suggest that the two sets of estimates bracket the true density per 1 km^2. We see that our assumptions about the size of the effective sample area have a substantial effect on the estimated density of the jay in Switzerland (Figure 7.8).

```
elev.mean <- attr(siteCovs(mhb.umf)$elev, "scaled:center")
elev.sd <- attr(siteCovs(mhb.umf)$elev, "scaled:scale")
elev.orig <- elev.sd*seq(-1.5, 2.42,,500) + elev.mean

# Remember length = 0 is saturation sampling because iLength = 1/L
newL <- data.frame(elev = seq(-1.5,2.42,,500), elev.orig, forest = 0,
  iLength = 1/5.1)          # 'Low' prediction
newH <- data.frame(elev = seq(-1.5,2.42,,500), elev.orig, forest = 0,
  iLength = 0)              # 'High' prediction
predL <- predict(fm17NB, type="lambda", newdata=newL, appendData=TRUE)
predH <- predict(fm17NB, type="lambda", newdata=newH, appendData=TRUE)
head(cbind(low = predL[,1:2], high = predH[,1:2]))
  low.Predicted    low.SE high.Predicted   high.SE
1      2.366013 0.4657475       5.394389  1.518786
2      2.380799 0.4626372       5.428100  1.521632
3      2.395534 0.4594896       5.461694  1.524521
4      2.410214 0.4563058       5.495166  1.527454
5      2.424840 0.4530872       5.528511  1.530434
6      2.439407 0.4498348       5.561724  1.533460

plot(Predicted ~ elev.orig, predL, type="l", lwd = 3, xlab="Elevation",
  ylab="Expected # territories", ylim=c(0, 13), frame=F, col = "red")
points(Predicted ~ elev.orig, predH, type="l", lwd = 3, col = "blue")
matlines(elev.orig, predL[,3:4], lty = 1, lwd = 1, col = "red")
matlines(elev.orig, predH[,3:4], lty = 1, lwd = 1, col = "blue")
```

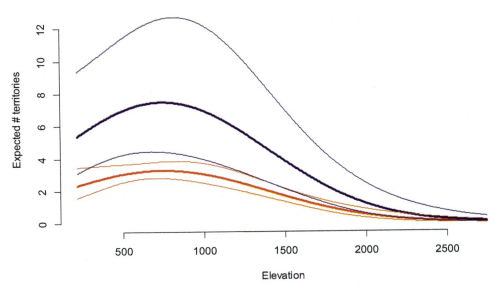

FIGURE 7.8

Expected density (number of territories per 1 km², with 95% CI) of the Eurasian jay (*Garrulus glandarius*) as a function of elevation in the analysis of the 2002 MHB territory detection/nondetection data in Switzerland. Red lines represent the assumption that the average survey route length (5.1 km) completely covers with search effort the nominal 1-km² sampling area. Blue lines represent the assumption that the nominal 1-km² sample area is represented by an infinite route length. Presumably, the truth lies somewhere in between. See text and Sections 6.9 and 6.10 for more detail.

Because we fitted a quadratic effect of elevation, depicted in Figure 7.8, we can compute directly the optimal elevation for the Eurasian jay, i.e., that which has the maximum density of territories, by employing a little bit of calculus. Denote the general quadratic response by the formula $y = a + b*x + c*x^2$, then we need to differentiate this expression with respect to x and set the result equal to 0 and solve: $dy/dx = b + 2*c*x = 0$ yields $x_{opt} = -b/(2c)$. Because elevation was standardized, we compute the optimal value and then back-transform the result.

```
b <- coef(fm17NB)[3]
c <- coef(fm17NB)[4]
elev.opt <- -b / (2*c)
```

```
(elev.opt <- elev.opt*elev.sd + elev.mean)
lambda(elev)
    740.3788
```

Hence, the maximum territory abundance for this species is estimated to occur at an elevation of about 740 m.

Perhaps we want to make the assumption that the minor lack of fit detected in Section 7.9.3 is simply some form of unstructured noise ("overdispersion"), which adds variance around the mean structure of the model rather than some structural deficiency in the mean structure. In that case, we may account for this overdispersion in the predictions using quasi-likelihood ideas with functions in the `modavgPred` package (Mazerolle, 2015). We can trick the function for model-averaged predictions into producing them for a single model, with or without overdispersion factor c-hat of 1.11 (as estimated for the full data set), by providing a model list with a single model only. We illustrate for the "low scenario" predictions.

```
require(AICcmodavg)
model.list <- list(fm17NB)   # candidate model list with single model
model.names <- c("AIC-best model")

# Compute model-averaged predictions of abundance for values of elevation, with
uncertainty (SE, CIs) adjusted for overdispersion (c.hat), with latter estimated from
bootstrapped Chisquare
pred.c.hatL <- modavgPred(cand.set = model.list, modnames = model.names, newdata = newL,
   parm.type = "lambda", type = "response", c.hat = 1.11)

# Compare predictions and SE without (cols 1-2) and with c.hat adjustment (cols 3-4)
head(cbind(predL[1:2], pred.c.hatL[1:2]), 10)
      Predicted       SE  mod.avg.pred  uncond.se
1      2.366013  0.4657475     2.366013  0.4906954
2      2.380799  0.4626372     2.380799  0.4874185
3      2.395534  0.4594896     2.395534  0.4841023
4      2.410214  0.4563058     2.410214  0.4807480
5      2.424840  0.4530872     2.424840  0.4773570
6      2.439407  0.4498348     2.439407  0.4739304
7      2.453915  0.4465501     2.453915  0.4704697
8      2.468361  0.4432342     2.468361  0.4669762
9      2.482743  0.4398885     2.482743  0.4634513
10     2.497058  0.4365144     2.497058  0.4598965
```

Hence, we see that the amount of overdispersion detected in the parametric bootstrap corresponds to a very slight upward adjustment of the SEs, which is probably not relevant in practice.

As in the previous chapter, we emphasize that our treatment of lack of fit as simple overdispersion (quantified by the magnitude of c.hat), to be adjusted for when computing SEs and possibly AIC scores via c-hat, is tentative. However, it makes sense to us and seems to be consistent with a very widely observed practice in the fields of capture-recapture, occupancy modeling, and distance sampling (e.g., MacKenzie and Bailey, 2004; Johnson et al., 2010; Cooch and White, 2014). We note, however, that it would be useful to conduct some simulation-based assessment of the validity of this approach. Moreover, detection and mitigation of lack of fit in hierarchical models seems to call for more research in general.

Next, we inspect the response of abundance to route length (see Section 6.9.4 for the analogous analysis but using the N-mixture model). We build a new data frame to use for predictions and then produce the response curve along with markers for the minimum, mean, and maximum route length (1.2, 5.135, and 9.4 km, respectively):

```
rlength <- seq(1, 30, 0.01)                # Vary route length from 1 to 30 kms
newData <- data.frame(elev=0, forest=0, iLength=1/rlength)
pred <- predict(fm17NB, type="lambda", newdata=newData, c.hat = 1.11)
par(mar = c(5,5,3,2), cex.lab = 1.5, cex.axis = 1.3)
plot(rlength, pred[[1]], type = "l", lwd = 3, col = "blue", frame = F, xlab = "Transect length
 (km)", ylab = "Exposed population (lambda)", ylim = c(0, 16), axes = F)
```

```
axis(1, at = seq(2,30,2))  ;  axis(2)
abline(v = c(1.2, 5.135, 9.4), lwd = 2)
matlines(rlength, cbind(pred[[1]]-pred[[2]], pred[[1]]+pred[[2]]), type = "l", lty = 1,
lwd = 2, col = "gray")

sat.pred <- predict(fm17NB, type="lambda", newdata= data.frame(elev=0, forest=0,
iLength=0), c.hat = 1.11)
abline(as.numeric(sat.pred[1]),0, lwd = 2, lty = 2)
```

To help interpret the response curve (Figure 7.9) we provide the estimates for the minimum, mean, and maximum route length (1.2, 5.135, 9.5 km, respectively) standardized by the saturation density of territories (evaluated at saturation sampling, or iLength = 0):

```
pred[round(rlength,2)==1.2,]/as.numeric(sat.pred[1])
     Predicted        SE       lower     upper
21 0.03011745 0.0244722 0.006125858 0.1480708

pred[round(rlength,2)==5.14,]/as.numeric(sat.pred[1])
     Predicted        SE       lower     upper
415 0.4414284 0.04940136 0.3544869 0.5496932

pred[round(rlength,2)==9.4,]/as.numeric(sat.pred[1])
     Predicted        SE       lower     upper
841   0.63945 0.1126562 0.4527356 0.903168
```

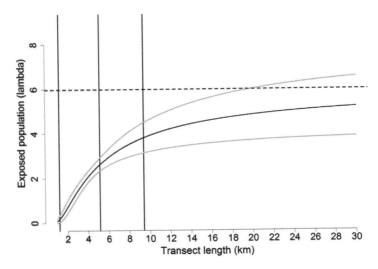

FIGURE 7.9

Estimated relationship (with 1 SE bounds) between the expected number of Jay territories (λ) that are exposed to detection by a 1 km² quadrat as a function of transect length. Vertical black lines show the minimum, mean, and maximum route lengths in the Swiss MHB (1.2, 5.135 and 9.4, respectively). The associated exposed population represents 3, 44 and 64% of the exposed population at saturation sampling (horizontal dashed line), when a hypothetical route of length infinity is squeezed into the 1 km² quadrat. See Figure 6.12 for the equivalent figure under the N-mixture model analysis of the Great Tit data.

Thus we see that about 3%, 44% and 64% of the territories are exposed to sampling by routes of length equal to the minimum, mean or maximum that were realized by the MHB survey.

7.9.5 SPATIAL PREDICTION

Our ultimate objective with the MHB jay data is to produce a distribution map to represent the spatial variation of the density of the species in Switzerland. To do this we first familiarize ourselves with the `raster` package function `rasterFromXYZ`, which we use to produce maps of the elevation and forest cover covariates over the whole of Switzerland (Figure 7.10).

```
library(raster)
library(rgdal)

# Swiss landscape data and shape files
data(Switzerland)         # Load Swiss landscape data from unmarked
CH <- Switzerland         # this is for 'confoederatio helvetica'
head(CH)
gelev <- CH[,"elevation"] # Median elevation of quadrat
gforest <- CH[,"forest"]
grid <- CH[,c("x", "y")]
lakes <- readOGR(".", "lakes")
rivers <- readOGR(".", "rivers")
border <- readOGR(".", "border")

# Draw two maps of Swiss elevation and forest cover   (Figure 7.10)
par(mfrow = c(1,2), mar = c(1,2,3,5))
mapPalette1 <- colorRampPalette(c("grey", "yellow", "orange", "red"))
mapPalette2 <- colorRampPalette(c("grey", "lightgreen", "darkgreen"))
r1 <- rasterFromXYZ(cbind(x = CH$x, y = CH$y, z = CH$elevation))
r2 <- rasterFromXYZ(cbind(x = CH$x, y = CH$y, z = CH$forest))
```

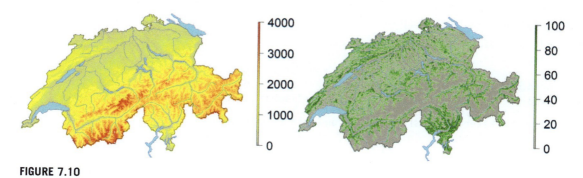

FIGURE 7.10

Swiss landscape: median elevation (left, in meters above sea level) and percent forest cover (right) in 1-km^2 pixels of Switzerland (total colored area is 42,275 km^2).

```r
plot(r1, col = mapPalette1(100), axes = FALSE, box = FALSE, main = "Elevation (m a.s.l.)",
    zlim = c(0, 4000))
plot(rivers, col = "dodgerblue", add = TRUE)
plot(border, col = "transparent", lwd = 1.5, add = TRUE)
plot(lakes, col = "skyblue", border = "royalblue", add = TRUE)

plot(r2, col = mapPalette2(100), axes = FALSE, box = FALSE, main = "Forest cover (%)",
    zlim = c(0, 100))
plot(rivers, col = "dodgerblue", add = TRUE)
plot(border, col = "transparent", lwd = 1.5, add = TRUE)
plot(lakes, col = "skyblue", border = "royalblue", add = TRUE)
```

Next, we apply this idea to producing a map of predictions of the expected abundance for each quadrat. To evaluate the predictions under the model we first have to get the landscape covariates on the same scale as the covariates used in fitting the model, i.e., standardize by the *same* mean and standard deviation that we used to standardize the *actual* data for the sample quadrats in our analysis. We can get the mean and standard deviation using the `attr` function applied to the `siteCovs` object as follows:

```r
# Standardize elevation for all grid cells using the mean at sample plots
elev.mean <- attr(siteCovs(mhb.umf)$elev, "scaled:center")
elev.sd <- attr(siteCovs(mhb.umf)$elev, "scaled:scale")
gelev <- (gelev - elev.mean) / elev.sd

# Standardize forest cover also using the mean at sample plots
forest.mean <- attr(siteCovs(mhb.umf)$forest, "scaled:center")
forest.sd <- attr(siteCovs(mhb.umf)$forest, "scaled:scale")
gforest <- (gforest - forest.mean) / forest.sd
```

Most of the approximately 42,000 quadrats on the landscape were not sampled, and so what do we do about route length for those quadrats? Well, we're interested in making a prediction of the true density that, as we just argued, by our parameterization of route length, should occur anywhere between a route length of 5.1 km and infinity. So, we will again predict density using the two values of the inverse route length covariate ($1/L = 1/5.1$ and $1/L = 0$). We plug the covariates into two data frames, invoke the `predict` function and plot the more conservative "low" predictions (Figure 7.11; the only difference with the "high" map would be the scale, while the spatial patterns are the same):

```r
# Form predictions for Swiss landscape
newL <- data.frame(elev=gelev, forest=gforest, iLength=1/5.1)
newH <- data.frame(elev=gelev, forest=gforest, iLength=0)
pred.mhb.NB.Low <- predict(fm17NB, type="lambda", newdata=newL, appendData=T)
pred.mhb.NB.High <- predict(fm17NB, type="lambda", newdata=newH, appendData=T)

# Create rasters and mask for elevation (mask areas > 2250 m)
r1 <- rasterFromXYZ(cbind(x = CH$x, y = CH$y, z = pred.mhb.NB.Low[,1]))
r2 <- rasterFromXYZ(cbind(x = CH$x, y = CH$y, z = pred.mhb.NB.Low[,2]))
elev <- rasterFromXYZ(cbind(CH$x, CH$y, gelev))
elev[elev > 2250] <- NA
r1 <- mask(r1, elev)
r2 <- mask(r2, elev)
```

7.9 EXAMPLE 3: JAYS IN THE SWISS MHB

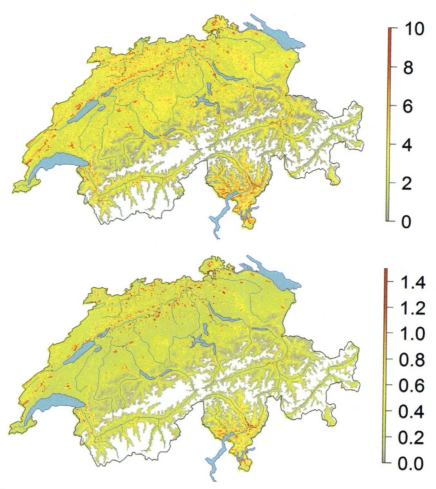

FIGURE 7.11

Abundance maps for the Eurasian Jay (*Garrulus glandarius*) in Switzerland in 2002 (territories per km^2) based on a multinomial mixture model for the detection/nondetection histories of the mapped territories in 238 one-km^2 quadrats in the Swiss breeding bird survey (MHB). Top: point estimates, bottom: standard errors.

```
# Draw maps of jay density and standard error of density   (Figure 7.11)
par(mfrow = c(1,2), mar = c(1,2,3,5))
plot(r1, col = mapPalette1(100), axes = FALSE, box = FALSE, main = "Density of European Jay",
  zlim = c(0, 10))
plot(rivers, col = "dodgerblue", add = TRUE)
plot(border, col = "transparent", lwd = 1.5, add = TRUE)
plot(lakes, col = "skyblue", border = "royalblue", add = TRUE)
```

```
plot(r2, col = mapPalette1(100), axes = FALSE, box = FALSE, main = "Standard errors of
density", zlim = c(0, 1.5))
plot(rivers, col = "dodgerblue", add = TRUE)
plot(border, col = "transparent", lwd = 1.5, add = TRUE)
plot(lakes, col = "skyblue", border = "royalblue", add = TRUE)
```

Since use of function `predict` can easily take several minutes for big prediction data sets such as the Swiss landscape, remember that we can get the point estimates very quickly in the old-fashioned way by building the linear predictor ourselves. Inspection of the `fm17NB` output object reveals that the required coefficients are the first four elements of the coefficients vector. Hence, we can build the design matrix, obtain the predictions, and produce a quick plot:

```
betavec <- coef(fm17NB)[1:4]     # yields predictions for saturation sampling
Xg <- cbind(rep(1, length(gforest)), gforest, gelev, gelev*gelev)
pred <- exp(Xg %*% betavec)
plot(rasterFromXYZ(cbind(grid, pred)))  # Quick and dirty version, not shown
```

7.9.6 POPULATION SIZE OF JAYS IN SWITZERLAND

Finally, we obtain a total population size estimate of breeding jay territories (presumed pairs) in Switzerland. This is a derived parameter that we can compute simply by summing up the predicted values (excluding quadrats with >50% water and >2250 m elevation) for the average route length of 5.1 km.

```
out <- which(CH$water > 50 | CH$elevation > 2250)
(N.jay <- sum(pred.mhb.NB.Low[-out,1]))    # 'low prediction'
[1] 88545.58
```

Or, for saturation sampling effort:

```
(N.jay <- sum(pred.mhb.NB.High[-out,1]))   # 'high prediction'
[1] 201879.4
```

We can embed the process of computing the predictions explicitly and summing them up in a function `Nhat`, which we can pass to `parboot` to obtain a bootstrapped confidence interval for the estimated total population size using the two quadrat sampling intensities corresponding to the average route length (`Nlow`) and an infinite route length (`Nhigh`):

```
Nhat <- function(fm) {
   betavec <- coef(fm)[1:5]
   Xg <- cbind(rep(1, length(gforest)), gforest, gelev, gelev*gelev, 1/5.1)
   predLow <- as.numeric(exp(Xg%*%(betavec)))
   predHigh <- as.numeric(exp(Xg[,-5]%*%(betavec[-5])))
   out <- which(CH$water > 50 | CH$elevation > 2250)
   Nlow <- sum(predLow[-out])
   Nhigh <- sum(predHigh[-out])
   return(c(Nlow = Nlow, Nhigh = Nhigh))
}
```

```
set.seed(2015)
(pb.N <- parboot(fm17NB, Nhat, nsim=100, report=1))

[....monitoring output deleted....]

Parametric Bootstrap Statistics:
           t0      mean(t0 - t_B)   StdDev(t0 - t_B)   Pr(t_B > t0)
Nlow    88546          -96.6              5593            0.485
Nhigh  201879       -13019.3             59753            0.535

t_B quantiles:
           0%     2.5%    25%      50%     75%    97.5%    100%
Nlow    78052   80211   84744   88445   92168   99105   110939
Nhigh  111480  126430  168324  208866  248581  349855  408421

t0 = Original statistic compuated from data
t_B = Vector of bootstrap samples
```

We obtain a 95% confidence interval of either 80,211–99,105 (for the low scenario) or of 126,430–349,855 jay territories (for the high scenario) in Switzerland. Of course we should exercise some caution in interpreting this as strict population size due to not being able to know how much "double counting" is going on around the edges of the quadrat or whether our prediction with 5.1 average survey transect length adjustment over- or undercorrects for this.

7.9.7 BAYESIAN ANALYSIS OF THE JAY DATA

We finish off this chapter by providing a Bayesian formulation of the multinomial mixture models used to analyze the MHB data in the BUGS language. One point of this illustration is to show how to include both site and observation covariates in the BUGS formulation of the multinomial N-mixture models. We note that the MHB data structure is somewhat complicated by the mixed sampling protocol, containing both two and three survey occasions, and so this is one additional challenge that we'll have to confront. We use the data augmentation (DA) approach first introduced for the removal model back in Section 7.8.4 for Chandler's ALFL data. We adapt that BUGS model formulation directly for the MHB data. To adapt this model we first have to convert the multinomial encounter frequencies for each site into *individual* encounter histories, a matrix with three columns and as many rows as there are individuals, and with an associated site ID variable that indicates in which MHB survey quadrat each individual encounter history was observed. The R commands to do this from the unmarked frame that we created previously are as follows:

```
# Extract data and do data augmentation up to M = 400
y <- as.matrix(getY(mhb.umf))

# Now we have to stretch out the encounter frequencies into individuals...
# There were 439 unique individuals observed during the survey
eh <- unlist(dimnames(y)[2])
ehid <- col(y)[y>0 & !is.na(y)]    # Column ids, index to encounter history
eh <- eh[ehid]
siteid <- row(y)[y>0 & !is.na(y)] # Site ids
y <- y[y>0 & !is.na(y)]            # Positive counts
```

```
eh <- rep(eh, y)
siteid <- rep(siteid, y)

eh.mat <- matrix(NA,nrow=length(eh),ncol=3)
for(i in 1:length(eh)){
    eh.mat[i,] <- as.numeric(unlist(strsplit(eh[i],split="")))
}
```

Note that one encounter history is "10x" (two samples at a high elevation site) and this produces a warning message about coercing the x to an NA. This is ok. The result is a 439×3 matrix of individual encounter histories. Now we proceed with the data augmentation as before. Here we augment up to 800 individuals and then we extract the site and observation covariates. Remember the obscovs are in row order so when we create the nsites $\times$ 3 matrix for a given observation covariate we need to stuff the vectorized version of the covariate into the matrix byrow (byrow=TRUE).

For illustration, we will pick one observation covariate, sampling intensity, for use in the model. We use all three of the site covariates (elevation, route length, and forest cover), the first and third being standardized. Recall that we fit the model using the inverse of route length, which we express directly in the BUGS model language below using the covariate length on the canonical scale.

```
# Define some things and do the data augmentation
nsites = nrow(sitecovs)
nind <- nrow(eh.mat)
M <- 800
y <- rbind(eh.mat, matrix(0, nrow=(M-nind), ncol=3))

# Augment site ID
site <- c(siteid, rep(NA, M-nind))
sitecovs <- siteCovs(mhb.umf)

obscovs <- obsCovs(mhb.umf)
Intensity <- matrix(obscovs[,"intensity"], nrow=nsites, ncol=3,byrow=TRUE)

# Bundle data for BUGS
data <- list(y = y, J = 3, M = M , nsites=nsites, X = as.matrix(sitecovs)[,1:3],
Intensity= Intensity, group=site)
str(data)

# Specify model in BUGS language
cat("
model {

# Prior distributions
p0 ~ dunif(0,1)
alpha0 <- log(p0 / (1-p0))
alpha1 ~ dnorm(0, 0.01)

beta0 ~ dnorm(0,0.01)
beta1 ~ dnorm(0,0.01)
beta2 ~ dnorm(0,0.01)
beta3 ~ dnorm(0,0.01)
psi <- sum(lambda[]) / M    # psi is a derived parameter
```

```
# Model for abundance: lambda depends on Elev, Length, Forest
for(s in 1:nsites){
  log(lambda[s]) <- beta0 + beta1 * X[s,1] + (beta2/X[s,2]) + beta3*X[s,3]
  probs[s] <- lambda[s] / sum(lambda[])
}

# Model for individual encounter histories
for(i in 1:M){
  group[i] ~ dcat(probs[])   # Group == site membership
  z[i] ~ dbern(psi)           # Data augmentation variables

  # Observation model: p depends on Intensity
  for(j in 1:J){
    logit(p[i,j]) <- alpha0 + alpha1 * Intensity[group[i],j]
    pz[i,j] <- p[i,j] * z[i]
    y[i,j] ~ dbern(pz[i,j])
  }
}
}
",fill=TRUE,file="model.txt")

# Parameters monitored
params <- c("p0", "alpha0", "alpha1", "beta0", "beta1", "beta2", "beta3", "psi")

# Initial values
zst <- rep(c(1,M-100), rep(0,100))
inits <- function(){
  list (p0 = runif(1), alpha1 = runif(1), beta0=runif(1),
    beta1=rnorm(1), z= zst ) }

# MCMC settings
ni <- 11000 ; nb <- 1000 ; nt <- 4 ; nc <- 3

# Call JAGS from R and summarize marginal posteriors
out <- jags(data, inits, params, "model.txt", n.thin = nt,
    n.chains = nc, n.burnin = nb, n.iter = ni)
print(out, digits = 2)
JAGS output for model 'model.txt', generated by jagsUI.
Estimates based on 3 chains of 11000 iterations,
burn-in = 1000 iterations and thin rate = 4,
yielding 7500 total samples from the joint posterior.
MCMC ran for 12.465 minutes at time 2015-07-01 12:37:12.
```

	mean	sd	2.5%	50%	97.5%	overlap0	f	Rhat	n.eff
p0	0.40	0.02	0.37	0.40	0.44	FALSE	1.00	1.00	963
alpha0	-0.39	0.07	-0.54	-0.39	-0.25	FALSE	1.00	1.00	952
alpha1	0.17	0.09	-0.01	0.17	0.36	TRUE	0.97	1.00	1944
beta0	1.16	0.21	0.75	1.16	1.58	FALSE	1.00	1.02	104
beta1(elev)	-0.67	0.07	-0.81	-0.67	-0.54	FALSE	1.00	1.00	783
beta2(ilen)	-3.09	1.00	-5.11	-3.07	-1.18	FALSE	1.00	1.02	101
beta3 (for)	0.47	0.05	0.38	0.47	0.57	FALSE	1.00	1.00	3297
psi	0.71	0.03	0.65	0.71	0.77	FALSE	1.00	1.00	2360
deviance	6857.13	58.13	6749.34	6855.07	6976.13	FALSE	1.00	1.00	938

We see some broad consistency between posterior means and MLEs from our best fitting Poisson model (which included more effects than here), which are shown below:

```
> fm17P       # AIC-best Poisson mixture model
Call:
gmultmix(lambdaformula = ~forest + elev + I(elev^2) + iLength,
    phiformula = ~1, pformula = ~day + I(day^2) + intensity +
        I(intensity^2), data = mhb.umf2, mixture = "P")

Abundance:
            Estimate    SE       z      P(>|z|)
(Intercept)  1.828   0.2505    7.30   2.92e-13
forest       0.370   0.0530    6.98   2.86e-12
elev        -0.674   0.0941   -7.16   7.89e-13
I(elev^2)   -0.480   0.1091   -4.40   1.09e-05
iLength     -4.392   1.1045   -3.98   6.98e-05

Detection:
                Estimate     SE       z      P(>|z|)
(Intercept)    -0.495411  0.1065  -4.6502  3.32e-06
day            -0.287148  0.0889  -3.2318  1.23e-03
I(day^2)        0.000859  0.0658   0.0131  9.90e-01
intensity       0.254052  0.0979   2.5949  9.46e-03
I(intensity^2) -0.047959  0.0798  -0.6009  5.48e-01
```

In the previous analyses of these data using `unmarked` we settled on a negative binomial model for abundance. As an exercise we have you fit overdispersion at the site level by adding a normal random effect to the mean of local abundance. You might also try fitting the negative binomial model directly and include some additional covariate effects.

7.10 SUMMARY AND OUTLOOK

A large number of multinomial sampling protocols are widely used in ecological studies focused on abundance, including "multiple observer" sampling, removal sampling, and essentially all classical capture-recapture models. In addition, typical studies of metacommunities (see Chapter 11) frequently produce replicated species detection/nondetection data, which can be treated in exactly the same way as were the individual detection histories in this chapter. Sampling based on such protocols produces spatially organized multinomial observations indexed by point count location, transect, or habitat patch. Frequently one objective of such "meta-population studies" of spatially structured populations is to investigate variation in abundance across those spatial units, and the hierarchical modeling framework, which leads to multinomial N-mixture models, is a natural approach for addressing such objectives.

You may wonder whether, given the choice, is it better to model detection histories by adopting a multinomial mixture model or by use of a binomial mixture model for the aggregated counts (per site and survey)? We have compared the two models for identical data sets (Kéry and Royle, 2010) and found that in that case studied (different sets of data from the Swiss MHB) the multinomial model produced slightly more precise estimates as expected (since it uses more information, i.e., that coming from individual identity). Hence, given the choice, one may think that it is best to always adopt a multinomial mixture model. However, this model class may also be more sensitive to two types of misclassification errors that must occur at least sometimes during territory mapping and also potentially other sampling protocols that require individual identity to be known over replicate occasions. With *splitting errors*, a single territory is erroneously divided up into two (or more), while with *lumping errors*, the converse is true: two (or more) territories are erroneously aggregated. Both types of errors are not unlike typical genotyping errors (Kalinowski et al., 2006; Wright et al., 2009) in noninvasive and other genetic sampling and would be expected to bias the estimates from a model. Interestingly, by ignoring individual identity across, though not within, occasions, the binomial N-mixture model is insensitive to these types of error and may therefore sometimes be favored. On the other hand, it may be that the multinomial N-mixture models are not (or less) affected by the 'good fit/bad prediction' dilemma which is sometimes found for binomial N-mixture models (see Section 6.9.3). More research may be warranted to compare these two types of models.

Most multinomial N-mixture models can be analyzed easily using either likelihood or Bayesian methods. To the best of our knowledge, the ability to do likelihood analysis of hierarchical models based on the multinomial observation model and explicit abundance models to describe variation in local abundance is unique to the `unmarked` package. And, `unmarked` has the capability to fit basic models for closed population sampling (e.g., within a season) and also certain types of open population models (Chapter 13). Multinomial N-mixture models are also easily described in the BUGS language, and as we have seen this permits considerable flexibility in how we analyze those models.

We spent a good deal of time in this chapter discussing the analysis of spatially stratified capture-recapture models (Converse and Royle, 2012; Royle and Converse, 2014; Chandler, 2015). We think this is an important topic because, despite the ubiquity of such data, these models cannot be fit directly in standard software packages for doing capture-recapture (e.g., MARK). This is because traditional software cannot explicitly handle the site-structured abundance component of the hierarchical model. Arguably, spatially stratified capture-recapture data have been collected for decades, but analysis of the spatial component has been largely if not entirely ignored. The readily available framework for analyzing multinomial N-mixture models should allow biologists to make more efficient use of spatially stratified ("meta-population") capture-recapture data and directly address inference problems that have to do with the spatial structure of abundance.

We cover some topics relevant to multinomial observation models later in this book. Certain manifestations of distance sampling give rise to multinomial observation models for which multinomial N-mixture models are directly applicable. Because distance sampling is such an important sampling protocol by itself, we devote two chapters (Chapters 8 and 9) exclusively to this sampling protocol. In the sequel to this volume, we also cover open population models using multinomial protocols and other types of spatially structured capture-recapture models in Chapter 22.

EXERCISES

1. In Section 7.6 we fitted a Poisson and lognormal mixture to the ovenbird data in BUGS. Using the NB construction mentioned at the end of that section, try to fit a negative binomial model with the same covariates in BUGS.

2. For the MHB data (Section 7.9.3) we were able to obtain a model with satisfactory fit by considering a negative binomial abundance distribution. Another strategy is to develop a more complex regression model for the mean, such as including higher-order interactions, spatial regression models, or polynomials. See if you can come up with a model that fits using a Poisson abundance model and more regression terms.
3. Use the function `simNmix` to generate a detection history data set under a model of your choice. Then, fit a binomial mixture model to the aggregated counts of individuals detected per survey. Compare the inferences with those from a multinomial mixture model fit to the detection histories directly.
4. In Section 7.9, first aggregate the detection histories to mere replicated counts (ignoring territory ID) and fit a binomial mixture model analogous to our multinomial mixture model. Second, use the total number of detected territories and fit a Poisson GLM with random site effect that is as close as possible to the original model in that section. Compare the inferences under the three models.
5. Extend the Bayesian analysis begun in Section 7.9.7 to include overdispersion in local abundance among sites by including a random site effect. In addition, carry out a Bayesian analysis of model fit using Bayesian p-values to investigate whether either model produces an adequate fit to the data.

CHAPTER 8

MODELING ABUNDANCE USING HIERARCHICAL DISTANCE SAMPLING

CHAPTER OUTLINE

- 8.1 Introduction 394
- 8.2 Conventional Distance Sampling 396
  - 8.2.1 The Full Likelihood 400
  - 8.2.2 Models of Detection Probability 400
  - 8.2.3 Simulating Distance Sampling Data 401
  - 8.2.4 Binned Data 403
    - 8.2.4.1 Conditional and Other Likelihoods 405
    - 8.2.4.2 Simulating Binned Distance Sampling Data 406
  - 8.2.5 Point Transect Data 408
    - 8.2.5.1 Simulating Point Transect Data 410
    - 8.2.5.2 Likelihood Analysis of Point Transect Data 412
  - 8.2.6 Sensitivity to Bin Width 415
  - 8.2.7 Spatial Sampling 416
- 8.3 Bayesian Conventional Distance Sampling 417
  - 8.3.1 Bayesian Analysis of Line Transect Data 418
  - 8.3.2 Other Formulations of the Distance Sampling Model 421
  - 8.3.3 A Treatise on the Integration of Mathematical Functions in One Dimension 421
  - 8.3.4 Bayesian Analysis of Point Transect Data 422
- 8.4 Hierarchical Distance Sampling (HDS) 426
  - 8.4.1 HDS Data Structure and Model 427
  - 8.4.2 HDS in unmarked 428
  - 8.4.3 Example: Estimating the Global Population Size of the Island Scrub-jay (ISSJ) 430
- 8.5 Bayesian HDS 444
  - 8.5.1 Simulating HDS Data 444
  - 8.5.2 Bayesian HDS Using Data Augmentation 445
  - 8.5.3 Bayesian HDS Using the Three-part Conditional Multinomial Model 452
  - 8.5.4 Point Transect HDS Using the Conditional Multinomial Formulation 455
  - 8.5.5 Bayesian HDS Analysis of the ISSJ Data 457
- 8.6 Summary 459
- Exercises 460

8.1 INTRODUCTION

Distance sampling (DS) is one of the most widely used statistical methods in ecology for estimating density or population size (Burnham et al., 1980; Buckland et al., 2001, 2004a; Williams et al., 2002). There is an enormous body of literature, which the two classics by Buckland et al. summarize succinctly. Clearly, our aim here is not to try and present an exhaustive overview of distance sampling but rather to summarize the salient features of this important methodology, in order to show how it fits into the larger picture of hierarchical modeling of spatially indexed abundance data.

Conventional distance sampling (CDS; Buckland et al., 2001) uses information on observed distances of animals from transect lines or observation points to characterize the detection probability of individuals. Under the eminently plausible hypothesis that detection probability is related to the distance between animals and the observer, one may obtain an estimate of absolute density. Often sampling is done from boats or planes in open environments and thus animals that are amenable to such sampling are widely studied using distance sampling methods, including ungulates, whales, and other marine mammals. More recently, CDS methods have become very popular in the sampling of birds using point counts (Buckland et al., 2001; Rosenstock et al., 2002). In point count surveys, distances are recorded from a point of observation (instead of along a transect), and this is usually referred to as point transect sampling. In this chapter, we use the terms 'point count' and 'point transect' synonymously. Distance sampling methods are attractive because they do not require that individuals be uniquely marked and recaptured (or resighted) through time. Furthermore, unlike most capture-recapture models (but not spatial capture-recapture models), distance sampling requires only a single sample of the population, making it "cheaper" in terms of logistics (because a single visit to a site is enough) and modeling (because no closure assumption is needed). Finally, distance sampling is one of the only methods, along with spatial capture-recapture (Royle et al., 2014), which accommodates the basic problem of unknown sample area (see also Section 6.10).

The main assumptions of the CDS method are: (1) animals are distributed uniformly in space, (2) detection probability is a function of distance and is equal to 1 at distance 0, (3) individuals are detected at their original location, i.e., there is no responsive movement, and (4) distances are measured without error. Usually assumption (1) is not stated explicitly but, instead, it is assumed that sample points or transects are distributed randomly (Buckland et al., 2001, p. 29). As far as the mechanics of distance sampling are concerned, these two are effectively equivalent assumptions about the system, the former being a model-based version of random sampling of individuals, the latter being more of a design-based argument.

Conventional distance sampling has been a popular sampling method for many decades. However, historically little attention has been paid to *modeling spatial variation* in abundance using distance sampling methods. While it has been standard practice to obtain distance sampling data at multiple sample units using essentially the type of meta-population design we have encountered in Chapters 1, 6 and 7, the data from such replicate samples have typically been pooled in order to estimate parameters of the detection probability model. Thus, information about, or explicit attention to, factors that influence abundance among sites has been neglected. This is unfortunate because spatial or spatiotemporal patterns in abundance are often the primary interest of ecological studies!

This is not to say that information from replicate sample units is not used at all in CDS—indeed, the spatial replicate sample units are used to estimate the *encounter rate variance*, i.e., the variance in n_s (number of encountered individuals) among replicate units $s = 1, 2, \ldots, S$. This provides a sort of nonparametric variance estimator while, on the other hand, in a formal hierarchical modeling framework, we use parametric models to describe variation in abundance among sample units. Instead of pooling data as in CDS, we think it makes sense, following the basic ideas of Chapters 6 and 7, to provide an explicit model for the variation of local population size N_s, the population size for spatial sampling unit s. We call this hierarchical distance sampling (HDS). By specifying a model for this latent variable, we can then build explicit models for distance sampling data that account for variation in population size (or local density) among sample units, thus facilitating inference about factors that influence spatial variation in abundance or the making of explicit spatial predictions of abundance.

While conventional distance sampling is very mature and established in ecology and wildlife science, HDS has only existed for a few years. There are two key conceptual papers that develop ideas of HDS. Hedley and Buckland (2004) adopt a two-stage estimation procedure where they use the usual distance sampling model for observed distances to estimate detection probability, pooling the data among sample units, and then in a second-stage procedure they fit a model (e.g., a Poisson GLM with an offset being a function of the probability of detection) to the observed count of individuals n_s for each of $s = 1, 2, ..., S$ sample units. Miller et al. (2013a) call this methodology *density surface modeling* and describe an R package named `dsm` and the implementation of this in the popular program Distance (Thomas et al., 2010). Royle et al. (2004) develop HDS as a formal hierarchical model in which the two components (detection and abundance) are simultaneously estimated in a single hierarchical model exactly analogous to the binomial or multinomial *N*-mixture model framework of Chapters 6 and 7. The HDS methodology is implemented in `unmarked` using the `distsamp` and `gdistsamp` functions and is the topic of this chapter and the next. We note that in this chapter we will temporarily deviate from our usual notation and for now index sites by s for $s = 1, ..., S$. The reason is that we also need to index individuals, for which we will use index i, with $i = 1, ..., M$, where M will usually be the number of individuals in an augmented data set; see below.

Since 2004, only a trickle of papers have appeared that develop HDS ideas or provide novel implementations. Chelgren et al. (2011b) and Moore and Barlow (2011) appear to be the first to do a Bayesian analysis of an HDS model in BUGS. Chelgren et al. (2011b) contains quite a few novel elements. They formulate an HDS model in continuous space (using the "ones trick" in BUGS) and provide a three-part hierarchical model with a binomial observation model for $n_s|N_s$, a Poisson model for N_s, and then the ordinary distance sampling model for the observed distances conditional on n_s. They also accommodate within-unit heterogeneity in density by zeroing out "nonhabitat" in the sampled region. Moore and Barlow (2011) have a temporal dimension and embed an exponential population model into their distance sampling observation model and also model group size (see also Pardo et al., 2015).

Shirk et al. (2014) adopt the Chelgren et al. formulation of the model and provide a nice application to sampling chameleons. Oedekoven et al. (2013) also use a variation of the three-part formulation of the HDS model but remove N from the model by summation to reduce this to a two-part model. Schmidt et al. (2012) and Schmidt and Rattenbury (2013) fit HDS models with variation in group size in BUGS using data augmentation (DA). They may be the first to use "S-fold data augmentation," i.e., doing DA for each (transect) population and then linking the different transects by modeling the data augmentation parameter ψ (see also Tenan et al., 2014b). Sillett et al. (2012) develop an application of likelihood-based HDS with covariates on detectability (e.g., the parameter σ of some detection function) and $E(N)$, and considered Poisson and negative binomial abundance models. This is the first paper using `unmarked`'s `gdistsamp` function. Chelgren et al. (2011b) and Shirk et al. (2014) included effects on σ in the context of Bayesian HDS models. Conn et al. (2012) develop a combined double-observer HDS model with group structure and a CAR formulation of spatial correlation, using a Bayesian analysis conducted with a custom MCMC algorithm implemented in R package `hierarchicalDS`. Amundson et al. (2014) develop an HDS model with time of removal and individual level effects (see Chapter 9). Finally, Niemi and Fernandez (2010) develop a spatial point process model for line transect data and Johnson et al. (2010), including their R package `Dspat`, develop a similar model (see also ver Hoef et al., 2014, and Pardo et al., 2015).

In this chapter we begin with a fairly detailed introduction to basic ideas of distance sampling models, absent the hierarchical structure of having multiple spatial sample units, i.e., conventional distance sampling, as it is covered in the classic textbook by Buckland et al. (2001) and implemented in the widely used Distance software (Thomas et al., 2010). We do this so that we can introduce the reader to the mechanics of formulating the distance sampling model, simulating data, and fitting the model for the two standard sampling contexts: (1) transects and (2) point counts ("point transects"). For both cases we consider both continuous and "binned" distance measurements. These are statistically equivalent models as the number of bins gets large, but the practical issues of their analysis, and especially their implementation in BUGS, are very different. Hence, it's useful to see and experiment with both formulations. In addition, there are technical distinctions having to do with whether we adopt a conditional or full likelihood formulation of the model, and also whether we analyze the data by classical likelihood or Bayesian analysis. We first cover all of these various manifestations of the conventional (nonhierarchical) distance sampling (CDS) model.

Only once these basic principles have been developed do we extend the ideas to hierarchical distance sampling, where we use hierarchical models to combine the data from sampling at multiple locations formally into a single joint model. The distance sampling protocol, combined with a model for abundance, produces what we'll call the hierarchical distance sampling (HDS) model. When continuous distance sampling measurements are binned into distance classes, a multinomial observation model is produced. Therefore, the multinomial mixture models of the previous chapter can be applied with only some minor technical modifications that we have to consider when computing the multinomial cell probabilities.

8.2 CONVENTIONAL DISTANCE SAMPLING

We first develop the basic concepts and technical details of "classic" distance sampling without thinking about spatial replication and hierarchical models. A simple way to motivate distance sampling is to think about our heuristic estimator of N derived by solving the relationship

$$E(n) = \bar{p}N.$$

Therefore, we can estimate N from a sample count n and an estimate of $\bar{p}$, the probability that an object (i.e., animal) appears in our sample of size n. The idea of distance sampling is to estimate $\bar{p}$ by modeling detection probability of objects as a function of distance x from the object to an observer recording data at a point or walking along a transect. This is done by specification of some function, the "detection (probability) function," $g(x; \theta)$, describing detection probability as a function of distance x and parameter(s) θ. That is, the detection function is a model for the probability of detection of an object conditional on its distance from the observer x, i.e., $g(x; \theta) = Pr(y = 1|x)$ in our usual notation of conditional probabilities, where y is a Bernoulli trial indicating detection ($y = 1$) or nondetection ($y = 0$). The traditional notation can be a little confusing because if we just write $g(x; \theta)$ then it looks like this could be a probability distribution for x, which it is not—rather it is the parameter of a Bernoulli probability mass function for a variable y, whether or not an object is detected conditional on x. This is why we will write this as $\Pr(y=1|x)$ when we want to be clear that it is a probability of an event, that is, of being detected.

How is $\bar{p}$ related to this detection function $g(x; \theta)$? It is the marginal or *average* detection probability (therefore we write $\bar{p}$ instead of $\hat{p}$), which is the probability that an individual in the population at large appears in the sample, and it is computed by averaging $g(x; \theta)$ over all possible values of x. Formally, the calculation is

$$\bar{p} \equiv \Pr(y=1) = \int_x g(x; \theta)[x]dx \tag{8.1}$$

Note that the averaging is being done with respect to a probability density for x, denoted here by [x] (using our established bracket notation), although we have yet to specify this quantity. Thus, the basic distance sampling model has two explicit and essential components:

1. The "observation model," which describes how individuals appear in the sample, characterized by the function $g(x; \theta)$.
2. The "process model," [x], which describes how objects in the population are distributed with respect to the observer or the transect.

Conventional distance sampling adopts an explicit and intensive focus on inference about component (1), typically considering many and fairly complex models for the detection function and choosing among those by AIC (e.g., Miller and Thomas, 2015). Historically, very little attention has been focused on modeling the "process," i.e., the probability distribution [x]. Conversely, HDS adopts an explicit focus on modeling [x] as we will see later in this chapter. It may seem like we're making a big deal about this because how can "distance from observer" be any kind of meaningful ecological process? Well, in and of itself it is not, but, in specific cases, the distribution of [x] is precisely equivalent to the distribution of objects in space, and models for such things are usually called *point process models* (Illian et al., 2008; Wiegand and Moloney, 2014). In a sense then, HDS is all about merging an "observation model" that describes the detection of individuals conditional on where they are located during sampling with a "process model" that describes where the individuals are located.

Before elaborating on that concept in more detail we discuss how to obtain $\bar{p}$ from a sample of distance data obtained by surveying a transect and recording distances to each of n objects that are detected. Conceptually, we think it is extremely helpful to think about distance sampling as a logistic regression capture-recapture problem by introducing a population of N individuals each characterized by a pair of random variables (y_i, x_i) where y_i is a binary indicator of whether we captured (or observed) that individual with $y_i = 1$ meaning "captured" and $y_i = 0$ "not captured," and x_i is the distance from the observer to the individual at the instantaneous time of sampling. Given the population of N (y, x) pairs, we only observe (y, x) for those n individuals having $y = 1$. So our "data" for a distance sampling study consist of the sample of distances $x_1,...,x_n$ that appear in our sample conditional on the event that $y = 1$ (i.e., that the individual was detected).

To obtain the likelihood for a sample of distances we need to identify the probability distribution of the *observed* distances x, which is to say the probability distribution of x conditional on the event $y = 1$. This can be calculated from a simple application of Bayes' rule. The observed data are the values of x for which $y = 1$, and thus we seek to identify the probability distribution $[x|y = 1]$. Bayes' rule tells us that

$$[x|y = 1] = \frac{[y = 1|x][x]}{[y = 1]}$$

where $[y = 1|x]$ is the function that we choose to model detection probability as a function of distance—the "detection function." The other two components require some specific discussion and analysis: (1) $[x]$ is the *population distribution of distances*. Therefore we must specify this probability distribution in order to compute $[x|y = 1]$. (2) Once we specify the probability distribution $[x]$, then we can compute the denominator directly as $[y = 1] = \int_0^B [y = 1|x][x]dx$. This is the average probability of detection over the interval $[0, B]$, where B represents some maximal distance out to which individuals are counted. So, the probability density for distance observation x_i is the following:

$$[x_i|y_i = 1] = \frac{g(x_i; \theta)[x]}{\int_x g(x; \theta)[x]dx} \tag{8.2}$$

for whatever distribution for $[x]$ we choose.

What sorts of distributions make sense for x? To gain some intuition about this we note that the distribution for x is essentially a prior distribution on "distance from observer," and it can be derived equivalently from a prior distribution on the location of individuals in the surveyed region. To make life simple here we first assume that the survey is done along a linear transect of length L so that we can imagine the surveyed region is a long rectangle with a line running down the middle. We will assume that individuals are only counted up to some maximum distance, say B, and so the surveyed region is a rectangle of dimension $L \times 2B$. Lacking specific knowledge to the contrary, it is sensible to assume that individuals are distributed uniformly over the sampled rectangle. Let's define the individual locations by the coordinates $\mathbf{u}_i$ for $i = 1, 2, \ldots, N$. As it turns out, if individuals are uniformly distributed in space, then their *distances* to a transect (but not to a point, see below) also have a uniform distribution on the interval $[0, B]$, i.e., the density $[x] = 1/B$.

One of the important concepts of distance sampling is that the observed distances are biased with respect to the population distribution $[x]$. The conditional density in Eq. (8.2) makes it clear that the density of observed x should be proportional to the detection function. So if we simulate data under a half-normal detection model (with scale parameter σ, see Figure 8.1 left) the distribution of the observed distances, represented by blue histogram shown in Figure 8.1 (right), is clearly not uniform.

Under the assumption that individuals are uniformly distributed in space, so that distances are uniformly distributed on the interval $[0, B]$, $[x]$ cancels from the numerator and denominator of the conditional distribution given above and therefore does not further influence the likelihood contribution of each x_i (for point count data, things are slightly different in the sense that x doesn't cancel from the likelihood; see below). The likelihood for n observed distances is therefore

$$L(\mathbf{x}; \theta) = \prod_{i=1}^{n} \frac{g(x_i; \theta)}{\int_x g(x; \theta)dx} \tag{8.3}$$

which we maximize to obtain $\widehat{\theta}$. It is worth pointing out that in order to evaluate the likelihood we have to do a numerical integration of the detection function over the support of x, the interval $[0, B]$. This is a key calculation because the integral in the denominator is also the average probability that an

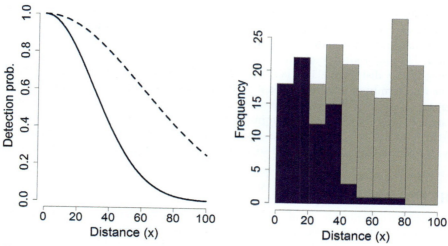

FIGURE 8.1

Half-normal detection function for two different values of σ (left: $\sigma = 30$ (solid) and $\sigma = 60$ (dashed)) and (right) histogram of a sample of true (gray) and observed distances (blue) for $\sigma = 30$; see Section 8.2.3 for R code (function `sim.1data`).

individual in the population is encountered, i.e., $\bar{p}$ from Eq. (8.1), and thus it is instrumental to "converting" n to $\widehat{N}$. Once we obtain MLEs of the model parameters $\widehat{\theta}$, we evaluate the expression $\bar{p} = [y = 1] = \int_x [y = 1|x][x]dx$, which is the marginal (or average) probability of detection, and we can get an estimate of N directly by $\widehat{N} = n/\widehat{p}$. It is customary in distance sampling not to estimate N but, rather, to estimate density D, which is related deterministically to N by dividing by the sampled area $\widehat{D} = \widehat{N}/(2*L*B) = n/(2*L*B*\widehat{p})$. The denominator here is the *effective sample area* (more commonly $B*\widehat{p}$ is called the effective strip half-width for transect sampling, see Buckland et al., 2001, p. 53).

We make four remarks here:

Remark 1: *Density and abundance.* Conventional distance sampling is talked about almost exclusively in terms of estimating density, $D = N/A$, where A is the area over which animals were counted. This area is not usually precisely defined; however, it is implicit in the estimation (i.e., buried under the hood) because formal bounds of integration for a distribution of distance must (usually) be specified, and this effectively implies an area A. We discuss this shortly. As a technical matter, estimation of N or D are statistically equivalent problems.

Remark 2: *Realized versus expected abundance.* In general, whenever an explicit model is placed on the unknown parameter N, this induces a distinction between realized and expected abundance and density (Efford and Fewster, 2013; Dorazio, 2013; Section 5.7.3 in Royle et al., 2014). Expected population size is $E(N)$ where the expectation is with respect to the distribution of N. Expected density is $E(N)/A$. The interpretation of these expected quantities is as the mean value of some hypothetical unit to which our model applies.

Remark 3: *Replicate transects.* Normally we have distances observed from multiple spatial units (transects, points), whereas we have so far only described the situation as if we had only a single spatial unit. Having multiple spatial sample units does not change the fundamental estimation problem nor the mechanics of how we achieve it in CDS. In this case, we simply pool all of the distances into one data set and do the analysis as just outlined. We show this with simulated data below.

Remark 4: *The uniformity assumption.* We stated an explicit assumption that distances from the transect are uniformly distributed. This is induced by the equivalent assumption that animal locations are distributed uniformly in two-dimensional space. However, sometimes formal distance sampling developments do *not* state such assumptions about distances or points or else they make general claims that they are not necessary. Instead, uniformity can be induced *by design* by randomly locating transects (Barry and Welsh, 2001). Our view is that the model assumption and sampling assumption yield equivalent statistical procedures, and so we're not too concerned with how you describe them.

8.2.1 THE FULL LIKELIHOOD

The previous section described estimation of N by first estimating the detection function parameter θ from the likelihood for the observed distances constructed for the n observations, a procedure that is naturally conditional on the event that $y = 1$. As a result, this is usually called the "conditional likelihood," and the estimator of N obtained by "adjusting" n is the *conditional estimator* of population size. However, it is also common in practice to use the so-called *full likelihood* (Borchers et al., 2002; p. 232 in Royle and Dorazio, 2008), which recognizes that n is also a stochastic outcome of the study and should be modeled. The distribution of n is

$$n \sim Binomial(N, \bar{p})$$

and to obtain the full likelihood we simply multiply the conditional likelihood by the binomial component for n. This yields (note: we leave the $[x]$ part in the conditional likelihood for a moment, instead of canceling it from both numerator and denominator):

$$L(\sigma, N) = \left\{ \prod_{i=1}^{n} \frac{g(x_i; \theta)[x]}{\int_x g(x; \theta)[x] dx} \right\} \frac{N!}{n!(N-n)!} \bar{p}^n (1-\bar{p})^{N-n}$$

where after some factorizing, canceling, and rearranging, we are left with:

$$L(\sigma, N) = \frac{N!}{n!(N-n)!} \left\{ \prod_{i=1}^{n} g(x_i; \theta)[x] \right\} (1-\bar{p})^{N-n}. \tag{8.4}$$

This resembles the usual full likelihood for every other capture-recapture type of model and, in particular, the individual covariate models (Borchers et al., 2002; Section 7.1 in Royle and Dorazio, 2008). Therefore, we can understand distance sampling as a special type of capture-recapture model where only a single ($J = 1$) sample is taken, and there is an individual covariate, distance x, measured on each observed individual. The full likelihood can be maximized to obtain the MLE of θ and N.

8.2.2 MODELS OF DETECTION PROBABILITY

So far we have just talked about the general concepts and mechanics of distance sampling and how to construct the likelihood of observed distances. However, much of practical distance sampling is focused on the encounter probability model $g(x; \theta)$ and, in particular, choosing among large classes of detection probability models to find models that fit the observed distance distribution (usually in the AIC sense). Here, we don't go into a catalog of types of models but simply mention that some common models include the following:

"half normal"	$g(x; \sigma) = \exp(-x^2/2\sigma^2)$
negative exponential	$g(x; \sigma) = \exp(-x/\sigma)$
hazard rate	$g(x; \sigma, b) = 1 - \exp(-h(x; \sigma, b))$ where $h(x; \sigma, b) = 1 - \exp(-(x/\sigma)^{-b})$

One key feature of these standard models is that they represent monotone decreasing functions of distance with one or two parameters. Another key feature is they all have a known intercept of 1, i.e., $g(0) = 1$, which is a requirement of conventional distance sampling models when we have no other ancillary data. We discuss generalizing this shortly (Alpizar-Jara and Pollock, 1996; Borchers et al., 1998). Why is it required that $g(0) = 1$? Think about the definition of $\bar{p}$:

$$\bar{p} = \Pr(y = 1) = \int_x [y = 1|x][x]dx$$

If our model for $[y = 1|x]$ had some arbitrary intercept, say $[y = 1|x] = \alpha * k(x; \sigma)$ where $k()$ was itself some function such that $k(0) = 1$, then the constant intercept α would be confounded with a level shift in density, i.e., a detection model with intercept α and density $[x] = 1/(2 * L * B)$ is equivalent to a detection model with intercept 1 and density $\alpha/(2 * L * B)$. The two are indistinguishable and, in fact, α just cancels from the conditional likelihood expression (Eq. (8.2)). In other words, an intercept in the detection function in CDS is not estimable using standard data for this design.

8.2.3 SIMULATING DISTANCE SAMPLING DATA

We demonstrate some of these basic distance sampling concepts by simulating an imaginary population of the extinct Chihuahuan musk oxen along a transect of length $L = 10$ km. We subject the individual musk oxen to detection by an observer traversing the transect and use a half-normal detection probability function (Figure 8.1). All of this goes according to the following:

```
strip.width <- 100    # one side of the transect, really half-width
sigma <- 30           # Scale parameter of half-normal detection function

# Define half-normal detection function
g <- function(x, sig) exp(-x^2/(2*sig^2)) # Function definition
g(30, sig=sigma)      # Detection probability at a distance of 30m

# Plot the detection function
par(mfrow=c(1,2))
curve(g(x, sig=30), 0, 100, xlab="Distance (x)", ylab="Detection prob.", lwd = 2, frame = F)
curve(g(x, sig=60), 0, 100, add=TRUE, lty = 2, lwd = 2)
```

```
# Define function to simulate non-hierarchical line transect data
sim.ldata <- function(N = 200, sigma = 30){
# Function to simulate line transect data under CDS.
# Function arguments:
#    N: number of individuals along transect with distance u(-100, 100)
#    sigma: scale parameter of half-normal detection function
# Function subjects N individuals to sampling, and then retains the value
# of x=distance only for individuals that are captured
par(mfrow = c(1,2))
# Plot the detection function
curve(exp(-x^2/(2*sigma^2)), 0, 100, xlab="Distance (x)", ylab="Detection prob.", lwd =
2, main = "Detection function", ylim=c(0,1)) # Plot detection function as function of sigma
text(80, 0.9, paste("sigma:", sigma))
xall <- runif(N, -100,100)    # Distances of all N individuals
hist(abs(xall), nclass=10, xlab = "Distance (x)", col = "grey", main = "True (grey) \nand
observed distances (blue)")  # Histogram of distances
g <- function(x, sig) exp(-x^2/(2*sig^2))
p <- g(xall, sig=sigma) # detection probability
y <- rbinom(N, 1, p)  # some inds. are detected and their distance measured
x <- xall[y==1]       # this has direction (right or left side of transect)
x <- abs(x)           # now it doesn't have direction
hist(x, col = "blue", add = TRUE)
return(list(N = N, sigma = sigma, xall = xall, x = x))
}

# Obtain a data set for analysis
set.seed(2015)              # If you want to get same results
tmp <- sim.ldata(sigma = 30) # Execute function and assign results to 'tmp'
attach(tmp)
```

We see that the blue histogram in Figure 8.1 vaguely resembles the half-normal detection probability function (you can increase the resemblance greatly by increasing N, e.g., to 10^6). Next, we will obtain the maximum likelihood estimates of the half-normal parameter σ (log-transformed to enforce a positive value) from the simulated data. To do this we define an R function that evaluates the conditional and full likelihoods and use `optim` to minimize the negative log-likelihood in each case.

```
# Conditional likelihood
Lcond <- function(lsigma){ # Define conditional nll
   sigma <- exp(lsigma)
   -1*sum(log(g(x,sig=sigma)/integrate(g, 0, 100, sig=sigma)$value/100))
}

# Call optim to maximize conditional likelihood
optim(log(30), Lcond, hessian=TRUE, method="Brent", lower=-5, upper=10)
$par
[1] 3.257716

$value
[1] 626.8964

[ ... output deleted ... ]
```

```
# Full likelihood
Lfull <- function(parm){    # Define full nll
    sigma <- exp(parm[1])
    n0 <- exp(parm[2])
    N <- length(x)+ n0
    pbar <- integrate(g, 0, 100, sig=sigma)$value/100
    -1*( lgamma(N+1) - lgamma(n0+1) + sum(log(g(x,sig=sigma)/100)) + n0*log(1-pbar) )
}

# Call optim to maximize full likelihood
optim(c(log(30), log(4)), Lfull, hessian=TRUE)
$par
[1] 3.259401 5.012220

$value
[1] 50.31808

[... output deleted ...]
```

In the first case, we get the MLE of $\log(\sigma)$, which we have to convert to the MLE of $\bar{p}$ and then compute $\widehat{N}_c = n/\widehat{\bar{p}}$. In the second case, we get the MLE of N directly by maximizing the full likelihood. Or rather, in this case, we estimate the logarithm of $n_0 = N - n$ and then have to back-transform and add back n to it for an estimate of N. To convert estimates of density we need simply divide the estimates of N by the area of the transect, which was 10 km long and 0.2 km wide (100 m on either side) $= 2$ km^2. This is all done as follows:

```
pbar <- integrate(g, 0, 100, sig=exp(3.26))$value/100
n <- length(tmp$x)

(Nhat.cond1 <- n/pbar)
[1] 223.6231
(Dhat.cond1 <- Nhat.cond1/(10*.2))
[1] 111.8115

n0hat <- exp(5.01)
(Nhat.full <- n + n0hat)
[1] 222.9047
(Dhat.full <- Nhat.full/(10*.2))
[1] 111.4524
```

We find that the densities of Chihuahuan musk oxen are quite respectable this year, to say the least (C. Amundson, pers. comm.), being on the order of 111 per km^2. Perhaps we will open a harvest season.

8.2.4 BINNED DATA

It is common in applications of distance sampling to produce observations in distance bands. In that case, the observation model is multinomial and therefore the multinomial N-mixture models, which we developed in Chapter 7, are directly relevant. We develop the likelihood for this case here. Suppose observations are recorded into $h = 1,2,...,H$ distance bands or strips on the intervals

$[b_0,b_1],(b_1,b_2],\ldots,(b_{H-1},b_H]$, where we define $b_0 = 0$ and $b_H = B$, the upper bound of recording distances. (Note that for pure notational convenience here we use H for the number of *observed* multinomial categories, while in Chapter 7 it was for *all* categories.) Let y_h be the frequency of encounters in distance interval h, and let $\mathbf{y} = (y_1,\ldots,y_H)$ denote the vector of frequencies with $n = \sum_h y_h$. We assert that the vector of observations $\mathbf{y}$ has a multinomial distribution:

$$\mathbf{y}|N \sim Multinomial(N; \{\pi_h\})$$

with parameters N and cell probabilities $\{\pi_h\}$. The number of individuals not detected will be denoted by $n_0 = N - n$, and the corresponding cell probability for these undetected individuals is $\pi_{H+1} = 1 - \sum_{h=1}^{H} \pi_h$.

It remains to define the cell probabilities π_h. These are, in words, "the probability that an individual occurs and is detected in distance class h" (see Buckland et al., 2001, p. 52), which is also "the probability that an individual is detected given that it occurs in class h *times* the probability that it occurs in class h." This is, using a formula,

$$\Pr(y = 1 \text{ and } x \in h) = \Pr(y = 1|x \in h)\Pr(x \in h)$$

which we'll simplify by writing

$$\pi_h = \bar{p}_h \psi_h$$

where ψ_h is the probability that x is located in distance interval h, which is, for a line transect, $\psi_h = (b_{h+1} - b_h)/B$, i.e., just the interval width over the transect half-width, and is implied by the uniform distribution assumption. But what is $\Pr(y = 1|x \in h)$? It is the integral over the distance band of the detection function multiplied by the conditional probability of x, given that x is in distance band h:

$$\bar{p}_h = \int_{x \in h} \Pr(y = 1|x, x \in h)\Pr(x|x \in h)dx$$

Under the uniformity assumption x is also uniformly distributed in each interval, and so the conditional pdf of x is $[x|x \in h] = 1/(b_{h+1} - b_h)$. Putting this all together then, we just integrate the detection function over the interval with an adjustment for area

$$\pi_h = (1/B) * \int_{x \in h} \Pr(y = 1|x, x \in h).$$

In our on-going line transect example with the muskoxen suppose we use 10-m distance bands for distances between 0 and 100 m. Then the conditional probability density of x is $1/10$ for each 10-m distance band, and we have to do the calculation

$$\bar{p}_h = \int_{b_{h-1}}^{b_h} \Pr(y = 1|x)/10\, dx$$

and then the multinomial cell probabilities are

$$\pi_h = \psi_h \bar{p}_h = (1/10)\bar{p}_h$$

The last cell probability, π_0, is 1 minus the sum of the rest: $\pi_0 = 1 - \sum \pi_h$. Thus, binned distance data have a multinomial distribution with these cell probabilities and we can obtain the full likelihood directly:

$$L(\sigma, n_0; \mathbf{y}) = \frac{(n+n_0)!}{n! n_0!} \pi_1^{y_1} \pi_2^{y_2} \ldots \pi_H^{y_H} \pi_0^{n_0} \tag{8.5}$$

where $n = \sum y_h$.

8.2.4.1 Conditional and Other Likelihoods

The conditional likelihood is easily derived by noting that the distribution of detections among the H distance classes, conditional on n, is also multinomial but with multinomial index n instead of N and conditional probabilities $\pi_h^c = \pi_h/(1-\pi_0)$. To uncondition on n, in order to obtain the full likelihood again, we note that $n \sim Binomial(N, 1-\pi_0)$ and so the full likelihood is the product of the conditional likelihood and this binomial piece, which leads us back to Eq. (8.5). The point is, the conditional multinomial and the binomial for n together are exactly equivalent specifications to the multinomial likelihood in Eq. (8.5) (see Section 5.1.2 in Royle and Dorazio, 2008). In practice, some may analyze the full likelihood by retaining the two individual pieces, although there is no need to do this in most cases.

As a final point, we might think like a Bayesian here and assume that N is not a fixed number to estimate but, rather, is itself the realization of a random variable. If $N \sim Poisson(\lambda)$, then this implies precisely that $n \sim Poisson((1-\pi_0)\lambda)$, and we can estimate the parameter λ instead of N. This "Poisson integrated full likelihood" has the following form (see Royle et al., 2014, p. 192):

$$L(\boldsymbol{\theta}, \lambda) = \left\{ \prod_{i=1}^{n} g(x_i; \boldsymbol{\theta}) \right\} \lambda^n \exp(-\lambda(1-\pi_0))$$

(see also Borchers and Efford, 2008). Instead of a Poisson prior for N we can consider a $Binomial(M, \psi)$ where M is prescribed, similar in spirit to the model for N in data augmentation (DA; Royle and Dorazio, 2012). For large M this approximates the Poisson prior but it yields a different likelihood, having the form (see Royle and Dorazio, 2008, p. 238):

$$\left(\prod_{i=1}^{n} \psi g(x_i; \boldsymbol{\theta}) \right) \frac{M!}{n!(M-n)!} (1 - \psi(1-\pi_0))^{M-n}.$$

This binomial integrated form of the full likelihood is equivalent to the model we would analyze using data augmentation (see below). These various considerations give us a number of essentially equivalent ways to analyze distance sampling models with binned data. It is worth knowing of these different formulations because one or another may have certain advantages in a given instance. For example, in Bayesian analysis of the distance sampling model we use a method of DA that is easily implemented in the BUGS language. The model implied by DA is the binomial integrated likelihood just shown.

8.2.4.2 Simulating Binned Distance Sampling Data

There are two ways to go about simulating binned distance sampling data, and we show both here. First, we can simulate continuous-space data exactly as we have done before with function sim.1data and then aggregate into distance intervals. Second, we can simulate directly multinomial observations with cell probabilities π_h. Here is a script that does it both ways using a half-normal detection function. (To verify that the same cell probabilities are produced, you could execute it with very large population size N).

```
set.seed(2015)
# Design settings and truth (population size N and detection function g)
interval.width <- 10
strip.width <- 100     # half-width really (one side of transect)
nbins <- strip.width%/%interval.width
sigma <- 30            # Scale parameter of half-normal detection function
g <- function(x, sig) exp(-x^2/(2*sig^2)) # Half-normal detection function
N <- 200               # Population size

# Method 1: simulate continuous distances and put into intervals
x <- runif(N, -strip.width, strip.width) # Distance all animals
p <- g(x, sig=sigma)   # Detection probability
y <- rbinom(N, 1, p)   # only individuals with y=1 are detected
x <- x[y==1]           # this has direction (right or left side of transect)
x <- abs(x)            # now it doesn't have direction

# Compute the distance category of each observation
xbin <- x %/% interval.width + 1  # note integer division function %/%

# Multinomial frequencies, may have missing levels
y.obs <- table(xbin)

# Pad the frequencies to include those with 0 detections
y.padded <- rep(0,nbins)
names(y.padded) <- 1:nbins
y.padded[names(y.obs)] <- y.obs
y.obs <- y.padded
y.true <- c(y.obs, N-length(xbin)) # Last category is "Not detected"

# Relative frequencies by binning continuous data (pi). These should compare
# with the cell probabilities computed below when N is very large
(y.rel <- y.true/N)   # Last category is pi(0) from above
(pi0.v1 <- y.rel[nbins+1])
0.635

# Compute detection probability in each distance interval
dist.breaks <- seq(0, strip.width, by=interval.width)
p <- rep(NA, length(dist.breaks)-1)
for(j in 1:length(p)){
  p[j] <- integrate(g, dist.breaks[j], dist.breaks[j+1],
      sig=sigma)$value / (dist.breaks[j+1]-dist.breaks[j])
}
```

```
round(p, 2)
[1] 0.98 0.88 0.71 0.51 0.33 0.19 0.10 0.05 0.02 0.01

# Compute the multinomial cell probabilities analytically. These are exact.
# psi = probability of occurring in each interval
interval.width <- diff(dist.breaks)
psi <- interval.width/strip.width
pi <- p * psi
sum(pi)                        # This is 1 - pi(0) from above
[1] 0.3756716
(pi0.exact <- 1-sum(pi))
[1] 0.6243284                  # Compare with 0.635 above

# Method 2: Use rmultinom to simulate binned observations directly
# This includes 0 cells AND n0
pi[length(p)+1] <- 1 - sum(pi)
(y.obs2 <- as.vector(rmultinom(1, N, prob=pi)))
(y.obs2 <- y.obs2[1:nbins]) # Discard last cell for n0 (because not observed)
```

We see that, under this model, we expect to encounter about 38% of the individuals along the transect.

Now let's take our simulated data and obtain the MLEs of the model parameters. Keep in mind that we have only simulated a single multinomial sample, which we could think of as sampling one transect of a certain length or multiple transects but then pooling the resulting data. Shortly we will get on to the meta-population sampling context and consider having spatial replicates, but, for now we continue our focus on the basic analysis of distance sampling data. The likelihood is just a multinomial, so if we package up most of the previous simulation R code into a function that computes the likelihood, given the parameter values, a multinomial data vector, and the distance breaks, then we can use `optim` or `nlm` to obtain the MLEs. Note that the multinomial full likelihood must include the combinatorial term in N and, as before, we parameterize the model in terms of the number of uncaptured individual n_0 so that $N = n + n_0$. Note also that the observed multinomial frequencies may include zero counts in some distance bands, and we must be sure to pad the observed data vector when appropriate.

```
Lik.binned <- function(parm, data, dist.breaks){
# Note that the parameters are parm[1] = log(sigma), parm[2] = log(n0)

sigma <- exp(parm[1])
n0 <- exp(parm[2])
p <- rep(NA, length(dist.breaks)-1)
for( j in 1:length(p)) {
   p[j] <- integrate(g, dist.breaks[j], dist.breaks[j+1],
       sig=sigma)$value / (dist.breaks[j+1]-dist.breaks[j])
}
psi <- interval.width/strip.width
pi <- p * psi
pi0 <- 1-sum(pi)

N <- sum(data) + n0
-1*(lgamma(N+1)-lgamma(n0+1) + sum(c(data,n0)*log(c(pi,pi0))))
}
```

```
# Evaluate likelihood for some particular value of the parameters
Lik.binned(c(2,0), data=y.obs, dist.breaks=dist.breaks)
[1] 335.1482

# Obtain the MLEs for the simulated data
optim(c(2,0), Lik.binned, data=y.obs, dist.breaks=dist.breaks)
$par
[1] 3.263681 5.006211

$value
[1] -117.0331
```

The MLE of N is $\widehat{N} = \widehat{n}_0 + n = \exp(5.006) + n$ where $n = \sum y_h$ for the observed distance categories; this yields $\widehat{N} = \exp(5.006) + 73 = 222.3063$, not too far from the true data-generating value of $N = 200$ (and compare to the estimates of musk oxen abundance we obtained back in Section 8.2.3). (Note the object y.obs2 simulated using the rmultinom function produces a different number of observed individuals due to randomness of random number generation.) The point of this was to build basic tools of simulating and fitting distance sampling data, for use later when we develop hierarchical distance sampling models, and so we don't do anything else with this for right now. At the same time, simulating data sets using R provides another and, to some, perhaps more intuitive, description of the basic distance sampling model than algebra.

8.2.5 POINT TRANSECT DATA

So-called "point transect data" are distance sampling data collected from circular point counts where an observer stands at a point and records distance to detected individuals within some radius B. Formulation of the likelihood for such data follows the same logic as for transect data, but this time the natural probability density for distance x is *not* uniform. Recall that, in the transect case, if we assume a uniform distribution of individuals, then distance is also uniform. But in the case of a circular sample unit, the uniform distribution of individuals implies a triangular distribution for distance. We can understand this by computing the cumulative probability distribution directly, noting that the probability that distance is less than any value x should be proportional to the area of a circle of radius x relative to a circle of radius B. That is, the cumulative distribution function F of distance x is (note that for clarity we use the uppercase X for the variable and x to represent a specific value):

$$F(x) = \Pr(X \le x) = \frac{\pi x^2}{\pi B^2}$$

The probability density is then obtained by differentiating with respect to x, i.e., $f(x) = \partial F(x)/\partial x$, which produces:

$$f(x) = \frac{2x}{B^2}$$

This triangular distribution has increasing probability density with distance from the center of the circle, to account for the increasing area of successive annuli. Recall our general expression for the conditional likelihood:

$$[x_i | y_i = 1] = \frac{g(x_i; \theta)[x]}{\int_x [y=1|x][x]dx}$$

8.2 CONVENTIONAL DISTANCE SAMPLING

In the case of point transects $[x]$ is not constant and so it does *not* cancel out from the numerator and denominator and the likelihood has to retain the $[x] \equiv f(x)$ bit. Using the half-normal detection function model, the contribution of each observed distance x_i to the conditional likelihood looks like this (note that $1/B^2$ cancels from numerator and denominator):

$$L(\sigma|x_i) = \frac{\exp(x_i^2/2\sigma^2)2x_i}{\int_x \exp(x^2/2\sigma^2)2x\,dx}$$

Shortly we will simulate some data and fit the model using the conditional likelihood expressed as an R function.

For binned point count data the vector of frequencies of encounters in each distance class, including the cell "not encountered," has a multinomial distribution with cell probabilities π_h for $h = 1,2,\ldots,H$ distance classes, and the last cell, $H+1$, corresponds to "not encountered." To compute these cell probabilities we have to do the integrations over successive annuli of the circle, and we have to make a smallish bit of math argumentation following our development in Section 8.2.4 above to do this. The multinomial cell probability π_h is, in words, "the probability that an individual is detected *and* in distance class h," which is the same as saying, "the probability that an individual is detected, given that it occurs in class h, times the probability that it occurs in class h." This is, using a formula,

$$\Pr(y = 1 \text{ and } x \in h) = \Pr(y = 1|x \in h)\Pr(x \in h)$$

which we write simply as: $\pi_h = \bar{p}_h \psi_h$ where ψ_h is the probability that x is in distance interval h, which is $\psi_h = (A_{h+1} - A_h)/\pi B^2$, where $A_h = \pi r_h^2$ is the area of a circle having radius r_h. So, ψ_h here is just the area of the annulus over the whole area of the point count circle. But what is $\Pr(y = 1|x \in h)$? Our expression from before had us compute this integral:

$$\bar{p}_h = \int_{x \in h} \Pr(y = 1|x, x \in h)\Pr(x|x \in h)dx.$$

Under the assumption that individuals are uniformly distributed in space, x has the triangular distribution on $[0, B]$ as noted above. But for the interval it has a slightly different form; we have to compute $f(x|x \in h)$, which is $f(x|x \in h) = f(x)/\Pr(x \in h)$, and it works out that the pdf of x is $[x|x \in h] = 2x/(b_{h+1}^2 - b_h^2)$. Also, $\psi_h = (b_{h+1}^2 - b_h^2)/B^2$. Putting this all together, we just integrate the detection function over the interval with an adjustment for area:

$$\pi_h = \psi_h \int_{b_h}^{b_{h+1}} \Pr(y = 1|x, x \in h) \times 2x \Big/ \left(b_{h+1}^2 - b_h^2\right)$$

This might seem a bit conceptual, but let's see how this looks in the form of an R function, which we then apply to our point count situation:

```
# Define function to compute cell probs for binned distance sampling
cp.ri <-function(radius1, radius2, sigma){
    Pi <- 3.141593
    a <- Pi*radius2^2 - Pi*radius1^2
```

```
      integrate(function(x, s=sigma) exp(-x^2 / (2 * s^2)) * x, radius1,
        radius2)$value *(2*Pi/a)
}
# Define distance intervals and compute multinomial probabilities
delta <- 0.5                    # Width of distance bins
B <- 3                          # Max count distance
dist.breaks <-seq(0, B, delta)  # Make the interval cut points
nD <-length(dist.breaks)-1
sigma <- 1
p.x <-rep(NA,nD)                # Conditional detection probabilities
for(i in 1:nD){
    p.x[i] <- cp.ri(dist.breaks[i], dist.breaks[i+1], sigma =1)
}
area <- 3.141593 * dist.breaks[-1]^2
ring.area <- diff(c(0, area))
# Pr(detection| in ring)*Pr(in ring)
cp <- p.x* ring.area/sum(ring.area)
```

These cell probabilities are used below to simulate data using the `rmultinom` function or to construct the multinomial likelihood, which was given previously (for line transects) as,

$$L(\sigma, n_0; \mathbf{y}) = \frac{(n+n_0)!}{n!n_0!} \pi_1^{y_1} \pi_2^{y_2} \ldots \pi_H^{y_H} \pi_0^{n_0}.$$

Here, $n = \sum y_h$, but we have to go through the gyrations of computing the cell probabilities π_h for the case of a circle instead of a nice rectangular transect.

8.2.5.1 Simulating Point Transect Data

To simulate point transect data we can simulate individuals uniformly on a $2B \times 2B$ square and then toss out those individuals located $>B$ from the center point of the square. This produces continuous distance data, which we can then bin into distance classes. Alternatively, we could compute the multinomial cell probabilities and simulate multinomial (i.e., grouped) observations directly using `rmultinom`. We show both in this section. First, we define a function that will simulate a population of individuals on the square and return the required data objects and give summary plots (Figure 8.2):

```
sim.pdata <- function(N=1000, sigma=1, B=3, keep.all=FALSE) {
# Function simulates coordinates of individuals on a square
# Square is [0,2*B] x[0,2*B], with a count location on the center
# point (B,B)
# Function arguments:
#     N: total population size in the square
#     sigma: scale of half-normal detection function
#     B: circle radias
#     keep.all: return the data for y = 0 individuals or not
```

```
# Plot the detection function
par(mfrow = c(1,2))
curve(exp(-x^2/(2*sigma^2)), 0, B, xlab="Distance (x)", ylab="Detection prob.", lwd = 2,
main = "Detection function", ylim = c(0,1))
text(0.8*B, 0.9, paste("sigma:", sigma))

# Simulate and plot simulated data
library(plotrix)
u1 <- runif(N, 0, 2*B)          # (u1,u2) coordinates of N individuals
u2 <- runif(N, 0, 2*B)
d <- sqrt((u1 - B)^2 + (u2 - B)^2) # distance to center point of square
plot(u1, u2, asp = 1, pch = 1, main = "Point transect")
N.real <- sum(d<= B)            # Population size inside of count circle

# Can only count indidivuals in the circle, so set to zero detection probability of
individuals in the corners (thereby truncating them):
p <- ifelse(d < B, 1, 0) * exp(-d*d/(2*(sigma^2)))
# Now we decide whether each individual is detected or not
y <- rbinom(N, 1, p)
points(u1[d <= B], u2[d <= B], pch = 16, col = "black")
points(u1[y==1], u2[y==1], pch = 16, col = "blue")
points(B, B, pch = "+", cex = 3, col = "red")
draw.circle(B, B, B)

# Put all of the data in a matrix:
#       (note we don't care about y, u, or v normally)
if(!keep.all){
   u1 <- u1[y==1]
   u2 <- u2[y==1]
   d <- d[y==1]
}
return(list(N=N, sigma=sigma, B=B, u1=u1, u2=u2, d=d, y=y, N.real=N.real))
}

# obtain a data set by distance sampling a population of N=1000 out to a distance of B=3
set.seed(1234)
tmp <- sim.pdata(N=1000, sigma=1, keep.all=FALSE, B=3) # produces Figure 8.2
attach(tmp)
```

Here we simulated a complete data set (Figure 8.2) but returned only the location coordinates of each individual (u_1, u_2) and the observed distances for captured ($y = 1$) individuals. We will use these locations later in the chapter. For now, we develop likelihood analyses of the distance data. We start by taking the data just simulated and "bin" them by using the integer division function %/%. To apply this to our simulated distance data, we do the following (and note we must *always* make sure that we have a vector of encounter frequencies that includes the zeros, i.e., the distance bins where nobody was detected!).

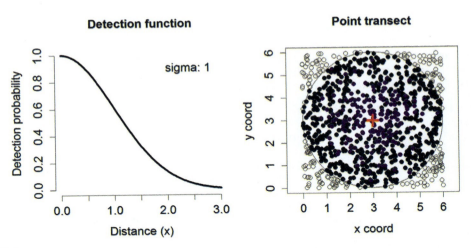

FIGURE 8.2

Plot of the simulation of point transect data (see R code sim.pdata). (left) Form of detection function for chosen value of sigma. (Right) Map of simulated point pattern. Locations of individuals inside of the maximum detection distance (circle) are black, individuals detected are blue, and the point count location is at the red cross.

```
# Bin the data and tabulate the bin frequencies. Be sure to pad the 0s!
delta <- 0.5                          # Width of distance bins
dist.breaks <-seq(0, B, delta)        # Make the interval cut points
dclass <- tmp$d %/% delta +1          # Convert distances to categorical distances
nD <-length(dist.breaks)-1            # How many intervals do we have ?
y.obs <- table(dclass)                # Next pad the frequency vector
y.padded <- rep(0, nD)
names(y.padded) <- 1:nD
y.padded[names(y.obs)] <- y.obs
y.obs <- y.padded
```

Next, we will simulate binned distance data with the rmultinom function which uses numerical construction of the multinomial cell probabilities based on the cp.ri function defined in the previous section:

```
cp <- c(cp, 1-sum(cp))  # Compute the last cell and add it to the vector

as.vector(rmultinom(n=1, size=1000, prob=cp))
[1]  25  59  74  53  25   9 755
```

We can check that these cell probabilities are in agreement with what we get when we simulate continuous distance data and then bin them as follows: we simulate an extremely large data set (e.g., $N = 10^5$) and compute the relative frequencies in each distance class. These should then be very close to the multinomial cell probabilities obtained by numerical integration.

8.2.5.2 Likelihood Analysis of Point Transect Data

To do likelihood analysis of the point transect data, we define an R function that evaluates the likelihood for a particular value of the parameter(s) and other arguments, such as the observed distance data and the upper distance bound of counting B. Here we provide three versions of the likelihood: (1) the multinomial full likelihood for binned data; (2) the full likelihood for continuous distance data; (3) the conditional likelihood for continuous data. (We omit the conditional likelihood based on the binned data.) In the following block of code, we define the three likelihood functions and then optimize each to obtain the estimated population size for the circular sample unit (note: you may want to re-create the data set from above in case you overwrote stuff in your R workspace).

```
# (1) Define multinomial likelihood for binned data
Lik.binned.point <- function(parm, data, dist.breaks){
    sigma <- exp(parm[1])
    n0 <- exp(parm[2])
    p.x <-rep(NA, nD)
    for(i in 1:nD){
        p.x[i] <- cp.ri(dist.breaks[i], dist.breaks[i+1], sigma =sigma)
    }
    area <- 3.141593 * dist.breaks[-1]^2
    ring.area <- diff(c(0, area))
    cp <- p.x* ring.area/sum(ring.area) # Pr(detection| in ring)*Pr(in ring)
    pi0 <- 1-sum(cp)
    N <- sum(data) + n0
    negLL <- -1*(lgamma(N+1)-lgamma(n0+1) + sum(c(data,n0)*log(c(cp,pi0))))
    return(negLL)
}

# Fit model
mle1 <- optim(c(2,0), Lik.binned.point, data=y.obs, dist.breaks=dist.breaks)

# (2) Define full likelihood for continuous data
Lik.cont.point <- function(parm, data, B){
    sigma <- exp(parm[1])
    n0 <- exp(parm[2])
    n <- length(data)
    N <- n + n0
    p <- exp(-data*data/(2*sigma*sigma))
    f <- 2*data/(B^2)
    pbar <- integrate(function(r, s=sigma) exp(-r^2 / (2 * s^2)) * r, 0, B)$value*2/(B^2)
    negLL <- -1*sum( log(p*f/pbar )) -1*(lgamma(N+1) - lgamma(n0+1) +
        n * log(pbar) + n0*log(1-pbar))
    return(negLL)
}

# Fit model
mle2 <- optim(c(0, 5), Lik.cont.point, data=tmp$d, B=B, hessian=TRUE)

# Compare two solutions and with realized true value of N
(Nhat.binned <- length(tmp$d) + exp(mle1$par[2]))
[1] 792.9483
(Nhat.cont <- length(tmp$d) + exp(mle2$par[2]))
[1] 799.0507
tmp$N.real
[1] 797
```

These are different by about 1%! It is tempting to regard the estimate under the continuous distance model as being better, and, indeed, it is both closer to the truth and also the data were simulated in that way, so in this case, it is. However, in practice, we don't know the truth, and there is no theoretical reason to prefer the continuous distance estimator to the binned estimator. They are alternative models both being used as approximations to the truth, which we don't know.

Finally, we provide the conditional likelihood for the continuous distance data, which only has a single parameter σ (for the half-normal model) to be estimated:

```
# (3) Define conditional likelihood for continuous data
Lik.cond.point <- function(parm, data, B){
   sigma <- exp(parm)
   p <- exp(-data*data/(2*sigma*sigma))
   f <- 2*data/(B^2)
   pbar <- integrate(function(x, s=sigma) exp(-x^2 / (2 * s^2)) * x, 0, B)$value*2/(B^2)
   negLL <- -1*sum( log(p*f/pbar ))
   return(negLL)
}

# Fit the model
mle3 <- optim(c(0), Lik.cond.point, data=tmp$d, B=B, method="Brent", hessian=TRUE,
lower=-10, upper=10)

# Inspect the output
mle3
$par
[1] 0.01523024

$value
[1] 160.8509

[ . . . output truncated . . . ]

$hessian
          [,1]
[1,] 544.0608

# Estimated sigma
(sigma.hat <- exp(mle3$par))
[1] 1.015347
```

We see that $\hat{\sigma} = \exp(0.0152) = 1.015$, which is pretty close to the true value of 1, and so we surmise that our likelihood implementation is likely correct. With the conditional estimator we don't obtain directly an estimate of density or population size. Instead we have to use the MLE of σ after the fact and compute $\bar{p}$, which we do as follows, finding that the conditional estimator of $N (= n/\bar{p})$ is in the same ballpark as the other two, and yet slightly different.

```
# Estimated average detection probability and conditional estimator of N
pbar <- integrate(
    function(x, s=sigma.hat) exp(-x^2 / (2 * s^2)) * x, 0, B)$value*2/(B^2)

(Nhat.cond1   <- length(d) / pbar)
[1] 800.2388
```

8.2.6 SENSITIVITY TO BIN WIDTH

When you write a paper that uses distance sampling with binned data, one criticism raised by a referee undoubtedly will be that it would be better to use a continuous data model instead of an "approximation." However, both the continuous distance model and the model for binned data are mere approximations to the actual data-generating process, which we don't know. Nevertheless, it is interesting to see how similar a discrete distance model is to a particular continuous data-generating model, which we can handily know if we happen to be simulating data.

To evaluate the effect of binning data that are generated from a truly continuous model, we show a small simulation study here that you can easily repeat for your own situation. We simulate data as above (all of the code is repeated here) using bin widths of $\delta = 0.5$, and we fit both the continuous and discrete distance models as in the previous section. We do 1000 Monte Carlo (simulation) replicates of each bin width scenario and, at the end, we compute the mean of $\widehat{N}$ and also the standard deviation. All of this goes as follows:

```
set.seed(1234)
simrep <- 1000                  # Number of sim reps
simout <- matrix(NA, nrow=simrep, ncol=3)
colnames(simout) <- c("N.real", "N.binned", "N.continuous")
delta <- 0.5                    # Set width of bins

# Begin simulation loop
for(sim in 1:simrep){
   tmp <- sim.pdata(N=1000, sigma=1, keep.all=FALSE, B=3)
   B <- tmp$B
   d <- tmp$d
   N.real <- tmp$N.real

   # Bin data, tabulate frequencies and pad 0s if necessary
   dist.breaks <- seq(0, B, delta)
   dclass <- d%/%delta + 1       # Convert distances to categorical distances
   nD <- length(dist.breaks) -1  # How many intervals do we have ?
   y.obs <- table(dclass)        # Next pad the frequency vector
   y.padded <- rep(0, nD)
   names(y.padded) <- 1:nD
   y.padded[names(y.obs)] <- y.obs
   y.obs <- y.padded

   # Obtain the MLEs using both models
   binned.est <- optim(c(2,0), Lik.binned.point, data=y.obs,
      dist.breaks=dist.breaks)
   cont.est <- optim(c(1,6), Lik.cont.point, data=d, B=B, hessian=TRUE)
   Nhat.binned <- length(d) + exp(binned.est$par[2])
   Nhat.cont <- length(d) + exp(cont.est$par[2])

   # Store results in a matrix
   simout[sim,] <- c(N.real, Nhat.binned, Nhat.cont)
}
```

```
# Now summarize the output
apply(simout, 2, mean)
     N.real    N.binned  N.continuous
    785.0580   782.5183    782.7314

sqrt(apply(simout, 2, var))
     N.real    N.binned  N.continuous
    12.84647   84.25812   82.70834
```

What we see here is essentially the same expected value of both estimators and an only very slightly increased standard deviation of the binned estimator. This is a general truth: binning has essentially no effect on bias and only negligibly decreases precision of the estimator *compared to fitting the correct continuous distance model*. Of course, in practice, we will not know the true model.

To gauge the sensitivity to bin width, we repeated the analyses for the same 1000 data sets using bin widths of 0.1, 0.2, 0.3, 0.5, 0.6, and 1.0 (all of these produce equal-width bins for $B = 3$). The results are tabulated as follows:

```
Width   Truth    Nhat.binned   SD.binned
 0.1    785.06    782.82        82.80
 0.2    785.06    782.90        82.82
 0.3    785.06    782.45        83.49
 0.5    785.06    782.52        84.26
 0.6    785.06    782.19        84.38
 1.0    785.06    783.51        89.27
```

What we see here is negligible bias in the estimated population size (much less than 1%), although we do see a systematic increase in the standard deviation of the estimator. For a bin width of 0.1 the SD is about the same as the continuous distance model, whereas for the bin width of 1.0 (one-third of the total count radius!) the SD increases by about 8%. What all of this means is that you'll suffer a small cost in terms of precision by using a bin width that is extremely coarse but no practical effect at all for bin widths that are roughly <10% of the count radius. Note that the mean MLE for all six cases is systematically less than the true average of 785.06. This is because the same 1000 data sets were used for each simulation and a different 1000 data sets will produce a different (higher or lower) discrepancy, on average. As always, we encourage you to play around with simulations to gain intuition and understanding of the effects of data collecting and analysis decisions.

8.2.7 SPATIAL SAMPLING

In practice we virtually always have more than a single sample unit. What do we do in this situation if we're doing conventional distance sampling? Let's say we have S transects, which we imagine to be S multinomial samples with size (= local population size) N_s, then the conventional distance sampling approach is to just pool all of the distance data and fit a single conditional likelihood to it (remember our change of notation for site index; see Section 8.1). So the N_s are not involved in this at all. Spatial sampling is ignored. This is surprising, since probably most applications of CDS have an interest in assessing hypotheses about spatiotemporal variation in N_s, and yet when doing CDS we usually ignore the problem almost completely by using the conditional likelihood of the pooled data.

We say "almost" since spatial sampling is not entirely ignored. The conditional estimator of density is

$$\widehat{D} = n/(\widehat{p} * 2 * L * B)$$

To obtain the variance of this estimator we need to estimate the quantity *Var(n)*, which CDS estimates using the variance of n_s among sample units (Buckland et al., 2001, p. 79). The latter is usually called the "encounter rate variance" and can be thought of as a nonparametric estimator of the variance among spatial units. That is the only way in which spatial sampling is dealt with in CDS.

8.3 BAYESIAN CONVENTIONAL DISTANCE SAMPLING

While we have not yet analyzed a hierarchical distance sampling (HDS) model, we have learned the four basic operations: simulation and analysis of continuous and binned data for line transects and for point transects. For point transects, this is only slightly more complicated than for transects due to the different geometry. We have shown how to write out the likelihood and obtain MLEs. We now cover how to analyze these models using Bayesian methods, which will come in handy when we finally get to the analysis of hierarchical distance sampling models. Part of the reason for building up this material in such a leisurely way is that there is not "one way" to analyze HDS models, just as we saw with CDS models. The various ways of analyzing these models will all be useful in analyzing HDS models in different situations or using different BUGS engines. So, we will go through the various formulations of the models above (line/point transects, continuous/binned distances, conditional/full likelihood) using Bayesian methods implemented in BUGS. To implement continuous distance models in BUGS often takes a little bit of trickery because the probability distribution of the observed distances is not usually a standard form. It is easy to resolve this by using distance bins, in which case we can use a categorical or multinomial distribution, where we build the cell probabilities explicitly. And, as we saw previously, there is almost no statistical cost for using a discrete distance model, even when we happen to know the correct continuous distance model.

For analyzing the full likelihood in BUGS we use the idea of *parameter-expanded data augmentation* (PX-DA or DA for short; Royle et al., 2007a; Royle and Dorazio, 2012), which we also discussed briefly in Section 7.8.4 and will encounter again in Chapters 9, 11, and later. The idea of DA is that we take our data set of *n* observed encounters (and distances) and augment it with a large number of $M - n$ "not encountered" individuals, which necessarily have missing distance data. We further expand our model by introducing a set of binary latent variables (the data augmentation variables) z_i, which are indicators of whether an individual in the larger data set of size *M* is a "real" individual, so that the observation of 0 is a stochastic (sampling) zero, or whether it is a fixed zero, which is to say $y = 0$ with probability 1. We assume $z_i \sim Bernoulli(\psi)$, where ψ is the data augmentation parameter. (There is a sense in which DA transforms a capture-recapture type of model into an occupancy type of model, and our notation with z and ψ is intended to reflect this.) This formulation of the model is equivalent to putting a *Binomial(M, ψ)* prior distribution on population size *N* (see Section 8.2.4.1 above) and a *Uniform*(0,1) prior on ψ. Those two priors together imply that the marginal (induced) prior distribution for *N* is *Discrete Uniform*(0, *M*). Of course the binomial prior is roughly equivalent to a Poisson prior distribution when *M* is large, but even when it is not, it is not clearly a better or worse prior than the Poisson, just different. Next, we demonstrate the use of DA for line transect data with continuous and binned data measurements.

8.3.1 BAYESIAN ANALYSIS OF LINE TRANSECT DATA

We illustrate a Bayesian analysis of distance sampling data from a transect using the famous impala data set from Burnham et al. (1980; analysis modified from Royle and Dorazio 2008, p. 235). In this study, distance data were collected along a 60 km transect. If we use a transect width of 1000 m, the total area is 60 km^2, which we'll use to convert estimated N to estimated density, D. The line transect situation is especially easy to deal with in BUGS because we can specify the uniform distribution for distance explicitly and then, conditional on the distances, the observation model is specified as a simple Bernoulli trial, like in a logistic regression. The Bayesian formulation of the distance sampling model therefore makes clear the elegant hierarchical structure of distance sampling as involving a process model (the distribution of individuals) and an observation model (the detection or nondetection of individuals; for this we use the half-normal model throughout). Next, we input the data directly into the R workspace, package things up, and run BUGS as follows:

```
# Get data and do data-augmentation
# Observed distances (meters) in the impala data set
x <- c(71.93, 26.05, 58.47, 92.35, 163.83, 84.52, 163.83, 157.33,
22.27, 72.11, 86.99, 50.8, 0, 73.14, 0, 128.56, 163.83, 71.85,
30.47, 71.07, 150.96, 68.83, 90, 64.98, 165.69, 38.01, 378.21,
78.15, 42.13, 0, 400, 175.39, 30.47, 35.07, 86.04, 31.69, 200,
271.89, 26.05, 76.6, 41.04, 200, 86.04, 0, 93.97, 55.13, 10.46,
84.52, 0, 77.65, 0, 96.42, 0, 64.28, 187.94, 0, 160.7, 150.45,
63.6, 193.19, 106.07, 114.91, 143.39, 128.56, 245.75, 123.13,
123.13, 153.21, 143.39, 34.2, 96.42, 259.81, 8.72)

B <- 500 # Strip half-width. Larger than max observed distance
nind <- length(x)

# Analysis of continuous data using data augmentation (DA)
nz <- 200 # Augment observed data with nz = 200 zeroes
y <- c(rep(1, nind), rep(0, nz)) # Augmented inds. have y=0 by definition
x <- c(x, rep(NA, nz)) # Value of distance are missing for the augmented

# Bundle and summarize data set
str( win.data <- list(nind=nind, nz=nz, x=x, y=y, B=B) )

# Save text file with BUGS model
cat("
model {

# Priors
sigma ~ dunif(0,1000)  # Half-normal scale
psi ~ dunif(0,1)       # DA parameter

# Likelihood
for(i in 1:(nind+nz)){
   # Process model
   z[i] ~ dbern(psi)   # DA variables
   x[i] ~ dunif(0, B)  # Distribution of distances
```

8.3 BAYESIAN CONVENTIONAL DISTANCE SAMPLING

```
# Observation model
    logp[i] <- -((x[i]*x[i])/(2*sigma*sigma)) # Half-normal detection fct.
    p[i] <- exp(logp[i])
    mu[i] <- z[i] * p[i]
    y[i] ~ dbern(mu[i])    # Simple Bernoulli measurement error process
}
# Derived quantities
N <- sum(z[1:(nind + nz)]) # Population size
D <- N / 60                # Density, with A = 60 km^2 when B = 500
}
",fill=TRUE,file="model1.txt")

# Inits
zst <- y
inits <- function(){ list (psi=runif(1), z=zst, sigma=runif(1,40,200)) }

# Params to save
params <- c("N", "sigma", "D")

# Experience the raw power of BUGS and summarize marginal posteriors
library(R2WinBUGS)
bd <- "c:/Program Files/WinBUGS14/"    # May have to adapt for your computer
out1 <- bugs(win.data, inits, params, "model1.txt", n.thin=2,n.chains=3,
    n.burnin=1000, n.iter=11000, debug=TRUE, DIC=FALSE, bugs.dir=bd)
print(out1, 3)
        mean     sd   2.5%   25%   50%   75%  97.5% Rhat n.eff
N     221.468 24.746 174.0 204.0 222.0 240.0 267.00 1.004   600
sigma 131.466 10.798 112.8 123.8 130.6 138.2 155.20 1.003  1100
D       3.691  0.412   2.9   3.4   3.7   4.0   4.45 1.004   600
```

Next, we provide an analysis of the impala data but using binned data to demonstrate the BUGS implementation using data augmentation. We first need to convert the distance data into distance bins, which we define here to be 50 m bins. Then we specify the model in BUGS using the dcat distribution for individual distance class observations. In BUGS we have to define detection probability for each interval, which we do by evaluating the half-normal detection probability function at the midpoint of each interval (input as data), which will look like this:

```
log(p[g]) <- -midpt[g] * midpt[g] / (2 * sigma * sigma)
```

We also have to compute the probability mass for each distance interval:

```
pi[g] <- delta / B   # probability of x in each interval
```

```
# Analysis of binned data using data augmentation
delta <- 50              # Width of distance bins
xg <- seq(0, B, delta)   # Make the interval cut points
dclass <- x %/% delta + 1 # Convert distances to distance category
nD <- length(xg) -1      # N intervals = length(xg) if max(x) = B
```

```
# Bundle data
# Note data changed to include dclass, nD, bin-width delta and midpt
midpt <- xg[-1] - delta/2  # Interval mid-points
str( win.data <- list (nind=nind, nz=nz, dclass=dclass, y=y, B=B,
    delta=delta, nD=nD, midpt=midpt) )   # Bundle and summarize

# BUGS model specification
cat("
model{
# Priors
psi ~ dunif(0, 1)
sigma ~ dunif(0, 1000)

# Likelihood
# Construct conditional detection probability and Pr(x) for each bin
for(g in 1:nD){          # midpt = mid point of each cell
   log(p[g]) <- -midpt[g] * midpt[g] / (2 * sigma * sigma)     # half-normal model
   pi[g] <- delta / B    # probability of x in each interval
}

for(i in 1:(nind+nz)){
   z[i] ~ dbern(psi)              # model for individual covariates
   dclass[i] ~ dcat(pi[])         # population distribution of distance class
   mu[i] <- z[i] * p[dclass[i]]   # p depends on distance class
   y[i] ~ dbern(mu[i])
}
# Derived quantities: Population size and density
N <- sum(z[])
D <- N / 60
}
",fill=TRUE, file = "model2.txt")

# Inits function
zst <- y # DA variables start at observed value of y
inits <- function(){ list (psi=runif(1), z=zst, sigma=runif(1,40,200)) }

# Parameters to save
params <- c("N", "sigma", "D")

# Unleash WinBUGS and summarize posteriors
out2 <- bugs(win.data, inits, params, "model2.txt", n.thin=2, n.chains=3,
   n.burnin=1000, n.iter=11000, debug=TRUE, DIC=FALSE, bugs.dir = bd)
print(out2, 2)
         mean     sd    2.5%     25%     50%     75%    97.5%  Rhat  n.eff
N      218.48  25.27  170.00  200.00  218.00  236.00  267.00  1.01   320
sigma  134.24  11.16  114.90  126.30  133.30  141.20  158.50  1.00   840
D        3.64   0.42    2.83    3.33    3.63    3.93    4.45  1.01   320
```

These are similar to those obtained previously using the continuous distance model, and they should become more similar as we decrease the bin width, and perhaps also by increasing the number of MCMC iterations so as to reduce Monte Carlo error.

8.3.2 OTHER FORMULATIONS OF THE DISTANCE SAMPLING MODEL

We have shown transect models with binned and continuous distances analyzed in BUGS using data augmentation. But there are many other exciting formulations of distance sampling models. For example, in Chapter 9 we will provide a formulation of the model not in terms of distance but in terms of location of encounter. This is one of our favorites. Another type of model that might be useful to develop is that in which we have binned data but parameterize the model in terms of latent continuous observations. The observed bin data are a "cut" of the continuous data that have the standard DS model for continuous data. This is an interesting idea because it allows us to "downscale" the observations, or make predictions, to a finer scale than the available observations. We think that this can be done directly in JAGS using its function for interval censoring. Finally, it is possible to formulate conditional likelihood models in BUGS. This is somewhat more complicated because the conditional distribution of the distances cannot be specified directly. However, we can formulate the model for binned data and then compute the conditional cell probabilities explicitly in the BUGS code. Then we can use either a multinomial distribution for distance bin frequencies or the categorical distribution for individual distance bin observations. We will show this in the next section in the context of point transect data, but we leave it to you as an exercise for the transect case.

8.3.3 A TREATISE ON THE INTEGRATION OF MATHEMATICAL FUNCTIONS IN ONE DIMENSION

Before we proceed with a development of Bayesian analysis of point transect data in BUGS, we first discuss the basic concept of *integration using the "rectangular rule"* whereby, to compute the integral of some function $f(x)$, we approximate the function by a bunch of rectangles centered at points x_i and then sum up the area of those rectangles. This is precisely the approach we used in the analysis of binned transect data in the previous section. The virtue of being able to do this is that it gives us a way of parameterizing *any* distribution in BUGS if we just know a formula for its pdf. Instead of the "ones trick" or the "zeros trick" (Section 5.8, and p. 204–206 in Lunn et al., 2013), we just compute the area under chunks of the curve and use `dcat` as a model for a binned version of the variable. The error in making a discrete approximation to any continuous distribution is usually negligible compared to the MC error in our MCMC analysis, provided we use enough bins.

To demonstrate this we show, in Figure 8.3, the right side of a normal kernel, and we ask, "what is the area under this curve?" Of course, we can use the `integrate` function to compute that directly (which we do below). Or, we can line up a bunch of rectangles as shown in Figure 8.3 and sum up the area of those rectangles. All of the code for doing this together is as follows:

```
sigma <- 2                          # normal scale (standard deviation)
curve(exp(-x^2 / (2*sigma^2)), 0, 10, frame = F)

delta <- 1                          # bin width
mid <- seq(0.5, 9.5, delta)         # 10 rectangles
f.mid <- exp(-mid^2 / (2*sigma^2))
barplot(f.mid, add=T, space=0, col="grey", width=delta)
curve(exp(-x^2 / (2*sigma^2)), 0, 10, add = TRUE, col = "blue", lwd = 3)
```

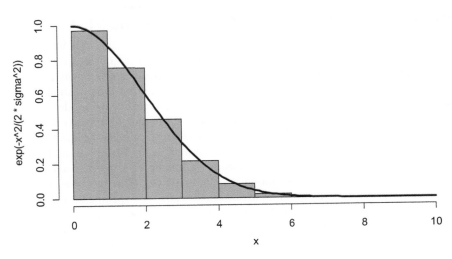

FIGURE 8.3
All you need to know about integration in one figure. The area under the blue curve is approximated by the area of the gray rectangles.

```
# Integral done using the integrate function
integrate( function(x){ exp(-x^2/(2*sigma^2)) }, lower=0, upper=100)
2.506628 with absolute error < 8.1e-07

# Summing up the 10 rectangular areas:
areas <- f.mid * delta
sum(areas)
[1] 2.506627
```

At the end of the day we see no practical difference between these two results, and therefore it stands to reason that if we model continuous distributions in BUGS using even a moderate number of rectangles, we would not expect to be badly led astray. Moreover, as we've said a few times already in this chapter and will continue to say some more, discrete distributions are perfectly reasonable models of random variables, without even having to think about them as approximations to any continuous thing.

8.3.4 BAYESIAN ANALYSIS OF POINT TRANSECT DATA

As with the formulation of the conditional likelihood for line transect data, the problem with analyzing point transect data in BUGS is that there is no built-in distribution for the distances, which we noted, in Section 8.2.5, has a triangular distribution. This can be dealt with in several ways by using the "zeros trick" or the "ones trick" (see Chelgren et al., 2011b, for a neat distance sampling application) or, alternatively, we can analyze the model for binned data on the circle and use the dcat or multinomial model. We show that here. This is sufficient in practice because of course we can always use a huge number of distance intervals to obtain what essentially is a continuous distance model if that was necessary. But also, as we discussed in Section 8.2.6, the continuous distance model is not any more correct than an a similar step function model. The key thing is to identify, for each individual, which distance category it belongs in and then compute the probabilities for that categorical random variable. We can use these to specify a model based either on the conditional likelihood or we can use specify an 'unconditional' model based on data augmentation. We show both of these.

The mathematical argumentation to define the distance class probabilities goes like this: The probability density of detections is the product of the detection function (here, a half-normal) and the density of x:

$$\Pr(\text{detection in } x) = \Pr(\text{detection}|x)\Pr(x) = \exp\left(-\frac{1}{2\sigma^2}x^2\right)f(x),$$

where $f(x)$ = pdf of radial distance from a point (for a transect $f(x)$ is constant). The probability distribution of radial distance x on a circle of radius B is:

$$f(x) = \frac{2x}{B^2}$$

(there is more mass in a distance band as you move far away from the point). As we showed back in Section 8.2.5, we need to integrate $\exp\left(-\frac{1}{2\sigma^2}x^2\right)f(x)$ over distance bands to get multinomial cell probabilities (this is what unmarked does, see Section 8.4.3). But from our treatise on the integration of one-dimensional functions, we know that, approximately, the multinomial cell probabilities should be "width times height" of a rectangle centered at x_h and therefore:

$$\pi(x_h) = \Pr(x_h - \delta/2 \leq x \leq x_h + \delta/2) \approx \exp\left(-\frac{1}{2\sigma^2}x_h^2\right)f(x_h)\delta$$

(this is the rectangular approximation to an integral). So we can choose x_h to be the mid-points of our intervals or we can use many very narrow intervals and then add them up into coarser bins. To implement a conditional model we need to compute conditional distance class probabilities:

$$\pi^c(x_h) = \frac{\pi(x_h)}{1 - \pi_0},$$

where the denominator: $1 - \pi_0 = \Pr(capture) = \sum_h \pi(x_h)$

Note that the conditional distance class probabilities are used when we analyze the conditional likelihood version of the model in BUGS, in which case N is a derived parameter. We can also analyze the full likelihood version of the model by data augmentation, which we also show below. The full likelihood DA version uses the probabilities $f(x_h)$ as the distribution for the population of true distances and then also models detection/nondetection of each individual, y_i.

We can simulate binned distance sampling data either directly by simulating categorical random variables, or we can simulate continuous distance data and bin the data as we did previously. We will simulate continuous data here, bin the data into classes, and then use the categorical distribution in BUGS to fit the point transect model.

```
### Version 1: Point count data in BUGS (conditional likelihood)
# Simulate a data set and harvest the output
set.seed(1234)
tmp <- sim.pdata(N=200, sigma=1, keep.all=FALSE, B=3)
attach(tmp)
```

```
# Chop the data into bins
delta <- 0.1              # width of distance bins for approximation
xg <- seq(0, B, delta)    # Make the mid points and chop up the data
midpt <- xg[-1] - delta/2

# Convert distances to categorical distances (which bin?)
dclass <- d %/% delta + 1
nD <- length(midpt) # how many intervals
nind <- length(dclass)

# Bundle and summarize data set
str( win.data <- list(midpt=midpt, delta=delta, B=B, nind=nind, nD=nD, dclass=dclass) )

# BUGS model specification, conditional version
cat("
model{

# Prior for single parameter
sigma ~ dunif(0, 10)

# Construct cell probabilities for nD cells (rectangle approximation)
for(g in 1:nD){    # midpt[g] = midpoint of each distance band
   log(p[g]) <- -midpt[g] * midpt[g] / (2*sigma*sigma)
   pi[g] <- (( 2 * midpt[g] ) / (B*B)) * delta
   f[g] <- p[g] * pi[g]
   fc[g] <- f[g] / pcap
}
pcap <- sum(f[]) # capture prob. is the sum of all rectangular areas

# Categorical observation model
for(i in 1:nind){
   dclass[i] ~ dcat(fc[])
}

# Derived quantities: population size and density
N <- nind / pcap
D <- N/(3.141*B*B)
}
",fill=TRUE, file="model3.txt")

# Inits function
inits <- function(){list (sigma=runif(1, 1, 10)) }

# Params to save
params <- c("sigma", "N","D")

# MCMC settings
ni <- 62000 ; nb <- 2000 ; nt <- 2 ; nc <- 3

# Run BUGS and summarize posteriors
bd <- "c:/Program Files/WinBUGS14/"    # May have to adapt this to your computer
```

```
out3 <- bugs(win.data, inits, params, "model3.txt", n.thin=nt,
n.chains=nc, n.burnin=nb, n.iter=ni, debug=FALSE, bugs.dir = bd)

## Version 2: point count data (full likelihood with data augmentation)
# Do data augmentation (for same simulated data set)
M <- 400
nz <- M - nind
y <- c(rep(1, nind), rep(0, nz))
dclass <- c(dclass, rep(NA, nz))

# Bundle and summarize data set
str( win.data <- list(midpt=midpt, delta=delta, B=B, nind=nind, nD=nD, dclass=dclass,
y=y, nz=nz) )

# BUGS model
cat("
model{

# Priors
sigma ~ dunif(0, 10)
psi ~ dunif(0, 1)

# Construct cell probabilities for nD cells (rectangle approximation)
for(g in 1:nD){            # midpt[g] = midpoint of each distance band
    log(p[g]) <- -midpt[g] * midpt[g] / (2*sigma*sigma)
    pi[g] <- ((2 * midpt[g]) / (B * B)) * delta
    pi.probs[g] <- pi[g] / norm
    f[g] <- p[g] * pi[g]
    fc[g] <- f[g] / pcap    # conditional probabilities
}
pcap <- sum(f[])# capture prob. is the sum of all rectangular areas
norm <- sum(pi[])

# Categorical observation model
for(i in 1:(nind+nz)){
    z[i] ~ dbern(psi)
    dclass[i] ~ dcat(pi.probs[])
    mu[i] <- p[dclass[i]] * z[i]
    y[i] ~ dbern(mu[i])
}

# Derived quantities: population size and density
N <- sum(z[])
D <- N/(3.141*B*B)

}
",fill=TRUE,file="model4.txt")

# Inits
inits <- function(){list (sigma=runif(1,1,10), psi=runif(1) ) }
```

```
# Parameters to save
params <- c("sigma","N","D","psi")

# MCMC settings
ni <- 62000 ; nb <- 2000 ; nt <- 2 ; nc <- 3

# Run BUGS and summarize posteriors
out4 <- bugs(win.data, inits, params, "model4.txt", n.thin=nt,
    n.chains=nc, n.burnin=nb, n.iter=ni, debug=FALSE, bugs.dir = bd)

# Compare posterior summaries
print(out3,2)    # Conditional likelihood
```
Inference for Bugs model at "model3.txt", fit using WinBUGS,
3 chains, each with 62000 iterations (first 2000 discarded), n.thin = 2
n.sims = 90000 iterations saved

	mean	sd	2.5%	25%	50%	75%	97.5%	Rhat	n.eff
sigma	1.11	0.13	0.91	1.02	1.10	1.18	1.41	1	90000
N	142.96	27.20	93.71	123.80	141.60	160.60	200.00	1	90000
D	5.06	0.96	3.31	4.38	5.01	5.68	7.08	1	90000
deviance	246.92	1.59	245.80	245.90	246.30	247.30	251.40	1	90000

```
print(out4,2)    # Full likelihood
```
Inference for Bugs model at "model4.txt", fit using WinBUGS,
3 chains, each with 62000 iterations (first 2000 discarded), n.thin = 2
n.sims = 90000 iterations saved

	mean	sd	2.5%	25%	50%	75%	97.5%	Rhat	n.eff
sigma	1.09	0.12	0.90	1.01	1.08	1.16	1.37	1	28000
N	150.16	35.14	91.00	125.00	147.00	172.00	229.00	1	51000
D	5.31	1.24	3.22	4.42	5.20	6.08	8.10	1	51000
psi	0.38	0.09	0.22	0.31	0.37	0.43	0.58	1	59000
deviance	396.00	18.01	361.80	383.70	395.70	407.90	432.40	1	44000

We see a slight inconsistency between the two analyses, both producing posterior means of N slightly less than the true value of $N = 152$ (=sum(tmp$N.real)). However, relative to the uncertainty of these estimates (quantified by the posterior standard deviation), the discrepancy between the two estimates is fairly small.

8.4 HIERARCHICAL DISTANCE SAMPLING (HDS)

Now we transition from the basic elements of conventional distance sampling models to situations where we have distance sampling data collected at S spatial locations, usually either transects or point count locations, but we could also have a mixture of both or even strange shapes or irregular transects. (Remember our change of notation for sites, which now have index s, which runs from 1 to S.) As we've noted a few times before, the traditional way to deal with this in distance sampling is to pool the distance data from all S spatial locations and estimate the parameter(s) of the detection function, e.g., σ for a half-normal. This is used to obtain an estimate of density, and then the variance is based on the encounter rate variance, which does use some information from among the sample units. However, conventional distance sampling does not directly address problems of spatial inference either in the form of modeling variation in N_s or local density across sample units s or making explicit predictions at other transects or point locations. We would argue that modeling variation in N among sample units is critically important and, indeed, often the primary interest in studies that use distance sampling. Therefore, HDS models should be in every ecologist's toolbox.

The models we deal with here assume that N_s is the population size of spatial sample unit s, and they don't make any explicit assumptions about "*within sample unit*" variation in density. Rather, they assume that the average covariate value defined for the sample unit is meaningful for explaining among sample unit variation. Thus, when we assume that $N_s \sim Poisson(\lambda_s)$ the parameter λ_s is constant for the sample location s and represents the mean for the sample unit. This is *not* to say that HDS models assume that density is constant within a sample unit, just that the aggregate density is adequately modeled by the covariates defined for the sample unit. We discuss this more in Chapter 9.

8.4.1 HDS DATA STRUCTURE AND MODEL

To develop distance sampling in an explicit meta-population setting, we suppose that S distinct spatial units are sampled using the distance sampling protocol. These might be transects or point counts for birds, distributed in some region (e.g., a park or forest). Distance x is naturally viewed as a continuous measurement, but for now we jump right into the discrete distance class formulation (we discuss continuous measurements shortly). Thus, we consider binned data here, wherein distances are recorded in discrete intervals from the central point of observation for each site. Let $h = 1, 2, ..., H$ index the distance classes, with end points, or *distance breaks* $(c_1, c_2), (c_2, c_3), ..., (c_H, c_{H+1})$. Here, c_{H+1} is the maximum distance at which birds were counted, or the radius of the point count (which we called B previously). Let y_{sh} be the observed count of individuals in distance class h for site $s = 1, 2, ..., S$. The data structure is summarized in Table 8.1.

We follow the basic ideas of the binomial and multinomial mixture models of Chapters 6 and 7, and assume that sample unit s has local abundance N_s, which is a random variable having a suitable distribution. For now we assume:

$$N_s \sim Poisson(\lambda_s)$$

Table 8.1 Typical hierarchical distance sampling data structure. For each of S transects we have encounter frequencies in each of a number of distance classes (three illustrated here). In addition, we may have one or more site-level covariates (v).

Transect	dclass 1 (0–50 m)	dclass 2 (50–100 m)	dclass 3 (100–200 m)	Covariate1	Covariate2
Transect 1	2	0	1	v_{11}	v_{12}
Transect 2	3	0	0	v_{21}	v_{22}
Transect 3	2	1	1	v_{31}	v_{32}
⋮	⋮	⋮	⋮	⋮	
Transect S	4	2	0	v_{S1}	v_{S2}

where one or more covariates (v) may influence the expected abundance, λ_s, on a suitable scale:

$$\log(\lambda_s) = \beta_0 + \beta_1 v_s.$$

In addition, we assume the detection frequencies in each of the H distance classes have, conditional on the population size N_s, a multinomial distribution:

$$(y_{s1}, \ldots, y_{sH}) \sim Multinomial(N_s, \pi_s)$$

where π_{sh} is the multinomial cell probability for distance class h and sample unit s—these depend on detection-function parameter(s) σ. These are computed exactly as we've described previously for either line or point transects. If there are no site covariates then there are no additional considerations. If, on the other hand, we also have covariates that influence detection probability and vary across sites, then we have to compute the multinomial cell probabilities separately for each site. It would be natural to model such covariates on the parameter σ, allowing this parameter to vary as a function of covariates that may be site specific (Marques et al., 2007). For example:

$$\log(\sigma_s) = \alpha_0 + \alpha_1 v_s$$

The scale parameter (σ) is a continuous, nonnegative number, hence, it is natural to apply a linear model of covariates on a transformed scale, typically the log, as for the expected count (λ) in a Poisson GLM.

8.4.2 HDS IN unmarked

The unmarked package has two specific functions for fitting HDS models. The older function distsamp assumes a basic closed population model (i.e., for one sample occasion) and allows only for a Poisson abundance model: $N_s \sim Poisson(\lambda_s)$. The more general (and newer) function gdistamp allows for a type of simple open population structure (see Chapter 9) and also for a negative binomial abundance distribution. Both of these functions accept only binned distance data, i.e., multinomial distance class frequencies.

In this section we mainly consider the distsamp function, which works about the same way as multinomPois (Chapter 7), where the abundance parameters N_s are marginalized out of the multinomial likelihood according to

$$[\mathbf{y}_s | \alpha, \beta] = \sum_{N_s=0}^{\infty} [\mathbf{y}_s | N_s, \alpha][N_s | \beta]$$

In practice, we truncate the upper bound of summation (called K in unmarked). As do other N-mixture model fitting functions in unmarked, distsamp uses as a default for K the maximum observed count at a site plus 100. For the Poisson abundance model this likelihood reduces to the product of independent Poisson components (as in Section 7.4), which is very efficient to compute with. As with other unmarked functions, distsamp and gdistsamp have helper functions, called unmarkedFrameDS and unmarkedFrameGDS, for packaging up the data for use by either fitting function. These functions take the basic data and some metadata and set it all up in an unmarkedFrame for analysis by either fitting function and certain summary functions. We demonstrate their use shortly. The distsamp function itself is used roughly like the multinomial *N*-mixture functions of Chapter 7, and it has a few critical arguments as follows (not all arguments shown):

```
distsamp(formula, data, keyfun=c("halfnorm", "exp", "hazard", "uniform"),
    output=c("density", "abund"), unitsOut=c("ha", "kmsq"), starts, ...)
  formula: Double right-hand formula describing detection covariates followed by abundance
           covariates. ~1~1 would be an intercepts-only model.
     data: object of class unmarkedFrameDS, containing response matrix, covariates,
           distance interval cut points, survey type ("line" or "point"), transect lengths
           (for survey = "line"), and units ("m" or "km") for cut points and transect lengths.
           See example for set up.
   keyfun: One of the following detection functions: "halfnorm", "hazard", "exp", or "
           uniform." See details.
   output: Model either "density" or "abund"
 unitsOut: Units of density. Either "ha" or "kmsq" for hectares and square kilometers,
           respectively.
   starts: Vector of starting values for parameters.
```

A more versatile function that allows the fitting of negative binomial abundance models is the gdistsamp function, which has the following structure:

```
gdistsamp(lambdaformula, phiformula, pformula, data,
    keyfun =c("halfnorm", "exp", "hazard", "uniform"),
    output = c("abund","density"), unitsOut = c("ha", "kmsq"),
    mixture = c("P", "NB"), K, starts, method = "BFGS",
    se = TRUE, rel.tol=1e-4, ...)
```

This function will also handle a type of open population structure, allowing for random availability to sampling with a parameter ϕ (corresponding to a temporary emigration probability $1 - \phi$). We cover such models in Chapter 9. The temporal structure is accommodated via a third formula argument, "phiformula," in addition to formulas for the expected abundance ($E(N)$, λ) and p. Note that the three formulas are separated in a gdistsamp call by a comma while there is no comma between the two components of the hierarchical model in distsamp. **And, importantly, the order of the formulas is "lambda, phi, p,"** whereas the order in the double formula in distsamp is **"p, lambda."** Because there are two "state" parameters (λ and ϕ), certain summary functions such as predict, which previously required type=state or type=det, require that you now specify which state parameter to predict (type=lambda, type=phi or type=det).

The abundance distribution (Poisson or negative binomial) is specified by the "mixture" argument. As with the pcount (Chapter 6) and the gmultmix function (see Section 7.5.3), the negative binomial parameterization used in gdistsamp contains the mean, λ, and logarithm of the negative binomial "size" parameter, say $log(\tau)$, with a variance of $\lambda + \lambda^2/\tau$. Therefore, as $1/\tau \rightarrow 0$ or $\tau \rightarrow \infty$, the negative binomial tends to the Poisson (i.e., no overdispersion is indicated). The gdistsamp function relies on the same basic technology as gmultmix. In general it computes the marginal likelihood by summing over possible values of N from $N = 0$ up to some finite value $N = K$. Thus K has to be specified either by the user or it defaults to 100 plus the maximum count at a site. Sensitivity of the estimates to the choice of K, beyond some large number such as this default, may indicate problems with parameter estimability (Couturier et al., 2013; Dennis et al., 2015a). The main arguments to the `gdistsamp` function are defined as follows:

> `lambdaformula`: A right-hand side formula describing the abundance covariates.
> `phiformula`: A right-hand side formula describing the availability covariates.
> `pformula`: A right-hand side formula describing the detection function covariates.
> `data`: An object of class 'unmarkedFrameGDS'.
> `keyfun`: One of the following detection functions: "halfnorm", "hazard", "exp", or "uniform." See details.
> `output`: Model either "density" or "abund"
> `unitsOut`: Units of density. Either "ha" or "kmsq" for hectares and square kilometers, respectively.
> `mixture`: Either "P" or "NB" for the Poisson and negative binomial models of abundance.
> `K`: An integer value specifying the upper bound used in the integration.

In the following section, we apply `distsamp` and `gdistsamp` to the analysis of a distance sampling data set on the island scrub-jay.

8.4.3 EXAMPLE: ESTIMATING THE GLOBAL POPULATION SIZE OF THE ISLAND SCRUB-JAY (ISSJ)

The island scrub-jay (*Aphelocoma insularis*; Figure 8.4) is a species that is endemic to Santa Cruz Island, California (Figures 8.5 and 8.6), and of some conservation interest to the National Park Service (NPS) and other organizations due to the extremely local distribution of the species and previous reports of low and declining population sizes. Our esteemed colleague T. S. Sillett and others initiated an island-wide survey in 2008 to obtain a statistical estimate of population size. The study was reported in Sillett et al. (2012), and we reproduce some of the analyses here. The island scrub-jay data are available in unmarked by typing `data(issj)`.

The data are distance sampling point count data from 307 point count locations (Figure 8.5) with counts made out to 300 m. For analysis, the raw distance data were binned into three 100-m distance classes because nearby birds were responding to the observer (by moving closer, representing responsive movement), so it was believed that the large distance classes should mitigate that affect. The objectives were to (1) estimate the global population size; (2) produce a map of the distribution of the population (i.e., $E(N)$), as a function of local habitat conditions; and (3) make predictions of $E(N)$ under alternative/historical landscapes. Until recently, the island had been heavily grazed by livestock, and an intense removal effort successfully eradicated the livestock causing vegetation to return to historical conditions. But we have the vegetation map for the state of the island under heavy grazing, and so we want make a hypothetical statement about how many jays there may have been.

FIGURE 8.4

A proud island scrub-jay (*Aphelocoma insularis*). *(Photo credit: Melanie Klein.)*

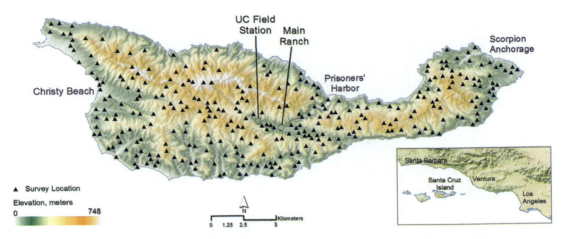

FIGURE 8.5

Santa Cruz Island, Channel Islands, California. The 307 distance sampling point count locations are shown as solid triangles.

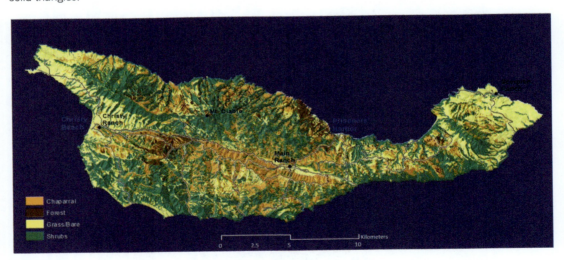

FIGURE 8.6

Current habitat and topography of Santa Cruz island.

To do the analysis in unmarked we load the data and do a few other bookkeeping things such as computing the area of the point count circle to use as an offset so that density in ha is reported, and we build the unmarkedFrameDS.

```
# Load, view and format the ISSJ data
library(unmarked)
data(issj)
```

```
round(head(issj), 2)
  issj[0-100] issj(100-200] issj(200-300]    x         y    elevation forest chaparral
1      0            0             2      234870.1  3767154    51.39    0.02    0.24
2      0            0             0      237083.0  3766804   156.88    0.01    0.47
3      0            0             0      235732.0  3766717   144.81    0.02    0.77
4      0            0             0      237605.0  3766719   184.27    0.26    0.21
5      0            0             0      234239.1  3766570   111.35    0.00    0.00
6      0            0             0      235005.1  3766420   204.13    0.16    0.34

# Package things up into an unmarkedFrame
covs <- issj[,c("elevation", "forest", "chaparral")]
area <- pi*300^2 / 100^2            # Area in ha
jayumf <- unmarkedFrameDS(y=as.matrix(issj[,1:3]),
    siteCovs=data.frame( covs, area),
    dist.breaks=c(0, 100, 200, 300),
    unitsIn="m", survey="point")
```

We note that the island scrub-jay (ISSJ) data comes with site covariates, which are elevation of the point, and cover types forest and chaparral. We input these along with the point count area (constant for all points) using the siteCovs argument to the unmarkedFrameDS. We also specify the distance breaks and the units of distance intervals, and declaring survey="point", that the sample unit is a point count circle (clearly, distances and distance breaks must be in the same units). Now we're ready to fit a few models, which we do like this, first with chaparral as a covariate on both the detection scale σ and on expected abundance λ and also elevation as a covariate on λ, and then, the second model has a constant σ:

```
# Fit model 1
(fm1 <- distsamp(~chaparral ~chaparral + elevation + offset(log(area)),
    jayumf, keyfun="halfnorm", output="abund"))

Call:
distsamp(formula = ~chaparral ~ chaparral + elevation + offset(log(area)),
    data = jayumf, keyfun = "halfnorm", output = "abund")

Abundance:
            Estimate      SE       z    P(>|z|)
(Intercept) -3.50982  0.31261  -11.23  2.99e-29
chaparral    4.11503  0.62458    6.59  4.44e-11
elevation   -0.00216  0.00073   -2.96  3.11e-03

Detection:
                 Estimate     SE      z    P(>|z|)
sigma(Intercept)    5.02   0.161  31.15  5.65e-213
sigmachaparral     -1.07   0.319  -3.36   7.73e-04

AIC: 964.7203

# Fit model 2
(fm2 <- distsamp(~1 ~chaparral + elevation + offset(log(area)),
    jayumf, keyfun="halfnorm", output="abund"))
```

```
Call:
distsamp(formula = ~1 ~ chaparral + elevation + offset(log(area)),
  data = jayumf, keyfun = "halfnorm", output = "abund")

Abundance:
             Estimate       SE        z    P(>|z|)
(Intercept)  -2.71972  0.200946  -13.53  9.77e-42
chaparral     2.12760  0.309172    6.88  5.92e-12
elevation    -0.00212  0.000728   -2.91  3.59e-03

Detection:
Estimate    SE      z    P(>|z|)
    4.58  0.0488  93.9      0

AIC: 976.2306
```

We see the model with chaparral on both σ and λ is favored by a wide margin according to AIC. We check the goodness-of-fit of this model by bootstrapping the `fitstats` function first introduced in Section 7.5.4 (see also Section 6.8).

```
(pb <- parboot(fm1, fitstats, nsim=1000, report=5))
(c.hat <- pb@t0[2] / mean(pb@t.star[,2])) # c-hat as ratio of observed
                            # and mean of expected value of Chi2 (under H0)
                            # (see, e.g., Johnson et al., Biometrics, 2010)

   Chisq
2.590553

residuals(fm1)             # Can inspect residuals
plot(pb)                   # Not shown
print(pb)

Call: parboot(object = fm1, statistic = fitstats, nsim = 1000, report = 5)

Parametric Bootstrap Statistics:
                 t0   mean(t0 - t_B)  StdDev(t0 - t_B)  Pr(t_B > t0)
SSE             421           262.7              16.5             0
Chisq          2357          1447.1              66.1             0
freemanTukey    210            42.9              10.1             0

t_B quantiles:
                0%  2.5%  25%  50%  75%  97.5%  100%
SSE            110   130  147  158  169    193   213
Chisq          739   806  864  903  950   1056  1364
freemanTukey   131   147  161  167  174    187   202

t0 = Original statistic compuated from data
t_B = Vector of bootstrap samples
```

The bootstrap analysis shows that the model does not fit at all, with not a single bootstrap sample falling to the right of the observed value, for any of the three fit statistics. That is, under the Null hypothesis of a fitting model, we don't expect to see any more extreme values of the fit statistics than their value for the observed data set. The "c-hat" statistic (Johnson et al., 2010) indicates a fairly high degree of overdispersion (2.59). This suggests that there is more unexplained variation in the data than allowed for by the distributional assumptions of the model. In the worst case this could mean that the main inference, e.g., regarding covariate effects, is wrong (i.e., that the model is structurally wrong), while in the much less dramatic case it could simply mean that we have unstructured noise, which would make the SEs too small and CIs too narrow. To mitigate that, we go through a more detailed process of model fitting, evaluation, and prediction by expanding the covariate structure of the model. First, however, we will standardize the covariates in the unmarkedFrame because this generally causes the fitting and analysis functions to perform more smoothly (i.e., often it avoids various types of numerical errors or errors due to bad starting values). You can repeat the analysis below without standardizing the covariates to see what happens.

```
# Standardize the covariates
sc <- siteCovs(jayumf)
sc.s <- scale(sc)
sc.s[,"area"] <- pi*300^2 / 10000  # Don't standardize area
siteCovs(jayumf) <- sc.s
summary(jayumf)

unmarkedFrameDS Object

point-transect survey design
Distance class cutpoints (m): 0 100 200 300

307 sites
Maximum number of distance classes per site: 3
Mean number of distance classes per site: 3
Sites with at least one detection: 76

Tabulation of y observations:
  0   1   2   3   4   5   6   9 <NA>
833  53  19   9   1   2   3   1    0

Site-level covariates:
   elevation            forest            chaparral             area
 Min.   :-1.4884   Min.   :-0.49215   Min.   :-1.1562    Min.   :28.27
 1st Qu.:-0.7974   1st Qu.:-0.49215   1st Qu.:-0.8721    1st Qu.:28.27
 Median :-0.1687   Median :-0.44295   Median :-0.2014    Median :28.27
 Mean   : 0.0000   Mean   : 0.00000   Mean   : 0.0000    Mean   :28.27
 3rd Qu.: 0.6650   3rd Qu.:-0.06982   3rd Qu.: 0.6872    3rd Qu.:28.27
 Max.   : 3.5731   Max.   : 5.42362   Max.   : 2.8809    Max.   :28.27

# Fit a bunch of models and produce a model selection table.
fall <- list()   # make a list to store the models

# With the offset output=abund is the same as output = density
fall$Null <- distsamp(~1 ~offset(log(area)), jayumf, output="abund")
fall$Chap. <- distsamp(~1 ~chaparral + offset(log(area)), jayumf,
    output="abund")
```

```
fall$Chap2. <- distsamp(~1 ~chaparral+I(chaparral^2)+offset(log(area)),
    jayumf, output="abund")
fall$Elev. <- distsamp(~1 ~ elevation+offset(log(area)), jayumf,
    output="abund")
fall$Elev2. <- distsamp(~1 ~ elevation+I(elevation^2)+offset(log(area)),
    jayumf, output="abund")
fall$Forest. <- distsamp(~1 ~forest+offset(log(area)), jayumf,
    output="abund")
fall$Forest2. <- distsamp(~1 ~forest+I(forest^2)+offset(log(area)),
    jayumf, output="abund")
fall$.Forest <- distsamp(~forest ~offset(log(area)), jayumf,
    output="abund")
fall$.Chap <- distsamp(~chaparral ~offset(log(area)), jayumf,
    output="abund")
fall$C2E. <- distsamp(~1 ~ chaparral + I(chaparral^2) + elevation +
    offset(log(area)),jayumf, output="abund")
fall$C2F2. <- distsamp(~1 ~chaparral + I(chaparral^2) + forest +
    I(forest^2)+offset(log(area)), jayumf, output="abund")
fall$C2E.F <- distsamp(~forest ~chaparral+I(chaparral^2)+elevation+
    offset(log(area)), jayumf, output="abund")
fall$C2E.C <- distsamp(~chaparral ~chaparral + I(chaparral^2) + elevation +
    offset(log(area)), jayumf, output="abund")

# Create a fitList and a model selection table
(msFall <- modSel(fitList(fits=fall)))

         nPars     AIC    delta   AICwt    cumltvWt
C2E.C        6  951.35     0.00  9.9e-01       0.99
C2E.         5  961.01     9.66  7.9e-03       1.00
C2E.F        6  962.95    11.60  3.0e-03       1.00
Chap2.       4  965.95    14.60  6.7e-04       1.00
C2F2.        6  968.13    16.78  2.2e-04       1.00
Chap.        3  981.39    30.04  3.0e-07       1.00
.Chap        3 1007.02    55.67  8.1e-13       1.00
Forest2.     4 1015.07    63.72  1.4e-14       1.00
Elev2.       4 1017.33    65.98  4.7e-15       1.00
Elev.        3 1018.10    66.75  3.2e-15       1.00
Null_D       2 1018.12    66.77  3.1e-15       1.00
Null         2 1018.12    66.77  3.1e-15       1.00
Forest.      3 1019.65    68.30  1.5e-15       1.00
.Forest      3 1020.08    68.73  1.2e-15       1.00

# Check out the best model
fall$C2E.C

Call:
distsamp(formula = ~chaparral ~ chaparral + I(chaparral^2) +
    elevation + offset(log(area)), data = jayumf, output = "abund")
```

```
Abundance:
                 Estimate    SE        z    P(>|z|)
(Intercept)       -2.562  0.1589   -16.12  1.75e-58
chaparral          1.230  0.1602     7.68  1.64e-14
I(chaparral^2)    -0.282  0.0775    -3.64  2.68e-04
elevation         -0.238  0.0926    -2.57  1.02e-02

Detection:
                 Estimate    SE        z    P(>|z|)
sigma(Intercept)   4.686  0.0682    68.75  0.000000
sigmachaparral    -0.208  0.0626    -3.32  0.000892

AIC: 951.3504

# Check out the goodness-of-fit of this model
(pb.try2 <- parboot(fall$C2E.C, fitstats, nsim=1000, report=5))
Call: parboot(object = fall$C2E.C, statistic = fitstats, nsim = 1000, report = 5)

Parametric Bootstrap Statistics:
                t0   mean(t0 - t_B)   StdDev(t0 - t_B)   Pr(t_B > t0)
SSE            425            267.7              16.13              0
Chisq         2197           1285.9              70.54              0
freemanTukey   207             43.4               9.76              0

t_B quantiles:
               0%   2.5%   25%   50%   75%  97.5%  100%
SSE           112    128   147   156   168    190   233
Chisq         752    794   864   904   946   1057  1419
freemanTukey  134    144   157   164   170    183   195

# Express the magnitude of lack of fit by an overdispersion factor
(c.hat <- pb.try2@t0[2] / mean(pb.try2@t.star[,2]))  #   Chisq
2.411948
```

Once again we see the fit is pretty bad, even considering the more complex covariate structures, and the overdispersion ratio is only negligibly smaller. We could think about trying to improve on this by considering more complex covariate models. However, there may be excess Poisson variation that simply cannot be explained by the available covariates. For example, jays are not uniformly distributed, and there is some amount of aggregation that might be explained by overdispersion (i.e., unstructured additional noise to the Poisson variation). So next, we try fitting a negative binomial model using the gdistsamp function. To do that we have to create a new unmarkedFrame using the unmarkedFrameGDS constructor function, which takes at a minimum one new argument called numPrimary, which is the number of sampling occasions within which it is reasonable to assume a closed population was sampled. In a normal distance sampling survey, we view the sampling as instantaneous and so we specify numPrimary=1. However, if we did a survey of the same points separated in time by days, weeks, or even years, then numPrimary would be the number of such temporal surveys. We don't discuss modeling temporal structure here (see Sections 9.5–9.7 in this and Chapter 14 in volume 2).

```
covs <- issj[,c("elevation", "forest", "chaparral")]
area <- pi*300^2 / 100^2                 # Area in ha
```

```
jayumf <- unmarkedFrameGDS(y=as.matrix(issj[,1:3]),
    siteCovs=data.frame(covs, area), numPrimary=1,
    dist.breaks=c(0, 100, 200, 300),
    unitsIn="m", survey="point")
sc <- siteCovs(jayumf)
sc.s <- scale(sc)
sc.s[,"area"] <- pi*300^2 / 10000  # Don't standardize area
siteCovs(jayumf) <- sc.s
summary(jayumf)

# Fit the model using gdistsamp and look at the fit summary
(nb.C2E.C <- gdistsamp( ~chaparral + I(chaparral^2) + elevation +
    offset(log(area)), ~1, ~chaparral, data =jayumf, output="abund",
    mixture="NB", K = 150))

gdistsamp(lambdaformula = ~chaparral + I(chaparral^2) + elevation +
    offset(log(area)), phiformula = ~1, pformula = ~chaparral,
    data = jayumf, output = "abund", mixture = "NB", K = 150)

Abundance:
                Estimate    SE      z     P(>|z|)
(Intercept)     -2.516  0.198  -12.73  4.17e-37
chaparral        1.432  0.229    6.25  4.01e-10
I(chaparral^2)  -0.376  0.114   -3.28  1.04e-03
elevation       -0.227  0.146   -1.55  1.20e-01

Detection:
                Estimate    SE      z     P(>|z|)
(Intercept)      4.679  0.0658  71.14  0.000000
chaparral       -0.199  0.0600  -3.32  0.000905

Dispersion:                       # Note the NB dispersion parameter
 Estimate    SE     z    P(>|z|)  #  scale is log(tau)
    -1.02  0.215  -4.73  2.23e-06

AIC: 695.4445
```

This produces a long list of warnings of this sort:

```
42: In log(cp[J + 1]) : NaNs produced
```

These are related to having near 0 probability in the very last cell (individuals > 300 m away) and in general are not a problem. The size parameter is $\exp(-1.02) = 0.36$, which, as we noted above, is the τ parameter in the negative binomial distribution. The AIC of `gdistsamp` is not comparable to that of `distsamp`. If you run the same Poisson model using both functions, you get a different AIC! This is because the likelihood construction is completely different. However, we will use this model here to carry out some further analysis. We check the model fit using our parametric bootstrap procedure:

```
(pb.try3 <- parboot(nb.C2E.C, fitstats, nsim=1000, report=5))

Call: parboot(object = nb.C2E.C, statistic = fitstats, nsim = 1000, report = 5)
```

8.4 HIERARCHICAL DISTANCE SAMPLING (HDS)

```
Parametric Bootstrap Statistics:
             t0   mean(t0 - t_B) StdDev(t0 - t_B) Pr(t_B > t0)
SSE          430            97.4             99.7     0.143856
Chisq       2200           868.1            161.2     0.000999
freemanTukey 211            20.8             20.9     0.159840

t_B quantiles:
              0%  2.5%   25%   50%   75%  97.5%  100%
SSE          117   188   265   315   379   576  1133
Chisq        921  1073  1225  1313  1425  1706  2556
freemanTukey 113   150   177   190   205   234   260

There were 50 or more warnings (use warnings() to see the first 50)

(c.hat <- pb.try3@t0[2] / mean(pb.try3@t.star[,2])) #
    Chisq
1.65186
```

This also produces many warnings of the previously mentioned variety. But, on the brighter side, this model does fit in a slightly more satisfactory way according to two out of three of our fit statistics. And, the overdispersion ratio is reduced by nearly 50% and so things appear to be more tolerable for this model. We could also produce predictions that are corrected for overdispersion as we did for the multinomial mixture models (see Section 7.9.4).

Next we use the results to produce an estimate of population size. We first define a function getN, which computes the sum of the predicted values for a given model object, and then we can apply it to any model we wish, and we can also use it as an input to the parboot function to produce uncertainty measures (SEs, CIs). For comparison, we also compute the predictions the old-fashioned way by constructing the model matrix and doing the linear algebra "by hand." In addition, we compute the Best Unbiased Predictor (BUP) of local abundance.

```
# *Expected* population size for the sample points
getN <- function(fm, newdata=NULL)
    sum(predict(fm, type="lambda", newdata=newdata)[,1])

getN(nb.C2E.C)
[1] 889.6142

# This does the same thing as the following commands
X <- model.matrix(~chaparral+I(chaparral^2)+elevation+log(offset(area)),
        siteCovs(jayumf))
head(X) # The design matrix
  (Intercept)  chaparral I(chaparral^2)    elevation log(offset(area))
1           1 -0.1218243     0.01484117  -1.20607849          3.341954
2           1  0.8384709     0.70303345  -0.36132054          3.341954
3           1  2.1319298     4.54512461  -0.45797193          3.341954
4           1 -0.2737078     0.07491594  -0.14196112          3.341954
5           1 -1.1562240     1.33685389  -0.72588125          3.341954
6           1  0.2989173     0.08935153   0.01704522          3.341954
```

```
# Prediction of total expected population size at the sample points
sum(exp(X %*% c(coef(nb.C2E.C, type="lambda"), 1)))
[1] 889.6142

# Empirical Bayes estimates of posterior distribution:
# Pr(N=x|y, lambda, sigma) for x=0,1,...,K
re.jay <- ranef(nb.C2E.C, K = 150)

# *Realized* population size
sum(bup(re.jay, "mean"))
[1] 827.4331
```

So there are about 889 ISSJs on the total area sampled by the 307 point counts, based on the fitted mean of the Poisson model. On the other hand, if we use the best unbiased predictor we have only about 827 ISSJs on the 307 point counts. In general, the two predictions should not be the same because the BUP "adjusts" the predictions toward the data (the observed counts) and so uses some additional information. The BUP is conditional on the particular sample at hand.

Next, we do two further summary analyses of the ISSJ models. First, we produce a graphical display of the effect of chaparral on expected local population size, and then we show a predictive map of expected density over the whole island (Figure 8.7).

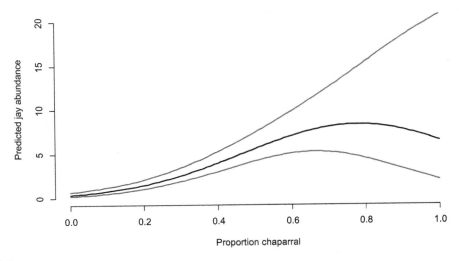

FIGURE 8.7

Response curve of the expected abundance, E(N), of island scrub-jays per 28 ha pixel to the covariate chaparral (with 95% CI limits).

8.4 HIERARCHICAL DISTANCE SAMPLING (HDS)

```
summary(jayumf) # Note the range of chaparral which we need to know
[...output shortened...]

Site-level covariates:
   elevation          forest            chaparral          area
 Min.   :-1.4884  Min.   :-0.49215  Min.   :-1.1562  Min.   :28.27
 1st Qu.:-0.7974  1st Qu.:-0.49215  1st Qu.:-0.8721  1st Qu.:28.27
 Median :-0.1687  Median :-0.44295  Median :-0.2014  Median :28.27
 Mean   : 0.0000  Mean   : 0.00000  Mean   : 0.0000  Mean   :28.27
 3rd Qu.: 0.6650  3rd Qu.:-0.06982  3rd Qu.: 0.6872  3rd Qu.:28.27
 Max.   : 3.5731  Max.   : 5.42362  Max.   : 2.8809  Max.   :28.27

# Create a new data frame with area 28.27 ha, the area of a 300 m circle
chap.orig <- seq(0, 1, 0.01)    # Values from 0 to 1 prop. chaparral
chap.pred <- (chap.orig - mean(issj$chaparral)) / sd(issj$chaparral)
newdat <- data.frame(chaparral = chap.pred, elevation = 0, area=28.27)

# Expected values of N for covariate values in "newdat"
E.N <- predict(fall$C2E.C, type="state", newdata=newdat, appendData=TRUE)
head(E.N)
  Predicted        SE     lower     upper chaparral elevation  area
1 0.3606945 0.1108475 0.1974927 0.6587612 -1.1562240         0 28.27
2 0.3907380 0.1163551 0.2179779 0.7004205 -1.1134578         0 28.27
3 0.4228468 0.1220244 0.2401844 0.7444256 -1.0706916         0 28.27
4 0.4571217 0.1278586 0.2642091 0.7908898 -1.0279254         0 28.27
5 0.4936646 0.1338616 0.2901490 0.8399297 -0.9851591         0 28.27
6 0.5325784 0.1400385 0.3181008 0.8916661 -0.9423929         0 28.27

# Make a plot of the response curve for the grid of chaparral values
plot(chap.orig, E.N[,"Predicted"], xlab="Proportion chaparral", ylab="Predicted jay
 abundance", type="l", ylim = c(0, 20), frame = F, lwd = 2)
matlines(chap.orig, E.N[,3:4], lty = 1, col = "grey", lwd = 1)
```

Finally, now we take the habitat map for the whole island (Figure 8.6) and we predict the expected abundance, $E(N)$, on every pixel of the map. These pixels are 9 ha pixels instead of 28 ha sample units, and so we have to account for that area change. In addition, because models were fitted with standardized covariates, we need to appropriately standardize the landscape variables by exactly the same mean and SD used for the data in the analysis. To do that we first look at the attributes of the scaled site covariates, for which we computed several pages of R code previously:

```
attributes(sc.s) # means are "scaled:center". SDs are "scaled:scale"
$dim
[1] 307   4

$dimnames
$dimnames[[1]]
NULL
```

```
$dimnames[[2]]
[1] "elevation" "forest"   "chaparral" "area"

$'scaled:center'
  elevation      forest  chaparral        area
202.0023616  0.0673357  0.2703592  28.2743339

$'scaled:scale'
  elevation      forest  chaparral        area
124.8818069  0.1368199  0.2338295   0.0000000
```

And now we can apply these values of the mean and SD to the grid variables and then predict for each pixel of the Santa Cruz landscape.

```
cruz.s <- cruz    # Created a new data set for the scaled variables
cruz.s$elevation <- (cruz$elevation*0.3048-202)/125
cruz.s$chaparral <- (cruz$chaparral-0.270)/0.234
cruz.s$area <- (300*300)/10000 # The grid cells are 300x300m=9ha
EN <- predict(nb.C2E.C, type="lambda", newdata=cruz.s)

# Total population size (by summing predictions for all pixels)
getN(nb.C2E.C, newdata=cruz.s)
[1] 2282.039

# Parametric bootstrap for CI
# A much faster function could be written to doing the sum
set.seed(2015)
(EN.B <- parboot(nb.C2E.C, stat=getN, nsim=1000, report=5))

Call: parboot(object = nb.C2E.C, statistic = getN, nsim = 1000, report = 5)

Parametric Bootstrap Statistics:
    t0  mean(t0 - t_B)  StdDev(t0 - t_B)  Pr(t_B > t0)
1  890        -13.1              160           0.535

t_B quantiles:
     0%  2.5%  25%  50%  75%  97.5%  100%
t*1  481  607  788  903  1004  1237  1425

t0 = Original statistic compuated from data
t_B = Vector of bootstrap samples
```

So we have a population size estimate and a 95% confidence interval, and now let's make a map of the predictions (i.e., create a species distribution map in terms of the expected abundance). To do this we use the `raster` package to create a raster stack using the land cover variables (standardized) that are provided when the ISSJ data are loaded. Then we use the `predict` function with this raster stack.

```
library(raster)
cruz.raster <- stack(rasterFromXYZ(cruz.s[,c("x","y","elevation")]),
    rasterFromXYZ(cruz.s[,c("x","y","chaparral")]),
    rasterFromXYZ(cruz.s[,c("x","y","area")]))
```

8.4 HIERARCHICAL DISTANCE SAMPLING (HDS)

```
names(cruz.raster) # These should match the names in the formula
[1] "elevation" "chaparral" "area"
plot(cruz.raster)            # not shown
# Elevation map on the original scale (not shown)
plot(cruz.raster[["elevation"]]*125 + 202, col=topo.colors(20),
main="Elevation (in feet) and Survey Locations", asp = 1)
points(issj[,c("x","y")], cex=0.8, pch = 16)
```

The `predict` function will use a raster stack having the appropriate covariates and produce raster output of predictions, SEs, and lower and upper confidence limits (Figure 8.8). Where else can you get such goodness from one function call? Here we show the Holy Grail of population ecology: a spatial

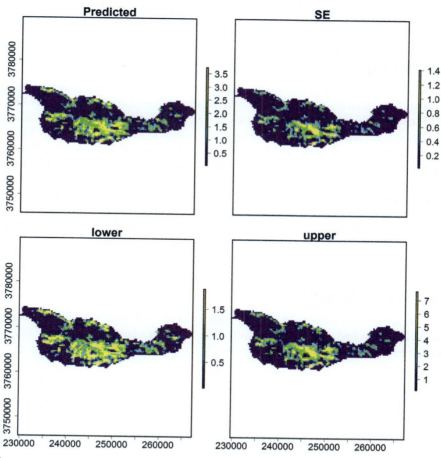

FIGURE 8.8

Global species distribution map of the island scrub-jay (*Aphelocoma insularis*), an endemic on the island of Santa Cruz, California, based on the best model in the model set. Predictions show the expected abundance (λ_s) under the AIC-best negative binomial model for every pixel. Three maps are shown to depict uncertainty in these predictions: the prediction SE and the lower and upper limit of a 95% prediction interval.

map depicting global population size predictions from the distance sampling model applied at the landscape scale:

```
EN.raster <- predict(nb.C2E.C, type="lambda", newdata=cruz.raster)
   doing row 1000 of 5625
   doing row 2000 of 5625
   doing row 3000 of 5625
   doing row 4000 of 5625
   doing row 5000 of 5625
plot(EN.raster, col = topo.colors(20), asp = 1)   # See Figure 8.8
```

While we conclude our analysis here merely by showing an estimate of the global population distribution of this important species, we note that the ultimate objective of this analysis was to use the model to make predictions of population size and distribution using the 1985 land cover of the island (pre-sheep cull; see Sillett et al., 2012, for more detail).

8.5 BAYESIAN HDS

Bayesian hierarchical distance sampling can be implemented in a number of different ways (following our developments of Section 8.3). There are two basic formulations that we demonstrate here: (1) The conditional (three-part) formulation of the model using either continuous or discrete data, which is similar to that which we outlined for the multinomial mixture model in Section 8.3. This three-part formulation of the model is similar to Chelgren et al. (2011b) and Shirk et al. (2014) and also similar to Hedley and Buckland (2004; although, they didn't do a joint estimation of the parameters from the different model components); and (2) formulation of the model for either discrete or continuous data using data augmentation. We should note before getting into the details that for some problems it might be perfectly reasonable to just pool all of the data and analyze one big data set having a single parameter N, the population size among all sampled populations. This may be reasonable to do if estimating overall abundance or mean density was the primary objective and the investigation of patterns in the variation among sample units was not important.

8.5.1 SIMULATING HDS DATA

We start by developing some familiarity with the data structure and processing by defining a function for simulating HDS data and fitting models to it. The function simHDS (with its default arguments shown) is called as follows:

```
simHDS(type="line", nsites = 100, mean.lambda = 2,
   beta.lam = 1, mean.sigma = 1, beta.sig = -0.5, B = 3, discard0=TRUE)
```

The function arguments mean the following:

- type lets you choose between either a line (type = "line") or a point (type = "point") transect protocol.
- nsites is the number of sites
- alpha.lam (= log(mean.lambda)) and beta.lam are the intercept and the slope of a log-linear regression of expected abundance per site on a habitat covariate
- alpha.sig (= log(mean.sigma)) and beta.sig are the intercept and the slope of a log-linear regression of scale parameter σ of the half-normal detection function on wind speed

- `B` is the strip half width
- `discard0=TRUE` subsets to sites at which >0 individuals were captured. You may or may not want to do this depending on how the model is formulated, so be careful.

Calling the function produces a visualization of the generated data set (see Figure 8.9 for type = "line" and Figure 8.10 for type = "point").

By default we simulate line transect data for 100 sites, with abundance N_s for transect s having a Poisson distribution with a mean that depends on some simulated site covariate, "habitat." We also incorporate an effect of another site-specific covariate, wind speed, which we assume affects the observation model via the detection function (specifically parameter σ; Marques et al., 2007). We now execute the function to obtain a point or a line transect data set (with default arguments).

```
set.seed(1234)
tmp1 <- simHDS("point")      # Point transect
tmp2 <- simHDS()             # Line transect (this is the default)
str(tmp1)                    # Look at function output
List of 14
 $ type        : chr "point"
 $ nsites      : num 100
 $ mean.lambda : num 2
 $ beta.lam    : num 1
 $ mean.sigma  : num 1
 $ beta.sig    : num -0.5
 $ B           : num 3
 $ data        : num [1:76, 1:5] 2 3 6 13 21 22 24 29 31 31 ...
  ..- attr(*, "dimnames")=List of 2
  .. ..$ : NULL
  .. ..$ : chr [1:5] "" "y" "u" "v" ...
 $ B           : num 3
 $ nsites      : num 100
 $ habitat     : num [1:100] -1.207 0.277 1.084 -2.346 0.429 ...
 $ wind        : num [1:100] 0.643 0.113 -0.73 1.071 0.105 ...
 $ N           : int [1:100] 0 6 6 0 6 3 0 4 1 1 ...
 $ N.true      : int [1:100] 0 5 5 0 3 2 0 3 1 1 ...
```

Note that `N.true` is the number of animals with distance $\leq$ B, so for a line transect, `N = N.true`, while for a point transect, `N.true` $\leq$ `N` because we are simulating on a square.

Now we have a nice set of distance sampling data collected at 100 sites and with two site-specific covariates called `habitat` (affecting local abundance) and `wind` (affecting the distance out to which individuals are detected).

As always, it is extremely useful to play around with data simulation functions with changed arguments to train your intuition about a certain modeled process and also about the statistical model we use to make an inference about the parameters in this process; see Exercise 7.

8.5.2 BAYESIAN HDS USING DATA AUGMENTATION

We have discussed data augmentation (Royle et al., 2007a) several times in previous chapters, and we analyzed the distance sampling model using the Impala data with DA in Section 8.3 for both binned and continuous distance measurements. Here we apply these ideas to HDS models. In general, every capture-recapture model can be analyzed using data augmentation. And, distance sampling can be regarded as

446 CHAPTER 8 MODELING ABUNDANCE USING HIERARCHICAL DISTANCE SAMPLING

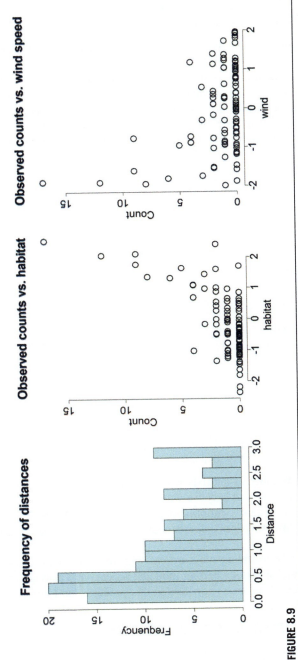

FIGURE 8.9

Visualization produced when the function simHDS is run for line transects. Histogram shows *observed* distances.

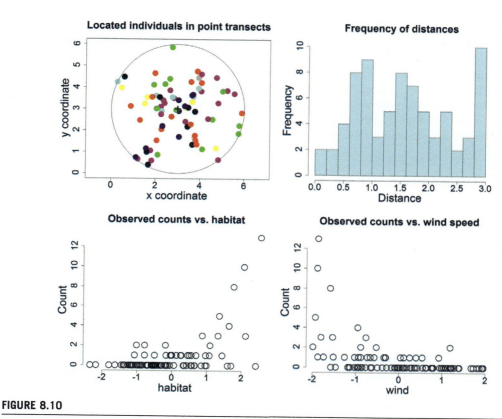

FIGURE 8.10

Visualization produced when the function simHDS is run for point transects. Colors in the top-left panel denote different sites. Histogram shows *observed* distances.

just another capture-recapture model with an individual covariate "distance," which affects p (also without the "recapture"). To apply DA to distance sampling, we augment a data set of captured individuals ($y = 1$) with a large number of uncaptured individuals (i.e., $y = 0$), and we recognize that the resulting "augmented data set" is a zero-inflated version of the known-N data set (similar to the relationship between an occupancy model and a logistic regression of detections at sites that are known to be occupied). That is, some of the added zeros are sampling zeros and some of them are structural zeros (fixed zeros that are not missed individuals). We can express this zero-inflated Bernoulli model directly in BUGS. It essentially recasts the capture-recapture (here distance sampling) model as a site occupancy model (see Chapter 10). In distance sampling, the only nuance is that p_i depends on distance and the distance "data" must be input as missing values for the augmented individuals, i.e., we have a site-occupancy model with a partially missing, site-level covariate, which are estimated as part of the model.

Now for site-structured data, i.e., HDS, where we have distance sampling data from a number S of sites, we have to consider how to get the multisite structure integrated into this DA formulation of the model. Such a framework for accomplishing this was described in Converse and Royle (2012), Royle et al. (2012), and Royle and Converse (2014). The main idea is simply to pool the data into one big data set having rows $i = 1, 2, \ldots, M$, which includes the n observed individuals (from among all sites) and a large number, $M - n$, of unobserved individuals. In addition, we add an additional individual covariate, which is the site membership of each individual, say $site_i$. The observed site membership of individuals is treated as a *categorical individual covariate*, which, for the augmented individuals, is missing data that can be estimated. In this formulation of the model we have both sites and individuals (as in Section 7.8.4) and so we need to be careful with our indexing, which will be slightly different from other places in the book. As in Chapter 7, we will use the index i for individual and $s = 1, 2, \ldots, S$ for sites. We require a prior distribution for the categorical individual covariate, which we specify as follows in the BUGS language:

```
site[i] ~ dcat( site.probs[]),
```

where `site.probs` is a vector of length S defined as

$$\text{site.probs}[s] = \frac{\lambda_s}{\sum_s \lambda_s}.$$

Here, λ_s is the expected abundance at site s, which may of course depend on site-specific covariates. This model is *implied* by the assumption that

$$N_s \sim \text{Poisson}(\lambda_s)$$

with

$$\log(\lambda_s) = \beta_0 + \beta_1 x_s.$$

The data augmentation part of the model includes a set of latent variables z_i which are Bernoulli trials taking on the value $z_i = 1$ when an *individual* observation corresponds to a real individual and $z_i = 0$ when it corresponds to a structural zero. As before, we assume $z_i \sim \text{Bernoulli}(\psi)$, where ψ is the data augmentation parameter. **One extremely important caveat** is that in this use of DA for such site-structured or stratified models, the intercept parameter β_0 of the abundance model is confounded with the DA parameter ψ (Royle and Converse 2014). They are equivalent parameters, and the model must be fitted by imposing a constraint, either by setting $\beta_0 = 0$ or setting $\psi = \sum_{s=1}^{S} \lambda_s / M$, where M is the total number of individuals in the augmented data set (see Royle et al., 2014, p. 314).

To summarize, we can do hierarchical distance sampling by: (1) including individual "site membership" as a categorical covariate; (2) specifying the site membership covariate as a categorical random variable with cell probabilities proportional to the Poisson mean parameter λ_s; and (3) specifying covariate effects directly on λ_s as in our usual binomial and multinomial N-mixture models.

For illustration, we also show how covariates can be introduced into the detection model, which we typically do by specifying a linear model for $\log(\sigma)$. We assume that we had measured a covariate wind speed, which affects detection probability via its effect on the *detection function*. Having measured wind speed at the time of our survey at every site, it is a site covariate (if we had temporal replicate measurements at a site, it would be an *observational covariate*). Thus, the observation part of our HDS model for every individual i in the augmented data set is this (note that sites are now indexed s and the scale parameter σ varies by site according to the log-linear model on wind speed):

$$y_i \sim \text{Bernoulli}(p_i)$$
$$p_i = \exp\left(- d_i * d_i \big/ \left(2 * \sigma_{s(i)}^2\right)\right)$$
$$\log(\sigma_s) = \alpha_0 + \alpha_1 * wind_s$$

Note that the first line specifies a relationship for individuals, the second one between individuals and an individual-specific covariate (d_i) and a site-specific parameter σ_s, and the third line a relationship purely at the site level. In line two, we emphasize the relationship between individual and site by use of the double subscript in $\sigma_{s(i)}$, which specifies membership of individual i to site s.

We now demonstrate using our simulated data from 100 line transect surveys.

```r
# Recreate line transect data set
set.seed(1234)
tmp <- simHDS()            # Line transect (default)
attach(tmp)

# Data augmentation: add a bunch of "pseudo-individuals"
nz <- 500                              # Augment by 500
nind <- nrow(data)
y <- c(data[,2], rep(0, nz))           # Augmented detection indicator y
site <- c(data[,1], rep(NA, nz))       # Augmented site indicator,
                                       #   unknown (i.e., NA) for augmented inds.
d <- c(data[,5], rep(NA,nz))           # Augmented distance data (with NAs)

# Bundle and summarize data set
str( win.data <- list(nsites=nsites, habitat=habitat, wind=wind, B=B, nind=nind, nz=nz,
y=y, d=d, site=site) )
win.data$site                          # Unknown site cov. for augmented inds.

# BUGS model for line transect HDS (NOT point transects!)
cat("
model{
# Prior distributions
beta0 ~ dunif(-10,10)      # Intercept of lambda-habitat regression
beta1 ~ dunif(-10,10)      # Slope of log(lambda) on habitat
alpha0 ~ dunif(-10,10)     # Intercept of log(sigma) (half-normal scale)
alpha1 ~ dunif(-10,10)     # Slope of log(sigma) on wind

# psi is a derived parameter under DA for stratified populations
psi <- sum(lambda[]) / (nind+nz)

# 'Likelihood' (sort of...)
for(i in 1:(nind+nz)){                 # i is index for individuals
  z[i] ~ dbern(psi)                    # Data augmentation variables
  d[i] ~ dunif(0, B)                   # Distance uniformly distributed
  p[i] <- exp(-d[i]*d[i]/(2*sigma[site[i]]*sigma[site[i]])) # Det. function
  mu[i] <- z[i]* p[i]                  # 'straw man' for WinBUGS
  y[i] ~ dbern(mu[i])                  # Bernoulli random variable
  site[i] ~ dcat(site.probs[1:nsites]) # Population distribution among sites
}

# Linear models for abundance and for detection
for(s in 1:nsites){                    # s is index for sites
  # Model for abundance
  # next line not necessary, but allows to make predictions
  N[s] ~ dpois(lambda[s])              # Realized abundance at site s
  log(lambda[s]) <- beta0 + beta1*habitat[s] # Linear model abundance
  site.probs[s] <- lambda[s] / sum(lambda[])

 # Linear model for detection
  log(sigma[s]) <- alpha0 + alpha1*wind[s]
}
```

```
# Derived parameters: total population size and average density across all sites
Ntotal <- sum(z[])
area <- nsites*1*2*B                    # Unit length == 1, half-width = B
D <- Ntotal/area
}
",fill=TRUE , file = "model1.txt")

# Inits
zst <- c(rep(1, sum(y)), rep(0, nz)) # ... and for DA variables
inits <- function(){list(beta0=0, beta1=0, alpha0=0, alpha1=0, z=zst)}

# Parameters to save
params <- c("alpha0", "alpha1", "beta0", "beta1", "psi", "Ntotal", "D")

# MCMC settings
ni <- 12000 ; nb <- 2000 ; nt <- 2 ; nc <- 3

# Call BUGS (ART 33 min) ...
bd <- "c:/Program Files/WinBUGS14/" # Never forget this for WinBUGS
out1 <- bugs(win.data, inits, params, "model1.txt", n.thin=nt,
  n.chains=nc, n.burnin=nb, n.iter=ni, debug=TRUE, bugs.dir = bd)

# ... or try JAGS for a change (ART 6 min)
library(jagsUI)     # never forget to load jagsUI
out1 <- jags(win.data, inits, params, "model1.txt", n.thin=nt,
    n.chains=nc, n.burnin=nb, n.iter=ni)
```

We note that JAGS completes the analysis about five times faster than WinBUGS (and would be faster still with argument parallel = TRUE). This is a good example of why testing things out in both BUGS engines can be helpful and efficient (you might also want to try out OpenBUGS).

```
# Summarize posterior output
print(out1, 2)
          mean    sd    2.5%    50%    97.5% overlap0    f  Rhat n.eff
alpha0    0.02  0.11   -0.17   0.02    0.28  TRUE   0.56   1 14232
alpha1   -0.66  0.10   -0.88  -0.65   -0.48  FALSE  1.00   1  4213
beta0     0.58  0.15    0.29   0.58    0.85  FALSE  1.00   1  1818
beta1     0.94  0.08    0.79   0.94    1.10  FALSE  1.00   1  1077
psi       0.41  0.04    0.33   0.40    0.50  FALSE  1.00   1  9092
Ntotal  259.11 24.40  215.00 258.00  310.00  FALSE  1.00   1 15000
D         0.43  0.04    0.36   0.43    0.52  FALSE  1.00   1 15000

# Truth in data simulation (note alpha0 and beta0 are log transformed)
 $ mean.lambda: num 2            # exp(beta0) above
 $ beta.lam   : num 1            # beta1
 $ mean.sigma : num 1            # exp(alpha0)
 $ beta.sig   : num -0.5         # alpha1
```

The posterior means of all parameters are not too far from their data-generating values. In addition, the posterior mean of Ntotal is 259.11 (CRI 215–310), which agrees quite well with the true total population size obtained by summing N_s over all S sites (note: only 136 individuals were detected).

```
sum(tmp$N.true)
[1] 305
```

Next, we consider the same analysis, but as if we had binned distance data. For this model we use the categorical distribution in BUGS. While there is nothing technically novel here, note that when we have the covariate on σ (wind) we must define a large matrix of site and distance-class specific detection probabilities, and this will slow things down substantially. This takes a few extra lines of code to make use of categorical data. We use the same simulated data set from earlier in this Section (which you may have to re-create in case you wrote over things in your R workspace). We first convert the distance data into categorical distance classes as we've done before, and then write out the BUGS model into a file and set things up for a BUGS run.

```
# Prepare data
delta <- 0.1                        # width of distance bins for approx.
midpt <- seq(delta/2, B, delta)     # make mid-points and chop up data
dclass <- d %/% delta + 1           # convert distances to cat. distances
nD <- length(midpt)                 # Number of distance intervals

# Bundle and summarize data set
str( win.data <- list (y=y, dclass=dclass, site=site, midpt=midpt, delta=delta, B=B,
nind=nind, nz=nz, nsites=nsites, nD=nD, habitat=habitat, wind=wind) )

# BUGS model specification for line-transect HDS (NOT point transects!)
cat("
model{
# Prior distributions
alpha0 ~ dunif(-10,10)
alpha1 ~ dunif(-10,10)
beta0  ~ dunif(-10,10)
beta1  ~ dunif(-10,10)

psi <- sum(lambda[])/(nind+nz)      # psi is a derived parameter

for(i in 1:(nind+nz)){               # Loop over individuals
  z[i] ~ dbern(psi)                  # DA variables
  dclass[i] ~ dcat(pi[site[i],])     # Population distribution of dist class
  mu[i] <- z[i] * p[site[i],dclass[i]]  # p depends on site AND dist class
  y[i] ~ dbern(mu[i])                # Basic Bernoulli response in DS model
  site[i] ~ dcat(site.probs[1:nsites])  # Site membership of inds
}

for(s in 1:nsites){                  # Loop over sites
# Construct cell probabilities for nD cells
for(g in 1:nD){                      # midpt = mid point of each cell
  log(p[s,g]) <- -midpt[g]*midpt[g]/(2*sigma[s]*sigma[s])
  pi[s,g] <- delta/B                 # Probability of x per interval
  f[s,g] <- p[s,g]*pi[s,g]           # pdf of observed distances
}

# not necessary  N[s]~dpois(lambda[s]) except for prediction
N[s] ~ dpois(lambda[s])              # Predict abundance at each site
```

```
   log(lambda[s]) <- beta0 + beta1 * habitat[s] # Linear model for N
   site.probs[s] <- lambda[s]/sum(lambda[])
   log(sigma[s]) <- alpha0 + alpha1*wind[s] # Linear model for sigma
}

# Derived parameters: total abundance and mean density across all sites
Ntotal <- sum(z[])       # Also sum(N[]) which is size of a new population
area <- nsites*1*2*B     # Unit length == 1, half-width = B
D <- Ntotal/area
}
",fill=TRUE, file = "model2.txt")

# Inits
zst <- c(rep(1, sum(y)), rep(0, nz))
inits <- function(){list (alpha0=0, alpha1=0, beta0=0, beta1=0, z=zst) }

# Params to save
params <- c("alpha0", "alpha1", "beta0", "beta1", "psi", "Ntotal","D")

# MCMC settings
ni <- 12000  ;  nb <- 2000  ;  nt <- 2  ;  nc <- 3

# Run JAGS with parallel processing (ART 1 min)
library(jagsUI)
out2 <- jags(win.data, inits, params, "model2.txt", n.thin=nt,
   n.chains=nc, n.burnin=nb, n.iter=ni, parallel = FALSE)
```

We summarize the posterior samples for each parameter (output truncated), the means of which, as before, are similar to the data-generating values.

```
print(out2,2)
          mean     sd    2.5%    50%    97.5%  overlap0    f  Rhat  n.eff
alpha0    0.02   0.11   -0.17   0.01    0.26     TRUE   0.55  1.00    838
alpha1   -0.66   0.10   -0.88  -0.65   -0.48    FALSE   1.00  1.00   6469
beta0     0.57   0.14    0.29   0.58    0.85    FALSE   1.00  1.01    266
beta1     0.94   0.08    0.79   0.94    1.11    FALSE   1.00  1.01    224
psi       0.41   0.04    0.33   0.40    0.49    FALSE   1.00  1.00    959
Ntotal  258.89  23.45  217.00 258.00  309.00    FALSE   1.00  1.00    834
D         0.43   0.04    0.36   0.43    0.52    FALSE   1.00  1.00    834
```

The results are, not surprisingly, numerically very similar to the analysis based on data augmentation of the previous section.

8.5.3 BAYESIAN HDS USING THE THREE-PART CONDITIONAL MULTINOMIAL MODEL

It is possible to specify HDS models in BUGS without using data augmentation. In Section 7.6 we discussed a specific formulation of multinomial models for implementation in BUGS to get around our inability to specify a random variable as a multinomial index. We will use that same formulation here. The basic idea is that we deconstruct the multinomial observation model by first conditioning on n_s for each site *s*, so that instead of having a multinomial/Poisson mixture model, we have the three-part multinomial/binomial/Poisson mixture model as follows:

$$y_s | n_s \sim Multinomial(n_s, \pi_s^c) \tag{1}$$

where $\pi_k^c = \pi_k/(1 - \pi_0)$, the index k here representing the kth element of the vector π_s^c,

$$n_s | N_s \sim Binomial(N_s, 1 - \pi_0) \tag{2}$$

$$N_s \sim Poisson(\lambda_s) \tag{3}$$

The first component is the model for distance class of the observed n_s individuals, the second describes imperfect detection of the N_s individuals leading to count n_s, and the third is our usual model for spatial variation in local abundance N_s. The key thing is that the multinomial of the first component has index n_s, which is observed, and so we're conditioning on observed data, not on the latent variable N_s. This three-part hierarchical model is easily implemented in BUGS. Chelgren et al. (2011b) may have been the first to do this, but they also formulated the model for continuous distance point count data using the "zeros trick" in BUGS. We prefer to use a fine binning if a nearly-continuous model is desired. In practice, we also usually specify the first stage in BUGS using a categorical observation model for individual observations, instead of the multinomial model for the distance class frequencies (as described above). This is a much more versatile formulation, which allows considerable flexibility to expand the model (see Chapter 9).

We simulate another data set using the same function as before, but this time *do not discard* the data from the sites where no animals were detected (discard0=F) so that those sites are carried as observed zeros in the data set and we can analyze the zero-filled data.

```
# Simulate line transect data set
set.seed(1234)
tmp <- simHDS(type="line", discard0=FALSE)
attach(tmp)

# Get number of individuals detected per site
# ncap = 1 plus number of detected individuals per site
ncap <- table(data[,1])                    # ncap = 1 if no individuals captured
sites0 <- data[is.na(data[,2]),][,1]       # sites where nothing detected
ncap[as.character(sites0)] <- 0            # Fill in 0 for sites with no detections
ncap <- as.vector(ncap)

# Prepare other data
site <- data[!is.na(data[,2]),1]           # site ID of each observation
delta <- 0.1                               # distance bin width for rect. approx.
midpt <- seq(delta/2, B, delta)            # make mid-points and chop up data
dclass <- data[,5] %/% delta + 1           # convert distances to cat. distances
nD <- length(midpt)                        # Number of distance intervals
dclass <- dclass[!is.na(data[,2])]         # Observed categorical observations
nind <- length(dclass)                     # Total number of individuals detected

# Bundle and summarize data set
str( win.data <- list(nsites=nsites, nind=nind, B=B, nD=nD, midpt=midpt, delta=delta,
  ncap=ncap, habitat=habitat, wind=wind, dclass=dclass, site=site) )
```

```
# BUGS model specification for line-transect HDS (NOT point transects!)
cat("
model{
# Priors
alpha0 ~ dunif(-10,10)
alpha1 ~ dunif(-10,10)
beta0 ~ dunif(-10,10)
beta1 ~ dunif(-10,10)

for(i in 1:nind){
   dclass[i] ~ dcat(fc[site[i],])  # Part 1 of HM
}

for(s in 1:nsites){
# Construct cell probabilities for nD multinomial cells
  for(g in 1:nD){                    # midpt = mid-point of each cell
    log(p[s,g]) <- -midpt[g] * midpt[g] / (2*sigma[s]*sigma[s])
    pi[s,g] <- delta / B             # Probability per interval
    f[s,g] <- p[s,g] * pi[s,g]
    fc[s,g] <- f[s,g] / pcap[s]
  }
  pcap[s] <- sum(f[s,])              # Pr(capture): sum of rectangular areas
  ncap[s] ~ dbin(pcap[s], N[s])      # Part 2 of HM
  N[s] ~ dpois(lambda[s])            # Part 3 of HM
  log(lambda[s]) <- beta0 + beta1 * habitat[s]  # Linear model abundance
  log(sigma[s])<- alpha0 + alpha1*wind[s]       # Linear model detection
}
# Derived parameters
Ntotal <- sum(N[])
area <- nsites*1*2*B   # Unit length == 1, half-width = B
D <- Ntotal/area
}
",fill=TRUE, file = "model3.txt")

# Inits
Nst <- ncap + 1
inits <- function(){list(alpha0=0, alpha1=0, beta0=0, beta1=0, N=Nst)}

# Params to save
params <- c("alpha0", "alpha1", "beta0", "beta1", "Ntotal","D")

# MCMC settings
ni <- 12000   ;   nb <- 2000   ;   nt <- 1   ;   nc <- 3

# Run JAGS (ART 1 min) and summarize posteriors
library(jagsUI)
out3 <- jags(win.data, inits, params, "model3.txt", n.thin=nt,
    n.chains=nc, n.burnin=nb, n.iter=ni)
print(out3, 2)
```

	mean	sd	2.5%	50%	97.5%	overlap0	f	Rhat	n.eff
alpha0	0.00	0.11	-0.19	-0.01	0.23	TRUE	0.54	1	11116
alpha1	-0.66	0.10	-0.87	-0.66	-0.49	FALSE	1.00	1	4359
beta0	0.64	0.14	0.35	0.64	0.90	FALSE	1.00	1	4330
beta1	0.91	0.08	0.76	0.91	1.07	FALSE	1.00	1	10802
Ntotal	266.98	24.52	223.00	266.00	318.00	FALSE	1.00	1	3085
D	0.44	0.04	0.37	0.44	0.53	FALSE	1.00	1	3085

This is not too bad (true $N = 305$) and the chains seem to be mixing well, so we're satisfied that this formulation of the model is viable. We could work up the three-part model for continuous distance data, but, as in Chelgren et al. (2011b) and Shirk et al. (2014), we would have to use the ones or zeros trick to implement this (but see Section 9.8.1 for an alternative formulation that avoids having to use the ones or zeros trick). However, there would be no practical difference between doing that and just using a large number of relatively narrow distance intervals (remember our treatise on integration in Section 8.3.2).

8.5.4 POINT TRANSECT HDS USING THE CONDITIONAL MULTINOMIAL FORMULATION

We now show how to apply the three-part conditional multinomial model to point transect data. As before, we simulate a data set but now keep the nondetection sites in the data set so that when we process the data we have a record of sites with $n = 0$ (where no individual was detected), and be sure to specify `type = "point"`.

```
# Simulate a point count data set using our simHDS function
set.seed(1234)
tmp <- simHDS(type="point", discard0=FALSE)
attach(tmp)

# Prepare data
# Number of individuals detected per site
ncap <- table(data[,1])                    # ncap = 1 if no individuals captured
sites0 <- data[is.na(data[,2]),][,1]       # Sites where nothing was seen
ncap[as.character(sites0)] <- 0            # Fill in 0 for sites with no detections
ncap <- as.vector(ncap)                    # Number of individuals detected per site

# Other data
site <- data[!is.na(data[,2]),1]           # Site ID of each observation
delta <- 0.1                               # Distance bin width for rect. approx.
midpt <- seq(delta/2, B, delta)            # Make mid-points and chop up data
dclass <- data[,5] %/% delta + 1           # Convert distance to distance category
nD <- length(midpt)                        # Number of distance intervals
dclass <- dclass[!is.na(data[,2])]         # Observed categorical observations
nind <- length(dclass)                     # Total number of individuals detected

# Bundle and summarize data set
str( win.data <- list(nsites=nsites, nind=nind, B=B, nD=nD, midpt=midpt,
  delta=delta, ncap=ncap, habitat=habitat, wind=wind, dclass=dclass,site=site) )
```

```
# BUGS model specification for point transect data
cat("
model{
# Priors
alpha0 ~ dunif(-10,10)
alpha1 ~ dunif(-10,10)
beta0 ~ dunif(-10,10)
beta1 ~ dunif(-10,10)

for(i in 1:nind){
   dclass[i] ~ dcat(fc[site[i],])  # Part 1 of HM
}
for(s in 1:nsites){
  # Construct cell probabilities for nD distance bands
  for(g in 1:nD){                   # midpt = mid-point of each band
    log(p[s,g]) <- -midpt[g] * midpt[g] / (2 * sigma[s] * sigma[s])
    pi[s,g] <- ((2 * midpt[g] ) / (B * B)) * delta # prob. per interval
    f[s,g] <- p[s,g] * pi[s,g]
    fc[s,g] <- f[s,g] / pcap[s]
  }
  pcap[s] <- sum(f[s,])              # Pr(capture): sum of rectangular areas

  ncap[s] ~ dbin(pcap[s], N[s])     # Part 2 of HM
  N[s] ~ dpois(lambda[s])            # Part 3 of HM
  log(lambda[s]) <- beta0 + beta1 * habitat[s] # Linear model abundance
  log(sigma[s]) <- alpha0 + alpha1*wind[s]     # Linear model detection
}

# Derived parameters
Ntotal <- sum(N[])
area <- nsites*3.141*B*B
D <- Ntotal/area
}
",fill=TRUE, file="model4.txt")

# Inits
Nst <- ncap + 1
inits <- function(){list(alpha0=0, alpha1=0, beta0=0, beta1=0, N=Nst)}

# Params to save
params <- c("alpha0", "alpha1", "beta0", "beta1", "Ntotal","D")

# MCMC settings
ni <- 12000 ; nb <- 2000 ; nt <- 1 ; nc <- 3

# Run BUGS (ART 2.3 min) and summarize posteriors
out4 <- bugs(win.data, inits, params, "model4.txt", n.thin=nt,
   n.chains=nc, n.burnin=nb, n.iter=ni, debug=TRUE, bugs.dir = bd)

print(out4, 2)
Inference for Bugs model at "model4.txt", fit using WinBUGS,
 3 chains, each with 12000 iterations (first 2000 discarded)
 n.sims = 30000 iterations saved
```

	mean	sd	2.5%	25%	50%	75%	97.5%	Rhat	n.eff
alpha0	-0.06	0.11	-0.26	-0.13	-0.06	0.01	0.18	1	7500
alpha1	-0.56	0.11	-0.80	-0.62	-0.55	-0.49	-0.38	1	30000
beta0	0.41	0.24	-0.09	0.24	0.41	0.57	0.87	1	3100
beta1	0.97	0.11	0.76	0.90	0.97	1.05	1.19	1	4800
Ntotal	227.92	40.85	159.00	198.00	224.00	253.00	317.00	1	4900
D	0.08	0.01	0.06	0.07	0.08	0.09	0.11	1	4900

Once again we obtain results very similar to the data generating values – we have only one individual difference between the posterior mean and the actual realized population size (and note that only 76 individuals were detected, so this estimation problem was not trivial). Of course density is estimated much lower in this case because the sample units are circles of radius $B = 3$ and not rectangles as before, and therefore the total area is much larger.

```
sum(tmp$N.true)                  # True realized population size
[1] 227
sum(!is.na(tmp$data[,"y"]))      # Observed index for population size (Johnson, 2008)
[1] 76
```

8.5.5 BAYESIAN HDS ANALYSIS OF THE ISSJ DATA

Finally, we illustrate an application of HDS using the ISSJ data. We could use either the data augmentation or the three-part model formulations of HDS for the ISSJ data. We'll use the three-part model formulation here and leave as an exercise for you to figure out how to implement the DA version of the model. From Section 8.4.3 we found that a negative binomial abundance model with quadratic effect of chaparral and linear effect of elevation on mean abundance, and a linear effect of chaparral on the distance function parameter σ was our preferred model, and provided a reasonable fit to the data. So we fit a model that is similar here, but minus the chaparral effect on σ (we have you do this as an exercise). And, instead of a negative binomial abundance model, we will illustrate the fitting of a Poisson lognormal model of overdispersion, where we add a site effect with standard deviation σ_{site} to the linear predictor of the expected abundance. First, we have to convert the vector of frequencies for each site to individual distance class observations.

```
# Load the ISSJ data
library(unmarked)
data(issj)

# Prepare some data
nD <- 3                          # Number of intervals
delta <- 100                     # Interval width
B <- 300                         # Upper bound (max. distance)
midpt <- c(50, 150, 250)         # mid points

# Convert vector frequencies to individual distance class
H <- as.matrix(issj[,1:3])
nsites <- nrow(H)
ncap <- apply(H, 1, sum)         # Number of individuals detected per site
dclass <- rep(col(H), H)         # Distance class of each individual
```

```
nind <- length(dclass)                              # Number of individuals detected
elevation <- as.vector(scale(issj[,c("elevation")])) # Prepare covariates
forest <- as.vector(scale(issj[,"forest"]))
chaparral <- as.vector(scale(issj[,"chaparral"]))

# Bundle and summarize data set
str( win.data <- list(nsites=nsites, nind=nind, B=B, nD=nD, midpt=midpt,delta=delta,
ncap=ncap, chaparral=chaparral, elevation=elevation, dclass=dclass) )

# BUGS model specification
cat("
model{
# Priors
sigma ~ dunif(0,1000)
beta0 ~ dunif(-10,10)
beta1 ~ dunif(-10,10)
beta2 ~ dunif(-10,10)
beta3 ~ dunif(-10,10)
sigma.site ~ dunif(0,10)
tau <- 1/(sigma.site*sigma.site)
# Specify hierarchical model
for(i in 1:nind){
    dclass[i] ~ dcat(fc[])                          # Part 1 of HM
}

# Construct cell probabilities for nD cells
for(g in 1:nD){                                     # midpt = mid-point of each cell
   log(p[g]) <- -midpt[g] * midpt[g] / (2 * sigma * sigma)
   pi[g] <- ((2 * midpt[g]) / (B * B)) * delta      # prob. per interval
   f[g] <- p[g] * pi[g]
   fc[g] <- f[g] / pcap
}
pcap <- sum(f[])                                    # Pr(capture): sum of rectangular areas
for(s in 1:nsites){
   ncap[s] ~ dbin(pcap, N[s])                       # Part 2 of HM
   N[s] ~ dpois(lambda[s])                          # Part 3 of HM
     log(lambda[s]) <- beta0 + beta1*elevation[s] + beta2*chaparral[s] +
     beta3*chaparral[s]*chaparral[s] + site.eff[s]
                                                    # Linear model for abundance
  site.eff[s] ~ dnorm(0, tau)                       # Site log normal 'residuals'
}
# Derived params
Ntotal <- sum(N[])
area <- nsites*3.141*300*300/10000                  # Total area sampled, ha
D <- Ntotal/area
}
",fill=TRUE, file="model5.txt")

# Inits
Nst <- ncap + 1
inits <- function(){list (sigma = runif(1, 30, 100), beta0 = 0, beta1 = 0, beta2 = 0,
beta3 = 0, N = Nst, sigma.site = 0.2)}
```

```
# Params to save
params <- c("sigma", "beta0", "beta1", "beta2", "beta3", "sigma.site", "Ntotal","D")

# MCMC settings
ni <- 52000 ; nb <- 2000 ; nt <- 2 ; nc <- 3
```

When we run WinBUGS, we may sometimes get an undefined real result error, which appears to be related to a bad choice of initial values for `sigma`. Simply try again until the algorithm works (also remember to define the object `bd`, which gives the WinBUGS Windows address)....

```
# Run BUGS (ART 0.9 min) and summarize posteriors
out5 <- bugs(win.data, inits, params, "model5.txt", n.thin=nt,
    n.chains=nc, n.burnin=nb, n.iter=ni, debug=TRUE, bugs.dir = bd)
```

... or else you can run JAGS, which we never observed to crash for numerical over/underflow. However, you may get the `'Observed node inconsistent with unobserved parent'` error if you use a prior for `sigma` that is too far away from the bulk of the posterior mass.

```
out5 <- jags(win.data, inits, params, "model5.txt", n.thin=nt,
    n.chains=nc, n.burnin=nb, n.iter=ni)
```

```
# Run JAGS (ART 0.5 min) and summarize posteriors
print(out5, 3)
              mean      sd    2.5%     50%   97.5%  overlap0    f  Rhat  n.eff
sigma       102.31    4.69   93.75  102.08  112.13     FALSE 1.00     1   9031
beta0        -0.10    0.25   -0.62   -0.09    0.36      TRUE 0.65     1   6307
beta1        -0.24    0.16   -0.56   -0.24    0.07      TRUE 0.94     1   1405
beta2         1.20    0.23    0.77    1.20    1.67     FALSE 1.00     1   3031
beta3        -0.52    0.15   -0.83   -0.52   -0.24     FALSE 1.00     1   1224
sigma.site    1.53    0.18    1.20    1.52    1.92     FALSE 1.00     1   2859
Ntotal      664.84   71.48  535.00  661.00  814.00     FALSE 1.00     1   4885
D             0.08    0.01    0.06    0.08    0.09     FALSE 1.00     1   4885
```

This is only one model, not exactly the best model from our analysis with `unmarked`, but the estimated effects are reasonably consistent. Under our Poisson lognormal (PLN) model for abundance we obtain an estimated `Ntotal` which is a bit less than the BUP (of about 827) from `unmarked`. Since we fitted the negative binomial model in the latter, which is similar but not identical to the PLN, some difference might be expected there.

8.6 SUMMARY

Distance sampling is an extremely important methodology in wildlife ecology and management. There is a very well worked-out theory and a truly huge body of literature (e.g., Buckland et al., 2001, 2004a), along with comprehensive software (Distance), R packages, user groups etc.; see, e.g., Thomas et al. (2010) and `distancesampling.org`. In this chapter, we did of course not aim to summarize all of distance sampling, but rather, we gave an introductory view of how distance sampling fits into the concept of hierarchical models, as we present them in this book. We gave a broad overview of various formulations of the distance sampling model, including conditional and full likelihood, point and line transects, and likelihood and Bayesian analysis. There are many formulations of distance sampling! We showed this vast scope of distance sampling implementations in BUGS because each one may be advantageous in a given situation.

Almost all studies that employ distance sampling collect data on multiple spatial sample units, either line transects or point counts (sometimes called point transects in the distance sampling literature). We call models that formally account for the variation in abundance N or in density among sample units hierarchical distance sampling (HDS) models, and we discussed analysis of HDS models in unmarked and in the BUGS language.

The R package unmarked implements models for data that are recorded or summarized into discrete distance intervals so that the resulting models are a variation of the multinomial N-mixture models of Chapter 7. We have argued in this chapter that there is no practical reason to favor continuous distance data and models over distance interval (i.e., binned) data models in principle, since a discrete distance model is no more or less an approximation to truth than is a continuous distance model. Both are just alternative approximations to the true but unknown detection function. And, they may or may not be close approximations to each other depending on the continuous model being approximated and the discreteness of the distance bins. If you really must have a continuous model, you can achieve this without any appreciable loss of precision simply by using many narrow distance bins.

We have introduced two general formulations of HDS in BUGS that can be useful: the conditional multinomial or "three-part" hierarchical formulation of the model, and the formulation based on data augmentation. The use of DA gives us an individual formulation of the model, and so this formulation should be advantageous for situations where we need to model individual covariates such as sex or other characteristics or when the observation unit is a group or cluster of individuals. We discuss some more advanced HDS models in the next chapter and then in chapters 14 and 24 in volume 2. One challenge to implementing point transect HDS models in BUGS is that nonstandard distributions are involved. In this case, we use binned data and approximate the cell probabilities using rectangular approximations to the area under the curve. An alternative would be to use the zeros or ones trick in BUGS (Chelgren et al., 2011b; Shirk et al., 2014).

Hierarchical distance sampling is a relatively recent advance that shows great promise to addressing fundamental problems related to the modeling of spatial variation in abundance or density. The HDS framework is not only flexible but, because it is so easy to implement in the BUGS language, ecologists can easily extend the ideas we have provided here to solve their own problems. Moreover, unmarked contains novel HDS modeling capabilities in a user-friendly and standardized analysis framework and will likely contain additional capabilities in the future. Given the accessibility of HDS models and the importance of modeling spatial variation in ecology, we think HDS models will become the de facto standard for the analysis of distance sampling data in the near future. HDS models are so important that we address several additional extensions in the next chapter (and more in volume 2!).

EXERCISES

1. The estimator of density as a function of the conditional MLE of $\bar{p}$ is: $\widehat{D} = n/(\widehat{\bar{p}} * L * 2 * B)$. Using basic statistical arguments, what is the variance of this estimator?
2. "Prove" by simulation that if individuals are randomly distributed on a rectangle then the distance to the center line has a uniform distribution. "Prove" by simulation that if individuals are randomly distributed about a point in space then the distance to the point has a triangular distribution. We put quotes around the verb 'to prove' because of course this does not represent a mathematical proof which you may also be able to derive.

3. Play around with the data simulation function (i.e., vary *N* and *sigma*) for line transects (this is function `sim.ldata` in Section 8.2.3) to train your intuition about line transect distance sampling.
4. Play around with the data simulation function for point transects (function `sim.pdata` in Section 8.2.5) to train your intuition about point transect distance sampling.
5. In Section 8.3.4, run a simulation study to "prove" that the two estimators (conditional likelihood and full likelihood) are about unbiased in the frequentist sense of the word.
6. In Section 8.4.3 (the ISSJ analysis using `unmarked`) see by how much the density estimate would be biased if the habitat effect of chaparral on *p* had NOT been taken account of. And by how much would we have erred in our global population estimate (assuming that the model in the section is the correct one of course) by assuming that chaparral does not affect the measurement error of density?
7. Play around with the data simulation function for hierarchical line and point transect sampling (function `simHDS` in Section 8.5.1) to train your intuition about HDS. Vary everything you can, especially, type, number of sites, average abundance, average half-normal scale parameter, and also strip half width (B).
8. Implement the HDS model using data augmentation for the ISSJ data (we used the three-part hierarchical model in Section 8.5.5).
9. For the ISSJ data in BUGS using the three-part model (Section 8.5.5), figure out how to model an effect of chaparral on the detection scale parameter σ.
10. For the analysis developed in question (9), modify the model to have a negative binomial abundance model instead of a Poisson-lognormal abundance model, and compare the inferences under the two alternatives of "overdispersed Poisson" models.

CHAPTER 9

ADVANCED HIERARCHICAL DISTANCE SAMPLING

CHAPTER OUTLINE

9.1 Introduction	464
9.2 Distance Sampling (DS) with Clusters, Groups, or Other Individual Covariates	464
9.2.1 Simulating HDS Data with Group Size	466
9.2.2 Analysis in BUGS	469
9.2.3 Imperfect Observation of Cluster Size	471
9.3 Time-Removal and DS Combined	472
9.3.1 The Four-Part Hierarchical Model	473
9.3.2 Simulating and Analyzing Time-Removal/DS Data	474
9.4 Mark-Recapture/Double-Observer DS	479
9.4.1 Simulating DO/DS Data	479
9.4.2 Analysis in BUGS	480
9.4.3 Remarks	483
9.5 Open HDS Models: Temporary Emigration	483
9.5.1 Data and Model Structure	484
9.5.2 Cautionary Note on Temporary Emigration Processes	485
9.5.3 Modeling Temporary Emigration with Distance Sampling in `unmarked` Using the Function `gdistsamp`	485
9.5.4 Fitting Temporary Emigration HDS Models in BUGS	498
9.5.4.1 Simulating a Temporary Emigration System	499
9.5.4.2 Bayesian Analysis of the Wagtail Data	503
9.5.4.3 Robust Design: Replicates within and among Years	509
9.6 Open HDS Models: Implicit Dynamics	510
9.7 Open HDS Models: Modeling Population Dynamics	513
9.7.1 Simulating the ISSJ Data Over Multiple Years	515
9.7.2 Fitting a Bunch of Open-Population Models	518
9.7.2.1 The Independence Model	518
9.7.2.2 The Reduced-Dynamics Model with Simple Exponential Growth	520
9.7.2.3 The Glorious Integrated HDS/DM Model	522
9.7.2.4 Summary Remarks on Modeling Populations Over Time	526
9.8 Spatial Distance Sampling: Modeling Within-Unit Variation in Density	526
9.8.1 DS with Location of Encounter	528
9.8.2 The Line Transect Case	532

9.8.3 Modeling Spatial Covariates 532
9.8.4 Spatial HDS Models in `unmarked` Using the `pcount` Function..........................539
9.8.5 Hierarchical Spatial DS...542
9.9 Summary 548
Exercises 549

9.1 INTRODUCTION

We call this chapter advanced hierarchical distance sampling (HDS), because it covers a number of topics that require more technical or conceptual development to understand and implement than do the conventional HDS models in Chapter 8. We address HDS models with individual covariates that include group or cluster size, and we discuss models for combined time-removal and distance-sampling (DS) data (Solymos et al., 2013; Amundson et al., 2014), which we show to be simply different types of the individual covariate model. The time-removal model leads naturally to a model in which we combine DS with double-observer (DO) sampling or any other multinomial sampling protocol. In using two observers to record distance data, there is also the possibility to obtain independent measurements of cluster or group size when individuals are counted in groups. Then it's possible to deal with imperfect observations of cluster size. We also discuss models that allow for modeling of within-sample unit variation in density. The HDS models developed in the previous chapter only specify a model for the aggregate population size, N_s, of each sample unit s, and sometimes we might wish to model within-unit variation in abundance or density which we discuss in this chapter. Further, we discuss variations of open-population HDS models in some detail. This goes a little against the main organization of our book, which covers static models for closed populations in volume 1, and then dynamic models for open populations in volume 2 (along with various types of more advanced models). However, we feel comfortable with this slight breach in organization, because it keeps both closed and open distance sampling material within two book chapters rather than spreading it out over two volumes.

Historically, DS has only been applied in the context of closed populations. However, as hierarchical models have increased in popularity, several formalizations of HDS for open populations have been described recently. In `unmarked`, we can fit a prototype of an open HDS model, one that allows for temporary emigration. This is based on the models of Chandler et al. (2011) for multinomial N-mixture models, allowing for temporary emigration (see Sections 6.14 and 7.5.3). These models are suitable when a DS protocol is repeated in time for each spatial unit, such as on multiple days within a season, and they provide a coherent framework for combining such replicated DS data. Another recent innovation in the area of open-population HDS models was described by Sollmann et al. (2015), who developed basically an equivalent of the Dail-Madsen (DM) model (see Chapters 12—13) with a DS observation model. This highlights the possibility of obtaining explicit estimates of population dynamics parameters (e.g., survival, recruitment, and growth rate) from simple DS data repeated over time.

9.2 DISTANCE SAMPLING (DS) WITH CLUSTERS, GROUPS, OR INDIVIDUAL COVARIATES

In wildlife surveys, individuals are seldom distributed in space independent of one another. Indeed, many species form aggregations or groups, and therefore if one individual in the group is detected, then

9.2 DS WITH CLUSTERS, GROUPS, OR OTHER INDIVIDUAL COVARIATES

normally others in the group, or the whole group, will be detected as well. This violates a key assumption of the methods in Chapter 8, which is that individuals are detected independently of one another. In this section, we consider generalizing the HDS models to allow for this group structure. Such models redefine the "unit" of the analysis to be the group rather than the individual, and a key attribute of the group is group size. There are several extremely common survey situations where grouped individuals are observed, including surveys of waterfowl (Koneff et al., 2008), large mammals (Koenen et al., 2002; Schmidt et al., 2012), upland game birds, and marine mammals (Hammond et al., 2002), as well as transect surveys for scat of various species (Marques et al., 2001). HDS models with group size as a covariate are a special case of more general models containing covariates at the level of the observation. In classical capture-recapture, these are called individual covariate models or model M_x (Chapter 6 in Kéry and Schaub, 2012). Besides group size, they can include things like sex or age class, or even continuous covariates like length or mass (though the latter are probably not relevant to HDS). As we will show, another type of individual covariate can be "time of removal," which is useful in combining HDS with time-removal data—i.e., a type of capture-recapture protocol (Amundson et al., 2014). We cover this last class of models separately in Section 9.3.

To formulate a model for this scenario, we adopt the basic individual-level formulation of the multinomial mixture model using data augmentation (DA) (Royle et al., 2007a; Royle and Dorazio, 2012), which we first introduced in Section 7.8.1.4. The idea in this formulation of the model is to construct the model for individual encounter data in stratified populations by introducing a categorical individual covariate that assigns each individual to a stratum (in this case, a site; note that in Section 7.8.1.4 we used the term "group structured" where group = stratum, not as used in the present context of counting groups of individuals). Thus we construct the individual-level data set with variable `site[i] ~ dcat(site.probs[])`, where the elements of `site.probs[]` depend on λ_s, the Poisson mean of the abundance N_s at site s. In BUGS, this will look as follows:

```
for(s in 1:nsites){
  N[s] ~ dpois(lambda[s])
  log(lambda[s]) <- beta0 + beta1*habitat[s]
  site.probs[s] <- lambda[s]/sum(lambda[])
}
```

The important idea is this: using DA, a Poisson abundance model implies a categorical model for population (site) membership of each individual observation in the data set. In such stratified models in which the site membership is derived from the model for Poisson abundance, the DA parameter ψ is confounded with the intercept of the Poisson mean, and thus we must introduce a constraint (see Chapter 14 in Royle et al., 2014) that appears like this in the BUGS model:

```
psi <- sum(lambda[])/(nind+nz).
```

We applied DA for stratified populations to a basic HDS situation in Section 8.5.2, and here we extend that formulation to include cluster or group size. There is actually very little to this extension. The basic idea with any type of individual covariate such as group size is to extend the HDS model to accommodate an individual-specific σ parameter that depends in some manner on the covariate in question. For this chapter, we will use a customary log-linear type of model for the covariate (Marques and Buckland, 2003):

$$log(\sigma_i) = \alpha_0 + \alpha_1 * Covariate_i$$

The value of the group size covariate is unknown for unobserved groups of individuals, and hence we need an additional model for the individual covariate to account for the fact that it is not observed for all individuals in the population. It seems natural to think of unobserved group sizes as having a Poisson (or negative binomial—NB) distribution truncated at 1, or shifted to the right, so that no mass occurs at value 0. It is somewhat more convenient in practice (and in BUGS) to simply subtract 1 from the observed group sizes so that the possible values are the integers 0, 1, 2,..., which is the approach we take here.

We note that while formulating models to account for group size is straightforward in BUGS, it is not presently possible to accommodate individual covariates in the unmarked package, and this is a capability that may be developed in the future.

9.2.1 SIMULATING HDS DATA WITH GROUP SIZE

We first define a new simulator function based on that of the previous chapter, called here simHDSg (with g for group), that will simulate HDS data of standard structure but with a group size covariate affecting the half-normal parameter σ for each observation. Recall that previously we had a site-level covariate, wind, that affected σ, but now σ varies at the level of the observation within the site (i.e., for each). We simulate Poisson group size and add 1 to it so that simulated data resemble "real" group sizes that might be observable (i.e., observed group size does not equal zero). In modeling the relationship between $log(\sigma)$ and group size, we put (groupsize-1) in the linear predictor so that the intercept is the value of σ for a real group size of 1.

```
# Function to simulate data under HDS protocol with groups
simHDSg <- function(type = "line", nsites = 100, lambda.group = 0.75, alpha0 = 0, alpha1 =
0.5, beta0 = 1, beta1 = 0.5, B = 4, discard0 = TRUE){
#
# Function simulates hierarchical distance sampling (HDS) data for groups under
#   either a line (type = "line") or a point (type = "point") transect protocol
#   and using a half-normal detection function (Buckland et al. 2001).
#   Other function arguments:
#     nsites: Number of sites (spatial replication)
#     lambda.group: Poisson mean of group size
#     alpha0, alpha1: intercept and slope of log-linear model relating sigma of
#        half-normal detection function to group size
#     beta0, beta1: intercept and slope of log-linear model relating the Poisson
#        mean of the number of groups per unit area to habitat
#     B: strip half width
#
# Get covariates
habitat <- rnorm(nsites)              # Simulated covariate

# Simulate abundance model for groups (Poisson GLM for N)
lambda <- exp(beta0 + beta1*habitat)  # Density of groups per "square"
N <- rpois(nsites, lambda)            # site-specific number of groups
N.true <- N                           # for point: inside of B
```

9.2 DS WITH CLUSTERS, GROUPS, OR OTHER INDIVIDUAL COVARIATES

```r
# Simulate observation model
data <- groupsize <- NULL

for(i in 1:nsites){
  if(N[i]==0){
    data <- rbind(data,c(i,NA,NA,NA,NA,NA)) # save site, y=1, u, v, d
  next
}

if(type=="line"){
  # Simulation of distances, uniformly, for each individual in the population
  d <- runif(N[i], 0, B)
  gs <- rpois(N[i],lambda.group) +1   # Observable group sizes >= 1
  groupsize <-c(groupsize,gs)
  sigma.vec <- exp(alpha0 + alpha1*(gs-1))   # Subtract 1 for interpretation
  # Detection probability for each group
  p <- exp(-d*d/(2*(sigma.vec^2)))
  # Determine if individuals are captured or not
  y <- rbinom(N[i], 1, p)
  u1 <- u2 <- rep(NA,N[i])
  # Subset to "captured" individuals only
  d <- d[y==1] ; u1 <- u1[y==1] ; u2 <- u2[y==1] ; gs <- gs[y==1] ; y <- y[y==1]
}

if(type=="point"){
  # Simulation of data on a circle of radius B (algorithm of Wallin)
  angle <- runif(N[i], 0, 2*π)
  r2 <- runif(N[i], 0, 1)
  r <-  B*sqrt(r2)
  u1 <-  r*cos(angle) + B
  u2 <-  r*sin(angle) + B

  d <- sqrt((u1 - B)^2 + (u2-B)^2)
  N.true[i] <- sum(d<= B)  # Population size inside of count circle, should be N[i] here.
  gs <- rpois(N[i], lambda.group) +1
  groupsize <-c(groupsize,gs)
  sigma.vec <- exp(alpha0 + alpha1*(gs-1))
  # For counting individuals on a circle so we truncate p here
  p <- ifelse(d<(B), 1, 0)*exp(-d*d/(2*(sigma.vec^2)))
  y <- rbinom(N[i], 1, p)
  # Subset to "captured" individuals only
  d <- d[y==1] ; u1 <- u1[y==1] ; u2 <- u2[y==1] ; gs <- gs[y==1] ; y <- y[y==1]
}
# Now compile things into a matrix and insert NA if no individuals were
# captured at site i. Coordinates (u,v) are preserved.
if(sum(y) > 0)
  data <- rbind(data,cbind(rep(i, sum(y)), y, u1, u2, d, gs))
else
  data <- rbind(data,c(i,NA,NA,NA,NA,NA)) # make a row of missing data
}
```

```
# Subset to sites at which individuals were captured. You may or may not
# do this depending on how the model is formulated so be careful.
if(discard0)
    data <- data[!is.na(data[,2]),]

# Visualization
 if(type=="line"){       # For line transect
  par(mfrow = c(1, 3))
  hist(data[,"d"], col = "lightblue", breaks = 20, main =
      "Frequency of distances to groups", xlab = "Distance")
  ttt <- table(data[,1])
  n <- rep(0, nsites)
  n[as.numeric(rownames(ttt))] <- ttt
  plot(habitat, n, main = "Observed group counts (n) vs. habitat", frame = F)
  plot(table(data[,"gs"]), main = "Observed group sizes", ylab = "Frequency", frame = F)
 }

 if(type=="point"){      # For point transect
  par(mfrow = c(2,2))
  plot(data[,"u1"], data[,"u2"], pch = 16, main =
      "Located groups in point transects", xlim = c(0, 2*B),
      ylim = c(0, 2*B), col = data[,1], asp = 1)
  points(B, B, pch = "+", cex = 3)
  library(plotrix)
  draw.circle(B, B, B)
  hist(data[,"d"], col = "lightblue", breaks = 20, main =
      "Frequency of distances to groups", xlab = "Distance")
  ttt <- table(data[,1])
  n <- rep(0, nsites)
  n[as.numeric(rownames(ttt))] <- ttt
  plot(habitat, n, main = "Observed group counts (n) vs. habitat", frame = F)
  plot(table(data[,"gs"]), main = "Observed group sizes", ylab = "Frequency", frame = F)
 }

# Output
list(type = type, nsites = nsites, lambda.group = lambda.group, alpha0 = alpha0, alpha1 =
alpha1, beta0 = beta0, beta1 = beta1, B = B, data=data, habitat=habitat, N = N, N.true =
N.true, groupsize=groupsize)
}
```

As always, we encourage you to execute this function many times and "play DS with groups" with your family or friends. It is instructive to observe how the simulated data change, either due to sampling variability (i.e., from one realization of the process to the next) or in response to your changes to the values of the function arguments. This will help you obtain an intuitive understanding of this type of data, which will then help you understand the model fit to it. Here are some possible settings in which you can use the function, in each case saving the function results in an R object data.

```
data <- simHDSg(type = "line")    # Defaults for line transect data
data <- simHDSg(type = "point")   # Default for point transect data
```

9.2 DS WITH CLUSTERS, GROUPS, OR OTHER INDIVIDUAL COVARIATES

```
data <- simHDSg(lambda.group = 5)   # Much larger groups
data <- simHDSg(lambda.group = 5, alpha1 = 0)  # No effect of groups size on sigma
```

Of course, you could also modify the function in different ways to suit your needs. For example, you could also add site-level covariates on σ as we did in the previous chapter. However, for now we focus on having the single covariate `habitat`, affecting the abundance of groups, for clarity in the exposition.

Note that in the previous chapter, we simulated DS data for point count sample units by simulating data on a square and then thinning the point pattern to remove points in the corners by truncating distances beyond radius B from the center point. However, in this function (and the remainder of this chapter) we adopt a more streamlined simulation method by simulating directly N_s points on the circle. To simulate a fixed number of spatial locations on a circle, we could use function `runifdisc` in package `spatstat` (Baddeley and Turner, 2005), but we use a different approach here, based on the following algorithm (see anderswallin.net/2009/05/uniform-random-points-in-a-circle-using-polar-coordinates/) which uses basic trigonometry for distributing points randomly in a circle.

- Simulate `angle ~ runif(0, 2*`π`)`
- Simulate `r2 ~ runif(0,1)`, where `r2` is the square of the radius of a unit circle
- Set: `r = B*sqrt(r2)`, where `B` is the point count radius
- Set: `u1 = r*cos(angle)`, where `u1` is the 'easting'
- Set: `u2 = r*sin(angle)`, where `u2` is the 'northing'

This algorithm is implemented in most of the simulation functions of this chapter.

9.2.2 ANALYSIS IN BUGS

Now we'll test out our function by simulating a trial data set, and fitting the model having a group size effect on σ using an extended version of the HDS model from Section 8.5.2. But before we do that, we have to go through our usual process of augmenting the data set to account for undetected individuals (refer back to Section 7.8.1.4 for an introduction to DA).

We start by simulating a line transect data set with the default function arguments, using our new favorite R function `simHDSg`, and then we harvest some data objects.

```
set.seed(1234)                      # We all create same data set
temp <- simHDSg(type="line")        # Execute function
data <- temp$data                   # harvest data
B <- temp$B                         # Get strip half width
habitat <- temp$habitat             # habitat covariate
nsites <- temp$nsites               # Number of spatial replicates
groupsize <- data[,"gs"] -1         # Input groupsize-1 as data
```

We augment all data by allowing for a bunch of "pseudogroups": this includes the distance data and the individual covariate data.

```
M <- 400                            # Size of augmented data set is M
nz <- M-nrow(data)                  # Number of "pseudo-groups" added
y <- c(data[,2],rep(0,nz))          # Indicator of capture (== 1 for all obs. groups)
nind <- nrow(data)                  # Number of observed groups
```

```
site <- c(data[,1], rep(NA,nz))   # Site they belong to is unknown
d <- c(data[,5], rep(NA,nz))      # Their distance data are missing ...
groupsize <- c(groupsize, rep(NA,nz)) # .... as is their size
zst <- y                          # Starting values for data augmentation variable

# Bundle data and produce summary
str(bugs.data <- list (y=y, B=B, nind=nind, nsites=nsites, d=d, habitat=habitat,
   site=site, nz=nz, groupsize=groupsize))
```

Here we demonstrate the analysis of models with group size using JAGS (again using the `jagsUI` package). In the BUGS model specification, we use the half-normal detection function as usual, and also show how to estimate total groups and total population size of individuals (for all groups combined) as derived parameters. This happens by multiplying the DA variables z_i by 1 plus the group size variable—so within the BUGS model specification we have this line of code:

```
zg[i] <- z[i]*(groupsize[i] + 1).
```

```
# Define model in BUGS langauge
cat("
model{

# Prior distributions for model parameters
alpha0 ~ dunif(-10,10)
alpha1 ~ dunif(-10,10)
beta0 ~ dunif(-10,10)
beta1 ~ dunif(-10,10)
lambda.group ~ dgamma(0.1, 0.1)
# psi is a derived parameter
psi <- sum(lambda[])/(nind+nz)

# Individual level model: observations and process
for(i in 1:(nind+nz)){
  z[i] ~ dbern(psi)                   # Data augmentation variables
  d[i] ~ dunif(0, B)                  # Distance is uniformly distributed
  groupsize[i] ~ dpois(lambda.group)  # Group size is Poisson

 log(sigma[i]) <- alpha0 + alpha1*groupsize[i]
 p[i] <- z[i]*exp(-d[i]*d[i]/(2*sigma[i]*sigma[i])) # p dep on distance
 # here using the half normal detection function
 y[i] ~ dbern(p[i])

 site[i] ~ dcat(site.probs[1:nsites])  # Population distribution among sites
 zg[i]<- z[i]*(groupsize[i] + 1)       # Number of individuals in that group
}

for(s in 1:nsites){
   # Model for population size of groups
   N[s] ~ dpois(lambda[s])
   log(lambda[s])<- beta0 + beta1*habitat[s]
   site.probs[s]<- lambda[s]/sum(lambda[])
}
```

```
# Derived quantities
G <- sum(z[])          # Total number of groups
Ntotal <- sum(zg[])    # Total population size (all groups combined)
}
",fill=TRUE, file="model1.txt")

# Load some libraries, define MCMC settings, inits function and parameters to save
library("R2WinBUGS")
library("jagsUI")
ni <- 6000 ; nb <- 2000 ; nt <- 2 ; nc <- 3
inits <- function(){list(alpha0=0, alpha1=0.5, beta0=0, beta1=0, z=zst)}
params <- c("alpha0", "alpha1", "beta0", "beta1", "psi", "Ntotal", "G",
   "lambda.group")

# Call JAGS (ART 1.4 min), check convergence and summarize posterior distributions
out1 <- jags(bugs.data, inits, params, "model1.txt", n.thin=nt,
    n.chains=nc, n.burnin=nb,n.iter=ni)
traceplot(out1)   ;   print(out1, 3)
```

	mean	sd	2.5%	50%	97.5%	overlap0	f	Rhat	n.eff
alpha0	0.026	0.114	-0.181	0.021	0.265	TRUE	0.57	1.008	1809
alpha1	0.377	0.102	0.181	0.376	0.584	FALSE	1.00	1.006	1840
beta0	0.836	0.120	0.596	0.837	1.068	FALSE	1.00	1.012	883
beta1	0.424	0.086	0.252	0.423	0.589	FALSE	1.00	1.001	4049
psi	0.602	0.070	0.473	0.598	0.749	FALSE	1.00	1.012	958
Ntotal	463.894	46.290	380.000	462.000	564.000	FALSE	1.00	1.003	3403
G	241.143	26.319	193.000	240.000	298.000	FALSE	1.00	1.010	1389
lambda.group	0.927	0.099	0.747	0.921	1.136	FALSE	1.00	1.013	1050

We see that the parameter estimates are not too far from the data-generating values, and mixing is adequate. The estimated number of groups (241) and individuals (464) is somewhat smaller than the data that were generated (`sum(temp$N.true)` = 273 groups, `sum(temp$groupsize)` = 496 individuals), but both 95% CRIs contain the true values in the simulated data.

9.2.3 IMPERFECT OBSERVATION OF CLUSTER SIZE

The model just developed and analyzed assumes that the cluster size is observed perfectly when the cluster is detected. Of course in practice, this assumption may not hold in some applications. How do we deal with unknown cluster size? One idea is that you could combine the *N*-mixture model for independent counts (Chapter 6) with the HDS model. If you have two observers, for example, and can obtain some independent counts of cluster size, then you get information about individual (within cluster) detection probability. We see two ways to implement this idea:

1. Using a standard mark-recapture distance-sampling (MRDS) model (Alpizar-Jara and Pollock, 1996; Borchers et al., 1998), where observers independently record information about each observed group including its size. See the next two sections for how to implement this without observations of group size.
2. A second design idea is to have one observer doing the DS and a "checker" who simply records an independent assessment of group size when the primary observer directs him to.

We have not implemented either of these ideas, although it seems reasonable to expect that imperfect observation of group size might occur in practice.

9.3 TIME-REMOVAL AND DS COMBINED

A standard problem in wildlife surveys is that some individuals in the population of interest may be unavailable for detection during the survey. This is often posed as a problem of "temporary emigration" (Kendall et al., 1997; Chandler et al., 2011), which is to say that individuals have temporarily left the population of interest and are not exposed to detection. For example, in salamander surveys (Bailey et al., 2004), individuals may burrow into the soil and thus not be captureable during a survey of artificial cover objects. In bird surveys, individuals may be in hiding during the survey (or not displaying or calling) and thus be unavailable (Diefenbach et al., 2007). Such individuals have $p = 0$, instead of the operative p for the individuals that are available for detection. One conceptual model for this problem is the temporary emigration model, where we assume that whether an individual in the population of interest is available for sampling at some occasion k is a random Bernoulli outcome with "availability" probability ϕ. Thus, in this conceptual view of the problem, there is a distinction between "availability" and ordinary detectability (also called "perceptibility") (Marsh and Sinclair, 1989). In essence, we admit that detection probability really has two components: availability probability and then, conditional on an individual being available, detection probability proper. We can formalize this line of thinking as follows (see also Nichols et al., 2009): suppose there exists some superpopulation of individuals, having size M, that is the population of individuals that might theoretically be detected at a site under some hypothetical infinite expenditure of effort at a point. Suppose that at the time of sampling, each individual flips the coin of availability having probability ϕ to decide whether it will burrow down into the mud for the day or not. Then the remaining N individuals that are available for sampling at all represent a binomial outcome: $N \sim \text{Binomial}(M, \phi)$. Our sample of observed individuals is also a binomial draw with probability p: $y \sim \text{Binomial}(N, p)$. Then, to fully model this system we need a model that accounts for both p and ϕ. DS, which is regarded as an instantaneous sampling method, provides direct information about p, but does not adjust for ϕ, and thus DS estimates can suffer from "availability bias." For example, if DS is done to count desert tortoises, and some fraction $1 - \phi$ of the tortoises reside in burrows during the sampling so that they cannot possibly be detected, then DS will produce a biased estimate of the density associated with the true M individuals present in the superpopulation. It will only estimate the number of available individuals. On the other hand, if DS is used in the point counting of birds, and the temporary emigration consists of birds that are temporarily outside of the circle of counting during the time when the point count takes place, then DS is properly estimating N and not M. Hence, in any sampling problem, one should consider whether inference needs to be based on the available population size N or the superpopulation size M (or perhaps both), and how the temporal and spatial methods of sampling may affect the definition of the population that is sampled. This is a recurrent and often difficult problem in ecology. And it is one which is relevant regardless of whether we do or do not include a component for measurement error in our model for abundance.

To accommodate these distinct components of detectability—i.e., those of availability and perceptibility—we can use a combination of simultaneous time removal (or temporary removal as we called it in Chapter 7) and DS. Such models have been considered by Farnsworth et al. (2005),

Solymos et al. (2013), Amundson et al. (2014), and in other types of multinomial observation models by Chandler et al. (2011) (see Chapter 13 in volume 2). The data we obtain from such a combined sampling protocol include both the time (or time interval) elapsed from the start of the survey for each observation, and the data for standard DS: distance measurements for a collection of S points or sites. To formulate a hierarchical model for this combined sampling protocol when time is recorded in discrete intervals, we recognize that "time of removal" is simply a categorical covariate at the observation level, not unlike group size in the previous section, so we need to specify an additional model component for the "population" of the removal time covariate. To develop the model, we use the basic DS formulation of the conditional model from Section 8.3.4, and Section 8.5.4 for the hierarchical version. Recall that under this formulation, we introduce a categorical covariate "distance class," and formulate the model in terms of the conditional probability distribution for distance class and then a separate model component for the number of detected individuals on a sample unit n_s conditional on N_s. For the extension of this model to the combined protocol, we include a second categorical covariate, time of removal. The cell probabilities for both categorical variables are constructed explicitly in BUGS. We provide a demonstration here using R and BUGS code modified from Amundson et al. (2014).

9.3.1 THE FOUR-PART HIERARCHICAL MODEL

We let M_s be the local population size at location (sample unit) s and, as before, we use a Poisson model as our null model of abundance (noting that the usual alternatives could be employed as well, such as NB, ZIP, and PLN; see Chapters 6–8):

$$M_s \sim Poisson(\lambda_s)$$

with variation among sites (e.g., covariates) modeled on the site-specific mean λ_s in the usual way via a log link. We model availability using the random temporary emigration model with probability of availability ϕ, so that the number of individuals available to be detected at point count location s is a binomial draw with parameters M_s and ϕ:

$$N_s \sim Binomial(M_s, \phi)$$

Then, N_s individuals are available to be detected by DS during the short DS survey. In the context of time-removal sampling, the parameter ϕ is the probability that an individual is available during the repeated removal samples. This parameter is related to a per-period (or interval) availability parameter p_a, which is the probability that an individual is available in any subsampling interval j, such as a three-minute period of a nine-minute point count. Then $\phi = 1 - (1 - p_a)^3$, where p_a is estimated from the data, the time interval in which each individual is first detected.

Now we need to describe the observation model, which has several components: a model for n_s, the number of individuals detected at point s during the time-removal sampling, and models for the distance and time-of-removal variables. We assume that n_s is binomially distributed conditional on the number available to be detected, N_s:

$$n_s \sim Binomial(N_s, \bar{p}_s)$$

where $\bar{p}_s$ is the net probability of an individual being detected (at all, in any distance class) for site s. Then, conditional on n_s, we specify the distribution of the two individual covariates $dclass_i$ (distance class) and *tint*$_i$ (time interval). These two covariates are both categorical individual covariates with cell probabilities for *dclass* that depend on the distance-based detection model being considered (see Section 8.2.4), and cell probabilities for *tint* that depend on the availability parameters p_a only.

Thus we have described a four-part hierarchical model defined by the following components: (1) a model for population size [*M*]; (2) a model for availability conditional on *M*, [*N*|*M*]; (3) a model for the observed count *n*, [*n*|*N*]; and finally (4) two component models for the individual categorical distance and time-of-removal observations given *n*. Shortly, we provide the BUGS model specification, which involves all four elements of the model. However, first, we note that very minor modifications to this model specification immediately yield several additional implementations that may work better, worse, or not at all across different BUGS implementations. Each additional model specification can be derived by collapsing (or combining) two or more components of the model. For example, we can combine components 1 and 2 and declare that $N_s \sim Poisson(\phi \lambda_s)$, thus producing a three-level hierarchical model. We can collapse components 2 and 3 into a combined binomial model, or we can combine all of the components 2, 3, and 4 into a Poisson model with mean $\bar{p}_s \phi \lambda_s$. Several published examples of combining one or more stages of the larger hierarchical model exist, including those by Chelgren et al. (2011b) and Oedekoven et al. (2014) (also see Section 7.4).

9.3.2 SIMULATING AND ANALYZING TIME-REMOVAL/DS DATA

We have written a function simHDStr (tr for time removal) for simulation of data combining a DS with either a time-removal or a double-observer sampling protocol. This code is based on Amundson et al. (2014). In addition to time-removal DS or double-observer DS data the function simulates on either a linear transect or a point count. Thus with this simulator, you can generate data under many different scenarios, although we do not demonstrate all possible situations in this chapter. We provide a basic version of the time-removal case here, and then in the following section, we discuss the double observer (DO) case (of course this is a specific version of a capture-recapture protocol). Internally, the function simulates a habitat covariate for each of the point count locations surveyed (the number of such point counts is specified by the user), and the resulting output returns the simulated covariate labeled habitat. Simulating the data and harvesting the output goes like this:

```
# Obtain a data set and harvest the results
set.seed(1235)                         # So we all create the same data set
temp <- simHDStr(type="point")         # Simulate point count-removal data set
data <- temp$data                      # Harvest data
B <- temp$B                            # Upper limit of counting (maximum count distance)
nsites <- temp$nsites                  # Number of sites
habitat <- temp$habitat                # Habitat covariate
K <- temp$K                            # Number of removal periods
```

Many of the parameters are set as defaults to the function in order to simulate a "basic" system. The ones of importance are the parameters of the model for $log(\sigma_s) = \alpha_0 + \alpha_1 * Habitat_s$, having default values of $\alpha_0 = 0$ and $\alpha_1 = 0$, and thus $\sigma = 1$ is constant among sites by default. For the model for local population size, we have $log(\lambda_s) = \beta_0 + \beta_1 * Habitat_s$, with $\beta_0 = 1$ and $\beta_1 = 0.5$. The variable $Habitat_s$ is simulated from a standard normal distribution. By default, the function uses $K = 3$ removal sample

9.3 TIME-REMOVAL AND DS COMBINED

periods of equal but unspecified length, with $p_a = 0.75$. When a double-observer data set is simulated, the default values of p are (0.40, 0.60) for observers 1 and 2, respectively. Detection decreases with distance according to the half-normal detection function (Buckland et al., 2001).

Now that we have simulated a data set, we need to do a little bit of processing. First, we subset the data to include encountered individuals only, and we create a vector **y** of the total encounters per site. We take care here to fill out the vector to its maximum extent with values of 0, so that sites where no individuals are counted get carried along as zeros in the analysis. Then we set up some things we need for the discrete version of the HDS model, including how many distance breaks to use, the interval width, and the midpoints of the intervals, and then we create a categorical distance class variable. Finally, we harvest the time interval of removal, called tint here, which is housed in the aux slot of the simulated data ("aux" for "auxiliary data").

```
# Create the observed encounter frequencies per site (include the zeros!)
data <- data[!is.na(data[,2]),]   # Sites where detections did occur
n <- rep(0,nsites)                # The full site vector
names(n) <- 1:nsites
n[names(table(data[,1]))] <- table(data[,1])  # Put in the counts
site <- data[,1]
nobs <- nrow(data)

# Create the distance class data
nD <- 10                          # Number of distance classes
delta <- B/nD                     # Bin size or width
mdpts <- seq(delta/2,B,delta)     # Midpoint distance of bins up to max distance
dclass <- data[,"d"]              # Distance class for each observation
dclass <- dclass%/%delta +1
tint <- data[,"aux"]

# Bundle data and summarize
str( win.data<-list(n=n, site=site, dclass=as.numeric(dclass),nsites=nsites,
nobs=nobs, delta=delta, nD=nD,mdpts=mdpts,B=B, K=K, tint=tint, habitat=habitat) )

List of 12
 $ n       : Named num [1:200] 0 0 2 0 1 1 1 1 1 1 ...
  ..- attr(*, "names")= chr [1:200] "1" "2" "3" "4" ...
 $ site    : num [1:98] 3 3 5 6 7 8 9 10 11 12 ...
 $ dclass  : num [1:98] 10 3 5 5 2 3 2 3 4 8 ...
 $ nsites  : num 200
 $ nobs    : int 98
 $ delta   : num 0.3
 $ nD      : num 10
 $ mdpts   : num [1:10] 0.15 0.45 0.75 1.05 1.35 1.65 1.95 2.25 2.55 2.85
 $ B       : num 3
 $ K       : num 3
 $ tint    : num [1:98] 1 2 1 1 2 1 1 1 1 2 ...
 $ habitat : num [1:200] -0.698 -1.285 0.99 0.112 0.114 ...
```

Next, we write out the BUGS model file for the four-part hierarchical model that includes the categorical individual covariates `dclass` and `tint`. We include two derived parameters: `Mtot`, which is the total population size among all 200 surveyed sites, and `Ntot`, which is the total population size of individuals that are *available* during the three time-removal sample intervals.

```
cat("
model {
# Prior distributions for basic parameters
# Intercepts
beta.a0 ~ dnorm(0,0.01)      # Intercept for availability
alpha0 ~ dnorm(0, 0.01)      # Intercept for sigma
alpha1 ~ dnorm(0,0.01)       # Slope on sigma covariate
# Coefficients
# beta.a1 ~ dnorm(0,0.01)    # Slope for availability covariate
beta0 ~ dnorm(0,0.01)        # Intercept for lambda
beta1 ~ dnorm(0,0.01)        # Slope for lambda covariate

for(s in 1:nsites){
  # Add covariates to scale parameter DISTANCE (perceptibility)
  log(sigma[s]) <- alpha0 + alpha1*habitat[s]
  # Add covariates for availability here TIME-REMOVAL (availability)
  p.a[s] <- exp(beta.a0) / (1+exp(beta.a0))
  # Optional covariates on availability
  # exp(beta.a0 + beta.a1*date[s])/(1+exp(beta.a0+beta.a1*date[s]))
  # Distance sampling detection probability model
  for(b in 1:nD){
    log(g[b,s]) <- -mdpts[b]*mdpts[b]/(2*sigma[s]*sigma[s]) # Half-normal
    f[b,s] <- ( 2*mdpts[b]*delta )/(B*B)  # Radial density function
    pi.pd[b,s] <- g[b,s]*f[b,s]   # Product Pr(detect)*Pr(distribution)
    pi.pd.c[b,s] <- pi.pd[b,s]/pdet[s] # Conditional probabilities
  }
  pdet[s] <- sum(pi.pd[,s])    # Probability of detection at all

  # Time-removal probabilities
  for (k in 1:K){
    pi.pa[k,s] <- p.a[s] * pow(1-p.a[s], (k-1))
    pi.pa.c[k,s] <- pi.pa[k,s]/phi[s] # Conditional probabilities of availability
  }
  phi[s] <- sum(pi.pa[,s]) # Probability of ever available
}
# Conditional observation model for categorical covariates
for(i in 1:nobs){
  dclass[i] ~ dcat(pi.pd.c[,site[i]])
  tint[i] ~ dcat(pi.pa.c[,site[i]])
}
# Abundance model
for(s in 1:nsites){
  # Binomial model for  # of captured individuals
```

```
# n[s] ~ dbin(pmarg[s], M[s])     # Formulation b, see text
# pmarg[s] <- pdet[s]*phi[s]
n[s] ~ dbin(pdet[s], N[s])        # Formulation a, see text
N[s] ~ dbin(phi[s],M[s])          # Number of available individuals
M[s] ~ dpois(lambda[s])           # Abundance per survey/site/point
# Add site-level covariates to lambda
log(lambda[s]) <- beta0 + beta1*habitat[s]
}
# Derived quantities
Mtot <- sum(M[])    # Total population size
Ntot <- sum(N[])    # Total available population size
PDETmean <- mean(pdet[]) # Mean perceptibility across sites
PHImean <- mean(phi[])   # Mean availability across sites
}
", fill=TRUE, file="tr-ds.txt")

# Create initial values (including for M and N) and list parameters to save
Mst <- Nst <- n + 1
inits <- function(){list(M=Mst, N=Nst, alpha0=1, beta0=runif(1,-1,1),
   beta.a1=runif(1,-1,1), beta1=runif(1,-1,1), alpha1=runif(1,-1,1),
   beta.a0=runif(1,-1,1))}
params <- c("beta.a0", "beta.a1", "alpha0", "alpha1", "beta0", "beta1", "PDETmean",
"PHImean", "Mtot", "Ntot")

# MCMC settings
ni <- 50000  ;  nb <- 10000  ;  nt <- 4  ;  nc <- 3

# Run JAGS in parallel (ART 7.3 min), check convergence and summarize posteriors
out2a <- jags(data=win.data, inits=inits, parameters=params,
   model.file ="tr-ds.txt",n.thin=nt, n.chains=nc, n.burnin=nb, n.iter=ni,
   parallel = TRUE)
traceplot(out2a)  ;     print(out2a, 3)
```

	mean	sd	2.5%	50%	97.5%	overlap0	f	Rhat	n.eff
beta.a0	-0.860	0.580	-2.490	-0.746	-0.059	FALSE	0.984	1.028	298
alpha0	0.068	0.069	-0.056	0.065	0.214	TRUE	0.840	1.004	629
alpha1	0.014	0.060	-0.099	0.013	0.135	TRUE	0.584	1.002	189
beta0	1.029	0.374	0.520	0.960	2.154	FALSE	1.000	1.043	181
beta1	0.382	0.141	0.111	0.382	0.653	FALSE	0.997	1.005	486
PDETmean	0.254	0.032	0.200	0.251	0.325	FALSE	1.000	1.004	591
PHImean	0.651	0.161	0.213	0.688	0.864	FALSE	1.000	1.018	402
Mtot	678.915	376.209	381.000	576.000	1930.075	FALSE	1.000	1.114	121
Ntot	391.799	56.581	290.000	389.000	511.000	FALSE	1.000	1.004	592
deviance	906.766	14.014	879.399	906.867	934.034	FALSE	1.000	1.003	793

We notice right away that the Markov chains mix only moderately well! In a moment, we will try an alternative formulation but, despite the fairly poor mixing, we remark that the posterior means are in the vicinity of the data-generating values. The true value of the intercept for the habitat model is 1.0 (posterior mean of beta0 = 1.029). The realized population size within a circle of radius 2B was:

```
sum(temp$M)
[1] 630
```

which is slightly lower than the posterior mean of 678.9 that we see here. The intercept for $log(\sigma)$ has a true value of 0.0, and the estimate is close to that; the coefficient for habitat in the $log(\sigma)$ model is 0 in the data simulation, and the posterior mean is near that. We monitored two other derived parameters: PDETmean and PHImean, the mean detectability and availability per site. The latter is an estimate of ϕ when there are no site covariates affecting availability. We see that it is slightly less than the data-generating value of 0.75.

However, because the MCMC algorithm is not very good in this case, we might run this for many more thousands of iterations in practice. Alternatively, we try a second formulation of the model that we produce by combining the two binomial models for the observed count n_s and the number of individuals available for sampling N_s. This is indicated in the BUGS model specification by the commented line "Formulation b" noted in red font. We see that this formulation of the model in BUGS produces better mixing (below), and the posterior summaries are numerically very similar.

```
print(out2b,3)

           mean      sd    2.5%     50%   97.5% overlap0     f  Rhat  n.eff
beta.a0  -0.820   0.469  -1.911  -0.755  -0.076    FALSE 0.987 1.004 10037
alpha0    0.066   0.071  -0.063   0.062   0.218     TRUE 0.828 1.002   880
alpha1    0.013   0.063  -0.107   0.012   0.141     TRUE 0.576 1.001  2152
beta0     1.003   0.301   0.483   0.975   1.694    FALSE 1.000 1.005  7323
beta1     0.384   0.152   0.086   0.385   0.680    FALSE 0.994 1.002  1017
PDETmean  0.253   0.033   0.198   0.249   0.326    FALSE 1.000 1.002   933
PHImean   0.661   0.136   0.339   0.685   0.860    FALSE 1.000 1.003 12617
Mtot    636.019 210.771 385.000 585.000 1197.000   FALSE 1.000 1.022  2419
Ntot    396.002  60.494 288.000 392.000  524.000   FALSE 1.000 1.002  1151
deviance 931.607  15.172 902.776 931.119  961.762   FALSE 1.000 1.001  3920
```

In summary, we have seen that mixing of the Markov chains for this class of models may not always be so good, and as a result you may have to do longer MCMC runs in practice, or else consider reparameterizations of the model as we have shown here. Because there are so many components to the hierarchical model, we can combine multiple components in many different ways to produce reparameterizations of the model. We showed two alternative versions here, but there are many others that may perform better or worse, and performance may vary across BUGS platforms. In practice, experimentation with different options may be helpful. As a general rule, MCMC in BUGS normally performs better when there are fewer completely latent states, especially if you can avoid latent variables that are nested between data and other latent variables. Formulation b in the above HDS/time-removal model gets rid of the intermediate state N_s; although we may still keep it in the model in order to make a prediction, it is not connected directly to the data.

Finally, we note that the model summarized here includes two categorical covariates for distance class and time of removal, which are assumed to be independent. We believe this is a sensible model for many applications, but there may be some interest in modeling nonindependence of these two covariates. While we have not experimented with nonindependent models. we imagine that specifying the joint distribution of the two categorical variables in BUGS would not be too difficult. One possibility is to allow the parameters of one categorical distribution to depend on the observed state of the other. Certainly, investigations into such models are worth considering.

9.4 MARK-RECAPTURE/DOUBLE-OBSERVER DS

The big idea of the previous section was that of a "combined protocol" where we integrated, in a single model, time-removal sampling with DS. The main technical aspect of these models is that the combined protocol involves a second *individual* covariate (in addition to distance), which is the time interval during which an individual was "removed." The two covariates (time interval and distance class) were assumed to be independent, so the model can be specified directly in the BUGS language without much difficulty. We can adapt that formulation of the model to consider other types of interesting combined protocols, including various types of capture-recapture models. These just amount to some other categorical covariate that is also independent of distance. Of course, the favorite of many ecologists is the double-observer (DO) protocol, which is a special case of capture-recapture/DS (or mark-recapture distance-sampling, MRDS: Alpizar-Jara and Pollock, 1996; Borchers et al., 1998; Laake et al., 2011). Conn et al. (2012) provide a Bayesian treatment of combined DS and DO sampling using hierarchical models. They do not provide a formulation of their model in BUGS, although they provide an R package `hierarchicalDS` for analysis of their model.

The combined DO/DS model closely follows our combined time-removal/DS model of the previous section. However, instead of "time of removal" as the categorical covariate, we have encounter history or "configuration of observers" as the categorical covariate, and the possible values for this covariate are 10, 01, or 11 for "first observer only," "second observer only," and "both observers." The structure of the model is fairly similar to the time-removal model, so we can modify the code very slightly to accommodate the double-observer situation.

9.4.1 SIMULATING DO/DS DATA

We use our handy `simHDStr` function and specify `method="double"`. The auxiliary variable for this method (called `aux` below) is a categorical indicator of which encounter history the observation goes with: detected by first observer only, by second observer only, or by both observers. We code these by 1, 2, or 3, respectively. Otherwise, the harvesting and organizing of the data goes pretty much like it does with the time-removal sampling that we looked at in the previous section:

```
# Simulate a double-observer distance sampling data set
set.seed(1235)
temp <- simHDStr(type="point", method="double") # Simulate double observer point count data set
data <- temp$data         # Harvest data
B <- temp$B               # Upper limit of counting (maximum count distance)
nsites <-temp$nsites      # Number of sites
habitat <-temp$habitat    # Habitat covariate
```

We have to pad the count vector with zeros, which is done below by creating a vector of zero counts for each site, giving the vector names that correspond to the number of sites, tabulating the observed data into frequencies, and then stuffing the observed frequencies into the vector of all-zero counts in site name order. In addition, we create the distance class information, and then bundle all of this together into a BUGS data list.

```
# Processing of the data: pad the count vector with 0s etc.
data <- data[!is.na(data[,2]),]
n <- rep(0,nsites)
names(n) <- 1:nsites
n[names(table(data[,1]))] <- table(data[,1])
site <- data[,1]
dclass <- data[,"d"]      # Categorical distance class for each observation
aux <- data[,"aux"]       # The auxiliary variable is capture history

# Create the categorical distance variable, use 10 classes here
nD <- 10
delta <- B/nD  # bin width
mdpts <-seq(delta/2,B,delta) # Midpoint of bins up to max distance
nobs <- nrow(data)
dclass <- dclass%/%delta +1

# Bundle data and look at overview of data
str( win.data <-list(n=n,site=site, dclass=as.numeric(dclass), nsites=nsites,
 nobs=nobs, delta=delta, nD=nD, mdpts=mdpts, B=B, aux=aux, habitat=habitat) )
```

9.4.2 ANALYSIS IN BUGS

Structurally, there is hardly any difference between the DO/DS and time-removal models. As before, we have to construct the multinomial cell probabilities explicitly, and they resemble standard capture-recapture probabilities instead of removal probabilities. We may find that the DO/DS model does not always mix very well, so long MCMC runs are often needed. Also, you can experiment with different formulations of the model by collapsing one or more levels of the hierarchical model. We write the model file here, and then produce posterior summaries for our simulated data set.

```
# Define model in BUGS langauge
cat("
model {

#Priors for fixed detection parameters
# 2 observer detection probability parameters
logitp1 ~ dnorm(0, 0.01)
logitp2 ~ dnorm(0, 0.01)
# Detection scale parameters
alpha0 ~ dnorm(0, 0.01)    # Intercept for sigma
alpha1 ~ dnorm(0, 0.01)    # Slope on sigma covariate
# Abundance parameters
beta0 ~ dnorm(0,0.01)      # Intercept for lambda
beta1 ~ dnorm(0,0.01)      # Slope for lambda covariate

# Detection scale parameter model
for(s in 1:nsites){
```

```
# Covariates on scale parameter (perceptibility)
log(sigma[s]) <- alpha0 + alpha1*habitat[s]
# Double observer cell probabilities here if there are covariates
logit(pobs[1,s]) <- logitp1 # + covariates
logit(pobs[2,s]) <- logitp2 # + covariates

# Distance sampling model and cell probabilities
  for(b in 1:nD){
    log(g[b,s]) <- -mdpts[b]*mdpts[b]/(2*sigma[s]*sigma[s])  # half-normal
    f[b,s] <- ( 2*mdpts[b]*delta )/(B*B)     # Scaled radial density function
    pi.pd[b,s] <- g[b,s]*f[b,s]              # Product Pr(detect)*Pr(distribution)
    pi.pd.c[b,s] <- pi.pd[b,s]/pdet[s]       # Conditional cell probabilities
  }
  pdet[s] <- sum(pi.pd[,s])                  # Marginal probability of detection

# Double observer cell probabilities and conditional probabilities
  doprobs[1,s] <- pobs[1,s]*(1-pobs[2,s])
  doprobs.cond1[1,s] <- doprobs[1,s]/sum(doprobs[,s])
  doprobs[2,s] <- (1-pobs[1,s])*pobs[2,s]
  doprobs.cond1[2,s] <- doprobs[2,s]/sum(doprobs[,s])
  doprobs[3,s] <- pobs[1,s]*pobs[2,s]
  doprobs.cond1[3,s] <- doprobs[3,s]/sum(doprobs[,s])
  pavail[s] <- sum(doprobs[,s]) # probability of availability AT ALL
 }

# Observation model for two categorical covariates
for(i in 1:nobs){
  dclass[i] ~ dcat(pi.pd.c[,site[i]])
  aux[i] ~ dcat(doprobs.cond1[,site[i]])
}

# Abundance model
for(s in 1:nsites){
  # Binomial model for # of captured individuals
  n[s] ~ dbin(pdet[s], N[s])
  N[s] ~ dbin(pavail[s], M[s])  # Binomial availability model
  # Abundance model
  M[s] ~ dpois(lambda[s])        # Predicted abundance per survey/site/point
  # Add site-level covariates to lambda
  log(lambda[s]) < - beta0 + beta1*habitat[s]
}
# Derived parameters
Mtot <- sum(M[])
Ntot <- sum(N[])
logit(p1) <- logitp1
logit(p2) <- logitp2
sigma0 <- exp(alpha0)            # Baseline sigma
}
",fill=TRUE,file="do_model.txt")
```

```
# Inits function
Nst <- n + 1              # inits for N
inits <- function(){list(M=Nst+1, N=Nst, alpha0=runif(1,1,2),
   beta0=runif(1,-1,1), beta1=runif(1,-1,1), alpha1=runif(1,-1,1),
   logitp1=0, logitp2=0)}

# Parameters to monitor
params <- c("alpha0", "alpha1", "beta0", "beta1", "Ntot", "Mtot", "logitp1",
   "logitp2", "p1", "p2", "sigma0")

# MCMC settings
ni <- 50000  ;  nb <- 10000  ;  nt <- 4  ;  nc <- 3

# Run JAGS in parallel (ART 6.8 min), check convergence and summarize the results
out3 <- jags(data=win.data, inits=inits, parameters.to.save=params,
      model.file="do_model.txt", n.thin=nt, n.chains=nc, n.burnin=nb, n.iter=ni,
      parallel = TRUE)
traceplot(out3)  ;  print(out3, 3)
```

	mean	sd	2.5%	50%	97.5%	overlap0	f	Rhat	n.eff
alpha0	-0.007	0.063	-0.126	-0.010	0.122	TRUE	0.556	1.005	515
alpha1	0.019	0.046	-0.070	0.018	0.111	TRUE	0.655	1.000	5148
beta0	0.993	0.180	0.653	0.990	1.350	FALSE	1.000	1.008	472
beta1	0.376	0.123	0.133	0.377	0.616	FALSE	0.999	1.001	8100
Ntot	439.573	63.443	329.000	435.000	576.000	FALSE	1.000	1.005	527
Mtot	605.797	101.860	436.000	596.000	832.000	FALSE	1.000	1.007	558
logitp1	-0.266	0.244	-0.746	-0.264	0.212	TRUE	0.862	1.001	3632
logitp2	0.106	0.265	-0.414	0.105	0.626	TRUE	0.656	1.001	5109
p1	0.435	0.059	0.322	0.434	0.553	FALSE	1.000	1.001	3623
p2	0.526	0.065	0.398	0.526	0.652	FALSE	1.000	1.001	5167
sigma0	0.995	0.063	0.882	0.990	1.130	FALSE	1.000	1.005	496
deviance	904.765	13.684	878.946	904.501	932.331	FALSE	1.000	1.003	840

Using the output from this run, we now summarize the posterior output in order to compare directly with data-generating quantities.

```
# Put true values into a vector
truth <- temp$parms
psi <- 1-(1-truth["p.double1"])*(1-truth["p.double2"])# Compute availability
truth <- c(truth[c("p.double1", "p.double2")], exp(truth["alpha0"]),
   truth["beta0"], "Mtot" = sum(temp$M),
   "Ntot" = sum(temp$M)*as.numeric(psi),
   truth[c("alpha0","alpha1","beta1")])

# Get posterior means and 2.5% and 97.5% percentiles (95% CRI)
post <- out3$summary[c("p1", "p2", "sigma0", "beta0", "Mtot", "Ntot",
   "alpha0", "alpha1", "beta1"), c(1,3,7)]

# Table compares truth with posterior mean and 95% CRI from JAGS
cbind(truth, posterior = round(post, 3))
```

	truth	mean	2.5%	97.5%
p.double1	0.40	0.435	0.322	0.553
p.double2	0.60	0.526	0.398	0.652

```
alpha0    1.00    0.995    0.882    1.130
beta0     1.00    0.993    0.653    1.350
Mtot    618.00  605.797  436.000  832.000
Ntot    469.68  439.573  329.000  576.000
alpha0    0.00   -0.007   -0.126    0.122
alpha1    0.00    0.019   -0.070    0.111
beta1     0.50    0.376    0.133    0.616
```

And we find that things are not too far off from the data-generating values, giving us faith that perhaps our model specification is properly done.

9.4.3 REMARKS

While the double-observer (DO) protocol is widely used in conjunction with DS, in which case the observers are making simultaneous observations, the formulation of the model is somewhat more versatile than this situation in the sense that general capture-recapture sampling could be used. For example, suppose you repeatedly sample a transect over time (e.g., on different days but within a closed period) and mark the individuals so that you obtain encounter histories. The individuals are not likely to be at the same location on each sample visit, so you also obtain, in addition to encounter histories, a location during each occasion on which the animal is observed. In this case, augmenting the model with an additional model component describing the variation (over time) in the observed locations of individuals is precisely a type of spatial capture-recapture (SCR) model (Royle et al., 2011), which can also be regarded as a DS model with measurement error, where the repeated observations of individual location are measurements of home range center made with error.

DO/DS with error in group size measurements: a great benefit of the DO/DS model is that we should be able to accommodate group size data from two observers and therefore account for imperfect observation of group size. This type of model, a combined DO/DS model with multiple measurements of group size, is formally a type of combined DO/DS/N-mixture protocol: two observers detecting groups, recording distance measurements, and doing independent counts of the group sizes—a triplicate sampling protocol, perhaps the only one ever conceived! From what we've just done, it should be obvious how you could implement this model. Each observer, in addition to detecting a group or not, also reports a group size. Therefore, we have additional data for each observation in the form of the pair of group size measurements y[i,1] and y[i,2] whenever both observers detected the group, and otherwise we have missing data for one or both counts (if not detected by the observer). We leave it as an exercise for you to invent this new statistical method and publish it in some journal.

9.5 OPEN HDS MODELS: TEMPORARY EMIGRATION

Many studies that use DS have a time dimension, either within a biological season or across seasons (years), so that multiple DS surveys are conducted on the same spatial units, and it may be unreasonable to make the closure assumption. Despite this, we would like to be able to integrate all the data into a single model and we might even want to model population dynamics from our DS data. Can this be done? In Section 9.7 below, we discuss fully dynamic models that allow us to directly estimate and model survival and recruitment using DS data, but first we consider a simpler type of open-population model, one that accommodates temporal variation in N using a temporary emigration type of model (Chandler et al., 2011; Sillet et al., 2012). This model is implemented in the function gdistsamp, found in the unmarked package. We demonstrate that here, and we also show how to fit these models in BUGS.

9.5.1 DATA AND MODEL STRUCTURE

Here we assume that DS data were obtained at each of $s = 1, 2, ..., S$ sites (points, transects) and for multiple survey occasions $k = 1, 2, ..., K$. The occasions may be separated by short time intervals such as a day or a week, or they may be longer time periods such as a season or a year. Or we may have both scales of sampling (also called a "robust design" in capture-recapture): within and across years (see below), or what are more usually called secondary (within) and primary (across) samples. For now we suppose that only one primary period (year) was sampled, and that a number of secondary samples are made within that primary period. Let M_s be the population size available to be detected at point s, for which we assume a distribution such as the Poisson: $M_s \sim Poisson(\lambda_s)$. The very simplest type of open-population model assumes that individuals within the population are randomly available for sampling at sample occasion k, say according to a binomial distribution. Then let $N_{s,k}$ be the population size of individuals available for sampling at occasion k:

$$N_{s,k} \sim Binomial(M_s, \phi)$$

This is a very simple type of open-population model that allows the population size to change during each sample period, but it does not specify explicit population dynamics. Rather, all of the dynamics are subsumed into the parameter ϕ, allowing temporal variation in N over the replicate samples. As the time interval between replicate samples increases, we then expect a lower value of ϕ and more variability in $N_{s,k}$ at a point. In the context of DS, we now suppose that distance measurements are obtained on individuals at each point and sample occasion, and let $y_{s,k}$ be the vector of distance class frequency observations at point s and occasion k. As before, this is a multinomial observation, with "sample size" $N_{s,k}$ and cell probabilities that depend on the DS detection probability model being considered. The three-level hierarchical model for repeated HDS data with temporary emigration is shown in Figure 9.1.

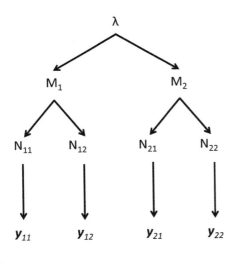

FIGURE 9.1

Schematic of the generalized distance sampling (DS) model in gdistsamp for estimating superpopulation abundance, availability, and detection, shown for two imaginary sites 1 and 2, with two DS surveys at each. SLM denotes 'some linear model': it is here where covariate effects are plugged into the model (and you could specify linear or non-linear models of course).

This model is an interesting one in part for its practical appeal since increasing numbers of studies or monitoring programs have adopted the combination of two common approaches to the estimation of detection probability: using replicated counts and collecting distance information at each time point. However, until recently, no readily available software existed to fit a unified DS model to the resulting data. With `gdistsamp`, you can fit such models using DS data, where the distance information allows you to model the parameter(s) of a detection function, and the (short-term) temporal replication (the point count aspect of the field protocol) allows you to model temporary emigration, or equivalently, availability (here in relation to superpopulation size).

9.5.2 CAUTIONARY NOTE ON TEMPORARY EMIGRATION PROCESSES

Distance sampling is a spatially explicit model in the sense that at an elemental level, the observations are spatial locations (see Section 9.8), and also the model involves an explicit (point process) model of individual locations. Therefore, there are two clearly distinct processes that might lead to temporary emigration in the context of DS and that affect the way we develop and interpret models of temporary emigration. First, you might have individuals moving about their territory and sometimes be outside of the fixed radius point count circle, or off the transect or plot being sampled (Chandler et al., 2011). We call this "spatial temporary emigration." Alternatively, especially for an "instantaneous" count, individuals might be hidden or unavailable for detection even if they are within the sampled spatial unit (e.g., tortoises in their burrow, or birds on a nest). We call this "random" temporary emigration, and it is not a spatially explicit process. However, the temporary emigration model described so far is actually adequate for either of these two processes (Chandler et al., 2011), although mechanistically it describes only the random temporary emigration model (and is merely a good approximation of spatial temporary emigration).

One consequence in developing simulations is that we have to simulate a fixed number of points on a circle, instead of on a square and thinning the corners as we did in Chapter 8, because otherwise the simulation mixes spatial temporary emigration and random temporary emigration. This is easy enough to do, as we described above in Section 9.2, and we implement this algorithm in our function `simHDSopen` below. Another consequence of these different temporary-emigration processes has to do with the interpretation: if temporary emigration is due to individuals leaving because they move about their home range, then density should be computed based on $N_{s,k}$. On the other hand, if temporary emigration is primarily due to "hiding" (e.g., burrowing for turtles), then the density should be estimated based on M_s. In practice, however, there may be a combination of both sources of temporary emigration, and this obviously muddles the interpretation of density estimates. We believe that it might be possible to build a model in JAGS (or writing your own MCMC) that accommodates both mechanisms of temporary emigration explicitly. This would be a good research project for a graduate student.

9.5.3 MODELING TEMPORARY EMIGRATION WITH DISTANCE SAMPLING IN unmarked USING THE FUNCTION gdistsamp

Program `unmarked` allows fitting temporary emigration DS models using the `gdistsamp` function. We briefly introduced the `gdistamp` function in Chapter 8, where we used it to fit NB abundance models to ordinary unreplicated DS data. However, the function accommodates the replicate samples of the temporary emigration model with the `numPrimary` option in the `unmarkedFrameGDS` constructor function. The data structure used to create the `unmarkedFrame` should be a matrix of `nsites` rows, and the number columns should be the number of distance classes, replicated K times. That is, if data are recorded in six distance classes and four samples are done over time, the input matrix should have 24 columns (see below). The `gdistsamp` function is based on the multinomial observation model for binned DS data (Chandler et al., 2011). In cases where we have continuous distance data, it will be sufficient to use a fine bin width (e.g., 1 or 2 m) without compromising any numerical precision. Here we demonstrate the use of `gdistsamp` for modeling spatial and temporal variation in superpopulation size from DS data, using a bird monitoring case study.

Dutch bird data: The Dutch Centre for Field Ornithology Sovon has provided the following data from their grassland bird monitoring program in Flevoland, where 235 points were surveyed four times between April and mid-July 2011, and the distance to each individual bird of a list of grassland bird species was measured based on mapped observations to a maximum of 300 m. Here we use data on the yellow wagtail (Figure 9.2), for which 913 individuals were detected at 191 sites. The number of wagtails detected per distance class and occasion is shown in Figure 9.3. Three site-level and two observation-level covariates are available:

- POTATO fields (% area): 0—100
- PERManent grassland (% area): 0—100
- Landscape SCALE (index for whether landscape is open (0) or closed (100)): 13—73

FIGURE 9.2

The yellow wagtail (*Motacilla flava*). (Photo: Michael Radloff)

9.5 OPEN HDS MODELS: TEMPORARY EMIGRATION

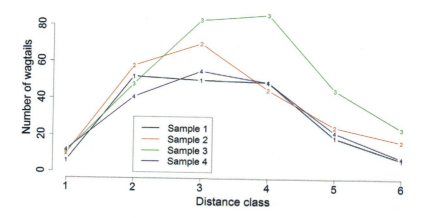

FIGURE 9.3

Number of yellow wagtails detected per distance class (50 m bins out to 300 m) and occasion (sample) in the Dutch grassland monitoring at 221 point count sites in Flevoland. Fourteen sites were dropped due to incomplete data. Note how there are fewer birds in closer distance classes, as expected with a point count distance sampling protocol.

In addition, we have two observation covariates that might affect detection probability or availability:

- Julian DATE of survey: 96–201
- HOUR of survey: 5.5–16

For the analysis here, we drop 14 sites that have 6 or more missing values, corresponding to one to three surveys that did not take place. We note however that removal of missing values is not necessary in this analysis but was done here to resolve a bug in the previous release of unmarked.

```
# Load the wagtail data, investigate NA patterns in detection data
load("wagtail.RData")
Y <- wagtail$Y
table(n.missing <- apply(Y, 1, function(x) sum(is.na(x))))   # Frequency distribution of
number of NAs per site
n.missing
   0   6  12  18
 221   5   3   6
keep <- which(n.missing == 0)      # Sites with complete distance data
Y <- Y[keep,]                      # restrict analysis to those

# Harvest other data for sites with complete distance data
potato <- wagtail$potato[keep]  ;  grass <- wagtail$grass[keep]
lscale <- wagtail$lscale[keep]  ;  hour <- wagtail$hour[keep,]
date <- wagtail$date[keep,]     ;  rep <- wagtail$rep[keep,]
breaks <- wagtail$breaks

# Look at the distance data
str(Y)
 num [1:235, 1:24] 0 0 0 0 0 0 1 0 0 0 ...
tmp <- apply(Y, 2, sum, na.rm = T)
```

```
matplot(1:6, t(matrix(tmp, nrow=4, byrow=T)), type="b", ylim=c(0,90), xlab="Distance
class", ylab="Number of wagtails", frame=F, lwd=3, lty=1)
# Standardize all continuous covariates
mn.potato <- mean(potato)   ;   sd.potato <- sd(potato)
mn.grass <- mean(grass)     ;   sd.grass <- sd(grass)
mn.lscale <- mean(lscale)   ;   sd.lscale <- sd(lscale)
mn.date <- mean(date)       ;   sd.date <- sd(c(date))
mn.hour <- mean(hour)       ;   sd.hour <- sd(c(hour))
POTATO <- (potato - mn.potato) / sd.potato
GRASS <- (grass - mn.grass) / sd.grass
LSCALE <- (lscale - mn.lscale) / sd.lscale
DATE <- (date - mn.date) / sd.date
HOUR <- (hour - mn.hour) / sd.hour

# Package into unmarked GDS data frame and inspect the data
umf <- unmarkedFrameGDS(y = Y[,1:24], survey="point", unitsIn="m",
   dist.breaks=breaks, numPrimary = 4,
   siteCovs = data.frame(POTATO, GRASS, LSCALE),
   yearlySiteCovs=list(rep = rep, DATE = DATE, HOUR = HOUR))
str(umf)
summary(umf)

Formal class 'unmarkedFrameGDS' [package "unmarked"] with 11 slots
  ..@ dist.breaks    : num [1:7] 0 50 100 150 200 250 300
  ..@ tlength        : num [1:221] 1 1 1 1 1 1 1 1 1 1 ...
  ..@ survey         : chr "point"
  ..@ unitsIn        : chr "m"
  ..@ numPrimary     : num 4
  ..@ yearlySiteCovs:'data.frame':    884 obs. of  3 variables:
  .. ..$ rep : Factor w/ 4 levels "1","2","3","4": 1 2 3 4 1 2 3 4 1 2 ...
  .. ..$ DATE: num [1:884] -1.446 -0.545 0.235 1.273 -1.293 ...
  .. ..$ HOUR: num [1:884] 1.717 0.155 -0.509 -1.073 2.291 ...
  ..@ y              : num [1:221, 1:24] 0 0 0 0 0 0 1 0 0 0 ...
  ..@ obsCovs        : NULL
  ..@ siteCovs       :'data.frame':    221 obs. of  3 variables:
  .. ..$ POTATO: num [1:221(1d)] 0.864 0.742 -0.762 0.969 -0.619 ...
  .. .. ..- attr(*, "dimnames")=List of 1
  .. .. .. ..$ : chr [1:221] "1" "2" "3" "4" ...
  .. ..$ GRASS : num [1:221(1d)] -0.389 -0.389 1.34 -0.389 -0.389 ...
  .. .. ..- attr(*, "dimnames")=List of 1
  .. .. .. ..$ : chr [1:221] "1" "2" "3" "4" ...
  .. ..$ LSCALE: num [1:221(1d)] -0.4592 -0.9044 0.5372 -0.0934 -0.6942 ...
  .. .. ..- attr(*, "dimnames")=List of 1
  .. .. .. ..$ : chr [1:221] "1" "2" "3" "4" ...
  ..@ mapInfo        : NULL
  ..@ obsToY         : num [1:4, 1:24] 1 0 0 0 1 0 0 0 1 0 ...

summary(umf)
unmarkedFrame Object
```

9.5 OPEN HDS MODELS: TEMPORARY EMIGRATION

```
221 sites
Maximum number of observations per site: 24
Mean number of observations per site: 24
Number of primary survey periods: 4
Number of secondary survey periods: 1
Sites with at least one detection: 185

Tabulation of y observations:
    0    1    2    3    4 <NA>
 4555  627  103   17    2    0

Site-level covariates:
     POTATO              GRASS              LSCALE
 Min.   :-0.7623    Min.   :-0.3893    Min.   :-1.7947
 1st Qu.:-0.7623    1st Qu.:-0.3893    1st Qu.:-0.6973
 Median :-0.5916    Median :-0.3893    Median :-0.1769
 Mean   : 0.0000    Mean   : 0.0000    Mean   : 0.0000
 3rd Qu.: 0.6977    3rd Qu.:-0.1487    3rd Qu.: 0.4909
 Max.   : 3.1970    Max.   : 3.6690    Max.   : 4.3437

Yearly-site-level covariates:
 rep          DATE                HOUR
 1:221   Min.   :-1.5066    Min.   :-1.89639
 2:221   1st Qu.:-0.8575    1st Qu.:-0.77184
 3:221   Median :-0.1014    Median :-0.03012
 4:221   Mean   : 0.0000    Mean   : 0.00000
         3rd Qu.: 0.7081    3rd Qu.: 0.66375
         Max.   : 1.7009    Max.   : 3.04683
```

We start as always by fitting simple models. We fit models with no covariates, and compare the fit of different detection functions (we are not going to fit the uniform detection function).

```
# Model fitting: Null models fm0
# exponential detection function
summary(fm0.exp <- gdistsamp(lambdaformula = ~1, phiformula = ~1, pformula = ~1,
    keyfun = "exp", output = "density", unitsOut = "ha",
    mixture = "P", K = 100, se = TRUE, data = umf) )

Abundance (log-scale):
 Estimate    SE    z   P(>|z|)
    -1.75 0.073  -24  5.33e-127

Availability (logit-scale):
 Estimate    SE    z  P(>|z|)
     7.57  8.51 0.89    0.374

Detection (log-scale):
 Estimate    SE    z P(>|z|)
     4.69 0.0455  103       0

AIC: 4618.906
```

```
# hazard detection function
summary(fm0.haz <- gdistsamp(lambdaformula = ~1, phiformula = ~1, pformula = ~1,
    keyfun = "haz", output = "density", unitsOut = "ha",
    mixture = "P", K = 100, se = TRUE, data = umf ) )
```

Abundance (log-scale):
 Estimate   SE   z   P(>|z|)
    -2.06 0.0889 -23.1 2.01e-118

Availability (logit-scale):
 Estimate   SE   z P(>|z|)
    0.581 0.266 2.18 0.0289

Detection (log-scale):
 Estimate   SE   z P(>|z|)
     5.13 0.0334 154   0

Hazard-rate(scale) (log-scale):
 Estimate   SE   z P(>|z|)
     1.59 0.0883 18 4.18e-72

AIC: 4508.873

```
# half-normal detection function
summary(fm0.hn <- gdistsamp(lambdaformula = ~1, phiformula = ~1, pformula = ~1,
    keyfun = "halfnorm", output = "density", unitsOut = "ha",
    mixture = "P", K = 100, se = TRUE, data = umf,control=list(trace=TRUE, REPORT=1)) )
```
Abundance (log-scale):
 Estimate   SE   z   P(>|z|)
    -2.06 0.0889 -23.1 1.72e-118

Availability (logit-scale):
 Estimate   SE   z P(>|z|)
     2.38 1.12 2.12 0.0343

Detection (log-scale):
 Estimate   SE   z P(>|z|)
     4.79 0.0235 204   0

AIC: 4530.66

```
# Compare AIC scores for 3 detection functions
rbind('AIC exp' = fm0.exp@AIC, 'AIC haz' = fm0.haz@AIC, 'AIC hn' = fm0.hn@AIC)
            [,1]
AIC exp 4618.906
AIC haz 4508.873
AIC hn  4530.660
```

We see that among these null models, the hazard function encounter probability model is much preferred by AIC. Hence, we will keep the hazard detection function and develop models that are more complex. We can produce estimates of the parameters on the canonical scale, and standard errors using the `backTransform` function:

```
backTransform(fm0.haz, type="lambda")
Backtransformed linear combination(s) of Abundance estimate(s)

 Estimate      SE LinComb (Intercept)
    0.128  0.0114   -2.06           1

Transformation: exp

backTransform(fm0.haz, type="phi")
Backtransformed linear combination(s) of Availability estimate(s)

 Estimate      SE LinComb (Intercept)
    0.641  0.0611   0.581           1

Transformation: logistic

backTransform(fm0.haz, type="det")
Backtransformed linear combination(s) of Detection estimate(s)

 Estimate    SE LinComb (Intercept)
      168  5.62    5.13           1

Transformation: exp
```

Mean abundance (per point) is 0.128, which, because we specified `output = "density"`, `unitsOut = "ha"` in the function call, represents 0.128 individuals per ha or, for a 300 m point count circle of 28.27 ha, about 3.62 birds per point count circle. You can compare this with the average total count of 1.27 individuals in the third sampling period, which yielded the highest counts on average. We see also that the detection shape parameter (it is called the shape parameter for the hazard model) is 168 m. The fitted hazard function model looks like Figure 9.4, and shows detection near 1.0 out to about 75 m, and then detection declines rapidly with distance.

```
plot(1:300, gxhaz(1:300, shape = exp(5.13), scale=1.59), frame = F, type = "l", xlab =
"Distance Wagtail-Observer (metres)", ylab = "Detection probability", lwd=3)
```

Next, we go on fitting some more-complex models that allow for occasion-specific temporary-emigration parameters and include covariates on various parts of the model. One thing you will notice, as the models get more and more complex, is *the use of estimates from simpler models as starting values for more-complex models*. This can be extremely important in practice, and it is something you should always try before seeking help for models that do not converge. Always start with simpler models and build up the analysis step-by-step! This vital modeling rule holds for both maximum likelihood and Bayesian analyses (see also Section 5.12).

```
# Model with time-dependent availability (phi)
fm1 <- gdistsamp(lambdaformula = ~1, phiformula = ~rep-1,
    pformula = ~1, keyfun = "haz", output = "density", unitsOut = "ha",
    mixture = "P", K = 100, se = TRUE, data = umf)
```

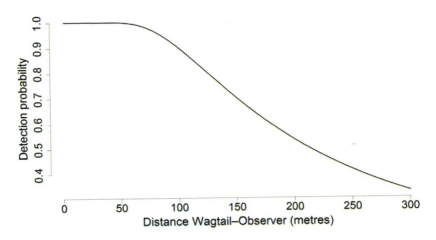

FIGURE 9.4

Fitted hazard function detection probability model for the Dutch wagtail data.

```
# Compare AIC for models with phi constant and phi time-dependent
rbind('AIC phi constant' = fm0.haz@AIC, 'AIC phi time-dep' = fm1@AIC)

                       [,1]
AIC phi constant    4508.873
AIC phi time-dep    4463.166    # Time-dependent phi gives much better AIC

# Add covariates on abundance/density (lambda): 2-phase fitting to assist convergence using
#  first K = 20, then K = 100

summary( fm2.init <- gdistsamp(lambdaformula = ~ POTATO+GRASS+LSCALE,
    phiformula = ~rep-1, pformula = ~1, keyfun = "haz", output = "density",
    unitsOut = "ha", control=list(trace=TRUE, REPORT=1),
    mixture = "P", K = 20, se = TRUE, data = umf))

starts <- coef(fm2.init)
summary( fm2 <- gdistsamp(lambdaformula = ~ POTATO+GRASS+LSCALE,
    phiformula = ~rep-1, pformula = ~1, keyfun = "haz", output = "density",
    unitsOut = "ha", starts=starts, control=list(trace=TRUE,
    REPORT=1), mixture = "P", K = 100, se = TRUE, data = umf))

Call:
gdistsamp(lambdaformula = ~POTATO + GRASS + LSCALE, phiformula = ~rep -
    1, pformula = ~1, data = umf, keyfun = "haz", output = "density",
    unitsOut = "ha", mixture = "P", K = 100, starts = starts,
    se = TRUE, control = list(trace = TRUE, REPORT = 1))

Abundance (log-scale):
            Estimate      SE       z    P(>|z|)
(Intercept)  -2.0809  0.0942  -22.093  3.67e-108
POTATO        0.0365  0.0432    0.846  3.98e-01
GRASS        -0.1977  0.0624   -3.169  1.53e-03
LSCALE       -0.1727  0.0486   -3.557  3.75e-04
```

```
Availability (logit-scale):
     Estimate    SE     z   P(>|z|)
rep1   0.0803 0.243 0.330  0.7411
rep2   0.5724 0.311 1.841  0.0656
rep3   1.6892 0.688 2.455  0.0141
rep4   0.1259 0.248 0.508  0.6115

Detection (log-scale):
 Estimate    SE    z P(>|z|)
     5.13 0.0334 153      0

Hazard-rate(scale) (log-scale):
 Estimate    SE    z  P(>|z|)
     1.59 0.0884 17.9 5.5e-72

AIC: 4436.963
```

Add covariates on lambda: optimisation with default, no inits
```
summary( fm2a <- gdistsamp(lambdaformula = ~ POTATO+GRASS+LSCALE,
    phiformula = ~rep-1, pformula = ~1, keyfun = "haz", output = "density",
    unitsOut = "ha", control=list(trace=TRUE, REPORT=1),
    mixture = "P", K = 100, se = TRUE, data = umf))

Abundance (log-scale):
            Estimate     SE       z   P(>|z|)
(Intercept)  -2.1851 0.0603 -36.250 1.00e-287
POTATO        0.0375 0.0440   0.853  3.93e-01
GRASS        -0.1907 0.0626  -3.045  2.33e-03
LSCALE       -0.1678 0.0492  -3.407  6.56e-04

Availability (logit-scale):
     Estimate    SE    z  P(>|z|)
rep1    0.389 0.184 2.11 3.50e-02
rep2    1.009 0.252 4.01 6.19e-05
rep3    7.038   NaN  NaN      NaN          # Numerical failure here
rep4    0.443 0.188 2.35 1.88e-02

Detection (log-scale):
 Estimate    SE   z P(>|z|)
      5.1 0.033 155      0

Hazard-rate(scale) (log-scale):
 Estimate    SE    z  P(>|z|)
     1.55 0.0878 17.7 1.01e-69

AIC: 4439.715
```

Estimates of availability probabilities
```
plogis(fm2@estimates@estimates$phi@estimates)
     rep1      rep2      rep3      rep4
0.5200552 0.6393067 0.8441211 0.5314240
```

The availability parameters ϕ indicate temporal variation in abundance: higher available site-specific population sizes in the third period compared with the others, and relatively lower abundance in the first and fourth. We proceed with fitting models that are even more complex, using the basic time specificity of the ϕ model as a building block:

```
# Models with time-dependent phi, AIC-best key functions and
#    three covariates on phi. Use previous estimates as starting values.
(tmp <- coef(fm2))
starts <- c(tmp[1:4], tmp[5:8], 0,0,0, tmp[9], tmp[10])

summary(fm3 <- gdistsamp(lambdaformula = ~ POTATO+GRASS+LSCALE,
    phiformula = ~(rep-1)+ POTATO+GRASS+LSCALE, pformula = ~1,
    keyfun = "haz", output = "density", unitsOut = "ha",
    mixture = "P", K = 100, control=list(trace=TRUE, REPORT=1),
    se = TRUE, data = umf, starts = starts))

# Models with time-dependent phi, AIC-best key function, 3 covariates on phi
#    and, in addition, date and hour on detection
# linear effects on detection
tmp <- fm3@estimates@estimates
starts <- c(tmp$lambda@estimates, tmp$phi@estimates, tmp$det@estimates, 0, 0, tmp$scale@estimates)
summary(fm4A <- gdistsamp(lambdaformula = ~ POTATO+GRASS+LSCALE,
    phiformula = ~(rep-1)+ POTATO+GRASS+LSCALE, pformula = ~ DATE + HOUR,
    keyfun = "haz", output = "density", unitsOut = "ha",
    mixture = "P", K = 100, control=list(trace=TRUE, REPORT=1),
    se = TRUE, data = umf, starts = starts) )

# quadratic effects on detection
tmp <- fm4A@estimates@estimates
p.start <- tmp$det@estimates
p.start <- c(p.start[1:2], 0, p.start[3], 0)
starts <- c(tmp$lambda@estimates, tmp$phi@estimates, p.start, tmp$scale@estimates)

summary(fm4B <- gdistsamp(lambdaformula = ~ POTATO+GRASS+LSCALE,
    phiformula = ~(rep-1)+ POTATO+GRASS+LSCALE,
    pformula = ~ DATE + I(DATE^2) + HOUR + I(HOUR^2),
    keyfun = "haz", output = "density", unitsOut = "ha",
    mixture = "P", K = 100, control=list(trace=TRUE, REPORT=1),
    se = TRUE, data = umf, starts = starts) )
```

Clearly, the model with quadratic effects on detection has a lower AIC score, but looking at the parameter estimates and their standard errors, we see that this improvement appears to come from the quadratic effect in DATE only. So we fit a model with quadratic effect of DATE, but linear effect of HOUR.

```
starts <- coef(fm4B)[-16]  # Drop coef for HOUR^2,
summary(fm4C <- gdistsamp(~ POTATO+GRASS+LSCALE,
~(rep-1)+ POTATO+GRASS+LSCALE, ~ DATE + I(DATE^2) + HOUR,
    keyfun = "haz", output = "density", unitsOut = "ha",
    mixture = "P", K = 100, control=list(trace=TRUE, REPORT=1),
  se = TRUE, data = umf, starts = starts) )
```

9.5 OPEN HDS MODELS: TEMPORARY EMIGRATION

This is the AIC-best model, and we try it with an NB distribution next.

```
starts <- c(coef(fm4C), 0)
summary(fm5 <- gdistsamp(~ POTATO+GRASS+LSCALE,
    ~(rep-1)+ POTATO+GRASS+LSCALE, ~ DATE + I(DATE^2) + HOUR,
    keyfun = "haz", output = "density", unitsOut = "ha",
    mixture = "NB", K = 100, control=list(trace=TRUE, REPORT=1),
    se = TRUE, data = umf , starts = starts) )

# Now we create a model selection table of these various models
modSel(fitList(fm0.haz, fm1, fm2, fm3, fm4A, fm4B, fm4C, fm5) )
         nPars      AIC  delta    AICwt cumltvWt
fm5         17  4380.57   0.00  1.0e+00     1.00
fm4C        16  4424.08  43.51  3.6e-10     1.00
fm4B        17  4425.48  44.91  1.8e-10     1.00
fm4A        15  4430.72  50.15  1.3e-11     1.00
fm3         13  4435.58  55.01  1.1e-12     1.00
fm2         10  4436.96  56.39  5.7e-13     1.00
fm1          7  4463.17  82.60  1.2e-18     1.00
fm0.haz      4  4508.87 128.30  1.4e-28     1.00
```

Hence, the NB model (fm5) beats the analogous Poisson model (fm4C) by over 43 AIC units. We inspect the estimates from the best model.

```
summary(fm5)

Abundance (log-scale):
            Estimate    SE       z P(>|z|)
(Intercept)  -1.2611 0.293  -4.304 1.68e-05
POTATO        0.3255 0.195   1.670 9.49e-02
GRASS        -0.0861 0.234  -0.367 7.13e-01
LSCALE        0.4174 0.214   1.954 5.07e-02

Availability (logit-scale):
      Estimate    SE      z P(>|z|)
rep1    -0.822 0.466 -1.765 0.077576
rep2    -0.818 0.441 -1.856 0.063455
rep3    -0.535 0.480 -1.114 0.265198
rep4    -1.125 0.423 -2.658 0.007870
POTATO  -0.463 0.253 -1.827 0.067690
GRASS   -0.213 0.300 -0.711 0.476800
LSCALE  -0.888 0.266 -3.335 0.000852

Detection (log-scale):
            Estimate    SE       z P(>|z|)
(Intercept)   5.1593 0.0384 134.38  0.0000
DATE          0.0469 0.0230   2.04  0.0417
I(DATE^2)    -0.0549 0.0231  -2.38  0.0175
HOUR         -0.0254 0.0161  -1.58  0.1143
```

```
Hazard-rate(scale) (log-scale):
 Estimate    SE    z  P(>|z|)
     1.56 0.0925 16.9 8.32e-64

Dispersion (log-scale):
 Estimate    SE   z P(>|z|)
    0.851 0.203 4.2 2.63e-05
```

The best model is fairly complicated—we see a positive response of abundance to POTATO and LSCALE, but a negative response to GRASS. Hence, density is higher with a larger proportion of potato fields and with more open landscape, but is lower when there is a larger proportion of (Dutch) grassland. Availability is time-specific, but it also has negative effects of all three of those covariates that are less easily interpreted, possibly meaning that they are using such areas less as the percentage in each habitat increases. Detection has a positive response to linear DATE and a negative response to both quadratic DATE and HOUR. There is a fair amount of overdispersion relative to the Poisson, since the NB dispersion parameter is estimated at $\exp(0.851) = 2.341$.

We next do an assessment of model fit for the best model by AIC (the NB model fm5) using our fitstats function, which we have used extensively in previous chapters.

```
# Bootstrap Goodness-of-fit assessment: ART ~ 20 hours
set.seed(1234)
(pb <- parboot(fm5, fitstats, nsim=100, report=5))

Call: parboot(object = fm5, statistic = fitstats, nsim = 200, report = 5)

Parametric Bootstrap Statistics:
                t0 mean(t0 - t_B) StdDev(t0 - t_B) Pr(t_B > t0)
SSE           1209          12.44             92.9        0.438
Chisq       184517         986.30           5476.1        0.433
freemanTukey   779          -2.28             40.2        0.507

t_B quantiles:
                0%    2.5%    25%    50%    75%  97.5%   100%
SSE            955    1031   1139   1195   1256   1375   1514
Chisq       168660  173254 179961 183467 187256 195116 198536
freemanTukey   664     705    757    780    812    851    883

t0 = Original statistic compuated from data
t_B = Vector of bootstrap samples

# Compute magnitude of "overdispersion" c.hat as ratio of observed to expected
#   chisquare test statistic
(c.hat <- pb@t0[2] / mean(pb@t.star[,2]))  # c-hat as ratio of observed/expected
   Chisq
1.005374
```

We find that the model fits the data very well, and we might therefore be satisfied with estimates of density obtained under this model. Note that the c-hat ratio is very near to 1, as we would expect when the model provides an adequate fit. We now use the model to make some predictions of density, which we can do directly with the `predict` function. Because we specified `output = "density", unitsOut = "ha"` in the function call, we merely have to back-transform the linear predictor to get an estimate of birds per ha. We also make predictions of parameter `phi` and of the scale of the detection function `sigma`. For the former, we predict at one level of the factor `rep`—note the special code for doing this.

```
# Predictions of lambda for POTATO, GRASS and LSCALE
newdat1 <- data.frame(POTATO=0, GRASS=0, LSCALE = seq(-1.8,4.33,,100))
newdat2 <- data.frame(POTATO=seq(-0.75,3,,100), GRASS=0, LSCALE = 0)
newdat3 <- data.frame(POTATO=0, GRASS=seq(-0.4, 3.6,,100), LSCALE = 0)
pred1 <- predict(fm5, type="lambda", newdata=newdat1, append = T)
pred2 <- predict(fm5, type="lambda", newdata=newdat2, append = T)
pred3 <- predict(fm5, type="lambda", newdata=newdat3, append = T)

# Predictions of phi for POTATO, GRASS and LSCALE and for rep = 1
newdat4 <- data.frame(rep = factor('1', levels = c('1','2','3','4')), POTATO=0, GRASS=0,
LSCALE = seq(-1.8,4.33,,100))
newdat5 <- data.frame(rep = factor('1', levels = c('1','2','3','4')), POTATO=seq(-
0.75,3,,100), GRASS=0, LSCALE = 0)
newdat6 <- data.frame(rep = factor('1', levels = c('1','2','3','4')), POTATO=0,
GRASS=seq(-0.4, 3.6,,100), LSCALE = 0)
pred4 <- predict(fm5, type="phi", newdata=newdat4, append = T)
pred5 <- predict(fm5, type="phi", newdata=newdat5, append = T)
pred6 <- predict(fm5, type="phi", newdata=newdat6, append = T)

# Predictions of detection function sigma for DATE and HOUR
newdat7 <- data.frame(DATE = seq(-1.51,1.69,,100), HOUR = 0)
newdat8 <- data.frame(DATE=0, HOUR = seq(-1.92,3.1,,100))
pred7 <- predict(fm5, type="det", newdata=newdat7, append = T)
pred8 <- predict(fm5, type="det", newdata=newdat8, append = T)
```

Now we plot all the predictions for all three parameters and use a different color for each covariate. These plots can be seen graphically in Figure 9.5.

```
par(mfrow = c(1,3), mar = c(5,5,2,2), cex.lab = 1.5, cex.axis = 1.5)
plot(newdat1$LSCALE, pred1[,1], xlab="Standardized covariate", ylab="Density (birds/
ha)", lwd=3,type="l", frame = F)
lines(newdat2$POTATO, pred2[,1], lwd=3, col="red")
lines(newdat3$GRASS, pred3[,1], lwd=3, col="blue")
legend(-1.6, 1.65, c("LSCALE", "POTATO", "GRASS"), col=c("black", "red", "blue"), lty=1,
lwd=3, cex=1.2)

plot(newdat4$LSCALE, pred4[,1], xlab="Standardized covariate", ylab="Availability
(phi)", lwd=3,type="l", frame = F)
lines(newdat5$POTATO, pred5[,1], lwd=3, col="red")
lines(newdat6$GRASS, pred6[,1], lwd=3, col="blue")
legend(2, 0.65, c("LSCALE", "POTATO", "GRASS"), col=c("black", "red", "blue"), lty=1,
lwd=3, cex=1.2)
```

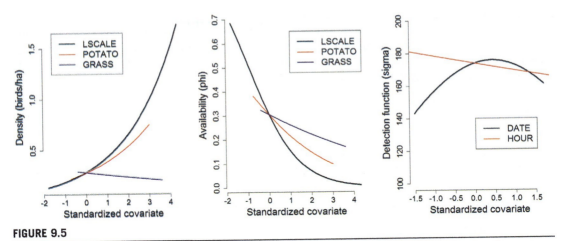

FIGURE 9.5

Fitted response curves of wagtail density (lambda; left), availability (phi; middle), and detection function scale (sigma; right) to standardized covariate ranges for LSCALE, POTATO, and GRASS for the first two, and DATE and HOUR for the last one (the standardized range for each response curve is the observed range in the data set).

```
plot(newdat7$DATE, pred7[,1], xlab="Standardized covariate", ylab="Detection function
 (sigma)", lwd=3,type="l", frame = F, ylim = c(100, 200))
lines(newdat8$HOUR, pred8[,1], lwd=3, col="red")
legend(0.5, 140, c("DATE", "HOUR"), col=c("black", "red"), lty=1, lwd=3, cex=1.2)
```

We see the above-mentioned effects, plus that the wagtails are less detectable (i.e., have a smaller hazard function shape parameter) early and late in the season, and toward the afternoon.

9.5.4 FITTING TEMPORARY EMIGRATION HDS MODELS IN BUGS

The data structure for the open HDS model with temporary emigration is similar to that of the hierarchical models that we considered earlier in this chapter (DO/DS and time-removal/DS), so the model is relatively easy to specify directly in the BUGS language. As with most HDS models, there are multiple approaches to formulating the model in terms of either individual observations (as in the time-removal and DO examples) or site-specific vectors of distance class frequencies. We take the latter approach here, because it is somewhat more straightforward with just ordinary distance data (not having additional auxiliary covariates).

As shown in Figure 9.1 above (and described first in Section 9.3.1), the basic temporary emigration model is a four-part hierarchical model with components: (1) a model for the superpopulation size M_s; (2) a model for the population size of availability of individuals $N_{s,k}$ conditional on M_s; (3) a model for the total number of detected individuals $n_{s,k}$ conditional on the number available for detection, $N_{s,k}$; and (4) an observation model for the distance class frequencies conditional on the number of detected individuals $n_{s,k}$. We can specify the four-part model directly as in Section 9.3.1 ("formulation a"), but as in that section, the formulation where we combine the second and third components of the model seems to provide better mixing ("formulation b"). A number of other forms of the model are possible by combining the different levels of the hierarchical model.

9.5.4.1 Simulating a Temporary Emigration System

We provide the function simHDSopen in the AHM package, which uses this algorithm. The default function arguments are:

```
simHDSopen(type="line", nsites = 100, mean.lam = 2, beta.lam = 0, mean.sig = 1,
beta.sig = 0, B = 3, discard0=TRUE, nreps=2, phi=0.7, nyears=5, beta.trend = 0)
```

where type ("line", "point") lets you choose the type of DS protocol, nsites determines the number of sampled sites, mean.lam and beta.lam specify the average density (on the natural scale) and the coefficient of log(density) on a site covariate (habitat), and mean.sig and beta.sig are the average detection function parameter (on the natural scale) and the coefficient of log(sigma) on an observational covariate. B is the maximum distance, discard0 determines whether you keep or discard sites where no individual is detected. nreps is the number of replicate DS surveys within a period of closure ("seasons" or years), nyears is the number of such (primary) periods, phi is the availability parameter, and beta.trend is the loglinear trend of annual population size or density.

The function will simulate general "4-D" data, including distance class observations for sites, replicates (secondary periods), and years (primary periods), although for the present purposes, we will only analyze data from a single primary period by subsetting a data set simulated for $T = 5$ years. To obtain a basic point count data set with $K = 7$ replicate samples during a season and $S = 100$ sites, we execute the function as follows:

```
# Obtain a temporary emigration data set
set.seed(1234)
str(tmp <- simHDSopen("point", nreps=7, nyears=5, nsites=100) )
attach(tmp)

[much output deleted]
 ...
 $ B          : num 3
 $ nsites     : num 100
 $ habitat    : num [1:100] -1.207 0.277 1.084 -2.346 0.429 ...
 $ wind       : num [1:100, 1:7, 1:5] -0.236 1.93 1.855 1.19 0.167 ...
 $ M.true     : int [1:100, 1:5] 2 2 1 3 2 3 1 1 1 6 ...
 $ K          : num 7
 $ nyears     : num 5
 $ Na         : int [1:100, 1:7, 1:5] 2 0 1 2 0 2 1 0 1 3 ...
 $ Na.real    : num [1:100, 1:7, 1:5] 2 0 1 2 0 2 1 0 1 3 ...
 $ mean.lam   : num 2
 $ beta.lam   : num 0
 $ mean.sig   : num 1
 $ beta.sig   : num 0
 $ phi        : num 0.7
 $ beta.trend : num 0
 $ parms      : Named num [1:6] 2 0 1 0 0.7 0
 ..- attr(*, "names")= chr [1:6] "mean.lam" "beta.lam" "mean.sig" "beta.sig" ...
```

There are quite a few default settings of the data simulator that you should be aware of. You can understand them by looking at the help file for the function, and also by inspecting the code (which is quite richly commented). For one thing, the function defaults to simulating five years of data, and we only need one year of data to fit a model of temporary emigration (since we have seven replicate samples per year) so to demonstrate the model, we will extract only the first year (but we could use any year), or we could extract the first survey for each of the five years, or construct a temporary emigration situation in a number of other ways. Note also that the model has a built-in trend in N across years, a set within-season temporary-emigration probability, built-in covariate effects, and so forth. For the default five-year data set, you can look at the data-generating values of total population size among the 100 sites (this is $\sum_{s=1}^{100} M_s$) as follows:

```
apply(tmp$M.true,2,sum)
[1] 195 192 196 203 186
```

You see that no trend in population size is evident over the five-year period, because we simulated the data with beta.trend=0. Using the output from the simulator function, we do a little bit of data processing to organize the default list format of the data into a 4-D array (sites × distance classes × replicates × years), and we use a distance class width of 0.5.

```
# Define distance class information
delta <- 0.5
nD <- B%/%delta                    # Number of distance classes
midpt <- seq(delta/2, B, delta)    # Mid-point of distance intervals

# Create the 4-d array
y4d <- array(0,dim=c(nsites, nD, K, nyears))
for(yr in 1:nyears){
  for(rep in 1:K){
    data <- tmp$data[[yr]][[rep]]
    site <- data[,1]
    dclass <- data[,"d"]%/%delta + 1
    ndclass <- B%/%delta
    dclass <- factor(dclass, levels= 1:ndclass)
    y4d[1:nsites,1:nD,rep,yr] <- table(site, dclass)
  }
}
```

For fitting a temporary-emigration model, we grab the first year of data for 100 sites, six distance classes, and seven within-year replicates.

```
y3d <- y4d[,,,1]
```

Next we set things up for analysis by JAGS, and fit the temporary emigration model to the replicated DS data we just simulated:

```
# Bundle and summarize the data set
nobs <- apply(y3d, c(1,3), sum) # Total detections per site and occasion
str( data <- list(y3d=y3d, nsites=nsites, K=K, nD=nD, midpt=midpt, delta=delta,
     habitat=habitat, B=B, nobs = nobs) )
```

9.5 OPEN HDS MODELS: TEMPORARY EMIGRATION

```
List of 9
 $ y3d     : num [1:100, 1:6, 1:7] 0 0 0 0 0 0 0 0 0 0 ...
 $ nsites  : num 100
 $ K       : num 7
 $ nD      : num 6
 $ midpt   : num [1:6] 0.25 0.75 1.25 1.75 2.25 2.75
 $ delta   : num 0.5
 $ habitat : num [1:100] -1.207 0.277 1.084 -2.346 0.429 ...
 $ B       : num 3
 $ nobs    : num [1:100, 1:7] 0 0 0 1 0 1 0 0 0 0 ...

# Define model in BUGS
cat("
model {
# Prior distributions
beta0 ~ dnorm(0, 0.01)    # Intercept for log(lambda)
mean.lam <- exp(beta0)
beta1 ~ dnorm(0, 0.01)    # Coefficient of lambda on habitat
phi ~ dunif(0,1)          # Probability of availability
sigma ~ dunif(0.01,5)     # Distance function parameter

# Detection probs for each distance interval and related things
for(b in 1:nD){
   log(g[b]) <- -midpt[b]*midpt[b]/(2*sigma*sigma)  # half-normal
   f[b] <- (2*midpt[b]*delta)/(B*B)     # radial density function
   cellprobs[b] <- g[b]*f[b]
   cellprobs.cond[b] <- cellprobs[b]/sum(cellprobs[1:nD])
}
cellprobs[nD+1] <- 1-sum(cellprobs[1:nD])

for (s in 1:nsites) {
   for (k in 1:K) {
      pdet[s,k] <- sum(cellprobs[1:nD])    # Distance class probabilities
      pmarg[s,k] <- pdet[s,k]*phi          # Marginal probability

      # Model part 4: distance class frequencies
      y3d[s,1:nD,k] ~ dmulti(cellprobs.cond[1:nD], nobs[s,k])
      # Model part 3: total number of detections:
      nobs[s,k] ~ dbin(pmarg[s,k], M[s])
      # nobs[s,k] ~ dbin(pdet[s,k], Navail[s,k]) # Alternative formulation
      # Model part 2: Availability. Not used in this model but simulated.
      Navail[s,k] ~ dbin(phi, M[s])
   }  # End k loop
   # Model part 1: Abundance model
   M[s] ~ dpois(lambda[s])
   log(lambda[s]) <- beta0 + beta1*habitat[s]
}  # End s loop
```

```
# Derived quantities
Mtot <- sum(M[])
for(k in 1:K){
  Ntot[k] <- sum(Navail[,k])
}
} # End model
",file="model.txt")

# Assemble the initial values and parameters to save for JAGS
Navail.st <- apply(y3d, c(1,3),sum)
Mst <- apply(Navail.st, c( 1), max) +2
inits <- function(){
  list(M=Mst, sigma = 1.0, phi=0.9, beta0=log(2), beta1=0.5)
}
params <- c("sigma", "phi", "beta0", "mean.lam", "beta1", "Mtot", "Ntot")

# MCMC settings
ni <- 60000 ; nb <- 10000 ; nt <- 5 ; nc <- 3

# Run WinBUGS or JAGS
library("R2WinBUGS")
library("jagsUI") # JAGS works but WinBUGS does not!
# bd <- "c:/WinBUGS14/"
# out1 <- bugs(data, inits, parameters, "model.txt", n.thin=nthin,
#       n.chains=nc, n.burnin=nb,n.iter=ni,debug=TRUE, bugs.dir = bd)
# We get this error: vector valued relation y3d must involve consecutive
# elements of variable

# Run JAGS: This fails quite often ('invalid parent node'), just keep trying
outTE1 <- jags(data, inits, params, "model.txt", n.thin=nt,n.chains=nc,
   n.burnin=nb,n.iter=ni, parallel = TRUE)
traceplot(outTE1) ; print(outTE1, 3)                       # ART 4 min
              mean      sd     2.5%     50%    97.5%  overlap0      f    Rhat  n.eff
sigma        0.997   0.041    0.921   0.996    1.084     FALSE  1.000  1.000  12226
phi          0.463   0.133    0.224   0.455    0.743     FALSE  1.000  1.006    371
beta0        1.020   0.296    0.528   0.996    1.686     FALSE  1.000  1.011    259
mean.lam     2.907   0.978    1.696   2.707    5.396     FALSE  1.000  1.019    228
beta1        0.029   0.089   -0.148   0.031    0.199      TRUE  0.634  1.001   1580
Mtot       290.784  95.815  175.000 270.000  536.000     FALSE  1.000  1.019    234
Ntot[1]    123.887  15.162   95.000 124.000  155.000     FALSE  1.000  1.000  18542
Ntot[2]    123.878  15.165   95.000 123.000  155.000     FALSE  1.000  1.000   9968
Ntot[3]    123.907  15.161   95.000 123.000  155.000     FALSE  1.000  1.000   8799
Ntot[4]    123.894  15.163   95.000 123.000  155.000     FALSE  1.000  1.000  30000
Ntot[5]    123.897  15.198   95.000 124.000  155.000     FALSE  1.000  1.000  30000
Ntot[6]    123.858  15.146   95.000 123.000  154.000     FALSE  1.000  1.000  10845
Ntot[7]    123.934  15.201   95.000 124.000  155.000     FALSE  1.000  1.000  13773
deviance  1402.895  18.445 1367.637 1402.711 1439.615     FALSE  1.000  1.004    497
```

Notice the prior distribution: sigma ~ dunif(0.01,5). In our original code, we had used a *Uniform*(0,5) prior on the DS scale parameter σ, and JAGS then often failed with an "invalid parent node" error, which was apparently due to a division by zero (Ken Kellner, pers. comm.). By keeping the mass just slightly away from zero, this error can be avoided. The Markov chains appear to have converged, and we have realized a fairly large posterior sample for each parameter. Now we will process the posterior output a little bit, and compare some quantities with the data-generating parameters that we extract from our simulated data object.

```
# Put true values into a vector
truth <- c(tmp$parms[c(1:3,5)], Mtot = sum(tmp$M[,1]),
    Ntot = (apply(tmp$Na.real[,,1],2,sum)))

# Get posterior means and 2.5% and 97.5% percentiles (95% CRI)
post <- outTE1$summary[c("mean.lam", "beta1", "sigma", "phi", "Mtot", "Ntot[1]",
"Ntot[2]", "Ntot[3]" ,"Ntot[4]", "Ntot[5]", "Ntot[6]", "Ntot[7]"), c(1,3,7)]

# Table compares truth with posterior mean and 95% CRI from JAGS
cbind(truth, posterior = round(post, 3))
           truth     mean      2.5%      97.5%
mean.lam    2.0     2.907     1.696      5.396
beta.lam    0.0     0.029    -0.148      0.199
mean.sig    1.0     0.997     0.921      1.084
phi         0.7     0.463     0.224      0.743
Mtot      195.0   290.784   175.000    536.000
Ntot1     130.0   123.869    95.000    154.000
Ntot2     128.0   123.916    95.000    155.000
Ntot3     122.0   123.914    95.000    155.000
Ntot4     136.0   123.835    95.000    154.000
Ntot5     133.0   123.854    95.000    155.000
Ntot6     144.0   123.908    95.000    155.000
Ntot7     136.0   123.847    95.000    155.000
```

We see some similarity between the estimates and the truth, and believe that the difference is attributable to Monte Carlo error. As always, for a full-blown study of the frequentist operating characteristics of the estimators of the model, you would have to simulate and analyze, say, 100 data sets, and then compare the distribution of estimates with the truth in the data-generating algorithm.

9.5.4.2 Bayesian Analysis of the Wagtail Data

We reanalyze the wagtail data, but this time we do it in BUGS using the temporary emigration model template from Section 9.5.3. This is based on the four-part hierarchical model "formulation b," which we met back in the time-removal HDS analysis. We first provide a basic model without all the covariates, using the half-normal distance function so you can see how it all works. Then we provide the analysis using the best model obtained from gdistsamp (model fm5). We compute density based on the expected value of M_s per site, and also using the expected value of available population size $N_{s,k}$. As we noted above, the former would be the correct density if temporary emigration were not due to movement, and the latter would be correct if all temporary emigration were due to movement out of the sample area. So, these provide something of a range under the limiting cases that might be causing temporary emigration.

To convert the wagtail data to the 3-D array format that we used in the BUGS analysis, we take the appropriate columns of the raw wagtail data and stuff them into the proper dimension of the 3-D array as follows.

```
y3d <- array(NA,dim=c(nrow(Y), 6, 4) )      # Create 3d array
y3d[,,1] <- Y[,1:6] ; y3d[,,2] <- Y[,7:12]  # Fill the array
y3d[,,3] <- Y[,13:18] ; y3d[,,4] <- Y[,19:24]

K <- 4                              # Number of primary occasions
nsites <- nrow(Y)                   # Number of sites
nD <- 6                             # Number of distance classes
midpt <- seq(25,275,50)             # Class midpoint distance
delta <- 50                         # Class width
B <- 300                            # Maximum distance
nobs <- apply(y3d, c(1,3), sum)     # Total detections per site and occasion

# Bundle and summarize data set
area <- pi*(300^2)/10000
str(data <- list(y3d=y3d, nsites=nsites, K=K, nD=nD, midpt=midpt, delta=delta,
   B=B, nobs=nobs, area=area))

# Write out the BUGS model file
cat("
model {

# Priors
# Abundance parameters
beta0 ~ dnorm(0, 0.01)
beta1 ~ dnorm(0, 0.01)

# Availability parameter
phi ~ dunif(0,1)

# Detection parameter
sigma ~ dunif(0,500)

# Multinomial cell probabilities
for(b in 1:nD){
  log(g[b]) <- -midpt[b]*midpt[b]/(2*sigma*sigma) # Half-normal model
  f[b] <- (2*midpt[b]*delta)/(B*B) # Scaled radial density function
  cellprobs[b] <- g[b]*f[b]
  cellprobs.cond[b] <- cellprobs[b]/sum(cellprobs[1:nD])
}
cellprobs[nD+1] <- 1-sum(cellprobs[1:nD])

for (s in 1:nsites) {
  for (k in 1:K) {
    # Conditional 4-part version of the model
    pdet[s,k] <- sum(cellprobs[1:nD])
    pmarg[s,k] <- pdet[s,k]*phi
    y3d[s,1:nD,k] ~ dmulti(cellprobs.cond[1:nD], nobs[s,k]) # Part 4: distance
    nobs[s,k] ~ dbin(pmarg[s,k], M[s])    # Part 3: number of detected individuals
    Navail[s,k] ~ dbin(phi, M[s])         # Part 2: Number of available individuals
  }   # end k loop
```

```
    M[s] ~ dpois(lambda[s])    # Part 1: Abundance model
    log(lambda[s]) <- beta0    # Habitat variables would go here
  } # end s loop
# Derived quantities
  for(k in 1:K){
    Davail[k] <- phi*exp(beta0)/area
  }
  Mtotal <- sum(M[])
  Dtotal<- exp(beta0)/area
} # end model
",fill=TRUE,file="wagtail.txt")

# Inits
Navail.st <- apply(y3d, c(1,3),sum)
Mst <- apply(Navail.st, c( 1), max,na.rm=TRUE) + 2
inits <- function() list(M=Mst, sigma = 100.0)

# Parameters to save
params <- c("sigma", "phi", "beta0", "beta1", "Mtotal", "Davail", "Dtotal")

# MCMC settings
ni <- 12000 ; nb <- 2000 ; nt <- 2 ; nc <- 3

# Run JAGS (ART 3 min)
library("jagsUI")
wag1 <- jags(data, inits, params, "wagtail.txt", n.thin=nt, n.chains=nc,
  n.burnin=nb, n.iter=ni, parallel = TRUE)
par(mfrow = c(3,3))    ;    traceplot(wag1)
summary(wag1)
```

```
              mean      sd      2.5%      50%     97.5% overlap0     f  Rhat  n.eff
sigma      121.161   2.756   116.007  121.061   126.824    FALSE  1.00 1.002   5401
phi          0.876   0.072     0.720    0.884     0.991    FALSE  1.00 1.001   2262
beta0        1.301   0.084     1.152    1.294     1.487    FALSE  1.00 1.001   2343
beta1        0.228  10.014   -19.385    0.233    19.648     TRUE  0.51 1.000  15000
Mtotal     814.971  64.398   715.000  806.000   969.025    FALSE  1.00 1.000   1886
Davail[1]    0.114   0.007     0.101    0.113     0.127    FALSE  1.00 1.000  15000
Davail[2]    0.114   0.007     0.101    0.113     0.127    FALSE  1.00 1.000  15000
Davail[3]    0.114   0.007     0.101    0.113     0.127    FALSE  1.00 1.000  15000
Davail[4]    0.114   0.007     0.101    0.113     0.127    FALSE  1.00 1.000  15000
Dtotal       0.130   0.011     0.112    0.129     0.156    FALSE  1.00 1.001   2038
```

Recall our best-fitting hazard model (without covariates) that produces a density estimate on the log scale of −2.06 (see `fm0.hn` from Section 9.5.2).

```
exp(-2.06)
[1] 0.127
```

This is not too different from what we have under the half-normal model when the calculation is based on $E(M_s)$ (we see a posterior mean of 0.130 birds/ha). Also note how easily we produced an estimate of the total number of wagtails (815 individuals) existing on the 221 point count sample units of ~28 ha each. We could also produce explicit estimates of $N_{s,k}$ by simply adding `Navail` to the list of saved parameters.

Next, we add covariates to the various parameters that appear in the best model obtained by our previous analysis, called `fm5`, and at the same time, we also consider the hazard detection model. This model has two parameters σ and θ, which in `unmarked` are called shape and scale, respectively. (Note: in the half-normal model, we usually refer to σ as the scale parameter, so this is different with the hazard model.) It is especially easy to write the hazard model in the BUGS language using the complementary log-log (cloglog) link function. Note that the hazard rate model has this form:

$$p = 1 - exp\left(-\left(x/shape\right)^{(-scale)}\right)$$

which after some rearrangement leads to:

$$cloglog(p) = scale * log(shape) - scale * log(x)$$

Therefore, we can use this model directly in BUGS by converting it to a linear model in log-distance, on the cloglog scale of detection probability.

We need to create a new data list to send to BUGS, which now includes all the covariates we've analyzed previously, along with area:

```
# Bundle and summmarize data set for BUGS
rep <- matrix(as.numeric(rep), ncol=4)
area <- pi*(300^2)/10000
str(data <- list(y3d=y3d, nsites=nsites, K=K, nD=nD, midpt = midpt, delta=delta,
   B=B, nobs=nobs, POTATO=POTATO, GRASS=GRASS, LSCALE=LSCALE, rep=rep, DATE=DATE,
   HOUR=HOUR, area=area))

# Define model in BUGS
cat("
model {

# Priors
# Abundance parameters
beta0 ~ dnorm(0, 0.01)
beta1 ~ dnorm(0, 0.01)
beta2 ~ dnorm(0, 0.01)
beta3 ~ dnorm(0, 0.01)

# Availability parameters
phi0 ~ dunif(0,1)
logit.phi0 <- log(phi0/(1-phi0))
for(k in 1:4){
   gamma1[k] ~ dunif(0, 1)   # Availability effects of surveys 1 - 4
   logit.gamma1[k] <- log(gamma1[k]/(1-gamma1[k]))
}
gamma2 ~ dnorm(0, 0.01)
gamma3 ~ dnorm(0, 0.01)
gamma4 ~ dnorm(0, 0.01)
```

```
# Detection parameters
sigma0 ~ dunif(0.1,500)    # Intercept
alpha2 ~ dnorm(0, 0.01)    # effect of DATE (linear)
alpha3 ~ dnorm(0, 0.01)    # effect of DATE (squared)
alpha4 ~ dnorm(0, 0.01)    # effect of HOUR
theta ~ dgamma(0.1, 0.1)
r ~ dunif(0, 10)

for (s in 1:nsites) {
  for (k in 1:K) {
    # Availability parameter
    logit.phi[s,k] <- logit.gamma1[k] + gamma2*POTATO[s] + gamma3*GRASS[s] + gamma4*LSCALE[s]
    phi[s,k] <- exp(logit.phi[s,k])/(1+ exp(logit.phi[s,k]))
    # Distance sampling parameter
    log(sigma[s,k]) <- log(sigma0) + alpha2*DATE[s,k] + alpha3*pow(HOUR[s,k],2) + alpha4*HOUR[s,k]
    # Multinomial cell probability construction
    for(b in 1:nD){
      #log(g[s,b,k]) <- -midpt[b]*midpt[b]/(2*sigma[s,k]*sigma[s,k]) # half-normal
      cloglog(g[s,b,k]) <- theta*log(sigma[s,k]) - theta*log(midpt[b])    # hazard
      f[s,b,k] <- (2*midpt[b]*delta)/(B*B)
      cellprobs[s,b,k] <- g[s,b,k]*f[s,b,k]
      cellprobs.cond[s,b,k] <- cellprobs[s,b,k]/sum(cellprobs[s,1:nD,k])
    }
    cellprobs[s,nD+1,k] <- 1-sum(cellprobs[s,1:nD,k])

    # Conditional 4-part hierarchical model
    pdet[s,k] <- sum(cellprobs[s,1:nD,k])
    pmarg[s,k] <- pdet[s,k]*phi[s,k]
    y3d[s,1:nD,k] ~ dmulti(cellprobs.cond[s,1:nD,k], nobs[s,k]) # Part 4
    nobs[s,k] ~ dbin(pmarg[s,k], M[s])       # Part 3: Number of detected individuals
    Navail[s,k] ~ dbin(phi[s,k],M[s])        # Part 2: Number of available individuals
  } # End k loop

  M[s] ~ dnegbin(prob[s], r)
  prob[s] <- r/(r+lambda[s])
  # M[s] ~ dpois(lambda[s])                  # Part 1: Abundance model
  log(lambda[s]) <- beta0 + beta1*POTATO[s] + beta2*GRASS[s] + beta3*LSCALE[s]
} # End s loop

# Derived quantities
for(k in 1:K){
  Davail[k] <- mean(phi[,k])*exp(beta0)/area
}
Mtotal <- sum(M[])
Dtotal <- exp(beta0)/area
} # End model
",fill=TRUE,file="wagtail2.txt")
```

Inits
```
Navail.st <- apply(y3d, c(1,3),sum)
Mst <- apply(Navail.st, c( 1), max,na.rm=TRUE) + 2
inits <- function() list(M=Mst, sigma0 = 100, alpha2=0, alpha3=0, alpha4=0, gamma2=0,
gamma3=0, gamma4=0, beta1=0,beta2=0,beta3=0, r = 1)
```

Parameters to save
```
params <- c("r","sigma0", "beta0", "beta1", "beta2", "beta3", "Mtotal", "alpha2",
"alpha3", "alpha4", "theta", "Dtotal", "Davail", "phi0", "gamma1", "gamma2", "gamma3",
"gamma4" ,"logit.gamma1")
```

MCMC settings
```
ni <- 32000 ; nb <- 2000 ; nt <- 2 ; nc <- 5
```

Run JAGS (ART 79 min), check convergence and summarize posteriors
```
wag2 <- jags(data, inits, params, "wagtail2.txt", n.thin=nt,n.chains=nc,
n.burnin=nb,n.iter=ni, parallel = TRUE)
```

	mean	sd	2.5%	50%	97.5%	overlap0	f	Rhat	n.eff
r	2.53	0.59	1.66	2.44	3.94	FALSE	1.00	1.00	9733
sigma0	165.22	6.28	152.52	165.40	176.99	FALSE	1.00	1.00	40722
beta0	1.93	0.28	1.47	1.90	2.57	FALSE	1.00	1.00	2046
beta1	0.27	0.21	-0.05	0.25	0.78	TRUE	0.95	1.01	514
beta2	0.09	0.27	-0.33	0.04	0.73	TRUE	0.57	1.03	129
beta3	0.12	0.28	-0.44	0.13	0.67	TRUE	0.68	1.02	138
Mtotal	1971.44	1158.54	1007.98	1661.00	4927.05	FALSE	1.00	1.02	656
alpha2	0.04	0.02	-0.01	0.03	0.08	TRUE	0.94	1.00	3626
alpha3	-0.01	0.01	-0.04	-0.01	0.02	TRUE	0.77	1.00	28143
alpha4	-0.03	0.02	-0.06	-0.03	0.00	TRUE	0.96	1.00	75000
theta	4.50	0.42	3.74	4.47	5.38	FALSE	1.00	1.00	75000
Dtotal	0.25	0.08	0.15	0.24	0.46	FALSE	1.00	1.01	1334
Davail[1]	0.09	0.01	0.07	0.09	0.12	FALSE	1.00	1.00	1649
Davail[2]	0.10	0.01	0.08	0.10	0.13	FALSE	1.00	1.00	1425
Davail[3]	0.12	0.01	0.09	0.11	0.15	FALSE	1.00	1.00	1326
Davail[4]	0.07	0.01	0.05	0.07	0.10	FALSE	1.00	1.00	1104
phi0	0.50	0.29	0.02	0.50	0.97	FALSE	1.00	1.00	75000
gamma1[1]	0.34	0.09	0.17	0.33	0.53	FALSE	1.00	1.00	3874
gamma1[2]	0.38	0.10	0.19	0.38	0.58	FALSE	1.00	1.00	3823
gamma1[3]	0.47	0.12	0.24	0.47	0.71	FALSE	1.00	1.00	4207
gamma1[4]	0.27	0.07	0.14	0.27	0.41	FALSE	1.00	1.00	3654
gamma2	-0.40	0.27	-0.99	-0.38	0.10	TRUE	0.94	1.01	519
gamma3	-0.43	0.33	-1.14	-0.40	0.13	TRUE	0.93	1.03	132
gamma4	-0.51	0.42	-1.21	-0.56	0.41	TRUE	0.87	1.02	147
logit.gamma1[1]	-0.71	0.44	-1.61	-0.70	0.13	TRUE	0.95	1.00	3104
logit.gamma1[2]	-0.51	0.44	-1.43	-0.50	0.33	TRUE	0.88	1.00	3134
logit.gamma1[3]	-0.13	0.50	-1.14	-0.13	0.87	TRUE	0.60	1.00	3681
logit.gamma1[4]	-1.04	0.38	-1.85	-1.02	-0.35	FALSE	1.00	1.00	2551

```
# Compare posterior means to MLEs obtained from unmarked
mle <- coef(fm5)
mle[12] <- exp(mle[12]) # convert to distance units
mle[16] <- exp(mle[16]) # back-transform the hazard parameter
mle[17] <- exp(mle[17]) # back-transform the NB dispersion parameter
bayes <- wag2$summary[,1]
bayes <- c(bayes[3:6],bayes[c(25:28,22:24)],bayes[c(2,8:10)],bayes[11],bayes[1])
bayes[1] <- log(exp(bayes[1])/area)# Convert from N per site to log(density) per ha

round( cbind(mle,bayes), 3)
                    mle     bayes
lambda(Int)      -1.261    -1.409
lambda(POTATO)    0.325     0.275
lambda(GRASS)    -0.086     0.086
lambda(LSCALE)    0.417     0.125
phi(rep1)        -0.822    -0.708
phi(rep2)        -0.818    -0.511
phi(rep3)        -0.535    -0.127
phi(rep4)        -1.125    -1.037
phi(POTATO)      -0.463    -0.398
phi(GRASS)       -0.213    -0.434
phi(LSCALE)      -0.888    -0.510
p(Int)          174.042   165.224
p(DATE)           0.047     0.035
p(I(DATE^2))     -0.055    -0.010
p(HOUR)          -0.025    -0.029
scale(scale)      4.761     4.496
alpha(alpha)      2.342     2.533
```

The posterior means are vaguely similar to the MLEs that we obtained from the best NB model (fm5) from gdistsamp. Note that you should also take into account the precision of an estimate (e.g., posterior SD, CRI) when judging whether or not it is similar to the data-generating values.

9.5.4.3 Robust Design: Replicates within and among Years

While many studies produce replicated DS data within a season that might be well described by a simple temporary emigration model, it is also common to produce replicated data in multiple years or seasons. That is, we might have $k = 1, 2,..., K$ DS surveys within $t = 1, 2,..., T$ years (or other longer periods). This is a classical robust design-type of data structure (Pollock, 1982), where we would normally use the terms "primary period" for the t samples and "secondary period" for the k samples. A simple extension of the temporary-emigration model easily accommodates this four-level hierarchical model, by changing the assumption on M_s to:

$$M_{s,t} \sim Poisson(\lambda_{s,t})$$

where now we can model variation across primary periods as well as across sites on the parameter $\lambda_{s,t}$. The unmarked package does not directly accommodate such models, although one could do close approximations by modeling variation in the temporary-emigration parameter across k and t, $\phi_{k,t}$. On the other hand, it is easy to specify the four-level hierarchical model in BUGS, which we do in the following Section 9.6. Note that we simulated a 4-D data set above with five years of data, each having seven replicate samples. For our analysis of the temporary emigration model, we used only one year of the simulated data set.

9.6 OPEN HDS MODELS: IMPLICIT DYNAMICS

We now show an example of true robust design data where we have sampled both within and across years—for example, multiple weekly samples in each of T years. Or more generally, the terms "primary" and "secondary" are used, but in the context of modeling animals, this will almost always be among- and within-year sample periods. We showed previously that our openHDS simulator provides such data by default, and as a result, previously we had to harvest the data from only one year in order to fit the basic temporary-emigration model of the previous section (and this is the type of model fitted in unmarked by the gdistsamp function). To describe a model for this true robust design situation, instead of M_s as in the previous section, we now have $M_{s,t}$ for years $t = 1, 2, \ldots, T$, and we wish to model variation across years in the superpopulation sizes $M_{s,t}$, which we naturally think of doing with a Poisson model:

$$M_{s,t} \sim Poisson(\lambda_{s,t})$$

where

$$log(\lambda_{s,t}) = \beta_0 + \beta_1 * Covariate_{s,t}$$

where $Covariate_{s,t}$ could be any covariate that depends on site, year, or both site and year, and of course, you could have multiple such covariates. Obviously, this could include things like "trend," a summary of weather or climate, or any other covariate that you care about. Then within a year, we allow for temporary emigration by letting $N_{s,k,t}$ be the population size available for detection by DS in secondary sample k. For each of these samples, we then obtain the vector of distance class frequencies $y_{s,k,t}$ for site s, replicate sample k, and year t. The basic model structure remains consistent with the four-part hierarchical model introduced in Section 9.3.1 with: (1) a model for superpopulation size $M_{s,t}$; (2) a model for the available population sizes $N_{s,k,t}$; (3) a model for the observed frequency of unique individuals $n_{s,k,t}$; and (4) a model for the distance class frequencies conditional on $n_{s,k,t}$ (note: we get a *lot* of mileage out of this four-part hierarchical model). You will see below that the BUGS model is only slightly extended to account for the fourth dimension of the data (that is, years).

As before, we simulate a data set with our handy simHDSopen function, but this time we simulate a trend in $E(N)$ across years by setting beta.trend = 0.2.

```
# Obtain a data set
set.seed(1236)
str(tmp <- simHDSopen("point", nreps=7, nyears=5, nsites=100, beta.trend=0.2) )
attach(tmp)

apply(tmp$M.true,2,sum)  # True population size per year
[1] 147 192 223 260 318
```

And we see that a strong trend in population size is evident over the five-year period. Using the output from the simulator function, we do a little bit of data processing to organize the data into a 4-D array (sites × distance classes × replicates × years), that, unlike in the previous analysis, we use directly in our BUGS data list.

```
# Define distance class information
delta <- 0.5
nD <- B%/%delta                  # Number of distance classes
midpt <- seq(delta/2, B, delta)  # Mid-point of distance intervals

# Create the 4-d array
y4d <- array(0, dim=c(nsites, nD, K, nyears))
for(yr in 1:nyears){
  for(rep in 1:K){
    data <- tmp$data[[yr]][[rep]]
    site <- data[,1]
    dclass <- data[,"d"]%/%delta + 1
    ndclass <- B%/%delta
    dclass <- factor(dclass, levels= 1:ndclass)
    y4d[1:nsites,1:nD,rep,yr] <- table(site, dclass)
  }
}
```

In the previous analysis of the one-year data set, we took the y4d[,,,1] sub-array from this larger array, but here we use the full 4-D data set to fit a model that includes all five years of data. In addition, the nobs data object has an extra dimension, because we summarize the number of observed individuals over sites, occasions, and years. We also add the number of years (T) to the data list for BUGS.

The model itself in BUGS does not require many modifications to deal with the four-level hierarchy, but we do have to define some quantities to depend on t. In this simple version of the model, we have the cell probabilities as constant across years. You should be able to make them depend on year if there is a year-specific effect on σ.

```
# Bundle and summarize the data set
nobs <- apply(y4d, c(1,3,4), sum)  # Total detections per site and occasion
str( data <- list(y4d=y4d, nsites=nsites, K=K, nD=nD, midpt=midpt, delta=delta,
habitat=habitat, B=B, nobs = nobs, T=tmp$nyears) )

# Define model in BUGS
cat("
model {

# Prior distributions
beta0 ~ dnorm(0, 0.01)   # Intercept for log(lambda)
mean.lam <- exp(beta0)
beta1 ~ dnorm(0, 0.01)   # Coefficient on habitat
phi ~ dunif(0,1)         # Probability of availability
sigma ~ dunif(0,5)       # Detection function parameter
beta.trend ~ dnorm(0, 0.01)

# Construct the multinomial cell probabilities
for(b in 1:nD){
  log(g[b]) <- -midpt[b]*midpt[b]/(2*sigma*sigma)   # half-normal
```

```
     f[b] <- (2*midpt[b]*delta)/(B*B)    # Radial density function
     cellprobs[b] <- g[b]*f[b]
     cellprobs.cond[b] <- cellprobs[b]/sum(cellprobs[1:nD])
  }
  cellprobs[nD+1] <- 1-sum(cellprobs[1:nD])
  for (s in 1:nsites) {
    for (k in 1:K) {
       pdet[s,k] <- sum(cellprobs[1:nD])  # Distance class probabilities
       pmarg[s,k] <- pdet[s,k]*phi        # Marginal probability
    }
  }

  for(t in 1:T){                          # Years
    for (s in 1:nsites) {                 # Sites
       for (k in 1:K) {                   # Replicates
         # Model part 4: distance class frequencies
         y4d[s,1:nD,k,t] ~ dmulti(cellprobs.cond[1:nD], nobs[s,k,t])
         # Model part 3: total number of detections:
         nobs[s,k,t] ~ dbin(pmarg[s,k], M[s,t])
         # Model part 2: Availability. Not used in this model but simulated.
         Navail[s,k,t] ~ dbin(phi, M[s,t])
       } # End k loop
       # Model part 1: Abundance model
       M[s,t] ~ dpois(lambda[s,t])
       log(lambda[s,t]) <- beta0 + beta1*habitat[s] + beta.trend*(t-2.5)
    } # End s loop
  } # End t loop

  # Derived quantities
  for(t in 1:T){
    Mtot[t] <- sum(M[,t])
      for(k in 1:K){
        Ntot[k,t] <- sum(Navail[,k,t])
   }
  }
} # End model
",file="tempemig4d.txt")

# Inits and parameters to save
Navail.st <- apply(y4d, c(1,3,4),sum)
Mst <- apply(Navail.st, c( 1,3), max) +2
inits <- function(){
   list(M=Mst, Navail = Navail.st, sigma = 1.0, phi=.9,beta0=log(2),beta1=.5)
}
params <- c("sigma", "phi", "beta0", "mean.lam", "beta.trend",
   "beta1", "Mtot", "Ntot")

# MCMC settings
ni <- 12000 ; nb <- 2000 ; nt <- 5 ; nc <- 3
```

```
# Run JAGS (ART 9 min), look at trace plots and summarize
outRD <- jags(data, inits, params, "tempemig4d.txt", n.thin=nt, n.chains=nc,
n.burnin=nb, n.iter=ni, parallel = FALSE)
par(mfrow = c(3,3)) ; traceplot(outRD)
summary(outRD)
```

	mean	sd	2.5%	50%	97.5%	overlap0	f	Rhat	n.eff
sigma	1.00	0.02	0.97	1.00	1.03	FALSE	1.00	1.00	4792
phi	0.71	0.06	0.60	0.71	0.84	FALSE	1.00	1.01	258
beta0	0.72	0.08	0.55	0.72	0.88	FALSE	1.00	1.01	286
mean.lam	2.05	0.17	1.74	2.04	2.42	FALSE	1.00	1.01	290
beta.trend	0.14	0.03	0.09	0.14	0.20	FALSE	1.00	1.00	4751
beta1	0.09	0.05	-0.01	0.09	0.18	TRUE	0.97	1.01	418
Mtot[1]	156.97	14.80	131.00	156.00	188.00	FALSE	1.00	1.00	426
Mtot[2]	203.43	17.51	171.00	203.00	241.00	FALSE	1.00	1.00	493
Mtot[3]	237.66	19.18	203.00	237.00	278.00	FALSE	1.00	1.01	346
Mtot[4]	250.75	21.73	212.00	250.00	297.00	FALSE	1.00	1.01	310
Mtot[5]	297.41	25.39	251.00	296.00	351.00	FALSE	1.00	1.01	307
Ntot[1,1]	110.94	9.42	93.00	111.00	130.00	FALSE	1.00	1.00	3735
Ntot[2,1]	110.95	9.36	93.00	111.00	129.00	FALSE	1.00	1.00	6000
Ntot[3,1]	110.75	9.31	93.00	111.00	129.00	FALSE	1.00	1.00	2280
[output truncated]									
Ntot[5,5]	210.19	13.93	183.00	210.00	238.00	FALSE	1.00	1.00	6000
Ntot[6,5]	210.03	13.75	184.00	210.00	238.00	FALSE	1.00	1.00	4334
Ntot[7,5]	210.25	13.82	183.00	210.00	237.00	FALSE	1.00	1.00	6000
deviance	8509.01	41.04	8429.53	8508.22	8588.80	FALSE	1.00	1.00	599

We see that the model accurately estimates the basic parameters, including the trend and the year-specific total population sizes. We think that this robust design-type model with temporary emigration and independent yearly population sizes is very versatile, because it allows for flexible variation in abundance within and among years, and it does not force you to make explicit Markovian assumptions about entries and exits from the population (as we will in the next section). Given the template just provided, it should be possible for you to develop your own extensions of the model for your own specific problem. For instance, if you have very long time-series of DS data (many years), you may want to fit a spline-type model for the annual variation in expected abundance (Fewster et al., 2000; see BUGS code in Section 10.14).

9.7 OPEN HDS MODELS: MODELING POPULATION DYNAMICS

In the previous section, we discussed a temporary emigration model that can be regarded as a primitive form of a population dynamics model in which survival and recruitment parameters are subsumed into a single "availability" parameter. In this section, we extend HDS models even further by considering fully dynamic HDS models. That is, these are models in which DS data are obtained over time on a population that is changing due to survival and recruitment, and we wish to use the observed DS data to estimate population vital rates (apparent survival and recruitment). Very little attention has been given to the formal integration of population dynamics models with

DS data. One idea has been to allow for gains or losses from the population over time by modeling estimates of N obtained by applying DS models to each time slice (Thomas et al., 2005; Buckland et al., 2004b, 2007)—i.e., a two-step analysis or "doing statistics on statistics" (Link, 1999). But formally embedding the population dynamics model for N within the DS model for the observation process has the advantage of efficiently using *all* of the data to provide information about parameters that are shared over time—either parameters of the observation model, or local site-specific parameters that affect abundance. Moreover, uncertainty is properly accounted for in every estimand; this is not so easy in a two-step analysis (although the bootstrap is quite versatile in this respect; and see Section 11.6.4 for a Bayesian solution to two-step modeling). The formulations of the model here (and the R/JAGS scripts) come from Sollmann et al. (2015), who developed open-population HDS models by merging the HDS framework with a basic survival/recruitment population dynamic model due to Dail and Madsen (2011)—i.e., the dynamic N-mixture (or DM) models that we discuss in Chapters 12–14).

As in the previous section, we assume that DS data are collected at $s = 1, 2,..., S$ sites and over $t = 1, 2,..., T$ periods, which we think of here as years, so that from one period to the next, individuals are being recruited and experiencing mortality. You can think of these occasions t as "primary periods" in the usual terminology of the robust design. While we do not explicitly address subsamples within each primary period (i.e., we have no secondary sampling occasions) as they were addressed in the temporary emigration models of the previous sections, the models developed here might be extended to include secondary periods in order to realize a type of integration of the two ideas, where we have a temporary emigration model to describe within-primary-period variation with a fully dynamic population model for among-year variation. For similar ideas for population dynamics models having within- and among-season dynamics, but without explicit measurement error components, see Matechou et al. (2014), Crewe et al. (2015), and Dennis et al. (2015b,c).

Let $N_{s,t}$ be the population size of individuals in the vicinity of site s and during year t. We assume that at some initial period $t = 1$, the population sizes have some distribution such as a Poisson:

$$N_{s,1} \sim Poisson(\lambda_{s,1})$$

where

$$log(\lambda_{s,1}) = \beta_0 + \beta_1 * x_s$$

and where x_s is some covariate that depends on site (and may also depend on time). To describe the population dynamics, we assume that $N_{s,t}$ for $t = 2,..., T$ is the sum of individuals that survive from the previous time, $S_{s,t}$, and new recruits $R_{s,t}$; this is exactly the process model of the DM model. We assume that the number of surviving individuals, $S_{s,t}$, is binomial, and we suppose that the number of new recruits, $R_{s,t}$, is Poisson (other recruitment models are possible; see Dail and Madsen, 2011):

$$S_{s,t} \sim Binomial(N_{s,t-1}, \phi)$$

$$R_{s,t} \sim Poisson(N_{s,t-1}\gamma)$$

Here, ϕ is apparent survival probability (i.e., true survival times the fidelity rate, which is 1 minus the probability of permanent emigration), and γ is the per capita recruitment rate. The population size at time t is then the sum of these two components:

$$N_{s,t} = S_{s,t} + R_{s,t}$$

We note some departures from previously adopted notation here. For example, we use $S_{s,t}$ to be the number of survivors for clarity and consistency with Sollmann et al. (2015), but note also that S has been used for the total number of sites elsewhere in this book. Given the specific context, there should be no confusion here. This fully dynamic model allows explicit estimation and modeling (e.g., as a function of covariates) of both apparent survival and recruitment—for example, by indexing them by s or t and using logit models such as:

$$logit(\phi_{s,t}) = a_0 + a_1 * Covariate_{s,t}$$

In principle, there are no novel technical considerations to building such models, and we do not demonstrate them here.

Reduced dynamics model: A more parsimonious model than the fully dynamic model, and one that is sometimes easier to fit to real (i.e., messy) data sets, is a model expressed in terms of a rate of change θ:

$$N_{s,t} \sim Poisson(N_{s,t-1}\theta)$$

This is then a simple exponential population growth model with 'trend parameter' θ; other population dynamics models—for instance, including density dependence—have been developed by Hostetler and Chandler (2015) in the context of the DM model.

Independence model: An even simpler "open" model is possible, one consistent with the independent population sizes model of the previous section. For this model, we remove the Markovian dependence in the abundance model altogether, and model $N_{s,t}$ simply as a function of site or year-specific effects:

$$N_{s,t} \sim Poisson(\lambda_{s,t})$$

where $log(\lambda_{s,t}) = \beta_0 + \beta_1 * Covariate_{s,t}$ as before, and where the covariate may be "Year" to produce a trend model (see p. 5 in Royle and Dorazio, 2008; Kéry et al., 2009).

In addition to any of these three basic demographic models, we can add the HDS observation model, which here we assume is the multinomial version for binned distance data because that is convenient to work with.

$$\mathbf{y}_{s,t} | N_{s,t} \sim Multinomial(N_{s,t}, \pi_{s,t})$$

where $\pi_{s,t}$ are the usual multinomial distance class probabilities that depend on the specific distance function chosen. As in the HDS/time-removal model from Section 9.3, we formulate this model in terms of individual observations that have a categorical distribution. While we do not show this here, likelihood formulations of such models can also be derived (as in Dail and Madsen, 2011), and the open-population (DM-type) HDS model should be available in `unmarked` sometime in the future.

We note that it would be possible to develop general models that include within-year replication, such as in the temporary emigration models of the previous section, combining both temporary emigration and population dynamics, but we have not yet done so. And as a final remark, we note that for some reason that we don't know we have only been able to fit Dail-Madsen models with JAGS, but not with WinBUGS and OpenBUGS. When fit in the latter two, the Markov chains appeared to converge, but the solutions never matched up the input values in simulated data.

9.7.1 SIMULATING THE ISSJ DATA OVER MULTIPLE YEARS

We simulate data on the Island Scrub-Jay (ISSJ) here and fit the models in JAGS. To do this we follow Sollmann et al. (2015, Suppl. 1) and use the ISSJ data from Sillett et al. (2012; see Section 8.4.3) as the basis for the simulation. We load the data here and refit one of the models from Chapter 8, and then use the estimated model parameters to simulate a *multiyear* data set as if we were repeating the survey each year for, in this case, $T = 6$ years. Then we can evaluate the utility of the open HDS models for characterizing population change. In this case, we will be simulating population dynamics according to the survival/recruitment model, but then we will fit three different models to a simulated data set: the data-generating model (the most complex), and two simpler models including the reduced dynamics and independence models introduced above. One difference between our analysis here and that of Sollmann et al. (2015) is that we use six (50 m) distance classes when we analyze the resulting data set, whereas Sollmann et al. (2015) used three. The rectangular area under the curve is a bad approximation to the integral for radial point counts with three distance classes (and the parameter values from the ISSJ data), and the induced heterogeneity causes bias in estimates of N simulated under the half-normal model that seems to be negligible with six distance classes, but not three.

First we load the ISSJ data, do some processing and organizing of the data into an `unmarkedFrame`, and then we fit one of the NB models to obtain parameter values to use in the simulation study.

```
# We load the ISSJ data analyzed in chapter 8, package into an unmarked frame
library(unmarked)
library(rjags)
data(issj)
covs <- issj[,c("elevation","forest","chaparral")]
area <- pi*300^2 / 100^2          # Area in ha
jayumf <- unmarkedFrameGDS(y=as.matrix(issj[,1:3]),
    siteCovs=data.frame(covs, area), numPrimary=1,
    dist.breaks=c(0, 100, 200, 300),
    unitsIn="m", survey="point")
sc <- siteCovs(jayumf)
sc.s <- scale(sc)
sc.s[,"area"] <- pi*300^2 / 10000 # Don't standardize area
covs <- siteCovs(jayumf) <- sc.s
summary(jayumf)

# Fit the model using gdistsamp and look at the fit summary
(nb.C2E.C <- gdistsamp( ~chaparral + I(chaparral^2) + elevation , ~1, ~chaparral,
    data =jayumf, output="abund", mixture="NB", K = 150))

# Get coefficient estimates to be used in data simulation
beta <- coef(nb.C2E.C)
betaFall <- beta[c("lambda(Int)", "lambda(chaparral)",
    "lambda(elevation)", "lambda(I(chaparral^2))")]

# Predict expected abundance per point count on log-scale for simulation
Xmat <- cbind(rep(1,307),covs[,3],covs[,3]^2,covs[,1]) # Order: chap, chap^2, elev
loglam <- Xmat%*%(betaFall)
lamnew <- exp(loglam)

# Parameters of the detection function
dparm <- beta[c("p(Int)", "p(chaparral)")]
sigma <- exp(Xmat[,c(1, 2)]%*%dparm)
nsites <- 307 # number of sampling points
```

9.7 OPEN HDS MODELS: MODELING POPULATION DYNAMICS

```
# Number of years
nyrs <- 6

# Set dynamics parameters to achieve a target growth rate of 0.95
phi <- 0.6       # Survival probability
gamma <- 0.35    # Recruitment rate

# Distance category info
db <- c(0, 50, 100, 150, 200, 250, 300)
midpt <- c(25, 75, 125, 175, 225, 275)
nD <- length(midpt)
delta <- 50      # Distance interval width
B <- 300
```

We have a simulator function (issj.sim), which we borrowed and modified from Sollmann et al. (2015), that simulates an ISSJ-like data set for a hypothetical ISSJ survey. The inputs are the distance information (B, db) defined above, vectors of the site-specific means and detection parameter values (lamnew, sigma), values of the dynamics parameters phi and gamma, and then the number of points (npoints; maximum 307 corresponding to the real ISSJ survey) and years (nyrs) to do the survey. The data harvested here are the categorical variables distance class (dclass) and year for each observation across all sites, and the matrix y containing the distance class frequencies for each site.

```
# Simulate an ISSJ data set and harvest the data objects
set.seed(2015)
dat <- issj.sim(B=300, db = db, lam=lamnew, sigma=sigma, phi=phi, gamma=gamma,
   npoints=nsites, nyrs=nyrs)

y <- dat$y
dclass <- dat$dclass
site <- dat$site

# Bundle and summarize the data set
str(data1 <-list(nsites=nsites, chap=as.vector(covs[,"chaparral"])[dat$cell],
   chap2=as.vector(covs[,"chaparral"]^2)[dat$cell],
   elev=as.vector(covs[,"elevation"])[dat$cell], T=nyrs, nD=nD, midpt=midpt,
   B=B, delta=delta, y=y, dclass=dclass, site=site, nind=sum(y)) )
List of 13
 $ nsites : num 307
 $ chap   : num [1:307] -0.122 0.838 2.132 -0.274 -1.156 ...
 $ chap2  : num [1:307] 0.0148 0.703 4.5451 0.0749 1.3369 ...
 $ elev   : num [1:307] -1.206 -0.361 -0.458 -0.142 -0.726 ...
 $ T      : num 6
 $ nD     : int 6
 $ midpt  : num [1:6] 25 75 125 175 225 275
 $ B      : num 300
 $ delta  : num 50
 $ y      : num [1:307, 1:6] 0 2 3 0 0 0 2 0 0 0 ...
  ..- attr(*, "dimnames")=List of 2
  .. ..$ : chr [1:307] "det" "det" "det" "det" ...
```

```
....$ : NULL
$ dclass : int [1:1108] 2 5 2 2 4 2 4 2 2 3 ...
$ site   : int [1:1108] 2 2 3 3 3 7 7 15 15 15 ...
$ nind   : num 1108
```

Now we're ready to experience once more the power of BUGS to fit fancy models.

9.7.2 FITTING A BUNCH OF OPEN-POPULATION MODELS

We fit in succession three models to the simulated ISSJ data: first the independence model, then the reduced dynamics model, and then the full-blown DM/HDS model. After fitting these models, we will process the output and compare the estimates of trend for each one. For each model, we fit the proper covariate effects on abundance and the DS scale parameter σ.

9.7.2.1 The Independence Model

The first model is the independent $N_{s,t}$ model, but with a linear trend across years in the expected population size at a site. This appears as a covariate on the abundance model in addition to the existing habitat covariates. Note that the independence model here is essentially the one covered in Section 9.6 (the robust design model), but *without* the intermediate availability stage.

```
# Write out the BUGS model file
cat("
model{

# Prior distributions
# Regression parameters
alpha0 ~ dunif(0,20)
alpha1 ~ dunif(-10,10)
beta0  ~ dunif(-20,20)
beta1  ~ dunif(-20,20)
beta2  ~ dunif(-20,20)
beta3  ~ dunif(-20,20)
beta4  ~ dunif(-20,20)    # Population trend parameter
r ~ dunif(0,5)            # NegBin dispersion parameter
rout <- log(r)

# 'Likelihood'
for (s in 1:nsites){
  # Linear model for detection function scale
  log(sigma[s]) <- alpha0+alpha1*chap[s]
  # Compute multinomial cell probabilities
  for(k in 1:nD){
     pi[k,s] <- (2*midpt[k]*delta )/(B*B)
     log(p[k,s]) <- -midpt[k]*midpt[k]/(2*sigma[s]*sigma[s])
     f[k,s] <- p[k,s]*pi[k,s]
     fc[k,s] <- f[k,s]/pcap[s]
     fct[k,s] <- fc[k,s]/sum(fc[1:nD,s])
  }
  pcap[s] <- sum(f[1:nD,s])  # Overall detection probability
```

9.7 OPEN HDS MODELS: MODELING POPULATION DYNAMICS

```
# Process model
  for (t in 1:T){
    log(lambda[s,t]) <- beta0 + beta1*chap[s] + beta2*chap2[s] + beta3*elev[s] +
beta4*(t - t/2)  # Note trend parameter here
    y[s,t] ~ dbin(pcap[s], N[s,t])
    N[s,t] ~ dnegbin(prob[s,t], r)
    prob[s,t] <- r/(r+lambda[s,t])
  } # End loop over years
} # End loop over sites

# Distance sampling observation model for observed (binned) distance data
for(i in 1:nind){
  dclass[i] ~ dcat(fct[1:nD,site[i]])
}
# Derived parameters
for(t in 1:6){
  Ntot[t] <- sum(N[,t])
  D[t] <- Ntot[t] / (28.27*nsites)   # 300 m point = 28.27 ha
}
}
", file="Sollmann1.txt")

# Set up initial values, parameters vector and MCMC settings
Nst <- y+1  # this is for trend model
inits <- function(){list(N=Nst, beta0=runif(1), beta1=runif(1), beta2=runif(1),
  beta3=runif(1), beta4=runif(1), alpha0=runif(1,3,5), alpha1=runif(1), r=1)}
params <-c('beta0', 'beta1', 'beta2', 'beta3', 'beta4', 'alpha0', 'alpha1', 'Ntot',
'D', 'r')
ni <- 22000 ; nb <- 2000 ; nt <- 1 ; nc <- 3

# JAGS setting b/c otherwise JAGS cannot build a sampler, rec. by M. Plummer
library(jagsUI)
set.factory("bugs::Conjugate", FALSE, type="sampler")

# Execute JAGS, look at convergence and summarize the results
open1 <- jags (data1, inits, params, "Sollmann1.txt", n.thin=nt, n.chains=nc,
  n.burnin=nb, n.iter=ni)
par(mfrow = c(3,3)) ; traceplot(open1) ; print(open1, 2)
```

JAGS output for model 'Sollmann1.txt', generated by jagsUI.
Estimates based on 3 chains of 22000 iterations, burn-in = 2000 and thin rate = 1,
yielding 60000 total samples from the joint posterior.
MCMC ran for 81.61 minutes at time 2015-07-12 21:27:49.

	mean	sd	2.5%	50%	97.5%	overlap0	f	Rhat	n.eff
beta0	0.99	0.14	0.72	0.99	1.26	FALSE	1.00	1	34306
beta1	1.75	0.09	1.58	1.75	1.94	FALSE	1.00	1	3476
beta2	-0.51	0.05	-0.61	-0.51	-0.41	FALSE	1.00	1	3640
beta3	-0.34	0.06	-0.47	-0.34	-0.22	FALSE	1.00	1	21284
beta4	-0.08	0.07	-0.21	-0.08	0.05	TRUE	0.88	1	60000
alpha0	4.68	0.02	4.64	4.68	4.73	FALSE	1.00	1	13522

alpha1	-0.19	0.02	-0.24	-0.19	-0.15	FALSE	1.00	1	4130
Ntot[1]	1094.84	76.19	953.00	1093.00	1251.00	FALSE	1.00	1	20103
Ntot[2]	972.49	70.50	841.00	971.00	1116.00	FALSE	1.00	1	56576
Ntot[3]	916.65	66.74	792.00	914.00	1054.00	FALSE	1.00	1	49904
Ntot[4]	934.19	67.10	809.00	932.00	1073.00	FALSE	1.00	1	40490
Ntot[5]	968.11	69.17	840.00	965.00	1110.00	FALSE	1.00	1	6407
Ntot[6]	854.38	65.43	732.00	852.00	989.00	FALSE	1.00	1	13167
D[1]	0.13	0.01	0.11	0.13	0.14	FALSE	1.00	1	20103
D[2]	0.11	0.01	0.10	0.11	0.13	FALSE	1.00	1	56576
D[3]	0.11	0.01	0.09	0.11	0.12	FALSE	1.00	1	49904
D[4]	0.11	0.01	0.09	0.11	0.12	FALSE	1.00	1	40490
D[5]	0.11	0.01	0.10	0.11	0.13	FALSE	1.00	1	6407
D[6]	0.10	0.01	0.08	0.10	0.11	FALSE	1.00	1	13167
r	0.35	0.03	0.29	0.35	0.41	FALSE	1.00	1	2109

9.7.2.2 The Reduced-Dynamics Model with Simple Exponential Growth

Next, we fit the reduced dynamics model. We see that the BUGS model specification looks essentially identical to that of the independence model above, except that the Poisson mean for $N_{s,t}$ now involves the previous population size.

```
# Write out the BUGS model file
cat("
model{

# Prior distributions
# Regression parameters
alpha0 ~ dunif(0,20)
alpha1 ~ dunif(-10,10)
beta0 ~ dunif(-20,20)
beta1 ~ dunif(-20,20)
beta2 ~ dunif(-20,20)
beta3 ~ dunif(-20,20)
theta ~ dunif(0,5)
# NegBin dispersion parameter
r ~ dunif(0,5)
rout <- log(r)

# 'Likelihood'
for (s in 1:nsites){
  # Linear model for detection function scale
  log(sigma[s]) <- alpha0+alpha1*chap[s]
  # Compute multinomial cell probabilities
  for(k in 1:nD){
    log(p[k,s]) <- -midpt[k]*midpt[k]/(2*sigma[s]*sigma[s])
    f[k,s] <- p[k,s]*pi[k,s]
    fc[k,s] <- f[k,s]/pcap[s]
    fct[k,s] <- fc[k,s]/sum(fc[1:nD,s])
    pi[k,s] <- (2*midpt[k]*delta )/(B*B)
  }
  pcap[s] <- sum(f[1:nD,s])    # Overall detection probability
```

```
# Process model
# Abundance model for Yr1 as in Sillett et al, 2012
log(lambda[s,1]) <- beta0 + beta1*chap[s] + beta2*chap2[s] + beta3*elev[s]
y[s,1] ~ dbin(pcap[s], N[s,1])
N[s,1] ~ dnegbin(prob[s,1], r)
prob[s,1] <- r/(r+lambda[s,1])

# Population dynamics model for subsequent years
  for (t in 2:T){
    N[s,t] ~ dpois(N[s, t-1] * theta)
    y[s,t] ~ dbin(pcap[s], N[s,t])
  }
}
# Distance sampling observation model for observed (binned) distance data
for(i in 1:nind){
    dclass[i] ~ dcat(fct[1:nD,site[i]])
}

# Derived parameters
for(t in 1:6){
   Ntot[t] <- sum(N[,t])
   D[t] <- Ntot[t] / (28.27*nsites) # 300 m point = 28.27 ha
}
}
", file="Sollmann2.txt")

# Set up initial values, parameter vector and MCMC settings
Nst <- y+1 # this is for trend model
inits <- function(){list(N=Nst, beta0=runif(1), beta1=runif(1), beta2=runif(1),
    beta3=runif(1), alpha0=runif(1,3,5), alpha1=runif(1), theta=runif(1,0.6,0.99))}
params <- c('beta0', 'beta1', 'beta2', 'beta3', 'alpha0', 'alpha1', 'theta',
    'rout', 'Ntot', 'D', 'r')
ni <- 22000 ; nb <- 2000 ; nt <- 1 ; nc <- 3

# Execute JAGS, look at convergence and summarize the results
open2 <- jags (data1, inits, params, "Sollmann2.txt", n.thin=nt, n.chains=nc,
    n.burnin=nb, n.iter=ni)
par(mfrow = c(3,3)) ; traceplot(open2) ; print(open2, 2)
```

JAGS output for model 'Sollmann2.txt', generated by jagsUI.
Estimates based on 3 chains of 22000 iterations, burn-in = 2000 and thin rate = 1,
yielding 60000 total samples from the joint posterior.
MCMC ran for 24.335 minutes at time 2015-07-12 19:05:01.

	mean	sd	2.5%	50%	97.5%	overlap0	f	Rhat	n.eff
beta0	0.97	0.14	0.70	0.97	1.25	FALSE	1.00	1	3706
beta1	1.78	0.18	1.44	1.78	2.13	FALSE	1.00	1	9749
beta2	-0.52	0.10	-0.71	-0.52	-0.33	FALSE	1.00	1	3305
beta3	-0.32	0.13	-0.58	-0.32	-0.06	FALSE	0.99	1	3715
alpha0	4.67	0.02	4.63	4.67	4.72	FALSE	1.00	1	60000
alpha1	-0.19	0.02	-0.23	-0.19	-0.15	FALSE	1.00	1	60000

theta	0.96	0.02	0.92	0.96	1.01	FALSE	1.00	1	8126
rout	-0.89	0.15	-1.18	-0.89	-0.59	FALSE	1.00	1	12607
Ntot[1]	1058.40	62.01	941.00	1057.00	1184.00	FALSE	1.00	1	8791
Ntot[2]	1009.30	50.46	913.00	1008.00	1110.00	FALSE	1.00	1	14850
Ntot[3]	974.86	45.87	887.00	974.00	1066.00	FALSE	1.00	1	14442
Ntot[4]	949.90	45.13	864.00	949.00	1040.00	FALSE	1.00	1	6549
Ntot[5]	926.22	47.53	837.00	925.00	1021.03	FALSE	1.00	1	5471
Ntot[6]	884.32	55.50	780.00	883.00	997.00	FALSE	1.00	1	6917
D[1]	0.12	0.01	0.11	0.12	0.14	FALSE	1.00	1	8791
D[2]	0.12	0.01	0.11	0.12	0.13	FALSE	1.00	1	14850
D[3]	0.11	0.01	0.10	0.11	0.12	FALSE	1.00	1	14442
D[4]	0.11	0.01	0.10	0.11	0.12	FALSE	1.00	1	6549
D[5]	0.11	0.01	0.10	0.11	0.12	FALSE	1.00	1	5471
D[6]	0.10	0.01	0.09	0.10	0.11	FALSE	1.00	1	6917
r	0.42	0.06	0.31	0.41	0.56	FALSE	1.00	1	12811

9.7.2.3 The Glorious Integrated HDS/DM Model

We specify a basic HDS/DM model here that has constant dynamics parameters ϕ and γ. For this model, some care has to be taken in setting up the starting values for the latent number of survivors and recruits.

```
# Write out the BUGS model file
cat("
model{
# Prior distributions
# Regression parameters
alpha0 ~ dunif(0,20)
alpha1 ~ dunif(-10,10)
beta0 ~ dunif(-20,20)
beta1 ~ dunif(-20,20)
beta2 ~ dunif(-20,20)
beta3 ~ dunif(-20,20)

# Priors for dynamics parameters: here they are constant across years
# We could add covariate models for logit(phi) and log(gamma)
phi ~ dunif(0,1)
gamma ~ dunif(0,5)

# NegBin dispersion parameter
r ~ dunif(0,5)
rout <- log(r)

# 'Likelihood'
for (s in 1:nsites){
  # Linear model for detection function scale
  log(sigma[s]) <- alpha0+alpha1*chap[s]

  # Compute multinomial cell probabilities
  for(k in 1:nD){
    log(p[k,s]) <- -midpt[k]*midpt[k]/(2*sigma[s]*sigma[s])
```

```
      f[k,s] <- p[k,s]*pi[k,s]
      fc[k,s] <- f[k,s]/pcap[s]
      fct[k,s] <- fc[k,s]/sum(fc[1:nD,s])
      pi[k,s] <- (2*midpt[k]*delta )/(B*B)
   }
   pcap[s] <- sum(f[1:nD,s]) # Overall detection probability

   # Process model
   # Abundance model for year 1
   log(lambda[s,1]) <- beta0 + beta1*chap[s] + beta2*chap2[s] + beta3*elev[s]
   y[s,1] ~ dbin(pcap[s], N[s,1])
   N[s,1] ~ dnegbin(prob[s,1], r)
   prob[s,1] <- r/(r+lambda[s,1])

   # Population dynamics model for subsequent years
   for (t in 2:T){                          # Loop over years
       S[s,t] ~ dbinom(phi, N[s, t-1])      # Survivors
       R[s,t] ~ dpois(gamma * N[s, t-1])    # Recruits
       N[s,t] <- S[s,t] + R[s,t]            # N = Survivors + Recruits
       y[s,t]~ dbin(pcap[s],N[s,t])         # Measurement error
   }
}

# Distance sampling observation model for observed (binned) distance data
for(i in 1:nind){
   dclass[i] ~ dcat(fct[1:nD,site[i]])
}

# Derived parameters
for(t in 1:6){
   Ntot[t] <- sum(N[,t])
   D[t] <- Ntot[t] / (28.27*nsites)    # 300 m point = 28.27 ha
}
}
", file="Sollmann3.txt")
```

With complex models such as this, finding initial values for the Markov chains that do not result in a conflict with the data and the model can be a major headache with JAGS, and indeed this is perhaps the major disadvantage of using this otherwise fantastic software.

```
# Set up some sensible starting values for S and R
yin <- y+1
yin[,2:6] <- NA
Sin <- Rin <- matrix(NA, nrow=nsites, ncol=nyrs)
y1 <- y + 1
for(s in 1:nsites){
   for (t in 2:6){
      Sin[s,t] <- rbinom(1,y1[s,t-1], phi )
      Rin[s,t] <- ifelse((y1[s,t]-Sin[s,t])>0, y1[s,t]-Sin[s,t], 0)
   }
}
```

```
# Set up initial values, parameters vector and MCMC settings
inits <-function(){list(N=yin, beta0=runif(1), beta1=runif(1), beta2=runif(1),
    beta3=runif(1), alpha0=runif(1,3,5), alpha1=runif(1), phi=0.6, gamma=0.3,
    R=Rin, S=Sin) }
params <- c('beta0', 'beta1', 'beta2', 'beta3', 'alpha0', 'alpha1', 'phi',
    'gamma', 'Ntot', 'D', 'r')
ni <- 152000 ;  nb <- 2000 ;  nt <- 10 ;   nc <- 5

# Run JAGS, look at convergence and summarize the results
library(jagsUI)
open3 <- jags (data1, inits, params, "Sollmann3.txt", n.thin=nt, n.chains=nc,
n.burnin=nb, n.iter=ni, parallel=TRUE)
par(mfrow = c(3,3)) ;  traceplot(open3) ;  print(open3, 2)
```

JAGS output for model 'Sollmann3.txt', generated by jagsUI.
Estimates based on 5 chains of 152000 iterations, burn-in = 2000 and thin rate = 10,
yielding 75000 total samples from the joint posterior.
MCMC ran for 388.514 minutes at time 2015-07-12 22:55:32.

	mean	sd	2.5%	50%	97.5%	overlap0	f	Rhat	n.eff
beta0	0.95	0.15	0.67	0.95	1.24	FALSE	1.00	1.00	13991
beta1	1.81	0.18	1.46	1.80	2.16	FALSE	1.00	1.00	61883
beta2	-0.52	0.10	-0.72	-0.52	-0.33	FALSE	1.00	1.00	75000
beta3	-0.32	0.13	-0.59	-0.32	-0.07	FALSE	0.99	1.00	40446
alpha0	4.68	0.02	4.64	4.68	4.73	FALSE	1.00	1.00	7042
alpha1	-0.19	0.02	-0.23	-0.19	-0.15	FALSE	1.00	1.00	14271
phi	0.47	0.19	0.04	0.52	0.73	FALSE	1.00	1.05	654
gamma	0.49	0.19	0.24	0.45	0.93	FALSE	1.00	1.05	658
Ntot[1]	1046.05	61.88	929.00	1044.00	1173.00	FALSE	1.00	1.00	9374
Ntot[2]	999.83	50.51	904.00	999.00	1102.00	FALSE	1.00	1.00	12246
Ntot[3]	965.78	45.50	879.00	965.00	1057.00	FALSE	1.00	1.00	10598
Ntot[4]	939.74	44.68	855.00	939.00	1030.00	FALSE	1.00	1.00	9358
Ntot[5]	915.14	47.45	826.00	914.00	1012.00	FALSE	1.00	1.00	7753
Ntot[6]	876.87	54.87	774.00	875.00	988.00	FALSE	1.00	1.00	10618
D[1]	0.12	0.01	0.11	0.12	0.14	FALSE	1.00	1.00	9374
D[2]	0.12	0.01	0.10	0.12	0.13	FALSE	1.00	1.00	12246
D[3]	0.11	0.01	0.10	0.11	0.12	FALSE	1.00	1.00	10598
D[4]	0.11	0.01	0.10	0.11	0.12	FALSE	1.00	1.00	9358
D[5]	0.11	0.01	0.10	0.11	0.12	FALSE	1.00	1.00	7753
D[6]	0.10	0.01	0.09	0.10	0.11	FALSE	1.00	1.00	10618
r	0.40	0.06	0.29	0.40	0.53	FALSE	1.00	1.00	17108

Technically, we have convergence here, but slow mixing of the dynamics parameters, which seems typical for this model. We might run the model for many more iterations in order to improve our inferences about the parameters. Further, the posterior means for some parameters are not so close to the data-generating values, although we did a small simulation study that confirmed generally good behavior of the model for estimating these parameters.

Now we plot the three abundance trajectories along with the realized values of N_t for the simulation, and we see how well they agree (see Figure 9.6). A noticeable feature is that the partial and fully

9.7 OPEN HDS MODELS: MODELING POPULATION DYNAMICS

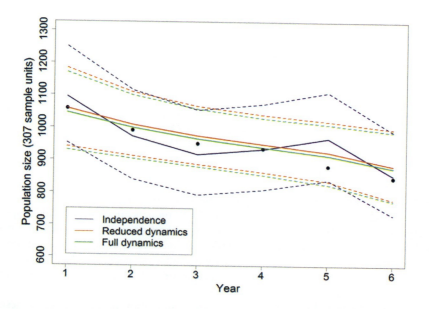

FIGURE 9.6

True population sizes (solid circles) and fitted trajectories under the three open-population HDS models: independence of population size in each year (blue); reduced dynamics with Markovian dependence of N_t on N_{t-1} (exponential population growth model, red); and full dynamics including explicit recruitment and survival (green).

dynamic models do quite a bit of smoothing compared with the independence model, and the estimates of each year-specific population size (shown in the output above) are somewhat more precise for these models, because they are borrowing information about population size from previous (and subsequent) years. That trend estimates can be improved by adding biological realism (i.e., population dynamics) into models for trends has also been noted by de Valpine (2003).

```
# Compare inferences in graph .... (Figure 9.6)
plot(apply(dat$N,2,sum),ylim=c(600,1300),xlab="Year",ylab="Population size (307
sample units)")
lines(apply(open1$sims.list$Ntot,2,mean), lty=1, col="blue", lwd=2)
lines(apply(open2$sims.list$Ntot,2,mean), lty=1, col="red", lwd=2)
lines(apply(open3$sims.list$Ntot,2,mean), lty=1, col="green", lwd=2)
open1.95cri <- apply(open1$sims.list$N,2,function(x) quantile(x, c(0.025,0.975)))
open2.95cri <- apply(open2$sims.list$N,2,function(x) quantile(x, c(0.025,0.975)))
open3.95cri <- apply(open3$sims.list$N,2,function(x) quantile(x, c(0.025,0.975)))
legend(1,750,legend=c("Independence","Reduced dynamics","Full dynamics"), lty=1,
col=c("blue","red","green"))

matlines(1:6, t(open1.95cri), type="l", lty=2, lwd=2, col="blue")
matlines(1:6, t(open2.95cri), type="l", lty=2, lwd=2, col="red")
matlines(1:6, t(open3.95cri), type="l", lty=2, lwd=2, col="green")
```

```
# .... and table
parms <- c(betaFall, dparm)
round(post <- cbind(parms, Independent=open1$summary[c(1:4,6,7),1],
    Partial=open2$summary[1:6,1], Full=open3$summary[1:6,1]), 3)

                      parms  Independent  Partial   Full
lambda(Int)           0.827      0.990     0.971   0.949
lambda(chaparral)     1.432      1.755     1.783   1.806
lambda(elevation)    -0.227     -0.509    -0.519  -0.523
lambda(I(chaparral^2)) -0.376   -0.343    -0.317  -0.323
p(Int)                4.679      4.682     4.672   4.680
p(chaparral)         -0.199     -0.194    -0.187  -0.192
```

9.7.2.4 Summary Remarks on Modeling Populations Over Time

We think the scope and relevance of open-population HDS models are enormous, and expect that they will be widely used in the future, given the ease with which they are implemented in the BUGS language and the accessibility of BUGS to practitioners for developing extensions of the model. We noted some limitations of our coverage earlier that we emphasize here: (1) The models cannot yet be fitted in unmarked, although similar models for *N*-mixtures can be fitted (see Chapters 12 and 14). We hope to implement HDS/DM models consistent with the Sollmann et al. (2015) framework in the near future; (2) We have not covered models with temporary emigration in the context of these multiyear dynamic models, although we think you can achieve that extension by working with the code we provided in the previous section and this one.

9.8 SPATIAL DISTANCE SAMPLING: MODELING WITHIN-UNIT VARIATION IN DENSITY

One of the basic assumptions of distance sampling is that density is constant—this is the "uniformity" assumption. This assumption is implicit in the mathematics upon which the DS estimator is based. We've relaxed this assumption a little bit, by allowing for variation among sample units using the HDS models of the previous chapter (and this chapter). The HDS models we dealt with so far assumed that N_s is the population size of spatial sample unit s, and they accommodate variation in population size among sample units by specifying covariates or random effects in a model for N_s. However, they don't make any explicit assumptions about *"within sample unit"* variation in density, but rather assume that the average covariate value defined for the sample unit is meaningful for explaining among-sample-unit variation. Thus, when we assume that $N_s \sim Poisson(\lambda_s)$, the parameter λ_s is constant for the sample location s and represents the aggregate mean for the sample unit. This is normally a sensible view when the sample unit is relatively small, such as a bird point count location. From a purely pragmatic standpoint, for small spatial units, we seldom have much within-sample-unit habitat information to work with (it's not available or not collected), so sample unit summaries of both counts of animals and of habitats "in the vicinity" of the count location are very practical to work with. In addition, the within-unit heterogeneity in density may not matter for the objective of obtaining a scalar density estimate, because of course DS is likely to be at least approximately design-unbiased anyhow. This just means in the case of within-unit heterogeneity, if the center point is random with respect to the habitat structure, then you probably won't get hurt too badly from a statistical standpoint.

While not modeling within-sample-unit heterogeneity will not usually be a critical misspecification of a model, there may be some cases where we want to model such heterogeneity. For example, if we are doing transects or point counts of a typical tree-dwelling passerine bird, and some transects fall over ponds, roads, or other habitats not suitable for the bird, we would expect within-unit variation in density that we may want to account for, or we may wish to exclude some locations as possible locations for individuals. Or we may have within-unit habitat measurements that are central to some biological or management-oriented hypothesis that we care about. And when sample units are very large (e.g., long transects), there are liable to be strong habitat gradients that we might consider (Hedley and Buckland, 2004). For these situations, we need spatially explicit DS models, which we describe in the following sections.

Spatially explicit models of density—that is to say, models that accommodate within-sample-unit variation—require that we come up with an alternative formulation of DS in terms of the locations of individuals. In statistical terminology, we formulate the model in terms of a point process that describes the distribution of individuals in space, and then we focus our model on the intensity function of the point process (Johnson et al., 2010; Miller et al., 2013a). These models are very similar to SCR models or spatially explicit capture-recapture models (Efford, 2004; Borchers and Efford, 2008; Royle and Young, 2008; Royle et al., 2011, 2014; Borchers et al., 2015), which also describe individual encounters conditional on an explicit spatial point process, and model the probability of detection as a function of distance from the observer (trap) to the object (in SCR, this is a latent variable).

As an example, consider a single point count location of radius 300 m shown in Figure 9.7, which is located in a habitat that is decidedly not homogeneous. In the present case, we'll say this is a shrubby field habitat, and that the blue patches are shrub cover. For illustration, we simulated a point pattern of $N = 200$ individuals in this landscape nonrandomly, so that the probability of an individual occurring in any given pixel is proportional to the metric value of the pixel and the species prefers open habitat over

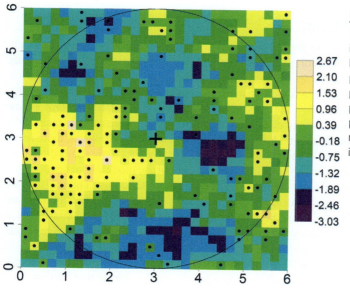

FIGURE 9.7

Point pattern generated in an inhomogeneous habitat with 200 bird locations simulated such that their probability of occurring in a pixel is proportional to the habitat value for that pixel. Plus sign shows the location of the point count sampler and the circle is the maximum distance out to which individuals are recorded.

shrubs. There are 900 pixels here, each being 10×10 m in dimension. It's hard to imagine that we would typically have such fine resolution data for most bird surveys (nor such high densities of birds), yet conceptually, the basic problem of having a nonuniform distribution of individuals should be obviously relevant. In the extreme case, we might have a binary mask "suitable or not," or "water or nonwater," or "road, nonroad," that precludes the possible observation of individuals at some locations. This is a direct way of describing constraints in the potential distribution of distances—i.e., nonuniformity. An example of this is in the work of Chelgren et al. (2011b), where they defined a covariate to be the percentage of suitable habitat multiplied by the $E(N)$ component of the model (see their Appendix 2). Similarly, the use of a habitat mask might have particular relevance to roadside sampling of birds, where one could construct DS models that account for density gradients in the vicinity of the road. Marques et al. (2010, 2013) discuss this particular case and propose a solution where auxiliary information about density is available. However, one does not need auxiliary information about density, provided one records individual locations rather than distance (as noted in Marques et al., 2010 in the context of point counts).

In the following section, we will use this simplified system of a nonuniform distribution (Figure 9.7) to extend our basic Bayesian DS model from Section 8.2.5. We also demonstrate a novel likelihood-based analysis framework in the unmarked package, using a customized version of the pcount function for N-mixture models. And we provide a simulator function to simulate a landscape and a point pattern to use for illustration. Of course, working with a single point count to illustrate fitting models requires that we simulate a fairly large sample size—e.g., $N = 200$ individuals—that we would never have in practice at a single point. Therefore, in Section 9.8.4 below, we'll consider having multiple points with a separate habitat raster for each point. Obviously, the size of the data sets that we must contend with expands pretty dramatically as the number of points increases, especially if we have increasing numbers of habitat variables. This will be something to keep in mind in practical applications of these spatial HDS models.

9.8.1 DS WITH LOCATION OF ENCOUNTER

It is relatively straightforward to reformulate the DS model to accommodate within-sample-unit heterogeneity. To do this, we first provide a formulation of DS that is based on the *locations of individuals* and *not* their *distance* to the observer. In other words, previously we formulated the model for detection probability directly in terms of distance, but in reality, distance is only a summary of what is actually observed, which is the location of each individual. Under classical DS models and the uniformity assumption, location is not informative about any of the model parameters once distance is taken into account, so it has not typically been used in the models, or wasn't even recorded. However, there is no real need to reduce location to distance if we could just formulate DS models directly in terms of spatial location. Once we have this formulation, it is easy and direct to think about the modeling of small-grain covariates (i.e., within the sample unit) that influence the locations of individuals. This is then an explicit spatial model of density. This formulation of the DS model based on location of encounter can be thought of as a primitive type of SCR model (see Section 4.6 in Royle et al., 2014). And this makes it easy to accommodate sampling units of arbitrary geometry or configuration. For example, we can then easily clip out some parts of "nonhabitats" where animals cannot reside.

We first simulate a data set using our handy DS simulator function sim.pdata from back in Section 8.2.5. Recall that the function also returned the coordinates of each individual in addition to the distance, but we haven't previously used the former. Here, we need these coordinates. We provide a formulation of the model using data augmentation (DA), and so we also augment the coordinates of the simulated data set. In this case, the coordinates of each augmented individual are regarded as missing data:

```
# Simulate a data set and harvest the output
set.seed(1234)
str(tmp <- sim.pdata(N=200,sigma=1,keep.all=FALSE,B=3))

List of 8
 $ N      : num 200
 $ sigma  : num 1
 $ B      : num 3
 $ u1     : num [1:37] 3.73 3.66 3.27 1.72 3.32 ...
 $ u2     : num [1:37] 3.17 1.9 1.68 3.83 2.54 ...
 $ d      : num [1:37] 0.753 1.276 1.345 1.529 0.557 ...
 $ y      : int [1:200] 0 1 1 0 0 0 0 0 0 0 ...
 $ N.real : int 152

# Harvest some data objects
B <- tmp$B
d <- tmp$d
u1 <- tmp$u1
u2 <- tmp$u2
nind <- length(d)

# Data augmentation
M <- 400                            # Max of 400 individuals
nz <- M-nind                        # Augment by nz individuals
y <- c(rep(1,nind), rep(0,nz))      # Augmented data augmentation variable
u <- cbind(u1,u2)                   # Locations of individuals detected
u <- rbind(u, matrix(NA, nrow=nz, ncol=2))
```

Next, we need to construct the model in BUGS. The key difference between this formulation of the model and our previous implementations of DS is that now we work directly with the location data instead of distance data, and the location coordinate observations are a vector of two values (easting, northing; x, y; u, v; u1, u2; etc.). Thus, we need to specify a prior distribution for the coordinates instead of for distance. What is a sensible prior distribution for the coordinates in a 2-dimensional region? Well, in Chapter 8 we talked in some detail about the basic fundamental assumption of DS, which is the uniformity assumption. That is, before seeing the data, it seems sensible to assume that the individuals are uniformly distributed in space, which induces a uniform distribution of distances (for line transects) or a triangular distribution of distances (for point counts). Obviously then, this uniformity assumption can serve as the basic assumption for individual coordinates. To be precise, let $\mathbf{u}_i$ be the location of individual i; in a population of N individuals, their locations are $\mathbf{u}_1,\ldots,\mathbf{u}_N$. In statistical parlance, we regard this collection of N points as a realization of a statistical point process. A probabilistic description of the mechanism that gave rise to this collection of N points is a *point process model*, and one of the fundamental requirements of specifying a point process model is that we have to describe the region within which the points are distributed. We'll call this the *state-space* of the point process, which we denote by U. This is just the region over which the N points are distributed, or in fact it is the set of *possible values* of each of the individuals' coordinates **u** in our population. The uniformity assumption for the point locations is equivalent to saying that each **u** is uniformly distributed over the region U, which we'll denote by $u_i \sim Uniform(U)$. Thus, in formulating the DS model in terms of coordinate locations instead of distance, we replace our explicit assumption about the distribution of d_i with an explicit assumption about the distribution of the coordinates u_i. This requires that we are explicit about where those points might be located; i.e., we have to specify the state-space U. When the region U is a rectangle, we can express the assumption that each u_i is uniform over the region U by equivalently expressing each coordinate of u_i (in the x and y directions) as having a uniform distribution on the possible values of u_1 and u_2. For example, in our simulated data set which we just created, we simulated observations on a circle centered at (B, B), so we can define our state-space by the square area $[0, 2B] \times [0, 2B]$ and then impose the uniformity assumption for u_i by specifying that the coordinates be independently distributed on the interval $[0, 2B]$. We show this in the BUGS language momentarily. Of course, we did this throughout Chapter 8, because it is a convenient way to simulate points uniformly in space.

```
# Bundle and summarize the data set
str(data <- list(B=B, nind=nind, u=u, y=y, nz=nz))
List of 5
 $ B    : num 3
 $ nind : int 37
 $ u    : num [1:400, 1:2] 3.73 3.66 3.27 1.72 3.32 ...
  ..- attr(*, "dimnames")=List of 2
  .. ..$ : NULL
  .. ..$ : chr [1:2] "u1" "u2"
 $ y    : num [1:400] 1 1 1 1 1 1 1 1 1 ...
 $ nz   : num 363
```

One final point about specifying the model in terms of individual coordinates: we still describe the detection probability model in terms of *distance* between u_i and the center point of the circle, so we have to define distance as a derived quantity given the locations, instead of "data" as it is in all conventional applications of DS (and in Chapter 8). This is just one additional line of BUGS code. The BUGS model is as follows:

```
# Write out the BUGS model file
cat("
model{

# Priors
sigma ~ dunif(0,10)
psi ~ dunif(0,1)

# Categorical observation model
for(i in 1:(nind+nz)){
  z[i] ~ dbern(psi)
  u[i,1] ~ dunif(0, 2*B)   # Here is the uniformity assumption made explicit
  u[i,2] ~ dunif(0, 2*B)
```

```
# Compute distance as a derived quantity
d[i] <- pow( pow( u[i,1]-B,2) + pow(u[i,2]-B,2), 0.5) # Pythagoras
p[i] <- exp(-d[i]*d[i] / (2*sigma*sigma))
mu[i] <- p[i] * z[i]
y[i] ~ dbern(mu[i])
}
# Other derived quantities
N <- sum(z[])
D <- N / (B*B)
}
",fill=TRUE,file="model1.txt")
```

An important point: note that in this model specification, we do not impose the restriction of counting individuals only in the circle of radius B. We could do this by adding one extra bit of model specification that converts `p[i]` to 0 if `d[i] > B`, if this is necessary due to how the data were recorded. The model as written defines the state-space of possible locations is the square of dimension $2B \times 2B$, so the estimate of N will apply to that square. Thus, when we compute density as a derived parameter, we use an area of $2B \times 2B$ in the denominator. What this means in practice is that we need not truncate our observations at B or else, if we do, then we would have to slightly modify this model.

Next, we set up the MCMC settings, the data, and the initial values, we set everything running in JAGS, and then produce some posterior summaries from fitting the model to our simulated data set:

```
# Load libraries and specify MCMC settings
library("R2WinBUGS") ; library(jagsUI)
ni <- 22000 ; nb <- 2000 ; nthin <- 2 ; nc <- 3

# Inits and parameters
inits <- function(){
  list(sigma=runif(1,1,10), psi=runif(1),z = c(rep(1,nind),rep(0,nz)) ) }
params <- c("sigma", "N", "psi")

# Execute jags and summarize the posterior distributions
out1 <- jags (data, inits, parameters, "model1.txt", n.thin=nthin,
    n.chains=nc, n.burnin=nb,n.iter=ni, parallel = FALSE)
par(mfrow = c(2,2)) ; traceplot(out1)
print(out1, 2)
             mean    sd    2.5%    50%   97.5% overlap0  f Rhat n.eff
sigma        1.06  0.10    0.89   1.05    1.29    FALSE  1    1 12430
N          199.65 45.95  123.00 196.00  302.00    FALSE  1    1 13681
psi          0.50  0.12    0.30   0.49    0.76    FALSE  1    1 13035
deviance   385.65 17.45  353.05 385.04  421.23    FALSE  1    1 18978
```

There's not too much to say about the results here, other than to reiterate the main point, which is that formulation of the model in terms of location instead of distance requires only changing the prior distribution to be uniform over the two-dimensional region being sampled instead of the specific triangular distribution for distance induced by point count sampling. In general, no matter the shape of the sample region, the distribution for object location can be specified easily (uniform or, as we see in the next section, proportional to a covariate). That is what makes the DS model, specified in terms of object *location*, so versatile.

9.8.2 THE LINE TRANSECT CASE

When we sample by a line transect, which may be linear or a general curved path, the formulation of the model is not so straightforward, because there is no sample "point," but rather an infinity of points (along the line), and the state-space U may not be so obvious (as a square around a circular sample unit). To deal with samples that are based on lines, it is convenient in BUGS to represent a line as a sequence of points that are relatively close together. For example, if we have a line transect that is 1 km in length, then perhaps we represent this by 20 points that are 50 m apart, or 100 points that are 10 m apart. As long as the point spacing is not so distant relative to the scale parameter of the distance function, there will be little loss of precision compared with a continuous line. Of course, we could always use a point grid with 1 or 0.1 m spacing if we were worried about a loss of precision due to numerical approximation. The good thing about this approach of chopping the line up into points is that it works in complete generality no matter the form of the line, which may be as wiggly and convoluted as imaginable. For an example, see Royle et al. (2011), who show an example of a hierarchical capture-recapture DS model with a wiggly track line for data from the Swiss breeding bird survey MHB; see Figure 6.8 for an example of such an irregular survey route.

The other issue to confront when dealing with line transects or general wiggly lines is prescribing the state-space. As it turns out, we can still use a nice rectangular state-space, but we need to choose it so that the line sample unit is nested within it and the edge of the state-space is sufficiently far from any point on the line, so that the probability of detection at the edge is near 0. For example, if we use the half-normal encounter probability function with scale parameter σ, it is sufficient to choose a rectangular state-space enclosing the line that is a minimum of two or three times σ from any point on the line. Some discussion of choosing the state-space and sensitivity to the state-space are discussed in Section 5.3.1 of Royle et al. (2014).

We discuss HDS for linear and wiggly transects in volume 2 (Chapter 24), along with some related SCR models.

9.8.3 MODELING SPATIAL COVARIATES

Now we extend the model slightly to accommodate explicit covariates within a sample unit. The conceptual model is that N is the population size of the sample unit, but the distribution of the N individuals within the unit is no longer uniform, and rather depends on some covariate value x (of course there may be multiple covariates). In making the transition to general models that extend beyond those assuming a uniform distribution of individuals (as in all previous sections), we need to also rethink the prior distribution for the individual locations $\mathbf{u}_i$. Conceptually, we regard the points $\mathbf{u}_1, \mathbf{u}_2, \ldots, \mathbf{u}_N$—i.e., the locations of all N individuals—as a realization of a point process with intensity function $\lambda(\mathbf{u}) = \lambda_0 exp(\beta_1 x(\mathbf{u}))$, where λ_0 and β_1 are parameters to be estimated, and $x(\mathbf{u})$ is some covariate that affects the density of points. The uniformity assumption of conventional DS is realized if we set $\beta_1 = 0$. In general, however, the probability density of point location is proportional to the intensity function, which can depend on explicit covariates. In principle, it is not difficult to fit such models in continuous space—such models have been fitted by Johnson et al. (2010) and Niemi and Fernandez (2010). But for our development here, we deviate from this a little bit and consider discrete versions of these models for the convenience of implementing them in the BUGS language (although we could probably use the zeros or ones tricks to implement the continuous space models, as Mike

Meredith has done in SCR models; see his website on the topic here: mikemeredith.net/blog/1309_SECR_in_JAGS_patchy_habitat.htm). Here, we suppose that the covariate is defined by a raster of discrete units of some fine resolution, and we index the raster pixels by g, so x_g is the value of the covariate at pixel g. Not only is the discrete space version of these models convenient in the BUGS language, but also we often have covariates that are defined by a discrete raster (e.g., in a GIS system), so we would have to either accommodate that in a continuous space point process model or discretize the continuous space model. We think the latter is somewhat more convenient to deal with.

A particular type of point process model is a Poisson point process (PPP) model with intensity function $\lambda(\mathbf{u}) = \lambda_0 exp(\beta_1 x(\mathbf{u}))$, or equivalently $\log(\lambda_\mathbf{u}) = \beta_0 + \beta_1 x(\mathbf{u})$, with $\beta_0 = \log(\lambda_0)$. The PPP model assumes that the total number of points N in the state-space has a Poisson distribution with parameter Λ:

$$N \sim Poisson(\Lambda)$$

where

$$\Lambda = \int_\mathbf{u} \lambda_0 exp(\beta_1 x(\mathbf{u})) d\mathbf{u}$$

One implication of the PPP model is that the number of points in any region of the state-space is a Poisson outcome with expected value obtained by integrating the intensity function over the relevant region. For discrete space—e.g., when pixel density is of sufficiently fine resolution—we are satisfied with the numerical approximation of the integral by summation (see Section 8.3.3):

$$\Lambda = \sum_g \lambda_0 exp(\beta_1 x_g)$$

for the grid of all points $g = 1, 2, \ldots, G$ that index the raster coordinates (by convention we define them by their center points). In this case, the baseline intensity λ_0 includes the area of the pixel, and thus it is the expected value *per pixel* when $\beta_1 x_g = 0$.

To analyze this model in BUGS, it is convenient to apply DA (Royle et al., 2007a) to the model, conditional on the total population size N, which is then easily adapted to multiple spatial units (sites; see below). To implement DA for this problem, we make sure to augment the observed data set to size M, and we introduce a collection of M binary DA variables $z_i \sim Bernoulli(\psi)$. As we noted in Section 7.8.1.4, the parameter ψ takes the place of the intercept β_0 in the model formulation (see Royle and Converse, 2014), and we estimate ψ and β_1 instead of β_0 and β_1. Alternatively, we could express ψ as a deterministic function of β_0 (see Section 7.8.1.4), but for now we retain the $\{\psi, \beta_1\}$ formulation of the model. The practical consequence of using DA here is this: given the total population size N for a sample unit, the distribution of individuals among the $g = 1, 2, \ldots, G$ pixels is multinomial with cell probabilities

$$\pi_g = \frac{exp(\beta_1 x_g)}{\sum_{g=1}^{G} exp(\beta_1 x_g)}$$

In other words, the PPP model, when conditioned on N in discrete space, is equivalent to a multinomial model for locations having cell probabilities π_g. This model can be readily specified in BUGS.

We provide a simulator function sim.spatialDS (shown below) in the AHM package that will simulate basic spatial DS data with a raster covariate as in Figure 9.7. The key idea of this function is that it creates a spatially explicit covariate on some raster, having spacing "delta," that is chosen in this case to chop the square up into a matrix of 30 × 30 pixels. Then a length 900 covariate vector is generated as a correlated multivariate normal variate. The basic principle behind this is as follows: if **V** is a variance-covariance matrix, and **e** is a compatible vector of iid normal random variables with mean 0 and variance 1, then if we define the vector **x** by

$$\mathbf{x} = \mathbf{V}^{1/2}\mathbf{e}$$

then it is a multivariate normal random variable with variance-covariance matrix **V**. The square root of a matrix is its Cholesky decomposition, and this is implemented in the R language with the function chol. So this is what our simulator function does, and it makes an image of the covariate and superimposes the N individual locations onto the map. The N individuals are distributed over the map in proportion to the habitat covariate according to the probabilities, for pixel g,

$$\pi_g = \frac{exp(\beta_1 x_g)}{\sum_g exp(\beta_1 x_g)}$$

In our simulator, instead of generating the total number of points N according to a Poisson distribution, we have N fixed (specified by the user). The result is that the model is not exactly consistent with a PPP model, but rather it is a binomial point process model. Binomial and Poisson point processes are virtually indistinguishable from a practical standpoint, both being fundamentally models of "randomly" distributed points. In the simulator function that follows, we note that locations are recorded as data, and we do not restrict locations to be on the circle, despite thinking of this as a "point count" type of sample. Of course, if you're in the field recording locations of individuals, then the circle boundary may not be so relevant.

```
# Simulator function for spatial distance sampling data
sim.spatialDS <-
function(N=1000, beta = 1, sigma=1, keep.all=FALSE, B=B, model="halfnorm"){
# Function simulates coordinates of individuals on a square
# Square is [0,2B] x [0,2B], with a count location on the point (B, B)
#    N: total population size in the square
#    beta: coefficient of SOEMTHING on spatial covariate x
#    sigma: scale of half-normal detection function
#    B: circle radius
#    keep.all: return the data for y=0 individuals or not
library(raster)      # Load required packages
library(plotrix)

# Create coordinates for 30 x 30 grid
delta <- (2*B-0)/30          # '2D bin width'
grx <- seq(delta/2, 2*B - delta/2, delta) # mid-point coordinates
gr <- expand.grid(grx,grx)    # Create grid coordinates

# Create spatially correlated covariate x and plot it
V <- exp(-e2dist(gr,gr)/1)
x <- t(chol(V))%*%rnorm(900)
par(mar=c(3,3,3,6))
```

9.8 SPATIAL DISTANCE SAMPLING

```r
image(rasterFromXYZ(cbind(gr,x)), col=topo.colors(10))
draw.circle(3, 3, B)
points(3, 3, pch="+", cex=3)
image.scale(x, col=topo.colors(10))

# Simulate point locations as function of habitat covariate x
probs <- exp(beta*x)/sum(exp(beta*x)) # probability of point in pixel (sum = 1)
pixel.id <- sample(1:900, N, replace=TRUE, prob=probs)
# could simulate randomly within the pixel but it won't matter so place centrally
u1 <- gr[pixel.id,1]
u2 <- gr[pixel.id,2]
points(u1, u2, pch=20, col='black', cex = 0.8) # plot points
title("Extremely cool figure")       # express your appreciation of all this

d <- sqrt((u1 - B)^2 + (u2-B)^2)     # distance to center point of square
#plot(u1, u2, pch = 1, main = "Point transect")
N.real <- sum(d<= B)                 # Population size inside of count circle

# Can only count individuals in the circle, so set to zero detection probability of
individuals in the corners (thereby truncating them)
# p <- ifelse(d< B, 1, 0) * exp(-d*d/(2*(sigma^2)))
# We do away with the circle constraint here.
if(model=="hazard")
   p <- 1-exp(-exp(-d*d/(2*sigma*sigma)))
if(model=="halfnorm")
   p <- exp(-d*d/(2*sigma*sigma))
# Now we decide whether each individual is detected or not
y <- rbinom(N, 1, p)                                         # detected or not
points(u1[d<= B], u2[d<= B], pch = 16, col = "black", cex = 1) # not detected
points(u1[y==1], u2[y==1], pch = 16, col = "red", cex = 1)    # detected

# Put all of the data in a matrix
if(!keep.all){
   u1 <- u1[y==1]
   u2 <- u2[y==1]
   d <- d[y==1]   }
# Output
return(list(model=model, N=N, beta=beta, B=B, u1=u1, u2=u2, d=d, y=y,
N.real=N.real, Habitat=x, grid=gr))
}
```

Next, we simulate a data set here and harvest the required objects including our "Habitat" covariate (see Figure 9.8). We note that the raster coordinates ("grid") are also extracted. In our simulation function, these are defined as the center point of each grid cell, which is how they usually are defined in practice. In our setup below, we take the simulated coordinates of each detection and identify the raster pixel that is closest simply by taking the distance to each pixel center point. In assigning observations to pixels, there may be more than one observation going to the same pixel—this is OK, we just have a duplicate "pixel ID" in the data set (see below). After we harvest our data, we do the DA to bring the data set up to size M. As in previous applications, this requires that we add a bunch of undetected individuals (having value $y = 0$) to our observation vector. However, in this case, we also need to augment the discrete location of each individual.

FIGURE 9.8

Observed locations of detected individuals (red points), undetected individuals (black points), and associated habitats in the vicinity of the point count location.

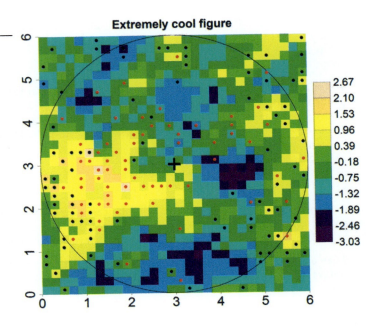

```
# Generate one data set and harvest the output
set.seed(1234)
str(tmp <- sim.spatialDS(N=200, beta=1, sigma=1.5, keep.all=FALSE, B=3))  # Figure 9.8

# Harvest data
B <- tmp$B
d <- tmp$d
u1 <- tmp$u1
u2 <- tmp$u2
Habitat <- as.vector(tmp$Habitat)
Habitat <- Habitat - mean(Habitat)
Habgrid <- tmp$grid
nind <- length(d)
G <- nrow(Habgrid)

# Do data augmentation, including for pixel ID
M <- 400
nz <- M-nind
pixel <- rep(NA, M)   # We use discrete "pixel ID" here instead of "s"
y <- c(rep(1,nind), rep(0,nz))

# Pick some starting values and figure out the pixel of each observation
s <- cbind(u1,u2)
s <- rbind(s, matrix(NA,nrow=nz,ncol=2))
D <- e2dist(s[1:nind,], Habgrid)
for(i in 1:nind){
```

```
    pixel[i] <- (1:ncol(D))[D[i,]==min(D[i,])] # pixel ID, there may be duplicates
}
# Bundle and summarize the data for BUGS
str(data <- list (B=B, nind=nind, y=y, nz=nz, Habitat=Habitat, Habgrid=Habgrid,
G=G, pixel=pixel))

List of 8
 $ B        : num 3
 $ nind     : int 76
 $ y        : num [1:400] 1 1 1 1 1 1 1 1 1 1 ...
 $ nz       : num 324
 $ Habitat  : num [1:900] -0.795 -0.417 0.356 -0.981 -0.482 ...
 $ Habgrid  :'data.frame':     900 obs. of 2 variables:
 ..$ Var1   : num [1:900] 0.1 0.3 0.5 0.7 0.9 1.1 1.3 1.5 1.7 1.9 ...
 ..$ Var2   : num [1:900] 0.1 0.1 0.1 0.1 0.1 0.1 0.1 0.1 0.1 0.1 ...
 ..- attr(*, "out.attrs")=List of 2
 .. ..$ dim       : int [1:2] 30 30
 .. ..$ dimnames:List of 2
 .. .. ..$ Var1: chr [1:30] "Var1=0.1" "Var1=0.3" "Var1=0.5" "Var1=0.7" ...
 .. .. ..$ Var2: chr [1:30] "Var2=0.1" "Var2=0.3" "Var2=0.5" "Var2=0.7" ...
 $ G        : int 900
 $ pixel    : int [1:400] 583 534 603 324 374 365 313 588 310 588 ...
```

Next, we write out the model file. We note that given the pixel location of each individual in the population, we define the coordinates of the individual by the center point of the pixel. We use that coordinate for computing the distance for use in the DS detection model (thus, there is implicit binning of distance in this model with the bin width related to the resolution of the habitat raster). Finally, we note that total population size for the $30 \times 30 = 900$-pixel raster (each pixel being 10×10 m, for a total size of 300×300 m, or 9 ha) is defined as a derived parameter related to the DA variables $N = \sum_{i=1}^{n+nz} z_i$, and density (per ha), D, is also a derived parameter related to N by $D = N/9$.

```
# Write BUGS model
cat("
model{

# Prior distributions
sigma ~ dunif(0,10)
psi ~ dunif(0,1)
beta ~ dnorm(0,0.01)

for(g in 1:G){   # g is the pixel index, there are G total pixels
   probs.num[g] <- exp(beta*Habitat[g])
   probs[g] <- probs.num[g]/sum(probs.num[])
}

# Models for DA variables and location (pixel)
for(i in 1:(nind+nz)){
   z[i] ~ dbern(psi)
```

```
   pixel[i] ~ dcat(probs[])
   s[i,1:2] <- Habgrid[pixel[i],]    # location = derived quantity
   # compute distance = derived quantity
   d[i] <- pow( pow( s[i,1]-B,2) + pow(s[i,2]-B,2), 0.5)
   p[i] <- exp(-d[i]*d[i]/(2*sigma*sigma))  # Half-normal detetion function
   mu[i] <- p[i]*z[i]
   y[i] ~ dbern(mu[i])                     # Observation model
}
# Derived parameters
N <- sum(z[])       # N is a derived parameter
D <- N/9            # area = 9 ha
}
",fill=TRUE, file="spatialDS.txt")

# Load libraries and specify MCMC settings
library("R2WinBUGS")
library(jagsUI)
ni <- 12000 ; nb <- 2000 ; nthin <- 2 ; nc <- 3

# Create inits and define parameters to monitor
inits <- function(){ list (sigma=runif(1,1,10),psi=runif(1),
     z = c(rep(1,nind), rep(0,nz)) ) }
params <- c("sigma", "N", "psi", "beta", "D")

# Run JAGS, check convergence and summarize posteriors
out2 <- jags (data, inits, params, "spatialDS.txt", n.thin=nthin,
    n.chains=nc, n.burnin=nb, n.iter=ni, parallel = TRUE)
par(mfrow = c(2,3) ; traceplot(out2)
print(out2, 2)

JAGS output for model 'spatialDS.txt', generated by jagsUI.
Estimates based on 3 chains of 12000 iterations,
burn-in = 2000 iterations and thin rate = 2,
yielding 15000 total samples from the joint posterior.
MCMC ran in parallel for 2.941 minutes at time 2015-07-11 18:41:35.

         mean     sd     2.5%     50%   97.5% overlap0 f Rhat n.eff
sigma    1.45   0.14    1.22    1.44    1.77   FALSE 1 1.00   472
N      214.00  37.32  151.00  211.00  298.00   FALSE 1 1.01   253
psi      0.53   0.10    0.37    0.53    0.75   FALSE 1 1.01   243
beta     1.00   0.13    0.73    1.00    1.26   FALSE 1 1.02   124
D       23.78   4.15   16.78   23.44   33.11   FALSE 1 1.01   253
```

While in this instance the Markov chain mixing appears adequate, the performance can be quite variable, and sometimes very long runs will be needed.

We can use the posterior output to create two different density maps (Figure 9.9). First, we can use the estimated habitat effect ("beta") and the observed habitat covariate to make a map of the *prior expected value* of N_g in pixel g (standardized by area this would be density, although we don't show that here since our spatial units are meaningless). In addition, by monitoring the "pixel ID" variable (not shown in the output above), we can make a map of the posterior means of N_g, which represents a

9.8 SPATIAL DISTANCE SAMPLING

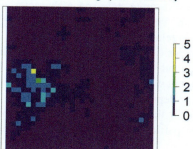
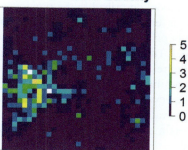

FIGURE 9.9

Estimated abundance maps: prior mean of N_g based on estimated habitat effect (left panel) and posterior mean of N_g (right panel). A stronger pattern would be evident if these were plotted on the log scale.

compromise between the estimated prior mean and the observed locations of individuals and thus will tend to be somewhat more speckled compared with the prior mean. This is analogous to the best unbiased predictor (BUP) that a frequentist would compute for a random effects model (see Section 6.4). We show the posterior means of N_s in Figure 9.9 (right). The R code to produce these figures is:

```
# Add pixel in order to make a density map
params <- c("sigma", "N", "psi", "beta", "D", "pixel")

# Run JAGS, check convergence and summarize posteriors
out2 <- jags (data, inits, params, "spatialDS.txt", n.thin=nthin,
   n.chains=nc, n.burnin=nb, n.iter=ni, parallel = FALSE)

# Plot density maps
library(raster)
par(mfrow=c(1,2))
pixel <- out2$sims.list$pixel
post <- table(pixel)/nrow(pixel)   # Average number of locations in each pixel
prior.mean <- mean(out2$sims.list$beta)*as.vector(Habitat)
prior.mean <- mean(out2$sims.list$psi)*M*exp(prior.mean)/sum(exp(prior.mean))
plot(rast.data <- rasterFromXYZ(cbind(Habgrid,prior.mean)), axes=FALSE,
     col=topo.colors(10) )
title("Prior mean density (estimated)")
plot(rast.post <- rasterFromXYZ(cbind(Habgrid,as.vector(post))),axes=FALSE,
     col=topo.colors(10) )
title("Posterior mean density")
```

9.8.4 SPATIAL HDS MODELS IN unmarked USING THE pcount FUNCTION

The unmarked package does not have any custom functions for fitting spatial HDS models. However, it is possible to hack the pcount function used for fitting *N*-mixture models and modify it for fitting spatial HDS models, and we describe that here.

To motivate the relevance of the binomial N-mixture model for this problem, we note that the spatial DS model described in the previous section, for a discrete state-space characterized by raster covariates, is easily formulated as a Poisson GLM in which we have data y_g = the number of detections in pixel g, and $N_g \sim Poisson(\lambda_g)$, where λ_g is related to the habitat variable x_g by the log-linear model. Then the model for the detection frequencies (in each pixel) is a type of hybrid Poisson GLM for the resulting data, in which $N_g \sim Poisson(\lambda_g)$ and $y_g|N_g \sim Binomial(N_g, p_g)$, which should look very familiar as the N-mixture model (Chapter 6), but with $K = 1$—that is, with a single visit. In this case, however, the detection probability parameter p_g depends on the distance of the pixel from the central observation point. Parameters are identifiable in this model without replication when we impose certain strict covariate relationships on p_g (in particular, the intercept must be 1.0 for the uncentered distance covariate). Thus, we can use the pcount function to fit this model if we can resolve the intercept problem. First note that pcount does accept the case in which there is a $K = 1$ replicate count, so there is no problem at all in taking spatial DS data structured as counts per pixel (in one vector) and inputting it into pcount (we show this in a minute). The main technical stumbling block is that pcount only allows logit-linear models for p, which are not standard DS models, and furthermore they do not have an intercept of 1.0. But we can trick pcount into fitting a DS-like model by removing the intercept from the detection model. For the logit model, this forces the intercept of p to be 0.50, which still isn't right, so we just edit the pcount function and multiply the inverse-logit transformation of the linear predictor by 2.0 to make the intercept equal to 1, and then we have the new function pcount.spHDS, which we demonstrate here (and provide with the AHM package and in the Web Supplement). Finally, we note that when we simulate a data set for analysis by this function, we will simulate it with a hazard encounter probability model of the form

$$p_g = 1 - exp\left(-exp\left(-d_g^2/2\sigma^2\right)\right)$$

Under this model, the complementary log-log link of p_g has only a quadratic term of distance:

$$cloglog(p_g) = -(1/2\sigma^2)d_g^2.$$

Our modified pcount.spHDS function only fits a logistic model for p, and that should be a better approximation to this hazard model (we may modify pcount.spHDS to fit either a log link or any other link to match standard DS models, but this had not yet been done at the time this book went into production). As a final note, we believe that the intercept of the logistic model could be estimated—that is, allowing for $p < 1$ at the center point—provided that replicate samples are available. This can be handled in pcount.spHDS as well, as it corresponds simply to the $K > 1$ situation. We note also that the multinomial N-mixture functions (Chapter 7) of unmarked could be hacked similarly to pcount.spHDS, to allow for the analysis of mark-recapture/DS (MRDS) models with (or without) spatial covariates.

To organize the data, we need to create a data set that consists of the pixel-specific counts and associated habitat variables. Using the data simulator we introduced previously, we demonstrate that here.

```
# Simulate a data set, N = 600 for the population size
set.seed(1234)
tmp <-sim.spatialDS(N=600, sigma=1.5, keep.all=FALSE, B=3, model= "hazard")
```

```
# Harvest the data
B <- tmp$B
d <- tmp$d
u1 <- tmp$u1
u2 <- tmp$u2
Habitat <- as.vector(tmp$Habitat)
Habitat <- Habitat - mean(Habitat)
Habgrid <- tmp$grid
nind <- length(d)
G <- nrow(Habgrid)

# Find which pixel each observation belongs to
s <- cbind(u1,u2)
D <- e2dist(s[1:nind,], Habgrid)
pixel <-rep(NA,nind)
for(i in 1:nind){
   pixel[i] <- (1:ncol(D))[D[i,]==min(D[i,])]
}

# Create a vector of counts in each pixel and pad it with zeros
pixel.count <- rep(0, G)
names(pixel.count) <- 1:G
pixel.count[names(table(pixel))] <- table(pixel)
# Create a covariate: distance between observer and pixel center
dist <- sqrt( (Habgrid[,1]-3)^2 + (Habgrid[,2]-3)^2 )
# Construct an unmarkedFrame
umf <- unmarkedFramePCount(y=matrix(pixel.count,ncol=1),
   siteCovs=data.frame(dist=dist,Habitat=Habitat))
summary(umf)

unmarkedFrame Object

900 sites
Maximum number of observations per site: 1
Mean number of observations per site: 1
Sites with at least one detection: 132

Tabulation of y observations:
   0   1   2   3   4   6  <NA>
 768 106  21   3   1   1    0

Site-level covariates:
      dist            Habitat
 Min.   :0.1414   Min.   :-2.61435
 1st Qu.:1.7029   1st Qu.:-0.67240
 Median :2.4042   Median :-0.03902
 Mean   :2.2946   Mean   : 0.00000
 3rd Qu.:2.9155   3rd Qu.: 0.68902
 Max.   :4.1012   Max.   : 3.08136
```

```
# Fit an N-mixture model with no intercept and distance squared using
#   the hacked function pcount.hds

(fm1 <- pcount.spHDS(~ -1 + I(dist^2) ~ Habitat, umf, K=20))

Call:
pcount.spHDS(formula = ~-1 + I(dist^2) ~ Habitat, data = umf, K = 20)

Abundance:
            Estimate    SE      z    P(>|z|)
(Intercept)  -0.875  0.1517  -5.77  8.07e-09
Habitat       1.038  0.0808  12.84  1.04e-37

Detection:
 Estimate    SE      z   P(>|z|)
  -0.202  0.0294  -6.88  5.99e-12

AIC: 730.7537
```

Note that the data were simulated with the hazard model with coefficient on the square of distance equal to $-1/2\sigma^2$, and with $\sigma = 1.5$ this works out to imply a coefficient value of -0.222, which is not too far from the estimated value of -0.202 that *applies to the logit model of detection*. The estimate of the habitat effect is very close to the data-generating value of 1.00. The intercept is harder to interpret directly because, although it is the intercept of a Poisson regression model, the data were simulated with a fixed number (600) for individuals, and not with a Poisson number of individuals per pixel. We can extract an estimate of the total population size by explicitly computing the estimates of $\hat{\lambda}$ for each pixel or by using the prediction function (or you could use the BUP function):

```
lam <- exp( coef(fm1)[1] + coef(fm1)[2]*Habitat )
pred <- predict(fm1, type='state')
sum(lam)
[1] 627.7792

sum(pred[,1])  # Same
```

We can see that our estimate of total population size is close to the simulated population size of $N = 600$. As with everything else in unmarked, we could use bootstrapping to get estimates of uncertainty for these and other quantities. To convince ourselves of the veracity of the spatial HDS capabilities just described, we should probably do a simulation study to confirm that things are working right.

We've done this analysis for data from a single point count having a population size much larger than we would ever realize in practice. However, the idea is fairly general—if we have a sample of point counts, we simply organize the data as one long vector of counts and one long vector of habitat variables (i.e., "stack" the data from multiple sites as if it were one big site). In this way, we could use the point count location as a blocking factor to model among-unit variation such as experimental treatment effects or other factors that affect variation "among points" in addition to variation "within points."

9.8.5 HIERARCHICAL SPATIAL DS

We want to extend the basic spatial DS model to an HDS situation where we have spatial replicates of the DS protocol using either point transects or line transects. There are many conceivable ways to formulate a spatial HDS model (we mentioned one approach to doing it in unmarked in the previous section), although we adopt here a fairly efficient implementation in BUGS using the basic DA approach that we've used many times. Here, however, we do an "S-fold" (S = number of sites) DA (Schmidt et al., 2012; Schmidt and Rattenbury, 2013; Tenan et al., 2014b; Sutherland et al., in review) where we independently augment each of the $s = 1, 2, \ldots, S$ sample units, bringing each data set up to a size of M. And now we introduce S DA parameters ψ_s, and the DA variables that before were z_i are now organized in a two-dimensional array $z_{i,s}$. Thus, there is one set of DA variables for each of the S data sets. A key idea in the formulation of the model here is that the ψ_s parameters are not estimated as independent parameters, but rather, there is a structure imposed on them by the underlying PPP assumption. From Section 7.8.1.4, we noted that in the context of stratified models, there is a deterministic relationship between ψ and the group means λ_g (Royle and Converse, 2014). In the present case, the "strata" are individual pixels within each of the S sample units, so:

$$\psi_s = \frac{\sum_{g=1}^{G} \lambda_{g,s}}{M}$$

where for our Poisson model,

$$log(\lambda_{g,s}) = \beta_0 + \beta_1 * Habitat_{g,s}$$

In essence, we formulate the spatial HDS model here as having S independent "group structured" (groups = pixels) DS models that we analyze as S copies of the basic model covered in Section 7.8.1.4. While there are S specific ψ_s parameters, note that these are all linked together deterministically by only two basic parameters β_0 and β_1, and thus the model is not at all highly parameterized. Strictly speaking, under DA, the resulting local population sizes then have a binomial distribution:

$$N_s \sim Binomial(M, \psi_s)$$

which is not exactly consistent with the PPP model, and indeed tends to the Poisson model only as M gets large and ψ_s is small. Thus, one should choose M to be much larger than typical values of N_s for the study being analyzed, if one wishes to approximate the PPP model well. However, in practice, there is no negative effect from simply asserting that the binomial model is the prior distribution for N_s instead of the Poisson, as the parameters of the intensity function λ_g will be well estimated in either case (we demonstrate this below). As in the simpler spatial DS model above, we parameterize the model in terms of discrete "pixels" having a categorical distribution, compute the location of detection as a derived parameter (the center of the pixel), and compute the distance between the point count location and the observation as a derived parameter as well.

To summarize the model for this spatial HDS model (when we use discrete space), we have the following key model components:

1. DA variables and parameters:

$$z_{i,s} \sim Bernoulli(\psi_s)$$

$$\psi_s = \frac{\sum_{g=1}^{G} \lambda_{g,s}}{M}$$

2a. Spatial location of each observation within each sample site:

$$pixel_{i,s} \sim Categorical(\pi_s)$$

where π_s is the vector of length G (number of pixels), which depends on the covariate raster $\mathbf{x}_s = (x_1, \ldots, x_G)$ specific to each sample unit s according to:

$$\pi_{g,s} = \frac{\lambda_{g,s}}{\sum_g \lambda_{g,s}}$$

2b. Location of individuals is derived from the *pixel* variable:

$$\mathbf{u}_{i,s} = raster.coordinate\left[pixel_{i,s},\right]$$

2c. Distance is a derived quantity:

$$d_{i,s} = \sqrt{(u_{i,s,1} - B)^2 + (u_{i,s,2} - B)^2}$$

2d. Detection probability is derived:

$$p_{i,s} = exp\left(-d_{i,s}^2/(2\sigma^2)\right)$$

3. Observation model: the binary encounter data have a Bernoulli distribution for each individual i in each population s:

$$y_{i,s} \sim Bernoulli(p_{i,s} z_{i,s})$$

As we have done many times in this book, we provide here a basic data simulator called sim.spatialHDS, which allows us to simulate data having an idealized structure so that we can manipulate it and then fit spatial HDS models to data where the data-generating parameter values are known. The function call has basic default arguments as follows:

```
sim.spatialHDS(lam0 = 4, sigma = 1.5, B = 3, nsites = 100)
# lam0 = expected population size per site
# nsites = number of point count locations
# B = count radius. Function simulates coordinates of individuals on a square
#      [0,2*B] x[0,2*B], with a count location on the point (B,B)
# sigma = scale of half-normal detection function
```

Internal to the function, a habitat raster is simulated for each point, which has dimension 20×20, and the habitat covariate is simulated as a spatially correlated multivariate normal random variable. In order to change the raster size or other features of the system or habitat covariate, you need to edit the function. A basic function call to simulate a data set for 100 bird point counts with a standardized radius of $B = 3$ and a baseline abundance of three birds per point (on average) goes like this:

```
library(raster)
# Simulate a data set and harvest the output
set.seed(1234)
str(tmp <-sim.spatialHDS(lam0 = 3, sigma = 1.5, B = 3, nsites = 100))

List of 6
 $ data   : num [1:462, 1:5] 1 1 1 2 2 2 3 3 3 3 ...
```

9.8 SPATIAL DISTANCE SAMPLING

```
  ..- attr(*, "dimnames")=List of 2
  .. ..$ : NULL
  .. ..$ : chr [1:5] "site" "u1" "u2" "d" ...
 $ B      : num 3
 $ Habitat: num [1:400, 1:100] -1.207 -0.708 0.204 -1.424 -0.767 ...
 $ grid   :'data.frame':    400 obs. of  2 variables:
  ..$ Var1 : num [1:400] 0.15 0.45 0.75 1.05 1.35 1.65 1.95 2.25 2.55 2.85 ...
  ..$ Var2 : num [1:400] 0.15 0.15 0.15 0.15 0.15 0.15 0.15 0.15 0.15 0.15 ...
  ..- attr(*, "out.attrs")=List of 2
  .. ..$ dim     : num [1:2] 20 20
  .. ..$ dimnames:List of 2
  .. .. ..$ Var1: chr [1:20] "Var1=0.15" "Var1=0.45" "Var1=0.75" "Var1=1.05" ...
  .. .. ..$ Var2: chr [1:20] "Var2=0.15" "Var2=0.45" "Var2=0.75" "Var2=1.05" ...
 $ N      : num [1:100] 3 3 4 6 6 5 4 6 7 7 ...
 $ nsites : num 100

# Process the simulated data set
data <- tmp$data
# To make it a 'real' data set:
data <- data[!is.na(data[,2]),] # Get rid of the 0 sites
data <- data[data[,"y"]==1,]    # Only keep detected individuals

# Now zero-pad the observed counts
nsites <- tmp$nsites
nobs <- rep(0, nsites)
names(nobs) <- 1:nsites
nobs[names(table(data[,1]))] <- table(data[,1])

# Extract data elements that we need
site <- data[,"site"]
s <- data[,c("u1","u2")]
B <- tmp$B
Habitat <- (tmp$Habitat) # Raster values
Habgrid <- tmp$grid      # Raster coordinates
nind <- nrow(data)
G <- nrow(Habgrid)

# We have to convert observed locations to pixels
pixel <- rep(NA,nrow(data))
D <- e2dist(s[1:nind,], Habgrid)
for(i in 1:nind){
   pixel[i] <- (1:ncol(D))[D[i,]==min(D[i,])]
}

# Do data augmentation of data from each site "S-fold" DA. Three objects need
# to have DA applied to them: Ymat, umat, pixmat
Msite <- 2*max(nobs) # Perhaps use a larger value
Ymat <- matrix(0,nrow=Msite,ncol=nsites)
umat <- array(NA, dim=c(Msite,nsites,2))
pixmat <- matrix(NA,nrow=Msite,ncol=nsites)
```

```
for(i in 1:nsites){
  if(nobs[i]==0) next
    Ymat[1:nobs[i],i]<- data[data[,1]==i,"y"]
    umat[1:nobs[i],i,1:2]<- data[data[,1]==i,c("u1","u2")]
    pixmat[1:nobs[i],i]<- pixel[data[,1]==i]
}
# Bundle the data for BUGS
str(data <- list (y=Ymat, pixel=pixmat, Habitat=Habitat, Habgrid=Habgrid, G = G,
 nsites=nsites, M = Msite, B = B))

# Write out the BUGS model file
cat("
model{

# Prior distributions
sigma ~ dunif(0,10)
beta1 ~ dnorm(0,0.01)
beta0 ~ dnorm(0,0.01)
lam0 <- exp(beta0)*G                      # Baseline lambda in terms of E(N) per sample unit

# For each site, construct the DA parameter as a function of lambda
for(s in 1:nsites){
  lamT[s] <- sum(lambda[,s])              # total abundance at a site
  psi[s] <- lamT[s]/M
    for(g in 1:G){                        # g is the pixel index, there are G total pixels
    lambda[g,s] <- exp(beta0 + beta1*Habitat[g,s])
    probs[g,s] <- lambda[g,s]/sum(lambda[,s])
  }
}

# DA variables and spatial location variables:
for(s in 1:nsites){
  for(i in 1:M){
    z[i,s] ~ dbern(psi[s])
    pixel[i,s] ~ dcat(probs[,s])
    u[i,s,1:2] <- Habgrid[pixel[i,s],]    # location = derived quantity
    # distance = derived quantity
    d[i,s] <- pow(  pow( u[i,s,1]-B,2) + pow(u[i,s,2]-B,2), 0.5)
    p[i,s] <- exp(-d[i,s]*d[i,s]/(2*sigma*sigma))  # Half-normal model
    mu[i,s] <- p[i,s]*z[i,s]
    y[i,s] ~ dbern(mu[i,s])               # Observation model
  }
# Derived parameters
  N[s] <- sum(z[,s])                      # Site specific abundance
}
  Ntotal <- sum(N[])                      # Total across all sites
  D <- Ntotal/(9*nsites)                  # Density: point area = 9 ha
}
```

9.8 SPATIAL DISTANCE SAMPLING

```
",fill=TRUE, file="spatialHDS.txt")

# Inits and parameters saved
zst <- Ymat
inits <- function(){ list (sigma=1.5, z = zst, beta0 = -5, beta1=1 ) }
params <- c("sigma", "Ntotal", "beta1", "beta0", "D", "lam0")

# MCMC settings
ni <- 12000  ;  nb <- 2000  ;  nthin <- 2  ;  nc <- 3

# Call JAGS, check convergence and summarize the results
out3 <- jags(data, inits, params, "spatialHDS.txt", n.thin=nthin, n.chains=nc,
n.burnin=nb, n.iter=ni, parallel = TRUE)
par(mfrow = c(2,3))  ;  traceplot(out3)
print(out3, 3)

JAGS output for model 'spatialHDS.txt', generated by jagsUI.
Estimates based on 3 chains of 12000 iterations, burn-in = 2000 and thin rate = 2,
yielding 15000 total samples from the joint posterior.
MCMC ran for 67.411 minutes at time 2015-07-07 18:39:52.

           mean      sd    2.5%     50%    97.5%  overlap0 f Rhat  n.eff
sigma      1.56    0.07    1.43    1.55    1.72    FALSE  1 1.00  10178
Ntotal   471.89   30.21  411.00  473.00  528.00    FALSE  1 1.01   1008
beta1      0.88    0.06    0.76    0.88    1.01    FALSE  1 1.01    325
beta0     -4.85    0.11   -5.07   -4.84   -4.65    FALSE  1 1.01    418
D          0.52    0.03    0.46    0.53    0.59    FALSE  1 1.01   1008
lam0       3.16    0.34    2.51    3.15    3.84    FALSE  1 1.01    427
```

Despite the large dimensions of some of the various data objects (400 rows, 100 columns), the model ran rather efficiently and seems to mix well. We find the estimate of total population size (posterior mean = 471.89) is not far from the simulated value (sum(tmp$N)=496), beta1 = 0.88 is near the data-generating value, and similarly for the posterior mean of beta0 for which truth is beta0 = log(3/400) = −4.89. The model was run a second time to save the site-specific N_s parameters, and a plot of the posterior means of each N_s as a function of the true site-specific population size is shown in Figure 9.10, produced by the following R commands:

```
Ntrue <- tmp$N
Nhat <- out3$summary[7:106,1]
plot(Ntrue, Nhat, xlab="Local population size, N, for each site", ylab="Posterior mean N
for each site", pch=20)
abline(0, 1, lwd=2)
```

We see the shrinkage behavior typical of Bayesian estimates, in which the estimated random effects tend to overpredict at the low end and underpredict at the high end, because they represent a compromise between the prior and the data (see also Section 6.4).

In summary, we imagine this specific type of spatial HDS model should prove to be very useful in practice for DS surveys or monitoring programs where developing explicit models of density is a research objective.

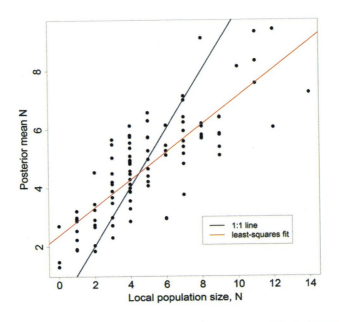

FIGURE 9.10

Posterior mean local population size at each of the 100 sites vs. the true local population size. The 1:1 line illustrates slight overprediction at the low end and slight underprediction at the high end, which is consistent with BUP-type estimators for random effects.

9.9 SUMMARY

In this chapter, we covered a number of extensions of HDS that allow for the inclusion of group as an observation unit, combining time-removal and double observer (DO) sampling with HDS, open models that include temporary emigration, partial- and full-population dynamics, and spatial HDS models. We think these topics, and their implementation in the versatile BUGS language, greatly extend the practical utility of DS for studying animal populations, because many population studies have broader interests beyond simply density estimation.

One of the big ideas that we emphasized early in the chapter is that a large number of different models, including models with group size, time removal, or double observers, can be formulated as simple HDS models (Chapter 8), but with discrete or categorical observation-level covariates ("individual covariates"). One remarkable thing about the formulation of this HDS model with group size under DA (Royle et al., 2007a) is that it essentially reduces all models to some kind of a multivariate GLM with partially latent random variables. For example, in the case of HDS models, the variables include the binary detection variable y_i, the distance or location variable d_i (or a categorical version of this), the DA variable z_i, and additional auxiliary variables such as group size, removal interval, or capture history.

A key aspect to building such "combined protocol models"—i.e., models that combine distance data with time-removal or multiple observers—is figuring out the specific probabilities that describe the categorical states of the additional covariate. Such combined protocols are another version of "integrated models," where we observed different kinds of data types that are informative about the same underlying process/state: in this case, both types of data are informative about abundance/density. These integrated models have exactly the same advantages as the more standard integrated population models (IPMs; Besbeas et al., 2002; Brooks et al., 2004; Schaub et al., 2007; Kéry and Schaub, 2012): (1) they are just eminently sensible, since you throw in all the data you have on some estimation problem; (2) thereby, you increase sample size and not surprisingly get more precise estimates (most of the time); and (3) sometimes you may be able to estimate additional parameters that you can't estimate from each data set alone.

A second major theme of this chapter had to do with formal integration of HDS with models of temporal variation in abundance. Drawing on two recent papers (Amundson et al., 2014; Sollmann et al., 2015, with the temporary-emigration models bridging these two papers), we provided a practical overview of how to fit models to different types of sampling designs involving spatially and temporally replicated DS data. These models are the first toward formally combining models of population structure with DS observation models, and we imagine this will be an important application area in the future.

Finally, we showed a number of novel formulations of spatial models based on PPPs (Johnson et al., 2010) that can be implemented in the BUGS language. Moreover, we argued that such models, when space is discretized (such as when covariates exist as GIS rasters), can be viewed as a type of binomial N-mixture model with $K = 1$ replicate, and with a special detection model. We therefore hacked the `unmarked` function `pcount` to create `pcount.spHDS`, which allows for the fitting of spatial DS models in a likelihood framework. We also showed how to fit spatial HDS models in BUGS using DA. The key idea is that we use a discrete version of observation location, where each observed animal is associated with the center point of a raster pixel cell. Then the latent variable, "pixel," has a categorical distribution. Our standard DA approach (Royle et al., 2007a) then allows us to formulate models for spatial DS.

EXERCISES

1. Temporary emigration, Section 9.5.4. Simulate 100 data sets, and fit the temporary-emigration model to assess the accuracy of the Bayesian procedure. For each data set, also obtain the MLEs using `unmarked`.
2. In Section 9.7.1, we used six distance classes compared with three in Sollmann et al. (2015), because three appears to induce bias in estimates of N. Do a small simulation study with the independence model to evaluate the degree of bias.
3. In Section 9.7.2.3, we fitted a HDS/DM model with constant dynamics parameters. As an exercise, make these specific to each time period. Think about how many of each parameter there should be.
4. For the full dynamic and partial dynamics models, we showed that posterior precision was improved compared with the independence model. Do a simulation study to characterize average precision and Bayesian credible interval coverage of the three models fitted to data generated under the fully dynamic model. This will likely take you all weekend, so don't plan to do anything fun (other than this simulation).

5. In Section 9.8.4, we described a hack of the pcount function that we called pcount.spHDS. Simulate data sets for multiple points (e.g., 10 points), supposing that five of the points are some kind of landscape treatment and five are a control, and simulate data so that the mean abundance between the groups represents a contrast. In addition to the treatment effect, be sure to simulate a habitat raster for each point. Then fit this model using pcount.spHDS, and estimate the treatment effect while controlling for the within-point heterogeneity in habitat. Refit the model without controlling for the habitat, and see what happens.
6. In Section 9.8.5, where we discussed the use of S-fold DA for HDS, we augmented the data from every site to a value of 2 * max(nobs). Use a smaller and larger value of M, and see if it makes a difference in the estimates. And, if so, which estimates are most affected?

CHAPTER 10

MODELING STATIC OCCURRENCE AND SPECIES DISTRIBUTIONS USING SITE-OCCUPANCY MODELS

CHAPTER OUTLINE

10.1 Introduction to the Modeling of Occurrence—Including Species Distributions 551
10.2 Another Exercise in Hierarchical Modeling: Derivation of the Site-Occupancy Model 557
10.3 Simulation and Analysis of the Simplest Possible Site-Occupancy Model 561
10.4 A Slightly More Complex Site-Occupancy Model with Covariates 564
10.5 A General Data Simulation Function for Static Occupancy Models: simOcc 577
10.6 A Model with Lots of Covariates: Use of R Function model.matrix with BUGS 581
10.7 Study Design, and Bias and Precision of Site-Occupancy Estimators 584
10.8 Goodness-of-Fit 589
10.9 Distribution Modeling and Mapping of Swiss Red Squirrels 590
10.10 Multiscale Occupancy Models 600
10.11 Space-for-Time Substitution 608
    10.11.1 A Magical Covariate 609
    10.11.2 No Magical Covariate Known: θ and p Are Confounded 611
10.12 Models for Data along Transects: Poisson, Exponential, Weibull, and Removal Observation Models 614
    10.12.1 Occupancy Models with "Survival Model" Observation Process: Exponential Time-to-Detection Model with Simulated Data 615
    10.12.2 Time-to-Detection Analysis with Real Data: Weibull Occupancy Model for the Peregrine Spring Survey 617
    10.12.3 Occupancy Models with Removal Design Observation Process 621
10.13 Occupancy Modeling of a Community of Species 621
10.14 Modeling Wiggly Covariate Relationships: Penalized Splines in Hierarchical Models 622
10.15 Summary and Outlook 626
Exercises 628

10.1 INTRODUCTION TO THE MODELING OF OCCURRENCE—INCLUDING SPECIES DISTRIBUTIONS

This chapter is about the joint modeling of occurrence and its ubiquitous false-negative measurement error. Occurrence means the presence or absence of some "thing" in some defined spatial and temporal unit. We have stressed many times that occurrence or presence/absence is a quantity that is directly derived from abundance, and that both abundance and occurrence are simple areal summaries of an underlying spatial point pattern. Thus, occurrence is exactly equivalent to the event that there is at least one "point" falling within a spatial unit or that the abundance of these "points" in a spatial unit is greater than zero. Despite being only a derived quantity, however, occurrence is hugely important in ecology and related sciences, such as wildlife management and conservation biology. Here, we usually deal with presence and absence of a species and "points" represent the individuals of that species. Reasons for the great importance of occurrence in ecology include the following (see also Chapter 2 in MacKenzie et al., 2006):

- Though only a reduced-information version of abundance, occurrence is typically positively related to abundance, and population changes typically are reflected by range changes (He and Gaston, 2000; Royle et al., 2005; see also Figure 1.2).
- Occurrence may be the only viable alternative for characterizing the state of a population if abundance cannot be reliably assessed for methodological or logistical reasons (MacKenzie et al., 2005), and the practical benefits relative to the measurement of the underlying spatial point pattern are much greater still.
- The parametric assumptions needed for modeling abundance (e.g., Poisson, negative binomial; see Chapter 6) may not be met in your data set. In contrast, the typical Bernoulli model for occurrence (see below) is likely pretty robust across a range of models for the underlying abundance distribution.
- Sometimes abundance may really not matter but occurrence is sufficient for the purpose at hand, e.g., for parasite infections where we may not worry about whether there are 10^5 or 10^6 parasites in a host (Lachish et al., 2012).
- Occurrence is identical to abundance when assessed at a spatial scale where a sample unit can be occupied by at most one individual, breeding pair, or family group. Examples include sites defined as territories of raptors or owls (MacKenzie et al., 2003; Martin et al., 2009). The number of occupied sites then corresponds to the number of breeding pairs, i.e., to the most widely used measure of the abundance of a population in avian ecology (Bibby et al., 2000).
- Occurrence is the basis for the most widely used biodiversity measure, species richness (see Chapter 11 in this book and Purvis and Hector, 2000). It is also the ingredient of an increasingly used index for species richness at top trophic levels computed from camera trap data, the Wildlife Picture Index (WPI; O'Brien et al., 2010).
- Occurrence is of great interest in the ecology of both invasive species (Rout et al., 2014) and diseases (McClintock et al., 2010b), both very popular fields of ecology.

Thus, occurrence is a very widely used state variable in ecology, and some of its subfields focus almost exclusively on it, such as metapopulation ecology (Hanski, 1998) or species distribution modeling (Elith and Leathwick, 2009). In addition, the occurrence of "things" in space is of interest in many scientific disciplines outside of ecology. Consequently, we believe that the potential scope of the models in this chapter may extend far beyond ecology, to the study of the occurrence of any kind of "thing" that is afflicted with false-negative measurement errors.

The basic approach in statistics to the modeling of occurrence is to treat presence and absence as a Bernoulli random variable governed by the "success probability" ψ, which in this context is known as *occupancy, or presence, probability*. Effects of covariates on ψ can be modeled on a link scale, in a logistic or related regression model, and many extensions are conceptually straightforward, including "wiggly" covariate relationships (GAMs, Hastie and Tibshirani, 1990; see Section 10.14) or the

10.1 INTRODUCTION TO THE MODELING OF OCCURRENCE—INCLUDING

modeling of spatial autocorrelation (Heikkinen and Högmander, 1994; Augustin et al., 1996; Wintle and Bardos, 2006; Bled et al., 2011a,b; Bardos et al., 2015; see also Chapters 21 and 22).

As for the underlying spatial point pattern and its areal summary of abundance, occurrence can rarely ever be assessed without error (Kéry, 2002; MacKenzie, 2005; Ferraz et al., 2007; Kéry and Schmidt, 2008; Kellner & Swihart, 2014; Lahoz-Monfort et al., 2014; Guillera-Arroita et al., 2014a, 2015). Instead, the *measurement of occurrence* often yields an observation $y = 0$ at an occupied site where $z = 1$, representing a false-negative error, or an observation of $y = 1$ where $z = 0$, representing a false-positive error. Both types of error usually lead to biased inferences about occupancy probability and its determinants, such as the strength of the relationships with environmental covariates. Occurrence measurement error has the potential to seriously mislead inferences from species distribution models (Kéry et al., 2010a,b, 2013; Kéry, 2011b; Dorazio, 2012; Guillera-Arroita et al., 2014a; Lahoz-Monfort et al., 2014). Thus, it would appear prudent to accommodate imperfect detection and false-positive errors in models of occurrence whenever possible. The dominant inference framework for the joint modeling of occurrence and its measurement error rate (especially for false-negatives) has the slightly odd name *site-occupancy model* (MacKenzie et al., 2002, 2006; Tyre et al., 2003). This chapter is the first in this book about this powerful class of models for presence/absence data; you will encounter more occupancy models in Chapter 11 (in this volume) and in many chapters in volume 2, especially 16—20. For now we only deal with false-negative measurement errors; see Chapter 19 for false-positive measurement errors in occupancy models.

In spite of its widespread use and regardless of the model used, the concept of presence/absence is widely misunderstood. First and foremost is the deterministic relationship between presence/absence and abundance: presence/absence (=occurrence) is simply a summary of local abundance, nothing more. When you have a good model of abundance you can explain both absences (=sites with abundance zero) and presences (=sites with any abundance greater than zero). Second is the importance to presence/absence of *your* definition of what constitutes a "presence". Whether you define an occurrence as the presence of a single individual, more than a single individual, a reproductive unit (e.g., pair, pack, etc.), or of a viable population will make a huge difference to the biological interpretation of "distribution" in your study. Third is the dependency of presence/absence on *your* choice of the size of the spatial and temporal scale, or grain, of the study. With increasing spatial grain and, if your species moves, also with increasing length of the observation period, occupancy probability increases monotonically, as has been rediscovered recently by Hayes and Monfils (2015). We can describe the spatial scale dependence under the assumption of a certain distribution of the underlying abundance, since occupancy probability is simply 1 minus the probability to get abundance zero under that distribution. Thus, for the simplest case of a Poisson abundance model, we can reconcile occupancy between different spatial scales by adopting a cloglog link for occupancy and treating the logarithm of quadrat size as an offset (see Section 3.3.6). Fourth, and finally, the issues around the unknown sampling area described in Section 6.10 in the context of an abundance model apply also to occupancy models (Efford and Dawson, 2012); see Chandler and Royle (2013), Chandler and Clark (2014) and Ramsey et al. (2015) for the modeling of a latent point process model based on observed counts or detection-nondetection data. Essentially, these models solve the problem of space as it relates to the definition of occupancy in occupancy models.

At least five types of data are used for the modeling of occurrence and species distributions: (1) point pattern data, (2) presence-only data, (3) "presence/absence" data, (4) "presence/absence" data replicated over time, and (5) count data, possibly replicated over time. The information content increases from 2 to 5. In theory, the information content of point pattern data (1) is greatest, but in practice inference based on this data type suffers from challenges for the proper modeling of measurement errors; see below. Here we briefly summarize these five data types, noting that the term "presence/absence" is misleading and actually wrong whenever there is occurrence measurement error; "detection/nondetection" data is then a better term. Nevertheless, we sometimes use "presence/absence" because it is used so widely, recognizing however that an "absence" may in fact represent an erroneous measurement of presence (unless the false-negative error rate is zero) and a "presence" may represent an erroneous measurement of absence (unless the false-positive error rate is zero).

1. *Point pattern data*: This can be viewed as the "mother" of all distribution and abundance data in ecology and beyond. Point pattern data typically arise when you exhaustively search some clearly defined area within some time frame and record the location of each object present (Illian et al., 2008; Wiegand and Moloney, 2014). Spatial point pattern models (PPMs) treat both the number *and* the locations of the objects in an area as the outcome of a random process, governed by an underlying intensity field. This field and its dependence on spatially indexed covariates can be modeled akin to a Poisson generalized linear model (GLM). When some objects fail to be recorded, we say that the point pattern is *thinned*. Typical data sets for which PPMs are adopted are produced by complete area searches for immobile and easily detectable objects, such as ore deposits, earthquakes, or, in ecology, trees, plants, or gopher mounds. Thus, these powerful methods have largely been developed in fields where measurement error could be assumed to be minor or perhaps even absent. Only relatively rarely have data on moving objects been analyzed using PPMs (Illian et al., 2012), but this use of PPMs is on the increase.

    Recently, there has been a spate of publications on PPMs for species distributions (Warton and Shepherd, 2010; Aarts et al., 2012; Dorazio, 2012, 2014; Fithian and Hastie, 2013; Renner and Warton, 2013; Renner et al., 2015). These authors argue that PPMs are the proper way of modeling species distribution data, in part because the scale dependence of gridded-data-based modeling methods is lost. We share the feeling of excitement about the opportunities offered by the application of PPMs to spatial data on distribution and abundance. However, we make three cautionary comments about PPMs for inference about distribution and abundance in ecology:

    a. The measurement error process underlying most species distribution and abundance data in ecology is totally different from that in classical applications of PPMs. Most ecological data sets to which PPMs are now being applied (e.g., Renner et al., 2015) are *not* the result of a complete area search for some easily detectable and unmistakable object. Rather, they will be the result of some very complicated thinning process induced by spatial sampling bias and additional bias induced by false-negative and false-positive measurement errors. For most large-scale ecological data sets we do *not* know where the observers went to produce their species records (i.e., the observed points). Thus, we cannot usually directly model the spatial sampling bias. In addition, animals move and can be overlooked and misidentified and so can plants (except for the movement; see e.g., Chen et al., 2013). Consequently, there will *always* be false-negative and false-positive errors in real-world point pattern data in ecology. False-negatives correspond to another thinning mechanism while false-positives represent the addition of some "ghost point pattern" that may bias the inference about the target species. Hence, all that is usually modeled with PPMs in ecology is some poorly defined index to density, i.e., the combination of real density, of spatial sampling bias, and of the effects of false-negative and false-positive errors.

        Currently, no PPM methods appear to be available for species distribution modeling that enable you to disentangle the true state from the measurement errors, as we do with most hierarchical models in this book (but see (c) below). We view this as a severe disadvantage of

10.1 INTRODUCTION TO THE MODELING OF OCCURRENCE—INCLUDING

this otherwise elegant and powerful class of models. For this reason, there is an urgent need for more work along the lines of Dorazio (2012, 2014), who confronts the challenges for PPMs posed by the measurement error process in real-world ecological data sets. Such work is particularly urgent since the typical species distribution data sets to which PPMs are likely to be applied are "dirty data" from citizen-science schemes, where there is often hardly any control over the sampling protocol. Consequently, the problems of spatial sampling bias and measurement errors will be particularly severe in these data sets. Naive application of PPMs in these cases would therefore seem to be risky. We fear that the excitement over a fancy novel modeling framework such as PPMs may easily let people forget about fundamental limitations of their data, which cannot be mended by most current PPMs.

b. The discretization of space for grid-based methods such as most of those described in this book engenders a scale dependence of the inferences. For instance, the larger your pixel, the larger will be both expected abundance and occupancy probability. Some authors have claimed that this is a serious disadvantage (e.g., Fithian and Hastie, 2013, Renner et al., 2015). However, gridding is not necessarily a disadvantage because it can greatly simplify the analysis. First, models for distribution and abundance for grid-based data are conceptually *much* easier than are PPMs. Second, almost all spatial sampling is either explicitly grid based or can be easily characterized by a grid-based framework, and therefore spatial sampling bias can be dealt with formally as part of the modeling and inference. Third, with grid-based models it is easy to incorporate a measurement error component, which we argue is often crucial for valid inference with ecological data sets. And fourth, adding a temporal dimension to a model for gridded data is straightforward, as you will see in several chapters in volume 2. In contrast, in the PPM framework adding time leads to rather complicated models.

c. There is one class of spatial PPM that has an integrated measurement error model: spatial capture-recapture (SCR) models (Efford, 2004; Borchers and Efford, 2008; Royle and Young, 2008; Royle et al., 2013, 2014; Borchers et al., 2015) and some variants of distance sampling models (see Section 9.8). They usually require repeated measurements of the locations of known individuals to jointly estimate the point pattern and its measurement error. Such data are not usually available for large-scale distribution modeling (but see Dorazio, 2014).

2. *Presence-only data*: This data type is typically represented by museum or other collections of locations where a species was identified, collected, or otherwise recorded. Without auxiliary information, we cannot estimate a Bernoulli parameter characterizing presence *and* absence (Pearce and Boyce, 2006; Boyce, 2010). However, typically such data are augmented with data on the environmental conditions at some or all of the larger number of sites from which the sites with recorded presences are sampled, hence, such data may be called "presence/background data" (Lahoz-Monfort et al., 2014). There has been some debate about whether the Bernoulli parameter can be estimated from such data or not, and several authors have developed methods to estimate logistic regression parameters from such data, including Lele and Keim (2006) and Royle et al. (2012). These methods critically depend on assumptions about the correct model structure, including the exact link function, and slight violations of these parametric assumptions may lead to badly biased estimators (Fithian and Hastie, 2013; Phillips and Elith, 2013; see also Section 10.11.1).

Hence, the prevailing opinion now is that with presence-only data we can estimate relative occupancy probability only, i.e., a regression slope but not the intercept. That this is a serious constraint for this data type is shown in Figure 10.1, where you see 21 regression lines all with an

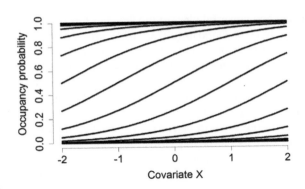

FIGURE 10.1

Twenty-one linear-logistic slopes equal to 1, or, what relative probability of presence means: in practice, it may mean hardly anything at all. All 21 linear-logistic regression lines have an identical slope of 1, but they differ in the intercept, which ranges from −10 to 10.

identical slope of 1 on the logit scale, but where the intercept ranges from −10 to +10. On the probability scale, depending on the intercept, a four-unit change in the covariate may correspond to a change of 0.0003 in occupancy probability, when the intercept is +10 or −10, or to a difference of 0.7616, when the intercept is zero. Hence, it appears to be difficult to obtain practically meaningful inferences from presence-only data.

```
alpha <- seq(-10, 10, by = 1)
curve(plogis(-10 + 1 * x), -2, 2, lwd = 3, ylim = c(0,1), xlab = "Covariate X",
   ylab = "Occupancy prob.", frame = F, main = "", cex.lab = 1.5, cex.axis = 1.5)
for(i in 2:21){ curve(plogis(alpha[i] + 1 * x), -2, 2, lwd = 3, add = T) }
(min <- (plogis(-10 + 1 * 2) - plogis(-10 + 1 * -2)) )
(max <- (plogis(0 + 1 * 2) - plogis(0 + 1 * -2)) )
```

An alternative is to use PPMs for presence-only data; indeed, Renner and Warton (2013) have shown that the statistical model underlying the use of the popular Maxent software for presence-only data (Phillips and Dudik, 2008) is equivalent to a Poisson PPM under ideal conditions, in which sampling of space is uniform so that the thinning is random or else one has explicit covariates that describe the thinning. One advantage of the PPM approach is that the dependence on the spatial scale is lost (though presumably not the dependence on temporal scale). A disadvantage is that the power and the elegance of PPMs for such data make it easy to forget what is really modeled in most cases: relative density. This is some unknown convolution of the true intensity of the underlying pattern and a potentially very complicated thinning process represented by the sampling processes that produced the data at hand; see above. To properly account for the sampling processes underlying observed point pattern data is a fundamental area in the PPM field where much more research is needed.

3. *Presence/absence data (more properly called detection/nondetection data)*: This kind of data is the classical input for logistic regression, or binomial generalized models and their extensions, such as generalized additive models (GAMs) or generalized linear mixed models (GLMMs). This data type contains more information about the logistic regression of probability of occupancy and

lets one estimate the intercept robustly, but does not allow one to separately estimate detection probability, except under very strong parametric assumptions (Dorazio, 2012; Lele et al., 2012; Knape and Korner-Nievergelt, 2015).
4. *Presence/absence (detection/nondetection) data replicated over time*: This data type contains the most information among the types 2–4 and is the type of data that we primarily consider in this chapter. As we will see, under the closure assumption, this data type allows one to jointly model probability of occupancy and of detection using the powerful site-occupancy models (MacKenzie et al., 2002; Tyre et al., 2003).
5. *Count data*: We have seen that presence/absence data are simply a summary of count data. Hence, it is clear that count data can also be used to model species distributions. Unreplicated counts may be modeled as a Poisson GLM and the probability of occupancy obtained as the probability that a count is greater than zero. Count data that are not replicated over time do not enable one to model abundance or occurrence jointly with detection probability except again under very restrictive assumptions (Solymos et al., 2013; Knape and Korner-Nievergelt, 2015). In contrast, count data that are replicated over time over a short time span, so that the closure assumption is met, may be modeled using the N-mixture model (Chapter 6), which can naturally also serve as a species distribution model (Royle et al., 2005; Dorazio, 2007; see Section 6.9) and so may data collected under any other protocol that provides the extra information to estimate measurement error, such as capture-recapture data (Section 7.9) or distance sampling (see Chapters 8 and 9).

You will note strong structural similarities between this chapter and Chapter 6 on the binomial mixture model. This is no accident, rather, it should underline the strong conceptual similarity between occupancy models and N-mixture models. Indeed, both are simply examples of a general form of two-level hierarchical models for distribution and occurrence or abundance, respectively.

10.2 ANOTHER EXERCISE IN HIERARCHICAL MODELING: DERIVATION OF THE SITE-OCCUPANCY MODEL

As for the N-mixture model (Section 6.2), we now want to derive the basic static site-occupancy model from first principles by thinking about the processes that underlie the observed detection/nondetection data. We ask three questions, and their answers will naturally lead to the basic site-occupancy model.

Question 1: Assume that 100 sites were inhabited each by a certain number of individuals of a study species of your choice, i.e., each site has some value of abundance N, such as $N = \{0, 1, 3, 1, 4, 0\}$ for the first six sites. However, perhaps you cannot measure abundance reliably, and only a summary of N, presence ($N > 0$) or absence ($N = 0$) is recorded. Hence, the true state of interest is then the occurrence, or the presence/absence state, which we denote by z, which would be $z = \{0, 1, 1, 1, 1, 0\}$ for these sites. If we want to model presence/absence, we want to treat z as the realization of a random variable, i.e., as the output from a named stochastic process, or statistical distribution. The process should accommodate both the randomness in the observed data as well as patterns that hold on average only and that we can describe in a GLM manner by introducing covariates into the expectation. So the first question is: **What is the customary statistical description of such presence/absence states?**

Arguably, the natural answer is a Bernoulli distribution for presence/absence at site i. We would write $z_i \sim Bernoulli(\psi)$, where ψ is the expected proportion of sites that are occupied, or the occupancy probability.

Question 2: Every naturalist and also every ecologist (even every statistical ecologist and perhaps even a statistician sometimes) who has *ever* set his foot into the field must know the ugly truth of presence/absence studies—that a species may sometimes be missed where it occurs. This induces a specific type of error in our *presence/absence measurement* (y)—sometimes, we measure an absence ($y = 0$) at a presence site (with $z = 1$); this represents a false-negative error. For instance, if we went to some occupied sites twice, we have the following presence/absence measurements that are possible: $\{0, 0\}$, $\{1, 0\}$, $\{0, 1\}$, and $\{1, 1\}$. So the second question is: **What is a sensible statistical model for the measurement error process at an occupied site?**

We think that the natural answer would again be a Bernoulli distribution. That is, we could write $(y_i|z_i = 1) \sim \text{Bernoulli}(p)$, where p is detection probability, i.e., the complement of the false-negative measurement error.

Question 3: And what about an absence site, where $z = 0$; what are possible presence/absence measurements and what statistical model might we choose for this process? Basically, we could again measure either a presence ($y = 1$) or an absence ($y = 0$) and a sensible model would be another Bernoulli, $(y_i|z_i = 0) \sim \text{Bernoulli}(q)$, where q would represent the probability of a false-positive error. However, in virtually all situations false-positives are far scarcer than false-negatives. Therefore, for the moment we ignore them and assume $q = 0$, i.e., that the false-positive error probability is zero. Standard occupancy models make this assumption, but in Chapter 19 we will see how we can relax the assumption and encounter occupancy models that enable estimation of both false-negative and false-positive measurement errors with an observation model that looks like this: $y_i \sim \text{Bernoulli}(z_i p + (1 - z_i)q)$ (Royle and Link, 2006; Miller et al., 2011, 2013b; Sutherland et al., 2013; Chambert et al., 2015).

If we combine these answers, we obtain *exactly* the basic site-occupancy model that was independently developed by MacKenzie et al. (2002) and Tyre et al. (2003); see Section 4.3 in MacKenzie et al. (2006) for some history. Thus, we have just reinvented the site-occupancy model from first principles by thinking about the processes that plausibly underlie the observed presence/absence data. This ability, to sequentially incorporate into a statistical model multiple, linked processes underlying an observed outcome, is one of the principal benefits of hierarchical modeling (Royle and Dorazio, 2008). Related to this is the benefit that hierarchical modeling almost enforces on us a more mechanistic thinking about the multiple processes that produce an observed data set (Kéry and Schaub, 2012).

In summary, here is the simplest site-occupancy model written in algebra:

1. State process: $z_i \sim \text{Bernoulli}(\psi)$
2. Observation process: $y_{ij}|z_i \sim \text{Bernoulli}(z_i p)$

The latent variable z_i is the true state of occurrence at site i ($i = 1...M$) and the Bernoulli parameter ψ is the expected value of z, called the probability of occupancy or of presence. The observed variable y_{ij} is our *measurement* of occurrence at site i during survey j ($j = 1...J$) and is conditional on z_i, and p is the detection probability of the study species at site i during survey j, i.e., the complement of the presence/absence (false-negative) measurement error. Detection probability here refers to *all* individuals inhabiting a site together and not to each individual singly as in the N-mixture model (Conceptually, the two are related as $P^* = 1 - (1 - p)^N$, where P^* is the per-site detection probability, p is the per-individual detection probability and N is the number of individuals at the site; see Section 6.13.1.). The outcome of the observation process is conditional on the outcome of the state process, because the parameter of the second Bernoulli distribution is the product of z_i and p. Thus, at unoccupied sites, this product is zero and only zero observations (absence measurements) can be made. It is here where our assumption about the absence of false-positives is manifest, i.e., that we assume that a species can be overlooked where it occurs but not erroneously recorded where it is absent.

Analogous to the N-mixture model, this hierarchical model can be described as consisting of two linked GLMs: a Bernoulli regression for the spatial variation in occurrence and another Bernoulli regression for the spatiotemporal variation of the observed detection/nondetection data at specific sites. The site-occupancy model is thus a hierarchical extension of a Bernoulli GLM or logistic regression. Logistic regression is the natural building block for models of occurrence (Royle and Dorazio, 2008) and is also the most widely used model for false-negative observation errors (i.e., imperfect detection; Kéry and Schaub, 2012). The site-occupancy model therefore combines the canonical model for species occurrence with the canonical model for imperfect detection. It is also a Bernoulli/Bernoulli mixture model. Recognizing the GLM character of the model, it becomes obvious that we can thus again start doing things that we do with GLMs, namely model structure in the parameters ψ and p, by first indexing them by site and site and time, respectively, and then expressing them as linear or other functions of covariates via some link function, or by the introduction of random effects to model hidden structure and correlations. We will see many examples of this in this chapter and throughout the book.

As for the N-mixture model (Chapter 6), we need repeated measurements of presence/absence for at least some sites; otherwise the parameters of the two parts of the model cannot be estimated separately. However, it is *not* required that we have the same number of repeated measurements at all sites, nor even that we have replicate observations for *all* sites! For instance, many site-occupancy models for species distributions will have replicate observations for only a minority of the sites; e.g., Kéry et al. (2010a,b), and Kéry (2011b). Nevertheless, the more replication the better (unless we risk violating the closure assumption; see below). If we do not have a balanced design with the same number of replicates at each site, it is best if the number of surveys per site is randomly allocated to a site. If it depends instead on some site characteristics, biased estimates may result. For instance, if multiple surveys are only undertaken at the "better" sites, where density and therefore detection probability (p) may be higher on average, the resulting estimate of p will be biased high with respect to all sites and therefore the occupancy estimator will be biased low.

Some authors have proposed variants of N-mixture (Solymos et al., 2012) and site-occupancy models for unreplicated surveys (Lele et al., 2012). Their models buy parameter identifiability by making very strong parametric assumptions about the covariates (Knape and Korner-Nievergelt, 2015). These assumptions are critical and they may well hold in some cases, but in others they may not, and it is not clear how they could be tested. Therefore, it appears risky to us to base inference about both ecological state and measurement error on unreplicated data alone. However, such data may of course be combined with data sets that *do* have replication in a form of integrated model (see Chapter 23).

You can fit a large array of occupancy models in a number of free software that use MLE. Most of all, program PRESENCE (Hines, 2006) has been developed specifically for occupancy models and allows you to fit a very large range of models for occurrence and also some for abundance. Then, MARK (White and Burnham, 1999) also contains a large number of occupancy models. Gimenez et al. (2014) have shown how E-SURGE (Choquet et al., 2009b), a powerful software for fitting hidden Markov models such as CJS and multistate models (see Chapter 15 in volume 2), can be tweaked to fit occupancy models as well. And of course, in this book we use `unmarked` and BUGS software.

The main assumptions of the basic site-occupancy model are the following. We will discuss them all in more detail later.

1. *Closure assumption*: We require that the presence/absence state z_i of site i does not change over the course of the study. This typically means that we will only use detection/nondetection data from a time period that is short relative to the dynamics of the modeled system. This may be hours or days if we model the occurrence of insects and years if we model the occurrence of trees. Certain violations of the closure assumption, corresponding to random temporary emigration, are usually not disastrous; they simply require one to interpret the occupancy parameter as *probability of use* sometime during the study period (i.e., of a site ever being occupied), rather than the probability of permanent occurrence. In addition, given the right kind of data (usually some form of temporal subsampling) one may estimate the probability of being temporarily absent formally in a multiscale occupancy model; see Section 10.10 for the occupancy case and Sections 6.14, 7.4, and 9.5 for related N-mixture models for abundance. Thus, if we have data collected under the so-called robust design (RD), closure is no problem. The RD denotes a sampling protocol with temporal replication at two scales, representing primary and nested secondary occasions (Williams et al., 2002). We assume that the system may change between primary occasions but is closed between secondary occasions within a primary occasion. With such data, we can simply fit the static model to each primary period separately or (and this results in identical parameter estimates) we fit a model to all data at once but fit separate parameters for every primary occasion. Alternatively, we can fit a dynamic model, where the change in occurrence is governed by parameters of persistence and colonization (see Chapter 16 in volume 2).

2. *No false positive errors*: This is an important assumption, since its violation can lead to strong bias in the occupancy estimator (Royle and Link, 2006; McClintock et al., 2010b; Miller et al., 2015). A common way to avoid false positives is to discard any observation with doubtful species identification or rather to treat it as a zero. This will lower the detection probability (if the record did in fact refer to the study species) but eliminate the deleterious false-positive errors (if it did not). There is important new work on joint estimation of both false-positive and false-negative measurement errors in occupancy models pioneered by Miller et al. (2011, 2013b) (see also Bailey et al., 2014, and Chambert et al., 2015). We review these models in Chapter 19 in volume 2.

3. *Independence of occurrence and independence of detection*: The former assumption means that occupancy probability at one site should be independent from the occupancy probability at another site, except insofar as we can explain such associations with covariates. The second assumption means that detection probability at a site should be independent across replicated visits. The most likely way in which the independence of occurrence assumption is violated is by mechanisms that lead to spatial autocorrelation and this can modeled; see Chapters 21 and 22. The most likely way in which the independence of detection assumption is violated is by "behavioral response" (Riddle et al., 2010), and this can also be modeled. A third case would be independence of the measurement error from the ecological state, i.e., the lack of density-dependent detection in a model for abundance (see Section 6.15), but this is not an issue in occupancy models, since for them detection is always conditional upon presence.

4. *Homogeneity of detection probability*: Unexplained heterogeneity in detection can greatly bias estimators (Miller et al., 2015). Specifically, unmodeled site-specific heterogeneity in detection

will lead to underestimates of occupancy (this is the second law of capture-recapture; Royle, 2006; Dorazio, 2007). We can try to eliminate such heterogeneity by modeling it via known covariates, by adopting the Royle-Nichols model (Section 6.13.1), using the N-mixture model if counts are available (Royle et al., 2005; Dorazio, 2007), or by modeling latent structure via finite or continuous mixture distributions (Royle, 2006). This can easily be achieved either in unmarked or else in BUGS.

5. *Parametric assumptions*: We assume that the two Bernoulli variables (typically with some covariates and potentially other structure in the mean) are a reasonable abstraction of reality in order to meet the objectives of the modeling. This and some of the other assumptions can be assessed with goodness-of-fit tests; see Section 10.8.

10.3 SIMULATION AND ANALYSIS OF THE SIMPLEST POSSIBLE SITE-OCCUPANCY MODEL

We illustrate and explain the simplest possible site-occupancy model using data simulation and analysis in unmarked and BUGS. We simulate data as described in 10.2, i.e., for a data set with constant occupancy (which we assume to be 0.8) and constant detection (assumed to be 0.5), collected at 100 sites with two presence/absence measurements each.

```
# Choose sample sizes and prepare observed data array y
set.seed(24)                               # So we all get same data set
M <- 100                                   # Number of sites
J <- 2                                     # Number of presence/absence measurements
y <- matrix(NA, nrow = M, ncol = J) # to contain the obs. data

# Parameter values
psi <- 0.8                                 # Probability of occupancy or presence
p <- 0.5                                   # Probability of detection

# Generate presence/absence data (the truth)
z <- rbinom(n = M, size = 1, prob = psi)   # R has no Bernoulli

# Generate detection/nondetection data (i.e. presence/absence measurements)
for(j in 1:J){
    y[,j] <- rbinom(n = M, size = 1, prob = z*p)
}

# Look at data
sum(z)                                     # True number of occupied sites
[1] 86

sum(apply(y, 1, max))                      # Observed number of occupied sites
[1] 61
```

Thus, in our simulation the species occurs at 86 sites and is detected at 61. The overall measurement error for the apparent number of occupied sites is thus $(86 - 61)/86 = -29\%$. Under our binomial model we'd expect a combined detection probability (over J surveys) of $1 - (1 - p)^J = 75\%$, i.e., a total measurement error of −25%. This difference between −29% and −25% is of course due to the sampling error inherent in the stochastic detection process. Now we inspect our data set:

```
head(cbind(z = z, y1 = y[,1], y2 = y[,2]))   # Truth and measurements for first 6 sites
     z y1 y2
[1,] 1  1  1
[2,] 1  1  1
[3,] 1  0  0
[4,] 1  0  1
[5,] 1  1  1
[6,] 0  0  0
```

Sites 1–5 are presence sites, while site 6 is unoccupied. Since we exclude false-positives, we will never observe the species at an absence site, but we may fail to detect it at a presence site. The first five sites illustrate three of the four possible detection histories at an occupied site: {1, 1} for sites 1, 2 and 5, {0, 1} at site 4, and {0,0} at site 3. You can look at the entire simulated, observed data y to see the fourth possible history, {1, 0}, first occurring at sites 25–27.

We now analyze the data with unmarked using function occu, where the linear model for detection is specified before that for occupancy.

```
library(unmarked)
umf <- unmarkedFrameOccu(y = y)    # Create unmarked data frame
summary(umf)                        # Summarize data frame
(fm1 <- occu(~1 ~1, data = umf))    # Fit model

Call:
occu(formula = ~1 ~ 1, data = umf)

Occupancy:
 Estimate    SE    z   P(>|z|)
     1.04 0.394 2.65  0.00807

Detection:
 Estimate   SE    z  P(>|z|)
    0.329 0.26 1.26   0.207

AIC: 270.2257

backTransform(fm1, "state")      # Get estimates on probability scale
backTransform(fm1, "det")

Backtransformed linear combination(s) of Occupancy estimate(s)
 Estimate     SE LinComb (Intercept)
     0.74 0.0759    1.04           1

Backtransformed linear combination(s) of Detection estimate(s)
 Estimate     SE LinComb (Intercept)
    0.581 0.0634   0.329           1
```

We observed the species at 61% of the sites, but we estimate that it really occurs at 74%, because detection probability is estimated at 58% for a single survey. Next, we conduct a Bayesian analysis of the model with JAGS.

10.3 SIMULATION AND ANALYSIS OF THE SIMPLEST POSSIBLE

```
# Bundle data and summarize data bundle
str( win.data <- list(y = y, M = nrow(y), J = ncol(y)) )

# Specify model in BUGS langauge
sink("model.txt")
cat("
model {
# Priors
   psi ~ dunif(0, 1)
   p ~ dunif(0, 1)
# Likelihood
  for (i in 1:M) {                    # Loop over sites
     z[i] ~ dbern(psi)                # State model
     for (j in 1:J) {                 # Loop over replicate surveys
        y[i,j] ~ dbern(z[i]*p)        # Observation model (only JAGS !)
#       y[i,j] ~ dbern(mu[i])         # For WinBUGS define 'straw man'
     }
#   mu[i] <- z[i]*p                   # Only WinBUGS
  }
}
",fill = TRUE)
sink()

# Initial values
zst <- apply(y, 1, max)               # Avoid data/model/inits conflict
inits <- function(){list(z = zst)}

# Parameters monitored
params <- c("psi", "p")

# MCMC settings
ni <- 5000 ; nt <- 1 ; nb <- 1000 ; nc <- 3

# Call JAGS and summarize posteriors
library(jagsUI)
fm2 <- jags(win.data, inits, params, "model.txt", n.chains = nc,
   n.thin = nt, n.iter = ni, n.burnin = nb)
print(fm2, dig = 3)
        mean    sd    2.5%   50%   97.5%  overlap0  f  Rhat  n.eff
psi    0.749  0.077  0.607  0.744  0.912    FALSE   1    1   12000
p      0.573  0.062  0.449  0.574  0.690    FALSE   1    1    4876
```

As usual, we get Bayesian estimates that are very similar to those using MLE (and we would get more similar ones still with a larger data set). As for the related N-mixture model (Section 6.3) we point out the striking similarity of a hierarchical model when written in algebra, in R when simulating the data, and in BUGS when fitting the model (Table 10.1). This illustrates well our frequent claims that once you know how to write a model in algebra, you're almost there at fitting it in BUGS, and that algebra, data simulation, and the BUGS language are similarly precise and useful ways of describing a hierarchical or, indeed, any model.

Table 10.1 Occupancy model descriptions in terms of algebra, data simulation code in R, and BUGS language. The latter is for JAGS only; for WinBUGS we have to define the success probability of the Bernoulli in the observation model outside as a "straw man" (see BUGS model code above).

	Algebraic Description	R Data Simulation	BUGS Model Statement
State model	$z_i \sim \text{Bernoulli}(\psi)$	`z <- rbinom(M,1,psi)`	`z[i] ~ dbern(psi)`
Obs. model	$y_{ij}|z_i \sim \text{Bernoulli}(z_i p)$	`y[,j] <- rbinom(M,1,z*p)`	`y[i,j] ~ dbern(z[i]*p)`

Finally, when no patterns over time (i.e., across the J replicate surveys) are modeled, the observation model can be simplified by fitting the model to the aggregated site-specific data, where y_i now is the detection frequency, i.e., the number of times over J surveys a species was detected.

$$y_i|z_i \sim \text{Binomial}(J, z_i p)$$

This is exactly the same model and will lead to exactly the same estimates, but is computationally more efficient (and may be worthwhile especially for large data sets or when the state model is very complex). If the number of replicates J is variable across sites, it must be made a vector; see Exercise 1.

This completes our simulation-based introduction to the simplest possible occupancy model, that with only an intercept in both parts of the model. In the next section, we illustrate a slightly more realistic and interesting model with one covariate in each model component.

10.4 A SLIGHTLY MORE COMPLEX SITE-OCCUPANCY MODEL WITH COVARIATES

We will hardly ever use the null/null site-occupancy model from the previous section but will typically be interested in effects of covariates, e.g., to model environmental effects on occupancy. In this section, we show covariate modeling and predictions of occupancy and detection, discuss the difference between the estimate of occupancy probability versus that of the realized occurrence state, and do a bootstrap assessment of uncertainty. We work with simulated data once again (note that in Section 10.9 you will see a similar analysis of a real data set). We mostly work with `unmarked`, but fit one model with BUGS as well.

We simulate data under the following model:

$$z_i \sim \text{Bernoulli}(\psi_i), \text{ with logit}(\psi_i) = \beta_0 + \beta_1 * \text{vegHt}_i$$

$$y_{ij}|z_i \sim \text{Bernoulli}(z_i p_{ij}), \text{ with logit}(p_{ij}) = \alpha_0 + \alpha_1 * \text{wind}_{ij}$$

Occupancy is affected by a site covariate (vegetation height) and detection is affected by a sampling, or observational covariate (wind speed).

10.4 A SLIGHTLY MORE COMPLEX SITE-OCCUPANCY MODEL

```
# Choose sample sizes and prepare obs. data array y
set.seed(1)                             # So we all get same data set
M <- 100                                # Number of sites
J <- 3                                  # Number of presence/absence measurements
y <- matrix(NA, nrow = M, ncol = J)     # to contain the obs. data

# Create a covariate called vegHt
vegHt <- sort(runif(M, -1, 1))          # Sort for graphical convenience

# Choose parameter values for occupancy model and compute occupancy
beta0 <- 0                              # Logit-scale intercept
beta1 <- 3                              # Logit-scale slope for vegHt
psi <- plogis(beta0 + beta1 * vegHt)    # Occupancy probability
# plot(vegHt, psi, ylim = c(0,1), type = "l", lwd = 3) # Plot psi relationship

# Now visit each site and observe presence/absence perfectly
z <- rbinom(M, 1, psi)                  # True presence/absence

# Look at data so far
table(z)
z
 0  1
49 51

# Plot the true system state
par(mfrow = c(1, 3), mar = c(5,5,2,2), cex.axis = 1.5, cex.lab = 1.5)
plot(vegHt, z, xlab="Vegetation height", ylab="True presence/absence (z)", frame = F,
cex = 1.5)
plot(function(x) plogis(beta0 + beta1*x), -1, 1, add=T, lwd=3, col = "red")
```

Figure 10.1 (left) shows how the relationship between occupancy probability and vegHt translates into a pattern of presence/absence. Of course, this is hardly ever the whole story behind "presence/absence data." Rather, there will almost always be false-negative measurement errors. Occurrence z becomes a latent state then, i.e., it will be only partially observable. We simulate this next and imagine that detection probability p is related to the covariate wind via a logit-linear regression with intercept -2 and slope -3 and that we make $J = 3$ presence/absence measurements at each site (Figure 10.1 middle).

```
# Create a covariate called wind
wind <- array(runif(M * J, -1, 1), dim = c(M, J))

# Choose parameter values for measurement error model and compute detectability
alpha0 <- -2                            # Logit-scale intercept
alpha1 <- -3                            # Logit-scale slope for wind
p <- plogis(alpha0 + alpha1 * wind)     # Detection probability
# plot(p ~ wind, ylim = c(0,1))         # Look at relationship

# Take J = 3 presence/absence measurements at each site
for(j in 1:J) {
    y[,j] <- rbinom(M, z, p[,j])
}

sum(apply(y, 1, max))                   # Number of sites with observed presences
[1] 32
```

```r
# Plot observed data and true effect of wind on detection probability
plot(wind, y, xlab="Wind", ylab="Observed det./nondetection data (y)", frame = F,
cex = 1.5)
plot(function(x) plogis(alpha0 + alpha1*x), -1, 1, add=T, lwd=3, col = "red")

# Look at the data: occupancy, true presence/absence (z), and measurements (y)
cbind(psi=round(psi,2), z=z, y1=y[,1], y2=y[,2], y3=y[,3])
        psi  z y1 y2 y3
 [1,]  0.05  0  0  0  0
 [2,]  0.05  0  0  0  0
 [3,]  0.07  0  0  0  0
 [4,]  0.07  1  0  0  0
 [5,]  0.07  0  0  0  0
    [ output truncated]
[91,]  0.90  1  1  0  0
[92,]  0.91  1  1  0  0
[93,]  0.91  1  0  0  0
[94,]  0.92  0  0  0  0
[95,]  0.92  1  0  1  1
    [ output truncated]
```

We suggest that you look at this table to make sure you *really* understand the relationships among ψ (psi), z, and y. Next, we use the site-occupancy model to analyze these data using unmarked and BUGS. We start with unmarked and will also illustrate the fitting of two factors that are unrelated to the data (because the response was not generated with their effects "built in"): time will index the first through the third survey, while hab will contrast three imaginary habitat types.

```r
# Create factors
time <- matrix(rep(as.character(1:J), M), ncol = J, byrow = TRUE)
hab <- c(rep("A", 33), rep("B", 33), rep("C", 34))   # Must have M = 100
```

To fit the model in unmarked, we package the data into an unmarked frame first. Note the difference between site covariates (indexed by site only) and sampling or observational covariates (indexed by site and survey). There is really a third possible type of covariate, for time, but in unmarked, this type has to be specified as an observational covariate, as we see for factor time, which codes for the first to the third survey.

```r
# Load unmarked, format data and summarize
library(unmarked)
umf <- unmarkedFrameOccu(
    y = y,                                             # Pres/Abs measurements
    siteCovs = data.frame(vegHt = vegHt, hab = hab),   # site-specific covs.
    obsCovs = list(wind = wind, time = time))          # obs-specific covs.
summary(umf)

unmarkedFrame Object

100 sites
Maximum number of observations per site: 3
Mean number of observations per site: 3
Sites with at least one detection: 32
```

10.4 A SLIGHTLY MORE COMPLEX SITE-OCCUPANCY MODEL

```
Tabulation of y observations:
   0    1 <NA>
 255   45    0

Site-level covariates:
     vegHt           hab
Min.   :-0.97322   A:33
1st Qu.:-0.35384   B:33
Median :-0.02438   C:34
Mean   : 0.03569
3rd Qu.: 0.53439
Max.   : 0.98381

Observation-level covariates:
     wind           time
Min.   :-0.99633   1:100
1st Qu.:-0.54111   2:100
Median :-0.10469   3:100
Mean   :-0.03803
3rd Qu.: 0.44824
Max.   : 0.99215
```

```
# Fit model and extract estimates
# Detection covariates follow first tilde, then occupancy covariates
summary(fm1.occ <- occu(~wind ~vegHt, data=umf))

Call:
occu(formula = ~wind ~ vegHt, data = umf)

Occupancy (logit-scale):
             Estimate    SE      z    P(>|z|)
(Intercept)   -0.136  0.449  -0.303  0.76177
vegHt          2.432  0.839   2.897  0.00377

Detection (logit-scale):
             Estimate    SE      z    P(>|z|)
(Intercept)   -1.70   0.398  -4.27  0.000019825
wind          -3.07   0.594  -5.16  0.000000243

AIC: 183.4468
Number of sites: 100
optim convergence code: 0
optim iterations: 36
Bootstrap iterations: 0
```

```
# Predict occupancy and detection as function of covs (with 95% CIs)
# Add truth from data simulation (below for full code to produce fig. 10-2)
newdat <- data.frame(vegHt=seq(-1, 1, 0.01))
pred.occ <- predict(fm1.occ, type="state", newdata=newdat)
newdat <- data.frame(wind=seq(-1, 1, 0.1))
pred.det <- predict(fm1.occ, type="det", newdata=newdat)
```

```
# Predictions for specified values of vegHt, say 0.2 and 2.1
newdat <- data.frame(vegHt=c(0.2, 2.1))
predict(fm1.occ, type="state", newdata=newdat, append = T)
  Predicted        SE       lower     upper   vegHt
1 0.5866972 0.12345081 0.3435511 0.7938296    0.2
2 0.9931116 0.01301558 0.7759108 0.9998334    2.1

# ... for values of wind of -1 to 1
newdat <- data.frame(wind=seq(-1, 1, , 5))
predict(fm1.occ, type="det", newdata=newdat, append = T)
  Predicted        SE       lower     upper   wind
1 0.797525949 0.081884645 0.593153464 0.9141025 -1.0
2 0.459433439 0.085800844 0.301588034 0.6258607 -0.5
3 0.154968761 0.052056383 0.077610026 0.2855637  0.0
4 0.038064095 0.022366542 0.011810009 0.1158403  0.5
5 0.008465936 0.007340634 0.001535816 0.0452499  1.0
```

We may summarize the analysis by plotting the observed data (the observed occurrence state of every site), the true data-generating values, and the estimated relationship between occupancy and covariate vegHt under the occupancy model (a "logistic regression" that does account for imperfect detection p) and under a simple logistic regression that does not account for p (Figure 10.2 right). We see that ignoring imperfect detection leads to (1) underestimation of the extent of occurrence and (2) to a bias toward zero (attenuation) of the regression coefficient of vegHt.

```
# Fit detection-naive GLM to observed occurrence and plot comparison
summary(fm.glm <- glm(apply(y, 1, max) ~ vegHt, family=binomial))
plot(vegHt, apply(y, 1, max), xlab="Vegetation height", ylab="Observed occurrence ('ever observed ?')", frame = F, cex = 1.5)
plot(function(x) plogis(beta0 + beta1*x), -1, 1, add=T, lwd=3, col = "red")
lines(vegHt, predict(fm.glm, ,"response"), type = "l", lwd = 3)
```

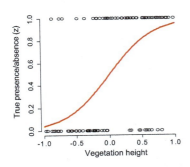 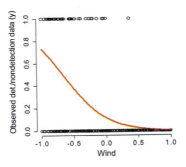 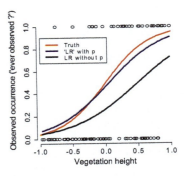

FIGURE 10.2

Left: The relationship between occurrence and vegetation height (red line); circles represent the true presence or absence (z) at each site. Middle: The relationship between detection probability and wind speed (red line); circles are the observed detection/nondetection (or "presence/absence") data. Right: Comparison between the true occupancy-covariate relationship (red) and its estimate under the occupancy model (blue) and a simple logistic regression (black); circles represent the observed occurrence status of each site.

```
lines(vegHt, predict(fm1.occ, type="state")[,1], col = "blue", lwd = 3)
legend(-1, 0.9, c("Truth", "'LR' with p", "LR without p"), col=c("red", "blue", "black"),
lty = 1, lwd=3, cex = 1.2)
```

Most of the data formatting in an unmarked frame, the model fitting, and the processing of the results in an occupancy analysis using function occu is very similar to what we did with the different N-mixture unmarked model-fitting functions in Chapters 6—9. Similarly, the binary random effects z_i can be estimated using the function ranef. These random effects have a very tangible meaning—they are the presence/absence state at each site, and their estimates represent our best guess of whether a particular site is occupied or not.

```
ranef(fm1.occ)
         Mean    Mode  2.5%  97.5%
[1,] 0.039231087   0    0     1
[2,] 0.005800029   0    0     0
[3,] 0.067784966   0    0     1
[4,] 0.044980764   0    0     1
[5,] 0.032297869   0    0     1
   [output truncated]
[91,] 1.000000000   1    1     1
[92,] 1.000000000   1    1     1
[93,] 0.776602771   1    0     1
[94,] 0.580214449   1    0     1
[95,] 1.000000000   1    1     1
     [ output truncated ]
```

These predictions of the random effects z are also called *conditional occupancy probability*, where conditional means "given the observed data at that site" (MacKenzie et al., 2006, pp. 97—98). When a species has been detected at least once at a site, under the usual assumption of no false-positives the site is occupied with certainty. This is why for sites 91, 92, and 95 in our example the conditional occupancy probability is equal to 1 with zero uncertainty. The case is more interesting for a site where a species was never detected during the J surveys, i.e., $\{y_i\} = 0$. The probability that site i is occupied then depends on three things: the expected occupancy probability for the site (ψ), detection probability for the site (p), and the number of surveys J:

$$\Pr(z_i = 1|\{y_i\} = 0) = \frac{\psi(1-p)^J}{(1-\psi) + \psi(1-p)^J}$$

This result follows directly from an application of Bayes' rule (Section 2.5.1) and makes sense intuitively—all else equal, given that the species was not observed at a site, we have higher confidence in its presence despite the negative survey results (1) when it is widespread overall (i.e., when occupancy probability ψ is high), (2) when it is elusive (i.e., when detection probability p is small), and (3) when the number of times we have looked for it (J) is small. When ψ is site- and p site- and survey-specific the equation changes to:

$$\Pr(z_i = 1|\{y_i\} = 0) = \frac{\psi_i \prod_{j=1}^{J}(1-p_{ij})}{(1-\psi_i) + \psi_i \prod_{j=1}^{J}(1-p_{ij})}$$

Let's double check this for site 1, where after three surveys the species was never detected. The probabilities of occupancy (1 value) and detection (1 value for each survey) for this site can be obtained from unmarked as follows:

```
(psi1 <- predict(fm1.occ, type="state")[1,1])
[1] 0.07565784
(p1 <- predict(fm1.occ, type="det")[c(1:3),1])
[1] 0.43290197 0.05942820 0.06472325
```

(Important note: The predictions of detection for the three surveys made at site 1 are in rows 1–3 and *not* in rows 1, 101, and 201, as you might perhaps think.) We can now calculate the conditional occupancy probability for site 1, given that all three surveys resulted in a negative result, as follows, and will find that it matches up the solution for site 1 obtained from the ranef function.

```
(z1 <- (psi1 * prod(1-p1)) / ((1 - psi1) + psi1 * prod(1-p1)))
[1] 0.03923109
```

One quantity that is frequently of interest is the finite-sample occupancy, i.e., the number of sites occupied in the sample of sites actually studied. In unmarked, we can obtain this quantity by summing over the estimates of the random effects z_i and for a confidence interval use a parametric bootstrap. For the latter, we first need to define a function that computes the finite-sample estimate of the number of sites occupied. Then, we use it for a large number of bootstrap samples to obtain uncertainty intervals around that estimate (Figure 10.3 left).

```
# Define function for finite-sample number and proportion of occupied sites
fs.fn <- function(fm){
    Nocc <- sum(ranef(fm)@post[,2,])
    psi.fs <- Nocc / nrow(fm@data@y)
    out <- c(Nocc = Nocc, psi.fs = psi.fs)
    return(out)
}
```

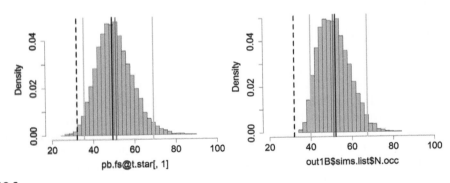

FIGURE 10.3

Bootstrap distribution (left) and posterior distribution (right) of the number of occupied sites in the actual sample of 100 surveyed sites (finite-sample occupancy). For the bootstrap, the point estimate (50.5) is the blue line and the 95% CI (35.4, 69.1) is shown in grey. The truth is 51 (red), and the species was observed at a total of 32 sites (dashed black line). For the Bayesian analysis, the point estimate is 52.0 and the 95% CRI is (40, 67.5).

10.4 A SLIGHTLY MORE COMPLEX SITE-OCCUPANCY MODEL

```
# Bootstrap the function
fs.hat <- fs.fn(fm1.occ)          # Point estimate
pb.fs <- parboot(fm1.occ, fs.fn, nsim=10000, report=2) # Takes a while (33 min)
# system.time(pb.fs <- parboot(fm.occ1, fs.fn, nsim=100, report=10)) # quicker

# Summarize bootstrap distributions
summary(pb.fs@t.star)
      Nocc           psi.fs
 Min.   :24.08   Min.   :0.2408
 1st Qu.:44.41   1st Qu.:0.4441
 Median :49.93   Median :0.4993
 Mean   :50.51   Mean   :0.5051
 3rd Qu.:55.87   3rd Qu.:0.5587
 Max.   :89.11   Max.   :0.8911

# Get 95% bootstrapped confidence intervals
(tmp1 <- quantile(pb.fs@t.star[,1], prob = c(0.025, 0.975)))
    2.5%     97.5%
35.44263 69.14740

(tmp2 <- quantile(pb.fs@t.star[,2], prob = c(0.025, 0.975)))
     2.5%      97.5%
0.3544263 0.6914740

# Plot bootstrap distribution of number of occupied sites (Fig. 10-3 left)
par(mfrow = c(1,2), mar = c(5,4,3,2))
hist(pb.fs@t.star[,1], col = "grey", breaks = 80, xlim = c(20, 100), main = "", freq = F)
abline(v = fs.hat[1], col = "blue", lwd = 3)              # add point estimate
abline(v = tmp1, col = "grey", lwd = 3)                   # add 95% CI
abline(v = sum(apply(y, 1, max)), lty = 2, lwd = 3)       # observed #occ sites
abline(v = sum(z), col = "red", lwd = 3)                  # true #occ sites
```

What is the difference between the estimates that you obtain with `predict` and those that you get from `ranef`? To understand this, look at the state model:

$$z_i \sim Bernoulli(\psi_i)$$

In short, `predict` yields estimates of the population parameter ψ, i.e., the *expected* presence/absence status for a site i that is drawn at random from the same statistical population of sites as the 100 we studied and has the given covariate values. In contrast, `ranef` yields estimates for z_i, i.e., the realized presence/absence status exactly of site i in the studied sample of sites, taking into account both the values of the modeled covariates *and* the data y_i observed at that site.

Next, we quickly illustrate the fitting of factors by first fitting what could be called a main-effects ANCOVA linear model for both model parts, i.e., a model with additive effects of a discrete (`hab` and `time`) and of a continuous covariate (`vegHt` and `wind`) for occurrence and detection, respectively. All linear models in `unmarked` are specified in exactly the same way as in other R functions such as `lm` or `glm`. So let's fit a model in the "means parameterizations," where the parameters for the factor levels directly have the meaning of the intercepts for each level. We can write these linear models in algebra as:

$$\text{logit}(\psi_i) = \beta 0_{hab(i)} + \beta 1 * \text{vegHt}_i$$

$$\text{logit}(p_{ij}) = \alpha 0_{time(i)} + \alpha 1 * \text{wind}_{ij}$$

```
# Fit model p(time+wind), psi(hab+vegHt)
summary(fm2.occ <- occu(~time+wind-1 ~hab+vegHt-1, data=umf))
Call:
occu(formula = ~time + wind - 1 ~ hab + vegHt - 1, data = umf)

Occupancy (logit-scale):
       Estimate   SE      z       P(>|z|)
habA   -0.570    1.191   -0.479   0.632
habB    0.476    0.648    0.735   0.462
habC   -1.055    1.276   -0.827   0.408
vegHt   2.869    1.829    1.569   0.117

Detection (logit-scale):
       Estimate  SE      z       P(>|z|)
time1  -1.37    0.500   -2.75    6.01e-03
time2  -2.17    0.530   -4.10    4.05e-05
time3  -1.51    0.522   -2.89    3.88e-03
wind   -3.17    0.619   -5.11    3.16e-07

# Predict occupancy for habitat factor levels at average covariate values
newdat <- data.frame(vegHt=0, hab = c("A", "B", "C"))
predict(fm2.occ, type="state", newdata = newdat, appendData = TRUE)
  Predicted    SE         lower        upper     vegHt hab
1 0.3611400  0.2747052  0.05195332  0.8536124    0    A
2 0.6168318  0.1531216  0.31138401  0.8514352    0    B
3 0.2582344  0.2445064  0.02773337  0.8094843    0    C

# Predict detection for time factor levels at average covariate values
newdat <- data.frame(wind=0, time = c("1", "2", "3"))
predict(fm2.occ, type="det", newdata=newdat, appendData = TRUE)
  Predicted    SE           lower        upper     wind time
1 0.2020662  0.08060966   0.08680271  0.4028639   0    1
2 0.1020261  0.04854268   0.03866499  0.2429754   0    2
3 0.1813491  0.07748170   0.07377254  0.3812293   0    3
```

See Section 6.4 for how we form predictions for one specific level of a factor. For the sake of exercise, let's now also fit a model with interaction effects in both the occupancy and the detection model and then use a likelihood ratio test to decide which is better supported by the data. In algebra, that model can be written as this:

$$\text{logit}(\psi_i) = \beta 0_{hab(i)} + \beta 1_{hab(i)} * \text{vegHt}_i$$

$$\text{logit}(p_{ij}) = \alpha 0_{time(i)} + \alpha 1_{time(i)} * \text{wind}_{ij}$$

The difference in this model is that now the slope parameters of vegHt and wind ($\beta 1$ and $\alpha 1$, respectively) are no longer a single number (scalar), but they are indexed and hence vary over the levels of the two factors hab and time. Hence, they are now vectors of length 3, corresponding to the three levels of the factors hab and time.

```
# Fit model p(time*wind), psi(hab*vegHt)
summary(fm3.occ <- occu(~time*wind-1-wind ~hab*vegHt-1-vegHt, data=umf))
```

10.4 A SLIGHTLY MORE COMPLEX SITE-OCCUPANCY MODEL

```
# Do likelihood ratio test
LRT(fm2.occ, fm3.occ)
      Chisq DF Pr(>Chisq)
1 6.233802   4   0.182355
```

The test says that interactive effects with the two continuous explanatory variables (i.e., model 3) are not preferred over a model with additive effects (model 2).

As a final part in this section, we illustrate the fitting of a simple occupancy model with covariates in BUGS and also show the forming of predictions (estimates of ψ), estimation of the realized presence/absence status of a site (estimates of z) and of the finite-sample occupancy, i.e., the number and proportion of occupied sites in the studied sample of 100 sites. As always in BUGS, knowing how to write a model in algebra gets us very close to the BUGS model description. We add three types of derived quantities: the number of occupied sites in the sample of 100 study sites (N.occ) and predictions of occupancy and of detection for a range of values of the covariates vegHt and wind, respectively. For the latter, we provide two sets of covariate values spaced evenly in the range over which predictions are desired (XvegHt, Xwind). We want to see the estimates of presence/absence at each site (z), so we add those in the list of parameters to be saved below.

```
# Bundle and summarize data set
str( win.data <- list(y = y, vegHt = vegHt, wind = wind, M = nrow(y), J = ncol(y), XvegHt =
seq(-1, 1, length.out=100), Xwind = seq(-1, 1, length.out=100)) )

# Specify model in BUGS language
sink("model.txt")
cat("
model {

# Priors
mean.p ~ dunif(0, 1)            # Detection intercept on prob. scale
alpha0 <- logit(mean.p)         # Detection intercept
alpha1 ~ dunif(-20, 20)         # Detection slope on wind
mean.psi ~ dunif(0, 1)          # Occupancy intercept on prob. scale
beta0 <- logit(mean.psi)        # Occupancy intercept
beta1 ~ dunif(-20, 20)          # Occupancy slope on vegHt

# Likelihood
for (i in 1:M) {
  # True state model for the partially observed true state
  z[i] ~ dbern(psi[i])          # True occupancy z at site i
  logit(psi[i]) <- beta0 + beta1 * vegHt[i]
  for (j in 1:J) {
    # Observation model for the actual observations
    y[i,j] ~ dbern(p.eff[i,j])  # Detection-nondetection at i and j
    p.eff[i,j] <- z[i] * p[i,j] # 'straw man' for WinBUGS
    logit(p[i,j]) <- alpha0 + alpha1 * wind[i,j]
  }
}
```

```
# Derived quantities
N.occ <- sum(z[])      # Number of occupied sites among sample of M
psi.fs <- N.occ/M      # Proportion of occupied sites among sample of M
for(k in 1:100){
    logit(psi.pred[k]) <- beta0 + beta1 * XvegHt[k] # psi predictions
    logit(p.pred[k]) <- alpha0 + alpha1 * Xwind[k]  # p predictions
}
}
",fill = TRUE)
sink()

# Initial values: must give for same quantities as priors given !
zst <- apply(y, 1, max)        # Avoid data/model/inits conflict
inits <- function(){list(z = zst, mean.p = runif(1), alpha1 = runif(1), mean.psi
= runif(1), beta1 = runif(1))}

# Parameters monitored
params <- c("alpha0", "alpha1", "beta0", "beta1", "N.occ", "psi.fs", "psi.pred",
"p.pred", "z") # Also estimate z = "conditional occ. prob."

# MCMC settings
ni <- 25000 ; nt <- 10 ; nb <- 2000 ; nc <- 3

# Call WinBUGS from R (ART 2 min) and summarize posteriors
out1B <- bugs(win.data, inits, params, "model.txt", n.chains = nc,
n.thin = nt, n.iter = ni, n.burnin = nb, debug = TRUE, bugs.directory = bugs.dir,
working.directory = getwd())
print(out1B, dig = 3)
          mean     sd    2.5%    25%    50%    75%   97.5% Rhat n.eff
alpha0  -1.739  0.385 -2.510 -1.992 -1.738 -1.479 -0.991 1.001  4900
alpha1  -3.047  0.586 -4.283 -3.419 -3.018 -2.644 -1.971 1.001  6900
beta0    0.080  0.585 -0.836 -0.325  0.001  0.386  1.496 1.001  4900
beta1    3.074  1.294  1.239  2.183  2.818  3.672  6.323 1.001  6900
N.occ   52.050  7.461 40.000 46.000 51.000 57.000 67.523 1.001  6900
psi.fs   0.520  0.075  0.400  0.460  0.510  0.570  0.675 1.001  6900
[   Output truncated  ]
```

We compare the truth with MLEs from unmarked and posterior inference from BUGS (Figure 10.4).

```
# Compare truth with MLEs and bayesian posterior inference in table ...
truth <- c(alpha0, alpha1, beta0, beta1, sum(z), sum(z)/M)
tmp <- summary(fm1.occ)
MLEs <- rbind(tmp[[2]][1:2,1:2], tmp[[1]][1:2,1:2], sumZ = c(mean(pb.fs@t.star[,1]),
sd(pb.fs@t.star[,1])), psi.fs = c(mean(pb.fs@t.star[,2]), sd(pb.fs@t.star[,2])))
print(cbind(truth, MLEs, out1B$summary[1:6, 1:2]))
              truth    Estimate         SE         mean         sd
(Intercept)   -2.00  -1.6961500 0.39751824  -1.73942300 0.38527615
wind          -3.00  -3.0670527 0.59406301  -3.04706739 0.58621034
(Intercept)1   0.00  -0.1360580 0.44880554   0.07996883 0.58483924
vegHt          3.00   2.4319320 0.83948163   3.07435855 1.29439358
sumZ          51.00  50.5097710 8.62495498  52.04971014 7.46061747
psi.fs         0.51   0.5050977 0.08624955   0.52049710 0.07460617
```

10.4 A SLIGHTLY MORE COMPLEX SITE-OCCUPANCY MODEL

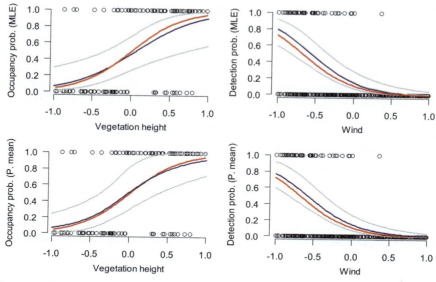

FIGURE 10.4

Estimated relationships (blue) between occurrence and vegetation height (left) and between detection probability and wind speed (right) from an occupancy model fit to the simulated data set. Top: maximum likelihood estimates (with 95% CIs in grey); bottom: Bayesian posterior means (with 95% CRIs in grey). Red lines represent the truth in the data simulation process. Circles show the true presence/absence (left) and the observed measurements of presence/absence (right).

```
# .... and in a graph (Fig. 10-4)
par(mfrow = c(2, 2), mar = c(4, 5, 2, 2), las = 1, cex.lab = 1, cex = 1.2)
plot(vegHt, z, xlab="Vegetation height", ylab="Occupancy prob. (MLE)", ylim = c(0, 1),
  frame = F)                      # True presence/absence
lines(seq(-1,1, 0.01), pred.occ[,1], col = "blue", lwd = 2)
matlines(seq(-1,1, 0.01), pred.occ[,3:4], col = "grey", lty = 1)
lines(vegHt, psi, lwd=3, col="red")    # True psi
plot(wind, y, xlab="Wind", ylab="Detection prob. (MLE)", ylim = c(0,1), frame=F)
lines(seq(-1, 1, 0.1), pred.det[,1], col = "blue", lwd = 2)
matlines(seq(-1, 1, 0.1), pred.det[,3:4], col = "grey", lty = 1)
plot(function(x) plogis(alpha0 + alpha1*x), -1, 1, add=T, lwd=3, col = "red")
plot(vegHt, z, xlab="Vegetation height", ylab="Occupancy prob. (P. mean)", las = 1,
  frame = F)                      # True presence/absence
lines(vegHt, psi, lwd=3, col="red")    # True psi
lines(win.data$XvegHt, out1B$summary[7:106,1], col="blue", lwd = 2)
matlines(win.data$XvegHt, out1B$summary[7:106,c(3,7)], col="grey", lty = 1)
plot(wind, y, xlab="Wind", ylab="Detection prob. (P. mean)", frame = F)
plot(function(x) plogis(alpha0 + alpha1*x), -1, 1, add=T, lwd=3, col = "red")
lines(win.data$Xwind, out1B$summary[107:206,1], col="blue", lwd = 2)
matlines(win.data$Xwind, out1B$summary[107:206,c(3,7)], col="grey", lty = 1)
```

The MLEs and Bayesian posterior means match fairly well in this realization of the simulated process (i.e., with a seed of 1 for the random number generator). However, during the development of this material we observed some cases where the posterior means did *not* match the MLEs so well, especially for the occupancy parameters. We believe that this was due to the relatively small sample size of only 100 sites. In this case, the priors have relatively more influence and the posterior will often be skewed. If we take the posterior mean as our point estimator, then by averaging over the whole posterior distribution, the Bayesian point estimate will differ from the MLE. This is presumably what McKann et al. (2013) called "small sample bias" in the case of the related dynamic occupancy model (see Chapter 16), where they observed that for truly extreme values of the probability parameters, the Bayesian estimates tended to be less extreme, i.e., pulled toward 0.5. Such a slight pulling in of extreme estimates may often be a good thing (Sauer and Link, 2002), for instance, it will completely prevent boundary estimates of 0 or 1 as they often occur in MLE (see Section 10.7).

Above, we discussed the finite-sample quantities, the number and the proportion of occupied sites among the sample of M studied sites. The estimate of the finite-sample occupancy is asymptotically equal to that of population occupancy. However, its uncertainty is smaller because one component of variation present in population occupancy is lacking: the binomial variance due to sampling M study sites from a hypothetical, infinite (statistical) population of sites. The only source of uncertainty in the variance estimate of finite-sample occupancy is due to imperfect detection and parameter uncertainty. These finite-sample quantities are frequently of more interest to practitioners than are the corresponding population quantities. Both are a function of the latent occurrence states z, which appear explicitly as latent variables in the model in BUGS, and you will obtain estimates of z simply by including them in the list of the estimated parameters.

Calculations on latent variables, such as z, in a Bayesian analysis is trivially easy and is conducted with a full propagation of all the involved uncertainties. In the Bayesian analysis, we directly estimate the quantities `N.occ` and `psi.fs`. Their estimates along with their uncertainties are quite comparable with their non-Bayesian counterparts when the variance is bootstrapped; see the table above and Figure 10.3 (right). However, the Bayesian posterior never extends to nonsensical values, i.e., the posterior does not extend to fewer occupied sites than were observed.

```
# Plot posterior distribution of number of occupied sites (see Fig. 10-3, right)
hist(out1B$sims.list$N.occ, col = "grey", breaks = 60, xlim = c(20, 100),
  main = "", freq = F)
abline(v = out1B$mean$N.occ, col = "blue", lwd = 3)    # add point estimate
abline(v = out1B$summary[5,c(3,7)], col = "grey", lwd = 3)  # add 95% CRI
abline(v = sum(apply(y, 1, max)), lty = 2, lwd = 3)    # observed #occ sites
abline(v = sum(z), col = "red", lwd = 2)               # true #occ sites
```

Perhaps the most powerful aspect of hierarchical models is their invitation to build custom models that are exactly tailored to your system and your questions. However, in applied work with hierarchical models, much of the power of hierarchical modeling simply stems from your ability to specify linear models in a smart way. Hence, being able to fit complex linear models is important for all of your hierarchical modeling. For this reason we want to illustrate an occupancy model with a slightly more complex linear predictor. But first we introduce a data simulation function that, among other things, will provide us with a data set for such a more complex occupancy model in Section 10.6.

10.5 A GENERAL DATA SIMULATION FUNCTION FOR STATIC OCCUPANCY MODELS: simOcc

In our AHM package, you find an R function simOcc that permits simulation of data sets under a very wide variety of static occupancy models. The function is similar to those in Chapter 4 and in Section 6.5, but for occurrence rather than for count data. We provide this function in the hope that it may be directly useful for you in one of the many ways in which data simulation can be valuable (see Section 4.4) and that it may serve as a starting point for adapting it to your more specific needs. Later, we use simOcc to validate BUGS code (Section 10.6), to investigate estimator quality in the occupancy model when the information content of the data set is low and variable (Section 10.7) and to study goodness-of-fit assessments (Section 10.8). Using the function, you can simulate data under the following most general model, where sites are indexed i and repeated presence/absence measurements j and the main notation (e.g., z, ψ, y, p) is standard in this chapter.

Ecological model for presence/absence (z):

$$z_i \sim Bernoulli(\psi_i)$$

$$logit(\psi_i) = \beta_0 + \beta_1 * elev_i + \beta_2 * forest_i + \beta_3 * elev_i * forest_i$$

Observation/measurement error model for detection/nondetection data (y):

$$y_{ij}|z_i \sim Bernoulli(z_i * p_{ij})$$

$$logit(p_{ij}) = \alpha_0 + \gamma_j + \alpha_1 * elev_i + \alpha_2 * wind_{ij} + \alpha_3 * elev_i * wind_{ij} + \varepsilon_i + b * y_{i,j-1}$$

Effects of three continuous covariates can be built into the simulated data set: elevation, forest cover (two site covariates), and wind speed (an observational covariate), as well as the interactions between elevation and forest cover and between elevation and wind speed. Elevation can affect both occupancy and detection, and the elevation-wind speed interaction is set to zero by default (see below). In addition, we may add the following:

γ_j: time (j)-specific effects on baseline detection probability (*time effects*); these are expressed as deviations from the logistic-linear detection intercept α_0

$\varepsilon_i \sim Normal(0, \sigma)$: site-specific random effects assumed to be draws from a normal distribution with standard deviation (σ, called sd.lp below: "*heterogeneity*" or *site random effects*)

b: a "*behavioral response*" term, which increases or lowers detection probability depending on whether the species was detected at site i at the preceding occasion ($j - 1$) (see Riddle et al., 2010); this effect can only occur from the second occasion onward (i.e., for $j > 1$).

Models with a single one of the last three effects are called model M_t, M_h, and M_b in the capture-recapture literature (Otis et al., 1978; Royle and Dorazio, 2008; Kéry and Schaub, 2012). The function is called as follows, with its default arguments shown and explained below.

```
simOcc(M = 267, J = 3, mean.occupancy = 0.6, beta1 = -2, beta2 = 2, beta3 = 1, mean.detection =
0.3, time.effects = c(-1, 1), alpha1 = -1, alpha2 = -3, alpha3 = 0, sd.lp = 0.5, b = 2,
show.plot = TRUE)
# Function arguments:
M:        Number of spatial replicates (sites)
J:        Number of temporal replicates (occasions)
```

```
mean.occupancy: Occupancy probability at value 0 of occ. covariates
beta1:          Main effect of elevation on occurrence
beta2:          Main effect of forest cover on occurrence
beta3:          Interaction effect on occurrence of elevation and forest cover
mean.detection: Detection probability at value 0 of detection covariates
time.effects: bounds for uniform distribution from which time effects gamma
                (on logit scale) will be drawn
alpha1:         Main effect of elevation on detection probability
alpha2:         Main effect of wind speed on detection probability
alpha3:         Interaction effect on detection of elevation and wind speed
sd.lp:          Standard deviation of random site effects (on logit scale)
b:              Constant value of 'behavioural response' leading to
                'trap-happiness' (if b > 0) or 'trap shyness' (if b < 0)
show.plot:      if TRUE, plots of the data will be displayed;
                should be set to FALSE if you are running simulations.
```

Executing the function produces two multipanel plots that visualize the simulated system (Figure 10.5 and 10.6). Here are some examples of the function's usage.

```
simOcc()                      # Execute function with default arguments
simOcc(show.plot = FALSE)  #    same, without plots
simOcc(M = 267, J = 3, mean.occupancy = 0.6, beta1 = -2, beta2 = 2, beta3 = 1, mean.detection =
0.3, time.effects = c(-1, 1), alpha1 = -1, alpha2 = -3, alpha3 = 0, sd.lp = 0.5, b = 2,
show.plot = TRUE)             # Explicit defaults
```

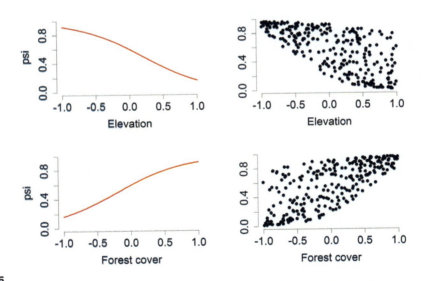

FIGURE 10.5

Visualization of the patterns in occupancy probability (psi) simulated by the function simOcc. Left panels show relationships between expected occupancy and one covariate when the other covariate is held constant, while the right panels show the same relationships at the observed values of the other covariate.

10.5 A GENERAL DATA SIMULATION FUNCTION FOR STATIC OCCUPANCY

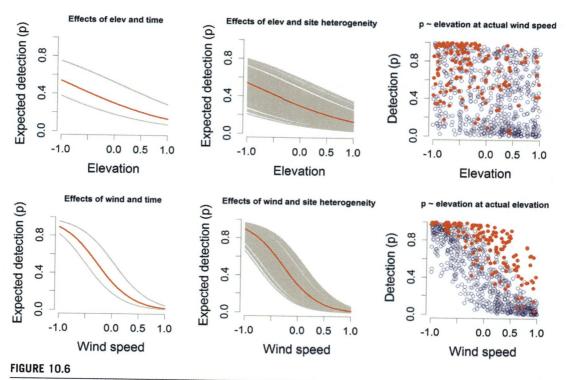

FIGURE 10.6

Visualization of the patterns in detection probability (*p*) simulated by the function `simOcc`. Left and middle panels show the relationships between the detection probability and one covariate when the other covariate is held constant (red line shows values with intercept). In the left panel, the effects of time, and in the middle panel, the effects of individual (site-specific) heterogeneity, are depicted in addition (these are the grey lines); each set of effects is also held constant in the other panel. In the right panel, these same relationships are shown at the observed values of the other covariate(s) and with red/blue color coding for the behavioral response effects (red: values of p when the species was detected at a site during the preceding survey; blue: same when the species was not detected during the preceding survey).

```
# Create a 'fix' data set and look at what we created
set.seed(24)
data <- simOcc()      # Assign results to an object called 'data'
str(data)
```

The output contains all argument settings employed plus all quantities created. Other than those that should be obvious from the preceding, we get occupancy probability (`psi`), presence/absence (`z`), the matrix of detection probability (`p`), the resulting data (`y`), the true number of occupied sites (`sumZ`), the number of sites at which at least one detection was made (`sumZ.obs`), the true proportion of occupied sites in the sample (`psi.fs.true`), and the observed proportion of occupied sites in the sample (`psi.fs.obs`). (The matrices `p0` and `p1` contain detection probability following a survey without detection (`p0`) or with detection (`p1`) and will not normally be of interest.).

You can use this function to generate data sets under various sampling designs (e.g., number of sites or surveys) and for a very large number of ecological and sampling "settings" (i.e., patterns in occurrence and detection). By setting to 0 or to 1 some arguments you can eliminate components from the simulation process, e.g., effects of covariates or patterns in detection probability (time, "heterogeneity" or "behavioral response") or the observation process altogether. It can be really useful for your understanding of occurrence data in general, and occupancy models specifically, to play around with this function with varying arguments to train your intuition about this important type of data and its sampling in the field under imperfect detection. Here are some illustrations of its use, with mostly only the relevant changes to the default arguments shown.

```
# Simplest possible occupancy model, with constant occupancy and detection
tmp <- simOcc(mean.occ=0.6, beta1=0, beta2=0, beta3=0, mean.det=0.3,
time.effects=c(0, 0), alpha1=0, alpha2=0, alpha3=0, sd.lp=0, b=0)
str(tmp)                    # give overview of results

# psi = 1 (i.e., species occurs at every site)
tmp <- simOcc(mean.occ=1) ; str(tmp)

# p = 1 (i.e., species is always detected when it occurs)
tmp <- simOcc(mean.det=1) ; str(tmp)
```

Other potentially interesting settings include these:

```
simOcc(J = 2)                              # Only 2 surveys
simOcc(M = 1, J = 100)                     # No spatial replicates, but 100 measurements
simOcc(beta3 = 1)                          # Including interaction elev-wind on p
simOcc(mean.occ = 0.96)                    # A really common species
simOcc(mean.occ = 0.05)                    # A really rare species
simOcc(mean.det = 0.96)                    # A really easy species
simOcc(mean.det = 0.05)                    # A really hard species
simOcc(mean.det = 0)                       # The dreaded invisible species
simOcc(alpha1=-2, beta1=2)                 # Opposing effects of elev on psi and p
simOcc(J = 10, time.effects = c(-5, 5))    # Huge time effects on p
simOcc(sd.lp = 10)                         # Huge (random) site effects on p
simOcc(J = 10, b = 0)                      # No behavioural response in p
simOcc(J = 10, b = 2)                      # Trap happiness
simOcc(J = 10, b = -2)                     # Trap shyness
```

You cannot simulate single-visit data (i.e., choose $J = 1$), but a simple workaround is to set J at any value greater than 1 and then discard everything except for one particular survey at each site (though you will have to compute "by hand" the observed number and proportion of occupied sites, `sumZ.obs` and `psi.fs.obs`). In addition, you cannot choose your own specific values for the time effects; rather, time effects on the logit scale are chosen randomly from a uniform distribution for which you specify the bounds.

We hope that this function is useful for you, either in its current version or as a template for modifications that you make to suit your needs. For instance, if you are interested in the question of how the quality of parameter estimates (e.g., bias, precision) varies as a function of sample size (number of sites and temporal replicates) and the number of covariate effects estimated in the model, then you could adapt the function to contain 20 or so covariate effects in both the occupancy and detection parts of the data-generating model (but vectorizing and using matrix-vector multiplication to build up the linear predictor would then be a good idea, see Section 3.2.1). Or if you're interested in how different patterns of missing values affect your estimates, then you could adapt the function by incorporating a missing value-generating process, which could be governed by the same or a different set of covariates. Finally, with three exceptions (the time, site, and behavioral response effects in p) the function contains continuous covariates only and no factors. Obviously, if you need more factors you could incorporate them.

10.6 A MODEL WITH LOTS OF COVARIATES: USE OF R FUNCTION model.matrix WITH BUGS

Next, we show how a fairly complex linear model can be fit as part of an occupancy model in BUGS and how we can simplify our life by using the powerful R linear modeling function model.matrix. We start by generating one data set with the simOcc function where we eliminate the effects of all three factors by setting to zero the arguments controlling them.

```
set.seed(148)
data <- simOcc(time.effects = c(0,0), sd.lp = 0, b = 0)
str(data)                              # Look at data object
```

To illustrate a fairly complex linear model with covariates and factors we invent a further factor that is unrelated to the response: habitat (hab), which divides the default 267 sites into three imaginary habitat types.

```
# Create habitat factor
hab <- c(rep("A", 90), rep("B", 90), rep("C", 87)) # must have M = 267 sites

# Load library, format data and summarize unmarked data frame
library(unmarked)
umf <- unmarkedFrameOccu(
    y = data$y,
    siteCovs = data.frame(elev = data$elev, forest = data$forest, hab = hab),
    obsCovs = list(wind = data$wind))
summary(umf)
```

For illustration, let's now use unmarked to fit a model with additive effects of elevation and wind speed in detection and fully interactive effects of elevation, forest cover, and habitat in occupancy probability.

```
summary(fm <- occu(~elev+wind ~elev*forest*hab, data=umf))
```

Looking at the long list of parameter estimates in the output, you'll probably agree that this is a fairly complicated linear model. We now fit the same model in BUGS in two ways: first, by writing out all the linear model terms explicitly, and second, by defining a design matrix for the linear model and

CHAPTER 10 MODELING STATIC OCCURRENCE AND SPECIES DISTRIBUTIONS

fitting this matrix in BUGS (see Section 6.11.1 for another example of this in an N-mixture model). First, the more difficult solution:

```
# Bundle and summarize data set
HAB <- as.numeric(as.factor(hab))   # Get numeric habitat factor
str( win.data <- list(y = data$y, M = nrow(data$y), J = ncol(data$y), elev = data$elev,
    forest = data$forest, wind = data$wind, HAB = HAB) )

# Specify model in BUGS language
sink("modelA.txt")
cat("
model {

# Priors
mean.p ~ dunif(0, 1)              # Detection intercept on prob. scale
alpha0 <- logit(mean.p)           #    same on logit scale
mean.psi ~ dunif(0, 1)            # Occupancy intercept on prob. scale
beta0 <- logit(mean.psi)          #    same on logit scale
for(k in 1:2){                    # 2 terms in detection model
    alpha[k] ~ dnorm(0, 0.1)      # Covariates on logit(detection)
}
for(k in 1:11){                   # 11 terms in occupancy model
    beta[k] ~ dnorm(0, 0.1)       # Covariates on logit(occupancy)
}
# Likelihood
for (i in 1:M) {                  # Loop over sites
  z[i] ~ dbern(psi[i])
  logit(psi[i]) <- beta0 +                           # occupancy (psi) intercept
    beta[1] * elev[i] +                              # effect of elev
    beta[2] * forest[i] +                            # effect of forest
    beta[3] * equals(HAB[i],2) +                     # effect of habitat 2 (= B)
    beta[4] * equals(HAB[i],3) +                     # effect of habitat 3 (= C)
    beta[5] * elev[i] * forest[i] +                          # elev:forest
    beta[6] * elev[i] * equals(HAB[i],2) +                   # elev:habB
    beta[7] * elev[i] * equals(HAB[i],3) +                   # elev:habC
    beta[8] * forest[i] * equals(HAB[i],2) +                 # forest:habB
    beta[9] * forest[i] * equals(HAB[i],3) +                 # forest:habC
    beta[10] * elev[i] * forest[i] * equals(HAB[i],2) +      # elev:forest:habB
    beta[11] * elev[i] * forest[i] * equals(HAB[i],3)        # elev:forest:habC
    for (j in 1:J) {              # Loop over replicates
      y[i,j] ~ dbern(z[i] * p[i,j])   # WinBUGS would need 'straw man' !
      logit(p[i,j]) <- alpha0 +       # detection (p) intercept
        alpha[1] * elev[i] +          # effect of elevation on p
        alpha[2] * wind[i,j]          # effect of wind on p
    }
  }
}
",fill = TRUE)
sink()
```

```r
# Inits
inits <- function(){list(z = apply(data$y, 1, max), mean.psi = runif(1), mean.p = runif(1),
alpha = rnorm(2), beta = rnorm(11))}

# Parameters monitored
params <- c("alpha0", "alpha", "beta0", "beta")

# MCMC settings
ni <- 50000  ;  nt <- 10  ;  nb <- 10000  ;  nc <- 3

# Run JAGS (ART 4 min), look at convergence and summarize posteriors
outA <- jags(win.data, inits, params, "modelA.txt", n.chains = nc, n.thin = nt,
n.iter = ni, n.burnin = nb, parallel = TRUE)
traceplot(outA)   ;   print(outA, 3)

# Compare MLEs and SEs with posterior means and sd's
tmp <- summary(fm)
cbind(rbind(tmp$state[1:2], tmp$det[1:2]), Post.mean = outA$summary[c(4:15, 1:3), 1],
Post.sd = outA$summary[c(4:15, 1:3), 2])
```

Second, the easy solution: in BUGS we simply fit a design matrix generated in R using model.matrix. We want to add the intercept in BUGS and hence create a design matrix without intercept and simply add the matrix to the data bundle. Then, in BUGS we define the linear predictor to be the matrix (or "inner") product of parameter vector and design matrix: inprod(beta[], occDM[i,]).

```r
# Create design matrix for occupancy covariates and look at it
occDM <- model.matrix(~ data$elev * data$forest * hab)[,-1] # Drop first col.
head(occDM)                   # Look at design matrix
str(occDM)

# Bundle and summarize data set
str( win.data <- list(y = data$y, M = nrow(data$y), J = ncol(data$y), elev = data$elev,
wind = data$wind, occDM = occDM) )

# Specify model in BUGS language
sink("modelB.txt")
cat("
model {

# Priors
mean.p ~ dunif(0, 1)          # Detection intercept on prob. scale
alpha0 <- logit(mean.p)       #   same on logit scale
mean.psi ~ dunif(0, 1)        # Occupancy intercept on prob. scale
beta0 <- logit(mean.psi)      #   same on logit scale
for(k in 1:2){                # 2 terms in detection model
   alpha[k] ~ dnorm(0, 0.1)   # Covariates on logit(detection)
}
for(k in 1:11){               # 11 terms in occupancy model
  beta[k] ~ dnorm(0, 0.1)     # Covariates on logit(occupancy)
}
```

```
# Likelihood
for (i in 1:M) {
  z[i] ~ dbern(psi[i])
  logit(psi[i]) <- beta0 + inprod(beta[], occDM[i,])  # slick !
  for (j in 1:J) {
    y[i,j] ~ dbern(z[i] * p[i,j])     # In WinBUGS need 'straw man'
    logit(p[i,j]) <- alpha0 +         # detection (p) intercept
      alpha[1] * elev[i] +            # effect of elevation on p
      alpha[2] * wind[i,j]            # effect of wind on p
  }
}
",fill = TRUE)
sink()
```

We can recycle all other parts of the code and directly launch JAGS.

```
# Call JAGS from R (ART 3.3 min) and summarize posteriors
outB <- jags(win.data, inits, params, "modelB.txt", n.chains = nc,
  n.thin = nt, n.iter = ni, n.burnin = nb, parallel = TRUE)
traceplot(outB)  ;   print(outB, 3)
```

Though the chains mix less well, up to MC error model B yields estimates that are identical to those under model A. Thus, you can use the model definition language in R to create the design matrix of a linear model and then fit that model, i.e., that design matrix, in BUGS directly. That may be a great simplification, because it may be much easier to specify a linear model in R, import its design matrix into BUGS, and fit it there, rather than constructing the model for every column in the design matrix, as we did in the previous section (for model A).

10.7 STUDY DESIGN, AND BIAS AND PRECISION OF SITE-OCCUPANCY ESTIMATORS

We next present a small simulation study with a basic occupancy model. We do this to emphasize the power of simulation to answer questions about study design and about the estimator quality from a model. Questions frequently heard about HMs like occupancy models are: "How many sites do I need?" or "How many replicate surveys are enough?" or "Is it better to visit more sites fewer times or fewer sites more frequently?" or (now this one is mean) "How come I get an NA or Inf standard error with my occupancy model with four sites?" These are important questions about study design and estimator quality, and these must represent one of the most neglected topics in all of ecological statistics. There are several important papers and rules of thumb about the design of occupancy studies (MacKenzie et al., 2002; Tyre et al., 2003; MacKenzie and Royle, 2005; Bailey et al., 2007; Guillera-Arroita et al., 2010, 2014b; Guillera-Arroita and Lahoz-Monfort, 2012; Ellis et al., 2015) and also software specifically designed for occupancy study design (GENPRES; Bailey et al., 2007; SODA; Guillera-Arroita et al., 2010, rSPACE; Ellis et al., 2015). However, by far the most powerful and most flexible way of answering such questions is by running your own custom simulations. For instance, the above questions may be tackled by running a factorial design where you vary both the number of sites

10.7 STUDY DESIGN, AND BIAS AND PRECISION OF SITE-OCCUPANCY

and the number of surveys, simulate and analyze 100—1000 data sets for each combination of this simulation design using the expected occupancy and detection probability of the species of your interest, and see which combination gives you the "best" estimates, where "best" would typically include statistical considerations such as the bias or the precision of the estimates as well as economical/logistical ones (e.g., how much do additional sites or additional surveys cost?; see MacKenzie and Royle, 2005). A simulation like the following can help you find the best trade-off between number of sites and number of visits specifically for *your* study.

Moreover, it is important to note that in principle, with enough data, all parameters of the site-occupancy model may be estimated. However, the quality of the estimates (i.e., whether there is bias and the magnitude of the precision of the estimates) will depend strongly on the sample size (number of sites, number of surveys) and on the magnitude of the parameters (ψ and p). For small sample sizes, the widely proclaimed unbiasedness of MLEs is typically lost (Le Cam, 1990) and solutions may become unstable (Welsh et al., 2013; Guillera-Arroita et al., 2015). Our little simulation draws attention to this basic fact of statistical inference and shows how the quality of estimators for particular scenarios (sample sizes, parameter values) can very easily be ascertained by simulation. In our example we do this for the null/null model without covariates, but the simulation could easily be extended to more complex models.

We study estimator quality in a design that varies the number of sites ($M = 20$, 120, or 250) and of surveys ($J = 2$, 5, or 10) and the magnitude of detection probability (covering almost the entire range between 0 and 1) for a species with occupancy equal to 0.5. For each combination of the two factors (sites, surveys) we repeat the following 1000 times: (1) randomly pick a value for detection probability from a *Uniform*(0.01, 0.99), (2) use simOcc to generate one data set, and (3) estimate parameters using MLE with unmarked. We don't need the SEs so we don't compute them, to avoid the simulation from breaking whenever a Hessian becomes singular.

```
# Do simulation with 1000 reps
simreps <- 1000
library(unmarked)

# Define arrays to hold the results
p <- array(dim = c(simreps,3,3))
estimates <- array(dim = c(2,simreps,3,3))

# Choose number and levels of simulation factors
nsites <- c(20, 120, 250)      # Number of sites
nsurveys <- c(2, 5, 10)         # Number of repeat surveys

# Start simulation
system.time(                    # Time whole thing
for(j in 1:3) {                 # Loop j over site factor
  for(k in 1:3) {               # Loop k over survey factor
    for(i in 1:simreps){        # Loop i over simreps
      # Counter
      cat("** nsites", j, "nsurveys", k, "simrep", i, "***\n")
      # Generate a data set: pick p and use p in simOcc()
      det.prob <- runif(1, 0.01, 0.99)
      data <- simOcc(M = nsites[j], J = nsurveys[k], mean.occupancy = 0.5,
        beta1 = 0, beta2 = 0, beta3 = 0, mean.detection = det.prob,
```

```
            time.effects = c(0, 0), alpha1 = 0, alpha2 = 0, alpha3 = 0,
            sd.lp = 0, b = 0, show.plot = F)
      # Fit model
      umf <- unmarkedFrameOccu(y = data$y)
      tmp <- occu(~1 ~1, umf, se = FALSE)      # Only get MLEs, not SEs
      # Save results (p and MLEs)
      p[i,j,k] <- data$mean.det
      estimates[,i,j,k] <- coef(tmp)
    }
  }
}
)

# Plot results
par(mfrow = c(3,3), mar = c(4,5,3,1), cex.main = 1.2)
for(j in 1:3){
  for(k in 1:3){
    lab <- paste(nsites[j],"sites,", nsurveys[k],"surveys")
    plot(p[,j,k], plogis(estimates[1,,j,k]), xlab = "Detection prob.",
        ylab = "Occupancy prob.", main = lab, ylim = c(0,1))
    abline(h = 0.5, col = "red", lwd = 2)
    lines(smooth.spline(plogis(estimates[1,,j,k])~ p[,j,k], df = 5),
        col = "blue", lwd = 2)
  }
}
```

Generation and analysis of 9,000 data sets takes less than four minutes on a moderate laptop! Figure 10.7 summarizes the results from the simulation for the nine scenarios that combine three levels each of the site and the survey factor and for the whole range of values of p between 1% and 99%. It shows that there can be substantial variation in the quality of the estimates under the site-occupancy model. When sample size (number of sites M or visits J) is small, the quality of the estimates can be fairly bad. Moreover, according to the first law of capture-recapture, the quality of the estimators becomes bad when p is low for any combination of M and J. Depending on the particular combination of number of sites and visits, the occupancy estimator is biased high when p is less than about 0.1 or 0.2, something that had already been noted by MacKenzie et al. (2002) and Guillera-Arroita et al. (2010).

We then repeated this simulation with WinBUGS, using the model in Section 10.3 with vague *Uniform*(0, 1) priors, and after waiting for about 15 hours, obtained the results in Figure 10.8. Using as the usual Bayesian point estimator the posterior mean rather than the posterior mode (which with vague priors would correspond to the MLE) always pulls estimates away from the boundary of the parameter space at 0 and 1 (McKann et al., 2013). Thus, the Bayesian estimates were much better than the MLEs because use of the posterior mean avoids the instabilities of the MLEs in situations with little information (small M, J, or p). Table 10.2 compares the Bayesian posterior means with the MLEs and shows that in every scenario the average estimation error (RMSE) was about halved in the Bayesian analysis with vague priors. Boundary estimates can be a serious problem with ML when

10.7 STUDY DESIGN, AND BIAS AND PRECISION OF SITE-OCCUPANCY

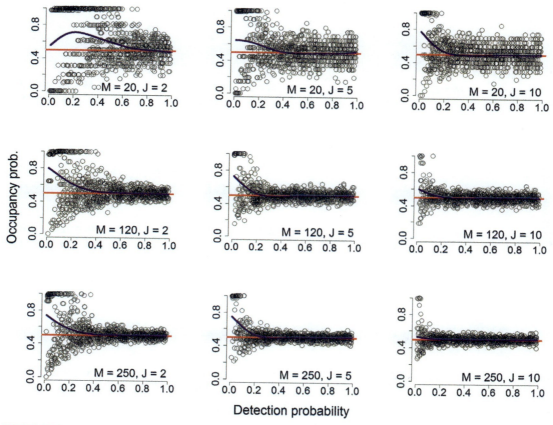

FIGURE 10.7

Results of a simulation study about the quality of MLEs of the site-occupancy estimator for occupancy probability in a grid spanned by three levels each of the two factors "number of sites" (M) and "number of repeat surveys" (J) and for a continuous range of values for detection probability ($0.01 \leq p \leq 0.99$). The red line shows the true occupancy probability (0.5) and the blue line is a spline smoother to show the average behavior of the estimator for a given value of p. The number of simulation replicates is 1000 for each plot. Look at Figure 10.8 for the Bayesian results of the same simulation exercise.

there is little information in the data. Using penalized likelihood can then stabilize the estimators. For static occupancy models, the function `occuPEN` has been incorporated in unmarked (Hutchinson et al., 2015).

The R code in this section can easily serve as a template for many other simulation studies, for instance, it is straightforward to gauge the effects of assumption violations, e.g., due to individual (site-specific) detection heterogeneity or behavioral response (this is Exercise 4).

588 CHAPTER 10 MODELING STATIC OCCURRENCE AND SPECIES DISTRIBUTIONS

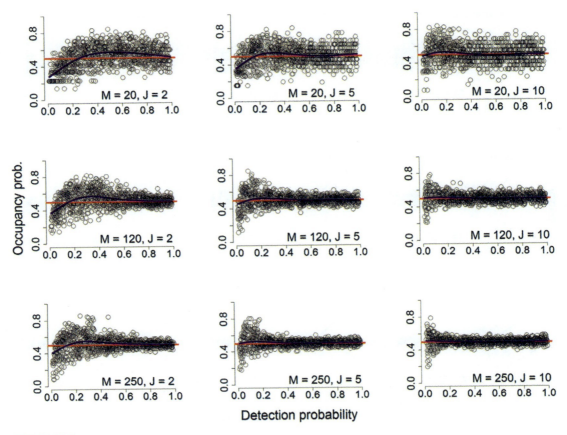

FIGURE 10.8

Results from the Bayesian version of the simulation. Figure 10.7 shows MLE results of the same simulation exercise; see it for explanations.

Table 10.2 Average estimation error (RMSE) for MLEs and for Bayesian posterior means under vague *Uniform*(0,1) priors for the nine scenarios (X / Y denotes root mean square error, in percent of the mean, for MLE / Bayes).

	2 Surveys	5 Surveys	10 Surveys
20 sites	59%/27%	42%/24%	35%/22%
150 sites	40%/20%	24%/15%	16%/12%
250 sites	36%/18%	22%/12%	12%/9%

10.8 GOODNESS-OF-FIT

Goodness-of-fit (GoF) implies a comparison of the observed data with the data expected under the model using some fit statistic, or discrepancy measure, such as residuals, Chi-square or deviance. With occupancy models, the data are binary unless aggregated to binomial counts (Section 10.3 and 11.6.1). Standard fit statistics are then a simple deterministic function of sample size and hence uninformative about model fit; see Section 4.4.5 in McCullagh and Nelder (1989) and Section 8.4.1.1 in Royle et al. (2014). In order to test GoF in binary response models such as the occupancy model, we always have to aggregate the binary response in some way. Clearly, many such aggregations are possible and it is not *a priori* clear which one is best in order to indicate a particular assumption violation nor how sensitive a test based on any such aggregation is. One particular way in which we can aggregate the binary detection/nondetection data is by unique detection history; we can then compare the observed with the expected number of sites that exhibit a certain detection history. This is what the GoF test by MacKenzie and Bailey (2004) does, which is implemented as function `mb.gof.test` in the R package `AICcmodavg` (Mazerolle, 2015). Another possible aggregation is to compute a fit statistic on row or column sums of the detection history matrix, i.e., aggregating over occasions or over sites (see also Section 8.4.2 in Royle et al., 2014).

To illustrate, we conducted a little simulation study and used function `simOcc` to simulate 20 data sets that each contained some effect that was not present in the data-analyzing model and that therefore represented an assumption violation of the analysis model. The basic function settings were the following, representing a data-generating model with forest cover effects in occupancy and elevation and wind speed effects in detection:

```
simOcc(M = 267, J = 3, mean.occupancy = 0.6, beta1 = 0, beta2 = 1, beta3 = 0, mean.detection = 0.3, time.effects = c(0, 0), alpha1 = 1, alpha2 = -1, alpha3 = 0, sd.lp = 0, b = 0)
```

We then either added an effect in the data simulation or dropped one in the analysis model, leading to a mismatch between data generation and data analysis, which we might hope to pick up with a GoF test. For each model fit we computed a Chi-square GoF test either directly on the cells of the detection history matrix or on the detection frequencies obtained by aggregating over columns or over rows. In addition, we calculated the GoF test by MacKenzie and Bailey (2004). To obtain a *p*-value for each, we bootstrapped 500 times (for Chi2) and 100 times (for the M&B test). Table 10.3 shows the median and the range of the *p*-values over the 20 data sets for each scenario.

We see that tests computed directly on the observed binary data (column "Chi2 on cells") are totally uninformative about lack of fit, but that tests on aggregated data have some power to detect to lack of fit, with aggregation by capture history (i.e., the MacKenzie-Bailey test) being the best among those tested. However, neither test was powerful in every case, and surprisingly, missing covariates remained undetected by all types of aggregation. We also see that test performance was highly variable among samples, with detection of lack of fit in some data sets and not in others.

Hence, diagnosing lack of fit in an occupancy model remains difficult and should perhaps not be relegated to the calculation of one single number such as a *p*-value from some GoF test. We could also plot residuals (which again must be computed on aggregated data) against modeled and unmodeled covariates or spatial coordinates to detect any systematic pattern, which could then be taken account of in an improved version of the model (see Section 6.9). Finally, there's the question of what to do if despite all our best efforts we don't succeed in identifying a model that passes our GoF test. The

Table 10.3 Median and range of *p*-values of Goodness-of-fit tests (for 20 simulated data sets for each type of assumption violation) from disaggregated data (Chi2 on cells) and from detection histories aggregated over columns (Chi2 on rows), over rows (Chi2 on columns) and per detection history type (MB test, MacKenzie and Bailey 2004) for a selection of seven scenarios of assumption violations.

Assumption Violation Type	Chi2 on Cells	Chi2 on Rows	Chi2 on Columns	MB Test
Strong behavioral response ($b = 2$)	0.78 (0.30–0.98)	0.18 (0.02–0.81)	0.00 (0.00–0.03)	0.00 (0.00–0.01)
Weak behavioral response ($b = 1$)	0.63 (0.11–0.89)	0.30 (0.01–0.69)	0.09 (0.00–0.99)	0.03 (0.00–0.31)
Detection heterogeneity (sd.lp = 1)	0.52 (0.22–0.95)	0.29 (0.00–0.65)	0.67 (0.02–0.99)	0.28 (0.01–0.86)
Missing site covariate in psi (forest)	0.61 (0.20–0.90)	0.44 (0.09–0.81)	0.35 (0.02–0.96)	0.45 (0.00–0.99)
Missing site covariate in p (elevation)	0.43 (0.06–0.90)	0.33 (0.09–0.91)	0.55 (0.10–0.93)	0.41 (0.07–0.90)
Missing observational covariate in p (wind)	0.52 (0.20–0.80)	0.55 (0.18–0.89)	0.63 (0.00–0.96)	0.44 (0.01–0.97)
Missing time effects in p (time.effects = c(−1,1))	0.57 (0.11–0.91)	0.51 (0.14–0.97)	0.00 (0.00–0.10)	0.00 (0.00–0.21)

discussions in Sections 6.9 and 8.4.3 are of course relevant here as well. That is, we could in theory throw away a data set as unanalyzable, but realistically this will rarely be done. More typically, we may simply stick to our analysis and acknowledge that we have more uncertainty about the inferences than what we formally account for in the SEs or CIs. Better still, we could inflate SEs and CIs by an estimate of the overdispersion parameter (c-hat) for the model at hand by dividing the observed Chi-square statistic by the mean of the statistics obtained from a bootstrap simulation. Seeing how little is changed in the SEs or CIs of predictions from the model may perhaps make us more comfortable in keeping an ill-fitting model.

Clearly all the same comments apply for a Bayesian model fit. That is, we must conduct posterior predictive checks on a response that is aggregated by summing over sites, occasions, or individual capture history. Computing it directly on the binary responses, as we erroneously did in Chapter 20 in Kéry (2010), will fail to indicate an ill-fitting model.

10.9 DISTRIBUTION MODELING AND MAPPING OF SWISS RED SQUIRRELS

At various places in this book we have emphasized that any model for abundance or occurrence with spatially indexed covariates can be used to produce a map of species abundance or occurrence, that is, a species distribution map. In particular, there is a sense in which site-occupancy models represent the most genuine species distribution model because they model true occupancy probability separately from false-negative detection error (Kéry et al., 2010a,b; 2013). This is different from any

10.9 DISTRIBUTION MODELING AND MAPPING OF SWISS RED SQUIRRELS

other species distribution modeling framework, which only model apparent occurrence, i.e., the product of occupancy and detection probability (Kéry, 2011b; Lahoz-Monfort et al., 2014; Guillera-Arroita et al., 2015). To emphasize the species distribution modeling role of site-occupancy models, and to finally show some real-data analysis with occupancy models, we next use `unmarked` to model the distribution of the European red squirrel (*Sciurus vulgaris*, Figure 10.9) in Switzerland. We base our analysis on data from the Swiss breeding bird survey MHB, where red squirrels are recorded as some sort of honorary avian species. Survey methods for the species are essentially identical to those for birds; see Section 6.9. The data set `SwissSquirrels.txt` contains detection/nondetection data for the red squirrel in 265 1 km^2 survey quadrats in Switzerland for 2007, along with some covariates. The goals of our analysis are threefold and exactly analogous to those for an analysis of great tit abundance in Section 6.9:

1. Identify environmental factors that affect the Swiss squirrel distribution
2. Produce a distribution map of the species
3. Estimate the Swiss range size of the species

We show a complete analysis that includes model selection, inference, GoF assessment, and prediction/mapping. We use two site (elevation, forest cover) and two observational covariates (survey

FIGURE 10.9

European red squirrel (*Sciurus vulgaris*), Cairngorms, Scotland, 2009 (*Photo by Aender Brepsom*).

date and duration) but do not use transect length now; see Sections 6.9 and 7.9.5 for how we could include route length into the analysis to accommodate coverage bias.

```
# Read in data set, select squirrels and harvest data
data <- read.table("SwissSquirrels.txt", header = TRUE)
str(data)
y <- as.matrix(data[,7:9])        # Grab 2007 squirrel det/nondet data
elev.orig <- data[,"ele"]         # Unstandardised, original values of covariates
forest.orig <- data[,"forest"]
time <- matrix(as.character(1:3), nrow=265, ncol = 3, byrow = T)
date.orig <- as.matrix(data[,10:12])
dur.orig <- as.matrix(data[,13:15])

# Overview of covariates
covs <- cbind(elev.orig, forest.orig, date.orig, dur.orig)
par(mfrow = c(3,3))
   for(i in 1:8){
   hist(covs[,i], breaks = 50, col = "grey", main = colnames(covs)[i])
}
pairs(cbind(elev.orig, forest.orig, date.orig, dur.orig))

# Standardise covariates and mean-impute date and duration
# Compute means and standard deviations
(means <- c(apply(cbind(elev.orig, forest.orig), 2, mean), date.orig =
mean(c(date.orig), na.rm = TRUE), dur.orig=mean(c(dur.orig), na.rm = TRUE)))
(sds <- c(apply(cbind(elev.orig, forest.orig), 2, sd), date.orig = sd(c(date.orig),
na.rm = TRUE), dur.orig=sd(c(dur.orig), na.rm = TRUE)))

# Scale covariates
elev <- (elev.orig - means[1]) / sds[1]
forest <- (forest.orig - means[2]) / sds[2]
date <- (date.orig - means[3]) / sds[3]
date[is.na(date)] <- 0
dur <- (dur.orig - means[4]) / sds[4]
dur[is.na(dur)] <- 0

# Load unmarked, format data and summarize
library(unmarked)
umf <- unmarkedFrameOccu(y = y, siteCovs = data.frame(elev = elev, forest = forest), obsCovs
= list(time = time, date = date, dur = dur))
summary(umf)
```

We want to identify a model that is useful for inference, specifically for prediction of squirrel distribution to the whole of Switzerland. We do some stepwise model selection first on the detection part, then on the occupancy part, while keeping the detection part as identified in the first step.

```
# Fit a series of models for detection first and do model selection
summary(fm1 <- occu(~1 ~1, data=umf))
summary(fm2 <- occu(~date ~1, data=umf))
summary(fm3 <- occu(~date+I(date^2) ~1, data=umf))
```

10.9 DISTRIBUTION MODELING AND MAPPING OF SWISS RED SQUIRRELS

```
summary(fm4 <- occu(~date+I(date^2)+I(date^3) ~1, data=umf))
summary(fm5 <- occu(~dur ~1, data=umf))
summary(fm6 <- occu(~date+dur ~1, data=umf))
summary(fm7 <- occu(~date+I(date^2)+dur ~1, data=umf))
summary(fm8 <- occu(~date+I(date^2)+I(date^3)+dur ~1, data=umf))
summary(fm9 <- occu(~dur+I(dur^2) ~1, data=umf))
summary(fm10 <- occu(~date+dur+I(dur^2) ~1, data=umf))
summary(fm11 <- occu(~date+I(date^2)+dur+I(dur^2) ~1, data=umf))
summary(fm12 <- occu(~date+I(date^2)+I(date^3)+dur+I(dur^2) ~1, data=umf))

# Put the fitted models in a "fitList" and rank them by AIC
fms <- fitList("p(.)psi(.)"                      = fm1,
    "p(date)psi(.)"                              = fm2,
    "p(date+date2)psi(.)"                        = fm3,
    "p(date+date2+date3)psi(.)"                  = fm4,
    "p(dur)psi(.)"                               = fm5,
    "p(date+dur)psi(.)"                          = fm6,
    "p(date+date2+dur)psi(.)"                    = fm7,
    "p(date+date2+date3+dur)psi(.)"              = fm8,
    "p(dur+dur2)psi(.)"                          = fm9,
    "p(date+dur+dur2)psi(.)"                     = fm10,
    "p(date+date2+dur+dur2)psi(.)"               = fm11,
    "p(date+date2+date3+dur+dur2)psi(.)"         = fm12)

(ms <- modSel(fms))
                                       nPars    AIC  delta  AICwt  cumltvWt
p(date+dur+dur2)psi(.)                     5 789.09   0.00 0.4612      0.46
p(date+date2+dur+dur2)psi(.)               6 790.94   1.85 0.1825      0.64
p(date+date2+date3+dur+dur2)psi(.)         7 791.59   2.50 0.1321      0.78
p(date+dur)psi(.)                          4 791.97   2.88 0.1091      0.88
p(date+date2+dur)psi(.)                    5 793.74   4.65 0.0451      0.93
p(date+date2+date3+dur)psi(.)              6 794.11   5.03 0.0373      0.97
p(date)psi(.)                              3 795.98   6.89 0.0147      0.98
p(date+date2+date3)psi(.)                  5 797.62   8.54 0.0065      0.99
p(date+date2)psi(.)                        4 797.67   8.58 0.0063      0.99
p(dur+dur2)psi(.)                          4 798.52   9.43 0.0041      1.00
p(dur)psi(.)                               3 801.29  12.20 0.0010      1.00
p(.)psi(.)                                 2 805.90  16.82 0.0001      1.00

# Continue with model fitting for occupancy, guided by AIC as we go
# Check effects of elevation
summary(fm13 <- occu(~date+dur+I(dur^2) ~elev, data=umf))
summary(fm14 <- occu(~date+dur+I(dur^2) ~elev+I(elev^2), data=umf))
summary(fm15 <- occu(~date+dur+I(dur^2) ~elev+I(elev^2)+ I(elev^3), data=umf))
cbind(fm13@AIC, fm14@AIC, fm15@AIC) # model 14 with elev2 best

# Check effects of forest and interactions
summary(fm16 <- occu(~date+dur+I(dur^2) ~elev+I(elev^2)+forest, data=umf))
summary(fm17<- occu(~date+dur+I(dur^2) ~elev+I(elev^2)+forest+I(forest^2), data=umf))
```

```
summary(fm18 <- occu(~date+dur+I(dur^2) ~elev+I(elev^2)+forest+I(forest^2)+elev:
forest, data=umf))
summary(fm19 <- occu(~date+dur+I(dur^2) ~elev+I(elev^2)+forest+I(forest^2)+elev:
forest+elev:I(forest^2), data=umf))
summary(fm20 <- occu(~date+dur+I(dur^2) ~elev+I(elev^2)+forest+I(forest^2)+elev:
forest+elev:I(forest^2)+I(elev^2):forest, data=umf))
summary(fm21 <- occu(~date+dur+I(dur^2) ~elev+I(elev^2)+forest+I(forest^2)+elev:
forest+elev:I(forest^2)+I(elev^2):forest+ I(elev^2):I(forest^2), data=umf))
cbind(fm16@AIC, fm17@AIC, fm18@AIC, fm19@AIC, fm20@AIC) # fm20 is best

# Check for some additional effects in detection
summary(fm22 <- occu(~date+dur+I(dur^2)+elev ~elev+I(elev^2)+forest+I(forest^2)+
elev:forest+elev:I(forest^2)+I(elev^2):forest, data=umf))
summary(fm23 <- occu(~dur+I(dur^2)+date*(elev+I(elev^2)) ~elev+I(elev^2)+
forest+I(forest^2)+elev:forest+elev:I(forest^2)+I(elev^2):forest, data=umf))
summary(fm24 <- occu(~dur+I(dur^2)+date*(elev+I(elev^2))+forest ~elev+I(elev^2)+
forest+I(forest^2)+elev:forest+elev:I(forest^2)+I(elev^2):forest, data=umf))
cbind(fm22@AIC, fm23@AIC, fm24@AIC) # None better, hence, stay with model 20
```

We do a bootstrapped GoF test on detection history frequencies (MacKenzie and Bailey, 2004), note that only the *observed* detection histories are shown in the table below.

```
library(AICcmodavg)
system.time(gof.boot <- mb.gof.test(fm20, nsim = 1000))
gof.boot

MacKenzie and Bailey goodness-of-fit for single-season occupancy model

Pearson chi-square table:

      Cohort  Observed  Expected  Chi-square
000      0       102     103.26      0.02
001      0        14      13.66      0.01
010      0        18      16.55      0.13
011      0        10      12.85      0.63
100      0        22      20.30      0.14
101      0        17      15.65      0.12
110      0        17      19.07      0.22
111      0        17      15.66      0.11
00NA     1        47      46.57      0.00
01NA     1         1       0.43      0.74

Chi-square statistic = 3.1134
Number of bootstrap samples = 1000
P-value = 0.878

Quantiles of bootstrapped statistics:
   0%    25%    50%    75%   100%
 0.67   4.16   6.12   8.60  25.64

Estimate of c-hat = 0.45
```

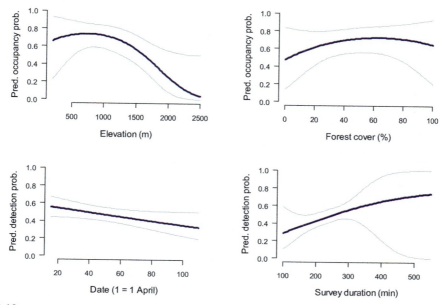

FIGURE 10.10

One-dimensional prediction: Estimated covariate relationships in the site-occupancy model for Swiss red squirrels in 2007. Grey lines show the 95% CIs.

Hence, the observed frequency of the squirrel site-level detection histories agrees reasonably well with that expected under the AIC-best model `fm20`. We conclude that this model is suitable to use for inference and to inspect covariate relationships and project them onto Swiss geographic space. First, we plot some one-dimensional covariate relationships. We use the `predict` function for the `unmarked` fitted object, which uses the delta rule to compute SEs and 95% CIs; the latter we plot as well (Figure 10.10). If our GoF analysis had detected some lack of fit, we could use it to compute an overdispersion factor c-hat and inflate the prediction variances by it, exactly analogous to what we did in Section 6.9.

```
# Create new covariates for prediction ('prediction covs')
orig.elev <- seq(200, 2500,,100)      # New covs for prediction
orig.forest <- seq(0, 100,,100)
orig.date <- seq(15, 110,,100)
orig.duration <- seq(100, 550,,100)
ep <- (orig.elev - means[1]) / sds[1]      # Standardize them like actual covs
fp <- (orig.forest - means[2]) / sds[2]
dp <- (orig.date - means[3]) / sds[3]
durp <- (orig.duration - means[4]) / sds[4]

# Obtain predictions
newData <- data.frame(elev=ep, forest=0)
pred.occ.elev <- predict(fm20, type="state", newdata=newData, appendData=TRUE)
```

```
newData <- data.frame(elev=0, forest=fp)
pred.occ.forest <- predict(fm20, type="state", newdata=newData, appendData=TRUE)
newData <- data.frame(date=dp, dur=0)
pred.det.date <- predict(fm20, type="det", newdata=newData, appendData=TRUE)
newData <- data.frame(date=0, dur=durp)
pred.det.dur <- predict(fm20, type="det", newdata=newData, appendData=TRUE)

# Plot predictions against unstandardized 'prediction covs'
par(mfrow = c(2,2), mar = c(5,5,2,3), cex.lab = 1.2)
plot(pred.occ.elev[[1]] ~ orig.elev, type = "l", lwd = 3, col = "blue", ylim = c(0,1),
las = 1, ylab = "Pred. occupancy prob.", xlab = "Elevation (m)", frame = F)
matlines(orig.elev, pred.occ.elev[,3:4], lty = 1, lwd = 1, col = "grey")
plot(pred.occ.forest[[1]] ~ orig.forest, type = "l", lwd = 3, col = "blue", ylim = c(0,1),
las = 1, ylab = "Pred. occupancy prob.", xlab = "Forest cover (%)", frame = F)
matlines(orig.forest, pred.occ.forest[,3:4], lty = 1, lwd = 1, col = "grey")
plot(pred.det.date[[1]] ~ orig.date, type = "l", lwd = 3, col = "blue", ylim = c(0,1),
las = 1, ylab = "Pred. detection prob.", xlab = "Date (1 = 1 April)", frame = F)
matlines(orig.date, pred.det.date[,3:4], lty = 1, lwd = 1, col = "grey")
plot(pred.det.dur[[1]] ~ orig.duration, type = "l", lwd = 3, col = "blue", ylim = c(0,1), las
= 1, ylab = "Pred. detection prob.", xlab = "Survey duration (min)", frame = F)
matlines(orig.duration, pred.det.dur[,3:4], lty = 1, lwd = 1, col = "grey")
```

We like to form predictions for two covariates simultaneously and therefore predict on a grid that combines each in a suitable range of values (Figure 10.11; see section 6.9.4 for similar R code to do this for predictions of expected abundance in an *N*-mixture model).

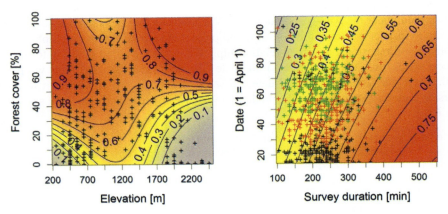

FIGURE 10.11

Two-dimensional predictions of the joint relationships of occupancy and detection probability, respectively, with two covariates under the site-occupancy model for Swiss red squirrels in 2007. Left: occupancy probability, right: detection probability. Plus signs denote the observed covariate values in the data set; in the right plot, their colors denote the first (black), second (red), and third survey (green).

10.9 DISTRIBUTION MODELING AND MAPPING OF SWISS RED SQUIRRELS

```
# Predict abundance and detection jointly along two separate covariate gradients
# abundance ~ (forest, elevation) and detection ~ (survey duration, date)
pred.matrix1 <- pred.matrix2 <- array(NA, dim = c(100, 100))  # Define arrays
for(i in 1:100){
    for(j in 1:100){
        newData1 <- data.frame(elev=ep[i], forest=fp[j])       # For abundance
        pred <- predict(fm20, type="state", newdata=newData1)
        pred.matrix1[i, j] <- pred$Predicted
        newData2 <- data.frame(dur=durp[i], date=dp[j])         # For detection
        pred <- predict(fm20, type="det", newdata=newData2)
        pred.matrix2[i, j] <- pred$Predicted
    }
}

par(mfrow = c(1,2), cex.lab = 1.2)
mapPalette <- colorRampPalette(c("grey", "yellow", "orange", "red"))
image(x=orig.elev, y=orig.forest, z=pred.matrix1, col = mapPalette(100), axes = FALSE,
xlab = "Elevation [m]", ylab = "Forest cover [%]")
contour(x=orig.elev, y=orig.forest, z=pred.matrix1, add = TRUE, lwd=1.5, col = "blue",
labcex = 1.3)
axis(1, at = seq(min(orig.elev), max(orig.elev), by = 250))
axis(2, at = seq(0, 100, by = 10))
box()
title(main = "Expected squirrel occurrence prob.", font.main = 1)
points(data$ele, data$forest, pch="+", cex=1)

image(x=orig.duration, y=orig.date, z=pred.matrix2, col = mapPalette(100), axes = FALSE,
xlab = "Survey duration [min]", ylab = "Date (1 = April 1)")
contour(x=orig.duration, y=orig.date, z=pred.matrix2, add = TRUE, lwd=1.5, col = "blue",
labcex = 1.3)
axis(1, at = seq(min(orig.duration), max(orig.duration), by = 50))
axis(2, at = seq(0, 100, by = 10))
box()
title(main = "Expected squirrel detection prob.", font.main = 1)
matpoints(as.matrix(data[, 13:15]), as.matrix(data[, 10:12]), pch="+", cex=1)
```

Next, we produce a Swiss distribution map for the red squirrel in 2007, along with a map of the uncertainty in these predictions at each 1-km^2 quadrat (Figure 10.12). As always, producing a map is simple if we have effects of spatially indexed covariates: we simply predict the response (occupancy or detection probability) for each quadrat in the area for which we want to produce the map and then we plot this. (If your computer has problems predicting at all $\sim$42,000 km pixels at once, do it in batches of $\sim$10,000 and then stack them afterwards.)

It is important to be able to gauge the uncertainty (SE, CI, etc.) in an estimate. For instance, we would never be able to publish the results of an analysis of variance (say, a histogram of group means) without indicating SEs or posterior standard deviations. However, in the species distribution modeling world, presenting estimates (i.e., maps) without showing the associated uncertainty is currently still the rule. Clearly, this is a state that can be improved. With a regression model as the site-occupancy model, it is easy to obtain uncertainty assessments for every estimate such as, here, a prediction of occupancy

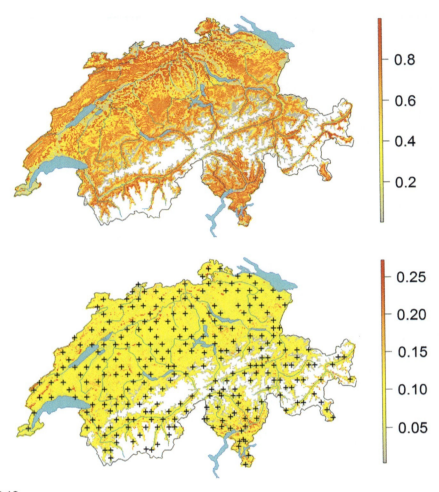

FIGURE 10.12

Species distribution map for red squirrels in Switzerland in 2007 under the best-fitting site-occupancy model (fm20) for data modeled at the 1-km² scale. The map on the top shows the expected probability of occupancy and that on the bottom the standard errors of those predictions, along with the locations of the sample locations (shown as plus signs). Areas with median elevation greater than 2250 m a.s.l. are masked (white).

or detection. So we next produce a map of the uncertainty in the preceding species distribution map, by also plotting the SEs of these predictions. (Why don't we also produce a map of the detection probability of Swiss red squirrels?)

```
# Load the Swiss landscape data from unmarked
data(Switzerland)      # Load Swiss landscape data in unmarked
CH <- Switzerland
```

10.9 DISTRIBUTION MODELING AND MAPPING OF SWISS RED SQUIRRELS

```
# Get predictions of occupancy prob for each 1km2 quadrat of Switzerland
newData <- data.frame(elev=(CH$elevation - means[1])/sds[1], forest=(CH$forest - means
[2])/sds[2])
predCH <- predict(fm20, type="state", newdata=newData)

# Prepare Swiss coordinates and produce map
library(raster)

# Define new data frame with coordinates and outcome to be plotted
PARAM <- data.frame(x = CH$x, y = CH$y, z = predCH$Predicted)
r1 <- rasterFromXYZ(PARAM)      # convert into raster object

# Mask quadrats with elevation greater than 2250
elev <- rasterFromXYZ(cbind(CH$x, CH$y, CH$elevation))
elev[elev > 2250] <- NA
r1 <- mask(r1, elev)

# Plot species distribution map (Fig. 10-14 left)
par(mfrow = c(1,2), mar = c(1,2,2,5))
mapPalette <- colorRampPalette(c("grey", "yellow", "orange", "red"))
plot(r1, col = mapPalette(100), axes = F, box = F, main = "Red squirrel distribution in 2007")

# Plot SE of the species distribution map (Fig. 10-14 right)
r2 <- rasterFromXYZ(data.frame(x = CH$x, y = CH$y, z = predCH$SE))
r2 <- mask(r2, elev)
plot(r2, col = mapPalette(100), axes = F, box = F, main = "Uncertainty map 2007")
```

Finally, we estimate the area of occurrence of red squirrels in Switzerland in 2007. For this, we make the assumption that each MHB route samples exactly 1 km^2, and, hence, we simply add up the occupancy probability for each quadrat. We do this both with and without the mask cutting out areas at elevation greater than 2250 m and see that our model predicts hardly any squirrels occurring at these very high elevations.

```
# Get extent of squirrel occurrence in 2007
sum(predCH$Predicted)                              # All quadrats
[1] 17354.57
sum(predCH$Predicted[CH$elevation < 2250])    # Only at elevations < 2250 m
[1] 17350.34
```

We also want to assess the uncertainty around this estimate via the bootstrap. We do the prediction "by hand" rather than using the predict function because we don't need the SEs and predict would take way too much time when used in a bootstrapped function.

```
# Standardise prediction covariate identical to those in analysis
pelev <- (CH$elevation - means[1]) / sds[1]
pforest <- (CH$forest - means[2]) / sds[2]

# Define function that predicts occupancy under model 20
Eocc <- function(fm) {
   betavec <- coef(fm)[1:8]         # Extract coefficients in psi
```

```
    DM <- cbind(rep(1,length(pelev)), pelev, pelev^2, pforest, pforest^2, pelev*pforest,
pelev*pforest^2, pelev^2*pforest)  # design matrix
    pred <- plogis(DM%*%(betavec))  # Prediction = DM * param. vector
    Eocc <- sum(pred)               # Sum over all Swiss quadrats (no mask)
    Eocc
}

(estimate.of.occurrence <- Eocc(fm20))  # Same as before, without mask
system.time(Eocc.boot <- parboot(fm20, Eocc, nsim=1000, report=10)) # 100 sec
plot(Eocc.boot)          # Plot bootstrap distribution of extent of occurrence
quantile(Eocc.boot@t.star, c(0.025, 0.975))
     2.5%    97.5%
15131.14 20185.21

# Convert these estimates to a proportion of the country
(c(point = 17354.57, quantile(Eocc.boot@t.star, c(0.025, 0.975))) / 42275
     point     2.5%    97.5%
 0.4104161 0.3579217 0.4774740
```

We conclude that in 2007 red squirrels occurred in Switzerland in about 41% of the 1-km^2 quadrats (95% CI 36—48%).

Importantly, whether we can interpret the estimated range size in terms of "permanent presence of at least one squirrel" or as "use by a squirrel sometimes during the study period" (see Section 10.2) depends on the assumption that the effective sampling area associated with an MHB sampling quadrat is exactly 1 km^2; see the discussion in Section 6.10 and also Efford and Dawson (2012). If the effective sampling area is smaller, then our estimate will be an underestimate with respect to the area of "permanent presence", and if it is bigger we will overestimate the range when we want to interpret the latter as the area of permanent presence. We think that to account for coverage bias and "sunflower effects" we could employ similar *ad hoc* methods as we did in Sections 6.9 and 6.10. To deal with the scaling of occupancy on area, we could perhaps model the underlying abundance rather than occupancy directly; see Section 3.3.6. Alternatively, a more formal treatment might be one along the lines of the seminal paper by Chandler and Clark (2014), who specify a model for binary detection/nondetection (i.e., occupancy) data that is formulated in terms of an underlying, i.e., latent point process model that accommodates movement of individuals within their home ranges. In an important paper, Ramsey et al. (2015) have reformulated the spatially explicit *N*-mixture model of Chandler and Royle (2013) for detection/nondetection data. That is, they estimate the parameters of a latent point process from the spatial correlation in the detection/nondetection data from of replicated surveys of adjacent sites (This means that their model could not be applied to MHB data as is, since no survey quadrat has any direct neighbor. Perhaps progress could be made by subdividing a 1 km^2 quadrat into four or 16 subquadrats and then modeling occurrence at that finer scale.). Making occupancy models spatially explicit is an important avenue for research.

10.10 MULTISCALE OCCUPANCY MODELS

The classical occupancy model has a single spatial scale (that for the chosen grid size) and two levels: one level for the latent occurrence state in a grid cell and another level for the repeated measurements. This two-level hierarchy can be extended to more than two levels in a straightforward fashion. For instance, we may repeatedly survey multiple spatial subunits that are each nested in multiple main units (representing

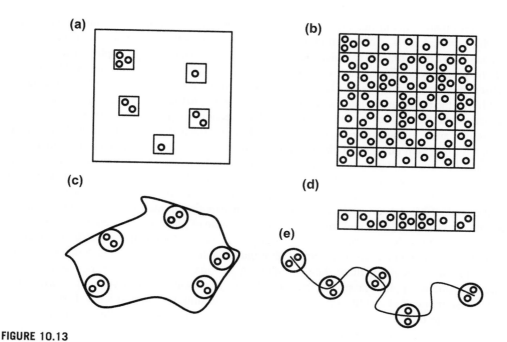

FIGURE 10.13

Five examples of multiscale designs, where smaller subunits are nested within a larger unit. A study will always comprise multiple such units, which represent the main spatial replicates that are required in occupancy modeling. Small circles represent the presence/absence measurements taken within a subunit, either at different locations or at different times at the same location. Often, we have imbalanced data, i.e., the number of measurements is not identical in every subunit; see (a), (b), and (d).

spatial subsampling); we may sample multiple units over several days, with each day subdivided into, say, three shorter time segments (representing temporal subsampling); or we may survey multiple spatial units using multiple detection methods over multiple occasions each. In disease surveillance, there may be a multilayered, nested sampling process, such as replicate PCRs run for each of a number of ducks examined at multiple ponds, which are nested within a collection of refuges (McClintock et al., 2010b). Hence, scales are defined spatially, temporally, spatiotemporally, or by different measurement methods, and, importantly, their definition will determine the precise meaning of the associated model parameters; see below. In reality, such multiscale occupancy designs are very common, but we believe that this has not yet been recognized widely enough. Formal analysis of multiscale occupancy designs started only fairly recently with Nichols et al. (2008), Aing et al. (2011) and Mordecai et al. (2011). In this section we focus on the simplest multiscale occupancy model, that with two scales and therefore three levels.

In a multiscale design, smaller units are nested within larger units (Figure 10.13). This scheme comes in a large number of variants and shapes: the larger sample units may have artificial (A, B, D) or "natural" shapes (e.g., C may be a pond and E may represent a river); only a fraction of the area (A, C, D) or the entire area of the larger unit may be sampled (B, D), and the nesting may be two-dimensional (A, B, C) or along a linear structure, such as a river or a transect; in the latter we typically have directionality in the sampling (D, E). Usually, the actual detection/nondetection data are collected in the subunits. However, there are cases where data may in addition be available at the unit level, for instance, we may have information about occurrence of a species *somewhere* within a unit, but the particular subunit in which the species was observed may be unavailable. Cases B and D represent a special case where all subunits in a main unit are sampled, i.e., each main unit is exhaustively sampled. In a sense, the relationship between occurrence at the two levels then becomes deterministic; see below. Combining presence/absence information at the unit level and covariate information at the subunit level in a three-level model forms the basis for the interesting work by Keil et al. (2014a,b) to downscale occupancy to form finer-scale maps from rougher-scale occurrence data; see also Dupuis et al. (2011).

Several issues arise in the analysis of multiscale designs. Basically, subunits provide replicate measurements for the main units and their proximity induces a similarity, or in other words, there is a dependency among subunits within the same unit that needs to be addressed in a model. At the same time, exactly this dependency can be used to make inferences about the units based on the subunits. To develop an intuition for a simple three-level occupancy model, let's first look at a nested sampling scheme in the context of experimental design outside of distribution modeling. Let's assume that we have measured plant mass (y_{ijk}) of multiple stems (k) in a sample of plants (j) collected in a number of populations (i). The basic *block structure* of the study is given by "population/plant/stem," that is, plants are nested within populations and stems within plants, and the resulting dependencies ought to be accounted for in a model for these data. In addition, we may have covariate measurements or experimental treatments applied at any of the three levels, representing a *treatment structure*. This experimental design is called a split-plot (Nelder, 1965a,b), leading to a nested analysis of variance (ANOVA). The basic block structure for this study can be described by the following hierarchical model for the individual plant mass measurements y_{ijk}.

Unit-level with population means γ_i: $\gamma_i \sim Normal(\mu, \sigma^2_{pop})$
Subunit-level with plant means a_{ij}: $a_{ij}|\gamma_i \sim Normal(\gamma_i, \sigma^2_{plant})$
Individual stem measurement y_{ijk}: $y_{ijk}|a_{ij} \sim Normal(a_{ij}, \sigma^2_{stem})$

Population means γ_i (note this is a Greek gamma, not y) cluster around the grand mean μ with variance σ^2_{pop}, plant means a_{ij} cluster around population mean γ_i with variance σ^2_{plant}, and the measurements y_{ijk} cluster around plant mean a_{ij} with residual variance σ^2_{stem}. Such nested ANOVA or split-plot designs can be analyzed in many software packages, including BUGS (Qian and Shen, 2007; Hector et al., 2011).

Now let's see how we can apply the same concepts to the case where instead of a continuous variable we measure binary "presence/absence" at three hierarchical levels and where that measurement may be affected by false-negative errors. Imagine that we had taken up to three presence/absence measurements (k) at each of five subunits (j) in a number of main units (i), as in Figure 10.13(a). One possible analysis inspired by the nested ANOVA treats the units as contributing a random block effect in the logit-linear model for occupancy ψ_{ij} of the subunits. That is, logit(ψ_{ij}) = $\gamma_i + \varepsilon_{ij}$, hence, the occupancy probability in a subunit (on the logit scale) is simply the sum of a contribution from unit (γ_i) and another one from subunit j in that unit (ε_{ij}).

Unit-level random effect γ_i: $\gamma_i \sim Normal(\mu, \sigma^2_{pop})$
Subunit-level level presence/absence a_{ij}: $a_{ij}|\gamma_i \sim Bernoulli(logit^{-1}(\gamma_i + \varepsilon_{ij}))$
Replicated pres/abs measurement y_{ijk}: $y_{ijk}|a_{ij} \sim Bernoulli(a_{ij} * p)$

In this "bottom-up" analysis (G. Guillera-Arroita, pers. comm.), the focus is on the subunit, and the nonindependence of subunits within the same unit is accounted for by a random unit effect (γ_i) in the logit of the occupancy probability at the subunit level. This random unit effect is the contribution of unit i to the occupancy probability at the smaller scale, making it a little higher or lower, on average, for subunits that are in the same unit. That is, γ_i is a typical random effect invoked to account for the correlation of grouped measurements, and it characterizes the unit. These random effects lack any specific biological meaning.

In an alternative model the unit-specific random effects *do* have a clear biological meaning—they are the usual "presence/absence" state z of the units. This leads to the following three-level occupancy model, which was nicely described by Aing et al. (2011) and Mordecai et al. (2011).

Unit-level presence/absence z_i: $z_i \sim Bernoulli(\psi)$
Subunit-level presence/absence a_{ij}: $a_{ij}|z_i \sim Bernoulli(z_i * \theta)$
Replicated pres/abs measurement y_{ijk}: $y_{ijk}|a_{ij} \sim Bernoulli(a_{ij} * p)$

In this "top-down" analysis (G. Guillera-Arroita, pers. comm.), we define a specific dependency in the presence/absence states between the unit and the subunit levels—a subunit can only be occupied ($a_{ij} = 1$) when the unit it belongs to is also occupied ($z_i = 1$). Random variables z_i and a_{ij} represent presence/absence in unit i and subunit ij, respectively, and y_{ijk} is the detection/nondetection datum in replicate k, subunit j and unit i. The parameters governing the three random variables are the unit-level (large-scale) occupancy probability ψ, the subunit-level (small-scale) occupancy probability θ, and the detection probability p associated with the actual replicated presence/absence measurement y taken at a subunit. These parameters can be indexed by i, ij, and ijk, respectively, though in practice we need to impose constraints, i.e., add covariates or random effects, to make the model identifiable.

The middle level has an important interpretation as the probability of availability (i.e., 1 minus temporal emigration) or small-scale, or temporary, occupancy, whereas the top level is the probability of permanent, or asymptotic, presence or large-scale occupancy probability (Efford and Dawson, 2012; Pavlacky et al., 2012). The middle level can either be viewed as a component of the state model (when we focus on different scales of occupancy) or as a component of the observation model (when temporary emigration is treated as a nuisance to be addressed to better estimate occupancy in the main units). Hence, if we are worried about the meaning of the occupancy parameter in a two-level occupancy model when there is temporary emigration (random in/out type of closure violation), we can simply collect data at an additional level and then estimate both probabilities of use (corresponding to ψ) and probability of temporary presence (corresponding to availability, θ). That is, to estimate parameters at each level, we need extra information, i.e., nested replicates at every level. Alternatively, we need extra information, e.g., an estimate of, say, p, from another study, which can be introduced as an informative prior in a Bayesian analysis, or perhaps make assumptions that some covariate contains information about variability in one of the parameters. There are two main ways in which we can get back to the usual two-level model: either by setting $\theta = 1$ or by setting $p = 1$. In the former, we have perfect availability, and whenever a unit is occupied, each single subunit is so, too. In the latter, whenever a subunit is occupied, we only observe detections, never nondetections.

In both the ANOVA-type and the triple-Bernoulli, three-level occupancy models just described, the similarity within the same unit is assumed to be identical for all subunits, and we assume there is no further spatial autocorrelation among subunits within the same unit. It is possible to account for spatial autocorrelation among subunits within a unit using methods presented in Chapters 21 and 22 (in volume 2), and analogous occupancy models have been developed for linear designs (such as D and E, representing sampling along trails or rivers; Hines et al., 2010, 2014; Aing et al., 2011). Another

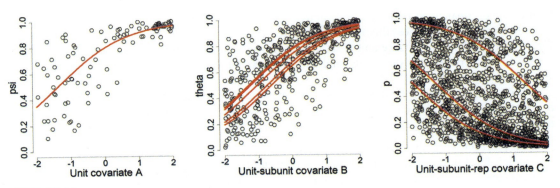

FIGURE 10.14

Example of the graphical output from the data simulation function sim3Occ (from model 4 in the text). Left: large-scale occupancy probability at 100 units (psi, red is expected value), middle: small-scale occupancy or availability probability for 500 subunits (theta, red shows expected value for five subunits which are best imagined to represent temporal variability; see Schmidt et al. (2013)), right: detection probability for 1500 individual measurements (p, red shows expected value for three replicates).

possibility to deal with temporal autocorrelation for multiscale data with temporal subsampling is to fit a dynamic occupancy model (MacKenzie et al., 2003; see Chapter 16 in volume 2); this is done by Rota et al. (2009). We present some simpler occupancy models for data collected along linear structures in Section 10.12.

Next, we illustrate a three-level occupancy model using simulated data and terminology inspired by a disease surveillance study reported in Schmidt et al. (2013). The function sim3Occ lets us generate three-level occupancy data sets with covariates, time effects, and unstructured random variability at every level possible (Figure 10.14). The function defaults are:

```
sim3Occ(nunit = 100, nsubunit = 5, nrep = 3, mean.psi = 0.8, beta.Xpsi = 1, sd.logit.psi = 0,
mean.theta = 0.6, theta.time.range = c(-1, 1), beta.Xtheta = 1, sd.logit.theta = 0, mean.p =
0.4, p.time.range = c(-2,2), beta.Xp = -1, sd.logit.p = 0)
```

This has us simulate data collected at 100 units, with five subunits in each and three presence/absence measurements on each subunit. The three main parameters of the model (large-scale occupancy ψ, small-scale occupancy θ, and detection p) are determined by their intercepts on the probability scale, mean.psi, mean.theta, and mean.p, respectively, and all three may depend on one unit, unit-subunit, and unit-subunit-replicate-specific, continuous covariate, with coefficients beta.Xpsi, beta.Xtheta, and beta.Xp, respectively. In addition, we can specify differences between the subunits (identical for all units) by the arguments theta.time.range and p.time.range and random noise/heterogeneity in all three parameters by setting to nonzero the arguments sd.logit.psi, sd.logit.theta, and sd.logit.p = 0. Executing the function also creates plots of how the three main parameters vary as a function of the covariates, time, and heterogeneity (Figure 10.14). We give a partially edited summary of the function output with some added comments.

```
data <- sim3Occ()      # Execute function with default args
str(data)              # Summary of output with some comments added
```

10.10 MULTISCALE OCCUPANCY MODELS

```
List of 28
[ ... output truncated ... ]
$ p          : num [1:100, 1:5, 1:3] 0.699    # p for each datum
$ z          : int [1:100(1d)] 0 1 1 0 1 0 1  # presence at unit
$ a          : int [1:100, 1:5] 0 0 1 0 0     # presence at subunit
$ y          : int [1:100, 1:5, 1:3] 0 0 1 0  # detection/nondetection
$ sum.z      : int 65                         # True number of units with presence
$ obs.sum.z  : int 63                         # Observed number of units with presence
$ sum.z.a    : int 63                         # see below
[ ... output truncated ... ]
```

Output `sum.z.a` is the true number of units with presence within the subunits actually sampled. Note that this will not always be identical to the number of occupied units, since a species may occur in a unit but happen to be absent in the finite number of the particular subunits surveyed. This will happen more often with small values of `nsubunit` and `mean.theta`. See Section 11.2 for the analogous problem of sampling a metacommunity of species and Adams et al. (2010) for that of estimating occurrence of a disease at a site (=unit) when individual amphibians are treated as subunits.

As always, we encourage you to play with this function with changed arguments to train your intuition about three-level occupancy designs, by looking at a summary of the output (as above) and the plots produced. Here is a sample of four possible settings. The simplest possible model has constant parameters and no other sources of variation.

```
# 'Null' model (model 1)
str(data <- sim3Occ(nunit = 100, nsubunit = 5, nrep = 3, mean.psi = 0.8, beta.Xpsi = 0,
sd.logit.psi = 0, mean.theta = 0.6, theta.time.range = c(0, 0), beta.Xtheta = 0,
sd.logit.theta = 0, mean.p = 0.4, p.time.range = c(0,0), beta.Xp = 0, sd.logit.p = 0))
```

We can let `theta` and `p` vary by "time," to make `theta` different among subunits and to make `p` different for each replicate. Such a design might be sensible if subunits denote different samples that are taken in time, as in the example described below (from Schmidt et al., 2013).

```
# No covariate effects, no random variability (model 2)
str(data <- sim3Occ(nunit = 100, nsubunit = 5, nrep = 3, mean.psi = 0.8, beta.Xpsi = 0,
sd.logit.psi = 0, mean.theta = 0.6, theta.time.range = c(-1, 1), beta.Xtheta = 0,
sd.logit.theta = 0, mean.p = 0.4, p.time.range = c(-2,2), beta.Xp = 0, sd.logit.p = 0))
```

We can let `psi`, `theta`, and `p` be affected by linear-logistic effects of three separate covariates.

```
# All covariate effects, but no random variability (model 3)
str(data <- sim3Occ(nunit = 100, nsubunit = 5, nrep = 3, mean.psi = 0.8, beta.Xpsi = 1,
sd.logit.psi = 0, mean.theta = 0.6, theta.time.range = c(-1, 1), beta.Xtheta = 1,
sd.logit.theta = 0, mean.p = 0.4, p.time.range = c(-2,2), beta.Xp = -1, sd.logit.p = 0))
```

And we can also add random unstructured noise at every level (Figure 10.14).

```
# Most complex model with all effects allowed for by sim function (model 4)
str(data <- sim3Occ(nunit = 100, nsubunit = 5, nrep = 3, mean.psi = 0.8, beta.Xpsi = 1,
sd.logit.psi = 1, mean.theta = 0.6, theta.time.range = c(-1, 1), beta.Xtheta = 1,
sd.logit.theta = 1, mean.p = 0.4, p.time.range = c(-2,2), beta.Xp = -1, sd.logit.p = 1))
```

Multiscale occupancy models cannot be fit in unmarked, but some can in PRESENCE and MARK. We here illustrate the fitting of model 3 in BUGS. We generate a data set and imagine a story using the scenario of disease sampling by environmental DNA (eDNA) described in Schmidt et al. (2013): five water samples were taken in each of 100 ponds and then analyzed twice with PCR for the presence or absence of a fungus that kills amphibians and that has an unpronounceable name that starts with letter B. Thus, we have the results of two PCR samples nested within five water samples (=subunits) nested within each of 100 ponds (=units).

```
set.seed(1)
str(data <- sim3Occ(nunit = 100, nsubunit = 5, nrep = 3, mean.psi = 0.8, beta.Xpsi = 1,
mean.theta = 0.3, theta.time.range = c(-1, 1), beta.Xtheta = 1, mean.p = 0.2, p.time.range =
c(-1,1), beta.Xp = -1))

Occupied units:                                      81
Units with >=1 occupied, surveyed subunit:           65
Observed number of occupied units:                   57
```

In this data set, among 100 ponds, 81 turned out to be occupied, but due to small-scale heterogeneity of occurrence, within the five water samples the B fungus actually occurred in only 65 of them. That is, it must have occurred elsewhere in 16 among the 81 occupied ponds. Due to imperfect detection of the PCR methods in the lab, the B fungus was only detected in 57 ponds among the 65 in which it did occur among the five water samples taken.

```
# Look at data
str(data$z)      # True quadrat (pond) occurrence state
str(data$a)      # True subquadrat (water sample) occurrence state
str(data$y)      # Observed data
cbind("pond"=data$z, "sample 1"= data$a[,1], "sample 2"= data$a[,2], "sample 3"=
data$a[,3], "sample 4"= data$a[,4], "sample 5"= data$a[,5])
which(data$z-apply(data$a, 1, max)==1) # Fungus present in pond, but not in examined
samples
```

We fit the data-generating model in BUGS.

```
# Bundle and summarize data set
y <- data$y
str( win.data <- list(y = y, n.pond = dim(y)[1], n.samples = dim(y)[2], n.pcr = dim(y)[3],
covA = data$covA, covB = data$covB, covC = data$covC) )

# Define model in BUGS language
sink("model.txt")
cat("
model {

# Priors and model for params
int.psi ~ dunif(0,1)          # Intercept of occupancy probability
for(t in 1:n.samples){
   int.theta[t] ~ dunif(0,1) # Intercepts of availability probability
}
```

```
  for(t in 1:n.pcr){
     int.p[t] ~ dunif(0,1)      # Intercepts of detection probability (1-PCR error)
  }
  beta.lpsi ~ dnorm(0, 0.1)     # Slopes of three covariates
  beta.ltheta ~ dnorm(0, 0.1)
  beta.lp ~ dnorm(0, 0.1)

  # "Likelihood" (or basic model structure)
  for (i in 1:n.pond){
     # Occurrence in pond i
     z[i] ~ dbern(psi[i])
     logit(psi[i]) <- logit(int.psi) + beta.lpsi * covA[i]
     for (j in 1:n.samples){
        # Occurrence in sample j
        a[i,j] ~ dbern(mu.a[i,j])
        mu.a[i,j] <- z[i] * theta[i,j]
        logit(theta[i,j]) <- logit(int.theta[j]) + beta.ltheta * covB[i,j]
        for (k in 1:n.pcr){
           # PCR detection error process in sample k
           y[i,j,k] ~ dbern(mu.y[i,j,k])
           mu.y[i,j,k] <- a[i,j] * p[i,j,k]
           logit(p[i,j,k]) <- logit(int.p[k]) + beta.lp * covC[i,j,k]
        }
     }
     tmp[i] <- step(sum(a[i,])-0.1)
  }

# Derived quantities
sum.z <- sum(z[])        # Total # of occupied ponds in sample
sum.a <- sum(tmp[])      # Total # of ponds with presence in >=1 of the 5 samples
} # end model
",fill=TRUE)
sink()

# Initial values
zst <- apply(y, 1, max)        # inits for presence (z)
ast <- apply(y, c(1,2), max)   # inits for availability (a)
inits <- function() list(z = zst, a = ast, int.psi = 0.5, beta.lpsi = 0)

# Parameters monitored
params <- c("int.psi", "int.theta", "int.p", "beta.lpsi", "beta.ltheta", "beta.lp",
"sum.z", "sum.a")

# MCMC setting
ni <- 5000 ; nt <- 2 ; nb <- 1000 ; nc <- 3

# Call WinBUGS and summarize posterior
out <- bugs(win.data, inits, params, "model.txt", n.chains = nc, n.thin = nt, n.iter = ni,
n.burnin = nb, debug = TRUE, bugs.dir = bd) # bd="c:/WinBUGS14/"
print(out, 3)
data$sum.z
```

Thus, in BUGS, it is straightforward to extend the basic two-level model to a three-level model and it would be equally easy to add more levels still (though JAR notes: "but with a concomitant decrease in our ability to understand the model"). Multiscale occupancy models are very powerful and occur surprisingly often in ecology and management, although only a mere handful of papers has used them so far. For related multiscale models of abundance, see Sections 6.14 and 9.5.

10.11 SPACE-FOR-TIME SUBSTITUTION

Typically, the information to estimate the detection parameters separately from the parameters in the occupancy submodel comes from temporal replicates, or else, from simultaneous deployment of several observers or detection devices. However, it has been argued that we can use *spatial* replicates as a surrogate, for instance, by dividing up a larger unit such as a 40-km route in the North American breeding bird survey into smaller, nested subunits of five 8-km segments or as in any of the cases depicted in Figure 10.13. Such a space-for-time substitution was used in some of the earlier capture-recapture literature on the estimation of species richness (Boulinier et al., 1998; Nichols et al., 1998a,b; Cam et al., 2002b,c; Doherty et al., 2003) and is sometimes also used in occupancy modeling (Royle and Kéry, 2007; Guillera-Arroita, 2011; Sadoti et al., 2013). Being able to use space for time in this way is very advantageous since it enables occupancy and detection probability to be estimated separately in an otherwise unreplicated design. This space-for-time substitution must be one of the least well understood topics in occupancy modeling. We include this section in part to motivate others to shed more light on this topic, but we would like to warn you that what we say here is quite tentative only.

One way to view the space-for-time design is as a restricted three-level occupancy model, where there is no replication at the bottom level. Therefore, when fitting the traditional two-level occupancy model, two parameters in the data-generation mechanism are lumped and only their product can be estimated. That is, we collapse our three-level model to two levels.

Large-scale occupancy at unit level (i): $z_i \sim Bernoulli(\psi)$
Detection model for unreplicated subunit surveys (ik): $y_{ik}|z_i \sim Bernoulli(z_i \times \theta \times p)$

As before, ψ is the (large-scale) occupancy probability, but what we estimate as "detection probability" in the two-level model now is the product of small-scale occupancy probability (θ) and detection probability proper (p). We have seen in the last section that with sufficient replication (i.e., at least some units have more than one subunit sampled and at least some subunits have more than a single survey), we can estimate all three parameters (ψ, θ, p) or functions thereof, such as regression coefficients for covariates. In contrast, when we only have a single observation per subunit (i.e., no replication at the lowest level), what we estimate when we feed these data into a traditional two-level occupancy model is ψ as "occupancy" and the product θp as "detection." An exception is the unlikely case in which we have measured covariates that exactly explain variability in θ and p. In this case, and under the strong assumption that the covariate model is known exactly, we can estimate all parameters in a restricted three-level design even without replication at the bottom level. Arguably, this special case is analogous to the restrictive conditions under which we can estimate parameters of a logistic regression from presence-only data (Lele and Keim, 2006; Royle et al., 2012) or those of a two-level occupancy model from unreplicated data (Lele et al., 2012; Knape and Korner-Nievergelt, 2015).

We think that one implicit assumption of space-for-time is that both unit- and subunit-level occupancy must be distinct random variables. This means that the subunits must not exhaustively sample the area represented by the unit because otherwise unit-level occupancy (z) is no longer a separate random variable but simply a deterministic function of the occurrence states (a) of the subunits comprising that unit. With exhaustive sampling, $z = \max(a)$ and ψ is

$$\psi_i = 1 - \prod_{j=1}^{J}(1 - \theta_{ij})$$

for the subunits in unit i, there is only one random process involved in the middle and the top level of the design. A different way of looking at this, but which leads to the same result, has to do with the movement of individuals among subunits. When the movement of individuals on and off a subunit lets a subunit be occupied or not in some random fashion, then unit-scale occupancy can be estimated in a space-for-time design, even when a unit is exhaustively sampled by subunits in the study. When subunits cover the entire unit (i.e., we sample exhaustively), we must then assume that individuals may be able to temporarily move off the unit, making the occurrence of the study species at the precise time at which surveys take place a random variable. In contrast, when individuals cannot temporarily move outside of a subunit, then we don't think that the third model in Section 10.10 makes sense and that instead the second model with an ANOVA-type of unit random effect ought to be used for inference. However, in this latter case, subunits don't contain information about detection probability in the absence of (temporal) replication at the bottom level.

If this argument makes sense, then in the case of the North American BBS (e.g., Royle and Kéry, 2007), we can adopt space-for-time *if* we assume that each route samples the species occurrence in some larger area, because the points sample species occurrence in some vaguely defined area around that route. A species may occur in the larger area without necessarily occurring at the sample points. In contrast, if we have regular quadrats (units) that are, say, divided up into four quadrants (subunits) each, then we are not sure whether space-for-time makes sense. Presumably it does, if we can assume that the units (and therefore also the subunits) sample some larger area around. Then, at any given time when surveyed, the subunit and the unit may be occupied by a species or not in some random fashion, because each individual present in the larger area happens to use a part of its home range that lies within the (sub)unit or it does not and this happens according to a Bernoulli process.

We illustrate space-for-time with two data sets generated with the `sim3Occ` function. First, we show the perhaps unlikely case where there is no replication at the bottom level, but where we have a covariate that explains detection at a known scale (here, the logistic). Only in this case, or if we have knowledge about one of the parameters from elsewhere, can we estimate all parameters of the three levels in the absence of replication at one of the levels. After that, we illustrate a more typical case where no such covariate is available.

10.11.1 A MAGICAL COVARIATE

In the first case, we assume that we have a magical covariate that explains variation in detection probability or in availability probability at a known scale. Here we illustrate the former and simulate additional random noise in all three parameters and set `nrep = 1`, thus, we have no replication at the bottom level.

```
set.seed(1)
data <- sim3Occ(nunit = 500, nsubunit = 5, nrep = 1, mean.psi = 0.8, beta.Xpsi = 1,
sd.logit.psi = 0.4, mean.theta = 0.6, theta.time.range = c(0, 0), beta.Xtheta = 1,
sd.logit.theta = 0.6, mean.p = 0.4, p.time.range = c(0,0), beta.Xp = -1, sd.logit.p = 0.8)

Occupied units:                                 367
Units with >=1 occupied, surveyed subunit: 361
Observed number of occupied units:          285
```

We use JAGS to fit the full three-level model with covariates, without estimating the magnitude of the random noise in the parameters.

```
# Bundle and summarize data set
y <- data$y
str( win.data <- list(y = y, nunit = dim(y)[1], nsubunit = dim(y)[2], nrep = dim(y)[3],
covA = data$covA, covB = data$covB, covC = data$covC) )

# Define model in BUGS langauge
sink("model.txt")
cat("
model {

# Priors
int.psi ~ dunif(0,1)      # Occupancy probability
int.theta ~ dunif(0,1)    # Availability probability
int.p ~ dunif(0,1)        # Detection probability
beta.lpsi ~ dnorm(0, 0.01)
beta.ltheta ~ dnorm(0, 0.01)
beta.lp ~ dnorm(0, 0.01)

# Likelihood
for (i in 1:nunit){
   # Occupancy model for quad i
   z[i] ~ dbern(psi[i])
   logit(psi[i]) <- logit(int.psi) + beta.lpsi * covA[i]
   for (j in 1:nsubunit){
      # Availability in subquad j
      a[i,j] ~ dbern(mu.a[i,j])
      mu.a[i,j] <- z[i] * theta[i,j]
      logit(theta[i,j]) <- logit(int.theta) + beta.ltheta * covB[i,j]
      for (k in 1:nrep){
         # PCR detection error process in replicate k
         y[i,j,k] ~ dbern(mu.y[i,j,k])
         mu.y[i,j,k] <- a[i,j] * p[i,j]
         logit(p[i,j]) <- logit(int.p) + beta.lp * covC[i,j,1]
      }
   }
   tmp[i] <- step(sum(a[i,])-0.1)
}
```

```
# Derived quantities
sum.z <- sum(z[])        # Total number of occupied quadrats
sum.a <- sum(tmp[])      # Total number of quads with presence in samples
p.theta <- int.p * int.theta   # What a 2-level model estimates as 'p'
}
",fill=TRUE)
sink()

# Initial values
inits <- function() list(z = array(1, dim = data$nunit), a = array(1, dim =c(data$nunit,
data$nsubunit)) )      # Set all to 1 to avoid conflict

# Parameters monitored
params <- c("int.psi", "int.theta", "int.p", "beta.lpsi", "beta.ltheta",
"beta.lp","p.theta", "sum.z", "sum.a")

# MCMC settings
ni <- 25000 ;  nt <- 2 ;  nb <- 2000 ;   nc <- 3

# Call JAGS (ART 15 min) and summarize posterior
out <- jags(win.data, inits, params, "model.txt", n.chains = nc, n.thin = nt, n.iter = ni,
n.burnin = nb, parallel = T)
traceplot(out) ;  print(out, 3)

# Compare truth and estimate in table
tmp <- cbind(rbind(mean.psi = data$mean.psi, mean.theta = data$mean.theta,
mean.p = data$mean.p, beta.lpsi = data$beta.Xpsi, beta.ltheta = data$beta.Xtheta, beta.lp
= data$beta.Xp, product.theta.p = data$mean.theta* data$mean.p, sum.z = data$sum.z),
out$summary[c(1:8),c(1, 3, 7)])
colnames(tmp) <- c("Truth", "Post.mean", "LCRL", "UCRL")
print(tmp, 3)
```

	Truth	Post.mean	LCRL	UCRL
mean.psi	0.80	0.783	0.708	0.863
mean.theta	0.60	0.615	0.466	0.776
mean.p	0.40	0.421	0.326	0.552
beta.lpsi	1.00	0.879	0.532	1.298
beta.ltheta	1.00	0.845	0.557	1.243
beta.lp	-1.00	-0.904	-1.168	-0.706
product.theta.p	0.24	0.255	0.226	0.287
sum.z	367.00	370.350	348.000	394.000

Note that the species was observed at 285 quadrats.

10.11.2 NO MAGICAL COVARIATE KNOWN: θ AND p ARE CONFOUNDED

By far the more typical case is when we *don't* have a magical covariate to explain variation at the unreplicated level and where θ and p are confounded; this is the typical application of the space-for-time design. Here, we demonstrate via simulation that a traditional (two-level) occupancy model fitted to space-for-time data appears to produce unbiased estimates of ψ and a covariate affecting ψ

and yields an estimate of the product of θ and p as its "probability of detection." Notably, we add unexplained noise in all three levels of the model—"time variation" in θ and random noise in all three levels. This represents the typical case where there is unexplained variation in θ and p as we would expect in the real world. When discussing the space-for-time occupancy design with colleagues, we had quite often sensed an uneasiness about a possible hidden assumption that availability probability (θ) must be constant within a unit. In our simulation, availability probability is *not* constant at all and neither is ψ and p. So let's see whether we can still estimate features of the model for ψ. We define a function to fit the model in unmarked, so that we can use the R function try to prevent a crash of the simulation whenever estimation fails in function occu.

```
# Load unmarked and define a function to fit the model with unmarked
library(unmarked)
occUM.fn <- function(data = data, inits = c(1, 1, -1)){
    umf <- unmarkedFrameOccu(y = data$y[,,1], siteCovs =
        data.frame(covA = data$covA))
    tmp1 <- summary(fm <- occu(~1 ~covA, data=umf, starts=inits))
    tmp <- matrix(unlist(c(tmp1$state[,1], tmp1$det[,1], tmp1$state[,2],
        tmp1$det[,2])), ncol = 2, byrow = F)
    dimnames(tmp) <- list(c("Occ_Int", "Occ_A", "Det_Int"), c("MLE", "SE"))
    return(MLE = tmp)
}

# Choose number of simulations and create structures to hold results
simreps <- 10000              # takes about 30 min
obs.stats <- array(NA, dim = c(simreps, 3))
dimnames(obs.stats) <- list(NULL, c("sum.z", "obs.sum.z", "sum.z.x"))
MLEs <- array(NA, dim = c(3,2,simreps))
dimnames(MLEs) <- list(c("Occ_Int", "Occ_A", "Det_Int"), c("MLE", "SE"), NULL)

# Set timer and launch simulation
system.time(
for(i in 1:simreps){
    cat("\n\n*** Simrep Number:", i, "***\n\n")
    # Generate data set
    data <- sim3Occ(nunit = 500, nsubunit = 5, nrep = 1, mean.psi = 0.8,
        beta.Xpsi = 1, sd.logit.psi = 0.4, mean.theta = 0.6,
        theta.time.range = c(-1, 1), beta.Xtheta = 0, sd.logit.theta = 0.6,
        mean.p = 0.4, p.time.range = c(0,0), beta.Xp = 0, sd.logit.p = 0.8)
    # Save stats
    obs.stats[i,] <- unlist(data[23:25])
    # Get MLEs of occupancy model and save them
    UMmle <- try(occUM.fn(data = data, inits = c(1,1,-1)))
    if (class(UMmle) == "try-error") {v<-1} else {
      MLEs[,,i] <- UMmle
    }
    rm(data, UMmle)
    }
)
```

10.11 SPACE-FOR-TIME SUBSTITUTION

```
# Visualize results
par(mfrow = c(1,3), mar = c(5,5,3,2), cex.lab = 1.5, cex.axis = 1.5, cex.main = 1.5)
# Estimate of occupancy (psi)
hist(plogis(MLEs[1,1,]), breaks = 40, col = "grey", main = "Quadrat occupancy (psi)")
abline(v = mean(plogis(MLEs[1,1,]), na.rm = T), col = "blue", lwd = 3)
abline(v = 0.8, col = "red", lwd = 3)
# Estimate of occupancy covariate (A)
hist(MLEs[2,1,], breaks = 40, col = "grey", main = "Quadrat occupancy covariate (covA)")
abline(v = mean(MLEs[2,1,], na.rm = T), col = "blue", lwd = 3)
abline(v = data$beta.Xpsi, col = "red", lwd = 3)
# Estimate of "detection": product of theta and p
hist(plogis(MLEs[3,1,]), breaks = 40, col = "grey", main = "'Detection probability' = \n theta * p")
abline(v = mean(plogis(MLEs[3,1,]), na.rm = T), col = "blue", lwd = 3)
abline(v = data$mean.theta * data$mean.p, col = "red", lwd = 3, lty = 2)
```

The means of all estimates in Figure 10.15 agree very well with the values used to simulate the data. In particular, what the two-level occupancy models calls "detection probability" (right) matches well the product of the probabilities of availability (θ) and detection (p) in the three-level data-generating occupancy model. Hence, it seems that valid inferences about occupancy, including covariate relationships, at the unit scale can be obtained with a space-for-time substitution design, even when there is plenty of unexplained variation, including nonconstant availability probability.

In spite of this, we think that a much better understanding of the space-for-time design is needed to apply it with confidence. Simulation studies that vary different parameters and that include an explicit description of the movements of individuals in their home range appear to be important in this endeavor.

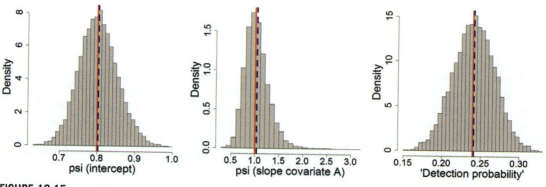

FIGURE 10.15

Simulation study with a space-for-time occupancy design. Left and middle: Estimates of intercept (on probability scale) and slope of a site covariate in the model for ψ. Right: Estimate of constant "detection probability"; the red line right shows the product of the probabilities of availability (θ) and detection (p) in the three-level data-generating occupancy model. Blue lines show the mean of 10,000 simulation replicates, and red lines show the truth in data generation.

10.12 MODELS FOR DATA ALONG TRANSECTS: POISSON, EXPONENTIAL, WEIBULL, AND REMOVAL OBSERVATION MODELS

Frequently, presence/absence data are collected along linear structures, which may be natural (e.g., surveys along rivers or coastlines) or artificial, such as any line transect in continuous habitat (e.g., in the North American BBS or in the Swiss MHB; see Figure 6.8). Issues that arise include the following: whether to discretize space, model serial autocorrelation, aggregate multiple detections, or use only part of the data to avoid dependency issues. Interestingly, the same issues arise when modeling occupancy data collected over time (e.g., with camera traps). In this sense, space and time can be thought of as more or less exchangeable in this section.

If you discretize occupancy data collected along a transect and analyze using a standard occupancy model, you lose some information whenever more than a single detection per occasion is subsumed into a single "1" in the resulting detection history. Hence, it might seem that it would be best to discretize so finely that you never have more than a single detection per occasion. On the other hand, this may lead to very large data sets (lots of occasions), and especially, there will likely be serial dependence, i.e., adjacent occasions may not be independent samples from the underlying process. You would have to deal with this problem by adopting more complex occupancy models, e.g., which assume Markovian dependence among neighboring segments (Hines et al., 2010, 2014; Aing et al., 2011). Such models have been implemented in PRESENCE and MARK, but not in `unmarked`.

Some might argue that discretization of measurements made on a continuous process is always a bad idea and that you should directly model a continuous detection process, in space or in time. This typically leads to Poisson process models for occupancy designs (Guillera-Arroita et al., 2011, 2012). The simplest such models assume independence in space or time—individual detections can be aggregated into a detection frequency per transect (or camera) and some total time interval, say C_i for the number of detections at trap or transect i of length L_i. We can then specify the following variant of an occupancy model, which simply has a Poisson instead of a binomial observation process, but is otherwise identical to the standard model.

$$z_i \sim Bernoulli(\psi)$$

$$C_i|z_i \sim Poisson(z_i * L_i * \lambda_i)$$

Here, L_i is the length of the spatial or temporal "observation window," λ_i is the detection rate per unit time or transect length, i.e., the expected number of detections per unit-length transect, given that a transect is occupied ($z_i = 1$). Patterns in occupancy can be modeled in the usual manner and those in the observation process via a log link, exactly as in a Poisson GLM. This model can easily be fitted in BUGS. Note that a single replicate (i.e., visit) is sufficient since there is no index j for replicates (though there would be if you have also temporal replicates, and more visits typically means more information and more precise estimates).

If independence can be assumed, this is a useful model. However, with dependence, e.g., aggregation of occurrence in time or space, we need to account for it, and one approach in continuous time/space is then to model this serial autocorrelation at the level of the individual detections. That is, we imagine two latent processes underlying the measured detections, between which there are switches with some probability that can be estimated. Guillera-Arroita et al. (2012) developed a two-state Markov-modulated point process model for this, and Murray Efford (pers. comm.) has written R code to fit the model, which is available from him on request.

The need for a more complex model for detections in continuous space or time in the case of serial dependence is one motivation for discretization and modeling of the aggregated data. In that way one may get rid of the serial dependence and be able to use a simpler model. Another way to avoid autocorrelation is to throw away some information and only model the first detection along every transect segment. In continuous space/time, this leads to an occupancy model that has an observation process like a continuous-time survival model (Garrard et al., 2008), where the time to first detection is modeled as a random variable with an exponential or Weibull distribution. For discretized data, we obtain a model with removal design (MacKenzie and Royle, 2005, p. 102 in MacKenzie et al., 2006; Rota et al., 2009; Guillera-Arroita and Lahoz-Monfort, in press). We cover all three in the remainder of this section.

10.12.1 OCCUPANCY MODELS WITH "SURVIVAL MODEL" OBSERVATION PROCESS: EXPONENTIAL TIME-TO-DETECTION MODEL WITH SIMULATED DATA

In this occupancy model with a continuous-time observation process, known as time-to-detection (TTD) protocol, "survival analysis" (in the medical or engineering sense) or time-to-event analysis, we obtain separate information about occurrence and detection probability from a single visit at each site (Garrard et al., 2008, 2013, 2015; also see McCarthy et al., 2013, and Bornand et al., 2014). It appears a very natural modeling approach for the types of search behavior that includes long and not necessarily very standardized surveys. The response variable in this model is the time (or possibly the transect length) until the first individual of the target species is detected (or the first cue, e.g., if you are searching for distinctive feces). A natural place to start modeling is the adoption of an exponential distribution, which describes the time between events in a Poisson process, i.e., where events occur independently with a constant rate in continuous time. This includes the time until the first event is observed. For a Poisson rate parameter λ, the expected (i.e., mean) time between events is $1/\lambda$.

One complication is that whenever the species is not detected after the maximum search time at a site, $Tmax_i$, there is uncertainty about the state of site, therefore, we must model time to detection (y_i) as a censored exponential random variable. By $d(Tmax_i)$ we indicate the response at a site to be censored at the value $Tmax_i$, i.e., when the species is not detected after $Tmax_i$, $d = 1$ and we treat the response y_i as missing. The model can be written as follows:

$$\text{Model for presence/absence: } z_i \sim Bernoulli(\psi)$$

$$\text{Model for censoring of data: } d(Tmax_i) = z_i * I(y_i > Tmax_i) + (1 - z_i)$$

$$\text{Model for the observed data: } y_i | z_i \sim Exponential(\lambda_i) \quad \text{if } d_i = 0$$

$$y_i = NA \quad \text{if } d_i = 1$$

Hence, response y_i is censored and will be set at NA either if time to detection happens to be greater than the maximum search time (but the site is occupied) or if the site is unoccupied. Detection probability (p) until time t is a function of the detection rate λ and the search time t: $p = 1 - \exp(-\lambda_i t)$ (Garrard et al., 2008). We next use a function (`simOccttd`) to simulate data under this model, where we may specify effects of one continuous covariate in each of occupancy and the Poisson rate parameter, specified via the usual logit and log links, respectively. The function arguments with their defaults are the following.

```
simOccttd(M = 250, mean.psi = 0.4, mean.lambda = 0.3, beta1 = 1, alpha1 = -1, Tmax = 10)
# Function arguments:
# M:              Number of sites
# mean.psi:       Intercept of occupancy probability
# mean.lambda:    Intercept of Poisson rate parameter
# beta1:          Slope of continuous covariate B on logit(psi)
# alpha1:         Slope of continuous covariate A on log(lambda)
# Tmax:           Maximum search time (in same units as response)
#                      (response will be censored at Tmax)
```

We execute the function once with default arguments to generate one data set.

```
set.seed(1)
data <- simOccttd()
str(data)
   Number of occupied sites ( among 250 ): 102
   Number of sites at which detected: 81
   Number of times censored: 169
```

We plot the response and then fit the data-generating exponential model in BUGS.

```
# Plot response (not shown)
hist(data$ttd, breaks = 50, col = "grey", main = "Observed distribution of time
to detection", xlim = c(0, data$Tmax), xlab = "Measured time to detection")
abline(v = data$Tmax, col = "grey", lwd = 3)

# Bundle data
str( win.data <- list(ttd = data$ttd, d = data$d, covA = data$covA,
   covB = data$covB, nobs = data$M, Tmax = data$Tmax) )

# Define occupancy model with exponential observation process
cat(file = "model1.txt", "
model {

# Priors
int.psi ~ dunif(0, 1)                    # Intercept occupancy on prob. scale
beta1 ~ dnorm(0, 0.001)                  # Slope coefficient in logit(occupancy)
int.lambda ~ dgamma(0.0001, 0.0001)      # Poisson rate parameter
alpha1 ~ dnorm(0, 0.001)                 # Slope coefficient in log(rate)

# Likelihood
for (i in 1:nobs){
# Model for occurrence
   z[i] ~ dbern(psi[i])
   logit(psi[i]) <- logit(int.psi) + beta1 * covB[i]

   # Observation model
   # Exponential model for time to detection ignoring censoring
   ttd[i] ~ dexp(lambda[i])
   log(lambda[i]) <- log(int.lambda) + alpha1 * covA[i]
```

```
    # Model for censoring due to species absence and ttd>=Tmax
    d[i] ~ dbern(theta[i])
    theta[i] <- z[i] * step(ttd[i] - Tmax) + (1 - z[i])
  }
  # Derived quantities
  n.occ <- sum(z[])     # Number of occupied sites among M
}
")

# Inits function for some params
# Initialize with z = 1 throughout and
#    all NA's due to censoring, rather than non-occurrence
zst <- rep(1, length(win.data$ttd))
ttdst <-rep(win.data$Tmax+1, data$M)
ttdst[win.data$d == 0] <- NA
inits <- function(){list(z =zst, ttd = ttdst, int.psi = runif(1), int.lambda = runif(1))}

# Parameters to estimate
params <- c("int.psi", "beta1", "int.lambda", "alpha1", "n.occ")

# MCMC settings
ni <- 12000 ; nt <- 2 ; nb <- 2000 ; nc <- 3

# Call WinBUGS from R (ART 1.3 min) and summarize posteriors
out1 <- bugs(win.data, inits, params, "model1.txt", n.chains=nc, n.iter=ni, n.burn = nb,
n.thin=nt, debug = TRUE, bugs.directory = bd)
print(out1, dig = 3)
```

	mean	sd	2.5%	25%	50%	75%	97.5%	Rhat	n.eff
int.psi	0.369	0.046	0.284	0.337	0.367	0.398	0.463	1.002	2200
beta1	1.408	0.243	0.963	1.240	1.396	1.564	1.910	1.001	15000
int.lambda	0.249	0.040	0.177	0.221	0.248	0.275	0.333	1.002	1600
alpha1	-1.248	0.157	-1.555	-1.355	-1.247	-1.141	-0.943	1.002	1900
n.occ	102.300	5.679	92.000	98.000	102.000	106.000	114.000	1.002	1700

The estimates seem to agree with the truth in the data simulation. You could run a simulation to check whether any discrepancy between the truth in the data simulation and the estimates is simply sampling variance or represents bias. As often with occupancy models, the number of occupied sites is estimated with staggering accuracy.

10.12.2 TIME-TO-DETECTION ANALYSIS WITH REAL DATA: WEIBULL OCCUPANCY MODEL FOR THE PEREGRINE SPRING SURVEY

To refine our understanding of a TTD occupancy model, we now analyze a small data set produced by a peregrine survey in the French Jura mountains during March 7–9, 2015, where *L'Equipe de choc helveto-britannique* (D. Parish, M. Kéry) visited 38 breeding cliffs for a total of 45 times and saw 57 peregrines (30 females, 27 males). Observation duration ranged from 3–95 mins and time to first detection from 0.1–48 mins. We saw birds in a total of 28 territories; hence, the observed proportion occupied was 0.74. We recorded time to detection separately for every bird present in a territory (typically, a male and a female, but sometimes including a visiting immature bird as well). The aims of our analyses are:

1. to estimate the proportion of occupied territories among the 38 visited sites,
2. to study effects on detection probability of time of day, sex, and duration of observation, and
3. to distinguish between a constant or a time-varying hazard (i.e., choose between an exponential or a Weibull distribution).

The Weibull distribution is a generalization of the exponential. In the exponential, the hazard rate (i.e., the instantaneous event rate) is constant over time, whereas the Weibull allows the hazard to vary over time, leading to an "accelerated failure time" survival model if the hazard increases over time. Conversely, in our case it could be that we lost faith with increasing time of not seeing a bird. The Weibull allows us to test and therefore adjust for this by way of an additional parameter, the shape, which governs the change of the hazard. We have an exponential with constant hazard when shape = 1, while 0 < shape < 1 means that the hazard declines over time and so does detection probability, whereas with shape > 1 the hazard and detection probability increase over time.

```
# Read in data
data <- read.table("ttdPeregrine.txt", header = T, sep = "\t")
str(data)   # Will be part of the AHM package later

# Manage data and standardize time of day
nobs <- length(data$SiteNumber)      # Number of observations
d <- as.numeric(is.na(data$ttd))     # Censoring indicator
mean.tod <- mean(data$MinOfDay)
sd.tod <- sd(data$MinOfDay)
tod <- (data$MinOfDay -mean.tod) / sd.tod

# Bundle and summarize data set
str( win.data <- list(M = max(data$SiteNumber), site = data$SiteNumber,
    tod = tod, male = as.numeric(data$sex)-1, ttd = data$ttd, d = d, nobs = nobs,
    Tmax = data$Tmax) )
```

Compared with Section 10.12.1, this analysis contains the following new elements: it assumes a Weibull instead of an exponential response, survey duration is not constant but a vector, there are multiple observations per site, and we have an observational-level covariate, the sex of each bird. We parameterize sex as an indicator for males. Its value is not known when no bird is observed at a site, therefore, we have to specify a model for sex to estimate it for the missed birds. We do this by simply specifying a Bernoulli model with success probability being the sex ratio (specified as the proportion of males). This is a case of a missing individual covariate that is exactly analogous to the case of a size covariate affecting detection probability in a closed model to estimate population size (see Section 6.4 in Kéry and Schaub 2012 and also Chapter 9 in this book, where we model group size in distance sampling).

```
# Define model
cat(file = "model2.txt", "
model {

# Priors
psi ~ dunif(0, 1)                       # Occupancy intercept
lambda.int[1] ~ dgamma(0.001, 0.001)    # Poisson rate parameter for females
lambda.int[2] ~ dgamma(0.001, 0.001)    # Poisson rate parameter for males
```

10.12 MODELS FOR DATA ALONG TRANSECTS: POISSON, EXPONENTIAL

```
alpha1 ~ dnorm(0, 0.001)         # Coefficient of time of day (linear)
alpha2 ~ dnorm(0, 0.001)         # Coefficient of time of day (squared)
shape ~ dgamma(0.001,0.001)      # Weibull shape
sexratio ~ dunif(0,1)            # Sex ratio (proportion males)
# Likelihood
for (i in 1:M){                  # Model for occurrence at site level
   z[i] ~ dbern(psi)
}

for (i in 1:nobs){               # Observation model at observation level
   # Weibull model for time to detection ignoring censoring
   ttd[i] ~ dweib(shape, lambda[i])
   log(lambda[i]) <- (1-male[i])*log(lambda.int[1])+male[i]*log(lambda.int[2])+alpha1
 * tod[i] + alpha2 * pow(tod[i],2)
   # Model for censoring due to species absence and ttd>=Tmax
   d[i] ~ dbern(theta[i])
   theta[i] <- z[site[i]] * step(ttd[i] - Tmax[i]) + (1 - z[site[i]])
   # Model for sex of unobserved individuals
   male[i] ~ dbern(sexratio)     # Will impute sex for unobserved individuals
}
# Derived quantities
n.occ <- sum(z[])                # Number of occupied sites among M
}
")

# Inits function
zst <- rep(1, win.data$M)
ttdst <-rep(win.data$Tmax+1)
ttdst[win.data$d == 0] <- NA
inits <- function(){list(z =zst, ttd = ttdst, psi = runif(1), lambda.int = runif(2),
alpha1 = rnorm(1), alpha2 = rnorm(1), shape = runif(1))}

# Parameters to estimate
params <- c("psi", "lambda.int", "alpha1", "alpha2", "n.occ", "z", "sexratio", "shape")

# MCMC settings
ni <- 15000 ; nt <- 2 ; nb <- 2000 ; nc <- 3

# Call JAGS from R (ART 0.6 min) and summarize posteriors
out2 <- bugs(win.data, inits, params, "model2.txt", n.chains=nc, n.iter=ni, n.burn = nb,
n.thin=nt, debug = T, bugs.directory = bd)
print(out2, dig = 3)
```

	mean	sd	2.5%	25%	50%	75%	97.5%	Rhat	n.eff
psi	0.923	0.062	0.771	0.889	0.938	0.972	0.998	1.002	2200
lambda.int[1]	0.129	0.045	0.061	0.098	0.123	0.154	0.236	1.002	2000
lambda.int[2]	0.109	0.042	0.048	0.079	0.102	0.132	0.208	1.001	5000
alpha1	-0.115	0.141	-0.392	-0.211	-0.115	-0.020	0.162	1.002	2800
alpha2	0.382	0.155	0.076	0.278	0.382	0.487	0.683	1.001	6900
n.occ	35.929	1.871	32.000	35.000	36.000	37.000	38.000	1.002	1300
z[1]	1.000	0.000	1.000	1.000	1.000	1.000	1.000	1.000	1
z[2]	1.000	0.000	1.000	1.000	1.000	1.000	1.000	1.000	1
z[3]	1.000	0.000	1.000	1.000	1.000	1.000	1.000	1.000	1

```
[ output truncated ]
z[36]           1.000 0.000 1.000 1.000 1.000 1.000 1.000 1.000      1
z[37]           0.632 0.482 0.000 0.000 1.000 1.000 1.000 1.002   1500
z[38]           1.000 0.000 1.000 1.000 1.000 1.000 1.000 1.000      1
sexratio        0.483 0.066 0.356 0.438 0.482 0.528 0.612 1.001  15000
shape           0.734 0.079 0.585 0.681 0.731 0.786 0.895 1.003   1100
```

We see that the sexes don't differ in their Weibull base rate, since the lambda intercepts for females (lambda.int[1]) and for males (lambda.int[2]) are very similar. The effect of time of day is not immediately clear, so we will form predictions below. We estimate that about 36 territories were occupied and, as always with a Bayesian fit of an occupancy model, we get the estimated presence/absence state of every site for free (these are the z's). Based on the observed birds, the sex ratio is estimated to be roughly even. Finally, interestingly, the Weibull shape parameter is estimated at a value clearly less than 1, indicating that the hazard for peregrine detection *declines* with increasing observation time. Perhaps we became increasingly pessimistic about the possible presence of a peregrine with increasing time without seeing it and then actually became more likely to miss a peregrine when it did appear? An alternative explanation, that the birds are disturbed and behave less conspicuous in our presence, does not apply here, because the distances between observer and the birds are simply too great.

We summarize the results from the analysis with Figure 10.16. To visualize the effects of time of day and duration on the detection probability, we average over the (very small) sex effect, and conduct

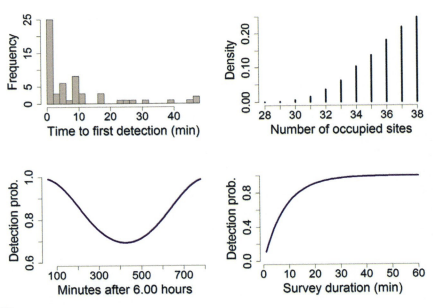

FIGURE 10.16

Results from fitting a Weibull time-to-detection occupancy model to part of the data from the peregrine spring survey 2015. Top left: the observed response: time to detection of the first bird for an individual visit; top right: posterior distribution of the number of occupied sites among the 38 sites visited; bottom: relationships between detection probability and time of day (left) and survey duration (right). Time of day is expressed in minutes after 6.00 h, so the x-axis extends from 7.00 to 19.00 hours.

the necessary calculations first. Time of day varies from 86—757 mins after 6.00 h, so we will predict for 60—780 mins and the x-axis extends from 7.00 to 19.00 hours.

```
# Predict detection over time of day: prediction cov. runs from 7h to 19h
minutes <- 60:780
pred.tod <- (minutes - mean.tod) / sd.tod  # Standardize as real data

# Predict p over time of day, averaging over sex, and for duration of 10 min
sex.mean <- apply(out2$sims.list$lambda.int, 1, mean)
p.pred1 <- 1 - exp(-exp(log(mean(sex.mean)) + out2$mean$alpha1 * pred.tod +
out2$mean$alpha2 * pred.tod^2) * 10)

# Predict p for durations of 1-60 min, averaging over time of day and sex
duration <- 1:60
p.pred2 <- 1 - exp(-exp(log(mean(sex.mean))) * duration)

# Visualize analysis
par(mfrow = c(2,2), mar = c(5,5,3,2), cex.lab = 1.5, cex.axis = 1.5)
hist(data$ttd, breaks = 40, col = "grey", xlab = "Time to first detection
(min)", main = "")
plot(table(out2$sims.list$n.occ)/length(out2$sims.list$n.occ), xlab = "Number of
occupied sites", ylab = "Density", frame = F)
plot(minutes, p.pred1, xlab = "Minutes after 6.00 hours (i.e., 7.00 - 19.00h)",
ylab = "Detection prob.", ylim = c(0.6, 1), type = "l", col = "blue", lwd = 3,
frame = F)
plot(duration, p.pred2, xlab = "Survey duration (min)", ylab = "Detection
prob.", ylim = c(0, 1), type = "l", col = "blue", lwd = 3, frame = F)
```

10.12.3 OCCUPANCY MODELS WITH REMOVAL DESIGN OBSERVATION PROCESS

A removal design is the discrete-time counterpart to what the time-to-detection design is in continuous time—we only model the events until the first detection, and a site needs only be surveyed until the species is first detected (or the data from later occasions is discarded). Although not always the most efficient occupancy design (MacKenzie and Royle, 2005; Bailey et al., 2007; Guillera-Arroita and Lahoz-Monfort, in press), the removal design tends to be popular among fieldworkers, who may not be motivated to revisit a site where a species' presence has already been ascertained. In individual capture-recapture (see Chapter 7), a removal design is coded using a categorical distribution for the type of observed detection history, and we can do the analogous thing also for an occupancy model with removal design. However, it is much easier if we simply turn all observations (if there are any) after the first detection at a site into NAs and fit the standard Bernoulli-Bernoulli occupancy model. This yields identical estimates to those obtained with the categorical parameterization and enables you to fit a removal design not only in BUGS but also in software such as unmarked, PRESENCE, MARK, or E-SURGE.

10.13 OCCUPANCY MODELING OF A COMMUNITY OF SPECIES

The occupancy framework can also be used to model a community, where species take the place of sites and the total number of "sites" is the list of species that can reasonably be assumed to occur in some area (MacKenzie et al., 2006). Such modeling makes sense especially in well-studied areas where the identity of all likely species in the regional pool is fairly well known, e.g., parts of

Europe and North America. The occupancy parameter might be interpreted as "relative species richness" or community integrity (Karr, 1990; Cam et al., 2002b,c), corrected for imperfect detection, i.e., species richness relative to some baseline list of species that may be thought to represent a regional pool of species present (Kéry, 2011a). Detection probability among animal or plant species in a community differs tremendously, hence, in this use of occupancy modeling for a community, adoption of a "heterogeneity model" (Royle, 2006) is imperative, that is we need a model that lets individual species differ in their detection probability. Otherwise, severe underestimation of the number of species, or the occupancy parameter, would result by unmodeled detection heterogeneity among species (this is again the second law of capture-recapture). Occupancy models with site heterogeneity in detection probability (beyond what can be explained by measured covariates) can only be fitted with PRESENCE and MARK and of course with BUGS. Typical specifications for such species-specific heterogeneity that cannot be explained by covariates include the logit-normal (see Section 6.11.2), beta-binomial, and finite mixture distributions (Dorazio and Royle, 2003; Royle, 2006). All of them can easily be specified in BUGS.

10.14 MODELING WIGGLY COVARIATE RELATIONSHIPS: PENALIZED SPLINES IN HIERARCHICAL MODELS

We end this chapter with a topic that is not specific to occupancy models and illustrate the modeling of highly irregular, "wiggly" covariate relationships using penalized splines (Ruppert et al., 2003; Crainiceanu et al., 2005; Gimenez et al. 2006a,b). Splines can be very useful in an exploratory analysis when nothing is known about the functional form of a covariate relationship; they may help suggesting a new parametric model. Penalized splines can be implemented as a form of GLMM, where the main features of the covariate relationship are described by fixed effects (e.g., linear and quadratic terms) and the wiggly part is accommodated by the random effects. We use function `wigglyOcc` to simulate static occupancy data with a really wild covariate relationship in both the occupancy and the detection part of the model (sample size is 240 sites and 3 replicate surveys). Executing the function draws two figures (Figure 10.17) and produces the simulated data and some summary output.

```
# Execute the function and inspect file produced
data <- wigglyOcc(seed = 1)
   True number of occupied sites: 141
   Observed number of occupied sites: 97
   Proportional underestimation of distribution: 0.31
str(data)
```

We use BUGS code due to Crainiceanu et al. (2005) to fit a model with penalized splines with truncated polynomial basis for each covariate in both parts of the model (i.e., for both occupancy and for detection). We also compare with a simpler model with two quadratic polynomials of the covariates. To spline-smooth a covariate in the detection model, we have to vectorize the analysis (see Chapter 21 in Kéry, 2010 and Chapter 16 in volume 2).

```
# Convert matrix data into vectors and prepare stuff
y <- c(data$y)                    # Detection/nondetection data (response)
Xsite <- data$Xsite               # Fine as is
Xsurvey <- c(data$Xsurvey)        # Survey covariate
site <- rep(1:data$M, data$J)     # Site index
```

10.14 MODELING WIGGLY COVARIATE RELATIONSHIPS

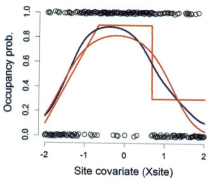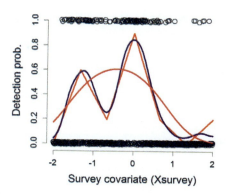

FIGURE 10.17

The crazy truth in our data set simulated using function wigglyOcc: red lines and circles show true occupancy probability and presence/absence z (left) and detection probability and the observed data y (right). Predictions under the spline occupancy model are blue and those under a simple model with linear and quadratic polynomial terms are brown. For both submodels in the spline model, 35 knots were evenly dispersed within the range of the covariates.

We then use our function spline.prep to create the two sets of fixed- and random-effects design matrices, based on code by Crainiceanu et al. (2005) and Zuur et al. (2012). The function call requires the name of a covariate and either the desired number of knots or else "NA," in which the latter is chosen using the rule of Ruppert (2002).

```
# tmp1 <-spline.prep(Xsite, 20)   # This would choose 20 knots for Xsite
tmp1 <-spline.prep(Xsite, NA)    # Choose variable default number of knots
tmp2 <-spline.prep(Xsurvey, NA)
Xocc <- tmp1$X         # Fixed-effects part of covariate Xsite in occ
Zocc <- tmp1$Z         # Random-effects part of covariate Xsite in occ
Xdet <- tmp2$X         # Fixed-effects part of covariate Xsite in det
Zdet <- tmp2$Z         # Random-effects part of covariate Xsite in det
nk.occ <- length(tmp1$knots)    # Number of knots in occupancy spline
nk.det <- length(tmp2$knots)    # Number of knots in detection spline
```

To compare estimates under the spline model with a simpler, parametric model with quadratic effects of both covariates, we fit both models inside of the same "hypermodel," by supplying a duplicate of the response data y.

```
# Bundle and summarize data set
win.data <- list(y1 = y, site = site, M = data$M, Xocc = Xocc, Zocc = Zocc, nk.occ = nk.occ,
Xdet = Xdet, Zdet = Zdet, nk.det = nk.det, nobs = data$M*data$J, y2 = y, onesSite = rep(1,
240), onesSurvey = rep(1, 720), Xsite = Xsite, Xsite2 = Xsite^2, Xsurvey = Xsurvey,
Xsurvey2 = Xsurvey^2)
str(win.data)   # onesSite and onesSurvey are for occ and det intercepts
```

```
# Specify two models in one in BUGS language
cat(file = "hypermodel.txt",
"model {

# *** Spline model for the data ***
# -------------------------------
# Priors
for(k in 1:3){                      # Regression coefficients
    alpha1[k] ~ dnorm(0, 0.1)       # Detection model
    beta1[k] ~ dnorm(0, 0.1)        # Occupancy model
}
for(k in 1:nk.occ){ # Random effects at specified knots (occupancy)
    b.occ[k] ~ dnorm(0, tau.b.occ)
}
for(k in 1:nk.det){ # Random effects at specified knots (detection)
    b.det[k] ~ dnorm(0, tau.b.det)
}
tau.b.occ ~ dgamma(0.01, 0.01)
tau.b.det ~ dgamma(0.01, 0.01)

# Likelihood
# Model for latent occupancy state
for (i in 1:M) {
    z1[i] ~ dbern(psi1[i])
    logit(psi1[i]) <- fix.terms.occ[i] + smooth.terms.occ[i]
    fix.terms.occ[i] <- beta1[1]*Xocc[i,1] + beta1[2]*Xocc[i,2] + beta1[3]*Xocc[i,2]
    smooth.terms.occ[i] <- inprod(b.occ[], Zocc[i,])
}
# Model for observations
for(i in 1:nobs){
    y1[i] ~ dbern(mu.y1[i])
    mu.y1[i] <- z1[site[i]] * p1[i]
    logit(p1[i]) <- fix.terms.det[i] + smooth.terms.det[i]
    fix.terms.det[i] <- alpha1[1]*Xdet[i,1] + alpha1[2]*Xdet[i,2] +
        alpha1[3]*Xdet[i,2]
    smooth.terms.det[i] <- inprod(b.det[], Zdet[i,])
}

# Derived quantities
sum.z1 <- sum(z1[])                 # Number of occupied sites in sample
sd.b.occ <- sqrt(1/tau.b.occ)       # SD of spline random effects variance Occ.
sd.b.det <- sqrt(1/tau.b.det)       # SD of spline random effects variance Det.

# *** Polynomial model for same data ***
# -------------------------------------
# Priors
for(k in 1:3){                      # Regression coefficients
    alpha2[k] ~ dnorm(0, 0.1)       # Detection model
    beta2[k] ~ dnorm(0, 0.1)        # Occupancy model
}
```

10.14 MODELING WIGGLY COVARIATE RELATIONSHIPS

```
# Likelihood
# Model for latent occupancy state
for (i in 1:M) {
    z2[i] ~ dbern(psi2[i])
    logit(psi2[i]) <- beta2[1]*onesSite[i] + beta2[2]*Xsite[i] + beta2[3]*Xsite2[i]
}
# Model for observations
for(i in 1:nobs){
    y2[i] ~ dbern(mu.y2[i])
    mu.y2[i] <- z2[site[i]] * p2[i]
    logit(p2[i]) <- alpha2[1]*onesSurvey[i] + alpha2[2]*Xsurvey[i] +
        alpha2[3] * Xsurvey2[i]
}

# Derived quantities
sum.z2 <- sum(z2[])              # Number of occupied sites in sample
}
")

# Initial values
zst <- apply(data$y, 1, max)
inits <- function(){list(z1=zst, alpha1=rnorm(3), beta1=rnorm(3), b.occ =
rnorm(nk.occ), b.det = rnorm(nk.det), tau.b.occ = runif(1), tau.b.det =
runif(1), z2=zst, alpha2=rnorm(3), beta2=rnorm(3))}

# Parameters monitored
params <- c("alpha1", "beta1", "psi1", "p1", "fix.terms.occ", "smooth.terms.occ",
"b.occ", "fix.terms.det", "smooth.terms.det", "b.det", "sum.z1", "sd.b.occ", "sd.b.det",
"alpha2", "beta2", "psi2", "p2", "sum.z2")

# MCMC settings
ni <- 100000  ;  nb <- 10000  ;  nt <- 90  ;  nc <- 3
```

We run JAGS in parallel, since this model takes a while.

```
# Call JAGS from R (ART 95 min) and summarize posteriors
system.time(fhm <- jags(win.data, inits, params, "hypermodel.txt", n.chains = nc, n.thin =
nt, n.iter = ni, n.burnin = nb, parallel = TRUE))
traceplot(fhm)  ;  print(fhm, 3)
print(fhm$summary[c(1:6, 2957:2965, 3926),], 3)   # Compare some key estimands

# Plot prediction of psi and p (Fig. 10-17)
par(mfrow = c(1,2), mar = c(5,4,3,2), cex.lab = 1.5, cex.axis = 1.5)
plot(Xsite, data$psi, main = "Occupancy probability", type = "l", ylim = c(-0.1, 1.1),
col = "red", xlab = "Site covariate (Xsite)", ylab = "", lwd = 3)
points(Xsite, jitter(data$z, amount = 0.02))
lines(Xsite, fhm$mean$psi1, col = "blue", lty = 1, lwd = 3)
lines(Xsite, fhm$mean$psi2, col = "brown", lty = 1, lwd = 3)

plot(Xsurvey[order(data$x.index)], data$p.ordered, main = "Detection probability
", type = "l", ylim = c(-0.1, 1.1), col = "red", xlab = "Survey covariate (Xsurvey)",
ylab = "", lwd = 3)
```

```
points(Xsurvey, jitter(y, amount = 0.02))
lines(Xsurvey[order(data$x.index)], fhm$mean$p1[order(data$x.index)], col = "blue",
lwd = 3)
lines(Xsurvey[order(data$x.index)], fhm$mean$p2[order(data$x.index)], col = "brown",
lwd = 3)
```

Thus, even with really wiggly covariate relationships in both model parts, we have a quite impressive ability to estimate these patterns (Figure 10.17). The inference was better about the detection model than for occupancy, where the inferred pattern with splines was not much different from that by a simple quadratic polynomial. This is generally the case for such HMs—features of the latent process are harder to estimate than those of the observed process. In spite of the better covariate fit of the spline model, there was only a slight improvement in the estimate of a key quantity, the number of occupied sites in the sample. The species occurred at 141 sites and was observed at 97. The simpler model with quadratic covariate effects produced an estimate of 130 occupied sites (95% CRI 115−148) and the spline model produced an estimate of 135 (95% CRI 120−152).

In general, when your interest focuses on the precise functional form of a covariate relationship (Strebel et al., 2014), you can add splines inside of your hierarchical model. While this offers great flexibility and power, it should probably also be applied with care since we can imagine that it may be easy to run into identifiability problems. Finally, as we have seen, the substantive conclusions may not always be much different from those under a much simpler, parametric model.

10.15 SUMMARY AND OUTLOOK

We have given an overview of a very powerful and flexible class of hierarchical models: occupancy models (MacKenzie et al., 2002, 2006; Tyre et al., 2003). The hierarchical definition of the model makes extremely transparent what these models are—they have one logistic regression for the incompletely observed presence/absence state, which is governed by probability of occupancy or presence, and then another, conditional model for the measurement of presence/absence, that accommodates false-negative observations. This measurement error model will typically be a binomial or Bernoulli distribution, but depending on the precise data collection protocol may be Poisson, exponential, or Weibull, or multinomial, or even another distribution, and we have given examples for most of those. As always, the occupancy model may be fit using likelihood or Bayesian methods. The occupancy model is applicable to any binary response with a binary measurement error (in the case of classical Bernoulli observation model). Therefore, depending on how you choose to define the event that is classified as a "presence," the same model may be applied to a vast number of different situations. In particular, it is the canonical species distribution model with explicit accommodation of the ubiquitous false-negative detection error (Kéry, 2011b; Lahoz-Monfort et al., 2014; Guillera-Arroita et al., 2015).

Nevertheless, we would like to remind you of a couple of caveats. First, there has been some debate recently about the usefulness of this model in practice (Welsh et al., 2013; also see McKann et al., 2013, and Hayes and Monfils, 2015). Most doubts expressed in the Welsh et al. paper were convincingly responded to by Guillera-Arroita et al. (2014a), and for McKann et al., see our comments in Section 10.7. (Hayes and Monfils, 2015, worry about the effects of temporary emigration (TE) on the meaning of the occupancy parameter. This interpretation can range from 'probability of permanent presence of the species' when there is no TE to 'probability of use sometimes during the study period' when there

10.15 SUMMARY AND OUTLOOK

is TE; see Section 10.2. Most people don't see this as a big problem. Moreover, we would claim that inferences under ANY model for occurrence may be affected by TE. So their main point really seems to be not specific to occupancy models at all.) But don't forget that to estimate parameters, you first need *data*. So, with extreme paucity of data, you cannot expect to obtain very good parameter estimates for any model (hierarchical or not) and with any inference method (e.g., Bayesian or not). Many such doubts could be avoided by conducting customized simulations, using a template like the one given in Section 10.7. With R, it is really easy to ascertain how good estimates under a model are likely to be for exactly *your* sample sizes, data collection protocol and model, and for hypothesized values of the parameters in your study. The second caveat is that the meaning of your occupancy parameters depends on three things: the definition of what constitutes a "presence," and the spatial and temporal scale of your sampling. Part of your job as an ecologist is to make a good choice here to ensure that you obtain parameters that are meaningful in the context of the biology of your species. And the third caveat is the relationship of occupancy with abundance—you cannot "escape" abundance (McCarthy et al., 2013), for instance, spatial variation in abundance will create spatial heterogeneity in detection probability when you study presence/absence, and this may bias your occupancy estimators unless modeled. And somewhat related to the last point, never forget that presence/absence is "the poor man's abundance." If you have counts you should always try to model those directly and only throw out information by modeling presence/absence instead if there are very good reasons for doing so (for instance, if you can't find a fitting abundance model).

A vast number of extensions are possible for the basic occupancy model. Many correspond to the relaxation of a specific assumption in the basic model (see Section 10.2). Lack of closure can be addressed by "multiseason" models (MacKenzie et al., 2003), especially dynamic occupancy models (see Chapter 16), and a specific lack of closure in the basic model has been addressed by Kendall et al. (2013). False-positives can be accommodated in addition to false-negatives; see Royle and Link (2006), Miller et al. (2011, 2013b), Chambert et al. (2015), and Chapter 19 (in volume 2). The assumption that sites are independent can be relaxed by modeling spatial dependence (autocorrelation); see Chapters 21 and 22. Finally, the assumption of homogeneous detection among sites can be relaxed by adopting a mixture model for detection probability (Royle, 2006), such as the "Royle-Nichols model" (Royle and Nichols, 2003; see Section 6.13) or, if replicated counts are available, an *N*-mixture model (see Chapter 6 in Dorazio, 2007). The Royle-Nichols model can be fit using `unmarked` with function `occuRN`. Other ways to address site-specific detection heterogeneity, including finite mixtures or the logit-normal model, are easy to implement in BUGS; for the latter, see the code for the random-effects models in Section 6.11.2.

Other extensions include multiple scales of occupancy, as we have shown in Section 10.10, and extensions to more than two scales (corresponding to four or more levels in the hierarchical model) are straightforward (McClintock et al., 2010b). We have also discussed the space-for-time substitution and suggested that it may be understood as a restricted multiscale occupancy model that lacks replication at the bottom level. Multiscale occupancy models are important because many occupancy sampling designs are in fact nested, leading to such models, but we think that this is not yet sufficiently widely recognized. We have also emphasized that the space-for-time substitution is perhaps one of the least understood topics in occupancy modeling that really warrants more research to help decide when it can and when it cannot be applied to model occupancy and detection probability with spatially instead of temporally replicated measurements.

We have also shown a variety of data collection protocols, which lead to different observation models (Section 10.12), as well as the move from a basic GLM-type occupancy model to a GAM-type occupancy model (Section 10.13). Fitting GAM-type models in BUGS is something that you can do with any flat or hierarchical model. Two-dimensional splines are one computationally efficient way of addressing spatial autocorrelation in N-mixture or occupancy models (Collier et al., 2012; Guélat and Kéry, in review). But there is more. Instead of two states (here, presence and absence), you may have multiple states, for instance absence, and presence with and without reproduction, leading to multistate occupancy models (Chapter 18), or instead of a single species, you may want to model multiple species jointly, with or without interactions (see Chapters 11, 17, and 20). Finally, Roth and Amrhein (2009) developed an occupancy model for territorial birds that yields estimates of what they call "local survival rates," somewhat bridging the gap between a site-level model for occurrence and an individual-level model.

A recurrent theme in ecological statistics is that of the integration of disparate information about the same state or process, so-called "integrated models" or IMs (Schaub and Abadi, 2011). Such models are very powerful and topical and, essentially, they are only available in practice to ecologists because of the accessible BUGS language. Integrated models typically link different types of data sets via a description of the most detailed, or "bottom," level; or we may also say that they simply link the modeling of different data sets that bear on the same process by *shared parameters* which simultaneously occur in two or more data sets. For the models in this chapter, in their most trivial form, an IM would fuse multiple data sets for the ecological state of occurrence that differ only in their observation process. For instance, it is very easy to combine in a single model occupancy data with observation models for capture-recapture (that is the standard model) and removal and time-to-detection designs. More complex IMs link data sets where the fundamental state about which the data are informative differs. For instance, Chandler and Clark (2014) integrate occupancy and telemetry data via a description of an underlying latent spatial point process for the individuals, and the occupancy data are simply an aggregation (or different observation process) of that "bottom" system description. The fusion of count and presence/absence data is straightforward conceptually (for instance, for a Royle-Nichols model for the latter) and should be done much more often. In an important paper, Conroy et al. (2008) show how occupancy and individual capture-recapture data can be integrated, again via a Royle-Nichols variant for the former. Dorazio (2014) demonstrated the fusion of spatial point pattern data and replicated counts, and a similar thing conceptually should be possible with detection/non-detection data. Such combinations lead to more precise estimates and in some cases to the compensation of the weaknesses of one data type by the information in the other data type. In Chapter 23 we illustrate some variants of such integrated models. They represent an extremely powerful and as yet mostly untapped framework for modeling multisite data on distribution and abundance.

EXERCISES

1. Referring to Section 10.1 and our classification of data used to model species distribution, we mention how occupancy probability scales with the area of the sampling unit. Write a little simulation that allows you to observe this relationship (hint: you may distribute individuals in a plane according to a Poisson process, overlay with grids of different size, and observe how the proportion of cells occupied changes with grid size).

2. In Section 10.3, extend the simulation such that false-positives are allowed with a probability of 0.01. Check how the traditional occupancy estimator in this section is biased when we conduct 2, 4, 6…20 surveys.
3. Also in Section 10.3, we claimed that the Bernoulli model illustrated was equivalent to a binomial model for the detection frequencies when no time-specific patterns in detection probability are present or modeled. (a) Prove this to yourself by fitting the binomial model to an example where you increase the number of sites 100-fold and the number of surveys 10-fold. How much faster in BUGS is the binomial model compared to the Bernoulli model? (b) Turn some of the final surveys into NAs and fit the binomial model when the binomial sample size varies and J has to be supplied in the analysis as a vector.
4. In Section 10.5, play around with function simOcc for at least half an hour to train your intuition about occurrence data, imperfect detection, and occupancy models.
5. Also in Section 10.5, use function simOcc to devise a simulation so see how much the true and the observed number of occupied sites vary by chance when using the default function arguments.
6. In Section 10.7, adapt the code to study the effects on the occupancy estimator of the standard model of assumption violations due to individual (site) detection heterogeneity and behavioral response.
7. In Section 10.9, produce a histogram that shows the elevation gradient of the distribution of the red squirrel in Switzerland.
8. In Section 10.14, fit the spline model with no fixed effect of the covariates at all (other than an intercept), so see how well the random parts of the spline model can recover the functional form of the two covariates and whether the resulting model can still estimate the number of occupied sites decently.

CHAPTER 11

HIERARCHICAL MODELS FOR COMMUNITIES

CHAPTER OUTLINE

11.1 Introduction 631
11.2 Simulation of a Metacommunity 633
11.3 Metacommunity Data from the Swiss Breeding Bird Survey MHB 643
11.4 Overview of Some Models for Metacommunities 646
11.5 Community Models That Ignore Species Identity.. 651
    11.5.1 Simple Poisson Regression for the Observed Community Size 651
    11.5.2 Poisson Random-Effects Model for the Observed Community Size 657
    11.5.3 N-mixture Model for Observed Community Size 658
11.6 Community Models that Fully Retain Species Identity 661
    11.6.1 Simplest Community Occupancy Model: n-fold Single Species Occupancy Model with Species Treated as Fixed Effects.. 662
    11.6.2 Community Occupancy Model with Bivariate Species-Specific Random Effects 667
    11.6.3 Modeling Species-Specific Effects in Community Occupancy Models 672
    11.6.4 Modeling Species Richness in a Two-Step Analysis.. 679
11.7 The Dorazio/Royle (DR) Community Occupancy Model with Data Augmentation (DA)....................... 682
    11.7.1 The Simplest DR Community Model with DA 683
    11.7.2 DR Community Model with Covariates 690
11.8 Inferences Based on the Estimated Z Matrix: Similarity among Sites and Species 709
11.9 Species Richness Maps and Species Accumulation Curves... 712
11.10 Community N-mixture (or Dorazio/Royle/Yamaura - DRY) Models 716
11.11 Summary and Outlook 724
Exercises 728

11.1 INTRODUCTION

In this chapter, we deal with the third main element in the subtitle of this book, species richness. But we do much more: we present a type of hierarchical model (HM) for communities, or more specifically for metacommunities, that was independently developed by Dorazio and Royle (2005) and by Gelfand et al. (2005). Of course, there is a vast body of models and methods for community analysis, and we won't even scratch the surface of community analysis. What we will do is showcase a single but extremely powerful approach for modeling metacommunities: the community occupancy model

(Dorazio and Royle, 2005) and the community abundance or community N-mixture model (Yamaura et al., 2012; Chandler et al., 2013). We call them the Dorazio/Royle (DR) and the Dorazio/Royle/Yamaura (DRY) models, respectively. Both are based on the eminently sensible idea of describing a metacommunity as a collection of individual species. In this framework, plenty of additional complexity can easily be built in as needed, or as the data—or your patience—allow, when fitting the models. Most importantly, we can accommodate environmental and other effects on distribution or abundance of the individual species, false-negative measurement error, temporal dynamics of species-level occurrence and abundance, and species interactions. We cover the former two in this chapter extensively and discuss interactions, but will cover temporal dynamics and interactions formally only in volume 2. We remind you that for the modeling of a single community, i.e., for a collection of species at a single place, you can use a static occupancy model where individual species take the place of a 'site'; see Section 10.13. In contrast, Chapter 11 is about the modeling of several or many such communities.

The DR/DRY community models enable you to estimate and model species richness at both the community and the whole metacommunity level (Brown and Maurer, 1989; Holyoak et al., 2005). A community is an ensemble of species occurring at one site, and a metacommunity is therefore a collection of such communities. In keeping with one of the main themes of this book, which is the analysis of ecological data on distribution and abundance that are *site-structured*, we strictly deal with metacommunities only. However, we quite often use the terms community and metacommunity exchangeably when the meaning should be clear. (Another slight vagueness in this chapter is the use of N to denote the number of species, although sometimes N will be used for local abundance of individuals, as throughout the book. Again, the meaning of N should be clear from the context.)

In a sense, the DR/DRY community models represent the culmination of all the other models for occurrence and abundance in this book: the DR is simply a community site-occupancy model, and the DRY is a community N-mixture model. While the basic HMs in Chapters 6–10 have two levels, one for the true but latent ecological state and another for the measurement error, the basic DR/DRY community models have three levels; we add a third level on top that describes the sampling of each species from the metacommunity. Thus, the community model is a "hypermodel" for abundance or distribution for a collection of species. The parameters for each species are treated as random effects endowed with prior distributions, and the hyperparameters of those priors describe the community.

DR/DRY community models have several key advantages. First, they permit inference at the level of the whole community and at that of each individual species, and second, our usual binomial measurement error model ideally eliminates (or at least reduces) the bias in our inferences due to false-negative measurement errors. In these models, we will encounter data augmentation (DA) in a novel setting that allows us not to estimate how many *individuals* we miss in a population (as in Chapters 7–9), but instead now DA allows us to make a formal guess about how many *individual species* we miss in a metacommunity. In that way, both community models permit inference at three hierarchical levels: metacommunity, community, and individual species. This, combined with the explicit description of the ubiquitous false-negative measurement error, is a unique feature of this modeling framework. We will show how inferences about alpha (local scale), beta (landscape scale), and gamma (macroscale) diversity are naturally made as a function of the estimated presence/absence matrix, where *estimation* means that we have corrected for false-negative measurement errors. Many important generalizations of these models, including the dependence of occurrence or abundance among species (i.e., species interactions), are fairly straightforward, at least conceptually.

Ten years after its initial development, the DR community occupancy model (but not the DRY community abundance model; see Section 11.10) has been applied quite widely and in an increasing number of papers, including the following: Kéry and Royle (2008), Kéry et al. (2008), Kéry and Royle (2009), Russell et al. (2009), Zipkin et al. (2009, 2010, 2012), DeWan and Zipkin (2010), Ruiz-Gutierrez et al. (2010), Holtrop et al. (2010), Kéry (2011a), Ruiz-Gutierrez and Zipkin (2011), Wells et al. (2011), Burton et al. (2012), Dorazio (2012), Jones et al. (2012), Chen et al. (2013), Giovanini et al. (2013), Guzy et al. (2013), Henden et al. (2013), Holt et al. (2013), Hunt et al. (2013), Linden and Roloff (2013), Mattsson et al. (2013), Sauer et al. (2013), Tingley and Beissinger (2013), White et al. (2013a,b), Carrillo-Rubio et al. (2014), Gilroy et al. (2014a,b, 2015), Homyack et al. (2014), Iknayan et al. (2014), Karanth et al. (2014), Kroll et al. (2014), Mata et al. (2014), McManamay et al. (2014), Pacifici et al. (2014), Sanderlin et al. (2014), Higa et al. (2015), Lewis et al. (2015), McNew and Handel (2015), Mihaljevic et al. (2015), Russell et al. (2015) and Sutherland et al. (in review).

Other authors have independently developed community models that are based on the notion of collecting together component models for individual species. Notably, first of all, the work by Gelfand and colleagues (e.g., Gelfand et al., 2005, 2006; Latimer et al., 2006) has important conceptual similarities to the models described in this chapter, though they lack an explicit description of presence/absence measurement error and thus cannot make explicit inferences about the community or estimates of species richness at any scale. Next, the principle of building community or joint species distribution models (JSDMs) from individual-species models is being reinvented quite regularly. Recent instances include Ovaskainen et al. (2010), Ovaskainen and Soininen (2011), Clark et al. (2014), and Pollock et al. (2014). However, we note that these authors all add interspecific dependency of occurrence in their models, which the basic DR/DRY models in this chapter do not contain.

The plan of this chapter is as follows: we use two broad strategies for describing the DR and the DRY models. The first is the simulation of a data set under these models, as we usually do at the start of a chapter. We provide a data simulation function for metacommunity presence/absence and count data, which hopefully provides you with deeper insight into some fundamentals of communities, such as the relationships between features of the individual species and emerging properties of the community, and the relationship between features of the observation process and the observed data. As always, when you really understand the simulation code for a data set under a specific model, you are well prepared for understanding the model as well. In addition, we hope that the simulation function will help you conduct simulation studies for investigation of topics about this type of model, or about communities in general, that are important for your work (for a recent example of such a simulation study see McNew and Handel, 2015). The second main feature of the chapter is that we approach the DR/DRY models by starting at a very basic and not even hierarchical model for a community, and then progressively change that until we are at the community model we desire. We hope that this *model progression* will help you understand our main hierarchical community models and clarify the ways in which they are different from other approaches.

11.2 SIMULATION OF A METACOMMUNITY

We use data simulation to clarify our thinking about two things. First, we emphasize how a metacommunity can be thought of as the result of a superposition of presence/absence or abundance patterns over a number of sites for a whole collection of species. Patterns at community (one site) or metacommunity (multiple sites) levels emerge as a function of community mean and among-species variance of species-specific traits that affect both the occupancy and the detection probability of all occurring species. Second, you will not be surprised that we emphasize the observation process underlying all real-world community data. Therefore, we explicitly describe two processes that underlie an observed presence/absence or abundance pattern in a metacommunity: the first is the *ecological process* governing the true occurrence or abundance, and the second is the *measurement process* (we allow for various patterns in false-negative detection errors).

Our function simComm permits you to simulate metacommunity presence/absence or count data that are constructed by combining these two processes. Beyond any association created by a similar response to covariates, the function assumes that species occur and are detected independently from one another. The default function arguments are the following:

```
simComm(type="det/nondet", nsite=30, nrep=3, nspec=100,
  mean.psi=0.25, sig.lpsi=1, mu.beta.lpsi=0, sig.beta.lpsi=0,
  mean.lambda=2, sig.loglam=1, mu.beta.loglam=1, sig.beta.loglam=1,
  mean.p=0.25, sig.lp=1, mu.beta.lp=0, sig.beta.lp=0, show.plot = TRUE)
```

Function simComm permits generation of replicated presence/absence (for type="det/nondet") or abundance data (for type="counts") with possible false-negative measurement error for a specified number of sites (nsite), replicates (or occasions, nrep), and size of a regional species pool (nspec) under the following models for occupancy or abundance, and detection for site i, replicate j, and species k. We generate the true presence/absence (z_{ik}) or abundance (N_{ik}) of species k ($k = 1...$nspec) at site i ($i = 1...$nsite) as follows. For now, we focus on the default function setting type = "det/nondet"—i.e., on the simulation of presence/absence measurements in a metacommunity.

Here is the model for presence/absence and occupancy probability for species k at site i:

$z_{ik} \sim Bernoulli(\psi_{ik})$ # True presence/absence
$logit(\psi_{ik}) = beta0_k + beta1_k * habitat_i$ # Occupancy probability affected by habitat
$beta0_k \sim Normal(logit(mean.psi), sig.lpsi^2)$ # Species heterogeneity in the intercept
$beta1_k \sim Normal(mu.beta.lpsi, sig.beta.lpsi^2)$ # Species heterogeneity in the slope

Hence, presence/absence z_{ik} of species k at site i is simulated as a Bernoulli trial with occupancy probability ψ_{ik}. A single site covariate, *habitat*, may affect the occupancy probability of a species via a logit-linear model. Both the intercept (the mean occupancy probability at a *habitat* value of zero) and the slope of the relationship are indexed by k; hence, they can differ by species. The average community response is $logit(\psi_i) = logit(mean.psi) + mu.beta.lpsi * habitat_i$, and the magnitude of the among-species variability around that community mean is governed by the standard deviations of the two normal distributions for the intercepts and the slopes. When a standard deviation is set to zero, the metacommunity becomes homogeneous in terms of that parameter. In contrast, when you set the standard deviation to a nonzero value, individual species will differ in their intercepts or their responses to the habitat. Depending on the magnitude of the variability among species, slopes may easily have opposing signs among the species in the community (see below).

All species occur and are detected independently; i.e., after accounting for covariate effects, the presence or the detection of one species does not have any effect on the likelihood of presence or detection of another species. If you want to relax that assumption, you could modify the function such that, for instance, $beta0_k$ is a draw from a multivariate normal distribution with a specified variance-covariance matrix that describes the residual associations among species. Nonzero off-diagonal elements in that matrix specify positive or negative residual association of the species; see Ovaskainen et al. (2010), Ovaskainen and Soininen (2011), and Dorazio and Connor (2014) for examples, as well as Chapter 20 in volume 2.

We generate the observed presence/absence measurements, or detection/nondetection data, (y_{ijk}) of species k at site i during occasion j as follows.

$y_{ijk} \sim Bernoulli(z_{ik} * p_{ijk})$ # Observed detection/nondetection data
$logit(p_{ijk}) = alpha0_k + alpha1_k * wind_{ij}$ # Detection probability affected by *wind*
$alpha0_k \sim Normal(logit(mean.p), sig.lp^2)$ # Species heterogeneity in the intercept
$alpha1_k \sim Normal(mu.beta.lp, sig.beta.lp^2)$ # Species heterogeneity in the slope

Thus, the observed data y_{ijk} are simulated as a Bernoulli trial with a success probability that is the product of the presence/absence of species k at site i (z_{ik}) and detection probability p_{ijk} for species k during occasion j at the site i. Our method of data simulation excludes false-positive errors. Detection probability on the logit scale may be affected by an observation covariate, *wind* speed, and the intercepts and slopes of this relationship for each species (on the logit scale) may again vary according to two independent normal distributions with means and standard deviations that describe the community mean response and the among-species variability in that response, respectively.

With `type="counts"`, abundance data from a metacommunity are simulated under an analogous model:

$N_{ik} \sim Poisson(\lambda_{ik})$ # True abundance
$log(\lambda_{ik}) = beta0_k + beta1_k * habitat_i$ # Expected abundance affected by *habitat*
$y_{ijk} \sim Binomial(N_{ik}, p_{ijk})$ # Observed count data
$logit(p_{ijk}) = alpha0_k + alpha1_k * wind_{ij}$ # Detection probability affected by *wind*

There are again normal prior distributions for all species-specific parameters, with hyperparameters that describe the community to which these species belong. Most of what we say about the simulation of metacommunity presence/absence data (for `type = "det/nondet"`) has a simple analogy when the function is used to simulate metacommunity count data (with `type = "counts"`); therefore, we don't explain the latter specifically. By default, the function generates data where each species has its own constant value of occupancy probability and detection probability, but there are no covariate effects.

```
# Execute function with default arguments
set.seed(1234)
data <- simComm(type="det/nondet", nsite=30, nrep=3, nspec=100,
mean.psi=0.25, sig.lpsi=1, mu.beta.lpsi=0, sig.beta.lpsi=0,
mean.lambda=2, sig.loglam=1, mu.beta.loglam=1, sig.beta.loglam=1,
mean.p=0.25, sig.lp=1, mu.beta.lp=0, sig.beta.lp=0, show.plot = TRUE)
# data <- simComm() # same
```

Executing the function produces two plots that visualize the simulated system. The first plot shows the relationships between occupancy probability and the *habitat* covariate and between detection probability and the *wind* covariate (Figure 11.1). The function's default arguments specify zero effects of both covariates, and hence all lines are horizontal. If you specify nonzero standard deviations for the

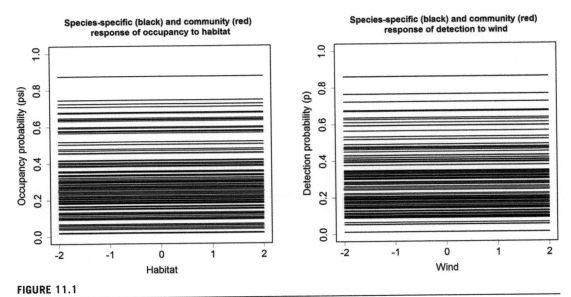

FIGURE 11.1

Species-specific and community-level responses of the probability of occupancy to the *habitat* covariate (left), and of the probability of detection to the *wind* covariate, in data simulated using function simComm with default settings. Black lines show the value of occupancy and detection probability for each of the 100 species and red lines show the community mean response. This is the first visualization produced by function simComm.

two normal distributions that govern species heterogeneity in the two covariate relationships, you will generate nonparallel lines (Figure 11.2), where you can see two important features of a community: (1) the community average of the response to some covariate may be very different from the response of the individual species; and (2) individual species may differ greatly in their response to a covariate. For instance, the community response of occupancy to the habitat is zero (the red line is horizontal), but most individual species are actually responding very clearly to that covariate. Moreover, there are many species with a positive response, but about just as many with a negative response.

The second plot produced when executing function simComm (Figure 11.3) depicts three matrices that show the true presence/absence pattern (z_{ik}) generated (top left), the observed detection frequency (i.e., the number of times that species k is detected at site i; top right), and the sites i where species k occurred but was never detected—that is, the combined presence/absence measurement error (bottom left). The frequency distributions of the true and observed numbers of species occurring per site are depicted in red and blue, respectively (bottom right). Finally, two interesting statistics are shown in the titles of the two plots on the left of this figure: one is the finite-sample number of occurring species (96 in this realization of the process), and the other is the observed number of species (92 in this realization). It is important to recognize that even though we specified 100 species in the regional pool in the area represented by our spatial sample of 30 sites, the 30 sampled sites only contained 96 species and failed to contain the remaining 4 species in the (meta)community. Thus, there is a sense in which there are two values for the "true species richness," one being 100 and the other (here) 96.

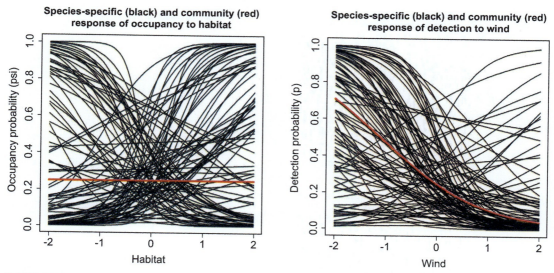

FIGURE 11.2

Species-specific and community-level response of the probability of occupancy to the *habitat* covariate (left) and of the probability of detection to the *wind* covariate in data simulated using function `simComm` with the following arguments: `mu.beta.lpsi=0`, `sig.beta.lpsi=2`, `mu.beta.lp=-1`, `sig.beta.lp=1` (other settings were at default values).

Dupuis et al. (2011) call them the *unconditional* and the *conditional* species richness, where conditional means "conditional on occurrence anywhere in the studied area." You will see below that the finite-sample (or conditional) total number of species (relative to the unconditional total) depends on the number of sites sampled, and on the mean and the among-species variability of occupancy probability.

Apart from the two plots, executing the function produces a large list of output that we here explain interspersed into an example. The first 14 elements simply repeat the function arguments that were operative during the function call. The remaining elements represent things produced by the function.

```
str(data)
List of 26
 $ type           : chr "det/nondet"
 $ nsite          : num 30
 $ nrep           : num 3
 $ nspec          : num 100
 $ mean.psi       : num 0.25
 $ mu.lpsi        : num -1.1
 $ sig.lpsi       : num 1
 $ mu.beta.lpsi   : num 0
 $ sig.beta.lpsi  : num 0
 $ mean.p         : num 0.25
 $ mu.lp          : num -1.1
```

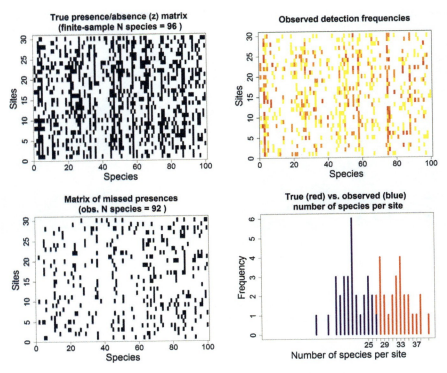

FIGURE 11.3

Top left: true presence and absence (z_{ik}) of species k at site i. Top right: number of times that species k was detected at site i (detection frequency; redder is more). Bottom left: Sites i where species k occurred, but was missed—i.e., never detected (this represents presence/absence measurement error). Bottom right: true (red) and observed (blue) frequency distribution of the number of species per site. This is the second visualization produced by function simComm.

```
 $ sig.lp       : num 1
 $ mu.beta.lp   : num 0
 $ sig.beta.lp  : num 0
# psi[i,k]: occupancy probability for site i and species k
 $ psi          : num [1:30, 1:100] 0.216 0.216 0.216 0.216 0.216 ...
  ..- attr(*, "dimnames")=List of 2
  .. ..$ : chr [1:30] "site1" "site2" "site3" "site4" ...
  .. ..$ : chr [1:100] "sp1" "sp2" "sp3" "sp4" ...
# p[i,j,k]: detection probability for site i, occasion j and species k
 $ p            : num  [1:30, 1:3, 1:100] 0.449 0.449 0.449 0.449 0.449 ...
  ..- attr(*, "dimnames")=List of 3
  .. ..$ : chr [1:30] "site1" "site2" "site3" "site4" ...
  .. ..$ : chr [1:3] "rep1" "rep2" "rep3"
  .. ..$ : chr [1:100] "sp1" "sp2" "sp3" "sp4" ...
```

```
# z[i,k]: true presence/absence for site i and species k
 $ z            : int [1:30, 1:100] 0 0 0 1 0 0 0 1 0 1 ...
 ..- attr(*, "dimnames")=List of 2
 .. ..$ : chr [1:30] "site1" "site2" "site3" "site4" ...
 .. ..$ : chr [1:100] "sp1" "sp2" "sp3" "sp4" ...
# z.obs[i,k]: observed presence/absence for site i and species k (i.e., an indicator for
whether species k was ever detected at site i during J surveys)
 $ z.obs        : int [1:30, 1:100] 0 0 0 1 0 0 0 1 0 1 ...
 ..- attr(*, "dimnames")=List of 2
 .. ..$ : chr [1:30] "site1" "site2" "site3" "site4" ...
 .. ..$ : chr [1:100] "sp1" "sp2" "sp3" "sp4" ...
# y[i,j,k]: detection/nondetection data for site i, occasion j and species k (for all 100
species)
 $ y.all        : int [1:30, 1:3, 1:100] 0 0 0 0 0 0 0 1 0 1 ...
 ..- attr(*, "dimnames")=List of 3
 .. ..$ : chr [1:30] "site1" "site2" "site3" "site4" ...
 .. ..$ : chr [1:3] "rep1" "rep2" "rep3"
 .. ..$ : chr [1:100] "sp1" "sp2" "sp3" "sp4" ...
# y[i,j,k]: detection/nondetection data for site i, occasion j and species k (only for the 92
species that were detected at least once)
 $ y.obs        : int [1:30, 1:3, 1:92] 0 0 0 0 0 0 0 1 0 1 ...
 ..- attr(*, "dimnames")=List of 3
 .. ..$ : chr [1:30] "site1" "site2" "site3" "site4" ...
 .. ..$ : chr [1:3] "rep1" "rep2" "rep3"
 .. ..$ : chr [1:92] "sp1" "sp2" "sp3" "sp4" ...
# detection frequency for all 100 species
 $ y.sum.all    : int [1:30, 1:100] 0 0 1 0 0 0 1 0 1 ...
 ..- attr(*, "dimnames")=List of 2
 .. ..$ : chr [1:30] "site1" "site2" "site3" "site4" ...
 .. ..$ : chr [1:100] "sp1" "sp2" "sp3" "sp4" ...
# detection frequency only for the 92 species detected at least once
 $ y.sum.obs    : int [1:30, 1:92] 0 0 1 0 0 0 1 0 1 ...
 ..- attr(*, "dimnames")=List of 2
 .. ..$ : chr [1:30] "site1" "site2" "site3" "site4" ...
 .. ..$ : chr [1:92] "sp1" "sp2" "sp3" "sp4" ...
# Finite sample (or conditional) species richness: number of species that occur in the 30
sampled sites
 $ Ntotal.fs    : int 96
# The observed version of the same (difference being due to imperfect detection)
 $ Ntotal.obs   : int 92
# The true number of species occurring at each of the 30 sites
 $ S.true       : Named int [1:30] 27 26 30 25 35 33 33 33 34 38 ...
 ..- attr(*, "names")= chr [1:30] "site1" "site2" "site3" "site4" ...
# The number of species observed at each of the 30 sites
 $ S.obs        : Named int [1:30] 17 17 16 11 21 20 19 25 20 24 ...
 ..- attr(*, "names")= chr [1:30] "site1" "site2" "site3" "site4" ...
```

As always, we would like to warmly encourage you to *play community*—i.e., execute this function many times, with changed function arguments, and observe how the output is affected by your choices and be astonished at how one realization from a given stochastic process can differ strikingly from another one. This will be a big help for your general understanding of some fundamentals of community ecology as well as of their measurement when the latter is contaminated by false-negative errors. Here we show a couple of settings that may be interesting. We note that you must have >1 site and replicate and usually >2 species, otherwise the function breaks. If you want to simulate data for a single site, replicate, or species, you must run the function for multiple sites, replicates, or species and then pull out a subset of the data representing a single site, replicate, or species. Don't specify no occurring species (mean.psi = 0) or invisible species (mean.p = 0)—both cause the function to crash—but the opposite (all species occurring everywhere, mean.psi = 1, or perfect detectability, mean.p = 1) works, though the plots don't look very nice anymore.

```
# Some possibly interesting settings of the function
data <- simComm(nsite = 267, nspec = 190, mean.psi = 0.25, sig.lpsi = 2,
mean.p = 0.12, sig.lp = 2) # similar to Swiss MHB; see Section 11.3
data <- simComm(mean.psi = 1)          # all species occur at every site
data <- simComm(mean.p = 1)            # no measurement error (perfect detection)

# Effect of spatial sample size (nsite) on species richness in sample (Ntotal.fs)
data <- simComm(nsite=50, nspec = 200) # 1-3 are usually missed in sample
data <- simComm(nsite=30, nspec = 200) # 4-6 usually missed
data <- simComm(nsite=10, nspec = 200) # around 30 typically missed
```

You should always repeatedly execute the function to study the average and the variation in the behavior of some feature in the metacommunity or of their measurement. Here is a simple simulation that tells you how many species in a metacommunity of 200 species will typically be missed when sampling 10 sites (using default function arguments otherwise): the answer is about 25 to 30 (under the assumptions about community assembly embodied by the function).

```
# Check for frequentist characteristics of such statistics
temp <- rep(NA, 100)
for(i in 1:100){
    cat("\nSimrep", i)
    temp[i] <- simComm(nsite=10, nspec = 200, show.plot = F)$Ntotal.fs
}
hist(200-temp, breaks = 30, main = "Number of species in the metacommunity
\nthat do not occur in the 10 sampled sites", col = "gold")
```

We conduct two further small simulations to illustrate an important feature of a metacommunity that holds regardless of any measurement error, and then a feature of presence/absence *measurements* of a metacommunity (i.e., which is due to measurement error when studying a metacommunity). In both, we simulate metacommunities composed of 200 species that are sampled at 50 sites. With the default function arguments, about 95% of the realizations will have at least 196 species actually occurring at the 50 studied sites. In the first simulation, we look at the effects of community mean occupancy probability (ψ) and among-species variability in logit(ψ) on the proportion of these 200 species in the metacommunity that actually occur in the sampled 50 sites. We generate 50 data sets for each combination of two factors with 10 levels, the first being the community mean occupancy probability (mean.psi, which we vary between 0.01 and 0.25) and the second being the interspecific variability of occupancy (sig.lpsi, which we vary between 0.1 and 5).

```r
# Simulation 1: effects of psi and sd(logit(psi)) on number of species actually occurring in
the 50 sampled sites
simrep <- 50             # Run 50 simulation reps
mpsi <- seq(0.01, 0.25,,10)
slpsi <- seq(0.1, 5,,10)
results1 <- array(NA, dim = c(10, 10, simrep))
for(i in 1:10){    # Loop over levels of factor mean.psi (mpsi)
  for(j in 1:10){  # Loop over levels of factor sig.lpsi (slpsi)
    for(k in 1:simrep){
      cat("\nDim 1:",i, ", Dim 2:", j, ", Simrep", k)
      tmp <- simComm(nsite=50, nspec = 200, show.plot = F, mean.psi = mpsi[i],
         sig.lpsi = slpsi[j])
      results1[i,j,k] <- tmp$Ntotal.fs
    }
  }
}
```

In the second simulation, we look at the effects of community mean detection probability (*p*) and among-species variability in logit(*p*) on the proportion of species occurring in the 50 sample sites that are detected at least once—i.e., on the converse of the measurement error in conditional species richness (Dupuis et al., 2011). We again generate 50 data sets for each combination of two factors with 10 levels, the first being the community mean detection probability (mean.p, which we vary between 0.01 and 0.25), and the second being the interspecific variability of detection (sig.lp, which we vary between 0.1 and 5).

```r
# Simulation 2: effects of p and sd(logit(p)) on the proportion of the species occurring in
the 50 sampled sites that are detected at least once
simrep <- 50             # Run 50 simulation reps again
mp <- seq(0.01, 0.25,,10)
slp <- seq(0.1, 5,,10)
results2 <- array(NA, dim = c(10, 10, simrep, 2))
for(i in 1:10){    # Loop over levels of factor mean.p (mp)
  for(j in 1:10){  # Loop over levels of factor sig.lp (slp)
    for(k in 1:simrep){
      cat("\nDim 1:",i, ", Dim 2:", j, ", Simrep", k)
      tmp <- simComm(nsite=50, nspec = 200, show.plot = F, mean.p = mp[i],
         sig.lp = slp[j])
      results2[i,j,k,] <- c(tmp$Ntotal.fs, tmp$Ntotal.obs)
    }
  }
}
```

We visualize the results in two image plots that show the mean (over the 50 simulated data sets) as a function of the 10 levels of each simulation factor (Figure 11.4).

```r
# Plot these two prediction matrices
par(mfrow = c(1, 2), mar = c(5,5,2,2), cex.lab = 1.5, cex.axis = 1.5)
mapPalette <- colorRampPalette(c("grey", "yellow", "orange", "red"))
```

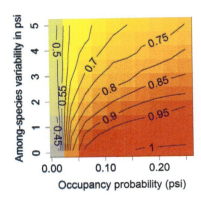 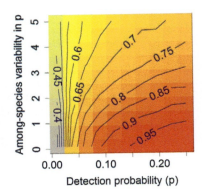

FIGURE 11.4

Left: Effects of community mean occupancy probability (ψ) and among-species variability in occupancy probability (logit(ψ)) on the proportion of 200 species in the wider community that actually *occur* somewhere within the 50 sampled sites (this is a true pattern in the ecological process). Right: Effects of community mean detection probability (p) and among-species variability in detection probability (logit(p)) on the proportion of *detected* species among those that occur in the sampled 50 sites (this is a pattern in the measurement error process). Each cell is the mean of 50 simulated data sets.

```
# Plot proportion of species occurring in sampled sites (Fig. 11-4 left)
z1 <- apply(results1/200, c(1,2), mean)   # Prop species occurring
image(x=mpsi, y=slpsi, z=z1, col = mapPalette(100), axes = T, xlab = "Occupancy probability
(psi)", ylab = "Among-species variability in psi")
contour(x=mpsi, y=slpsi, z=z1, add = T, col = "blue", labcex = 1.5, lwd = 1.5)

# Plot proportion of species detected in sampled sites (Fig. 11-4 right)
z2 <- apply(results2[,,,2] / results2[,,,1], c(1,2), mean)
image(x=mp, y=slp, z=z2, col = mapPalette(100), axes = T, xlab = "Detection probability
(p)", ylab = "Among-species variability in p")
contour(x=mp, y=slp, z=z2, add = T, col = "blue", labcex = 1.5, lwd = 1.5)
```

This concludes our introduction to metacommunity studies based on the occurrence of individual species. We have seen that the average metacommunity may behave quite differently from the behaviour of the individual species. We have also seen that we can distinguish two characterizations of the size of a metacommunity: one unconditional, which is the number of species that *would* be sampled at saturation effort in a region (in terms of the spatial replicates, the number of sites), and the other the realized or conditional on occurrence in at least one of the sampled sites (Dupuis et al., 2011). The former is also the asymptote of a species accumulation curve (Dorazio et al., 2006); see also Section 11.9. Whether one of them is relevant for you, and if so which, is something each ecologist must decide independently. In addition to metacommunity size, we have the size of local communities (local species richness); see Figure 11.3 (bottom left). Other terms for these descriptors of the richness of a metacommunity include gamma and alpha diversity (Whittaker et al., 2001), where the term beta diversity is used to characterize the dissimilarity of the communities among different sites—i.e., the spatial variability in species occurrence within a metacommunity.

The other important topic in the analysis of a metacommunity that we have emphasized is that of the measurement error induced by imperfect detection. Measurement error in community studies may bias *any* descriptor of a community or metacommunity, including all three types of diversity and community and individual-species responses to covariates.

11.3 METACOMMUNITY DATA FROM THE SWISS BREEDING BIRD SURVEY MHB

To illustrate HMs for communities we use data collected in the Swiss breeding bird survey MHB (Monitoring Häufige Brutvögel; Schmid et al., 2004). The most complex models that we will fit will be constructed exactly analogous to the concept of our data simulation in Section 11.2. That is, we will assume independence among species in both occurrence and detection. Independence of occurrence may seem like a strong assumption, but whether it holds is likely to be scale-dependent: for relatively large sites such as the 1-km^2 quadrats in the Swiss breeding bird survey MHB, it appears unlikely that the presence or absence or the abundance of one species has any effect on that of another beyond what we can explain by the habitat (and it is this *conditional independence* that we mean when we make this assumption).

In the MHB, all individuals of all breeding bird species are surveyed using territory mapping during three surveys in the breeding season (approx. mid-April through the end of June) in 267 1-km^2 sampling units laid out as a grid over Switzerland. Only two surveys are conducted in quadrats above the tree line. Surveys are conducted along irregular and quadrat-specific transects of length ranging from 1.2 to 9.4 km (mean 5.1). Thus, in the MHB counts and territory ID data are collected, but we will first reduce the information content of the original data and simply model the species-specific detection/nondetection data—i.e., the replicated presence/absence measurements. In Section 11.10 we will model the counts directly. We chose data from 2014, when 266 quadrats were surveyed and 145 species detected. The variables in our data set are explained in the following interspersed R output.

```
# Read in data set and look at data first
data <- read.csv2("MHB_2014.csv", header = T, sep = ";", dec= ".")
str(data)      # (Later read in as a system file of R package AHM)

'data.frame':   42186 obs. of 24 variables:
# record number
 $ id           : int  1 2 3 4 5 6 7 8 9 10 ...
# species ID: numeric, English name, Latin abbreviation and Latin name
 $ specid       : int  50 50 50 50 50 50 50 50 50 50 ...
 $ engname      : Factor w/ 158 levels "Alpine Accentor",..: 98 98 98 98 98 98 98 ...
 $ latabb       : Factor w/ 158 levels "ACCGEN","ACCNIS",..: 145 145 145 145 145 145 ...
 $ latname      : Factor w/ 158 levels "Accipiter gentilis",..: 146 146 146 146 146 146 ...
# species traits: body length (cm), wing span (cm), body mass (g)
 $ body.length  : int  27 27 27 27 27 27 27 27 27 27 ...
 $ wing.span    : int  43 43 43 43 43 43 43 43 43 43 ...
 $ body.mass    : int  150 150 150 150 150 150 150 150 150 150 ...
# x and y coordinates of sample 1km² quadrat
 $ coordx       : int  922942 928942 928942 934942 934942 946942 946942 952942 ...
 $ coordy       : int  63276 79276 103276 95276 111276 95276 111276 119276 111276 ...
```

```
# length of survey route and number of surveys per breeding season
 $ rlength    : num  6.4 5.5 4.3 4.5 5.4 3.6 3.9 6.1 5.8 4.5 ...
 $ nsurvey    : int  3 3 3 3 3 3 3 3 3 3 ...
# mean quadrat elevation (in metres) and forest cover (as a percentage)
 $ elev       : int  450 450 1050 950 1150 550 750 650 550 550 ...
 $ forest     : int  3 21 32 9 35 2 6 60 5 13 ...
# number of birds counted for 1st through 3rd survey
 $ count141   : int  0 0 0 0 0 0 0 0 0 0 ...
 $ count142   : int  0 0 0 0 0 0 0 0 0 0 ...
 $ count143   : int  0 0 0 0 0 0 0 0 0 0 ...
# survey date (1 = 1 April 2014)
 $ date141    : int  21 26 25 40 16 52 18 17 18 25 ...
 $ date142    : int  52 47 52 55 38 61 40 39 45 50 ...
 $ date143    : int  70 59 73 65 62 69 60 61 59 76 ...
# survey duration (in minutes)
 $ dur141     : int  215 195 210 310 240 180 180 195 190 195 ...
 $ dur142     : int  220 175 270 300 240 145 195 225 180 203 ...
 $ dur143     : int  240 185 210 285 210 140 180 210 205 215 ...
# id for surveyor who did the 3 surveys in 2014
 $ obs14      : int  386 147 77 293 77 361 77 77 179 165 ...

# Create various species lists (based on English names and systematic order)
(species.list <- levels(data$engname))   # alphabetic list
(spec.name.list <- tapply(data$specid, data$engname, mean))  # species ID
(spec.id.list <- unique(data$specid))    # ID list
(ordered.spec.name.list <- spec.name.list[order(spec.name.list)])  # ID-order list
```

We first grab the three counts in an array and convert them into simple replicated presence/absence measurements or detection/nondetection data. Note how this illustrates once again the (one-way) deterministic relationship between these two types of data.

```
COUNTS <- cbind(data$count141, data$count142, data$count143)  # Counts 2014
DET <- COUNTS
DET[DET > 1] <- 1          # now turned into detection/nondetection data
```

In BUGS, it is convenient to fit the model to the data formatted in a three-dimensional array, because the model description is much neater if we can use the dimensions of a multidimensional array to convey the information about factors such as site, species, and replicate survey. We do this next. There are different ways in which you may organize this array. We organize it such that the third dimension of the array indexes species, and we name it using the appropriate species list. This array is a direct generalization of the typical data array for single-species occupancy models, where we usually model a site-by-replicate matrix of species detections (Chapter 10). Here, we stack the species-specific matrices, and the species represent the slices of the resulting 3-D array.

```
# Put detection data into 3D array: site x rep x species
nsite <- 267                      # number of sites in Swiss MHB
nrep <- 3                         # number of replicate surveys per season
nspec <- length(species.list)     # 158 species occur in the 2014 data
y <- array(NA, dim = c(nsite, nrep, nspec))
```

11.3 METACOMMUNITY DATA FROM THE SWISS BREEDING BIRD

```
for(i in 1:nspec){
  y[,,i] <- DET[((i-1)*nsite+1):(i*nsite),]
}
dimnames(y) <- list(NULL, NULL, names(ordered.spec.name.list))

# Check data for one species, here chaffinch, and pull them out from 3D array
which(names(ordered.spec.name.list) == "Common Chaffinch")
(tmp <- y[,,which(names(ordered.spec.name.list) == "Common Chaffinch")])

# Frequency distribution of number of surveys per site in chosen year
table(nsurveys <- apply(y[,,1], 1, function(x) sum(!is.na(x))))
  0   2   3
  1  47 219

# Which site has NA data in 2014 ?
(NAsites <- which(nsurveys == 0) )
[1] 30
```

Hence, 219 sites were surveyed three times, 47 twice, and one site was not surveyed in 2014. We next look at the observed number of occupied sites among the 266 with nonmissing values in 2014.

```
# Observed number of occupied sites
tmp <- apply(y, c(1,3), max, na.rm = TRUE)
tmp[tmp == -Inf] <- NA    # Only 266 quadrats surveyed in 2014
sort(obs.occ <- apply(tmp, 2, sum, na.rm = TRUE))
           Little Bittern      Little Ringed Plover                   Barn Owl
                        0                         0                          0
        Eurasian Eagle-Owl                Little Owl             Eurasian Hoopoe
                        0                         0                          0
   White-backed Woodpecker                Bluethroat         Great Reed Warbler
                        0                         0                          0
     Red-breasted Flycatcher           Woodchat Shrike            Common Rosefinch
                        0                         0                          0
            Ortolan Bunting               White Stork                Greylag Goose
                        0                         1                          1
[ ... Output truncated ..... ]
             Mistle Thrush                 Great Tit                    Coal Tit
                      190                       198                         199
              Song Thrush          Common Chiffchaff           Eurasian Blackcap
                      210                       213                         217
            European Robin           Common Blackbird                 Winter Wren
                      219                       223                         234
           Common Chaffinch             Black Redstart
                      243                       244

# Plot species 'occurrence frequency' distribution (not shown)
plot(sort(obs.occ), xlab = "Species number", ylab = "Number of quads with detections")
```

```
# Drop data from species that were not observed in 2014
toss.out <- which(obs.occ == 0)
y <- y[,,-toss.out]
obs.occ <- obs.occ[-toss.out]
# Redefine nspec as the observed # species: 145
( nspec <- dim(y)[3] )
```

We are now left with binary detection/nondetection data from 267 sites (of which one has missing data) and two to three replicate surveys for 145 species observed during the 2014 surveys. We could toss out the data from site 30 with the missing data, but we keep the site to illustrate the ability of some models to estimate things for "new" or unsurveyed sites.

```
str(y)
 num [1:267, 1:3, 1:145] 0 0 0 0 0 0 0 0 0 0 ...
 - attr(*, "dimnames")=List of 3
  ..$ : NULL
  ..$ : NULL
  ..$ : chr [1:145] "Little Grebe" "Great Crested Grebe" "Grey Heron" "White Stork" ...
```

We continue our descriptive overview of the data and compute the observed number of species per quadrat ("observed species richness").

```
# Get observed number of species per site
tmp <- apply(y, c(1,3), max, na.rm = TRUE)
tmp[tmp == "-Inf"] <- NA
sort(C <- apply(tmp, 1, sum))   # Compute and print sorted species counts
   [1]  6  6  6  7  7  7  8  8  9  9  9 10 10 10 11 12 12 12 13 14 14 14 15 15 15
  [26] 15 16 17 17 18 18 19 19 19 19 19 20 21 24 24 24 24 25 25 25 25 25 25 25 26
  [51] 26 26 26 27 27 27 28 28 28 28 29 29 29 29 29 29 30 30 30 30 30 30 30 30 30
  [76] 30 31 31 31 31 31 31 32 32 32 32 32 32 32 32 32 32 32 33 33 33 33 33 33 33
 [101] 33 33 33 33 33 34 34 34 34 34 34 34 34 34 34 34 34 34 35 35 35 35 35 35 35
 [126] 35 35 35 35 36 36 36 36 36 36 37 37 37 37 37 37 37 37 37 37 38 38 38 38 38
 [151] 38 38 38 38 38 38 38 38 39 39 39 39 39 39 39 39 39 39 39 39 39
 [176] 40 40 40 40 40 40 40 40 40 40 40 41 41 41 41 41 41 41 41 41 41 42 42 42
 [201] 42 42 42 42 42 42 42 42 43 43 43 43 43 43 43 43 43 44 44 44 44 44 44 44
 [226] 44 44 44 44 44 45 45 45 45 45 45 45 45 45 45 46 46 46 46 46 46 46 47 47 47
 [251] 47 47 47 47 48 48 48 48 48 48 49 50 51 51 56 59
```

So the *observed* species richness, or *observed* community size, in 266 surveyed 1-km^2 quadrats in Switzerland varied from 6 to 59, with an average of 34. Let's plot that (Figure 11.5).

```
plot(table(C), xlim = c(0, 60), xlab = "Observed number of species", ylab = "Number of quadrats", frame = F)
abline(v = mean(C, na.rm = TRUE), col = "blue", lwd = 3)
```

11.4 OVERVIEW OF SOME MODELS FOR METACOMMUNITIES

We use the MHB 2014 data set to fit a series of models for species richness and other features of a metacommunity. These models will form a progression of 11 models that range from very simplistic to increasingly realistic/mechanistic; see overview in Table 11.1. We believe that such a progression will help you to both *understand* the differences between different models for communities and better grasp

11.4 OVERVIEW OF SOME MODELS FOR METACOMMUNITIES

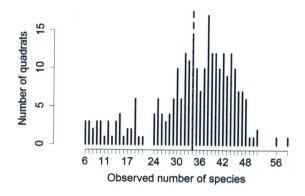

FIGURE 11.5

Frequency distribution of the observed number of breeding bird species per 1-km² quadrat (that is, observed species richness) in the Swiss MHB survey in 2014 (n = 267 quadrats, with data from one quadrat missing). Blue line shows the mean of 34 species.

the difficulties when fitting them in BUGS. The original MHB data can be denoted y_{ijk} and contain the count of species k ($k = 1...145$) at site i ($i = 1...267$) during replicate survey (or occasion) j ($j = 1...3$). In models 1–10, we quantize these data to become binary detection/nondetection data, so that y_{ijk} contains the detection (1) or nondetection (0) of species k at site i during occasion j. In addition, in models 1–4 we summarize these data one more step and treat as a response the number of detected species. In model 11 we will model the counts directly.

The 11 different modeling frameworks can be described as follows:

1. Our simplest models (1–3) for the community consist of regression models that relate the *number of observed species* to covariates. Two drawbacks of such an analysis are that species richness measurement error is not allowed for, and that the species identities are lost—i.e., individual species are not distinguished across sites nor even across replicate surveys of each site.
2. A slightly more advanced approach (model 4) to community modeling is the adoption of a straightforward *N*-mixture model for the replicated counts of observed species (model 4). Under this approach, imperfect detection of species is accounted for, but species heterogeneity in detection cannot be modeled apart from differences among sites and surveys, leading to underestimation of local species richness due to the second law of capture-recapture (which is that unmodelled detection heterogeneity induces a negative bias in abundance and occupancy estimators). Furthermore, species identities are again not retained. In terms of the use of information about species identities, there is also an intermediate model between the N-mixture and the occupancy models (in 3., below): the multinomial mixture model (see Chapter 7). When applied to *species* detection/nondetection data, it would only ignore species identity *across* sites, but retain species identity *within* a site across replicate surveys. Indeed, we could use unmarked to fit a multinomial mixture model of the M_h type to estimate site-specific species richness in a heterogeneity model that would allow each species to have its own detection probability, exactly as shown in Section 7.8.3 for site-stratified estimation of population abundance. One interesting side effect of this formulation of the species richness estimation problem would be that we could then directly model local species richness as a function of covariates. See Exercise 2 for an example of this idea applied to the MHB data set using both unmarked and BUGS.

Table 11.1 Overview of hierarchical (and some "flat") models for (meta)community inference in this chapter.

Nr.	Model Name/ Description	Modeled State	Inference on Community	Inference on Individual Species	Serial Autocorrelation	Measurement Error Model (Detection Probability)	Allows Detection Heterogeneity	Inference About Unseen Species	Inference About Community Composition
1	Poisson GLM for maximum species count	Species count	yes	no	NA	no, only covs.	NA	no	no
2	Poisson GLM for replicated species counts	Species count	yes	no	no	only covs.	no	no	no
3	Random-effects Poisson GLM for replicated species counts	Species count	yes	no	yes	only covs.	no	no	no
4	N-mixture model for replicated species counts	Species count	yes	no	yes	yes	no	no	no
5	Fixed-effects community occupancy model	Pres/Abs	yes	yes	yes	yes	yes	no	yes
6	Random-effects community occupancy model with (ψ, p) correlation	Pres/Abs	yes	yes	yes	yes	yes	no	yes
7	Random-effects community occupancy model with categorical covariate to explain species effects (simple DR model without DA)	Pres/Abs	yes	yes	yes	yes	yes	no	yes

8	Random-effects community occupancy model with continuous covariate to explain species effects (simple DR model, no DA)	Pres/Abs	yes	yes	yes	yes	no	yes
9	Full DR community occupancy model with DA, no covariates	Pres/Abs	yes	yes	yes	yes	yes	yes
10	Full DR community occupancy model with DA and with covariates	Pres/Abs	yes	yes	yes	yes	yes	yes
11	DRY community N-mixture model with covariates	Abund.	yes	yes	yes	yes	yes[a]	yes

[a]*The DRY community N-mixture model (model 11) can be specified with or without DA.*

3. Model 5 is a straightforward site-occupancy model with one component model per observed species. Hence, *species identity is retained*, so the model accommodates features of the individual member species of a (meta)community as well as emerging characteristics of the metacommunity. Furthermore, heterogeneity among species in occurrence and detection can be fully accounted for. The simplest such model consists of fitting *separate occupancy model parameters* to each observed species, but inside a single model. No relationship among species will be imposed on the parameters. In ANOVA terms, species will be treated as fixed effects; the estimates for one species will not be affected by the data from another species. This analysis allows one to estimate (correcting for imperfect detection) the number of species occurring at each site among the total number of species ever detected in the metacommunity, but not the total number of species occurring in the entire metacommunity, because some species may have been missed everywhere. This is the approach taken in calculating the popular wildlife picture index (O'Brien et al., 2010).

4. In models 6–8, we treat each species as a random sample from the studied community; i.e., each species is assigned a random effect. These analyses allow us to *formally estimate characteristics of the observed community*—e.g., the mean probability of occupancy or detection, or the mean response of occupancy probability to a covariate for some environmental conditions. This approach also allows us to estimate the number of species occurring at each site among all the species that were ever detected anywhere in the metacommunity, but not the total size of the metacommunity. That is, no inference is made about those species that were never observed anywhere. These models are quite similar to the models developed by Gelfand et al. (2005, 2006) and Latimer et al. (2006).

5. The final two models for detection/nondetection data (models 9 and 10) accommodate the fact that some species in the metacommunity may never be observed anywhere at all. We can extend the random-effects multispecies model to those unseen species in an attempt to make an inference about the entire community. This analysis uses parameter-expanded DA (Tanner and Wong, 1987; Royle et al., 2007a; Royle and Dorazio, 2012)—i.e., the fitting of a more complex HM (with one additional hierarchical layer) to a modified data set, which includes an added portion of data to accommodate potential unseen species. This is the full Dorazio/Royle (DR) community occupancy model with data-augmentation (Dorazio and Royle, 2005; Dorazio et al., 2006).

6. The final model, 11, is the abundance version of the DR community, community N-mixture, or Dorazio/Royle/Yamaura (DRY) model (Yamaura et al., 2012).

In community occupancy models, in the absence of covariates that vary by replicate survey j, it is convenient to aggregate binary detection/nondetection data y_{ijk} into site- and occasion-specific counts by summing over replicates, and then model detection frequency $ysum_{ik}$—i.e., the number of detections of species k at site i—as a binomial random variable with a binomial index given by the number of surveys (which may or may not be the same for all sites and/or species). Binomial versions of the models are computationally much more efficient to analyze in BUGS.

Importantly, neither the community occupancy nor the community N-mixture model contains a structural parameter for species richness, since they model the occurrence or abundance of each individual species in the community. Species richness is then a derived quantity based on the occurrence of individual species. One simple way to model species richness as a function of other spatially or temporally indexed covariates is to do a two-step analysis and plug species-richness estimates into a regression analysis to model the effects of those covariates (e.g., see Tingley and Beissinger, 2013), but in the second analysis you must account for the estimation uncertainty coming from the first analysis. Section 11.6.4 gives an example of how to do this in the context of estimating an elevation profile of avian species richness in Switzerland. A more formal way of modeling species richness is by adopting a logit-linear model for the data augmentation parameter omega in Section 11.7 (Sutherland et al., in review) or else by the construction shown in Section 7.8.4. We may show this in volume 2.

11.5 COMMUNITY MODELS THAT IGNORE SPECIES IDENTITY

We start by fitting a few fairly simplistic models to the Swiss MHB 2014 data set that focus on spatial or spatiotemporal patterns of species richness only. These models directly describe the number of observed species, but do not keep track of *which* species was observed where and when. That is, these models do not keep track of species identity across replicate surveys at each site and especially, they don't keep track of species identity across replicate sites. Thus, any inferences about individual species are impossible, and at the community level, we risk missing relationships with environmental covariates if a similar number of species show a negative and a positive response (see Figure 11.2, left). Finally, accommodation of false-negative measurement error of species richness is only possible in a very rudimentary way (and in practice is hardly ever done in the type of models in this section).

11.5.1 SIMPLE POISSON REGRESSION FOR THE OBSERVED COMMUNITY SIZE

Probably the most common approach to inference about species richness in much of current ecology would be to fit some kind of regression model to the observed numbers of species. Here, we relate this number to elevation (linear and squared) and forest cover (linear). In the simplest approach with this model, we might discard the information coming from repeated measurements of presence/absence and simply model the total number of species detected across all survey occasions combined. If we want to account for any effects of survey date or survey duration, we then have to summarize these over the surveys, typically by taking the mean for each site (see end of this section for a version of this model where we don't need to aggregate the data). We fit the following model to the observed number of species C at site i:

$$C_i \sim Poisson(\lambda_i)$$
$$\log(\lambda_i) = \gamma_0 + \gamma_1 * elev_i + \gamma_2 * elev_i^2 + \gamma_3 * forest_i +$$
$$\gamma_4 * mean.date_i + \gamma_5 * mean.date_i^2 + \gamma_6 * mean.duration_i$$

Presumably, the first three covariates would be assumed to act on the true species richness at each site, while the last three might be seen to adjust for seasonal and survey-effort-related differences in the detectability of the average species. However, there is no way to explicitly specify such mechanisms, since in this model there is no distinction between a true ecological state (referring to the true number of species at each site) and the measurement process that relates such a state to our observed data which is only an index to true species richness. To emphasize that we don't know the nature of these effects in this model, we call the coefficients `gamma` rather than `beta` or `alpha` as we do elsewhere throughout the book to denote covariate effects for the true state and its measurement, respectively.

```
# Get covariates and standardize them
# Quadrat elevation and forest cover
orig.ele <- data$elev[1:nsite]
(mean.ele <- mean(orig.ele, na.rm = TRUE))
(sd.ele <- sd(orig.ele, na.rm = TRUE))
ele <- (orig.ele - mean.ele) / sd.ele
orig.forest <- data$forest[1:nsite]
```

```
(mean.forest <- mean(orig.forest, na.rm = TRUE))
(sd.forest <- sd(orig.forest, na.rm = TRUE))
forest <- (orig.forest - mean.forest) / sd.forest

# Average date and duration of survey
tmp <- cbind(data$date141, data$date142, data$date143)[1:nsite,]
orig.mdate <- apply(tmp, 1, mean, na.rm = TRUE)
(mean.mdate <- mean(orig.mdate[-NAsites]))  # drop unsurveyed site
(sd.mdate <- sd(orig.mdate[-NAsites]))
mdate <- (orig.mdate - mean.mdate) / sd.mdate
mdate[NAsites] <- 0              # impute mean for missing

tmp <- cbind(data$dur141, data$dur142, data$dur143)[1:nsite,]
orig.mdur <- apply(tmp, 1, mean, na.rm = TRUE)
(mean.mdur <- mean(orig.mdur[-NAsites]))
(sd.mdur <- sd(orig.mdur[-NAsites]))
mdur <- (orig.mdur - mean.mdur) / sd.mdur
mdur[NAsites] <- 0               # impute mean for missing

# Bundle data and summarize input data for BUGS
str( win.data <- list(C = C, nsite = length(C), ele = ele, forest = forest, mdate = mdate,
mdur = mdur) )
List of 6
 $ C      : num [1:267] 30 32 51 40 44 41 34 35 42 35 ...
 $ nsite  : int 267
 $ ele    : num [1:267] -1.1539 -1.1539 -0.2175 -0.3735 -0.0614 ...
 $ forest : num [1:267] -1.1471 -0.4967 -0.0992 -0.9303 0.0092 ...
 $ mdate  : num [1:267] -0.3814 -0.6011 -0.2415 -0.0418 -0.9207 ...
 $ mdur   : num [1:267] -0.274 -0.994 -0.184 1.047 -0.184 ...

# Specify model in BUGS language
sink("model1.txt")
cat("
model {

# Priors
gamma0 ~ dnorm(0, 0.001)              # Regression intercept
for(v in 1:6){                        # Loop over regression coef's
   gamma[v] ~ dnorm(0, 0.001)
}

# Likelihood for Poisson GLM
for (i in 1:nsite){
   C[i] ~ dpois(lambda[i])
   log(lambda[i]) <- gamma0 + gamma[1] * ele[i] + gamma[2] * pow(ele[i],2) +
gamma[3] * forest[i] + gamma[4] * mdate[i] + gamma[5] * pow(mdate[i],2) +
gamma[6] * mdur[i]
}
}
",fill = TRUE)
sink()
```

11.5 COMMUNITY MODELS THAT IGNORE SPECIES IDENTITY

```
# Initial values
inits <- function() list(gamma0 = rnorm(1), gamma = rnorm(6))

# Parameters monitored
params <- c("gamma0", "gamma")

# MCMC settings
ni <- 6000 ; nt <- 4 ; nb <- 2000 ; nc <- 3

# Call WinBUGS from R (ART <1 min)
out1 <- bugs(win.data, inits, params, "model1.txt", n.chains = nc,
n.thin = nt, n.iter = ni, n.burnin = nb, debug = TRUE, bugs.directory = bugs.dir,
working.directory = getwd())

# Call JAGS from R (ART <1 min), check convergence and summarize posteriors
library(jagsUI)
out1J <- jags(win.data, inits, params, "model1.txt", n.chains = nc, n.thin = nt,
n.iter = ni, n.burnin = nb)
traceplot(out1J)    ;    print(out1J, dig = 3)
```

	mean	sd	2.5%	50%	97.5%	overlap0	f	Rhat	n.eff
gamma0	3.701	0.020	3.662	3.701	3.739	FALSE	1.000	1.001	1790
gamma[1]	-0.169	0.021	-0.209	-0.169	-0.130	FALSE	1.000	1.000	2738
gamma[2]	-0.152	0.018	-0.187	-0.152	-0.118	FALSE	1.000	1.002	1734
gamma[3]	-0.015	0.012	-0.038	-0.015	0.010	TRUE	0.882	1.000	3000
gamma[4]	-0.003	0.024	-0.049	-0.003	0.044	TRUE	0.543	1.001	1901
gamma[5]	-0.071	0.014	-0.099	-0.071	-0.043	FALSE	1.000	1.000	3000
gamma[6]	0.090	0.012	0.067	0.090	0.113	FALSE	1.000	1.000	3000

```
# Plot posterior distributions for potentially 'ecological' parameters
par(mfrow = c(1,3))
hist(out1J$sims.list$gamma[,1], breaks = 50, col = "grey", main = "", xlab =
"Slope of elevation (linear)")
hist(out1J$sims.list$gamma[,2], breaks = 50, col = "grey", main = "", xlab =
"Slope of elevation (squared)")
hist(out1J$sims.list$gamma[,3], breaks = 50, col = "grey", main = "", xlab =
"Slope of forest")
```

The evidence for linear and quadratic effects of elevation is strong, while that for an effect of forest cover is much less so. Let's inspect these relationships by plotting them (Figure 11.6, left and middle).

```
# Get covariate values for prediction
orig.pred.ele <- seq(250, 2750,, 500) # 500 vals spread between 250 and 2750
p.ele <- (orig.pred.ele - mean.ele) / sd.ele
orig.pred.forest <- seq(1, 100,, 500)
p.forest <- (orig.pred.forest - mean.forest) / sd.forest

# Compute predictions
nsamp <- out1J$mcmc.info$n.samples
pred.ele <- pred.forest <- array(NA, dim = c(500, nsamp))
for(i in 1:nsamp){
   pred.ele[,i] <- exp(out1J$sims.list$gamma0[i] + out1J$sims.list$gamma[i,1] *
p.ele + out1J$sims.list$gamma[i,2]* p.ele^2)
```

CHAPTER 11 HIERARCHICAL MODELS FOR COMMUNITIES

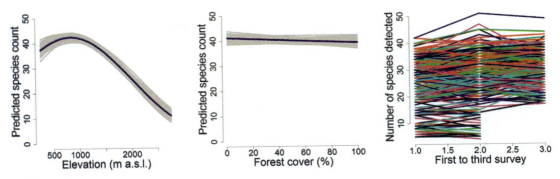

FIGURE 11.6

Left and middle: Inferences about the relationship between species richness and elevation and forest cover, respectively, in Switzerland in 2014 based on a simple Poisson GLM that does not retain species identity and ignores imperfect detection (blue line shows posterior mean and gray lines a random sample of size 100 of the posterior distribution of the regression equations). Right: Number of detected bird species per site and survey during the 2014 Swiss MHB survey.

```
   pred.forest[,i] <- exp(out1J$sims.list$gamma0[i] + out1J$sims.list$gamma[i,3]
   * p.forest)
}

# Plot posterior mean and a random sample of 100 from posterior of regression
selection <- sample(1:nsamp, 100)
par(mfrow = c(1,3))
matplot(orig.pred.ele, pred.ele[,selection], ylab = "Predicted species count", xlab =
   "Elevation (m a.s.l.)", type = "l", lty = 1, lwd = 1, col = "grey", ylim = c(0, 50), frame = F)
lines(orig.pred.ele, apply(pred.ele, 1, mean), lwd = 3, col = "blue")
matplot(orig.pred.forest, pred.forest[,selection], ylab = "Predicted species count",
   xlab = "Forest cover (%)", type = "l", lty = 1, lwd = 1, col = "grey",
   ylim = c(0, 50), frame = F)
lines(orig.pred.forest, apply(pred.forest, 1, mean), lwd = 3, col = "blue")
```

So, the response of the community to elevation seems clear: the largest communities are observed at low to medium elevations and the smallest at high elevations (Figure 11.6, left). There is only a very weak negative response of the community as a whole to forest cover in terms of the observed species richness (Figure 11.6, middle). Hence, we *know* that forest cover must be hugely significant on the occurrence patterns of many species and, yet, when we model the aggregate total we see nothing. That's all we have learned about the metacommunity of Swiss breeding birds, using a variant of what may be the most traditional approach to inference about the ecological determinants of the size of a community.

By simply taking the maximum species count over surveys, we obviously lost some information, especially about the relationship between counts and the two covariates that presumably may also affect the observed species richness via detection probability. So as an improvement over the current model, let's fit a similar model to the individual counts for each survey (Figure 11.6, right). This model is again a classical analysis that attempts to correct for measurement error (here associated with species richness N) by simply adding covariates that are believed, hoped, or claimed to be correlated with detection probability. We think that with longer survey duration, we see a larger proportion of the

11.5 COMMUNITY MODELS THAT IGNORE SPECIES IDENTITY

species present. Survey date is also expected to be related to detection probability (see, for instance, our earlier analyses of abundance and distribution of Swiss MHB data in this book). In principle, in the case of an open community, N could also be related to survey date; i.e., species might move in or out during the entire sampling period, and so covariate *date* might capture part of the resulting variation. However, in accordance with the usual closure assumption that we are willing to make for MHB data within a single two-to-three-month breeding season, we will interpret effects of *date* in terms of variation in detection probability rather than in terms of variation in the true community size. We will tally up the number of species detected per site and survey and then prepare the observational covariates for survey date (*DAT*) and survey duration (*DUR*), which vary by site and survey only (but not by species); we scale both.

```
# Get observed species richness per site and rep and plot
CC <- apply(y, c(1,2), sum, na.rm = TRUE)
CC[CC == 0] <- NA              # 0 means not surveyed
matplot(t(CC), type = 'l', lty = 1, lwd = 2, xlab = "First to third survey",
ylab = "Number of species detected", frame = F)# Fig. 11-6 right

# Get survey date and survey duration and standardise both
# Survey date (this is Julian date, with day 1 being April 1)
orig.DAT <- cbind(data$date141, data$date142, data$date143)[1:nsite,]
(mean.date <- mean(orig.DAT, na.rm = TRUE))
(sd.date <- sd(c(orig.DAT), na.rm = TRUE))
DAT <- (orig.DAT - mean.date) / sd.date       # scale
DAT[is.na(DAT)] <- 0                          # mean-impute missings
# Survey duration (in minutes)
orig.DUR <- cbind(data$dur141, data$dur142, data$dur143)[1:nsite,]
(mean.dur <- mean(orig.DUR, na.rm = TRUE))
(sd.dur <- sd(c(orig.DUR), na.rm = TRUE))
DUR <- (orig.DUR - mean.dur) / sd.dur         # scale
DUR[is.na(DUR)] <- 0                          # mean-impute missings

# Bundle data and summarize
str( win.data <- list(CC = CC, M = nrow(CC), J = ncol(CC), ele = ele, forest =
forest, DAT = DAT, DUR = DUR) )
List of 7
 $ CC     : num [1:267, 1:3] 21 28 32 29 29 31 24 32 30 30 ...
 ..- attr(*, "dimnames")=List of 2
 .. ..$ : NULL
 .. ..$ : NULL
 $ M      : int 267
 $ J      : int 3
 $ ele    : num [1:267] -1.1539 -1.1539 -0.2175 -0.3735 -0.0614 ...
 $ forest : num [1:267] -1.1471 -0.4967 -0.0992 -0.9303 0.0092 ...
 $ DAT    : num [1:267, 1:3] -1.415 -1.19 -1.235 -0.559 -1.64 ...
 $ DUR    : num [1:267, 1:3] -0.43511 -0.78764 -0.52324 1.23944 0.00556 ...

# Specify model in BUGS language
sink("model2.txt")
cat("
model {
```

```
# Priors
gamma0 ~ dnorm(0, 0.001)
for(v in 1:6){
  gamma[v] ~ dnorm(0, 0.001)
}

# Likelihood for Poisson GLM
for (i in 1:M){                        # Loop over sites
  for(j in 1:J){                       # Loop over occasions
    CC[i,j] ~ dpois(lambda[i,j])
    log(lambda[i,j]) <- gamma0 + gamma[1] * ele[i] + gamma[2] * pow(ele[i],2) + gamma[3] *
    forest[i] + gamma[4] * DAT[i,j] + gamma[5] * pow(DAT[i,j],2) + gamma[6] * DUR[i,j]
  }
}
}
",fill = TRUE)
sink()

# Initial values
inits <- function() list(gamma0 = rnorm(1), gamma = rnorm(6))

# Parameters monitored
params <- c("gamma0", "gamma")

# MCMC settings
ni <- 6000 ; nt <- 4 ; nb <- 2000 ; nc <- 3

# Call WinBUGS from R (ART 1.7 min)
out2 <- bugs(win.data, inits, params, "model2.txt", n.chains = nc,
n.thin = nt, n.iter = ni, n.burnin = nb, debug = TRUE, bugs.directory =
bugs.dir, working.directory = getwd())

# Call JAGS from R (ART 0.6 min)
out2J <- jags(win.data, inits, params, "model2.txt", n.chains = nc, n.thin = nt,
n.iter = ni, n.burnin = nb)
traceplot(out2J) ; print(out2J, dig = 2)
```

	mean	sd	2.5%	50%	97.5%	overlap0	f	Rhat	n.eff
gamma0	3.44	0.01	3.42	3.44	3.47	FALSE	1.00	1	3000
gamma[1]	-0.19	0.01	-0.21	-0.19	-0.17	FALSE	1.00	1	1336
gamma[2]	-0.15	0.01	-0.17	-0.15	-0.13	FALSE	1.00	1	3000
gamma[3]	-0.01	0.01	-0.03	-0.01	0.00	TRUE	0.93	1	3000
gamma[4]	0.00	0.01	-0.01	0.00	0.02	TRUE	0.67	1	2352
gamma[5]	-0.03	0.01	-0.05	-0.03	-0.02	FALSE	1.00	1	3000
gamma[6]	0.10	0.01	0.09	0.10	0.12	FALSE	1.00	1	856

Now we see "significant" effects of elevation (linear and squared—gamma[1] and gamma[2]), but the effect of forest (gamma[3]) is still not "significant." Two of the three effects that we think are related with the measurement error of species richness rather than with species richness itself (i.e., date squared and survey duration—gamma[5] and gamma[6]) are "significant"—i.e., have 95% CRIs that do not overlap zero.

11.5.2 POISSON RANDOM-EFFECTS MODEL FOR THE OBSERVED COMMUNITY SIZE

The next model in our progression is a random-effects Poisson (REP) model that accounts for possible dependence of repeated counts at the same site by adopting normal random site effects on the log scale. Thus, while the previous model treated repeated species counts at the same site as independent (given the covariate values), the improvement made by the REP model is that it accommodates a possible correlation of counts made at the same site.

$$C_{ij} \sim Poisson(\lambda_{ij})$$
$$\log(\lambda_{ij}) = \gamma_{0,i} + \text{covariates}_{ij}$$
$$\gamma_{0,i} \sim Normal(mu.gamma0, sd.gamma0^2)$$

This is a simplified version of the REP models developed by Bill Link and John Sauer (e.g., Link and Sauer, 2002) for counts of individual species. Note also that there is a relationship between this Poisson/Normal mixture model and the Binomial/Poisson N-mixture model (Chapter 6; see also Dennis et al., 2015a).

```
# Bundle and summarize data set
str( win.data <- list(CC = CC, M = nrow(CC), J = ncol(CC), ele = ele, forest =
forest, DAT = DAT, DUR = DUR) )

# Specify model in BUGS language
sink("model3.txt")
cat("
model {

# Priors
mugamma0 ~ dnorm(0, 0.001)              # Hyperparameters
taugamma0 <- pow(sd.gamma0,-2)
sd.gamma0 ~ dunif(0, 10)
for(v in 1:6){                           # Parameters
   gamma[v] ~ dnorm(0, 0.001)
}

# Likelihood for Poisson GLMM
for (i in 1:M){                          # Loop over sites
   gamma0[i] ~ dnorm(mugamma0, taugamma0) # Site intercepts random now
   for(j in 1:J){                        # Loop over repeated measurements
     CC[i,j] ~ dpois(lambda[i,j])
     log(lambda[i,j]) <- gamma0[i] + gamma[1]*ele[i] + gamma[2] * pow(ele[i],2) +gamma[3] *
     forest[i] + gamma[4] * DAT[i,j] + gamma[5] * pow(DAT[i,j],2) + gamma[6] * DUR[i,j]
   }
}
}
",fill = TRUE)
sink()

# Initial values
inits <- function() list(gamma0 = rnorm(nrow(CC)), gamma = rnorm(6))
```

```
# Parameters monitored
params <- c("mugamma0", "sd.gamma0", "gamma0", "gamma")

# MCMC settings
ni <- 6000 ; nt <- 4 ; nb <- 2000 ; nc <- 3

# Call WinBUGS from R (ART 2.9 min)
out3 <- bugs(win.data, inits, params, "model3.txt", n.chains = nc,
n.thin = nt, n.iter = ni, n.burnin = nb, debug = TRUE, bugs.directory =
bugs.dir, working.directory = getwd())

# Call JAGS from R (ART 0.7 min)
out3J <- jags(win.data, inits, params, "model3.txt", n.chains = nc, n.thin = nt,
n.iter = ni, n.burnin = nb)
traceplot(out3J, c('mugamma0', 'sd.gamma0', 'gamma'))      print(out3J, dig = 2)
```

	mean	sd	2.5%	50%	97.5%	overlap0	f	Rhat	n.eff
mugamma0	3.43	0.02	3.38	3.43	3.47	FALSE	1.00	1.01	189
sd.gamma0	0.18	0.01	0.16	0.18	0.21	FALSE	1.00	1.00	3000
gamma0[1]	3.31	0.10	3.10	3.31	3.51	FALSE	1.00	1.00	3000
gamma0[2]	3.44	0.10	3.25	3.44	3.62	FALSE	1.00	1.00	1021
gamma0[3]	3.58	0.08	3.41	3.58	3.75	FALSE	1.00	1.00	3000
[... output truncated ....]									
gamma0[265]	3.39	0.14	3.10	3.39	3.67	FALSE	1.00	1.00	3000
gamma0[266]	3.81	0.10	3.61	3.81	3.99	FALSE	1.00	1.00	3000
gamma0[267]	3.88	0.09	3.70	3.88	4.04	FALSE	1.00	1.00	608
gamma[1]	-0.21	0.02	-0.24	-0.21	-0.17	FALSE	1.00	1.00	415
gamma[2]	-0.18	0.02	-0.22	-0.18	-0.14	FALSE	1.00	1.01	278
gamma[3]	0.00	0.02	-0.03	0.00	0.03	TRUE	0.54	1.00	3000
gamma[4]	0.01	0.01	-0.01	0.01	0.03	TRUE	0.84	1.00	3000
gamma[5]	-0.02	0.01	-0.04	-0.02	-0.01	FALSE	1.00	1.00	3000
gamma[6]	0.10	0.01	0.08	0.10	0.13	FALSE	1.00	1.00	1454

We mostly present this model as one element in our progression to models that are more sophisticated in the context of metacommunities, so we don't say much about it here. But we note than when we account for the nonindependence of species counts from the same sites, the effect of forest (gamma[3]) is again clearly not "significant." We also note that JAGS is much faster than WinBUGS (0.7 vs. 2.9 minutes), so for the rest of the chapter we will fit models only in JAGS.

11.5.3 N-MIXTURE MODEL FOR OBSERVED COMMUNITY SIZE

A next step in our progression of increasingly sophisticated community models is a simple binomial N-mixture model. This model also accounts for the correlation of repeated measurements of species richness at a site. Moreover, it obtains an estimate of the average detection probability of a species from this correlation and thereby also produces an estimate of local community size: the number of species at each site. However, it does not allow for species heterogeneity in detection probability, nor for inference about individual species, because species identities are lost in this approach; we use only counts of "unmarked" species. As a consequence, we cannot estimate metacommunity size. Nevertheless, this model achieves an improved conceptual clarity by separating covariate effects on the community from covariates that affect its measurement only, and so we will return to our usual alpha-beta notation for the coefficients. We fit the following model to C_{ij}, the number of species recorded at site i during occasion j, where N_i is the community size (the number of species occurring) at site i.

11.5 COMMUNITY MODELS THAT IGNORE SPECIES IDENTITY

$$C_{ij} \sim Binomial(N_i, p_{ij})$$
$$\text{logit}(p_{ij}) = \alpha_0 + \text{covariates}_{ij}$$
$$N_i \sim Poisson(\lambda_i)$$
$$\log(\lambda_i) = \beta_0 + \text{covariates}_i$$

Here, p_{ij} is the average detection probability of a species at site i during occasion j, and λ_i is the expected community size (i.e., local species richness) at site i. Under the closure assumption, the number of species occurring at a site does not change among repeated measurements of community size at site i. We index the detection and community richness covariates by ij and by i to say that they can be observation or site covariates for the former, but only site covariates for the latter.

```
# Bundle and summarize data set
str( win.data <- list(CC = CC, M = nrow(CC), J = ncol(CC), ele = ele, forest =
forest, DAT = DAT, DUR = DUR) )

# Specify model in BUGS language
sink("model4.txt")
cat("
model {

# Priors
alpha0 ~ dnorm(0, 0.01)      # Base-line community detection probability
beta0 ~ dnorm(0, 0.01)       # Base-line community size (number of species)
for(v in 1:3){
    alpha[v] ~ dnorm(0, 0.01)  # Covariate effects on detection
    beta[v] ~ dnorm(0, 0.01)   # Covariate effects on community size
}

# Likelihood
# Ecological model for true community size
for (i in 1:M){               # Loop over sites
  N[i] ~ dpois(lambda[i])     # Community size
  lambda[i] <- exp(beta0 + beta[1] * ele[i] + beta[2] * pow(ele[i],2) +
  beta[3] * forest[i])

  # Observation model for repeated measurements
  for (j in 1:J){             # Loop over occasions
    CC[i,j] ~ dbin(p[i,j], N[i])
    p[i,j] <- 1 / (1 + exp(-lp[i,j]))
    lp[i,j] <- alpha0 + alpha[1] * DAT[i,j] + alpha[2] * pow(DAT[i,j],2) +
    alpha[3] * DUR[i,j]
  # logit(p) = ... causes undefined real result in WinBUGS (but not JAGS)
  }
}
}
",fill = TRUE)
sink()
```

The total over all sites, `Ntotal` in our previous applications of this model in Chapter 6, is no longer a meaningful quantity. We do not keep track of which species contribute to the species total at each site, and this prevents us from adding up the species at all sites for an estimate of the total number of species in the metacommunity.

```
# Define function to generate random initial values
Nst <- apply(CC, 1, max, na.rm = TRUE) + 1
Nst[Nst == -Inf] <- max(Nst, na.rm = T) # Some nonzero val. for unsurv. sites
inits <- function() list(N = Nst, alpha0 = rnorm(1), alpha = rnorm(3), beta0 =
rnorm(1), beta = rnorm(3))

# Parameters monitored
params <- c("alpha0", "alpha", "beta0", "beta","N")

# MCMC settings
ni <- 6000 ; nt <- 4 ; nb <- 2000 ; nc <- 3

# Run JAGS from R (ART 1.5 min) in parallel, look at traceplots
#    and summarize posteriors
out4 <- jags(win.data, inits, params, "model4.txt",
   n.chains = nc, n.thin = nt, n.iter = ni, n.burnin = nb, parallel = TRUE)
traceplot(out4, c('alpha0', 'alpha', 'beta0', 'beta')); print(out4, 3)

            mean    sd     2.5%    50%     97.5%  overlap0  f      Rhat   n.eff
alpha0      1.180   0.075  1.022   1.182   1.319  FALSE     1.000  1.006  503
alpha[1]    0.052   0.020  0.013   0.052   0.092  FALSE     0.994  1.000  3000
alpha[2]   -0.051   0.017 -0.083  -0.051  -0.017  FALSE     0.998  1.001  1565
alpha[3]    0.334   0.033  0.270   0.333   0.400  FALSE     1.000  1.000  3000
beta0       3.713   0.025  3.667   3.713   3.763  FALSE     1.000  1.000  3000
beta[1]    -0.200   0.013 -0.225  -0.199  -0.175  FALSE     1.000  1.004  491
beta[2]    -0.176   0.017 -0.208  -0.176  -0.143  FALSE     1.000  1.001  1413
beta[3]    -0.007   0.013 -0.032  -0.007   0.018  TRUE      0.720  1.001  1581
N[1]       34.904   2.072 31.000  35.000  39.000  FALSE     1.000  1.000  3000
N[2]       40.330   2.510 36.000  40.000  45.000  FALSE     1.000  1.000  3000
N[3]       50.092   2.247 46.000  50.000  55.000  FALSE     1.000  1.001  2803
[ ... output truncated ... ]
```

We can compare the main abundance parameter estimates from the REP regression with the N-mixture model.

```
print(tmp <- cbind(out3$summary[c(1, 270:272),c(1:3,7)], out4$summary[5:8,
c(1:3, 7)]), 3) # REP model: cols. 1-4; Nmix model: cols. 5-8
          mean     sd      2.5%     97.5%    mean     sd      2.5%     97.5%
mugamma0  3.43072  0.0234  3.3850   3.4770   3.71304  0.0249  3.6665   3.7633
gamma[1] -0.21030  0.0165 -0.2431  -0.1780  -0.19952  0.0129 -0.2252  -0.1746
gamma[2] -0.18101  0.0196 -0.2208  -0.1440  -0.17632  0.0167 -0.2078  -0.1430
gamma[3]  0.00196  0.0159 -0.0292   0.0328  -0.00709  0.0129 -0.0323   0.0184
```

The species richness intercept under the REP model is exp(3.43) = 30.88 species, while that under the N-mixture model is exactly 10 more—i.e., exp(3.71) = 40.85 species. This is surely not so surprising, since the REP model does not account for imperfect detection. Otherwise, we see qualitatively similar inferences about the abundance covariates. In particular, we don't see any effect of forest cover on estimated true species richness. We finish by plotting the estimates of local species richness and the observed species counts against elevation (Figure 11.7). Although we can't easily tell which estimate under the N-mixture model belongs to which observed value of species richness, we get the impression that the N-mixture model must underestimate species richness, since the observed number of species at a site is on average very similar to that estimated under the N-mixture model. We think it unlikely that we see all occurring species with only three surveys. Rather, it appears more likely that among-species heterogeneity in detection probability leads to an underestimation of the site-specific N in the N-mixture

11.6 COMMUNITY MODELS THAT FULLY RETAIN SPECIES IDENTITY

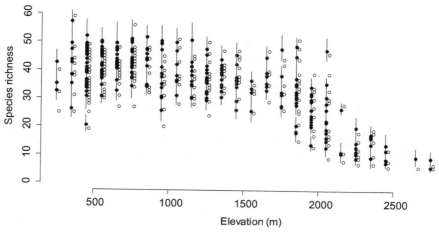

FIGURE 11.7

Estimated number of species at each of 267 sites in the Swiss breeding bird survey MHB under the binomial N-mixture model (filled circles, with 95% CRI) along the elevation profile from 200–2750 m a.s.l. The open circles show the *observed* number of species over the two or three surveys combined. Note counts are slightly offset for improved visibility.

model (per the second law of capture-recapture). In the N-mixture model, we can't accommodate such heterogeneity among species, and this is one of the main drawbacks of this approach.

```
plot(orig.ele, out4$summary[9:275, 1], pch = 16, xlab = "Elevation (m)", ylab =
"Species richness", ylim = c(0, 60), frame = F)
segments(orig.ele, out4$summary[9:275, 3], orig.ele, out4$summary[9:275, 7])
points(orig.ele+20, C)     # elevation slightly offset
```

Remember that instead of a binomial N-mixture model for site-specific species richness estimation, you could also adopt a multinomial N-mixture model (Chapter 7) for the *species* detection/nondetection data and model local species richness instead of local population abundance as in the more typical applications of this model in Chapter 7. In contrast to the binomial N-mixture model, species identity would only be lost across sites, but not across replicate surveys at each site. Hence, more information would be used, plus we could model individual detection heterogeneity, which is compulsory for species richness applications of closed population abundance estimators (Kéry 2011a). Importantly, we could then also model local species richness directly as a function of site covariates. A pure M_h type of such a model (i.e., with only species heterogeneity, but no other effect in detection probability) may even be fit in unmarked (see Section 7.8.3), where we could then test hypotheses about drivers of species richness using formal significance tests or AIC. When fit in BUGS, we could also model other structure into detection probability beyond individual (i.e., species-specific) heterogeneity, e.g., any of the detection covariates that we consider in this chapter. See Exercise 2 for an example of this.

11.6 COMMUNITY MODELS THAT FULLY RETAIN SPECIES IDENTITY

In the next set of models in our progression toward a powerful and satisfactory community modeling framework, we adopt a totally different approach and model the specific response to covariates of each member of the community. From now on, our community model will be a collection of component models for the presence and absence of individual species. This will allow us to get much deeper insights into the metacommunity as well as into the local communities, and the ways in which they respond to environmental and other covariates as a function of individual species' responses. The species-specific approach to modeling the metacommunity will permit us to look at the average response of both the (meta)community and every one of its members. Remember our binary detection/nondetection indicators y_{ijk} for species k at site i during survey j; how can we retain species identities in an analysis of these data?

11.6.1 SIMPLEST COMMUNITY OCCUPANCY MODEL: N-FOLD SINGLE SPECIES OCCUPANCY MODEL WITH SPECIES TREATED AS FIXED EFFECTS

The first and simplest, but already very powerful, approach is to simply fit a separate species distribution model (SDM) to every species. Of course, we will use our favorite SDM, the occupancy model from Chapter 10. We could do this in a loop by fitting a separate occupancy model to the data from each of the 145 species in turn. However, it is *much more efficient* to fit these 145 models to the data of *all* species at once. This can easily be done in BUGS by organizing the data as a three-dimensional array (which we have already done). For illustration, we simply fit a null model to each species, with a species-specific constant intercept for both the occupancy and the detection probability, though we could readily add covariates. One big advantage of this approach over one that fits 145 separate models is that in the combined model, we can easily sum or compare quantities across species. For instance, it is trivial to add up the estimated number of occurring species for each site with a full propagation of all uncertainties involved in producing such a sum. As usual, the ease with which we can do computations on latent variables (here, the indicators of occurrence z) is one of the great practical benefits of a Bayesian analysis using MCMC. Similarly, we could constrain parameters among species, for instance by making them constant for some or all species or modeling them as a function of species-specific covariates (as we will see in Section 11.6.3).

We start the series of multispecies or community occupancy models without accounting for any structure among the three replicate surveys, so that we can collapse the detection/nondetection data to detection frequencies (which tell us whether a species was detected 0, 1, 2, or 3 times). We fit the following two-level HM to our detection frequency data $ysum_{ik}$ for species k ($k = 1...145$) and site i ($i = 1...267$).

$$\text{Process model}: \quad z_{ik} \sim Bernoulli(\psi_k)$$
$$\text{Observation model}: \quad ysum_{ik}|z_{ik} \sim Binomial(J_i, z_{ik}p_k)$$

Here, J_i is the number of surveys and it is indexed by site, because at 47 high-elevation sites, there are two instead of the usual three surveys per breeding season. This corresponds *exactly* to the binomial version of the simplest possible occupancy model (see Section 10.3), except that we now stratify by species: every quantity is now also indexed by k, allowing it to be different for every species. No relationship among species is imposed on the species-specific parameters of occupancy (ψ_k) and detection probability (p_k), so their estimates will be identical to what you would get if you looped over all species and fitted a separate model to each.

We fit the binomial model to the aggregated data for illustrative purposes, mostly because this makes it easier to understand the basic structure of this model and the next few models in our progression of community models. It would be easy to fit the analogous models with a Bernoulli response for the disaggregated data y_{ijk}; see Exercise 3. You will see that the Bernoulli model takes about 50% longer to run for this data set.

11.6 COMMUNITY MODELS THAT FULLY RETAIN SPECIES IDENTITY

```
# Collapse 3D detection/nondetection data to 2D detection frequencies
ysum <- apply(y, c(1,3), sum, na.rm = T)     # Collapse to detection frequency
ysum[NAsites,] <- NA                          # Have to NA out sites with NA data

# Bundle and summarize data set
str( win.data <- list(ysum = ysum, M = nrow(ysum), J = data$nsurvey[1:nsite],
nspec = dim(ysum)[2]) )

# Specify model in BUGS language
sink("model5.txt")
cat("
model {

# Priors
for(k in 1:nspec){                # Loop over species
    psi[k] ~ dunif(0, 1)
    p[k] ~ dunif(0, 1)
}

# Ecological model for latent occurrence z (process model)
for(k in 1:nspec){                # Loop over species
  for (i in 1:M) {                # Loop over sites
      z[i,k] ~ dbern(psi[k])
  }
}

# Observation model for observed data ysum
for(k in 1:nspec){                # Loop over species
    for (i in 1:M) {
       mup[i,k] <- z[i,k] * p[k]
       ysum[i,k] ~ dbin(mup[i,k], J[i])
    }
}

# Derived quantities
for(k in 1:nspec){                # Loop over species
    Nocc.fs[k] <- sum(z[,k])      # Add up number of occupied sites among the 267
}
for (i in 1:M) {                  # Loop over sites
    Nsite[i] <- sum(z[i,])        # Add up number of occurring species at each site
}
}
",fill = TRUE)
sink()

# Initial values
zst <- apply(y, c(1,3), max)      # Observed occurrence as inits for z
zst[is.na(zst)] <- 1
inits <- function() list(z = zst, psi = rep(0.4, nspec), p = rep(0.4, nspec))

# Parameters monitored
params <- c("psi", "p", "Nsite", "Nocc.fs")
```

```
# MCMC settings
ni <- 2500 ; nt <- 2 ; nb <- 500 ; nc <- 3

# Call JAGS from R (ART 2.1 min)
out5 <- jags(win.data, inits, params, "model5.txt", n.chains = nc, n.thin = nt,
n.iter = ni, n.burnin = nb, parallel = TRUE)
par(mfrow = c(4,4)) ; traceplot(out5) ; print(out5, dig = 3)
```

	mean	sd	2.5%	50%	97.5%	overlap0	f	Rhat	n.eff
psi[1]	0.025	0.011	0.009	0.024	0.051	FALSE	1	1.000	3000
psi[2]	0.025	0.031	0.005	0.019	0.067	FALSE	1	1.211	159
psi[3]	0.022	0.011	0.007	0.020	0.047	FALSE	1	1.009	2555
psi[4]	0.008	0.007	0.001	0.007	0.025	FALSE	1	1.054	658
psi[5]	0.023	0.009	0.009	0.022	0.043	FALSE	1	1.000	3000
psi[6]	0.170	0.191	0.003	0.090	0.657	FALSE	1	1.059	69
[...]									
psi[143]	0.035	0.021	0.011	0.031	0.086	FALSE	1	1.012	3000
psi[144]	0.124	0.020	0.087	0.123	0.167	FALSE	1	1.000	3000
psi[145]	0.026	0.017	0.007	0.023	0.070	FALSE	1	1.005	3000
p[1]	0.592	0.134	0.317	0.597	0.840	FALSE	1	1.001	2551
p[2]	0.417	0.181	0.097	0.418	0.760	FALSE	1	1.002	1007
p[3]	0.578	0.152	0.268	0.583	0.840	FALSE	1	1.000	3000
[...]									
p[143]	0.392	0.149	0.115	0.386	0.687	FALSE	1	1.001	1353
p[144]	0.734	0.049	0.633	0.737	0.822	FALSE	1	1.001	2022
p[145]	0.463	0.167	0.142	0.469	0.767	FALSE	1	1.000	3000
Nsite[1]	35.684	2.189	32.000	35.000	40.000	FALSE	1	1.050	45
Nsite[2]	37.697	2.201	34.000	38.000	42.000	FALSE	1	1.041	53
Nsite[3]	56.031	1.972	52.000	56.000	60.000	FALSE	1	1.027	78
[...]									
Nsite[265]	20.881	2.741	16.000	21.000	27.000	FALSE	1	1.022	97
Nsite[266]	52.958	1.997	49.000	53.000	57.000	FALSE	1	1.066	35
Nsite[267]	55.543	1.939	52.000	55.000	60.000	FALSE	1	1.056	41
Nocc.fs[1]	5.851	1.363	5.000	5.000	10.000	FALSE	1	1.000	899
Nocc.fs[2]	5.710	7.773	3.000	4.000	16.000	FALSE	1	1.230	160
Nocc.fs[3]	4.898	1.782	4.000	4.000	9.000	FALSE	1	1.044	1032
[...]									
Nocc.fs[143]	8.249	4.749	5.000	7.000	21.000	FALSE	1	1.029	809
Nocc.fs[144]	32.172	1.212	31.000	32.000	35.000	FALSE	1	1.003	327
Nocc.fs[145]	6.038	3.510	4.000	5.000	15.000	FALSE	1	1.016	3000

In this analysis, species can be said to correspond to "fixed effects," so the parameter estimates for each species are exclusively determined by the data for that species. This means that the quality of estimates of species with very few observed presences will be pretty mediocre or even downright terrible; you can see this by the large posterior standard deviations and also the bad mixing of the chains for psi and p for some species. Nevertheless, one of the advantages of fitting this model with species strata is that we can compute derived quantities that are functions of the estimates for more than one species. One such example is the estimated number of species (among the list of 145 that were detected anywhere in 2014) occurring at a site, Nsite. Hence, as for the N-mixture model approach,

11.6 COMMUNITY MODELS THAT FULLY RETAIN SPECIES IDENTITY

this analysis accommodates imperfect detection, and yields an estimate of site-specific species richness conditional on the list of species that were detected at least once. But in contrast to the *N*-mixture model, the community occupancy approach does *not* assume all species are identical. Rather, every species is allowed to differ in terms of *both* occupancy *and* detection probability. Thus, we would expect the resulting estimates of *N* to be greater than those under the binomial mixture model, which we argued were probably underestimates due to unmodeled individual detection heterogeneity (Boulinier et al., 1998; Cam et al., 2002b,c). We check this in a plot (Figure 11.8), where we moreover distinguish between sites surveyed twice versus three times. We see that indeed, all estimates of *N* are now clearly greater than the observed number of species (they lie above the 1:1 line), presumably because our new model better accounts for species-specific detection heterogeneity than did the *N*-mixture model. The second striking observation is that there are two sets of sites, and in one of them, the underestimation of species richness by the raw data is greater than in the other. This is of course the distinction of sites surveyed twice only: it is expected that a larger proportion of occurring species is missed.

```
# Compare observed and estimated site species richness
par(cex = 1.3)
twice <- which(data$nsurvey[1:267] == 2)
plot(C[twice], out5$summary[291:557,1][twice], xlab = "Observed number of
species", ylab = "Estimated number of species", frame = F, xlim = c(0, 60), ylim = c(0, 70),
col = "red", pch = 16)
segments(C[twice], out5$summary[291:557,3][twice], C[twice],
out5$summary[291:557,7][twice], col = "red")
points(C[-twice], out5$summary[291:557,1][-twice], col = "blue", pch = 16)
segments(C[-twice], out5$summary[291:557,3][-twice],
C[-twice], out5$summary[291:557,7][-twice], col = "blue")
abline(0,1)
```

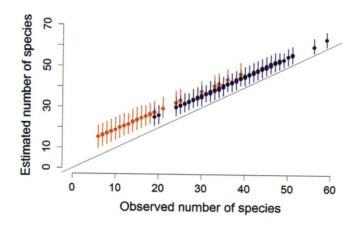

FIGURE 11.8

Comparison of observed and estimated number of species in 2014 at 267 sites under a community occupancy model with species treated as fixed effects (with 95% CRIs). The species richness estimate is conditional upon the list of 145 species that were detected at least once anywhere during the breeding season 2014 among the MHB sample quadrats. Blue: sites with three visits, red: sites with only two visits.

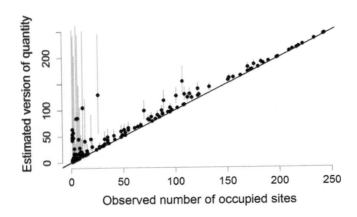

FIGURE 11.9

Observed and estimated number of sites (out of 267) where a species occurred in 2014 (with 95% CRIs).

We can also compare the observed and estimated number of sites where each species occurs (Figure 11.9): not surprisingly, when a species is detected at only few sites, the estimates become much more imprecise.

```
# Observed and estimated number of occupied sites for each species
# in a table
cbind(obs.occu = obs.occ, out5$summary[558:702, c(1,3,7)])
                       obs.occu      mean     2.5%    97.5%
Little Grebe                  5   5.850667    5.000   10.000
Great Crested Grebe           3   5.709667    3.000   16.000
Grey Heron                    4   4.898333    4.000    9.000
White Stork                   1   1.277667    1.000    3.025
Mute Swan                     5   5.145000    5.000    6.000
Greylag Goose                 1  44.787667    1.000  176.000
Mallard                      55  61.984000   56.975   70.000
Common Merganser              1   1.188333    1.000    3.000
European Honey Buzzard        5  83.990667    8.000  262.000
Red Kite                     95 103.973000   98.000  112.000
[ ... output truncated ... ]

# and in a plot
plot(obs.occ, out5$summary[558:702, 1], xlab = "Observed number of occupied sites",
ylab = "Estimated version of quantity", ylim = c(0, 267), frame = F, pch = 16)
abline(0,1)
segments(obs.occ, out5$summary[558:702,3], obs.occ, out5$summary[558:702,7], col
= "grey", lwd = 2)
```

Finally, we want to compare the estimates of detection probability with those for occupancy probability for all 145 species under model 5 (Figure 11.10).

```
# Estimated occupancy and detection probability for each species (model 5)
plot(out5$summary[1:145,1], out5$summary[146:290,1], xlab = "Occupancy estimate",
ylab = "Detection estimate", xlim = c(0,1), ylim = c(0,1), frame = F, pch = 16)
```

11.6 COMMUNITY MODELS THAT FULLY RETAIN SPECIES IDENTITY

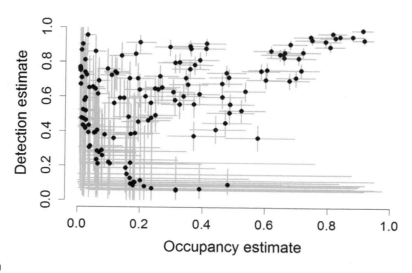

FIGURE 11.10

Estimates of detection and occupancy probabilities for the 145 species detected in the MHB in 2014 (with 95% CRIs).

```
segments(out5$summary[1:145,3], out5$summary[146:290,1], out5$summary[1:145,7],
  out5$summary[146:290,1], col = "grey", lwd = 2)
segments(out5$summary[1:145,1], out5$summary[146:290,3], out5$summary[1:145,1],
  out5$summary[146:290,7], col = "grey", lwd = 2)
```

Naturally, with small sample sizes, estimates become worse in the sense of having (much) wider 95% CRIs. Here, in what is exactly equivalent to independent single-species occupancy models fitted to 145 species, we see that when occupancy is smaller than about 0.1 and detection smaller than about 0.2, estimates become very imprecise and likely also biased, as is well known in the occupancy literature (MacKenzie et al., 2002; Guillera-Arroita et al., 2010). We will next see how we can improve by not estimating the parameters for all species independently (by treating species as a fixed effects factor), but rather treating the observed species as a sample from a larger statistical population of species—i.e., by treating species as random effects sharing hyperparameters that we estimate.

11.6.2 COMMUNITY OCCUPANCY MODEL WITH BIVARIATE SPECIES-SPECIFIC RANDOM EFFECTS

In this analysis, we treat the parameters of each species as random effects. This means that we assume that species-specific effects are drawn from a common distribution, called a prior distribution, with hyperparameters that we estimate. Under the assumption that the species are exchangeable ("similar but not identical"), the random-effects assumption will typically lead to improved estimates for individual species, in the sense of reducing prediction error or uncertainty intervals. The former is not something that we can prove or observe for real data because we don't know the truth; instead, we would have to check that with simulated data. However, the latter is easily shown (Kéry and Royle, 2008; Zipkin et al., 2009).

Our new model is almost exactly the same as before, except that now the species-specific parameters will be constrained by a common prior distribution. All other things, for instance the finite-sample number of occupied quadrats or species richness (among our list of 145) can be computed in exactly the same way as in the previous analysis where species were treated as fixed effects. In the previous analysis, there was no need to apply a link function to the parameters, because the $Uniform(0,1)$ priors for each parameter naturally enforced the usual range constraint for probability parameters. However, now that we make the random-effects assumption, we make the conventional assumption of normal priors for the parameters on the logit scale.

We could fit the following three-level HM, where the species random effects add another level to our HM.

$$\text{Process model}: \quad z_{ik} \sim Bernoulli(\psi_k)$$
$$\text{Observation model}: \quad ysum_{ik}|z_{ik} \sim Binomial(J_i, z_{ik}p_k)$$
$$\text{Models of species heterogeneity}: \quad logit(\psi_k) \sim Normal\left(\mu_{lpsi}, \sigma^2_{lpsi}\right)$$
$$logit(p_k) \sim Normal\left(\mu_{lp}, \sigma^2_{lp}\right)$$

The species-specific parameters are no longer estimated independently; rather, they are constrained by the assumption of a common normal prior distribution for the logits of the occupancy and detection probabilities. It is the last two equations that declare species effects to be random. Also note that the species index k still runs from 1 to 145, which is the number of observed species. In this model, the pairs of values $[logit(\psi_k), logit(p_k)]$ for a species are assumed to be independent; i.e., there is no *a priori* assumption of any relationship between the two. However, as Dorazio et al. (2006) have argued, we may well assume that there is such a relationship induced by the underlying abundance of a species: more-common species are expected to be both more widespread and more detectable. This would lead to a positive correlation between occupancy and detection probabilities. We can actually see something of this in Figure 11.10 (though perhaps a triangular relationship better describes the results): very widespread species always seem to have high values of p, while many species that are not widespread exhibit small and in other cases large values of detection probability.

If we want to quantify an association between two parameters, we can assume a multivariate normal distribution for them instead of two independent normal distributions, and this has been done in many applications of multispecies occupancy models (e.g., Dorazio et al., 2006; Kéry and Royle, 2008, 2009). The final two lines in the above model then become this:

$$\text{Models of species heterogeneity}: \quad (lpsi_k, lp_k) \sim MVN\left(\begin{pmatrix}\mu_{lpsi}\\ \mu_{lp}\end{pmatrix}, \begin{pmatrix}\sigma^2_{lpsi} & \sigma_{lp*lpsi}\\ \sigma_{lp*lpsi} & \sigma^2_{lp}\end{pmatrix}\right)$$

This model has one additional parameter, the covariance $\sigma_{lp*lpsi}$. The covariance can be expressed as $\rho\sigma_{lpsi}\sigma_{lp}$—i.e., as the product of the correlation coefficient ρ and the standard deviations of the logit-scale parameters *lpsi* and *lp*. This is the parameterization that we had chosen in Kéry and Royle (2008); see also Chapter 12 in Kéry (2010). However, here we adapt code for covariance parameters from Chapter 7 in Kéry and Schaub (2012); also see Cam et al. (2002a). For a version of the model with two univariate normals and therefore a zero covariance, see Section 11.6.3 (where we add "species group" effects).

```
# Bundle and summarize data set
str( win.data <- list(ysum = ysum, M = nrow(ysum), J = data$nsurvey[1:nsite],
nspec = dim(ysum)[2], R = matrix(c(5,0,0,1), ncol = 2), df = 3) )

# Specify model in BUGS language
sink("model6.txt")
cat("
model {
```

11.6 COMMUNITY MODELS THAT FULLY RETAIN SPECIES IDENTITY

```
# Priors
for(k in 1:nspec){    # Group lpsi and lp together in array eta
   lpsi[k] <- eta[k,1]
   lp[k] <- eta[k,2]
   eta[k, 1:2] ~ dmnorm(mu.eta[], Omega[,])
}
# Hyperpriors
# Priors for mu.lpsi=mu.eta[1] and mu.lp=mu.eta[2]
# probs = community means of occupancy and detection probability
for(v in 1:2){
   mu.eta[v] <- log(probs[v] / (1-probs[v]))
   probs[v] ~ dunif(0,1)
}
# Prior for variance-covariance matrix
Omega[1:2, 1:2] ~ dwish(R[,], df)
Sigma[1:2, 1:2] <- inverse(Omega[,])

# Ecological model for latent occurrence z (process model)
for(k in 1:nspec){
   logit(psi[k]) <- lpsi[k]      # Must take outside of i loop (b/c only indexed k)
   for (i in 1:M) {
      z[i,k] ~ dbern(psi[k])
   }
}

# Observation model for observed data ysum
for(k in 1:nspec){
   logit(p[k]) <- lp[k]          # Needs to be outside of i loop
   for (i in 1:M){
      mu.p[i,k] <- z[i,k] * p[k]
      ysum[i,k] ~ dbin(mu.p[i,k], J[i])
   }
}

# Derived quantities
rho <- Sigma[1,2] / sqrt(Sigma[1,1] * Sigma[2,2])     # Correlation coefficient
for(k in 1:nspec){
   Nocc.fs[k] <- sum(z[,k])      # Number of occupied sites among the 267
}
for (i in 1:M) {
   Nsite[i] <- sum(z[i,])        # Number of occurring species
}
}
",fill = TRUE)
sink()

# Initial values
zst <- apply(y, c(1,3), max)  # Observed occurrence as starting values for z
zst[is.na(zst)] <- 1
inits <- function() list(z = zst, Omega = matrix(c(1,0,0,1), ncol = 2), eta =
matrix(0, nrow = nspec, ncol = 2))
```

```
# Parameters monitored
params <- c("mu.eta", "probs", "psi", "p", "Nsite", "Nocc.fs", "Sigma", "rho")

# MCMC settings
ni <- 20000 ; nt <- 15 ; nb <- 5000 ; nc <- 3

# Call JAGS from R (ART 12 min), check traceplots and summarize posteriors
out6 <- jags(win.data, inits, params, "model6.txt", n.chains = nc, n.thin = nt, n.iter = ni,
n.burnin = nb, parallel = TRUE)
par(mfrow = c(3,3)) ; traceplot(out6, c('mu.eta', 'probs', 'Sigma', 'rho'))
print(out6, 3)
```

	mean	sd	2.5%	50%	97.5%	overlap0	f	Rhat	n.eff
mu.eta[1]	-1.889	0.176	-2.230	-1.891	-1.535	FALSE	1	1.000	3000
mu.eta[2]	0.536	0.116	0.305	0.536	0.760	FALSE	1	1.000	3000
probs[1]	0.133	0.020	0.097	0.131	0.177	FALSE	1	1.000	3000
probs[2]	0.630	0.027	0.576	0.631	0.681	FALSE	1	1.000	3000
psi[1]	0.024	0.010	0.009	0.023	0.047	FALSE	1	1.001	3000
psi[2]	0.018	0.010	0.005	0.016	0.045	FALSE	1	1.009	2854
psi[3]	0.020	0.009	0.007	0.018	0.042	FALSE	1	1.001	2228
[...]									
psi[143]	0.027	0.013	0.010	0.025	0.056	FALSE	1	1.009	1138
psi[144]	0.121	0.020	0.084	0.120	0.162	FALSE	1	1.000	3000
psi[145]	0.022	0.011	0.007	0.020	0.045	FALSE	1	1.001	3000
p[1]	0.570	0.125	0.320	0.574	0.802	FALSE	1	1.000	3000
p[2]	0.437	0.148	0.162	0.431	0.744	FALSE	1	1.001	1681
p[3]	0.562	0.132	0.305	0.568	0.806	FALSE	1	1.002	1012
[...]									
p[143]	0.431	0.128	0.188	0.429	0.683	FALSE	1	1.000	3000
p[144]	0.735	0.049	0.636	0.737	0.827	FALSE	1	1.001	1581
p[145]	0.474	0.137	0.214	0.472	0.737	FALSE	1	1.001	2019
Nsite[1]	32.007	1.393	30.000	32.000	35.000	FALSE	1	1.001	1705
Nsite[2]	34.023	1.359	32.000	34.000	37.000	FALSE	1	1.001	1064
Nsite[3]	52.689	1.229	51.000	53.000	55.000	FALSE	1	1.001	3000
[...]									
Nsite[265]	17.135	2.181	13.000	17.000	22.000	FALSE	1	1.000	3000
Nsite[266]	49.537	1.211	48.000	49.000	52.000	FALSE	1	1.000	3000
Nsite[267]	52.223	1.086	51.000	52.000	55.000	FALSE	1	1.001	3000
Nocc.fs[1]	5.829	1.230	5.000	5.000	9.000	FALSE	1	1.003	2650
Nocc.fs[2]	4.331	1.986	3.000	4.000	10.000	FALSE	1	1.007	3000
Nocc.fs[3]	4.759	1.222	4.000	4.000	8.000	FALSE	1	1.007	1294
[...]									
Nocc.fs[143]	6.894	2.450	5.000	6.000	13.000	FALSE	1	1.017	3000
Nocc.fs[144]	32.102	1.181	31.000	32.000	35.000	FALSE	1	1.000	3000
Nocc.fs[145]	5.292	1.870	4.000	5.000	10.000	FALSE	1	1.003	3000
Sigma[1,1]	4.309	0.550	3.350	4.258	5.548	FALSE	1	1.000	3000
Sigma[2,1]	1.462	0.285	0.937	1.450	2.056	FALSE	1	1.000	3000
Sigma[1,2]	1.462	0.285	0.937	1.450	2.056	FALSE	1	1.000	3000
Sigma[2,2]	1.522	0.224	1.143	1.499	2.021	FALSE	1	1.000	3000
rho	0.571	0.078	0.408	0.576	0.707	FALSE	1	1.000	3000

11.6 COMMUNITY MODELS THAT FULLY RETAIN SPECIES IDENTITY

We see that the species heterogeneity in the logit transform of detection probability (sd.lp) is estimated at 1.522, while for occupancy probability (sd.lpsi) it is estimated at a much larger value of 4.309, and the correlation between the two logit-scale parameters is estimated at 0.57.

```
# Graphically compare some estimates between fixed- and random-effects model
par(mfrow = c(2,2))      # not shown
# Species-specific occupancy (probability scale)
plot(out5$summary[1:145,1], out6$summary[5:149,1], main = "Species-specific occupancy
probability") ;             abline(0,1)
# Species-specific detection (probability scale)
plot(out5$summary[146:290,1], out6$summary[150:294,1], main = "Species-specific
detection probability") ;        abline(0,1)
# Site-specific species richness
plot(out5$summary[291:557,1], out6$summary[295:561,1], main = "Site-specific
species richness (conditional on list of 145 detected)") ;       abline(0,1)
# Species-specific number of presences
plot(out5$summary[558:702,1], out6$summary[562:706,1], main = "Species-specific number
of presences (in 267 sites)") ;        abline(0,1)
```

For the most part (i.e., for species with at least moderate sample size), point estimates are fairly similar under the two models. However, for species with relatively sparse data, the estimates are substantially more precise, and there can be considerable shifts in the posterior means; this is the "shrinkage" effect of Bayesian estimators at work. Finally, we want to compare the estimates of detection probability with those of occupancy probability for all 145 species under model 6 (Figure 11.11). The length of the 95% CRI appears reduced overall compared with those under model 5, which did not share information among the species via the random-effects assumption.

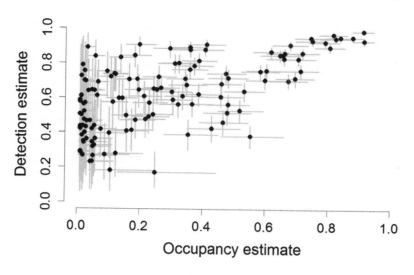

FIGURE 11.11

Estimates of detection probability and occupancy probability for the 145 species detected in the MHB in 2014 (with 95% CRI) under random-effects model 6.

```
# Estimated occupancy and detection probability for each species
plot(out6$summary[5:149,1], out6$summary[150:294,1], xlab = "Occupancy estimate",
ylab = "Detection estimate", xlim = c(0,1), ylim = c(0,1), frame = F, pch = 16)
segments(out6$summary[5:149,3], out6$summary[150:294,1], out6$summary[5:149,7],
out6$summary[150:294,1], col = "grey", lwd = 2)
segments(out6$summary[5:149,1], out6$summary[150:294,3], out6$summary[5:149,1],
out6$summary[150:294,7], col = "grey", lwd = 2)
```

11.6.3 MODELING SPECIES-SPECIFIC EFFECTS IN COMMUNITY OCCUPANCY MODELS

Frequently, we want to compare groups of species—for instance, guilds (e.g., herbivores vs carnivores), life-history categories (slow vs fast species), migration modes (resident vs migratory species), or other distinctions between species. We then treat species again as random effects, but specify a linear model for the hyperparameters that govern the species-specific occupancy and detection parameters. We illustrate this here. In our data file, there are data on three traits that characterize species: body length (cm), body mass (g), and wingspan (cm). As a first example for the modeling of species-specific effects, we will fit a community occupancy model with separate parameters for three species size groups that we define by body mass. We try to obtain three groups with roughly similar numbers of species, and take as cut points 1.3 and 2.6 for the log10 of body mass; this corresponds to about 20 and 398 g. One species with 11-kg body mass (the mute swan) is assigned to group 3, too.

Modeling species effects in such groups corresponds to an ANOVA-type of linear model for species-specific effects in the community model. In our second example below, we will illustrate the corresponding "regression-type" of linear model—i.e., one that models the effects of a continuous covariate. Clearly, when you understand how to fit linear models in the BUGS language, you could easily combine the two types of effects (see Chapters 3 and 5)—for instance, if you had effects of body size (continuous) and a binary factor such as whether species are sexually dimorphic (Doherty et al., 2003).

```
# Look at distribution of body mass among 136 observed species
mass <- tapply(data$body.mass, data$specid, mean)  # Get mean species mass
hist(log10(mass), breaks = 40, col = "grey")       # Look at log10
gmass <- as.numeric(log10(mass) %/% 1.3 + 1)       # Size groups 1, 2 and 3
gmass[gmass == 4] <- 3                             # Mute swan is group 3, too
```

We will index the three size groups as $g = 1, 2, 3$ and could adopt any one of the following three models, which differ only in the structure assumed for species heterogeneity:

Interspecific Comparison 1: Group Effects for the Means Only

Process model : $\quad z_{ik} \sim Bernoulli(\psi_k)$

Observation model : $\quad ysum_{ik}|z_{ik} \sim Binomial(J_i, z_{ik}p_k)$

Models of species heterogeneity : $\quad logit(\psi_k) \sim Normal\left(\mu_{lpsi}[g], \sigma^2_{lpsi}\right)$

$\quad logit(p_k) \sim Normal\left(\mu_{lp}[g], \sigma^2_{lp}\right)$

11.6 COMMUNITY MODELS THAT FULLY RETAIN SPECIES IDENTITY

Interspecific Comparison 2: Group Effects for Both Means and Variances
(same process and observation models)

Models of species heterogeneity: $\text{logit}(\psi_k) \sim Normal\left(\mu_{lpsi}[g], \sigma^2_{lpsi}[g]\right)$

$\text{logit}(p_k) \sim Normal\left(\mu_{lp}[g], \sigma^2_{lp}[g]\right)$

By the *g* within square brackets, we are saying that we estimate a different parameter for every group. Model 1 only distinguishes between the groups in terms of the mean of the intercepts for occupancy and detection probability, while model 2 also allows the *variances* to differ between the groups. So, this is another rare example where we explicitly model a variance parameter as a function of a covariate (see also Section 6.11.2.2 for an example in the context of a binomial mixture model). For illustration, we here fit model 2 (model 1 is a simple restriction of model 2). In this model, the only changes to a model without group effects (e.g., model 6) happen in the prior and hyperprior sections of the BUGS model description.

```
# Bundle and summarize data set
str( win.data <- list(ysum = ysum, g = gmass, M = nrow(ysum), J =
data$nsurvey[1:nsite], nspec = dim(ysum)[2]) )

# Specify model in BUGS language
sink("model7.txt")
cat("
model {

# Priors: note group effects specified in this section
for(k in 1:nspec){       # loop over species
   lpsi[k] ~ dnorm(mu.lpsi[g[k]], tau.lpsi[g[k]])   # note g-dependence now
   lp[k] ~ dnorm(mu.lp[g[k]], tau.lp[g[k]])
}

# Hyperpriors
for(g in 1:3){           # loop over 3 groups (g)
   mu.lpsi[g] <- logit(mu.psi[g])       # everything is indexed g now
   mu.lp[g] <- logit(mu.p[g])
   mu.psi[g] ~ dunif(0,1)
   mu.p[g] ~ dunif(0,1)
   tau.lpsi[g] <- pow(sd.lpsi[g], -2)
   sd.lpsi[g] ~ dunif(0,5)
   tau.lp[g] <- pow(sd.lp[g], -2)
   sd.lp[g] ~ dunif(0,5)
}

# Ecological model for latent occurrence z (process model)
for(k in 1:nspec){                   # no change at all down here in model
   logit(psi[k]) <- lpsi[k]
   for (i in 1:M) {
      z[i,k] ~ dbern(psi[k])
   }
}
```

```
# Observation model for observed data ysum
for(k in 1:nspec){        # Loop over species
   logit(p[k]) <- lp[k]
   for (i in 1:M) {
      mu.px[i,k] <- z[i,k] * p[k]   # call mu.px to avoid conflict with above
      ysum[i,k] ~ dbin(mu.px[i,k], J[i])
   }
}

# Derived quantities
for(k in 1:nspec){               # Loop over species
   Nocc.fs[k] <- sum(z[,k])      # Number of occupied sites among the 267
}
for (i in 1:M) {                 # Loop over sites
   Nsite[i] <- sum(z[i,])        # Number of occurring species at each site
}
}
",fill = TRUE)
sink()

# Initial values
zst <- apply(y, c(1,3), max)
zst[is.na(zst)] <- 1
inits <- function() list(z = zst)

# Parameters monitored
params <- c("mu.psi", "mu.lpsi", "sd.lpsi", "mu.p", "mu.lp", "sd.lp")

# MCMC settings
ni <- 6000 ; nt <- 2 ; nb <- 2000 ; nc <- 3

# Call JAGS from R (ART 6 min), look at convergence and summarize posteriors
out7 <- jags(win.data, inits, params, "model7.txt", n.chains = nc, n.thin = nt,
   n.iter = ni, n.burnin = nb, parallel = TRUE)
par(mfrow = c(3,3)) ; traceplot(out7)
print(out7, dig = 3)
```

	mean	sd	2.5%	50%	97.5%	overlap0	f	Rhat	n.eff
mu.psi[1]	0.240	0.063	0.134	0.235	0.375	FALSE	1.000	1.001	2417
mu.psi[2]	0.132	0.028	0.082	0.129	0.192	FALSE	1.000	1.000	6000
mu.psi[3]	0.080	0.031	0.035	0.074	0.151	FALSE	1.000	1.006	347
mu.lpsi[1]	-1.183	0.349	-1.865	-1.181	-0.511	FALSE	0.999	1.001	2291
mu.lpsi[2]	-1.909	0.246	-2.413	-1.905	-1.439	FALSE	1.000	1.000	6000
mu.lpsi[3]	-2.517	0.407	-3.305	-2.523	-1.725	FALSE	1.000	1.006	316
sd.lpsi[1]	2.200	0.275	1.734	2.174	2.820	FALSE	1.000	1.001	2656
sd.lpsi[2]	2.133	0.195	1.794	2.118	2.553	FALSE	1.000	1.000	6000
sd.lpsi[3]	1.871	0.323	1.350	1.835	2.613	FALSE	1.000	1.000	6000
mu.p[1]	0.750	0.033	0.680	0.752	0.810	FALSE	1.000	1.000	6000
mu.p[2]	0.666	0.036	0.591	0.667	0.735	FALSE	1.000	1.000	6000
mu.p[3]	0.412	0.095	0.237	0.410	0.598	FALSE	1.000	1.003	759

```
mu.lp[1]    1.108   0.178   0.753   1.110   1.453   FALSE   1.000   1.000   6000
mu.lp[2]    0.693   0.165   0.369   0.693   1.019   FALSE   1.000   1.000   6000
mu.lp[3]   -0.369   0.407  -1.169  -0.364   0.396    TRUE   0.822   1.003    723
sd.lp[1]    1.031   0.136   0.801   1.018   1.329   FALSE   1.000   1.001   2990
sd.lp[2]    1.292   0.147   1.041   1.280   1.617   FALSE   1.000   1.001   3446
sd.lp[3]    1.903   0.409   1.262   1.851   2.835   FALSE   1.000   1.004    471
```

We note some nice patterns along the body mass gradient in every hyperparameter. The average occupancy probability of small birds (in size class 1) is 0.24, and then drops to 0.13 and 0.08 for species in size classes 2 and 3, respectively. The interspecific variability in occupancy probability follows a similar pattern, as suggested by the estimates for sd.lpsi. The average detection probability decreases with increasing average body mass of a species (you see this in mu.p), while the interspecific variability in detection probability seems to increase with mean body mass (look at sd.lp).

Perhaps you feel that binning body mass is artificial. Of course, you can easily fit a model where body mass is treated as a continuous covariate, thus avoiding the categorizing of species. We next fit a linear regression of the hyperparameters governing species-specific values of occupancy and detection probability on the *continuous* mass covariate (which we log-transform). This is model 8, and it looks like this (the only change to the previous model is again in the part specifying the nature of species heterogeneity).

Interspecific Comparison 3: Linear Regressions for Means and Variances

Models of species heterogeneity:
$$\text{logit}(\psi_k) \sim Normal\left(\mu_{lpsi,k}, \sigma^2_{lpsi,k}\right)$$
$$\text{logit}(p_k) \sim Normal\left(\mu_{lp,k}, \sigma^2_{lp,k}\right)$$
$$\mu_{lpsi,k} = delta0.lpsi + delta1.lpsi * \log(mass_k)$$
$$\mu_{lp,k} = delta0.lp + delta1.lp * \log(mass_k)$$
$$\log\left(\sigma^2_{lpsi,k}\right) = phi0.lpsi + phi1.lpsi * \log(mass_k)$$
$$\log\left(\sigma^2_{lp,k}\right) = phi0.lp + phi1.lp * \log(mass_k)$$

Now, each of the four hyperparameters that govern the two species traits (occupancy and detection probability) is indexed by species (k), and we simply add to our model four linear regressions on the log10 of body mass. To enforce the range constraint for a variance, we apply the linear model for variances with a log link function. Now, the "top-level" parameters in our HM (the hyper-hyperparameters) are the four intercepts and the four slopes. To keep our notation for covariate effects tidy, we now call these hypercoefficients delta and phi.

```
# Bundle and summarize data set
logmass <- as.numeric(log10(mass))      # Take log10 of body mass
str( win.data <- list(ysum = ysum, logmass = logmass, M = nrow(ysum),
    J = data$nsurvey[1:nsite], nspec = dim(ysum)[2]) )

# Specify model in BUGS language
sink("model8.txt")
cat("
model {
```

```
# Priors
for(k in 1:nspec){                      # loop over species
    lpsi[k] ~ dnorm(mu.lpsi[k], tau.lpsi[k])    # now all indexed by k, not g
    tau.lpsi[k] <- 1/var.lpsi[k]
    lp[k] ~ dnorm(mu.lp[k], tau.lp[k])
    tau.lp[k] <- 1/var.lp[k]
    mu.lpsi[k] <- delta0.lpsi + delta1.lpsi * logmass[k]
    mu.lp[k] <- delta0.lp + delta1.lp * logmass[k]
    log(var.lpsi[k]) <- phi0.lpsi + phi1.lpsi * logmass[k]
    log(var.lp[k]) <- phi0.lp + phi1.lp * logmass[k]
}
# Priors for regression params for means
delta0.lpsi ~ dnorm(0, 0.01)
delta1.lpsi ~ dnorm(0, 0.01)
delta0.lp ~ dnorm(0, 0.01)
delta1.lp ~ dnorm(0, 0.01)
# Priors for regression params for variances
phi0.lpsi ~ dnorm(0, 0.01)
phi1.lpsi ~ dnorm(0, 0.01)
phi0.lp ~ dnorm(0, 0.01)
phi1.lp ~ dnorm(0, 0.01)

# Ecological model for latent occurrence z (process model)
for(k in 1:nspec){
    logit(psi[k]) <- lpsi[k]
    for (i in 1:M) {
        z[i,k] ~ dbern(psi[k])
    }
}

# Observation model for observed data ysum
for(k in 1:nspec){                      # Loop over species
    logit(p[k]) <- lp[k]
    for (i in 1:M) {
        mu.p[i,k] <- z[i,k] * p[k]
        ysum[i,k] ~ dbin(mu.p[i,k], J[i])
    }
}

# Derived quantities
for(k in 1:nspec){                      # Loop over species
    Nocc.fs[k] <- sum(z[,k])            # Number of occupied sites among the 267
}
for (i in 1:nsite) {                    # Loop over sites
    Nsite[i] <- sum(z[i,])              # Number of occurring species at each site
}
}
",fill = TRUE)
sink()
```

11.6 COMMUNITY MODELS THAT FULLY RETAIN SPECIES IDENTITY

```
# Initial values
zst <- apply(y, c(1,3), max)
zst[is.na(zst)] <- 1
inits <- function() list(z = zst, delta0.lpsi = rnorm(1), delta1.lpsi = rnorm(1),
   delta0.lp = rnorm(1), delta1.lp = rnorm(1), phi0.lpsi = rnorm(1),
   phi1.lpsi = rnorm(1), phi0.lp = rnorm(1), phi1.lp = rnorm(1))

# Parameters monitored
params <- c("delta0.lpsi", "delta1.lpsi", "delta0.lp", "delta1.lp", "phi0.lpsi",
"phi1.lpsi", "phi0.lp", "phi1.lp", "psi", "p", "Nocc.fs", "Nsite")

# MCMC settings
ni <- 12000 ; nt <- 2 ; nb <- 2000 ; nc <- 3

# Call JAGS from R (ART 12 min), look at convergence and summarize posteriors
out8 <- jags(win.data, inits, params, "model8.txt", n.chains = nc, n.thin = nt,
n.iter = ni, n.burnin = nb, parallel = TRUE)
par(mfrow = c(3,3))
traceplot(out8, c('delta0.lpsi', 'delta1.lpsi', 'delta0.lp', 'delta1.lp', 'phi0.lpsi',
'phi1.lpsi', 'phi0.lp', 'phi1.lp'))
print(out8, dig = 3)
```

	mean	sd	2.5%	50%	97.5%	overlap0	f	Rhat	n.eff
delta0.lpsi	-0.553	0.490	-1.530	-0.552	0.414	TRUE	0.871	1.001	3854
delta1.lpsi	-0.705	0.239	-1.177	-0.702	-0.238	FALSE	0.998	1.001	2538
delta0.lp	2.038	0.300	1.449	2.037	2.633	FALSE	1.000	1.000	15000
delta1.lp	-0.758	0.175	-1.105	-0.756	-0.424	FALSE	1.000	1.000	15000
phi0.lpsi	1.716	0.376	0.995	1.711	2.467	FALSE	1.000	1.001	1751
phi1.lpsi	-0.159	0.192	-0.533	-0.158	0.221	TRUE	0.799	1.001	1743
phi0.lp	-0.477	0.414	-1.260	-0.491	0.369	TRUE	0.873	1.002	1157
phi1.lp	0.480	0.219	0.044	0.484	0.897	FALSE	0.985	1.002	1936

This example highlights the considerable power that BUGS gives you for testing biological hypotheses even inside a fairly complex HM! We finish up this section by plotting the body mass relationships of parameters for the community mean and community heterogeneity (=variance) of the logits of occupancy and detection probabilities in Swiss birds (Figure 11.12). On average, larger species are less widespread and less detectable than smaller species. There is less unexplained variability among species in this relationship for occupancy in larger species, while for detection, larger species are more variable around this relationship than are smaller species. We note, though, that these relationships are less clear for the two variances, and the 95% CRI of phi1.lpsi covers 0.

```
# Get covariate values for prediction
predm <- seq(10, 10000, ,500)          # Predict for mass of 10g to 10 kg
pred.logm <- log10(predm)

# Compute predictions (all in one array)
tmp <- out8$sims.list                  # Grab simulation list
nsamp <- out8$mcmc.info$n.samples      # Number of MCMC samples
pred <- array(NA, dim = c(500, nsamp, 4))   # Array for predictions
```

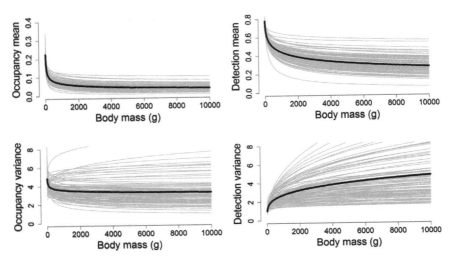

FIGURE 11.12

Estimates of the community relationships between occupancy and detection probability and body size in Swiss birds (top: community means, bottom: community variance, or interspecific variability). Gray lines show a random sample of size 100 of the posterior distribution of these relationships and the blue lines show the posterior mean of these relationships.

```
# [,,1] mu.psi, [,,2] mu.p, [,,3] var.lpsi, [,,4] var.lp
for(i in 1:nsamp){                          # Fill array
  pred[,i,1] <- plogis(tmp$delta0.lpsi[i] + tmp$delta1.lpsi[i] * pred.logm)
  pred[,i,2] <- plogis(tmp$delta0.lp[i] + tmp$delta1.lp[i] * pred.logm)
  pred[,i,3] <- exp(tmp$phi0.lpsi[i] + tmp$phi1.lpsi[i] * pred.logm)
  pred[,i,4] <- exp(tmp$phi0.lp[i] + tmp$phi1.lp[i] * pred.logm)
}

# Plot posterior mean and a random sample of 100 from posterior of regression
selection <- sample(1:nsamp, 100)         # Choose random sample of MCMC output
par(mfrow = c(2,2), mar = c(5,5,2,2))
matplot(predm, pred[,selection,1], ylab = "Occupancy mean", xlab = "Body mass (g)",
   type = "l", lty = 1, lwd = 1, col = "grey", ylim = c(0, 0.4), frame = F)
lines(predm, apply(pred[,,1], 1, mean), lwd = 3, col = "blue")
matplot(predm, pred[,selection,2], ylab = "Detection mean", xlab = "Body mass (g)",
   type = "l", lty = 1, lwd = 1, col = "grey", ylim = c(0, 0.8), frame = F)
lines(predm, apply(pred[,,2], 1, mean), lwd = 3, col = "blue")
matplot(predm, pred[,selection,3], ylab = "Occupancy variance", xlab = "Body mass (g)",
   type = "l", lty = 1, lwd = 1, col = "grey", ylim = c(0, 8), frame = F)
lines(predm, apply(pred[,,3], 1, mean), lwd = 3, col = "blue")
matplot(predm, pred[,selection,4], ylab = "Detection variance", xlab = "Body mass (g)",
   type = "l", lty = 1, lwd = 1, col = "grey", ylim = c(0, 8), frame = F)
lines(predm, apply(pred[,,4], 1, mean), lwd = 3, col = "blue")
```

We make two final comments, one on "phylogenetic independence," and the other on the inference about unseen species. First, our analysis treats every species as independent, but in an evolutionary sense, they are not: two closely related species may be similar purely because of common ancestry, and you may want to adjust for such phylogenetic nonindependence (Dorazio and Connor, 2014). If you want to weigh species by their phylogenetic "similarity," such that two closely related species contribute less information to the among-species model than do two species that are more distantly related, you could specify a multivariate normal distribution for the species effects, and model the effects of a relationship matrix (which expresses how closely every pair of species are related) into the variance-covariance matrix (Ives and Zhu, 2006; Waldmann, 2009; Papaïx et al., 2010). Second, the extension of this model with species-specific covariates to include unseen species in the metacommunity (see Section 11.7) is straightforward. Except that your model then must include a submodel to estimate the values of such missing "individual covariates" for the unseen species (Royle, 2009; Chapter 6 in Kéry and Schaub, 2012; and also see Chapters 7–8 in this book).

11.6.4 MODELING SPECIES RICHNESS IN A TWO-STEP ANALYSIS

In a community model composed of component models for the presence and absence of individual species, species richness is not a structural parameter that can be modeled (though it is in the multinomial mixture model; see our comments in Sections 11.4 and 11.5.3, as well as Exercise 2, and note that we could model species richness directly also by adopting a logit-linear model for the data-augmentation parameter (Sutherland et al., in review) or using the construction explained in Section 7.8.4.). Instead, species richness is a quantity computed from the matrix of the individual species presence indicators (the presence/absence matrix Z). Often, questions in community ecology revolve around relationships between species richness (N) and environmental conditions. How can these be addressed in a community occupancy framework?

Here is a simple two-step analysis, which has been called "doing statistics on statistics" (Link, 1999), but which may be a useful first exploratory step in hierarchical modeling (Murtaugh, 2007): we take estimates from one analysis (here, our community size estimates `Nsite`) and plug them into a second analysis to relate them to the environmental variables of our choice; see Tingley and Beissinger (2013) for an example in the context of a community model. The only problem is that these estimates from the first-step analysis are not independent (they typically have covariances), *and* they come with estimation uncertainty (standard errors). To do a two-step analysis properly, we must propagate the estimation uncertainty from the first analysis into the second analysis; otherwise, we will typically underestimate the uncertainty in the final analysis—i.e., get CIs that are too narrow and with too many significant test results. To do this right is not so trivial, and may often jeopardize the very reason for doing a two-step analysis of an HM, which is simplicity. For instance, it would be *wrong* to conduct a second-step analysis and simply weigh each estimate by the reciprocal of its squared standard error, since this would assume that the residuals in the second analysis are composed exclusively of estimation uncertainty from the first analysis (Jenni and Kéry, 2003).

Here we show how to properly do such a two-step analysis with propagation of the first-step estimation uncertainty into the second-step analysis (we ignore the covariances, but see below that they appear to be negligible). We illustrate by exploring the relationship between richness and elevation using species richness (`Nsite`) estimates from model 5, where we would assume that the lack of random species effects would minimize any covariances between the estimates of the Ns.

CHAPTER 11 HIERARCHICAL MODELS FOR COMMUNITIES

This analysis can also be called a meta-analysis, because it synthesizes multiple estimates into a single estimate. In a more typical meta-analysis, the former come from different studies, while here they are simply different estimates from a single model. The code in this section is partly taken from McCarthy and Masters (2005).

```
# Extract estimates of N from model 5
N.pm <- out5$summary[291:557, 1]       # Posterior means of Nsite
N.psd <- out5$summary[291:557, 2]      # ... posterior sd's of Nsite
N.cri <- out5$summary[291:557, c(3,7)] # ... CRL's of Nsite

# Plot estimates as a function of elevation
elev <- data$elev[1:267]
plot(elev, N.pm, xlab = "Altitude (m a.s.l.)", ylab = "Estimated avian species richness",
ylim = c(0, 70), frame = F)
segments(elev, N.cri[,1], elev, N.cri[,2], col = "grey")
lines(smooth.spline(N.pm ~ elev, w = 1 / N.psd), col = "grey", lwd = 3)
```

We fit a simple regression model with cubic polynomials of elevation, but one that has two residual components. The first is the estimation uncertainty coming from the first-step analysis, the magnitude of which is assumed to be known: this is the posterior standard deviation of the community size estimate Nsite. The second component is the usual lack of fit component, which allows the individual data point to lie off the modeled relationship. This component will be estimated from the data. In our model, we will compute predictions of species richness for a range of values for elevation.

```
# Bundle and summarize data set
pred.ele <- (seq(200, 2750,5) - mean.ele) / sd.ele      # elevation standardised
str(win.data <- list(ele = ele, N = N.pm, psd = N.psd, n = length(N.pm),
pred.ele = pred.ele, npred = length(pred.ele)))

# Define model in BUGS language
sink("meta.analysis.txt")
cat("
model{

# Priors
for(v in 1:4){      # Priors for intercept and polynomial coefficients
   beta[v] ~ dnorm(0, 0.0001)
}
tau.site <- pow(sd.site, -2)
sd.site ~ dunif(0,10)

# Likelihood
for (i in 1:n){
   N[i] ~ dnorm(muN[i], tau.psd[i])    # Measurement error model for estimated N
   tau.psd[i] <- pow(psd[i], -2)       # 'Known' part of residual: meas. error
   muN[i] <- beta[1] + beta[2] * ele[i] + beta[3] * pow(ele[i],2) +
   beta[4] * pow(ele[i],3) + eps.site[i]   # add another source of uncertainty
   eps.site[i] ~ dnorm(0, tau.site)    # this is the usual 'residual'
}
```

11.6 COMMUNITY MODELS THAT FULLY RETAIN SPECIES IDENTITY

```
# Get predictions for plot
for(i in 1:npred){
    Npred[i] <- beta[1] + beta[2] * pred.ele[i] + beta[3] * pow(pred.ele[i],2) +
beta[4] * pow(pred.ele[i],3)
}
} # end model
",fill=TRUE)
sink()

# Initial values, params monitored, and MCMC settings
inits <- function() list(beta = rnorm(4))
params <- c("beta", "sd.site", "Npred")
ni <- 12000 ; nt <- 10 ; nb <- 2000 ; nc <- 3

# Call JAGS and summarize posterior
out <- jags(win.data, inits, params, "meta.analysis.txt", n.chains = nc, n.thin
= nt, n.iter = ni, n.burnin = nb)
print(out, 3)
```

We add to our plot the predictions of species richness from our meta-analysis, with 95% CRI of the polynomial relationship.

```
lines(seq(200, 2750,5), out$mean$Npred, col = "blue", lwd = 3)
matlines(seq(200,2750,5), out$summary[6:516,c(3, 7)], col = "blue", lwd = 2,
lty= "dashed")
```

The analysis represented by the blue lines in Figure 11.13 (though not the gray spline smooth) properly propagates the uncertainty in the estimates of N from the first analysis into the estimates of the second analysis, but ignores covariances. We can see how bad this may be by plotting the joint posteriors of N for pairs of sites. They seem to be mostly uncorrelated (most clouds are fairly round), and we conclude that the covariances of the N estimates are small at best, and hence results from our two-step analysis are likely to be adequate.

```
par(mfrow = c(3, 3), mar = c(5,4,3,2))
for(i in 1:267){
    for(j in 1:267){
        plot(jitter(out5$sims.list$Nsite[,i]), jitter(out5$sims.list$Nsite[,j]),
        main = paste("Joint posterior sites", i, "and", j))
#       browser()
    }
}
```

Such a Bayesian meta-analysis is a very useful way of synthesizing in a single estimate or in several parameters (here, the coefficients of a polynomial regression) a collection of estimates while properly accommodating, and therefore weighing, the uncertainty in these estimates. In addition to formal model parameters, you can easily do this for quantities that are *not* formal parameters in your model and for which you therefore cannot directly specify a submodel, as for N here. Finally, this method can be very useful to model parameters in a two-step hierarchical procedure as a substitute for fitting a formal HM all at once. Sometimes, such a two-step analysis (which is a hierarchical procedure, but not

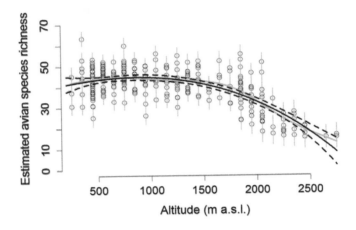

FIGURE 11.13

Relationship between elevation and estimated avian species richness in Switzerland at 1-km² sample sites, conditional on the 145 species detected at least once during the 2014 surveys. Symbols denote point estimates with 95% CRIs from community model 5. Gray line is a spline smooth with weights equal to the reciprocal of the squared posterior standard deviations. Blue line is the cubic regression line estimated in the meta-analysis that accounts for both estimation error (posterior standard deviations) and residual variation around the regression line. The dashed blue lines give the 95% CRI of the prediction.

a hierarchical model) may be easier to understand or to explain to somebody else (Murtaugh, 2007), or it may be necessary because your full HM is too big to be fit by your computer all at once (see Sauer and Link, 2002, 2011, for a good example of the latter).

11.7 THE DORAZIO/ROYLE (DR) COMMUNITY OCCUPANCY MODEL WITH DATA AUGMENTATION (DA)

In all previous models, our inferences were restricted to the set of those 145 species that were detected at least once during the MHB surveys in 2014. We did *not* explicitly extend the scope of inference in these models to any species that may have been exposed to sampling, but happened to be missed. Species may be missed because either they did occur in the general area (i.e., were part of the regional species pool), but happened to be absent from the particular 267 sites surveyed; or the species did occur in the particular set of surveyed sites, but simply failed to be detected during the repeated "measurements" of their presence/absence. In models 7 and 8, we already extended the scope of inference from the 145 observed species to some undefined larger statistical "population" or community of species, by treating species-specific parameters as random effects. Thereby, we were able to formally estimate certain quantities of that community: these were the hyperparameters governing species-specific random effects. In the final and most sophisticated set of models in this chapter, we will formally extend the scope of inference to *all* species that constitute the avian metacommunity sampled by the Swiss MHB. This model will allow us to also make inferences about occurring species that were missed by the surveys. That is, to species that were not detected in any survey of any sample quadrat.

11.7 THE DORAZIO/ROYLE (DR) COMMUNITY OCCUPANCY MODEL

By failing to account for such missed species, any estimator of a community quantity such as species richness or the average response to some covariate, under all previous models, was conditional on the list of 145 detected species. If there are any undetected species, the previous estimators also underestimate the true species richness at each site. Furthermore, with our previous models, we were unable to estimate the total size of the metacommunity. Finally, our previous models may give biased results with respect to the total metacommunity, because the 145 species that were detected at least once may be a biased sample from the total metacommunity. For instance, quite likely, the sample of detected species will be biased positively with respect to the average detection probability in the metacommunity. As a consequence, the sample of species detected may also be biased in other traits directly or indirectly related to detection probability of a species; probably, for instance, occupancy (see Section 11.6.2).

Our method of making an inference about species that were never detected is again based on *parameter-expanded data augmentation*, or *data augmentation* for short (PX-DA or DA) (Tanner and Wong, 1987; Liu and Wu, 1999; Royle et al., 2007a; Royle and Dorazio, 2012; see also Sections 7.8.1.4 and 8.3). Dorazio and Royle (2005) developed a multispecies occupancy community model that extends inferences to unseen species, and Dorazio et al. (2006) fit this model using DA with BUGS. In this section, we will fit two such Dorazio/Royle (DR) community occupancy models. We will see that they are conceptually fairly simple extensions of the models in Section 11.6; hence, everything that we said and did there remains relevant for the full DR community models with DA.

11.7.1 THE SIMPLEST DR COMMUNITY MODEL WITH DA

To formally extend our inferences to the unseen species in a metacommunity, we imagine a superpopulation of, say, M species (where M is much larger than the total number of species in the metacommunity N), and then we estimate which members of M are also members of N. That is, which of a number of M potentially occurring species are exposed to our spatial sample comprising our S sites. (In this section, we again deviate slightly from our standard notation, and use S instead of M for the number of study sites, as we did in Chapters 7–9.) We achieve this by PX-DA (Royle et al., 2007a). In a sense, "DA" comes before "PX": it means that we first add all-zero detection histories for a number of $nz = M - n$ additional potential species, where n is the number of observed species (here 145) and nz is the number of all-zero species added. The PX part means that we add another hierarchical layer (represented by another parameter) to our model, one that describes the random sampling of the N occurring "real" species from the total of M potential species. This sampling process is represented by a Bernoulli random variable w, which is an indicator for a species that is part of the studied metacommunity (we also call it the data augmentation variable). Consequently, our hierarchical community model now has three main levels and can be written in following conditional (i.e., linked) probability statements:

1. Superpopulation process : $w_k \sim Bernoulli(\Omega)$
2. State process (occurrence) : $z_{ik}|w_k \sim Bernoulli(w_k \psi_k)$
3. Observation process (detection) : $ysum_{ik}|z_{ik} \sim Binomial(J_i, z_{ik} p_k)$
4. Models of species heterogeneity : $logit(\psi_k) \sim Normal\left(\mu_{lpsi}, \sigma^2_{lpsi}\right)$

   $logit(p_k) \sim Normal\left(\mu_{lp}, \sigma^2_{lp}\right)$

The species index k now runs from 1 to M—i.e., up to the number of species in the full augmented data set rather than only up to the 145 observed species. But otherwise, the model of species heterogeneity is *exactly* the same as for model 6 (except for the trivial difference that here we choose not to specify a correlation parameter). In a sense, the "community occupancy parameter" Ω now takes the place of the metacommunity size N, since the expected value of metacommunity size N is $M\Omega$. You can think of this as exactly analogous to an occupancy problem, where the number of occupied sites (corresponding to N) can be estimated as the product of the occupancy parameter (corresponding to Ω) and the total number of surveyed sites (corresponding to M). Note, hence, that Ω has no meaning if you don't know M.

Table 11.2 shows a conceptual outline of the community occupancy model with PX-DA for the Swiss MHB 2014 data set. Purely for layout reasons, we transpose the matrices $ysum$ and $\mathbf{Z}$, but this doesn't change anything in the model. The observed data (the dark-gray shaded rectangle) is represented by $ysum_{ki}$, i.e., the detection frequencies of the 145 species observed in the 267 quadrats. To these we will add 150 species-detection histories containing only zeros (this is the data augmentation part represented by the medium-gray shaded rectangle below), bringing our total augmented data set to

Table 11.2 Conceptual outline of the DR community occupancy model for the Swiss MHB 2014 data set.

		Observed: $ysum$					Only partially observed: $\mathbf{Z}$ and w					
	Quadrat i	1	2	3	...	267	1	2	3	...	267	w_k
Species k	1	3	2	0	...	2	1	1	1	...	1	1
	2	0	0	1	...	2	0	1	1	...	1	1
	3	2	0	1	...	0	1	0	1	...	1	1
	...	...	...	...	...	...	...	...	...	...	...	...
n	145	0	0	1	...	0	0	1	1	...	1	1
n+1	146	0	0	0	...	0	0	1	0	...	0	1
	...	...	...	...	...	...	...	...	...	...	...	...
	...	0	0	0	...	0	0	0	1	...	0	1
N	?	0	0	0	...	0	0	0	0	0	0	0
N+1	?+1	0	0	0	...	0	0	0	0	...	0	0
	...	...	...	...	...	...	...	...	...	...	...	...
	...	...	...	...	...	...	...	...	...	...	...	...
M=n+nz	295	0	0	0	...	0	0	0	0	...	0	0

11.7 THE DORAZIO/ROYLE (DR) COMMUNITY OCCUPANCY MODEL

$M = 295$ "potential" species. The model permits inferences about two latent structures, the true (augmented) presence/absence matrix $\mathbf{Z}$ (i.e., the M-by-267 matrix containing the occurrence indicators z_{ki}) and the "metacommunity membership indicators" or DA variables w_k (both represented by yellow shading to indicate the studied metacommunity). Essentially, one main aim of the modeling is to estimate the unobserved values in the arrays $\mathbf{Z}$ and $\mathbf{w}$, which are shown in red. Note that we show the $\mathbf{Z}$ matrix for the studied metacommunity with fewer than N species to emphasize that the "conditional" metacommunity size (the number of species that occur in the actual sample of 267 survey sites) may well be less than the number of species that form the entire metacommunity (N) inhabiting the wider area sampled by these sites. The "unconditional" metacommunity size is the asymptote of the former—i.e., the number of species actually occurring anywhere in your sampled sites as the number of sample sites goes to infinity (Dorazio et al., 2006; Dupuis et al., 2011).

DA in models for abundance or related quantities can always be imagined as turning a sort of logistic regression model into an occupancy model where each "individual" appears on a row in the data set. We do this by adding to the data a large number of potential and undetected individuals with all-zero detection histories and by adding to the logistic regression model one hierarchical layer representing the presence/absence indicator z (which we here denote w). We have used DA extensively in the multinomial abundance models in Chapters 7—9. What perhaps makes DA more challenging to grasp in the context of the DR community model is that even the basic DR models without DA (i.e., the models in Section 11.6) are *already* site-occupancy models (we have one occupancy model for each observed species). So when we do DA to extend our inferences to the entire metacommunity, we add a second level of such an occupancy, or zero-inflated, type of model. If you find this concept challenging to grasp, then it is best to first try to fully understand DA in the simpler context of estimation of abundance in a closed population (e.g., see Chapter 6 in Kéry and Schaub, 2012; Royle et al., 2007a; Royle and Dorazio, 2012, and Chapters 7—9 in this book). Let's now look at PX-DA in the community models in practice.

```
# Augment data set (DA part)
nz <- 150                # Number of potential species in superpopulation
M <- nspec + nz          # Size of augmented data set ('superpopulation')
yaug <- cbind(ysum, array(0, dim=c(nsite, nz)))    # Add all zero histories

# Bundle and summarize data set
str( win.data <- list(yaug = yaug, nsite = nrow(ysum), nrep =
data$nsurvey[1:nsite], M = M, nspec = nspec, nz = nz) )

# Specify model in BUGS language
sink("model9.txt")
cat("
model {

# Priors to describe heterogeneity among species in community
for(k in 1:M){                 # Loop over all species in augmented list
   lpsi[k] ~ dnorm(mu.lpsi, tau.lpsi)
   lp[k] ~ dnorm(mu.lp, tau.lp)
}

# Hyperpriors to describe full community
omega ~ dunif(0,1)             # Data augmentation or 'occupancy' parameter
```

```
mu.lpsi ~ dnorm(0,0.001)              # Community mean of occupancy (logit)
mu.lp ~ dnorm(0,0.001)                # Community mean of detection (logit)
tau.lpsi <- pow(sd.lpsi, -2)
sd.lpsi ~ dunif(0,5)                  # Species heterogeneity in logit(psi)
tau.lp <- pow(sd.lp, -2)
sd.lp ~ dunif(0,5)                    # Species heterogeneity in logit(p)

# Superpopulation process: this is the 'paramater expansion' part of PX-DA
for(k in 1:M){
   w[k] ~ dbern(omega)                # Metacommunity membership indicator
}                                     # (or data augmentation variable)

# Ecological model for latent occurrence z (process model)
for(k in 1:M){
   mu.psi[k] <- w[k] * psi[k]         # species not part of community zeroed out for z
   logit(psi[k]) <- lpsi[k]
   for (i in 1:nsite) {
      z[i,k] ~ dbern(mu.psi[k])
   }
}

# Observation model for observed detection frequencies
for(k in 1:M){
   logit(p[k]) <- lp[k]
   for (i in 1:nsite) {
      mu.p[i,k] <- z[i,k] * p[k]      # non-occurring species are zeroed out for p
      yaug[i,k] ~ dbin(mu.p[i,k], nrep[i])
   }
}

# Derived quantities
for(k in 1:M){
   Nocc.fs[k] <- sum(z[,k])           # Number of occupied sites among the 267
}
for (i in 1:nsite) {
   Nsite[i] <- sum(z[i,])             # Number of occurring species at each site
}
n0 <- sum(w[(nspec+1):(nspec+nz)])    # Number of unseen species in metacommunity
Ntotal <- sum(w[])                    # Total metacommunity size (= nspec + n0)
}
",fill = TRUE)
sink()

# Initial values
wst <- rep(1, nspec+nz)                     # Simply set everybody at 'occurring'
zst <- array(1, dim = c(nsite, nspec+nz))   # ditto for z
inits <- function() list(z = zst, w = wst, lpsi = rnorm(n = nspec+nz), lp =
rnorm(n = nspec+nz))
```

11.7 THE DORAZIO/ROYLE (DR) COMMUNITY OCCUPANCY MODEL

```
# Parameters monitored
params <- c("mu.lpsi", "sd.lpsi", "mu.lp", "sd.lp", "psi", "p", "Nsite",
"Ntotal", "omega", "n0")

# MCMC settings
ni <- 22000 ; nt <- 2 ; nb <- 2000 ; nc <- 3

# Call JAGS from R (ART 62 min), check convergence and summarize posteriors
out9 <- jags(win.data, inits, params, "model9.txt", n.chains = nc, n.thin = nt,
n.iter = ni, n.burnin = nb, parallel = TRUE)
par(mfrow = c(2,2)) ; traceplot(out9, c('mu.lpsi', 'sd.lpsi', 'mu.lp', 'sd.lp'))
print(out9, dig = 3)
```

	mean	sd	2.5%	50%	97.5%	overlap0	f	Rhat	n.eff
mu.lpsi	-2.511	0.343	-3.285	-2.480	-1.937	FALSE	1	1.004	808
sd.lpsi	2.596	0.263	2.153	2.573	3.178	FALSE	1	1.005	530
mu.lp	0.639	0.127	0.382	0.642	0.881	FALSE	1	1.000	14319
sd.lp	1.332	0.114	1.129	1.325	1.576	FALSE	1	1.001	2335
psi[1]	0.022	0.010	0.007	0.020	0.044	FALSE	1	1.000	28192
psi[2]	0.016	0.010	0.004	0.014	0.041	FALSE	1	1.001	15194
psi[3]	0.018	0.009	0.005	0.016	0.039	FALSE	1	1.000	30000
[... output truncated ...]									
Nsite[265]	17.654	2.222	14.000	18.000	22.000	FALSE	1	1.000	10161
Nsite[266]	49.997	1.333	48.000	50.000	53.000	FALSE	1	1.000	8094
Nsite[267]	52.619	1.209	51.000	52.000	55.000	FALSE	1	1.000	4639
Ntotal	164.818	9.102	152.000	163.000	187.000	FALSE	1	1.006	570
omega	0.558	0.042	0.484	0.556	0.647	FALSE	1	1.003	978
n0	19.818	9.102	7.000	18.000	42.000	FALSE	1	1.006	570

Now, estimates of local species richness or community size, Nsite, *do* include a contribution from those n0 species that were never seen during the surveys in 2014. Consequently, we can now also estimate the total number of species in the larger area from which the 267 study quadrats were drawn as a random sample (this is quantity *N* or Ntotal in the model). We can look at the posterior distributions of the site-specific number of occurring species, and compare these estimates with the observed number of species (Figure 11.14). We see fairly different species-coverage rates in different quadrats (compare, for instance, quadrat 9 and 250, with a larger proportion of species estimated to have been missed in the latter). Quadrat 30 was not surveyed in 2014, and yet we can make a formal guess of the likely number of species occurring in it. In this trivial model, this may not be a very interesting estimate, but below we will see how we can refine such estimates for unsurveyed sites in a model with covariates on detection and occupancy.

```
# Plot posterior distribution of site-specific species richness (Nsite)
par(mfrow = c(3,3), mar = c(5,4,3,2))
for(i in 1:267){
  plot(table(out9$sims.list$Nsite[,i]), main = paste("Quadrat", i),
  xlab = "Local species richness", ylab = "", frame = F,
  xlim = c((min(C[i], out9$sims.list$Nsite[,i], na.rm = T)-2),
  max(out9$sims.list$Nsite[,i])) )
  abline(v = C[i], col = "grey", lwd = 4)
  browser()
}
```

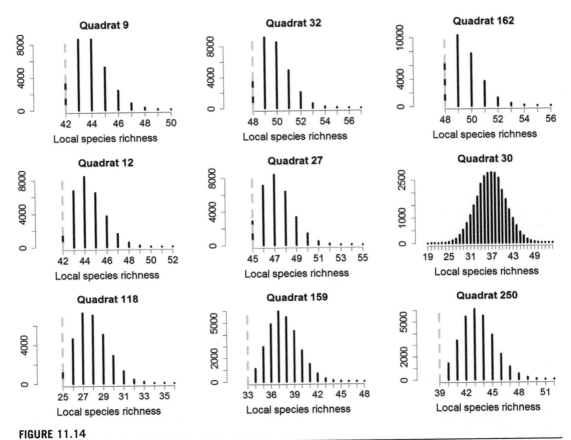

FIGURE 11.14

Posterior distributions of local species richness, or community size (Nsite), at a selection of nine 1-km^2 sample quadrats in the Swiss breeding bird survey MHB. Gray dashed lines show the observed number of species in that year. Site 30 was not surveyed in 2014.

```
# Plot it only for a selection of sites
par(mfrow = c(3,3), mar = c(5,4,3,2))
for(i in c(9, 32, 162, 12, 27, 30, 118, 159, 250)){
    plot(table(out9$sims.list$Nsite[,i]), main = paste("Quadrat", i),
    xlab = "Local species richness", ylab = "", frame = F,
    xlim = c((min(C[i], out9$sims.list$Nsite[,i], na.rm = T)-2),
    max(out9$sims.list$Nsite[,i])))
    abline(v = C[i], col = "grey", lwd = 4)
}
```

To finish up, we plot the posterior distribution of the metacommunity size (Ntotal)—i.e., the estimated number of species that occur in some wider area that is sampled by our 267 survey quadrats (Figure 11.15). This "regional pool" of species is a somewhat hypothetical value, and strictly speaking,

11.7 THE DORAZIO/ROYLE (DR) COMMUNITY OCCUPANCY MODEL

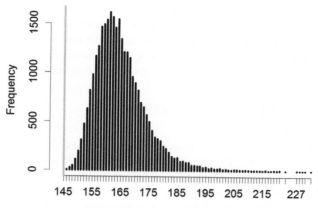

FIGURE 11.15

Posterior distribution of the metacommunity size in the Swiss MHB or, more precisely, the expected number of bird species occurring in some larger area that is sampled by the 267 quadrats in the hypothetical case that the number of sample quadrats goes to infinity. Gray line indicates the 145 species observed in the year 2014.

we don't know exactly to which area it refers. Given the nice spatial coverage of Switzerland by the MHB sample (see Figures 1.3 and 10.12), we can loosely think of Ntotal as an estimate of the total number of breeding bird species in Switzerland. Sometimes estimates of Ntotal agree quite well with the known number of species in some larger region such as Switzerland (e.g., in the analysis in Kéry and Royle, 2009), but there is no reason that they ought to do so always. Since our model is not spatially explicit, we do not know the effective sample area, and hence cannot say to which area our metapopulation size estimate refers.

```
# Plot posterior distribution of total species richness (Ntotal)
plot(table(out9$sims.list$Ntotal), main = "", ylab = "", xlab = "Avian
metacommunity size in Swiss MHB survey (267 1km2 quadrats)", frame = F, xlim =
c(144, 245))
abline(v = nspec, col = "grey", lwd = 4)
```

We estimate that there were 165 species (95% CRI 152–187), and therefore that 20 species (95% CRI 7–42) did either not occur in the 267 sample quadrats or they did occur in the surveyed quadrats, but they were missed by the observers surveying the quadrats. If we make the *ad hoc* interpretation that Ntotal is the number of Swiss breeding birds, then this is an underestimate. As of 2015, there were 215 known breeding bird species in Switzerland (179 regular, 20 irregular, 16 occasional; V. Keller, pers. comm.); hence, in any one year there might be 190–200 species breeding in Switzerland.

Our current model does not use all information about the occurrence of a species at a site where it was not detected. There are at least two further sources of such information: one comes from covariate relationships and the other from co-occurring species (Clark et al., 2014). (A third source of such information might be represented by "spatial relationships", i.e., if a species is known at a nearby site, it may also be more likely to occur at a given site, even if undetected; see Chapters 21–22 for possible ways in which such information could be introduced in the model.) Exploiting these relationships with the habitat and the occurrence of other bird species would allow us to further improve our estimates of how likely it is for a given species to occur at a site where it was not detected. Therefore, we might also improve our community and metacommunity size estimates. In the next section, we will add covariate information into the model. Hierarchical modeling of species co-occurrence patterns as well as the modeling of spatial autocorrelation in HMs will be dealt with in volume 2.

11.7.2 DR COMMUNITY MODEL WITH COVARIATES

When we incorporate covariate information into our DR community model, we can study the aggregate response of all species in a community to such covariates by looking at the mean and variance hyperparameters governing the species-specific occupancy and detection parameters. In addition, we can also investigate species-specific relationships. This is a great advantage over simpler approaches that directly model species richness as a response to the environment. To model time-specific covariates into the detection part of the DR community model, our current response (the detection frequency $ysum_{ik}$) must be disaggregated. That is, we must go back to the original detection/nondetection data y_{ijk} and treat them as a Bernoulli random variable. We can write the most complex occupancy model in this chapter as:

1. Superpopulation process : $\quad w_k \sim Bernoulli(\Omega)$
2. State process (occurrence) : $\quad z_{ik}|w_k \sim Bernoulli(w_k \psi_k)$
3. Observation process (detection) : $\quad y_{ijk}|z_{ik} \sim Bernoulli(z_{ik} p_{ijk})$
4. Models of species heterogeneity : $\quad logit(\psi_{ik}) = lpsi_k + betalpsi_k * elevation_i + \ldots$
   $\quad logit(p_{ijk}) = lp_k + betalp_k * date_{ij} + \ldots$

with

$$lpsi_k \sim Normal\left(\mu_{lpsi}, \sigma^2_{lpsi}\right)$$

$$betalpsi_k \sim Normal\left(\mu_{betalpsi}, \sigma^2_{betalpsi}\right)$$

$$lp_k \sim Normal\left(\mu_{lp}, \sigma^2_{lp}\right)$$

$$betalp_k \sim Normal\left(\mu_{betalp}, \sigma^2_{betalp}\right)$$

So, occupancy and detection probability are regressed on a number of covariates, and the intercepts and slopes of both regressions are species-specific random effects, with a common prior distribution as well as mean and variance hyperparameters that are estimated. As in the previous section the index for species (k) runs from 1 to $M = n + nz$—i.e., up to the number of species in the full augmented data set. For covariates on occupancy, we will fit elevation (linear and squared) and forest cover; for detection covariates, we will fit survey date (linear and squared) and survey duration.

We will see that with covariates, metacommunity size will often be estimated at a (much) higher value than it is under the more simplistic previous DR model that assumed both the occurrence and detection probabilities of a species were constant over all sites and all surveys. This makes intuitive sense: according to the second law of capture-recapture, unmodeled heterogeneity in detection will bias low population size estimators, and here arguably, occupancy probability and therefore species richness. Our new analysis will account for much more such heterogeneity among species, sites, and surveys, and therefore should be better able to accommodate such heterogeneity and thereby avoid negative bias.

How should we choose the amount of DA—i.e., the number of "potential" species nz by which we augment our data set? There are basically two approaches. The first is the one that we chose for

11.7 THE DORAZIO/ROYLE (DR) COMMUNITY OCCUPANCY MODEL

model 9: we made nz so large that the posterior of Ntotal was not affected by our choice. Our use of DA induces a discrete uniform prior on Ntotal—i.e., *Ntotal* ~ *DU(0, M)*. Hence, if we want a vague prior for Ntotal, we choose nz by trial and error: we start by adding a couple of *nz* all-zero detection histories, fit the model, and inspect the posterior distribution of Ntotal. If its mass is piling up against the chosen value of $M = n + nz$, we repeat the analysis with a larger value of *nz*, and do this until the posterior distribution of Ntotal is no longer right-truncated. In model 9, the posterior of Ntotal was not truncated by our choice of $M = n + nz$ (Figure 11.15), so our prior for metacommunity size was vague in that model. A second approach (Dorazio et al., 2010, 2011) is to fix *M* at a biologically plausible value, thus effectively choosing an informative prior on Ntotal by fixing it at a known value of species richness in the wider studied area. Currently, there are about 215 known breeding bird species in Switzerland, and hence this approach would let us augment by $nz = 215 - 145 = 70$ all-zero species. We fitted the following model using both approaches. First, after some trial and error (and a l-o-o-o-n-g waiting time each), we found that $nz = 250$, and hence $M = 145 + 250 = 395$, represented a fairly vague prior for Ntotal. Second, we fit the model with $M = 145 + 70 = 215$ species. Below, we briefly compare the results from the two approaches, but first we show how to fit the model. (Also see Link, 2013, on use of scale priors in models with data augmentation, including DR models.)

```
# Augment data set: choose one of two different priors on Ntotal
nz <- 250            # Use for vague prior on Ntotal: M = 395
nz <- 215 - nspec    # Use for informative prior on Ntotal: M = 215
yaug <- array(0, dim=c(nsite, nrep, nspec+nz)) # array with only zeroes
yaug[,,1:nspec] <- y     # copy into it the observed data

# Create same NA pattern in augmented species as in the observed species
missings <- is.na(yaug[,,1]) # e.g., third survey in high-elevation quads
for(k in (nspec+1):(nspec+nz)){
    yaug[,,k][missings] <- NA
}
```

In our experience, one of the main practical challenges in this kind of model (and also similar models where we deal with arrays rather than just vectors) is to come to grips with modeling the data in a multidimensional array (which you can imagine as some kind of an orderly "box"). Also, sometimes some lines of code must be moved around inside or outside of some of the two to three loops within which we define the elements of these arrays, to avoid defining a quantity repeatedly (this is the "multiple definition of…" trap in BUGS). This requires a *very* clear understanding of the "box" into which we place these quantities: you must know exactly which dimension stands for what index in an array. Note that we can loop over the dimensions of the array in the order we like (and some orders may result in better mixing—see tip 12 in Appendix 1 of Kéry and Schaub, 2012). However, each index of the array has a clearly defined meaning. In our case, one is for sites, another is for replicate surveys, and the last is for species. Which is which is defined by the way that we build this array in the first place.

For all but tiny communities, these are very parameter-rich models, and saving all species-specific parameters, including those for the *nz* potential species, will produce huge results files that will cost us a lot of memory. Moreover, the computation time to produce just the posterior summary when running BUGS from R will be huge. Therefore, to minimize both at the end of the BUGS program, we may save into new structures the parameter estimates for the observed species plus one potential species and save only these (though we do this only when we run the model with $nz = 250$).

CHAPTER 11 HIERARCHICAL MODELS FOR COMMUNITIES

We now package the data set and run the analysis.

```
# Bundle and summarize data
str(win.data <- list(y = yaug, nsite = dim(y)[1], nrep = dim(y)[2], nspec =
dim(y)[3], nz = nz, M = nspec + nz, ele = ele, forest = forest, DAT = DAT, DUR = DUR) )
List of 10
 $ y      : num [1:267, 1:3, 1:395] 0 0 0 0 0 0 0 0 0 0 ...
 $ nsite  : int 267
 $ nrep   : int 3
 $ nspec  : int 145
 $ nz     : num 250            # This is for nz = 250 and M = 395
 $ M      : num 395
 $ ele    : num [1:267] -1.1539 -1.1539 -0.2175 -0.3735 -0.0614 ...
 $ forest : num [1:267] -1.1471 -0.4967 -0.0992 -0.9303 0.0092 ...
 $ DAT    : num [1:267, 1:3] -1.415 -1.19 -1.235 -0.559 -1.64 ...
 $ DUR    : num [1:267, 1:3] -0.43511 -0.78764 -0.52324 1.23944 0.00556 ...

# Specify model in BUGS language
sink("model10.txt")
cat("
model {

# Priors
omega ~ dunif(0,1)
# Priors for species-specific effects in occupancy and detection
for(k in 1:M){
   lpsi[k] ~ dnorm(mu.lpsi, tau.lpsi)      # Hyperparams describe community
   betalpsi1[k] ~ dnorm(mu.betalpsi1, tau.betalpsi1)
   betalpsi2[k] ~ dnorm(mu.betalpsi2, tau.betalpsi2)
   betalpsi3[k] ~ dnorm(mu.betalpsi3, tau.betalpsi3)
   lp[k] ~ dnorm(mu.lp, tau.lp)
   betalp1[k] ~ dnorm(mu.betalp1, tau.betalp1)
   betalp2[k] ~ dnorm(mu.betalp2, tau.betalp2)
   betalp3[k] ~ dnorm(mu.betalp3, tau.betalp3)
}

# Hyperpriors
# For the model of occupancy
mu.lpsi ~ dnorm(0,0.01)
tau.lpsi <- pow(sd.lpsi, -2)
sd.lpsi ~ dunif(0,8)    # as always, bounds of uniform chosen by trial and error
mu.betalpsi1 ~ dnorm(0,0.1)
tau.betalpsi1 <- pow(sd.betalpsi1, -2)
sd.betalpsi1 ~ dunif(0, 4)
mu.betalpsi2 ~ dnorm(0,0.1)
tau.betalpsi2 <- pow(sd.betalpsi2, -2)
sd.betalpsi2 ~ dunif(0,2)
mu.betalpsi3 ~ dnorm(0,0.1)
tau.betalpsi3 <- pow(sd.betalpsi3, -2)
sd.betalpsi3 ~ dunif(0,2)
```

```
# For the model of detection
mu.lp ~ dnorm(0,0.1)
tau.lp <- pow(sd.lp, -2)
sd.lp ~ dunif(0, 2)
mu.betalp1 ~ dnorm(0,0.1)
tau.betalp1 <- pow(sd.betalp1, -2)
sd.betalp1 ~ dunif(0,1)
mu.betalp2 ~ dnorm(0,0.1)
tau.betalp2 <- pow(sd.betalp2, -2)
sd.betalp2 ~ dunif(0,1)
mu.betalp3 ~ dnorm(0,0.1)
tau.betalp3 <- pow(sd.betalp3, -2)
sd.betalp3 ~ dunif(0,1)

# Superpopulation process: Ntotal species sampled out of M available
for(k in 1:M){
   w[k] ~ dbern(omega)
}

# Ecological model for true occurrence (process model)
for(k in 1:M){
  for (i in 1:nsite) {
    logit(psi[i,k]) <- lpsi[k] + betalpsi1[k] * ele[i] + betalpsi2[k] * pow(ele[i],2) +
      betalpsi3[k] * forest[i]
    mu.psi[i,k] <- w[k] * psi[i,k]
    z[i,k] ~ dbern(mu.psi[i,k])
  }
}

# Observation model for replicated detection/nondetection observations
for(k in 1:M){
  for (i in 1:nsite){
    for(j in 1:nrep){
      logit(p[i,j,k]) <- lp[k] + betalp1[k] * DAT[i,j] + betalp2[k] * pow(DAT[i,j],2) +
        betalp3[k] * DUR[i,j]
      mu.p[i,j,k] <- z[i,k] * p[i,j,k]
      y[i,j,k] ~ dbern(mu.p[i,j,k])
    }
  }
}

# Derived quantities
#for(k in 1:M){
#   Nocc.fs[k] <- sum(z[,k])           # Number of occupied sites among the 267
#}
for (i in 1:nsite){
   Nsite[i] <- sum(z[i,])              # Number of occurring species at each site
}
n0 <- sum(w[(nspec+1):(nspec+nz)])     # Number of unseen species
Ntotal <- sum(w[])                     # Total metacommunity size
```

```
# Vectors to save (S for 'save'; discard posterior samples for
# all minus 1 of the potential species to save disk space)
# we do this for nz = 250 (i.e., M = 395)
lpsiS[1:(nspec+1)] <- lpsi[1:(nspec+1)]
betalpsi1S[1:(nspec+1)] <- betalpsi1[1:(nspec+1)]
betalpsi2S[1:(nspec+1)] <- betalpsi2[1:(nspec+1)]
betalpsi3S[1:(nspec+1)] <- betalpsi3[1:(nspec+1)]
lpS[1:(nspec+1)] <- lp[1:(nspec+1)]
betalp1S[1:(nspec+1)] <- betalp1[1:(nspec+1)]
betalp2S[1:(nspec+1)] <- betalp2[1:(nspec+1)]
betalp3S[1:(nspec+1)] <- betalp3[1:(nspec+1)]
}
",fill = TRUE)
sink()

# Initial values
wst <- rep(1, nspec+nz)                    # Simply set everybody at occurring
zst <- array(1, dim = c(nsite, nspec+nz))  # ditto
inits <- function() list(z = zst, w = wst, lpsi = rnorm(n = nspec+nz), betalpsi1 = rnorm(n =
nspec+nz), betalpsi2 = rnorm(n = nspec+nz), betalpsi3 = rnorm(n = nspec+nz), lp = rnorm(n =
nspec+nz), betalp1 = rnorm(n = nspec+nz), betalp2 = rnorm(n = nspec+nz), betalp3 = rnorm(n =
nspec+nz))
```

The number of estimated quantities ("parameters" or "estimands") in this analysis is truly enormous and may make your computer "explode," or you may have to wait for many days for models to be fit—literally! Hence, we may run the analyses in multiple steps, with a different set of estimands saved each time. We also choose different MCMC settings for each, and different modes of running JAGS. The first set is where we monitor convergence and save MCMC samples for the main structural parameters of the model, which are the hyperparameters describing the metacommunity. In addition, we want to look at the estimates of metacommunity and community size, or Ntotal and Nsite, respectively, yielding a total of $1 + 16 + 1 + 267 = 285$ "parameters." That may seem like a lot. But wait, you ain't seen nothing yet...

```
# Set 1
params1 <- c("omega", "mu.lpsi", "sd.lpsi", "mu.betalpsi1", "sd.betalpsi1",
"mu.betalpsi2", "sd.betalpsi2", "mu.betalpsi3", "sd.betalpsi3", "mu.lp", "sd.lp",
"mu.betalp1", "sd.betalp1", "mu.betalp2", "sd.betalp2", "mu.betalp3", "sd.betalp3",
"Ntotal", "Nsite")

# MCMC settings
ni <- 15000 ; nt <- 10 ; nb <- 5000 ; nc <- 3

# Run JAGS, check convergence and summarize posteriors
out101 <- jags(win.data, inits, params1, "model10.txt", n.chains = nc, n.thin =
nt, n.iter = ni, n.burnin = nb, parallel = TRUE)
par(mfrow = c(2, 2))
traceplot(out101, c(c("omega", "mu.lpsi", "sd.lpsi", "mu.betalpsi1",
"sd.betalpsi1", "mu.betalpsi2", "sd.betalpsi2", "mu.betalpsi3", "sd.betalpsi3",
"mu.lp", "sd.lp", "mu.betalp1", "sd.betalp1", "mu.betalp2",
"sd.betalp2", "mu.betalp3", "sd.betalp3", "Ntotal")) )
```

11.7 THE DORAZIO/ROYLE (DR) COMMUNITY OCCUPANCY MODEL

In the second set, we want to get posterior samples from species-specific effects in occupancy and detection, data augmentation variable w, and the presence/absence matrix $\mathbf{Z}$. For this, we simply need the MCMC samples and are not interested in any summaries (which take *many* days to compute in R on a laptop); hence, we use a more basic way of running JAGS from R, with function `jags.basic` in the `jagsUI` package, that will only return the MCMC samples. For DA of up to $M = 215$ species, we will estimate $215 * 8 + 215 * 267 + 215 = 59,340$ "parameters." Now, *that* is a big model!

```
# Set 2
params2 <- c("mu.lpsi", "sd.lpsi", "mu.betalpsi1", "sd.betalpsi1", "mu.betalpsi2",
"sd.betalpsi2", "mu.betalpsi3", "sd.betalpsi3", "lpsi", "betalpsi1", "betalpsi2",
"betalpsi3", "lp", "betalp1", "betalp2", "betalp3", "z", "w")
ni <- 12000 ; nt <- 20 ; nb <- 2000 ; nc <- 3
out102 <- jags.basic(win.data, inits, params2, "model10.txt", n.chains = nc,
n.thin = nt, n.iter = ni, n.burnin = nb, parallel = TRUE)
library(coda)
all10 <- as.matrix(out102)     # Put output from 3 chains into a matrix
summary(out102)                # May take a loooong time
gelman.diag(out102)            # ditto
```

Augmenting a data set up to size M induces a discrete uniform prior on the range $(0, M)$ for the metacommunity size Ntotal. We fitted this model with Ntotal ~ $DU(0, 215)$ and with Ntotal ~ $DU(0, 395)$. The former is a clearly informative prior, while the latter is only very weakly informative about metacommunity size. We needed much longer Markov chains to achieve convergence, and computation time was *very much* longer for the latter. Interestingly, however, in terms of inferences, the choice hardly mattered at all! The only exceptions were the estimates of metacommunity size Ntotal, and the mean and standard deviation hyperparameters for the species-specific intercepts of the occupancy model, which are μ_{lpsi} and σ^2_{lpsi}, respectively. With $M = 395$, we estimated a much greater metacommunity composed of species with lower mean occupancy probability, but more interspecific variability in the occupancy intercept. Estimates of all other hyperparameters were hardly changed, although posterior standard deviations in the occupancy model were just slightly larger for $M = 395$.

```
# Comparison of main hyperparameters when M = 215 and with M = 395
# (not all code to produce this output is shown)
print(cbind(out10.215$summary[1:17,c(1:3, 7)], out10.395$summary[1:17, c(1:3, 7)]), 2)
```

	mean	sd	2.5%	97.5%	mean	sd	2.5%	97.5%
mu.lpsi	-4.404	0.466	-5.309	-3.450	-6.32	1.197	-8.638	-4.135
sd.lpsi	4.785	0.357	4.126	5.520	5.66	0.634	4.451	6.960
mu.betalpsi1	-0.663	0.205	-1.065	-0.268	-0.65	0.198	-1.046	-0.270
sd.betalpsi1	2.443	0.178	2.115	2.809	2.41	0.171	2.091	2.760
mu.betalpsi2	-0.974	0.094	-1.159	-0.790	-1.00	0.098	-1.189	-0.812
sd.betalpsi2	0.804	0.085	0.649	0.979	0.80	0.082	0.653	0.976
mu.betalpsi3	-0.102	0.091	-0.279	0.075	-0.11	0.091	-0.286	0.066
sd.betalpsi3	0.997	0.078	0.854	1.155	0.99	0.074	0.853	1.141
mu.lp	0.729	0.137	0.451	0.991	0.71	0.140	0.428	0.981
sd.lp	1.417	0.127	1.192	1.696	1.45	0.129	1.220	1.727
mu.betalp1	0.068	0.052	-0.034	0.173	0.07	0.051	-0.026	0.170
sd.betalp1	0.461	0.048	0.374	0.563	0.46	0.048	0.377	0.564

mu.betalp2	-0.137	0.039	-0.216	-0.064	-0.14	0.039	-0.219	-0.064
sd.betalp2	0.333	0.039	0.260	0.414	0.33	0.040	0.261	0.416
mu.betalp3	0.210	0.031	0.148	0.272	0.21	0.031	0.149	0.272
sd.betalp3	0.227	0.033	0.167	0.296	0.23	0.033	0.168	0.294
Ntotal	205.763	8.238	184.975	215.000	273.50	42.635	200.000	361.000

In addition, community size estimates (`Nsite`) were essentially identical under both metacommunity size priors. Posterior means were on average greater by 0.16 species (corresponding to 0.4%) with $M = 395$ than with $M = 215$, and posterior standard deviations of `Nsite` were on average 1.2% greater. This suggests that the very rare species have only a negligible effect on the spatial patterns of local species richness. In summary then, in our example all of the important parameters were remarkably insensitive to the choice of prior on metacommunity size. Hence, we base our inference on the model with informative prior on the metacommunity size ($M = 215$), since this takes *much* less time to fit, plus we feel that it is motivated by existing knowledge of the community of birds in Switzerland.

```
out10 <- out101
```

Our model enables inferences at all three hierarchical levels—metacommunity, local community, and individual species—and we now give examples of each.

First the metacommunity, which is described by the following quantities in the model: `omega, mu.lpsi, sd.lpsi, mu.betalpsi1, sd.betalpsi1, mu.betalpsi2, sd.betalpsi2, mu.betalpsi3, sd.betalpsi3, mu.lp, sd.lp, sd.betalp1, mu.betalp2, sd.betalp2, mu.betalp3, sd.betalp3, n0, Ntotal`. There is perhaps not much point in estimating metacommunity size, since we used an informative prior on `Ntotal`. However, we can use the hyperparameters to describe the metacommunity—e.g., in terms of the species mean and among-species variability of occupancy and detection parameters, or the community average response to covariates. We can do this by simply sampling the respective normal distributions and then inverse-logit transforming the resulting values. Figure 11.16 suggests that there is plenty of among-species variability in both occupancy and detection probability, but that a minority of species are widespread, whereas many have an extremely limited

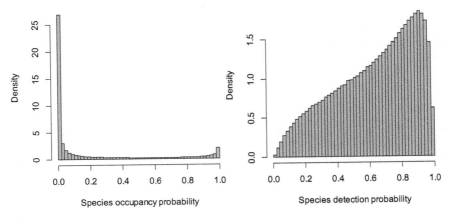

FIGURE 11.16

Community distribution of average occupancy and detection probability among Swiss breeding bird species based on estimates of the species intercept model parameters (`mu.lpsi, sd.lpsi, mu.lp, sd.lp`) for the spatiotemporal scale of 1 km^2 and the three-month-long breeding season in 2014.

11.7 THE DORAZIO/ROYLE (DR) COMMUNITY OCCUPANCY MODEL

distribution, and that a majority are relatively easy to detect. The average species occupancy probability for the spatiotemporal scale in the MHB survey was 0.19, and the average species detection probability at the mean survey conditions (date and duration) was 0.63.

```
par(mfrow = c(1,2))            # Fig. 11-16
psi.sample <- plogis(rnorm(10^6, mean = out10$mean$mu.lpsi, sd = out10$mean$sd.lpsi))
p.sample <- plogis(rnorm(10^6, mean = out10$mean$mu.lp, sd = out10$mean$sd.lp))
hist(psi.sample, freq = F, breaks = 50, col = "grey", xlab = "Species occupancy
probability", ylab = "Density", main = "")
hist(p.sample, freq = F, breaks = 50, col = "grey", xlab = "Species detection probability",
ylab = "Density", main = "")
summary(psi.sample) ; summary(p.sample)
```

We can also look at the interspecific variability in every parameter, which appears in our model as standard deviations of normal priors for species-specific effects. We simply plot histograms of the posteriors for each. Executing the next body of code, you will see that the difference among species in the Swiss breeding bird community varies a great deal among the different parameters of the occupancy and detection models (note different scales of abscissa).

```
par(mfrow = c(2,4))           # Among-species variability in parameters (not shown)
hist(out10$sims.list$sd.lpsi, breaks = 100, col = "grey", xlim = c(0,6), main =
"Occupancy: intercept")
abline(v = mean(out10$sims.list$sd.lpsi), col = "blue", lwd = 3)
hist(out10$sims.list$sd.betalpsi1, breaks = 100, col = "grey", xlim = c(0,3),
main = "Occupancy: linear effect of elevation")
abline(v = mean(out10$sims.list$sd.betalpsi1), col = "blue", lwd = 3)
hist(out10$sims.list$sd.betalpsi2, breaks = 100, col = "grey", xlim = c(0,3),
main = "Occupancy: quadratic effect of elevation")
abline(v = mean(out10$sims.list$sd.betalpsi2), col = "blue", lwd = 3)
hist(out10$sims.list$sd.betalpsi3, breaks = 100, col = "grey", xlim = c(0,3),
main = "Occupancy: linear effect of forest cover")
abline(v = mean(out10$sims.list$sd.betalpsi3), col = "blue", lwd = 3)
hist(out10$sims.list$sd.lp, breaks = 100, col = "grey", xlim = c(0,2),
main = "Detection: intercept")
abline(v = mean(out10$sims.list$sd.lp), col = "blue", lwd = 3)
hist(out10$sims.list$sd.betalp1, breaks = 100, col = "grey", xlim = c(0,1), main
= "Detection: linear effect of survey date")
abline(v = mean(out10$sims.list$sd.betalp1), col = "blue", lwd = 3)
hist(out10$sims.list$sd.betalp2, breaks = 100, col = "grey", xlim = c(0,1), main
= "Detection: quadratic linear effect of survey date")
abline(v = mean(out10$sims.list$sd.betalp2), col = "blue", lwd = 3)
hist(out10$sims.list$sd.betalp3, breaks = 100, col = "grey", xlim = c(0,1), main
= "Detection: linear effect of survey duration")
abline(v = mean(out10$sims.list$sd.betalp3), col = "blue", lwd = 3)
```

Next, we look at the community average response to modeled covariates, where we predict the mean community response of occupancy probability to elevation and forest cover, and of species detection probability to survey date and survey duration. As always, we create "original" covariates that cover the entire range of a covariate over which we want to make predictions of a modeled quantity. Then we apply the identical scaling that we used for the actual covariates used in the analysis, and next we apply the linear model parameters estimated in the analysis to get the predictions (we do this here for each of the 3000 posterior samples). Finally, we plot posterior summaries of these predictions against the values of the "original" (i.e., unscaled) covariates. This time we put all four sets of predictions into a single array.

```
# Visualize covariate mean relationships for the average species
o.ele <- seq(200, 2500,,500)              # Get covariate values for prediction
o.for <- seq(0, 100,,500)
o.dat <- seq(15, 120,,500)
o.dur <- seq(100, 420,,500)
ele.pred <- (o.ele - mean.ele) / sd.ele
for.pred <- (o.for - mean.forest) / sd.forest
dat.pred <- (o.dat - mean.date) / sd.date
dur.pred <- (o.dur - mean.dur) / sd.dur

# Predict occupancy for elevation and forest and detection for date and duration
# Put all four predictions into a single
str( tmp <- out10$sims.list )             # grab MCMC samples
nsamp <- length(tmp[[1]])                  # number of mcmc samples
predC <- array(NA, dim = c(500, nsamp, 4)) # "C" for 'community mean'
for(i in 1:nsamp){
    predC[,i,1] <- plogis(tmp$mu.lpsi[i] + tmp$mu.betalpsi1[i] * ele.pred +
    tmp$mu.betalpsi2[i] * ele.pred^2 )
    predC[,i,2] <- plogis(tmp$mu.lpsi[i] + tmp$mu.betalpsi3[i] * for.pred)
    predC[,i,3] <- plogis(tmp$mu.lp[i] + tmp$mu.betalp1[i] * dat.pred +
    tmp$mu.betalp2[i] * dat.pred^2 )
    predC[,i,4] <- plogis(tmp$mu.lp[i] + tmp$mu.betalp3[i] * dur.pred)
}

# Get posterior means and 95% CRIs and plot (Fig. 11-17)
pmC <- apply(predC, c(1,3), mean)
criC <- apply(predC, c(1,3), function(x) quantile(x, prob = c(0.025, 0.975)))

par(mfrow = c(2, 2))
plot(o.ele, pmC[,1], col = "blue", lwd = 3, type = 'l', lty = 1, frame = F,
ylim = c(0, 0.05), xlab = "Elevation (m a.s.l)", ylab = "Community mean occupancy")
matlines(o.ele, t(criC[,,1]), col = "grey", lty = 1)
plot(o.for, pmC[,2], col = "blue", lwd = 3, type = 'l', lty = 1, frame = F,
ylim = c(0, 0.05), xlab = "Forest cover", ylab = "Community mean occupancy")
matlines(o.for, t(criC[,,2]), col = "grey", lty = 1)
plot(o.dat, pmC[,3], col = "blue", lwd = 3, type = 'l', lty = 1, frame = F, ylim =
c(0.2, 0.8), xlab = "Survey date", ylab = "Community mean detection")
matlines(o.dat, t(criC[,,3]), col = "grey", lty = 1)
plot(o.dur, pmC[,4], col = "blue", lwd = 3, type = 'l', lty = 1, frame = F, ylim =
c(0.2, 0.8), xlab = "Survey duration", ylab = "Community mean detection")
matlines(o.dur, t(criC[,,4]), col = "grey", lty = 1)
```

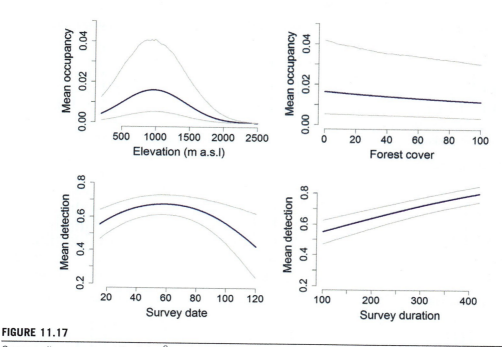

FIGURE 11.17

Community response at the 1-km² spatial scale of Swiss breeding bird occupancy probability to elevation and forest cover, and of species detection probability to survey date and survey duration.

The community average occupancy probability of Swiss breeding birds was greatest around 1000 m in elevation and essentially did not respond to forest cover (Figure 11.17, top). On average, Swiss breeding bird species were most detectable around day 60 (May 30), and increasing survey duration in 1-km² quadrats from 100 to 400 min pushed detection probability from 0.55 to 0.80 (Figure 11.17, bottom). You may wonder about the strange values of occupancy, whose community mean varies from just about nothing to slightly less than 1.5%. How can we ever observe more than just a handful of species with such low values? The reason we can do this is because we model species variability as random noise around a mean *on the logit scale*, and the community average of the logit occupancy probability (−4.351, corresponding to 0.012 on the probability scale) does not correspond to the average of the probability-scale occupancy intercept. We saw above that the expected value of the community distribution of occupancy (Figure 11.16, left) was about 19%.

Second, moving to inferences at the community (i.e., site-specific) level, we can again obtain the estimates of species richness at each site (Nsite), and then do so for the same sample of nine sites that we already plotted in Figure 11.14. The estimates under model 10 in Figure 11.18 are slightly larger, presumably because we accounted for more detection heterogeneity in this model than in model 9. Of course, you may plot estimates of Nsite against any site covariate in your model or even outside, to generate hypotheses about drivers of variation in community richness. When plotting the Nsite estimates against site elevation (Figure 11.19), we see that incorporating covariates increases Nsite estimates at lower elevations, but decreases them at higher elevations, compared with the

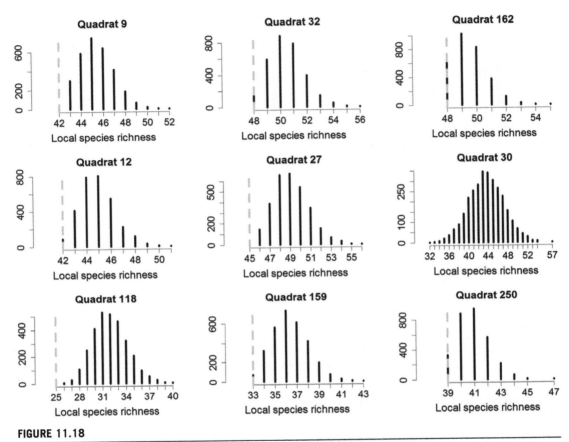

FIGURE 11.18

Sample of posterior distributions of local species richness (community size Nsite or alpha diversity) under model 10 at the same selection of nine 1-km² MHB sample quadrats as shown in Figure 11.14 (which are estimates of covariate-free model 9).

covariate-free estimates under model 9. This sensible pattern is due to elevation relationships for most species that are estimated in the model (we will see these species-specific estimates below).

```
# Plot posterior distribution of site-specific species richness (Nsite)
par(mfrow = c(3,3), mar = c(5,4,3,2))
for(i in 1:267){
    plot(table(out10$sims.list$Nsite[,i]), main = paste("Quadrat", i), xlab = "Local
    species richness", ylab = "", frame = F, xlim = c((min(C[i], out10$sims.list$Nsite[,i],
    na.rm = T)-2), max(out10$sims.list$Nsite[,i]) ))
    abline(v = C[i], col = "grey", lwd = 4)
    browser()
}
```

11.7 THE DORAZIO/ROYLE (DR) COMMUNITY OCCUPANCY MODEL

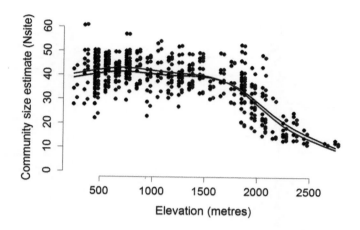

FIGURE 11.19

Comparison between DR models 9 (black, no covariates) and 10 (blue, with covariates) in terms of the relationship between community size (Nsite) and elevation. Note slight offset of blue in the x direction.

```
# Plot it only for a selection of sites (Fig. 11-18)
par(mfrow = c(3,3), mar = c(5,4,3,2))
for(i in c(9, 32, 162, 12, 27, 30, 118, 159, 250)){
   plot(table(out10$sims.list$Nsite[,i]), main = paste("Quadrat", i),
   xlab = "Local    species richness", ylab = "", frame = F,
   xlim = c((min(C[i], out10$sims.list$Nsite[,i], na.rm = T)-2),
   max(out10$sims.list$Nsite[,i])))
   abline(v = C[i], col = "grey", lwd = 4)
}

# Plot Nsite estimates under models 9 & 10 vs. elevation (Fig. 11-19)
offset <- 30      # Set off elevation for better visibility
plot(elev, out9$mean$Nsite, xlab = "Elevation (metres)", ylab = "Community size
estimate (Nsite)", frame = F, ylim = c(0,60), pch = 16) # black: model 9
lines(smooth.spline(out9$mean$Nsite ~ elev), lwd = 3)
points(elev+offset, out10$mean$Nsite, pch = 16, col = "blue") # blue: model 10
lines(smooth.spline(out10$mean$Nsite ~ elev), lwd = 3, col = "blue")
```

Lastly, we present inferences about individual species under our community occupancy model. To present the species-specific inferences and look at the parameter estimates of covariate effects, we use the second set of MCMC output and first compute posterior means and 95% CRIs of the species-specific parameters from the "naked" MCMC samples contained in out1o2.

```
str(all10)                    # Look at the MCMC output
pm <- apply(all10, 2, mean)    # Get posterior means and 95% CRIs
cri <- apply(all10, 2, function(x) quantile(x, prob = c(0.025, 0.975)))    # CRIs
```

Now we produce plots of the species-specific parameter estimates in the occupancy and detection models for all 145 observed species and compare them with the community average

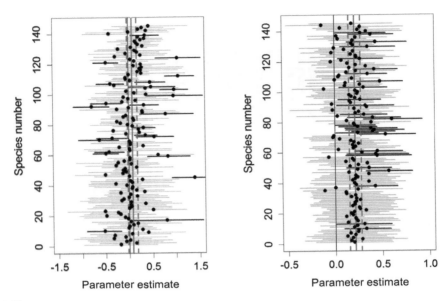

FIGURE 11.20

Comparison between community and individual species response of detection probability to survey date (linear, left) and survey duration (right). Red lines show posterior mean and 95% CRI of the community mean hyperparameter, and black dots and gray lines show the same thing for each individual species (for the 145 observed species), with species CRIs that do not overlap zero colored in blue.

(the hyperparameters). Beyond the interest of inspecting parameter estimates for each species, these plots should emphasize again that the community average may be quite different from that of an individual species. In case you know European birds and would like to inspect the parameter estimates for a particular species, remember that the 145 parameters are in the same order as in the object ordered.spec.name.list.

```
# Effects of date (linear and quadratic) and of duration on detection
#par(mfrow = c(1,3), cex.lab = 1.3, cex.axis = 1.3) # Can put all three in one
par(mfrow = c(1,2), cex.lab = 1.3, cex.axis = 1.3)
# Date linear (Fig. 11-20 left)
plot(pm[1:145], 1:145, xlim = c(-1.5, 1.5), xlab = "Parameter estimate", ylab
= "Species number", main = "Effect of date (linear) on detection", pch = 16)
abline(v = 0, lwd = 2, col = "black")
segments(cri[1, 1:145], 1:145, cri[2, 1:145], 1:145, col = "grey", lwd = 1)
sig1 <- (cri[1, 1:145] * cri[2, 1:145]) > 0
segments(cri[1, 1:145][sig1 == 1], (1:145)[sig1 == 1], cri[2, 1:145][sig1 == 1], (1:145)
[sig1 == 1], col = "blue", lwd = 2)
abline(v = out101$summary[11,1], lwd = 3, col = "red")
abline(v = out101$summary[11,c(3,7)], lwd = 2, col = "red", lty = 2)

# Date quadratic (not shown)
plot(pm[216:360], 1:145, xlim = c(-1.5, 1.5), xlab = "Parameter estimate", ylab = "Species
number", main = "Effect of date (quadratic) on detection", pch = 16)
```

11.7 THE DORAZIO/ROYLE (DR) COMMUNITY OCCUPANCY MODEL

```
abline(v = 0, lwd = 2, col = "black")
segments(cri[1, 216:360], 1:145, cri[2, 216:360], 1:145, col = "grey", lwd = 1)
sig2 <- (cri[1, 216:360] * cri[2, 216:360]) > 0
segments(cri[1, 216:360][sig2 == 1], (1:145)[sig2 == 1], cri[2, 216:360][sig2 == 1],
  (1:145)[sig2 == 1], col = "blue", lwd = 2)
abline(v = out101$summary[13,1], lwd = 3, col = "red")
abline(v = out101$summary[13, c(3,7)], lwd = 3, col = "red", lty = 2)
```

In the detection model, a total of 24 species among the 145 observed had a "significant" effect of date (linear, do sum(sig1)), in the sense that the 95% CRI of the parameter estimate did not cover zero (these are the blue lines in Figure 11.20). The community mean effect of date (linear), mu.betalp1, was estimated at 0.068 (95% CRI: −0.034, 0.173). Only 20 species had a "significant" effect of date (squared) on detection probability, although the community mean effect of date (squared), mu.betalp2, was "significantly" negative (posterior mean and 95% CRI: −0.137 (−0.216, −0.064)). Thirty-seven species had either a significant linear or a significant squared effect of date (or both) on detection probability (do sum((sig1 + sig2) > 0)). Finally, even though the community mean effect of survey duration on detection, mu.betalp3, was clearly positive (posterior mean and 95% CRI: 0.210 (0.148, 0.272), individually it was so for only 29 species (Figure 11.20, right).

```
# Survey duration (Fig. 11-20 right)
plot(pm[431:575], 1:145, xlim = c(-0.5, 1), xlab = "Parameter estimate", ylab =
  "Species number", main = "Effect of survey duration on detection", pch = 16)
abline(v = 0, lwd = 2, col = "black")
segments(cri[1, 431:575], 1:145, cri[2, 431:575], 1:145, col = "grey", lwd = 1)
sig3 <- (cri[1, 431:575] * cri[2, 431:575]) > 0
segments(cri[1, 431:575][sig3 == 1], (1:145)[sig3 == 1], cri[2, 431:575][sig3 == 1],
  (1:145)[sig3 == 1], col = "blue", lwd = 2)
abline(v = out101$summary[15,1], lwd = 3, col = "red")
abline(v = out101$summary[15, c(3,7)], lwd = 3, col = "red", lty = 2)
```

In the model for occupancy probability, 104 species had a "significant" linear and 68 a "significant" quadratic effect of elevation (Figures 11.21 and 11.22). The community mean response was negative for both, although the "significant" individual species responses are both negative and positive for the former, but always negative for the latter. This shows that even though species have different elevation preferences, ultimately their occupancy probability declines toward zero with increasing elevation. As we have seen before, there was no effect of forest cover at the community mean level, but for 66 species individually, we did find such an effect (35 negative and 31 positive; Figure 11.23). Of course, this is not surprising at all, since everybody knows (well, at least most birdwatchers do) that some species like forest and others don't like it at all. But it illustrates nicely the limitations for community analyses that do not keep track of individual species responses (e.g., the models in Section 11.5).

```
# Effect of elevation (linear) on occupancy probability (Fig. 11-21)
plot(pm[646:790], 1:145, xlim = c(-8, 8), xlab = "Parameter estimate", ylab =
  "Species number", main = "Effect of elevation (linear) on occupancy", pch = 16)
abline(v = 0, lwd = 2, col = "black")
segments(cri[1, 646:790], 1:145, cri[2, 646:790], 1:145, col = "grey", lwd = 1)
```

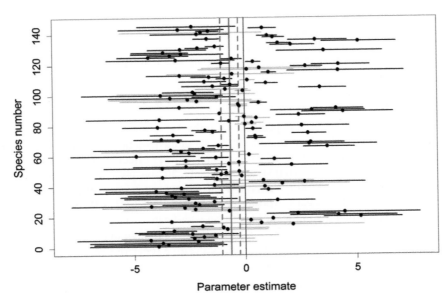

FIGURE 11.21

Comparison between community and individual species response of occupancy probability at the 1-km² scale to site elevation (linear) for 145 observed species. Color and symbols as in Figure 11.20.

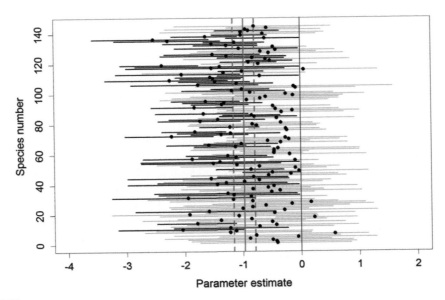

FIGURE 11.22

Comparison between community and individual species response of occupancy probability at the 1-km² scale to site elevation (squared) for 145 observed species. Color and symbols as in Figure 11.20.

11.7 THE DORAZIO/ROYLE (DR) COMMUNITY OCCUPANCY MODEL

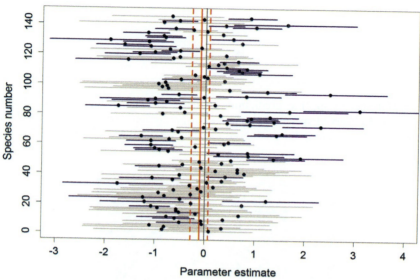

FIGURE 11.23

Comparison between community and individual species response of occupancy probability at the 1-km^2 scale to forest cover for 145 observed species. Color and symbols as in Figure 11.20.

```
sig4 <- (cri[1, 646:790] * cri[2, 646:790]) > 0
segments(cri[1, 646:790][sig4 == 1], (1:145)[sig4 == 1], cri[2, 646:790][sig4 == 1],
  (1:145)[sig4 == 1], col = "blue", lwd = 2)
abline(v = out101$summary[3,1], lwd = 3, col = "red")
abline(v = out101$summary[3,c(3,7)], lwd = 3, col = "red", lty = 2)

# Effect of elevation (quadratic) on occupancy probability (Fig. 11-22)
plot(pm[861:1005], 1:145, xlim = c(-4, 2), xlab = "Parameter estimate", ylab =
  "Species number", main = "Effect of elevation (quadratic) on occupancy", pch = 16)
abline(v = 0, lwd = 2, col = "black")
segments(cri[1, 861:1005], 1:145, cri[2, 861:1005], 1:145, col = "grey", lwd=1)
sig5 <- (cri[1, 861:1005] * cri[2, 861:1005]) > 0
segments(cri[1, 861:1005][sig5 == 1], (1:145)[sig5 == 1], cri[2, 861:1005][sig5 == 1],
  (1:145)[sig5 == 1], col = "blue", lwd = 2)
abline(v = out101$summary[5,1], lwd = 3, col = "red")
abline(v = out101$summary[5,c(3,7)], lwd = 3, col = "red", lty = 2)

# Effect of forest (linear) on occupancy probability (Fig. 11-23)
plot(pm[1076:1220], 1:145, xlim = c(-3, 4), xlab = "Parameter estimate", ylab = "Species
  number", main = "Effect of forest cover on occupancy", pch = 16)
abline(v = 0, lwd = 2, col = "black")
segments(cri[1, 1076:1220], 1:145, cri[2, 1076:1220],1:145, col = "grey", lwd=1)
sig6 <- (cri[1, 1076:1220] * cri[2, 1076:1220]) > 0
```

CHAPTER 11 HIERARCHICAL MODELS FOR COMMUNITIES

```
segments(cri[1, 1076:1220][sig6 == 1], (1:145)[sig6 == 1], cri[2, 1076:1220][sig6 == 1],
    (1:145)[sig6 == 1], col = "blue", lwd = 2)
abline(v = out101$summary[7,1], lwd = 3, col = "red")
abline(v = out101$summary[7,c(3,7)], lwd = 3, col = "red", lty = 2)
negsig6 <- (cri[1, 1076:1220] < 0 & cri[2, 1076:1220] < 0) == 1 # sig negative
possig6 <- (cri[1, 1076:1220] > 0 & cri[2, 1076:1220] > 0) == 1 # sig positive
```

Finally, we plot species-specific predictions of detection and occupancy as a function of the covariates for the 145 observed species, using code for prediction covariates used for the community mean predictions above and doing the analogous thing as we did there. Figures 11.24 and 11.25 show

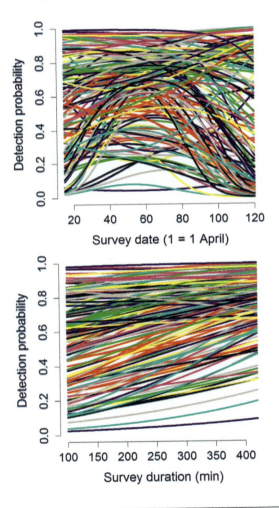

FIGURE 11.24

Species-specific predictions of detection probability as a function of survey date and survey duration in 1-km^2 quadrats in the Swiss breeding bird survey MHB in 2014 under community occupancy model number 10. Each line represents one of the 145 observed species. Note that one covariate is kept at the observed average for the computation of the prediction for the other covariate, and vice versa.

11.7 THE DORAZIO/ROYLE (DR) COMMUNITY OCCUPANCY MODEL

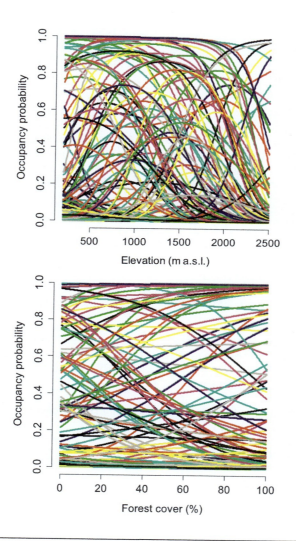

FIGURE 11.25

Species-specific predictions of occupancy probability as a function of elevation and forest cover in 1-km^2 quadrats in the Swiss breeding bird survey MHB in 2014 under community occupancy model number 10. Each line represents one of the 145 observed species. Note that one covariate is kept at the observed average for the computation of the prediction for the other covariate, and vice versa.

how the interspecies variability in the parameters that we have illustrated in the last few figures translates into variation among species in occupancy and detection probability.

```
# Predict detection for date and duration and occupancy for elevation and forest
# for each of the 145 observed species
predS <- array(NA, dim = c(500, nspec, 4))   # covariate value x species x
response, "S" for 'species'
```

```
p.coef <- cbind(lp=pm[1292:1436], betalp1 = pm[1:145], betalp2 = pm[216:360],
betalp3 = pm[431:575])
psi.coef <- cbind(lpsi=pm[1507:1651], betalpsi1 = pm[646:790], betalpsi2 = pm[861:1005],
betalpsi3 = pm[1076:1220])

for(i in 1:nspec){           # Loop over 145 observed species
    predS[,i,1] <- plogis(p.coef[i,1] + p.coef[i,2] * dat.pred +
       p.coef[i,3] * dat.pred^2 )            # p ~ date
    predS[,i,2] <- plogis(p.coef[i,1] + p.coef[i,4] * dur.pred) # p ~ duration
    predS[,i,3] <- plogis(psi.coef[i,1] + psi.coef[i,2] * ele.pred +
       psi.coef[i,3] * ele.pred^2 )          # psi ~ elevation
    predS[,i,4] <- plogis(psi.coef[i,1] + psi.coef[i,4] * for.pred) # psi ~ forest
}

# Plots for detection probability and survey date and duration (Fig. 11-24)
par(mfrow = c(1,2), cex.lab = 1.3, cex.axis = 1.3)
plot(o.dat, predS[,1,1], lwd = 3, type = 'l', lty = 1, frame = F,
    ylim = c(0, 1), xlab = "Survey date (1 = 1 April)",
    ylab = "Detection probability")
for(i in 2:145){
    lines(o.dat, predS[,i,1], col = i, lwd = 3)
}

plot(o.dur, predS[,1,2], lwd = 3, type = 'l', lty = 1, frame = F, ylim = c(0, 1), xlab = "Survey
duration (min)", ylab = "Detection probability")
for(i in 2:145){
    lines(o.dur, predS[,i,2], col = i, lwd = 3)
}

# Plots for occupancy probability and elevation and forest cover (Fig. 11-25)
par(mfrow = c(1,2), cex.lab = 1.3, cex.axis = 1.3)
plot(o.ele, predS[,1,3], lwd = 3, type = 'l', lty = 1, frame = F, ylim = c(0, 1),
    xlab = "Elevation (m a.s.l.)", ylab = "Occupancy probability")
for(i in 2:145){
    lines(o.ele, predS[,i,3], col = i, lwd = 3)
}

plot(o.for, predS[,1,4], lwd = 3, type = 'l', lty = 1, frame = F, ylim = c(0, 1), xlab = "Forest
    cover (%)", ylab = "Occupancy probability")
for(i in 2:145){
    lines(o.for, predS[,i,4], col = i, lwd = 3)
}
```

We see that the DR community occupancy models give us tremendous power to make inferences at all three levels involved in the sampling: metacommunity, community, and individual species. But there is more. One of the most exciting but perhaps underappreciated things about HMs such as this one is its estimate of the latent variable z, collected here in the presence/absence matrix **Z**. We have used this matrix already to estimate alpha diversity (local species richness `Nsite`). However, we can use it for much more: to compare sites in terms of their occurring species (species turnover or beta diversity), and to compare species in terms of the sites at which they co-occur. We illustrate this next.

11.8 INFERENCES BASED ON THE ESTIMATED Z MATRIX: SIMILARITY AMONG SITES AND SPECIES

Our model provides us with a detection-error-corrected estimate of the true presence/absence matrix **Z** for every site and species, and with an estimate of the latent variable w for every species in the superpopulation of size M. Both enable computation of any presence/absence-based classical measure of alpha, beta, and gamma diversity. Summing up the values of **Z** over species for each site yields species richness at each site, which is the traditional measure of alpha diversity. Conversely, summing up vector w yields the total species richness in the area sampled by the study sites (metacommunity size); this is a measure of gamma diversity. For beta diversity, a very large number of measures have been proposed (Whittaker et al., 2001). One of them is the Jaccard index (J), which expresses the similarity of two sites in terms of their occurring species (Dorazio et al., 2011). For two sites r and s, J is the proportion of species shared among those species that occur either at r or s (note the momentary abuse of capital J in this section only):

$$J_{r,s} = \frac{\sum z_r z_s}{\sum z_r + \sum z_s - \sum z_r z_s},$$

where summations run over species. Values of this similarity index range from 0 to 1, corresponding respectively to the extremes of no shared species, and all species occurring at both sites. Instead of the similarity between two sites, we can choose to express the *dissimilarity* between two sites by taking $1 - J_{r,s}$. Similarly, we can express the similarity *between species* in terms of the sites where they co-occur by simply switching the dimensions in the above expressions. That is, we can compute the similarity between species r and species s in terms of the sites where they both occur, and where they occur alone, by applying this expression to the rows of the **Z** matrix. The summations in the expression are then over sites rather than species. For our last model, 10, we first have to format the MCMC samples for the **Z** matrix into a 3-dimensional array. After some inspection of the MCMC output, we find out that the 267 × 215 elements of the **Z** matrix are in rows 1937 through 59,341 of the vector `all10`, which we have computed already.

```
# Plug MCMC samples for full z matrix into 3D array
str(all10)
nsite <- 267
nspec <- 215
nsamp <- dim(all10)[1]      # 1200 MCMC samples
z <- array(NA, dim = c(nsite, nspec, nsamp))
Jacc <- array(NA, dim = c(nsite, nspec, nsamp))
for(j in 1:nsamp){         # Fill z matrix by column (default)
   cat(paste("\nMCMC sample", j, "\n"))
   z[,,j] <- all10[j, 1937:59341]
}

# Restrict computations to observed species
zobs <- z[,1:145,]     # Species 1 to 145

# Compute Jaccard index for sites and for species
Jsite <- array(NA, dim = c(nsite, nsamp))
Jspec <- array(NA, dim = c(145, nsamp))
```

We now compute the values of the two Jaccard indices for all pair-wise comparisons for any reference site or species that we like. For illustration, we here chose site 1 and species 13 (which is the sparrowhawk *Accipiter nisus*).

```
# Choose reference site and species for Jaccard indices
ref.site <- 1          # Just choose first site
ref.species <- 13      # European Sparrowhawk (check object 'obs.occ')

# Get posterior distributions for Jsite and Jspec (for references)
for(k in 1:nsamp){
  for(i in 1:nsite){ # Jaccard index for sites (in terms of shared species)
    Jsite[i,k] <- sum(zobs[ref.site,,k] * zobs[i,,k]) / (sum(zobs[ref.site,,k]) +
      sum(zobs[i,,k]) - sum(zobs[ref.site,,k] * zobs[i,,k]))
  }
  for(i in 1:(nspec-nz)){ # Jacc. index for species (in terms of shared sites)
    Jspec[i,k] <- sum(zobs[,ref.species,k] * zobs[,i,k]) / (sum(zobs[,ref.species,k]) +
      sum(zobs[,i,k]) - sum(zobs[,ref.species,k] * zobs[,i,k]))
  }
}
# NA's arise when a site has no species or a species no sites

# Get posterior means, standard deviations and 95% CRI
# Jaccard index for sites, compared to reference site 1
pm <- apply(Jsite, 1, mean, na.rm = TRUE)      # Post. mean of Jsite wrt. site 1
psd <- apply(Jsite, 1, sd, na.rm = TRUE)       # Post. sd of Jsite wrt. site 1
cri <- apply(Jsite, 1, function(x) quantile(x, prob = c(0.025, 0.975), na.rm =
  TRUE)) # CRI
cbind('post. mean' = pm, 'post. sd' = psd, '2.5%' = cri[1,], '97.5%' = cri[2,])
        post. mean   post. sd      2.5%         97.5%
 [1,]   1.00000000   0.00000000    1.00000000   1.00000000
 [2,]   0.56538646   0.03273553    0.50000000   0.63414634
 [3,]   0.34548127   0.02181290    0.30882353   0.39070786
 [4,]   0.51197748   0.03099446    0.45454545   0.57446809
 [5,]   0.42943330   0.02526238    0.38596491   0.48214286
 [6,]   0.60248009   0.03155804    0.54000000   0.66666667
 [7,]   0.54569316   0.03233974    0.48914689   0.61363636
 [8,]   0.36845865   0.02617650    0.32075472   0.41822727
 [9,]   0.52875245   0.03116781    0.47169811   0.59183673
[10,]   0.40707376   0.02977571    0.35294118   0.47058824
[ output truncated ]
```

We can look at the table that expresses the similarity of the bird communities of all sites compared with site 1. As always with spatially referenced data, a map can be interesting, so we map the posterior means of these Jaccard indices (Figure 11.26). Looking at the Swiss geography in some other maps in this book (e.g., Figure 7.10), you will note that site 1 is most similar (in terms of the occurrence of bird species) to other lowland sites. And perhaps you will not be surprised, but still, it is nice to find a sensible result in this formal comparison of the bird communities among these sites.

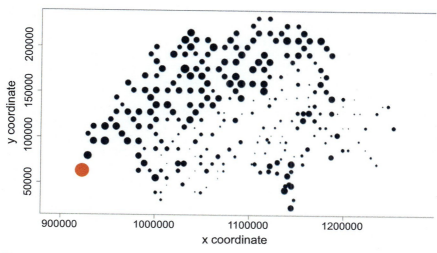

FIGURE 11.26

Map of the posterior means of the Jaccard site indices with respect to site 1 (shown in red; this is in the lowlands near Geneva). Circle size is proportional to the value of the Jaccard index, with red circle equal to 1.

```
# Make a map of Jaccard site indices (Fig. 11-26)
x <- 3          # size setting for plotting symbol
plot(data$coordx[1:267], data$coordy[1:267], xlab = "x coordinate", ylab = "y
coordinate", cex = x*pm, asp = 1, pch = 16)
points(data$coordx[which(pm == 1)], data$coordy[which(pm == 1)], cex = x*pm, col = "red",
pch = 16)
```

Turning to similarities among species in terms of presence/absence patterns, we compute posterior summaries for the 145 observed species and order them according to decreasing similarity in the next table.

```
# Jaccard index for species, compared with a reference species
# (species 13, European Sparrowhawk)
pm <- apply(Jspec, 1, mean, na.rm = TRUE)     # Post. mean of Jspec wrt. species 1
psd <- apply(Jspec, 1, sd, na.rm = TRUE)      # Post. sd of Jspec wrt. species 1
cri <- apply(Jspec, 1, function(x) quantile(x, prob = c(0.025, 0.975), na.rm =
TRUE)) # CRI
tmp <- cbind('post. mean' = pm, 'post. sd' = psd, '2.5%' = cri[1,], '97.5%' = cri[2,])
rownames(tmp) <- names(obs.occ)
print(tmp])              # print in systematic order
print(tmp[rev(order(tmp[,1])),]) # print in order of decreasing Jacc. values
```

	post. mean	post. sd	2.5%	97.5%
Eurasian Sparrowhawk	1.000000000	0.000000000	1.000000000	1.00000000
Common Chaffinch	0.671351689	0.227396146	0.235251929	0.93308394
Common Blackbird	0.651996602	0.214434239	0.230418594	0.90376980

Eurasian Blackcap	0.650924688	0.208590706	0.240886364	0.90043290
Winter Wren	0.646294658	0.217968460	0.232050947	0.89959878
European Robin	0.636539164	0.211018891	0.232380668	0.89270386
Black Redstart	0.631596397	0.210977120	0.227082359	0.91760300
Common Chiffchaff	0.627099631	0.199463573	0.237635548	0.86122449
Great Tit	0.616446917	0.187948719	0.232227488	0.84719465
[... output truncated ...]				
Common Grasshopper Warbler	0.004857411	0.004743370	0.000000000	0.01695646
White Stork	0.004850678	0.003766593	0.000000000	0.01408451
Common Pheasant	0.004783376	0.003576669	0.000000000	0.01398848
Savi's Warbler	0.004734415	0.003690696	0.000000000	0.01304634
Rook	0.004598997	0.003595877	0.000000000	0.01351351
Northern Lapwing	0.004587117	0.003802500	0.000000000	0.01360777
Tawny Pipit	0.004294793	0.004583141	0.000000000	0.01428571
Common Merganser	0.004178675	0.003455891	0.000000000	0.01265823
Golden Eagle	0.002372237	0.005180995	0.000000000	0.01105438

```
plot(1:145, tmp[rev(order(tmp[,1])),1]) # can also plot
```

There is a host of community analyses that are based on the presence/absence matrix and that you can now conduct for a corrected presence/absence matrix, thereby adjusting all your inferences for false-negative detection errors.

11.9 SPECIES RICHNESS MAPS AND SPECIES ACCUMULATION CURVES

We can further use the model output to make extrapolations in space. We show two such examples. First, we map the posterior predictive distribution of local species richness (always at the 1-km^2 scale) for all of Switzerland, and second, we compute and plot the species accumulation curve corrected for species presence/absence measurement error. This section is strongly based on code from Dorazio et al. (2006) and Dorazio et al. (2011). Remember our model for the occurrence of each species in superpopulation M (abbreviated slightly):

```
for(k in 1:M){              # Loop over 215 species
    w[k] ~ dbern(omega)     # Model for regional species pool
    for (i in 1:nsite) {    # Loop over sites: model for presence/absence Z
        logit(psi[i,k]) <- lpsi[k] + betalpsi1[k] * ele[i] + betalpsi2[k] * pow(ele[i],2) +
        betalpsi3[k] * forest[i]
        z[i,k] ~ dbern(w[k] * psi[i,k]) # Presence/absence Z
    }
}
```

We can use our estimates of the data augmentation variable (w) and of the community regression parameters (lpsi[k], etc.) to make predictions in space, using the values of the covariates for a site. That is, for each species in the list of length M, we draw values of w, lpsi, betalpsi1, betalpsi2, and betalpsi3 to obtain a sample from the posterior distribution of presence/absence z_{ik} of species k at every site i in a larger area, and then add up species and plot the result to produce a species richness map (as in Dorazio et al., 2011). We do this now for the Swiss landscape. To avoid a crash, we do this only for a sample of 50 posterior draws (of 1200).

11.9 SPECIES RICHNESS MAPS AND SPECIES ACCUMULATION CURVES

```
# Get Swiss landscape data and standardise covariates as for model 10
library(unmarked)
data(Switzerland)
ch <- Switzerland
ELE <- (ch$elevation - mean.ele) / sd.ele
FOREST <- (ch$forest - mean.forest) / sd.forest

nsamp <- nrow(all10)                        # 1200 ..... far too many
nkm2 <- length(ch[[1]])                     # 42275, that's a LOT!
select.samp <- sort(sample(1:nsamp, 50))    # Chose random sample of 50
nsamp <- length(select.samp)                # new sample size 50

# Create posterior predictive distribution for Z for Swiss landscape
str( zCH <- array(NA, dim = c(nkm2, 215, nsamp)) )    # BIG array!
W <- all10[,1722:1936]                      # Grab MCMC samples from w
LPSI <- all10[,1507:1721]                   # Grab MCMC samples from logit(psi)
BETALPSI1 <- all10[,646:860]                # Grab MCMC samples from betalpsi1
BETALPSI2 <- all10[,861:1075]               # Grab MCMC samples from betalpsi2
BETALPSI3 <- all10[,1076:1290]              # Grab MCMC samples from betalpsi3
for(i in 1:nkm2){                           # takes about 5 mins !
   cat(paste("\nQuadrat", i, "\n"))
   for(u in 1:length(select.samp)){
      psi <- W[select.samp[u],] * plogis(LPSI[select.samp[u],] +
         BETALPSI1[select.samp[u],] * ELE[i] +
         BETALPSI2[select.samp[u],] * ELE[i]^2 +
         BETALPSI3[select.samp[u],] * FOREST[i] )
      zCH[i,,u] <- rbinom(215, 1, psi)
   }
}

# Compute posterior distribution of species richness by collapsing z array
SR <- apply(zCH, c(1,3), sum)    # posterior distribution
pmSR <- apply(SR, 1, mean)       # posterior mean
sdSR <- apply(SR, 1, sd)         # posterior standard deviation
```

We can now plot the posterior mean of species richness at a 1-km^2 scale in all of Switzerland based on the values of elevation and forest cover in the Swiss landscape, and on the parameters relating each individual species' occurrence with these covariates under our model 10. We see that the map is strongly influenced by the severe elevation gradient in Switzerland, and that the estimates are fairly precise (Figure 11.27). See Dorazio et al. (2010) for another such example of a species richness map that is corrected for measurement error.

```
library(raster)
par(mfrow = c(1,2), mar = c(2,2,3,5))
# Posterior mean map (code for more rudimentary plots only shown)
r1 <- rasterFromXYZ(data.frame(x = ch$x, y = ch$y, z = pmSR))
elev <- rasterFromXYZ(cbind(ch$x, ch$y,ch$elevation))
elev[elev > 2250] <- NA          # Mask areas > 2250 m a.s.l.
r1 <- mask(r1, elev)
```

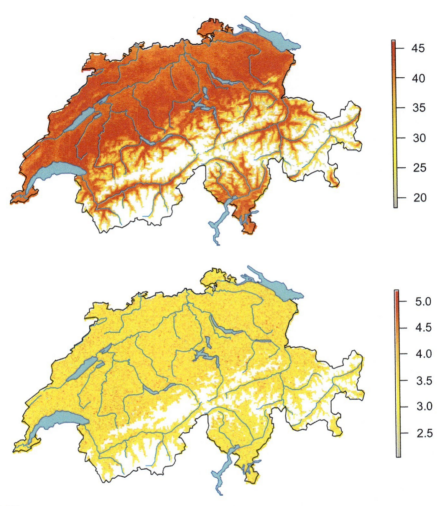

FIGURE 11.27

Breeding bird species richness at the 1-km^2 scale in Switzerland based on community occupancy model 10 for MHB survey data collected in 2014. Top: posterior mean; bottom: posterior standard deviation. Areas above 2250 m in elevation are masked (white).

```
mapPalette <- colorRampPalette(c("grey", "yellow", "orange", "red"))
plot(r1, col = mapPalette(100), axes = F, box = FALSE, main ="")

# Posterior standard deviation map (this code will only produce a more rudimentary map)
r2 <- rasterFromXYZ(data.frame(x = ch$x, y = ch$y, z = sdSR))
elev <- rasterFromXYZ(cbind(ch$x, ch$y,ch$elevation))
elev[elev > 2250] <- NA          # Mask areas > 2250 m a.s.l.
r2 <- mask(r2, elev)
plot(r2, col = mapPalette(100), axes = F, box = FALSE, main ="")
```

11.9 SPECIES RICHNESS MAPS AND SPECIES ACCUMULATION CURVES

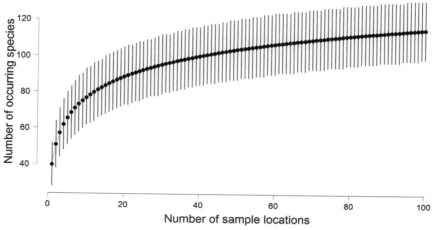

FIGURE 11.28

Species accumulation curve for Swiss breeding birds based on the analysis of the 2014 data harvested from the Swiss breeding bird survey MHB using a DR community occupancy model. These are predictions for 1-km^2 quadrats at the average values of the occupancy covariates elevation and forest cover. *(Code taken from Dorazio et al. (2006).)*

A final use of the parameter estimates from a DR community occupancy model that we illustrate is the computation of a detection-error-corrected species accumulation curve; that is, a curve showing the total number of occurring species in a sample of sites when the number of sites increases (Cam et al., 2002b,c). Such curves can be used, for instance, to standardize spatial sampling effort when comparing different sites. The following code is taken directly from the appendix in Dorazio et al. (2006). All computations are based on posterior samples of the data augmentation parameter omega, and on the mean and standard deviation of the prior distribution of the intercept of the occupancy model. We compute the theoretical relationship between the cumulative number of occurring species, and the number of sample sites for a random sample of sites with up to 100 sites. Since we do not account for covariate effects, our result applies at a covariate value of 0—i.e., for hypothetical 1-km^2 quadrats at 1189-m elevation and with 35% forest cover (Figure 11.28).

```
# Get 3,000 posterior samples of omega, and the mean and sd hyperparameters
omega <- out101$sims.list$omega
mu.lpsi <- out101$sims.list$mu.lpsi
str( sd.lpsi <- out101$sims.list$sd.lpsi )        # Confirms we have 3,000 draws

# compute posterior predictions of species occurrence probabilities
nsites <- 100
ndraws <- length(omega)
Nmax <- 215
psi <- matrix(NA, nrow=ndraws, ncol=Nmax)
for (i in 1:ndraws) {
  w <- rbinom(215, 1, omega[i])
  psi[i,] <- w * plogis(rnorm(Nmax, mean = mu.lpsi[i], sd=sd.lpsi[i]))
}
```

```
# compute posterior predictions of species presence at each site
z <- array(NA, dim=c(ndraws, Nmax, nsites))
for (i in 1:ndraws) {
  for (j in 1:Nmax) {
    z[i,j, ] <- rbinom(nsites, size=1, prob=psi[i,j])
  }
}

# compute posterior predictions of cumulative number of species present
Ntot <- matrix(NA, nrow=ndraws, ncol=nsites)
for (i in 1:ndraws) {
  for (j in 1:nsites) {
    zsum <- rep(NA, Nmax)
    if (j>1) {
      zsum <- apply(z[i, , 1:j], 1, sum)
    }
    else {
      zsum <- z[i, , 1]
    }
    Ntot[i,j] <- sum(zsum>0)
  }
}                           # takes about 4 min

# compute summary stats of species accumulation curve
nSpeciesPresent <- matrix(NA, nrow=3, ncol=nsites)
for (j in 1:nsites) {
  x <- Ntot[,j]
  nSpeciesPresent[1, j] <- mean(x)
  nSpeciesPresent[2:3, j] <- quantile(x, probs=c(0.05, 0.95))
}

# Plot species accumulation curve
ylim = c(min(nSpeciesPresent[2,]), max(nSpeciesPresent[3,]))
plot(1:nsites, nSpeciesPresent[1,], pch=16, ylim=ylim, type="b",
xlab="Number of sample locations", ylab="Number of occurring species",
las=1, cex.axis=1.2, cex.lab=1.5, cex=1.2, frame = F)
segments(1:nsites, nSpeciesPresent[2,], 1:nsites, nSpeciesPresent[3,])
```

11.10 COMMUNITY *N*-MIXTURE (OR DORAZIO/ROYLE/YAMAURA - DRY) MODELS

All community models in this chapter so far were "incidence-based," modeling the presence/absence of species or functions thereof, such as the number of observed species from detection/nondetection data. In this section, we extend the framework of the DR community occupancy model to become an "abundance-based" community *N*-mixture model. Interestingly, the history of this development has proceeded exactly along the same lines as that for single-species models, where the Royle-Nichols (RN) model in 2003 (Section 6.13) opened up the way from the simple Bernoulli-Bernoulli, or site-occupancy, mixture model toward more general HMs with abundance modeled as a latent state—i.e., the *N*-mixture models (Royle, 2004b). For community models, Yamaura et al. (2011) first extended the community occupancy models by adopting an RN formulation for the modeling of species detection/nondetections. Then, Yamaura et al. (2012) modeled counts and developed a full community N-mixture community model, which we here call Dorazio/Royle/Yamaura (DRY) model. Such abundance-based community models are still in their infancy, and only very few papers have been published on them so far, including those by Chandler et al. (2013), Barnagaud et al. (2014), Beesley et al. (2014), Dorazio et al. (2015), Tobler et al. (2015), Sollmann et al. (in press), and Yamaura et al. (in press). However, we expect these to become more common soon.

The basic community N-mixture model with DA is really just a couple of trivial steps away from the community occupancy model (Section 11.7.2). It describes the relationship between latent abundance N_{ik} of species k at site i and the observed response y_{ijk}, which is the *count* of species k at site i during replicate j as follows.

1. Superpopulation process: $w_k \sim Bernoulli(\Omega)$
2. State process (abundance): $N_{ik}|w_k \sim Poisson(w_k \lambda_k)$
3. Observation process (detection): $y_{ijk}|N_{ik} \sim Binomial(N_{ik}, p_{ijk})$
4. Models of species heterogeneity: $\log(\lambda_{ik}) = beta0_k + beta1_k * elevation_i + ...$
$logit(p_{ijk}) = alpha0_k + alpha1_k * date_{ij} + ...$

with
$beta0_k \sim Normal(\mu_{beta0}, \sigma^2_{beta0})$
$beta1_k \sim Normal(\mu_{beta1}, \sigma^2_{beta1})$
$alpha0_k \sim Normal(\mu_{alpha0}, \sigma^2_{alpha0})$
$alpha1_k \sim Normal(\mu_{alpha1}, \sigma^2_{alpha1})$

You can see that this is simply an N-mixture model for M species, where M may be either the total number of observed species or (as in the equations above) the number of species in the augmented data set, exactly analogous to what we have seen in Sections 11.6 and 11.7. Hence, in principle you can apply, inside a community model, everything you learned about N-mixture models in Chapters 6–9. For instance, Chandler et al. (2013) modeled removal counts, and Sollmann et al. (in press) modeled distance sampling counts, for a whole community of species. Below, we will be extending the model to contain a zero-inflation component. We start by organizing the 2014 MHB counts into a 3-D array that we call yc.

```
# Organize counts in 3D array: site x rep x species
COUNTS <- cbind(data$count141, data$count142, data$count143)     # Counts 2014
nsite <- 267                       # number of sites in Swiss MHB
nrep <- 3                          # number of replicate surveys per season
nspec <- length(species.list)      # 158 species occur in the 2014 data
yc <- array(NA, dim = c(nsite, nrep, nspec))
for(i in 1:nspec){
   yc[,,i] <- COUNTS[((i-1)*nsite+1):(i*nsite),]     # 'c' for counts
}
dimnames(yc) <- list(NULL, NULL, names(ordered.spec.name.list))
```

We very briefly look into the raw data to get an idea of the variability of these counts, among sites, among replicates, among species. You have to execute this and see yourself.

```
# Observed maximum and mean maximum count per species
tmp <- apply(yc, c(1,3), max, na.rm = TRUE)
```

```
tmp[tmp == -Inf] <- NA           # 1 quadrat with NA data in 2014
sort(round(meanmax <- apply(tmp, 2, mean, na.rm = TRUE), 3))   # mean of max
sort(obs.max.C <- apply(tmp, 2, max, na.rm = TRUE))            # max

# Plot observed species abundance distribution
plot(sort(meanmax), xlab = "Species number", ylab = "Mean maximum count")

# Spatio-temporal patterns in counts (mean over sites)
tmp <- apply(yc, c(2,3), mean, na.rm = TRUE)
matplot(log10(tmp+0.1), type = "l", lty = 1, lwd = 3, xlab = "MHB survey 1 - 3",
  ylab = "log10 of mean count over sites", frame = F, cex.lab = 1.3, cex.axis = 1.3)

# Drop data from 13 species not observed in 2014
toss.out <- which(obs.max.C == 0)    # list of species not seen
yc <- yc[,,-toss.out]                # toss them out
obs.max.C <- obs.max.C[-toss.out]
nspec <- dim(yc)[3]                  # Redefine nspec as 145

# So here are our data
str(yc)
int [1:267, 1:3, 1:145] 0 0 0 0 0 0 0 0 0 0 ...
- attr(*, "dimnames")=List of 3
  ..$ : NULL
  ..$ : NULL
  ..$ : chr [1:145] "Little Grebe" "Great Crested Grebe" "Grey Heron" ...
plot(table(yc))  # Extremely skewed distribution of observed counts
```

Hence, we model the observed 145 species only and do not data-augment. When developing the material here, we started with a model entirely analogous to the simplest community occupancy model in Section 11.6.1. This model did not pass a commonsense goodness-of-fit (CSGoF) test, since it yielded a flat rather than declining relationship between estimated community size (Nsite) and elevation (unlike in Figure 11.19). Adding in our standard set of covariates led to the desired decline of Nsite at higher elevations, but the overall level was about 14 species higher than were estimates from the comparable community occupancy model (model 10). Especially at high elevations, this does not seem plausible. So, we finally added zero-inflation for every species separately, and this model then did pass our CSGoF test. Compared with the algebra above, the ZIP-DRY community model has an additional hierarchical level that you can think of as the usual "suitability indicator," but it lacks the superpopulation process (the data augmentation variable w). We chose not to data augment to make the model simpler and quicker to converge. 'Quick' is relative, because it takes much longer to get this multispecies N-mixture model to converge than it does for the analogous occupancy model. However, the combination of DA and zero inflation would not pose any problems in principle.

2a. State process 1 (suitability, zero-inflation) : $a_{ik} \sim Bernoulli(\phi_k)$

2b. State process 2 (abundance) : $N_{ik}|a_{ik} \sim Poisson(a_{ik}\lambda_k)$

We made the model fully hierarchical by adopting community priors for all species-specific parameters except for the zero-inflation parameter ϕ_k.

```
# Bundle and summarize data set
str(win.data <- list(yc = yc, nsite = dim(yc)[1], nrep = dim(yc)[2],
nspec = dim(yc)[3], ele = ele, forest = forest, DAT = DAT, DUR = DUR))
List of 9
 $ yc    : int [1:267, 1:3, 1:145] 0 0 0 0 0 0 0 0 0 0 ...
  ..- attr(*, "dimnames")=List of 3
  .. ..$ : NULL
  .. ..$ : NULL
  .. ..$ : chr [1:145] "Little Grebe" "Great Crested Grebe" "Grey Heron" ...
 $ nsite : int 267
 $ nrep  : int 3
 $ nspec : int 145
 $ ele   : num [1:267] -1.1539 -1.1539 -0.2175 -0.3735 -0.0614 ...
 $ forest: num [1:267] -1.1471 -0.4967 -0.0992 -0.9303 0.0092 ...
 $ DAT   : num [1:267, 1:3] -1.415 -1.19 -1.235 -0.559 -1.64 ...
 $ DUR   : num [1:267, 1:3] -0.43511 -0.78764 -0.52324 1.23944 0.00556 ...
```

In the model, we adopted suitably wide normal priors for all mean parameters, and uniform priors for all variance parameters. The presence/absence matrix **Z** is computed as a derived quantity from the posterior samples of the latent abundance variable, and then species richness can be tallied up in exactly the same way as before (and we could also compute similarity etc).

```
# Specify model in BUGS language
sink("model11.txt")
cat("
model{

# Community priors (with hyperparameters) for species-specific parameters
for(k in 1:nspec){
   phi[k] ~ dunif(0,1)                              # Zero-inflation
   alpha0[k] ~ dnorm(mu.alpha0, tau.alpha0)         # Detection intercepts
   beta0[k] ~ dnorm(mu.beta0, tau.beta0)            # Abundance intercepts
   for(v in 1:3){
      alpha[k, v] ~ dnorm(mu.alpha[v], tau.alpha[v])  # Slopes detection
      beta[k, v] ~ dnorm(mu.beta[v], tau.beta[v])     # Slopes abundance
   }
}

# Hyperpriors for community hyperparameters
# abundance model
mu.beta0 ~ dunif(-1, 2)
tau.beta0 <- pow(sd.beta0, -2)
sd.beta0 ~ dunif(0, 3)
for(v in 1:3){
   mu.beta[v] ~ dunif(-1.5, 1)
   tau.beta[v] <- pow(sd.beta[v], -2)
}
sd.beta[1] ~ dunif(0, 3)
sd.beta[2] ~ dunif(0, 1.5)
sd.beta[3] ~ dunif(0, 1)
```

```
# detection model
mu.alpha0 ~ dunif(-2, 0)
tau.alpha0 <- pow(sd.alpha0, -2)
sd.alpha0 ~ dunif(0, 2)
for(v in 1:3){
  mu.alpha[v] ~ dunif(-0.5, 0.5)
  tau.alpha[v] <- pow(sd.alpha[v], -2)
}
sd.alpha[1] ~ dunif(0, 0.8)
sd.alpha[2] ~ dunif(0, 0.5)
sd.alpha[3] ~ dunif(0, 0.3)

# Ecological model for true abundance (process model)
for(k in 1:nspec){
  for (i in 1:nsite){
    a[i,k] ~ dbern(phi[k])              # zero-inflation
    N[i,k] ~ dpois(a[i,k] * lambda[i,k])
    log(lambda[i,k]) <- beta0[k] + beta[k,1] * ele[i] + beta[k,2] * pow(ele[i],2) +
      beta[k,3] * forest[i]
    # Compute presence/absence matrix z (for N > 0) from latent abundance
    z[i,k] <- step(N[i,k]-1)            # returns TRUE if N >= 1
  }
}

# Observation model for replicated counts
for(k in 1:nspec){
  for (i in 1:nsite){
    for (j in 1:nrep){
      yc[i,j,k] ~ dbin(p[i,j,k], N[i,k])
      logit(p[i,j,k]) <- alpha0[k] + alpha[k,1] * DAT[i,j] +
        alpha[k,2] * pow(DAT[i,j],2) + alpha[k,3] * DUR[i,j]
    }
  }
}

# Other derived quantities
for(k in 1:nspec){
  mlambda[k] <- phi[k] * exp(beta0[k])   # Expected abundance on natural scale
  logit(mp[k]) <- alpha0[k]              # Mean detection on natural scale
  Nocc.fs[k] <- sum(z[,k])               # Number of occupied sites among the 267
}
for (i in 1:nsite) {
  Nsite[i] <- sum(z[i,])                 # Number of occurring species at each site
}
}
",fill = TRUE)
sink()

# Initial values
ast <- matrix(rep(1, nspec*nsite), nrow = nsite)
some.more <- 5          # May have to play with this until JAGS is happy
Nst <- apply(yc, c(1,3), max, na.rm = T) + some.more
```

```
Nst[Nst == '-Inf'] <- 20            # May have to play with this, too
Nst <- Nst
inits <- function()list(a = ast, N = Nst)

# OR: use inits at earlier solutions (greatly speeds up convergence)
pm <- out11$mean     # Pull out posterior means from earlier run
inits <- function() list(a = ast, N = Nst, alpha0 = rnorm(nspec), beta0 =
rnorm(nspec), alpha = matrix(rnorm(n = nspec*3), ncol = 3), beta =
matrix(rnorm(n = nspec*3), ncol = 3), mu.beta0 = pm$mu.beta0, sd.beta0 = pm$sd.beta0,
mu.beta = pm$mu.beta, sd.beta = pm$sd.beta, mu.alpha0 = pm$mu.alpha0, sd.alpha0 =
pm$sd.alpha0, mu.alpha = pm$mu.alpha, sd.alpha = pm$sd.alpha )

# Parameters monitored
params <- c("phi", "mp", "mlambda", "alpha0", "beta0", "alpha", "beta", "mu.beta0",
"sd.beta0", "mu.beta", "sd.beta", "mu.alpha0", "sd.alpha0", "mu.alpha", "sd.alpha",
"Nsite")

# MCMC settings
ni <- 180000 ; nt <- 90 ; nb <- 90000 ; nc <- 3

# Call JAGS from R, check convergence and summarize posteriors
out11 <- jags(win.data, inits, params, "model11.txt", n.chains = nc,
n.thin = nt, n.iter = ni, n.burnin = nb, parallel = FALSE)
par(mfrow = c(3,3)) ;traceplot(out11, c("mu.beta0", "sd.beta0", "mu.beta", "sd.beta",
"mu.alpha0", "sd.alpha0", "mu.alpha", "sd.alpha") )
print(out11, 2)
```

	mean	sd	2.5%	50%	97.5%	overlap0	f	Rhat	n.eff
mu.beta0	0.48	0.21	0.07	0.48	0.89	FALSE	0.99	1.00	598
sd.beta0	2.14	0.18	1.80	2.14	2.50	FALSE	1.00	1.00	3000
mu.beta[1]	-0.47	0.21	-0.91	-0.46	-0.08	FALSE	0.99	1.00	2173
mu.beta[2]	-0.60	0.11	-0.83	-0.60	-0.38	FALSE	1.00	1.00	3000
mu.beta[3]	-0.18	0.06	-0.30	-0.18	-0.07	FALSE	1.00	1.00	526
sd.beta[1]	2.27	0.17	1.98	2.26	2.66	FALSE	1.00	1.01	181
sd.beta[2]	1.13	0.10	0.94	1.12	1.34	FALSE	1.00	1.02	92
sd.beta[3]	0.60	0.05	0.51	0.60	0.71	FALSE	1.00	1.00	1473
mu.alpha0	-1.17	0.11	-1.39	-1.16	-0.96	FALSE	1.00	1.02	136
sd.alpha0	0.98	0.10	0.80	0.97	1.19	FALSE	1.00	1.02	120
mu.alpha[1]	0.04	0.04	-0.04	0.04	0.13	TRUE	0.84	1.00	1885
mu.alpha[2]	-0.11	0.04	-0.18	-0.11	-0.04	FALSE	1.00	1.00	1051
mu.alpha[3]	0.26	0.02	0.22	0.26	0.29	FALSE	1.00	1.04	53
sd.alpha[1]	0.44	0.04	0.37	0.44	0.52	FALSE	1.00	1.00	1186
sd.alpha[2]	0.34	0.03	0.28	0.34	0.41	FALSE	1.00	1.00	1441
sd.alpha[3]	0.14	0.02	0.10	0.14	0.18	FALSE	1.00	1.35	10

Convergence has not formally been achieved for one hyperparameter, but we have been running the model for a really long time and its traceplots do not look so bad, so we're still going to summarize the results from this run for now. Basically, we are able to do everything we did when summarizing the analysis of model 10. For instance, we can find that the average per-individual detection probability is only 0.27.

```
summary(p.sample <- plogis(rnorm(10^6, mean = -1.170, sd = 0.980)) )
hist(p.sample, breaks = 50, col = "grey", xlab = "Per-individual detection probability",
freq = FALSE)
```

722　CHAPTER 11　HIERARCHICAL MODELS FOR COMMUNITIES

For illustration, we plot the species-specific covariate relationships in the detection (Figure 11.29) and abundance parts of the model (Figure 11.30). Note that for predicting expected abundance, we have to multiply with the zero-inflation parameter. We will make use of the prediction covariates generated in the last section.

```
# Predict detection for date and duration and expected abundance for elevation and forest
# for each of the 145 observed species
predI <- array(NA, dim = c(500, nspec, 4))    # covariate value x species x
response, "I" for 'individual' (as opposed to 'species' in model 10)
pm <- out11$mean        # Grab posterior means from model 11
```

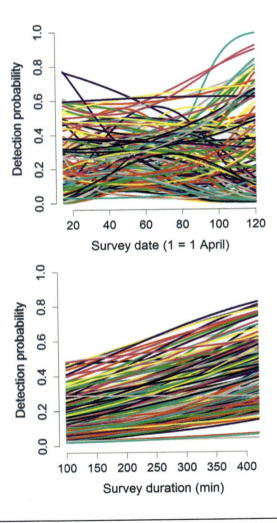

FIGURE 11.29

Species-specific predictions of detection probability (per individual) as a function of survey date and survey duration in 1-km² quadrats in the Swiss breeding bird survey MHB under community N-mixture model number 11. Each line represents one of the 145 species observed in 2014. Note that one covariate is kept at the observed average for the computation of the prediction for the other covariate, and vice versa.

11.10 COMMUNITY N-MIXTURE MODELS

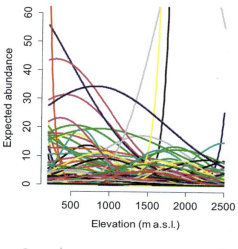

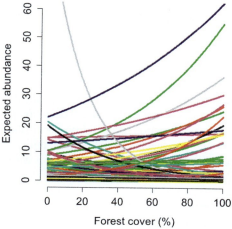

FIGURE 11.30

Species-specific predictions of expected abundance probability as a function of elevation and forest cover in 1-km^2 quadrats in the Swiss breeding bird survey MHB under community N-mixture model number 11. Each line represents one of the 145 species observed in 2014. Note that one covariate is kept at the observed average for the computation of the prediction for the other covariate, and vice versa.

```
for(i in 1:nspec){      # Loop over 145 observed species
    predI[,i,1] <- plogis(pm$alpha0[i] + pm$alpha[i,1] * dat.pred +
    pm$alpha[i,2] * dat.pred^2 )          # p ~ date
    predI[,i,2] <- plogis(pm$alpha0[i] + pm$alpha[i,3] * dur.pred)    # p ~ duration
    predI[,i,3] <- pm$phi[i] * exp(pm$beta0[i] + pm$beta[i,1] * ele.pred +
    pm$beta[i,2] * ele.pred^2 )           # E(N) ~ elevation
    predI[,i,4] <- pm$phi[i] * exp(pm$beta0[i] + pm$beta[i,3] * for.pred)   # E(N) ~ forest
}

# Plots for detection probability and survey date and duration (Fig. 11-29)
par(mfrow = c(1,2), cex.lab = 1.3, cex.axis = 1.3)
```

```
plot(o.dat, predI[,1,1], lwd = 3, type = 'l', lty = 1, frame = F, ylim = c(0, 1),
    xlab = "Survey date (1 = 1 April)", ylab = "Per-individual detection probability")
for(i in 2:145){
   lines(o.dat, predI[,i,1], col = i, lwd = 3)
}

plot(o.dur, predI[,1,2], lwd = 3, type = 'l', lty = 1, frame = F, ylim = c(0, 1),
    xlab = "Survey duration (min)", ylab = "Per-individual detection probability")
for(i in 2:145){
   lines(o.dur, predI[,i,2], col = i, lwd = 3)
}

# Plots for expected abundance and elevation and forest cover (Fig. 11-30)
par(mfrow = c(1,2), cex.lab = 1.3, cex.axis = 1.3)
plot(o.ele, predI[,1,3], lwd = 3, type = 'l', lty = 1, frame = F,
ylim = c(0, 60), xlab = "Elevation (m a.s.l.)",ylab = "Expected abundance")
for(i in 2:145){
   lines(o.ele, predI[,i,3], col = i, lwd = 3)
}

plot(o.for, predI[,1,4], lwd = 3, type = 'l', lty = 1, frame = F,
ylim = c(0, 60),    xlab = "Forest cover (%)", ylab = "Expected abundance")
for(i in 2:145){
   lines(o.for, predI[,i,4], col = i, lwd = 3)
}
```

Comparing these figures with those from the community occupancy model (Figures 11.24 and 11.25) we note broadly similar patterns. Obviously, the per-individual detection probability under the N-mixture model is lower than the per-species detection probability under the occupancy model (because $P^* = 1 - (1-p)^N$, where P^* is the per-species detection probability, p the per-individual detection probability and N is local abundance). Therefore, the increase of p with increasing survey duration is much more striking under the N-mixture model.

11.11 SUMMARY AND OUTLOOK

We have given an overview of the community versions of site-occupancy models and N-mixture models: that is, the Dorazio/Royle (DR) community occupancy model (Dorazio and Royle, 2005; Dorazio et al., 2006) and the Dorazio/Royle/Yamaura (DRY) community abundance model (Yamaura et al., 2012; Chandler et al., 2013). In both models, spatial patterns in the presence/absence or the abundance of a community of species are described by a collection of elemental models for each individual species. Thus, these models are straightforward extensions to a community of species of any of the single-species models for occurrence and abundance that we presented in Chapters 6—10. Both the DR and the DRY model meet our main requirements for a flexible and robust community modeling framework: they allow us to make inferences at all levels of a metacommunity (species, community, metacommunity), and they explicitly deal with measurement error so that inferences can be made beyond just the sample of observed species, including estimation of species richness at multiple scales. Naturally, this object of inference includes the number of observed species at a point in time as well as the number of unobserved species. Only the DR/DRY class of models allows for this generality and breadth of inference.

These powerful community models can describe patterns, such as covariate relationships, at the level of the individual species, the local community (all species occurring at one site), and the whole metacommunity—i.e., for the regional pool of species in the wider area of study. We have contrasted the DR/DRY community models with a much more simplistic approach that directly models emerging properties of a community, such as species richness, but that lacks the ability to make inferences about individual species, and we have seen that the DR community model was considerably more powerful. For instance, at the community level there was no measurable effect of forest cover on species richness, but at the level of individual species, almost half of the observed species were found to have a strong response to forest (31 species positive and 35 negative). In the Swiss breeding bird survey MHB, the average species in the metacommunity sampled had a detection probability of 0.63 at the level of a presence/absence measurement (i.e., in the occupancy model) and of 0.27 at the level of the abundance measurement for an individual (i.e., in the model for abundance). Therefore, by correcting for the resulting measurement error, arguably the DR and DRY community models had much reduced bias compared to other community models that do not contain a measurement error submodel.

We have also shown some further analyses that are possible after fitting a DR/DRY community model. Examples include the inspection of estimates for individual species, and extrapolation of species richness to a region such as the whole of Switzerland to produce a detection-corrected species richness map along with its associated uncertainty map (Dorazio et al., 2010). We also showed the computation of a species accumulation curve that corrects for imperfect and heterogeneous detection probability and that is insensitive to the ordering of sites (Dorazio et al., 2006). One estimand of particular interest is the presence/absence matrix Z, which has been called the fundamental unit of analysis in biogeography and community ecology (McCoy and Heck, 1987). Most ecologists mistake the *observed* presence/absence patterns for the *true* presence/absence matrix, and therefore will obtain biased inferences due to presence/absence measurement error. In contrast, in the DR/DRY community models, we obtain an *estimate* of Z that is *corrected* for detection error. We can now do any calculation we would like on this estimate of Z to describe the size and composition of the metacommunity. As an illustration, we have computed Jaccard indices to compare sites in terms of occurring species, and to compare species in terms of the sites where they occur.

The community models in this chapter are joint species distribution models (JSDMs) in the sense that they simultaneously describe, in a single model, the occurrence or abundance of all species in a community. This simultaneous modeling has great advantages; for instance, we can formally compare species or even fit models to explain differences among species; see Section 11.6.3. In addition, estimates for information-poor rare species benefit from the richer information that comes from more common species, resulting in improved estimates for such species (Kéry and Royle, 2008; Zipkin et al., 2009). Also, we can easily compute derived quantities such as the number of occurring species with full propagation of the associated uncertainty. However, these models are based on the assumption that all species occur and are detected independently from one another given the modeled covariates. Clearly, this assumption will not always hold, and several recent lines of research have developed community models that do not make the independence assumption (Ovaskainen et al., 2010; Ovaskainen and Soininen, 2011; Clark et al., 2014; Dorazio and Connor, 2014; Pollock et al., 2014). This is exciting work. However, whether the more complex dependence models should be favored over simpler independence models will depend on many things, such as the spatial and temporal scale of a study, the taxonomic group, or even the specific set of species and the richness or sparsity of the data. For instance, it seems likely that the larger the grain (or spatial scale) of study, the more independent will the species occurrence or abundance appear to be. Similar reasoning also seems possible for the temporal scale: when analyzing data over larger temporal "observation windows," it would appear that observed dependencies in occurrence among species would be dampened. Finally, we would surmise that dependence is stronger for plants (e.g., trees, Clark et al., 2014) than for animals, because plants compete very heavily along few environmental gradients (light, nutrients), while the ability to move would seem to greatly reduce competition among animal species by increasing the number of niches available at any given place.

In our analyses of breeding bird species data collected in 1-km^2 quadrats for a temporal window of two to three months, we think it unlikely that substantial dependencies would occur beyond what can be modeled by covariates. However, in principle, extending the DR/DRY models to contain species dependency appears simple: all we have to do is to add site-specific random effects to the linear predictor of occupancy or expected abundance for every species, and then specify a multivariate normal distribution as a prior for these "site residuals." The off-diagonals in the variance-covariance matrix then describe the positive or negative residual association for all pairs of observed species, where "residual" means "after all effects explained by the covariates in the model have been accounted for." We will cover multispecies models with (statistical) species interactions in Chapter 20 of volume 2.

We have presented two variants of community models, with and without data augmentation (DA). DA lets us make inferences about the whole metacommunity, including not only species that were detected but also species that were never detected. This allows one to estimate the full size of the metacommunity or the regional species pool. Also, it will reduce bias in inferences about the community that occur whenever a species trait is associated with detection probability. For instance, we found that there was a positive correlation between occupancy and detection probability in Swiss birds, and hence it is likely that the 145 observed species represent a biased sample from the entire Swiss avian metacommunity (because we are more likely to ever observe those with higher-than-average detection probability). Doing DA and modeling the entire metacommunity will ideally eliminate this bias. On the other hand, DR/DRY community models are naturally very parameter-rich and therefore costly in terms of computation time; DA usually *greatly* increases that cost. Moreover, we very rarely know what the metacommunity actually refers to in terms of an effective sample area, because none of these models is spatially explicit; see the discussion in Section 6.10. Hence, the estimated metacommunity size is often a largely hypothetical parameter and may therefore be of little practical interest. As a consequence, it may often be fine to apply a DR/DRY community model to the observed portion of a metacommunity only, as we have done for the DRY community N-mixture model.

Where will the development of DR/DRY and related community models go from here? First of all, we would like to think that the number of applications of these models will greatly increase. These models are just tremendously sensible mechanistically, and they offer powerful and multifaceted insights into communities at multiple scales. In addition, the required data are quite common, especially detection/nondetection data for the DR community model.

We have seen how easily we can move from a model for presence/absence data to a model for abundance, either for counts (Yamaura et al., 2012) or for detection/nondetection data combined with an RN model (Yamaura et al., 2011). Something in between might be a model with multiple states of occurrence—i.e., the community analog of a multistate occupancy model (Royle and Link, 2005; Nichols et al., 2007; MacKenzie et al., 2009)—see Fukaya and Royle (2013) for an example of such a model. In addition, in Chapters 6–10 we have presented a multitude of observation protocols for measuring presence/absence or abundance. All of these could be applied in a community context. For instance, it would seem to be easy to develop a community occupancy model for a removal or a time-to-detection protocol. Chandler et al. (2013) and Sollmann et al. (in press) have developed community N-mixture models for removal and distance sampling protocols, respectively, and the same could be done for any other multinomial mixture model discussed in Chapter 7.

One way of looking at a DR/DRY community model is that it simply represents the most sensible way of integrating the information about all species in a community or a subset thereof (Ovaskainen

11.11 SUMMARY AND OUTLOOK

and Soininen, 2011). Thus, whenever your aim is to characterize a community by some average over the species that occur in it, clustering the species together in a single analysis and specifying community hyperparameters to describe the mean and the variability among species seems a very sensible thing to do. This leads exactly to a DR/DRY type of community model, and not surprisingly, this has also been discovered independently (Gelfand et al., 2005) and rediscovered multiple times since 2005 (e.g., Ovaskainen et al., 2010; Ovaskainen and Soininen, 2011).

One area that is almost always underinvestigated is study design, and currently only Sanderlin et al. (2014) and Yamaura et al. (in press) have looked at DR and DRY community model design (also see McNew and Handel, 2015). Presumably, much may be deduced about community model study design from studies of single-species occupancy and N-mixture model study design (e.g., MacKenzie and Royle, 2005; Bailey et al., 2007; Guillera-Arroita et al., 2010, 2014b; McIntyre et al., 2012; Yamaura, 2013; Ellis et al., 2015), but there is certainly room for other and more focused studies. Perhaps the data simulation function and the analysis code given in this chapter may serve as a starting point.

One important difference between the DR/DRY community models and other otherwise similar community models recently developed (e.g., Gelfand et al., 2005; Ovaskainen et al., 2010; Clark et al., 2014; Pollock et al., 2014) is that the DR/DRY models contain an explicit description of the measurement error for occurrence or abundance. Unless this measurement error is small, all such inferences from community models will be biased to an unknown degree. However, DR/DRY community models only account for one of the two possible errors, false-negatives. False-negative errors are likely to be much more common than false-positive errors, especially in well-designed surveys like the Swiss MHB and for taxa such as birds, which are relatively species poor, comparatively easy to identify, and for which there are armies of well-trained volunteer observers in countries such as Switzerland. However, clearly there must be many situations in which species misidentifications can be a serious problem, especially in burgeoning Internet-based citizen-science programs that are now mushrooming all over the world, and where observer quality is extremely variable, generally unknown, and certainly often dubious. The consequences of false-positives on estimators of community models have not been studied. Some may be deduced from studies for single-species occupancy models (e.g., Miller et al., 2015), but the topic would warrant a thorough simulation study. Even more important would be the development of community models that can quantify and thereby correct for false-positives. However, this would appear to be a very difficult problem, because the basic DR/DRY model is already very parameter-rich and may be difficult to fit in practice. Adding complexity for false-positives would greatly increase the number of parameters that need to be estimated and exacerbate these problems.

It is intriguing to hypothesize about the interaction between false-positive errors in community studies and the perceived dependence in the *observed* occurrence of species (e.g., Clark et al., 2014; Pollock et al., 2014). We think it is possible that apparent dependency in the observed occurrence among species could be partly the result of misidentification. Falsely detecting a species A, which was in reality species B, is also the commission of a false-negative error for species B and, at the same time, should lead to understatement of occurrence probability for species B and vice versa for species A. This looks like a good research question: To investigate the mechanism underlying associations in the *observed* occurrence of species in a community; are they true (biological) or are they spurious, due to measurement error only, or perhaps both?

Spatial models for occurrence or abundance accommodate spatial autocorrelation and are increasingly used (e.g., Wintle and Bardos, 2006; Bled et al., 2011a,b; Yackulic et al., 2012; Broms et al., 2014; see also Chapters 21 and 22 in volume 2). Mattsson et al. (2013) developed a community model with autologistic formulation of spatial autocorrelation on occupancy, and presumably a conditional autoregressive (CAR) or related formulation would work as well (Johnson et al., 2013), or else a spline model (Collier et al., 2012, see also Section 10.14). For single-species models, accommodating spatial autocorrelation by use of splines or CAR random effects can produce *much* better species distribution maps than do models that ignore spatial autocorrelation (Guélat and Kéry, in review). Hence, presumably maps of species richness would benefit similarly. But while in principle it would be straightforward to add autocorrelation terms for each species to the model, in practice, as for the modeling of false-positives, such models may just become too data hungry to be fittable in practice. However, this is clearly something that ought to be investigated.

Finally, an obvious extension is to accommodate time. All models in this chapter assumed closure. Dorazio et al. (2010) have extended the community model to include colonization/extinction dynamics, and so have Fukaya and Royle (2013), Henden et al. (2013), and Tobler et al. (2015). This is a straightforward extension conceptually, starting with the basic dynamic occupancy model for a single species (MacKenzie et al., 2003; Royle and Kéry, 2007) and using the same ideas of treating species-level parameters as fixed or random effects and of accounting or not for unseen species via DA, as we saw in Sections 11.6 and 11.7. We cover dynamic community models in Chapter 17 in volume 2.

EXERCISES

1. "Play community" by running the data simulation function `simComm` with changed function arguments, and observe how the latent state of presence/absence or abundance, and the observed data, are affected by your choice. Ideally, run a little simulation study to improve your intuition about metacommunity studies. Perhaps you can come up with some idea to even morph such play into a project for a paper.

2. In Sections 11.4 and 11.5.3 we mentioned that a multinomial mixture model would be intermediate between an N-mixture and a community site-occupancy model in the sense of keeping track of species identity over replicate surveys (unlike the former), though not across sites (unlike the latter). Use `unmarked` to fit a multinomial mixture model to the Swiss MHB data that we used in this chapter. You will not be able to incorporate any covariates into p, but do add the same covariates into "abundance" (which is the model part for local species richness in this application of the model) that we do for species occupancy in later sections of this chapter. Also fit the same multinomial mixture model with BUGS; there, you will be able to add any covariates into the model for p that you like.

3. In Section 11.6.1, turn model 5 (which has a binomial response) into a model with Bernoulli data distribution by fitting the Bernoulli variant of the same model to the disaggregated 3-D data array. Compare the timings between this and the binomial model 5.

4. In model 10, add into the detection model the effects of elevation and its interaction with date. The reason for this is that a survey date of, say, April 15, does not mean the same at an elevation of 250 and one of 2500 m. In the lowlands, this is the middle of spring, while at higher elevations, this is late winter.

5. At the start of Section 11.7.2, we mentioned that unmodeled detection heterogeneity among sites would likely bias low species richness estimators. Conduct a simulation study to investigate whether any bias results from site heterogeneity in occupancy probability. (Hint: you can use function `simComm`.)

SUMMARY AND CONCLUSION

CHAPTER OUTLINE

A.1 Objectives of *Applied Hierarchical Modeling in Ecology (AHM)* 729
A.2 What's in the Two *AHM* Books? 730
    A.2.1 Hierarchical Models (HMs) 730
    A.2.2 Measurement Error Models 730
    A.2.3 Demographic Modeling of Populations and Communities 731
    A.2.4 Brief Chapter Overview of *AHM* Volume 1 731
    A.2.5 The HM as the "Grand Unifying Model" 733
A.3 Outlook: *AHM* Volume 2 734

A.1 OBJECTIVES OF *APPLIED HIERARCHICAL MODELING IN ECOLOGY (AHM)*

Our general goals in writing *AHM* were: to provide a comprehensive overview of hierarchical models (HMs) or state-space models for the demographic analysis of populations and communities using a "meta-population design" with an explicit model structure for the measurement error process; to showcase the R package unmarked for likelihood analysis and BUGS software for Bayesian analysis of HMs; and finally, to emphasize the powerful and unifying concept of hierarchical modeling as an overarching principle in all of these analyses.

More specifically, we wanted to make this synthesis as complete and yet as gentle and easy to understand as possible. We had two kinds of audiences in mind. On the one hand are ecologists and scientists, as well as managers working in related disciplines such as conservation biology, biogeography, geography, wildlife research, and biodiversity monitoring. On the other hand are statisticians, especially those first becoming acquainted with these classes of HMs, which they are unlikely to have encountered in standard methodology classes or classical applied statistics texts. For ecologists, we wanted to make the material as accessible as possible. This includes our novel use of data simulation as a method of explaining models, and the ample use of graphs to illustrate the models and the results of the analyses. Crucially, it also includes cookbook analyses, which we think are truly essential for letting most nonstatisticians really understand a new model, and enabling them to apply it to actual data in practice. The dual inference paradigm approach, with seamless integration of frequentist and Bayesian analyses, is also part of our recipe for maximum accessibility of the material presented. Very often, a likelihood-based analysis with a function coded up in R (which in this book typically means a function in the R package unmarked) will let you start fitting your models easily. It was our goal to enable you as a practitioner to immediately start off with your own data analyses after reading a specific chapter of this book.

At the same time, *AHM* is not simply a lab manual for the analysis of HMs. Instead, it contains a substantial amount of novel material, especially in the two "monographs" on distance sampling but also throughout the rest of the book. At several places, we emphasized less well-understood topics where research may be warranted. This includes the good-fit/bad-prediction dilemma in binomial N-mixture models (which may afflict multinomial mixture models less or perhaps not at all?), goodness-of-fit assessment for HMs, and the issue of spatial instead of temporal replication in occupancy models. Thus, by presenting what we think is a fairly comprehensive overview of the field of HMs for demographic analysis of populations and communities, we also hope to catalyze new research and new developments in this exciting and extremely useful class of models.

A.2 WHAT'S IN THE TWO *AHM* BOOKS?
A.2.1 HIERARCHICAL MODELS (HMs)

In the two *AHM* volumes, we give a modern synthesis of both frequentist and Bayesian treatments of HMs for the demographic analysis of populations, communities, and metacommunities studied at a collection of sites, or in what we have called a "meta-population design" (Royle, 2004a; Kéry and Royle, 2010). Hierarchical or state-space models decompose a complex statistical model into a sequence of simpler submodels, and this decomposition often helps in fitting a complex model (Berliner, 1996; Cressie et al., 2009; Cressie and Wikle, 2011; Hobbs and Hooten, 2015). In addition, the individual modules of HMs may naturally represent multiple processes, biological scales, or groupings in space, in time, or along other dimensions such as multiple species in a community. Hence, different modules in HMs may describe spatial vs. temporal variation of a studied quantity, subunits nested in larger spatial or temporal units (e.g., multiple scales of measurements), or multiple levels in the hierarchy of biological organization such as individuals nested within a population, species nested within a community, or communities nested within a metacommunity. In a sense, hierarchical modeling represents a natural way of keeping your statistical modeling tidy by organizing like processes, scales, and other things in an orderly manner. At the same time, HMs render your statistical modeling extremely versatile, since they allow you to simply switch, add, or drop individual modules as needed.

A.2.2 MEASUREMENT ERROR MODELS

One hallmark of all main HMs featured in the *AHM* books is that they have one module for the actual ecological process and its result—e.g., abundance or presence/absence—and another module to adjust the inference for errors in the measurement of that ecological process (and there may be additional modules, of course). We often term these two modules the *ecological (sub)model*, and the *observation or measurement error (sub)model*, whereas in the state-space model literature these are typically called the *process model* and the *observation model* (e.g., Buckland et al., 2004b, 2007; Newman et al., 2014).

The main parameters in the measurement error part of an HM typically represent the probability of false-negative errors or its complement, detection probability (p). Detection error is present in virtually all ecological data sets ever collected in the field, and often even in those from lab studies. If unaccounted for, detection error often leads to biased estimates of all population or community quantities of interest. This is radically different from perhaps better-known types of measurement error for continuous quantities, such as body size, where the mean of replicated measurements is often unbiased. By not sweeping under the rug this ugly reality of all ecological data, but rather confronting measurement error up front, *AHM* continues a tradition of other statistics books since the 1970s, including those by Otis et al. (1978), Seber (1982), Buckland et al. (2001, 2004a), Borchers et al.

(2002), Williams et al. (2002), Amstrup et al. (2005), MacKenzie et al. (2006), Royle and Dorazio (2008), King et al. (2009), Kéry and Schaub (2012), McCrea and Morgan (2014), and Royle et al. (2014).

By explicitly accounting for measurement error in our demographic population and community models, the HMs in *AHM* are *explicit hierarchical models* in the sense that the latent variables and parameters in the ecological part of the model have clear biological meanings. For instance, they may represent abundance N rather than *relative abundance*. The latter is a vaguely defined quantity that confounds true abundance with detection probability, and typically represents the product Np only (Johnson, 2008, Section 1.1.1 in Royle and Dorazio, 2008; Kéry, 2011b). Relative abundance and relative, or apparent, occurrence are the targets of inference of statistical models that do *not* have an explicit measurement error component. Adding an explicit measurement error component to our model may represent a cost in terms of model complexity and data requirements, but it buys us a much clearer interpretation of all ecological parameters (provided of course that the model assumptions are not violated too strongly). Scientific interpretability of model parameters is something that we value a lot in a model.

A.2.3 DEMOGRAPHIC MODELING OF POPULATIONS AND COMMUNITIES

Key parameters in the ecological part of the HMs in *AHM* describe state variables such as distribution, abundance, and species richness, rate parameters such as a trend (in abundance or distribution), or vital rates governing population dynamics such as survival probability or recruitment rates. We have organized the two *AHM* volumes so that we first set the scene for the development and use of HMs for inferences about populations, communities, and metacommunities by introducing the necessary context, concepts, and methods. We did this in the first five chapters that form Part 1 of *AHM* volume 1. In the six chapters in Part 2 of *AHM* volume 1, we provide six "monographs" for a number of key HMs for inferences about abundance, distribution, and species richness, as well as related demographic quantities in the demographic analysis of populations, metapopulations, communities, and metacommunities. We emphasized static models in volume 1—i.e., on distribution, abundance, species richness, and similar quantities at a single point in time. That is, we do not keep track of any temporal dynamics, which is a main topic of *AHM* volume 2, when we present dynamic versions of all the static HMs introduced in *AHM* volume 1. In addition, in volume 2 we cover a series of more "advanced" models; see below for more information, and also the table in the Preface of this volume. We have deviated a little bit from this organizing principle in the two distance-sampling monographs (Chapters 8 and 9), where we also cover distance-sampling models with population dynamics.

A.2.4 BRIEF CHAPTER OVERVIEW OF *AHM* VOLUME 1

The first five chapters in *AHM* volume 1 represented the introductory part, or Prelude, of the two-volume book project. They should teach you the concepts and techniques required to understand all the subsequent material in *AHM*. In Chapter 1, we described how we view abundance, occurrence/distribution, and species richness as simple summaries of an underlying point pattern within some spatial discretization scheme: if you know the point pattern—i.e., the locations of all animals or plants in one or more study areas (or "sites")—then you know exactly the abundance and occurrence of every species, and the species richness at every site. Occurrence/distribution is a simple summary of the abundance distribution across sites. Similarly, species richness is a simple summary of the occurrence of every species at a site. We have also argued in Chapter 1 that we typically view zero-inflated models as a mere modeling trick to account for incomplete knowledge of the factors governing abundance. We do *not* usually view these models as a useful representation of the ecological processes underlying spatial heterogeneity in abundance. Therefore, we would usually refrain from attributing much ecological meaning to the zero-inflation part of such models.

In Chapters 2 and 3, we formally introduced HMs and their analysis using the two dominant statistical inference paradigms: maximum likelihood estimation (MLE) and Bayesian posterior inference. We argued that you need both in order to be an effective ecological modeler, since in different situations, for different persons, or for different problems, MLE may be more practically useful than a Bayesian analysis, or vice versa. This dual inference paradigm approach is one of the main themes of *AHM*. To analyze HMs in *AHM*, we used the two dominant statistical software programs in ecology: R for frequentist inference and WinBUGS, OpenBUGS, and JAGS for Bayesian inference (see below). Note that for the latter we could also have used the interesting new R package NIMBLE (NIMBLE Development Team, 2015; de Valpine et al., in review). NIMBLE is a general model-fitting software that uses and extends the BUGS language for flexible specification of HMs, but allows a wide variety of numerical algorithms for fitting these models, including maximum likelihood and Bayesian posterior inference.

For maximum likelihood inference, we used mainly the R package unmarked (Fiske and Chandler, 2011) and often also unmarked-specific functions in the R package AICcmodavg (Mazerolle, 2015). The functions in these packages allow quick and easy fitting of a very wide range of HMs for abundance and distribution in "meta-population designs," all fully integrated in the R computing environment. Hence, data preparation and further analyses such as parametric bootstrap for goodness-of-fit assessment, significance testing, model selection, model averaging, and plotting (for instance, of predictions) are made extremely easy. *AHM* is the first book written on the unmarked package.

One of the main benefits of hierarchical modeling is the ease with which you can develop custom models that represent the processes that you think underlie an observed data set. These models may accommodate potentially complicated dependencies created through groupings in space, time or along multiple other dimensions, such as species in a community. To represent effects of covariates on individual modules of HMs, we typically use linear models for a parameter that may be link-transformed in order to enforce range constraints, exactly as in generalized linear models (GLMs; McCullagh and Nelder, 1989). Hence, in practice HMs can be viewed as a series of connected GLMs for a corresponding series of linked random variables, some of them observed (the data) and some of them not directly observable (the latent variables or random effects). So to use HMs in practice, you need a very good grasp of these three key concepts in applied statistical modeling (linear models, GLMs, and random effects), which are exactly the statistical concepts underlying some of the most commonly used R functions such as lm, glm, lmer, and glmer. Due to their great importance in applied work as part of HMs, we presented a very gentle and applied, yet comprehensive overview of linear models, GLMs, and random effects in Chapter 3.

Throughout *AHM*, we use data simulation frequently and for a large number of reasons. A completely unique and novel feature of *AHM* is that we explicitly *use data simulation in R to explain a statistical model* (though see Kéry (2010) and Kéry and Schaub (2012) for precursors of this approach). Parametric statistical models represent an abstraction of the processes that we think underlie our observed data set. Hence, by writing R code to simulate a data set generated by these very processes, we essentially represent the exact same model but simply in a different way. We believe that the explanation of a model by means of data simulation may often be one of the most accessible ways to explain a statistical model to a nonstatistician. Owing to the great importance of data simulation, we dedicated the entirety of Chapter 4 to it.

In Chapter 5, we introduced the current industry standard for applied Bayesian modeling, which is also our software of choice for the Bayesian analysis of HMs: BUGS (Gilks et al., 1994; Lunn et al., 2000, 2009, 2013; Plummer, 2003). BUGS is an extremely simple and yet astonishingly powerful model-definition language that essentially allows for specification of arbitrarily complex statistical models as a sequence of simple "local" relationships among modeled quantities such as data, parameters, and derived estimands. Hence using BUGS, we can easily specify even very complex likelihoods by decomposing (technically, factorizing) them into much simpler component parts. Over the first quarter century of its existence, the BUGS language has proven extremely intuitive and easy to learn for many ecologists and other nonstatisticians. The language is currently implemented in three "BUGS engines": WinBUGS (Lunn et al., 2000), OpenBUGS (Thomas et al., 2006), and JAGS (Plummer, 2003). In addition, it has recently been adopted as the model-definition language in the new and exciting NIMBLE project (NIMBLE Development Team, 2015), which is a generic modeling platform for analysis of HMs using a number of likelihood and Bayesian methods. There are now several good introductory books on BUGS for ecologists—e.g., by McCarthy (2007), Kéry (2010), and Korner-Nievergelt et al. (2015). In spite of this, we have chosen to write what in essence turned out to be yet another such introductory book, and we packaged it into Chapter 5. There, we showed some of the latest tricks and also focused on topics that are important for the HMs in the later chapters of this book.

After this Prelude, the main content of the two *AHM* books consists of three broad categories: static models in volume 1 and dynamic and advanced models in volume 2. Of particular note, the latter part of volume 1 contains six chapters or "monographs," of which the first four covered spatial modeling of abundance under widely different data collection protocols: replicated point counts, capture-recapture and related protocols, and distance sampling. A further monograph dealt with the modeling of distribution, that is, with "species distribution models" that have an explicit measurement error component. The final monograph in volume 1 covered the community versions of all the models covered previously in Part 2. These community models represent a joint model of abundance or distribution in a "meta-population design" not for a single species, but for multiple or potentially all species occurring in a metacommunity.

A.2.5 THE HM AS THE "GRAND UNIFYING MODEL"

The four chapters on abundance modeling covered very different data collection protocols: replicated point, transect or areal counts (leading to the adoption of binomial mixture models in Chapter 6), double-observer, removal, or related capture-recapture types of data collection protocols (leading to the adoption of a variant of the multinomial mixture models in Chapter 7), and distance sampling models in Chapters 8 and 9. One of the main themes that we emphasized throughout the book is the extreme similarity of the models for these very different data types, or indeed of all the HMs in this book (both volumes). HMs help you recognize the similarities, rather than being puzzled by their apparent differences, among a vast range of statistical models and procedures that are often thought to be entirely different. One typical example is provided by distance sampling and capture-recapture. As we have pointed out many times, when viewed as HMs with one module for the underlying latent abundance, or density, and another module for the measurement error process, distance sampling and capture-recapture models suddenly begin to look like rather minor variants of an HM that differ only in their specific submodel for the abundance measurement error. All that is different are the data we collect to model abundance and the specific measurement error model, which is based on replicated detections of individuals in the former model and on distance (or possibly location) measurements in the latter model.

Similarly, occupancy models for presence/absence and N-mixture models for abundance modeling are sometimes thought to represent quite different model classes. But when viewed as HMs, they are naturally seen to represent mere variants of the general theme of an HM: there is one module for the true underlying state (which is occurrence—or presence/absence—for the occupancy model, and abundance for the N-mixture model) and another module for the associated measurement error process for the two different kinds of data collected (replicated counts in the former and replicated detection/nondetection data for the latter). In addition, there are intermediate models, such as the Royle–Nichols (or Bernoulli/Poisson mixture), Poisson/Poisson mixture (Section 6.13), and Poisson/Bernoulli mixture models (Section 10.12), and these naturally fit into the larger picture of demographic HMs for meta-population designs and link the "typical" site-occupancy and binomial N-mixture models. Thus, HMs unify these and a large number of other demographic models in ecology.

Letting you recognize the astonishing similarities among a host of such demographic population and community models is one of the most important benefits of a description of these model classes as HMs. In this way, you perceive them not as widely separated and unrelated entities, but begin to see their family tree. This has a huge holistic advantage for your understanding of these and many other "estimationist" models (demographic models with an explicit model for the measurement error) such as those covered in treatises by Buckland et al. (2001, 2004a), Borchers et al. (2002), Williams et al. (2002), Amstrup et al. (2005), Royle and Dorazio (2008), King et al. (2009), and McCrea and Morgan (2014). In addition, it may help you develop your own variant of HM for your particular question or data set. We like to think of the HM, as represented by those in this book, as the "grand unifying model" in the demographic analysis of populations and communities.

A.3 OUTLOOK: *AHM* VOLUME 2

Part of volume 1 of *AHM* sets the scene for further HMs that we cover in volume 2; see Table 1 in the Preface for an overview of the contents of volume 2. Of course, the models in volume 2 are based on the same concepts as those in volume 1: they are HMs for data collected in meta-population designs, and they usually contain one module for the measurement error in the main modeled quantity such as abundance, distribution, species richness, survival probability, or recruitment rate. And we will present analyses of these models with both classical/frequentist (i.e., maximum likelihood) and Bayesian methods (i.e., BUGS).

Specifically, Part 3 in volume 2 will present the dynamic versions of the models in Chapters 6–11 in volume 1. First, we will dedicate a full three chapters (12–14) to the modeling of population dynamics—i.e., for various specifications of the change of abundance over time—for instance, in terms of a simple trend model or in a fully demographic model that explains the change in terms of the two main vital rates, apparent survival, and recruitment rate. Then, owing to its great practical importance in research and management, we will single out one of these vital rates and deal with the modeling of apparent survival in the context of the Cormack–Jolly–Seber model in Chapter 15.

Chapter 16 will represent a "monograph" on the dynamic occupancy model (MacKenzie et al., 2003), i.e., the dynamic version of the models in Chapter 10 in volume 1. Chapter 17 will do the same for community occupancy models by presenting the dynamic community models of Dorazio et al. (2010).

Part 4 in volume 2 will contain a number of more specialized HMs for populations and communities of various kinds and present plenty of material on occupancy models in particular. In Chapter 18, we will present the modeling of multiple occupancy states (Royle and Link, 2005, Nichols et al., 2007; MacKenzie et al., 2009). Chapter 19 deals with the modeling of false-positives in the occupancy context (Royle and Link, 2005; Miller et al., 2011; Chambert et al., 2015). In Chapter 20, we cover the modeling of species interactions, either for a small number of species with a specified direction in their interactions (Waddle et al., 2009; Richford et al., 2010) or for a whole community of species (Ovaskainen et al., 2010; Pollock et al., 2014; Clark et al., 2014). Chapters 21 and 22 cover spatial models, including the more phenomenological conditional-autoregressive and related models (e.g., Aing et al., 2011; Johnson et al., 2013), models with exponential spatial correlation (Royle et al., 2007b; Chelgren et al., 2011a; Post van den Burg et al., 2011), and the more mechanistic autologistic models (Bled et al., 2011a,b; Bardos et al., 2015). Chapter 23 covers the important topic of "integrated models," where two or more qualitatively different data types are informative about the same demographic quantities and are merged in a single model. Thereby, their combined information is utilized in a joint likelihood. Such models are becoming famous in the context of demographic analyses of the Leslie matrix type (e.g., Besbeas et al., 2002; Brooks et al., 2004; Schaub et al., 2007; Schaub and Abadi, 2011), but the issue of model-based combination of disparate sources of information is highly relevant in a much larger number of model classes (Schaub and Kéry, 2012). Chapter 24 revisits some distance sampling and related spatial capture–recapture models (Efford, 2004; Royle et al., 2014), and finally Chapter 25 provides a grand conclusion for the entire two volumes of the *AHM* book project and an outlook to where the field of hierarchical demographic models for populations and communities may steer in the future.

References

Aarts, G., Fieberg, J., Matthiopoulos, J., 2012. Comparative interpretation of count, presence-absence and point methods for species distribution models. Methods Ecol. Evol. 3, 177–187.

Abadi, F., Gimenez, O., Arlettaz, R., Schaub, M., 2010a. An assessment of integrated population models: bias, accuracy, and violation of the assumption of independence. Ecology 91, 7–14.

Abadi, F., Gimenez, O., Ullrich, B., Arlettaz, R., Schaub, M., 2010b. Estimation of immigration rate using integrated population modeling. J. Appl. Ecol. 47, 393–400.

Adams, M.J., Chelgren, N.D., Reinitz, D., Cole, R.A., Rachowicz, L.J., Galvan, S., McCreary, B., Pearl, C.A., Bailey, L.L., Bettaso, J., Bull, E.L., Leu, M., 2010. Using occupancy models to understand the distribution of an amphibian pathogen, *Batrachochytrium dendrobatidis*. Ecol. Appl. 20, 289–302.

Aing, C., Halls, S., Oken, K., Dobrow, R., Fieberg, J., 2011. A Bayesian hierarchical occupancy model for track surveys conducted in a series of linear, spatially correlated, sites. J. Appl. Ecol. 48, 1508–1517.

Aitkin, M., 1991. Posterior bayes factors. J. R. Stat. Soc. Series B (Methodol.) 53, 111–142.

Alldredge, M.W., Pollock, K.H., Simons, T.R., Collazo, J.A., Shriner, S.A., 2007. Time-of-detection method for estimating abundance from point-count surveys. Auk 124, 653–664.

Alpizar-Jara, R., Pollock, K.H., 1996. A combination line transect and capture-recapture sampling model for multiple observers in aerial surveys. Environ. Ecol. Stat. 3 (4), 311–327.

Amstrup, S.C., McDonald, T.L., Manly, B.F.J. (Eds.), 2005. Handbook of Capture-recapture Analysis. Princeton University Press, Princeton NJ.

Amundson, C.L., Royle, J.A., Handel, C.M., 2014. A hierarchical model combining distance sampling and time removal to estimate detection probability during avian point counts. Auk 131, 476–494.

Andrewartha, H.G., Birch, L.C., 1954. The Distribution and Abundance of Animals. University of Chicago Press, Chicago.

Arnason, A.N., 1972. Parameter estimates from mark-recapture experiments on two populations subject to migration and death. Res. Pop. Ecol. 13, 97–113.

Arnason, A.N., Schwarz, C.J., 1999. Using POPAN-5 to analyse banding data. Bird Study 46, 157–168.

Augustin, N.H., Mugglestone, M.A., Buckland, S.T., 1996. An autologistic model for the spatial distribution of wildlife. J. Appl. Ecol. 33, 339–347.

Baddeley, A., Berman, M., Fisher, N.I., Hardegen, A., Milne, R.K., Schuhmacher, D., Shah, R., Turner, R., 2010. Spatial logistic regression and change-of-support in Poisson point processes. Electron. J. Stat. 4, 1151–1201.

Baddeley, A., Turner, R., 2005. Spatstat: an R package for analyzing spatial point patterns. J. Stat. Softw. 12, 1–42.

Baillie, S.R., 1991. Integrated population monitoring of breeding birds in Britain and Irland. Ibis 132, 151–166.

Bailey, L.L., Hines, J.E., Nichols, J.D., MacKenzie, D.I., 2007. Sampling design trade-offs in occupancy studies with imperfect detection: examples and software. Ecol. Appl. 17, 281–290.

Bailey, L.L., MacKenzie, D.I., Nichols, J.D., 2014. Advances and applications of occupancy models. Methods Ecol. Evol. 5, 1269–1279.

Bailey, L.L., Simons, T.R., Pollock, K.H., 2004. Estimating site occupancy and species detection probability parameters for terrestrial salamanders. Ecol. Appl. 14, 692–702.

Balmford, A., Green, R.E., Jenkins, M., 2003. Measuring the changing state of nature. Trends Ecol. Evol. 18, 326–330.

Banerjee, S., Carlin, B.P., Gelfand, A.E., 2004. Hierarchical Modeling and Analysis for Spatial Data. Chapman & Hall/CRC Press.

Banks-Leite, C., Pardini, R., Boscolo, D., Cassano, C.R., Püttker, T., Barros, C.S., Barlow, J., 2014. Assessing the utility of statistical adjustments for imperfect detection in tropical conservation science. J. Appl. Ecol. 51, 849–859.

Bardos, D.C., Guillera-Arroita, G., Wintle, B.A., 2015. Valid auto-models for spatially autocorrelated occupancy and abundance data. Methods Ecol. Evol. http://dx.doi.org/10.1111/2041-210X.12402.

Barker, R.J., 1997. Joint modeling of live-recapture, tag-resight, and tag-recovery data. Biometrics 53, 666–677.

Barnagaud, J.Y., Barbaro, L., Papaix, J., Deconchat, M., Brockerhoff, E., 2014. Habitat filtering by landscape and local forest composition in native and exotic New Zealand birds. Ecology 95, 78–87.

Barry, S.C., Welsh, A.H., 2001. Distance sampling methodology. J. R. Stat. Soc. Series B 63, 31–53.

Bates, D., Mächler, M., Bolker, B.M., Walker, S.C., 2014. Fitting Linear Mixed-Effects Models Using lme4. http://arxiv.org/abs/1406.5823.

Bayes, T., 1763. An essay towards solving a problem in the doctrine of chances. Phil. Trans. R. Soc. A 53, 370–418.

Beesley, L.S., Gwinn, D.C., Price, A., King, A.J., Gawne, B., Koehn, J.D., Nielsen, D.L., 2014. Juvenile fish response to wetland inundation: how antecedent conditions can inform environmental flow policies for native fish. J. Appl. Ecol. 51, 1613–1621.

Begon, M., Harper, J.L., Townsend, C.R., 1986. Ecology: Individuals, Populations and Communities. Blackwell, Oxford.

Beissinger, S.R., 2002. Population viability analysis: past, present, future. In: Beissinger, S.R. (Ed.), Population Viability Analysis. The University of Chicago Press, Chicago, pp. 5–17.

Bellier, E., M. Kéry, M. Schaub. Dynamic N-mixture models with density–dependence in vital rates. In review.

Berliner, L.M., 1996. Hierarchical Bayesian time series models. In: Hanson, K., Silver, R. (Eds.), Maximum Entropy and Bayesian Methods. Kluwer Academic Publishers, Dordrecht, The Netherlands, pp. 15–22.

Besbeas, P., Freeman, S.N., Morgan, B.J.T., Catchpole, E.A., 2002. Integrating mark-recapture-recovery and census data to estimate animal abundance and demographic parameters. Biometrics 58, 540–547.

Bibby, C.J., Burgess, N.D., Hill, D.A., Mustoe, S., 2000. Bird Census Techniques, second ed. Academic Press.

Bled, F., Royle, J.A., Cam, E., 2011a. Assessing hypotheses about nesting site occupancy dynamics. Ecology 92, 938–951.

Bled, F., Royle, J.A., Cam, E., 2011b. Hierarchical modeling of an invasive spread: case of the Eurasian collared dove *Streptopelia decaocto* in the USA. Ecol. Appl. 21, 290–302.

Bolker, B.M., 2008. Ecological Models and Data in R. Princeton University Press, Princeton, New Jersey.

Bolker, B.M., Brooks, M.E., Clark, C.J., Geange, S.W., Poulsen, J.R., Stevens, M.H.H., White, J.S., 2009. Generalized linear mixed models: a practical guide for ecology and evolution. Trends Ecol. Evol. 24, 127–135.

Bonner, S.J., Schwarz, C.J., 2006. An extension of the Cormack-Jolly-Seber model for continuous covariates with application to *Microtus pennsylvanicus*. Biometrics 62, 142–149.

Borchers, D.L., Buckland, S.T., Zucchini, W., 2002. Estimating Animal Abundance. Springer, London.

Borchers, D.L., Efford, M.G., 2008. Spatially explicit maximum likelihood methods for capture-recapture studies. Biometrics 64, 377–385.

Borchers, D.L., Stevenson, B.C., Kidney, D., Thomas, L., Marques, T.A., 2015. A unifying model for capture–recapture and distance sampling surveys of wildlife populations. J. Am. Stat. Assoc. 110, 195–204.

Borchers, D.L., Zucchini, W., Fewster, R.M., 1998. Mark-recapture models for line transect surveys. Biometrics 1207–1220.

Bornand, C.N., Kéry, M., Bueche, L., Fischer, M., 2014. Hide-and-seek in vegetation: time-to-detection is an efficient design for estimating detectability and occurrence. Methods Ecol. Evol. 5, 433–442.

Boulinier, T., Nichols, J.D., Sauer, J.R., Hines, J.E., Pollock, K.H., 1998. Estimating species richness: the importance of heterogeneity in species detectability. Ecology 79, 1018–1028.

Boyce, M.S., 2010. Presence-only data, pseudo-absences, and other lies about habitat selection. Ideas Ecol. Evol. 3, 26–27.

Broms, K.M., Johnson, D.S., Altwegg, R., Conquest, L.L., 2014. Spatial occupancy models applied to atlas data show Southern Ground Hornbills strongly depend on protected areas. Ecol. Appl. 24, 363–374.

Brooks, S.P., Catchpole, E.A., Morgan, B.J.T., 2000. Bayesian animal survival estimation. Stat. Sci. 15, 357–376.

Brooks, S.P., Gelman, A., 1998. Alternative methods for monitoring convergence of iterative simulations. J. Comput. Gr. Stat. 7, 434–455.

Brooks, S.P., King, R., Morgan, B.J.T., 2004. A Bayesian approach to combining animal abundance and demographic data. Anim. Biodivers. Conserv. 27, 515–529.

Brown, J.H., Maurer, B.A., 1989. Macroecology: the division of food and space among species on continents. Science 243, 1145–1150.

Brownie, C., Anderson, D.R., Burnham, K.P., Robson, D.S., 1985. Statistical Inference from Band Recovery Data - a Handbook. In: US Fish and Wildlife Service, vol. 156. Resource Publication, Washington DC.

Brownie, C., Hines, J.E., Nichols, J.D., Pollock, K.H., Hestbeck, J.B., 1993. Capture-recapture studies for multiple strata including non-Markovian transitions. Biometrics 49, 1173–1187.

Buckland, S.T., Anderson, D.R., Burnham, K.P., Laake, J.L., Borchers, D.L., Thomas, L., 2001. Introduction to Distance Sampling. Oxford University Press, Oxford.

Buckland, S.T., Anderson, D.R., Burnham, K.P., Laake, J.L., Borchers, D.L., Thomas, L. (Eds.), 2004a. Advanced Distance Sampling. Oxford University Press, Oxford, 414 pp.

Buckland, S.T., Newman, K.B., Fernandez, C., Thomas, L., Harwood, J., 2007. Embedding population dynamics models in inference. Stat. Sci. 22, 44–58.

Buckland, S.T., Newman, K.B., Thomas, L., Koesters, N.B., 2004b. State-space models for the dynamics of wild animal populations. Ecol. Model. 171, 157–175.

Burnham, K.P., 1993. A theory for combined analysis of ring recovery and recapture data. In: Lebreton, J.D. (Ed.), Marked Individuals in the Study of Bird Populations. Birkhäuser, Basel, pp. 199–213.

Burnham, K.P., Anderson, D.R., 2002. Model Selection and Multimodel Inference: A Practical Information Theoretic Approach. Springer, New York.

Burnham, K.P., Anderson, D.R., Laake, J.L., 1980. Estimation of density from line transect sampling of biological populations. Wildl. Monogr. 72, 3–202.

Burton, A., Sam, M., Balangtaa, C., Brashares, J., 2012. Hierarchical multi-species modeling of carnivore responses to hunting, habitat and prey in a West African protected area. PLoS ONE 7 (5), e38007.

Cam, E., Link, W.A., Cooch, E.G., Monnat, J.Y., Danchin, E., 2002a. Individual covariation in life-history traits: seeing the trees despite the forest. Am. Nat. 159, 96–105.

Cam, E., Nichols, J.D., Hines, J.E., Sauer, J.R., Alpizar-Jara, R., Flather, C.H., 2002b. Disentangling sampling and ecological explanations underlying species-area relationships. Ecology 83, 1118–1130.

Cam, E., Nichols, J.D., Sauer, J.R., Hines, J.E., 2002c. On the estimation of species richness based on the accumulation of previously unrecorded species. Ecography 25, 102–108.

Carlin, B.P., Chib, S., 1995. Bayesian model choice via Markov chain Monte Carlo methods. J. R. Stat. Soc. Series B (Methodol.) 57, 473–484.

Carlin, B.P., Louis, T.A., 2009. Bayesian Methods for Data Analysis. CRC Press/Taylor & Francis Group, Boca Raton.

Carrillo-Rubio, E., Kéry, M., Morreale, S.J., Sullivan, P.J., Gardner, B., Cooch, E.G., Lassoie, J.P., 2014. Use of multispecies occupancy models to evaluate the response of bird communities to forest degradation associated with logging. Conserv. Biol. 28, 1034–1044.

Casella, G., Berger, R.L., 2002. Statistical inference (Vol. 2). Duxbury, Pacific Grove, CA.

Caswell, H., 2001. Matrix Population Models. Construction, Analysis, and Interpretation. Sinauer Associates, Sunderland, Massachusetts.

Catchpole, E.A., Kgosi, P.M., Morgan, B.J.T., 2001. On the near-singularity of models for animal recovery data. Biometrics 57, 720–726.
Catchpole, E.A., Morgan, B.J.T., 1997. Detecting parameter redundancy. Biometrika 84, 187–196.
Catchpole, E.A., Morgan, B.J.T., Viallefont, A., 2002. Solving problems in parameter redundancy using computer algebra. J. Appl. Stat. 29, 625–636.
Caughley, G., 1974. Bias in aerial survey. J. Wildl. Manage. 38, 921–933.
Chambert, T., Miller, D.A.W., Nichols, J.D., 2015. Modeling false positive detections in species occurrence data under different study designs. Ecology 96, 332–339.
Chandler, R.B., 2015. Modeling Variation in Abundance Using Capture-Recapture Data. Available at: cran.r-project.org/web/packages/unmarked/vignettes/cap-recap.pdf.
Chandler, R.B., Clark, J.D., 2014. Spatially explicit integrated population models. Methods Ecol. Evol. 5, 1351–1360.
Chandler, R.B., King, D.I., 2011. Habitat quality and habitat selection of golden-winged warbler in Costa Rica: application of hierarchical models for open populations. J. Appl. Ecol. 48, 1038–1047.
Chandler, R.B., King, D.I., Chandler, C.C., 2009a. Effects of management regime on the abundance and nest survival of shrubland birds in wildlife openings in northern New England, USA. For. Ecol. Manage. 258, 1669–1676.
Chandler, R.B., King, D.I., DeStefano, S., 2009b. Scrub-shrub bird habitat associations at multiple spatial scales in beaver meadows in Massachusetts. Auk 126, 186–197.
Chandler, R.B., King, D.I., Raudales, R., Trubey, R., Chandler, C., Chavez, V.J.A., 2013. A small-scale land-sparing approach to conserving biological diversity in tropical agricultural landscapes. Conserv. Biol. 27, 785–795.
Chandler, R.B., Royle, J.A., 2013. Spatially explicit models for inference about density in unmarked or partially marked populations. Ann. Appl. Stat. 7, 936–954.
Chandler, R.B., Royle, J.A., King, D.I., 2011. Inference about density and temporary emigration in unmarked populations. Ecology 92, 1429–1435.
Chelgren, N.D., Adams, M.J., Bailey, L.L., Bury, R.B., 2011a. Using multilevel spatial models to understand salamander site occupancy patterns after wildfire. Ecology 92, 408–421.
Chelgren, N.D., Samora, B., Adams, M.J., McCreary, B., 2011b. Using spatiotemporal models and distance sampling to map the space use and abundance of newly metamorphosed Western toads (*Anaxyrus boreas*). Herpetol. Conserv. Biol. 6, 175–190.
Chen, G., Kéry, M., Plattner, M., Ma, K., Gardner, B., 2013. Imperfect detection is the rule rather than the exception in plant distribution studies. J. Ecol. 101, 183–191.
Chen, G., Kéry, M., Zhang, J., Ma, K., 2009. Factors affecting detection probability in plant distribution studies. J. Ecol. 97, 1383–1389.
Choquet, R., Cole, D., 2012. A hybrid symbolic-numerical method for determining model structure. Math. Biosci. 236, 117–125.
Choquet, R., Lebreton, J.D., Gimenez, O., Reboulet, A.M., Pradel, R., 2009a. U-CARE: utilities for performing goodness of fit tests and manipulating CApture-REcapture data. Ecography 32, 1071–1074.
Choquet, R., Rouan, L., Pradel, R., 2009b. Program E-SURGE: a software application for fitting multievent models. In: Thomson, D.L., Cooch, E.G., Conroy, M.J. (Eds.), Modeling Demographic Processes in Marked Populations. Springer, New York, pp. 845–865.
Clark, J.S., 2007. Ecological Data Models with R. Princeton University Press, Princeton, NJ.
Clark, J.S., Gelfand, A.E. (Eds.), 2006. Hierarchical Modelling for the Environmental Sciences. Oxford University Press, Oxford, England.
Clark, J.S., Gelfand, A.E., Woodall, C.W., Zhu, K., 2014. More than the sum of the parts: forest climate response from joint species distribution models. Ecol. Appl. 24, 990–999.

Cole, D.J., 2012. Determining parameter redundancy of multi-state mark-recapture models for sea birds. J. Ornithol. 152 (Suppl. 2), 305—315.

Collier, B.A., Groce, J.E., Morrison, M.L., Newnam, J.C., Campomizzi, A.J., Farrell, S.L., Mathewson, H.A., Snelgrove, R.T., Carroll, R.J., Wilkins, R.N., 2012. Predicting patch occupancy in fragmented landscapes at the rangewide scale for an endangered species: an example of an American warbler. Divers. Distrib. 18, 158—167.

Conn, P.B., Laake, J.L., Johnson, D.S., 2012. A hierarchical modeling framework for multiple observer transect surveys. PLoS ONE 7, e42294.

Conroy, M.J., Runge, J.P., Barker, R.J., Schofield, M.R., Fonnesbeck, C.J., 2008. Efficient estimation of abundance for patchily distributed populations via two-phase, adaptive sampling. Ecology 89, 3362—3370.

Converse, S.J., Royle, J.A., 2012. Dealing with Incomplete and Variable Detectability in Multi-year, Multi-site Monitoring of Ecological Populations. Design and Analysis of Long-term Ecological Monitoring Studies. Cambridge University Press, Cambridge, England, United Kingdom, pp. 426—442.

Cooch, E., White, G., 2014. Program MARK: A Gentle Introduction. Available in pdf format for free download at http://www.phidot.org/software/mark/docs/book.

Cook, R.D., Jacobson, J.O., 1979. A design for estimating visibility bias in aerial surveys. Biometrics 735—742.

Cormack, R.M., 1964. Estimates of survival from the sighting of marked animals. Biometrika 51, 429—438.

Coull, B.A., Agresti, A., 1999. The use of mixed logit models to reflect heterogeneity in capture-recapture studies. Biometrics 55, 294—301.

Couturier, T., Cheylan, M., Bertolero, A., Astruc, G., Besnard, A., 2013. Estimating abundance and population trends when detection is low and highly variable: a comparison of three methods for the Hermann's tortoise. J. Wildl. Manage. 77, 454—462.

Crainiceanu, C.M., Ruppert, D., Wand, M.P., 2005. Bayesian analysis for penalized spline regression using WinBUGS. J. Stat. Softw. 14, 1—25.

Crewe, T.L., Taylor, P.D., Lepage, D., 2015. Modeling systematic change in stopover duration does not improve bias in trends estimated from migration counts. PLoS ONE 10 (6), e0130137. http://dx.doi.org/10.1371/journal.pone.0130137.

Cressie, N., Calder, C.A., Clark, J.S., Ver Hoef, J.M., Wikle, C.K., 2009. Accounting for uncertainty in ecological analysis: the strengths and limitations of hierarchical statistical modeling. Ecol. Appl. 19, 553—570.

Cressie, N., Wikle, C.K., 2011. Statistics for Spatio-Temporal Data. Wiley.

Cribari-Neto, F., Zeileis, A., 2010. Beta regression in R. J. Stat. Softw. 34, 1—24. http://www.jstatsoft.org/v34/i02/.

Dail, D., Madsen, L., 2011. Models for estimating abundance from repeated counts of an open population. Biometrics 67, 577—587.

Dellaportas, P., Forster, J.J., Ntzoufras, I., 2002. On Bayesian model and variable selection using MCMC. Stat. Comput. 12, 27—36.

Dénes, F.V., Silveira, L.F., Beissinger, S.R., 2015. Estimating abundance of unmarked animal populations accounting for imperfect detection and other sources of zero inflation. Methods Ecol. Evol. 6, 543—556.

Dennis, B., 1996. Discussion: should ecologists become Bayesians? Ecol. Appl. 6, 1095—1103.

Dennis, E.B., Morgan, B.J.T., Freeman, S.N., Brereton, T., Roy, D.B., 2015c. A Generalised Abundance Index for Seasonal Invertebrates. Technical report UKC/SMSAS/14/006. University of Kent.

Dennis, E.B., Morgan, B.J.T., Freeman, S.N., Roy, D.B., Brereton, T., 2015b. Dynamic models for longitudinal butterfly data. J. Agric Biol. Environ. Stat. http://dx.doi.org/10.1007/s13253-015-0216-3.

Dennis, E.B., Morgan, B.J.T., Ridout, M.S., 2015a. Computational aspects of N-mixture models. Biometrics 71, 237—246.

de Valpine, P., 2003. Better inferences from population-dynamics experiments using Monte Carlo state-space likelihood methods. Ecology 84, 3064—3077.

de Valpine, P., 2009. Shared challenges and common ground for Bayesian and classical analysis of hierarchical statistical models. Ecol. Appl. 19, 584–588.

de Valpine, P., Turek, D., Paciorek, C.J., Anderson-Bergman, C., Temple Lang, D., Bodik, R. Programming with models: writing statistical algorithms for general model structures with NIMBLE. J. Comput. Gr. Stat. (in review).

DeWan, A.A., Zipkin, E.F., 2010. An integrated sampling and analysis approach for improved biodiversity monitoring. Environ. Manage. 45, 1223–1230.

Denwood, M.J., 2015. runjags: An R package providing interface utilities, parallel computing methods and additional distributions for MCMC models in JAGS. J. Stat. Softw. http://cran.r-project.org/web/packages/runjags/.

Dice, L., 1938. Some census methods for mammals. J. Wildl. Manage. 2, 119–130.

Diefenbach, D.R., Marshall, M.R., Mattice, J.A., Brauning, D.W., Johnson, D.H., 2007. Incorporating availability for detection in estimates of bird abundance. Auk 124, 96–106.

Dixon, P.M., 2006. Bootstrap resampling. Encycl. Environmetrics (Online, Wiley). http://dx.doi.org/10.1002/9780470057339.vab028.

Dobson, A., Barnett, A., 2008. An Introduction to Generalized Linear Models. CRC/Chapmann & Hall, Boca Raton.

Dodd, C.K., Dorazio, R.M., 2004. Using counts to simultaneously estimate abundance and detection probabilities in salamander surveys. Herpetologica 60, 468–478.

Doherty Jr., P.F., Sorci, G., Royle, J.A., Hines, J.E., Nichols, J.D., Boulinier, T., 2003. Sexual selection affects local extinction and turnover in bird communities. Proc. Nat. Acad. Sci. 100, 5858–5862.

Dorazio, R.M., 2007. On the choice of statistical models for estimating occurrence and extinction from animal surveys. Ecology 88, 2773–2782.

Dorazio, R.M., 2012. Predicting the geographic distribution of a species from presence-only data subject to detection errors. Biometrics 68, 1303–1312.

Dorazio, R.M., 2013. Bayes and empirical Bayes estimators of abundance and density from spatial capture-recapture data. PLoS ONE 8, e84017.

Dorazio, R.M., 2014. Accounting for imperfect detection and survey bias in statistical analysis of presence-only data. Glob. Ecol. Biogeogr. 23, 1472–1484.

Dorazio, R.M., Connor, E.F., 2014. Estimating abundances of interacting species using morphological traits, foraging guilds, and habitat. PLoS ONE 9, e94323.

Dorazio, R.M., Connor, E.F., Askins, R.A., 2015. Estimating the effects of habitat and biological interactions in an avian community. PLoS ONE 10, e0135987. http://dx.doi.org/10.1371/journal.pone.0135987.

Dorazio, R.M., Gotelli, N.J., Ellison, A.M., 2011. Modern methods of estimating biodiversity from presence-absence surveys. In: Grillo, O., Venora, G. (Eds.), Biodiversity Loss in a Changing Planet. InTech, ISBN 978-953-307-707-9.

Dorazio, R.M., Jelks, H.L., Jordan, F., 2005. Improving removal-based estimates of abundance by sampling a population of spatially distinct subpopulations. Biometrics 61, 1093–1101.

Dorazio, R.M., Kéry, M., Royle, J.A., Plattner, M., 2010. Models for inference in dynamic metacommunity systems. Ecology 91, 2466–2475.

Dorazio, R.M., Martin, J., Edwards, H.H., 2013. Estimating abundance while accounting for rarity, correlated behavior, and other sources of variation in counts. Ecology 94, 1472–1478.

Dorazio, R.M., Mukherjee, B., Zhang, L., Ghosh, M., Jelks, H.L., Jordan, F., 2008. Modeling unobserved sources of heterogeneity in animal abundance using a Dirichlet process prior. Biometrics 64, 635–644.

Dorazio, R.M., Rodriguez, D.T., 2012. A Gibbs sampler for Bayesian analysis of site-occupancy data. Methods Ecol. Evol. 3, 1093–1098.

Dorazio, R.M., Royle, J.A., 2003. Mixture models for estimating the size of a closed population when capture rates vary among individuals. Biometrics 59, 351–364.

Dorazio, R.M., Royle, J.A., 2005. Estimating size and composition of biological communities by modeling the occurrence of species. J. Am. Stat. Assoc. 100, 389–398.

Dorazio, R.M., Royle, J.A., Söderström, B., Glimskär, A., 2006. Estimating species richness and accumulation by modeling species occurrence and detectability. Ecology 87, 842–854.

Dupuis, J.A., 1995. Bayesian estimation of movement and survival probabilities from capture-recapture data. Biometrika 82, 761–772.

Dupuis, J.A., Bled, F., Joachim, J., 2011. Estimating the occupancy rate of spatially rare or hard to detect species: a conditional approach. Biometrics 67, 290–298.

Edwards, A.W., 1992. Likelihood, expanded ed. Johns Hopkins, Baltimore.

Efford, M., 2004. Density estimation in live-trapping studies. Oikos 106, 598–610.

Efford, M.G., Borchers, D.L., Mowat, G., 2013. Varying effort in capture-recapture studies. Methods Ecol. Evol. 4, 629–636.

Efford, M.G., Dawson, D.K., 2009. Effect of distance-related heterogeneity on population size estimates from point counts. Auk 126, 100–111.

Efford, M.G., Dawson, D.K., 2012. Occupancy in continuous habitat. EcoSphere 3, 1–12. Article 12.

Efford, M.G., Fewster, R.M., 2013. Estimating population size by spatially explicit capture-recapture. Oikos 122, 918–928.

Efron, B., 1982. The jackknife, the bootstrap and other resampling plans, Vol. 38. Society for industrial and applied mathematics, Philadelphia.

Efron, B., Tibshirani, R.J., 1993. An Introduction to the Bootstrap. Chapman & Hall/CRC Press.

Elith, J., Leathwick, J.R., 2009. Species distribution models: ecological explanation and prediction across space and time. Annu. Rev. Ecol. Evol. Syst. 40, 677–697.

Ellis, M.M., Ivan, J.S., Tucker, J.M., Schwartz, M.K., 2015. rSPACE: spatially based power analysis for conservation and ecology. Methods Ecol. Evol. 6, 621–625.

Etterson, M.A., Niemi, G.J., Danz, N.P., 2009. Estimating the effects of detection heterogeneity and overdispersion on trends estimated from avian point counts. Ecol. Appl. 19, 2049–2066.

Farnsworth, G.L., Nichols, J.D., Sauer, J.R., Fancy, S.G., Pollock, K.H., Shriner, S.A., Simons, T.R., 2005. Statistical Approaches to the Analysis of Point Count Data: A Little Extra Information Can Go a Long Way. USDA Forest Service. Gen. Tech. Rep. PSW-GTR-191.

Farnsworth, G.L., Pollock, K.H., Nichols, J.D., Simons, T.R., Hines, J.E., Sauer, J.R., 2002. A removal model for estimating detection probabilities from point-count surveys. Auk 119, 414–425.

Ferraz, G., Nichols, J.D., Hines, J.E., Stouffer, P.C., Bierregaard Jr., R.O., Lovejoy, T.E., 2007. A large-scale deforestation experiment: effects of patch area and isolation on Amazon birds. Science 315, 238–241.

Fewster, R.M., Buckland, S.T., Siriwardena, G.M., Baillie, S.R., Wilson, J.D., 2000. Analysis of population trends for farmland birds using generalized additive models. Ecology 81, 1970–1984.

Fiske, I., Chandler, R., 2011. unmarked: an R package for fitting hierarchical models of wildlife occurrence and abundance. J. Stat. Softw. 43, 1–23.

Fithian, W., Elith, J., Hastie, T., Keith, D.A., 2014. Bias correction in species distribution models: pooling survey and collection data for multiple species. Methods Ecol. Evol. 6, 424–438.

Fithian, W., Hastie, T., 2013. Finite-sample equivalence in statistical models for presence-only data. Ann. Appl. Stat. 7, 1917–1939.

Fujisaki, I., Mazzotti, F.J., Dorazio, R.M., Rice, K.G., Cherkiss, M., Jeffery, B., 2011. Estimating trend in alligator populations from nightlight survey data. Wetlands 31, 147–155.

Fukaya, K., Royle, J.A., 2013. Markov models for community dynamics allowing for observation error. Ecology 94, 2670–2677.

Garrard, G.E., Bekessy, S.A., McCarthy, M.A., Wintle, B.A., 2008. When have we looked hard enough? A novel method for setting minimum survey effort protocols for flora surveys. Austral Ecol. 33, 986–998.

Garrard, G.E., Bekessy, S.A., McCarthy, M.A., Wintle, B.A., 2015. Incorporating detectability into environmental impact assessment for threatened species. Conserv. Biol. 29, 216–225.

Garrard, G.E., McCarthy, M.A., Williams, N.S.G., Bekessy, S.A., Wintle, B.A., 2013. A general model of detectability using species traits. Methods Ecol. Evol. 4, 45–52.

Gelfand, A.E., Smith, A.F., 1990. Sampling-based approaches to calculating marginal densities. Journal of the American Statistical Association 85, 398–409.

Gelfand, A.E., Hills, S.E., Racine-Poon, A., Smith, A.F., 1990. Illustration of Bayesian inference in normal data models using Gibbs sampling. Journal of the American Statistical Association 85 (412), 972–985.

Gelfand, A.E., Schmidt, A.E., Wu, S., Silander Jr., J.A., Latimer, A., Rebelo, A.G., 2005. Modelling species diversity through species level hierarchical modelling. Appl. Stat. 54, 1–20.

Gelfand, A.E., Silander Jr., J.A., Wu, S., Latimer, A., Lewis, P.O., Rebelo, A.G., Holder, M., 2006. Explaining species distribution patterns through hierarchical modeling. Bayesian Anal. 1, 41–92.

Gelman, A., 2005. Analysis of variance: why is it more important than ever (with discussion). Ann. Stat. 33, 1–53.

Gelman, A., 2006. Prior distributions for variance parameters in hierarchical models. Bayesian Anal. 1, 515–534.

Gelman, A., Carlin, J.B., Stern, H.S., Dunson, D.B., Vehtari, A., Rubin, D.B., 2014. Bayesian Data Analysis, third ed. CRC/Chapman & Hall, Boca Raton.

Gelman, A., Carlin, J.B., Stern, H.S., Rubin, D.B., 2003. Bayesian Data Analysis, second ed. Chapman and Hall, London.

Gelman, A., Hill, J., 2007. Data Analysis Using Regression and Multilevel/hierarchical Models. Cambridge University Press, Cambridge.

Gelman, A., Meng, X.-L., Stern, H.S., 1996. Posterior predictive assessment of model fitness via realized discrepancies (with discussion). Stat. Sinica 6, 733–807.

Gelman, A., Pardoe, I., 2006. Bayesian measures of explained variance and pooling in multilevel (hierarchical) models. Technometrics 48, 241–251.

Gelman, A., Rubin, D.B., 1992. Inference from iterative simulation using multiple sequences. Stat. Sci. 7, 457–511.

George, E.I., McCulloch, R.E., 1993. Variable selection via Gibbs sampling. J. Am. Stat. Assoc. 88, 881–889.

Ghosh, S., Gelfand, A.E., Zhu, K., Clark, J.S., 2012. The k-ZIG: flexible modeling for zero-inflated counts. Biometrics 68, 878–885.

Gilks, W.R., Thomas, A., Spiegelhalter, D.J., 1994. A language and program for complex Bayesian modelling. Statistician 43, 169–177.

Gilroy, J.J., Edwards, F.A., Uribe, C.A.M., Haugaasen, T., Edwards, D.P., 2014b. Surrounding habitats mediate the trade-off between land-sharing and land-sparing agriculture in the tropics. J. Appl. Ecol. 51, 1337–1346.

Gilroy, J.J., Prescott, G.W., Cardenas, J.S., Castaneda, P.G.D., Sanchez, A., Rojas-Murcia, L.E., Uribe, C.A.M., Haugaasen, T., Edwards, D.P., 2015. Minimizing the biodiversity impact of Neotropical oil palm development. Glob. Change Biol. 21, 1531–1540.

Gilroy, J.J., Woodcock, P., Edwards, F.A., Wheeler, C., Uribe, C.A.M., Haugaasen, T., Edwards, D.P., 2014a. Optimizing carbon storage and biodiversity protection in tropical agricultural landscapes. Glob. Change Biol. 20, 2162–2172.

Gimenez, O., Choquet, R., Lebreton, J.D., 2003. Parameter redundancy in multistate capture-recapture models. Biomet. J. 45, 704–722.

Gimenez, O., Covas, R., Brown, C.R., Anderson, M.D., Bomberger Brown, M., Lenormand, T., 2006a. Nonparametric estimation of natural selection on a quantitative trait using mark-recapture data. Evolution 60, 460–466.

Gimenez, O., Crainiceanu, C., Barbraud, C., Jenouvrier, S., Morgan, B.J.T., 2006b. Semiparametric regression in capture-recapture modeling. Biometrics 62, 691–698.

Gimenez, O., Viallefont, A., Catchpole, A.E., Choquet, R., Morgan, B.J.T., 2004. Methods for investigating parameter redundancy. Anim. Biodivers. Conserv. 27, 561–572.

Gimenez, O., Viallefont, A., Charmantier, A., Pradel, R., Cam, E., Brown, C.R., Anderson, M.D., Brown Bomberger, M., Covas, R., Gaillard, J.M., 2008. The risk of flawed inference in evolutionary studies when detectability is less than one. Am. Nat. 172, 441–448.

Gimenez, O., Blanc, L., Besnard, A., Pradel, R., Doherty, P.F., Marboutin, E., Choquet, R., 2014. Fitting occupancy models with E-SURGE: hidden Markov modelling of presence–absence data. Methods Ecol. Evol. 5, 592–597.

Giovanini, J., Kroll, A.J., Jones, J.E., Altman, B., Arnett, E.B., 2013. Effects of management intervention on post-disturbance community composition: an experimental analysis using Bayesian hierarchical models. PLoS ONE 8, e59900. http://dx.doi.org/10.1371/journal.pone.0059900.

Gopalaswamy, A.M., Royle, J.A., Delampady, M., Nichols, J.D., Karanth, K.U., Macdonald, D.W., 2012. Density estimation in tiger populations: combining information for strong inference. Ecology 93, 1741–1751.

Gotelli, N.J., McGill, B.J., 2006. Null versus neutral models: what's the difference? Ecography 29, 793–800.

Govindan, B.N., Swihart, R.K., 2015. Community structure of acorn weevils (Curculio): inferences from multispecies occupancy models. Can. J. Ecol. 93, 31–39.

Grimm, V., 1999. Ten years of individual-based modelling in ecology: what have we learned and what could we learn in the future? Ecol. Model. 115, 129–148.

Green, A.W., Hooten, M.B., Grant, E.H.C., Bailey, L.L., 2013. Evaluating breeding and metamorph occupancy and vernal pool management effects for wood frogs using a hierarchical model. J. Appl. Ecol. 50, 1116–1123.

Grosbois, V., Gimenez, O., Gaillard, J.M., Pradel, R., Barbraud, C., Clobert, J., Möller, A.P., Weimerskirch, H., 2008. Assessing the impact of climate variation on survival in vertebrate populations. Biol. Rev. 83, 357–399.

Gu, W., Swihart, R.K., 2004. Absent or undetected? Effects of non-detection of species occurrence on wildlife-habitat models. Biol. Conserv. 116, 195–203.

Guélat, J., Kéry, M. Effects of spatial autocorrelation and imperfect detection on large-scale maps of abundance (in review).

Guillera-Arroita, G., 2011. Impact of sampling with replacement in occupancy studies with spatial replication. Methods Ecol. Evol. 2, 401–406.

Guillera-Arroita, G., Lahoz-Monfort, J.J., 2012. Designing studies to detect changes in species occupancy: power analysis under imperfect detection. Methods Ecol. Evol. 3, 860–869.

Guillera-Arroita, G., Lahoz-Monfort, J.J. Species occupancy estimation and imperfect detection: shall surveys continue after the first detection? Adv. Stat. Anal. (in press).

Guillera-Arroita, G., Lahoz-Monfort, J.J., Elith, J., Gordon, A., Kujala, K., Lentini, P.E., McCarthy, M.A., Tingley, R., Wintle, B.A., 2015. Is my species distribution model fit for purpose? Matching data and models to applications. Glob. Ecol. Biogeogr. 24, 276–292.

Guillera-Arroita, G., Lahoz-Monfort, J.J., MacKenzie, D.I., Wintle, B.A., McCarthy, M.A., 2014a. Ignoring imperfect detection in biological surveys is dangerous: a response to 'Fitting and Interpreting Occupancy Models'. PLoS ONE 9, e99571.

Guillera-Arroita, G., Morgan, B.J.T., Ridout, M.S., Linkie, M., 2011. Species occupancy modelling for detection data collected along a transect. J. Agric. Biol. Environ. Stat. 3, 301–317.

Guillera-Arroita, G., Morgan, B.J.T., Ridout, M.S., Linkie, M., 2012. Models for species detection data collected along transects in presence of abundance-induced heterogeneity and clustering in the detection process. Methods Ecol. Evol. 3, 358–367.

Guillera-Arroita, G., Ridout, M.S., Morgan, B.J.T., 2010. Design of occupancy studies with imperfect detection. Methods Ecol. Evol. 1, 131–139.

Guillera-Arroita, G., Ridout, M.S., Morgan, B.J.T., 2014b. Two-stage Bayesian study design for species occupancy estimation. J. Agric. Biol. Environ. Stat. 19, 278–291.

Guzy, J.C., Price, S.J., Dorcas, M.E., 2013. The spatial configuration of greenspace affects semi-aquatic turtle occupancy and species richness in a suburban landscape. Landscape Urban Plan. 117, 46–56.

Hammond, P.S., Berggren, P., Benke, H., Borchers, D.L., Collet, A., Heide-Jørgensen, M.P., Heimlich, S., Hiby, A.R., Leopold, M.F., Øien, N., 2002. Abundance of harbour porpoise and other cetaceans in the North Sea and adjacent waters. J. Appl. Ecol. 39, 361–376.

Hanski, I., 1998. Metapopulation dynamics. Nature 396, 41–49.

Hastie, T.J., Tibshirani, R.J., 1990. Generalized Additive Models. Chapman & Hall/CRC, Boca Raton.

Hastings, W.K., 1970. Monte Carlo sampling methods using Markov chains and their applications. Biometrika 57, 97–109.

Hayne, D.W., 1949. Calculation of size of home range. J. Mammal. 30, 1–18.

Hayes, D.B., Monfils, M.J., 2015. Occupancy modeling of bird point counts: Implications of mobile animals. J. Wildl. Manage. http://dx.doi.org/10.1002/jwmg.943.

He, F., Gaston, K.J., 2000. Estimating species abundance from occurrence. Am. Nat. 156, 553–559.

Hector, A., Bell, T., Hautier, Y., Isbell, F., Kéry, M., Reich, P.B., van Ruijven, J., Schmid, B., 2011. Bugs in the analysis of biodiversity experiments: species richness and composition are of similar importance for grassland productivity. PLoS ONE 6, e17434.

Hedley, S.L., Buckland, S.T., 2004. Spatial models for line transect sampling. J. Agric. Biol. Environ. Stat. 9, 181–199.

Heikkinen, J., Högmander, H., 1994. Fully Bayesian approach to image restoration with an application in biogeography. Appl. Stat. 43, 569–582.

Henden, J.A., Yoccoz, N.G., Ims, R.A., Langeland, K., 2013. How spatial variation in areal extent and configuration of labile vegetation states affect the riparian bird community in Arctic tundra. PLoS ONE 8, e63312. http://dx.doi.org/10.1371/journal.pone.0063312.

Hestbeck, J.B., Nichols, J.D., Malecki, R.A., 1991. Estimates of movement and site fidelity using mark-resight data of wintering Canada Geese. Ecology 72, 523–533.

Higa, M., Yamaura, Y., Koizumi, I., Yabuhara, Y., Senzaki, M., Ono, S., 2015. Mapping large-scale bird distributions using occupancy models and citizen data with spatially biased sampling effort. Divers. Distrib. 21, 46–54.

Hines, J.E., 2006. PRESENCE – Software to Estimate Patch Occupancy and Related Parameters. USGS-PWRC. www.mbr-pwrc.usgs.gov/software/presence.shtml.

Hines, J.E., Nichols, J.D., Collazo, J.A., 2014. Multiseason occupancy models for correlated replicate surveys. Methods Ecol. Evol. 5, 583–591.

Hines, J.E., Nichols, J.D., Royle, J.A., MacKenzie, D.I., Gopalaswamy, A.M., Samba Kumar, N., Karanth, K.U., 2010. Tigers on trails: occupancy modeling for cluster sampling. Ecol. Appl. 20, 1456–1466.

Hostetler, J.A., Chandler, R.B., 2015. Improved state-space models for inference about spatial and temporal variation in abundance from count data. Ecology 96, 1713–1723.

Hobbs, N.T., Hooten, M.B., 2015. Bayesian Models: A Statistical Primer for Ecologists. Princeton University Press, Princeton, NJ.

Holt, B.G., Rioja-Nieto, R., MacNeil, M.A., Lupton, J., Rahbek, C., 2013. Comparing diversity data collected using a protocol designed for volunteers with results from a professional alternative. Methods Ecol. Evol. 4, 383–392.

Holtrop, A.M., Cao, Y., Dolan, C.R., 2010. Estimating sampling effort required for characterizing species richness and site-to-site similarity in fish assemblage surveys of wadeable Illinois streams. Trans. Am. Fish. Soc. 139, 1421–1435.

Holyoak, M., Leibold, M.A., Holt, R.D. (Eds.), 2005. Metacommunities: Spatial Dynamics and Ecological Communities. University of Chicago Press, Chicago, IL.

Homyack, J.A., O'Bryan, C.J., Thornton, J.E., Baldwin, R.F., 2014. Anuran assemblages associated with roadside ditches in a managed pine landscape. For. Ecol. Manage. 334, 217–231.

Hooten, M.B., Hobbs, N.T., 2015. A guide to Bayesian model selection for ecologists. Ecol. Monogr. 85, 3–28.

Hunt, S.D., Guzy, J.C., Price, S.J., Halstead, B.J., Eskew, E.A., Dorcas, M.E., 2013. Responses of riparian reptile communities to damming and urbanization. Biol. Conserv. 157, 277–284.

Hurlbert, S.H., 1984. Pseudoreplication and the design of ecological field experiments. Ecol. Monogr. 54, 187–211.

Hutchinson, R.A., Valente, J.V., Emerson, S.C., Betts, M.G., Dietterich, T.G., 2015. Penalized likelihood methods improve parameter estimates in occupancy models. Methods Ecol. Evol. http://dx.doi.org/10.1111/2041-210X.12368. Article first published online: March 31, 2015.

Illian, J.B., Penttinen, A., Stoyan, H., Stoyan, D., 2008. Statistical Analysis and Modelling of Spatial Point Patterns. Wiley, Chichester.

Illian, J.B., Sørbye, S.H., Rue, H., Hendrichsen, D.K., August 2012. Using INLA to fit a complex point process model with temporally varying effects – a case study. J. Environ. Stat. 3 (7). http://www.jenvstat.org/v03/i07.

Iknayan, K.J., Tingley, M.W., Furnas, B.J., Beissinger, S.R., 2014. Detecting diversity: emerging methods to estimate species diversity. Trends Ecol. Evol. 29, 97–106.

Ives, A.R., Zhu, J., 2006. Statistics for correlated data: phylogenies, space, and time. Ecol. Appl. 16, 20–32.

Jackman, S., 2009. Bayesian Analysis for the Social Sciences. John Wiley & Sons.

Jackman, S., 2012. pscl: Classes and Methods for R Developed in the Political Science Computational Laboratory, Stanford University. Department of Political Science, Stanford University, Stanford, California. R package version 1.04.4. http://pscl.stanford.edu.

Jenni, L., Kéry, M., 2003. Timing of autumn bird migration under climate change: advances in long-distance migrants, delays in short-distance migrants. Proc. R. Soc. Lond. Series B 270, 1467–1471.

Jolly, G.M., 1965. Explicit estimates from capture-recapture data with both death and immigration-stochastic model. Biometrika 52, 225–247.

Johnson, D.H., 2008. In defense of indices: the case of bird surveys. J. Wildl. Manage. 72, 857–868.

Johnson, D.S., Conn, P.B., Hooten, M.B., Ray, J.C., Pond, B.A., 2013. Spatial occupancy models for large data sets. Ecology 94, 801–808.

Johnson, D.S., Laake, J.L., Ver Hoef, J.M., 2010. A model-based approach for making ecological inference from distance sampling data. Biometrics 66, 310–318.

Johnson, F.A., Dorazio, R.M., Castellón, T.D., Martin, J., Garcia, J.O., Nichols, J.D., 2014. Tailoring point counts for inference about avian density: dealing with nondetection and availability. Nat. Res. Model. 27, 163–177.

Jones, J.E., Kroll, A.J., Giovanini, J., Duke, S.D., Ellis, T.M., Betts, M.G., 2012. Avian species richness in relation to intensive forest management practices in early seral tree plantations. PLoS ONE 7, e43290. http://dx.doi.org/10.1371/journal.pone.0043290.

Jonsen, I.D., Mills Flemming, J., Myers, R.A., 2005. Robust state-space modeling of animal movement data. Ecology 86, 2874–2880.

Joseph, L.N., Elkin, C., Martin, T.G., Possingham, H., 2009. Modeling abundance using N-mixture models: the importance of considering ecological mechanisms. Ecol. Appl. 19, 631–642.

Kadane, J.B., Lazar, N.A., 2004. Methods and criteria for model selection. J. Am. Stat. Assoc. 99, 279–290.

Kalinowski, S.T., Taper, M.L., Creel, S., 2006. Using DNA from non-invasive samples to identify individuals and census populations: an evidential approach tolerant of genotyping errors. Conserv. Genet. 7, 319–329.

Karanth, K.U., Nichols, J.D., 1998. Estimation of tiger densities in India using photographic captures and recaptures. Ecology 79, 2852–2862.

Karanth, K.U., Nichols, J.D., Kumar, N.S., Hines, J.E., 2006. Assessing tiger population dynamics using photographic capture-recapture sampling. Ecology 87, 2925–2937.

Karanth, K.K., Nichols, J.D., Sauer, J.R., Hines, J.E., Yackulic, C.B., 2014. Latitudinal gradients in North American avian species richness, turnover rates and extinction probabilities. Ecography 37, 626–636.

Karr, J.R., 1990. Biological integrity and the goal of environmental legislation: lessons for conservation biology. Conserv. Biol. 4, 244–250.

Kellner, K., 2015. jagsUI: A Wrapper Around 'rjags' to Streamline 'JAGS' Analyses. R package version 1.3.7. http://CRAN.R-project.org/package=jagsUI.

Kellner, K.F., Swihart, R.K., 2014. Accounting for Imperfect Detection in Ecology: A Quantitative Review. PLoS ONE 9, e111436.

Kendall, W.L., Hines, J.E., Nichols, J.D., 2003. Adjusting multistate capture-recapture models for misclassification bias: manatee breeding proportions. Ecology 84, 1058–1066.

Kendall, W.L., Hines, J.E., Nichols, J.D., Grant, E.H.C., 2013. Relaxing the closure assumption in occupancy models: staggered arrival and departure times. Ecology 94, 610–617.

Kendall, W.L., Nichols, J.D., Hines, J.E., 1997. Estimating temporary emigration using capture-recapture data with Pollock's robust design. Ecology 78, 563–578.

Kéry, M., 2002. Inferring the absence of a species - a case study of snakes. J. Wildl. Manage. 66, 330–338.

Kéry, M., 2004. Extinction rate estimates for plant populations in revisitation studies: importance of detectability. Conserv. Biol. 18, 570–574.

Kéry, M., 2008. Estimating abundance from bird counts: binomial mixture models uncover complex covariate relationships. Auk 125, 336–345.

Kéry, M., 2010. Introduction to WinBUGS for Ecologists - a Bayesian Approach to Regression, ANOVA, Mixed Models and Related Analyses. Academic Press, Burlington.

Kéry, M., 2011a. Species richness and community dynamics – a conceptual framework. In: O'Connell, A.F., Nichols, J.D., Karanth, K.U. (Eds.), Camera Traps in Animal Ecology: Methods and Analyses. Springer, Tokyo, pp. 207–231.

Kéry, M., 2011b. Towards the modeling of true species distributions. J. Biogeogr. 38, 617–618.

Kéry, M., Dorazio, R.M., Soldaat, L., van Strien, A., Zuiderwijk, A., Royle, J.A., 2009. Trend estimation in populations with imperfect detection. J. Appl. Ecol. 46, 1163–1172.

Kéry, M., Gardner, B., Monnerat, C., 2010a. Predicting species distributions from checklist data using site-occupancy models. J. Biogeogr. 37, 1851–1862.

Kéry, M., Gardner, B., Stoeckle, T., Weber, D., Royle, J.A., 2011. Use of spatial capture-recapture modeling and DNA data to estimate densities of elusive animals. Conserv. Biol. 25, 356–364.

Kéry, M., Gregg, K.B., 2003. Effects of life-state on detectablity in a demographic study of the terrestrial orchid *Cleistes bifaria*. J. Ecol. 91, 265–273.

Kéry, M., Gregg, K.B., 2004. Demographic analysis of dormancy and survival in the terrestrial orchid *Cypripedium reginae*. J. Ecol. 92, 686–695.

Kéry, M., Gregg, K.B., Schaub, M., 2005a. Demographic estimation methods for plants with unobservable life-states. Oikos 108, 307–320.

Kéry, M., Guillera-Arroita, G., Lahoz-Monfort, J.J., 2013. Analysing and mapping species range dynamics using dynamic occupancy models. J. Biogeogr. 40, 1463–1474.

Kéry, M., Madsen, J., Lebreton, J.D., 2006. Survival of Svalbard pink-footed geese *Anser brachyrhynchus* in relation to winter climate, density and land-use. J. Anim. Ecol. 75, 1172–1181.

Kéry, M., Royle, J.A., 2008. Hierarchical Bayes estimation of species richness and occupancy in spatially replicated surveys. J. Appl. Ecol. 45, 589–598.

Kéry, M., Royle, J.A., 2009. Inference about species richness and community structure using species-specific occupancy models in the national Swiss breeding bird survey MHB. In: Thomson, D.L., Cooch, E.G., Conroy, M.J. (Eds.), Modeling Demographic Processes in Marked Populations, pp. 639–656.

Kéry, M., Royle, J.A., 2010. Hierarchical modeling and estimation of abundance in metapopulation designs. J. Anim. Ecol. 79, 453–461.

Kéry, M., Royle, J.A., Schmid, H., 2005b. Modeling avian abundance from replicated counts using binomial mixture models. Ecol. Appl. 15, 1450–1461.

Kéry, M., Royle, J.A., Schmid, H., 2008. Importance of sampling design and analysis in animal population studies: a comment on Sergio et al. J. Appl. Ecol. 45, 981–986.

Kéry, M., Royle, J.A., Schmid, H., Schaub, M., Volet, B., Häfliger, G., Zbinden, N., 2010b. Site-ocupancy distribution modeling to correct population-trend estimates derived from opportunistic observations. Conserv. Biol. 24, 1388–1397.

Kéry, M., Schaub, M., 2012. Bayesian Population Analysis Using WinBUGS: A Hierarchical Perspective. Academic Press.

Kéry, M., Schmidt, B.R., 2008. Imperfect detection and its consequences for monitoring for conservation. Community Ecol. 9, 207–216.

Kéry, M., Spillmann, J.H., Truong, C., Holderegger, R., 2006. How biased are estimates of extinction probability in revisitation studies. J. Ecol. 94, 980–986.

King, R., Morgan, B.J.T., Gimenez, O., Brooks, S.P., 2009. Bayesian Analysis for Population Ecology. Chapmann & Hall, Boca Raton.

Knape, J., Korner-Nievergelt, F., 2015. Estimates from non-replicated population surveys rely on critical assumptions. Methods Ecol. Evol. 6, 298–306.

Koenen, K.K., DeStefano, S., Krausman, P.R., 2002. Using distance sampling to estimate seasonal densities of desert mule deer in a semidesert grassland. Wildl. Soc. Bull. 30, 53–63.

Koneff, M.D., Royle, J.A., Otto, M.C., Wortham, J.S., Bidwell, J.K., 2008. A double-observer method to estimate detection rate during aerial waterfowl surveys. J. Wildl. Manage. 72, 1641–1649.

Korner-Nievergelt, F., von Felten, S., Roth, T., Almasi, B., Guélat, J., Korner-Nievergelt, P., 2015. Bayesian Data Analysis in Ecology Using Linear Models with R, BUGS, and Stan, first ed. Academic Press.

Krebs, C.J., 2009. Ecology: The Experimental Analysis of Distribution and Abundance, sixth ed. Benjamin Cummings, San Francisco.

Kroll, A.J., Ren, Y., Jones, J.E., Giovanini, J., Perry, R.W., Thill, R.E., White, D., Wigley, T.B., 2014. Avian community composition associated with interactions between local and landscape habitat attributes. For. Ecol. Manage. 326, 46–57.

Kuo, L., Mallick, B., 1998. Variable selection for regression models. Sankhya 60B, 65–81.

Laake, J.L., Collier, B.A., Morrison, M.L., Wilkins, R.N., 2011. Point-based mark-recapture distance sampling. J. Agric. Biol. Environ. Stat. 16, 389–408.

Lachish, S., Gopalaswamy, A.M., Knowles, S.C.L., Sheldon, B.C., 2012. Site-occupancy modelling as a novel framework for assessing test sensitivity and estimating wildlife disease prevalence from imperfect diagnostic tests. Methods Ecol. Evol. 3, 339–348.

Lahoz-Monfort, J.J., Guillera-Arroita, G., Wintle, B.A., 2014. Imperfect detection impacts the performance of species distribution models. Glob. Ecol. Biogeogr. 23, 504–515.

Langtimm, C.A., Dorazio, R.M., Stith, B.M., Doyle, T.J., 2011. New aerial survey and hierarchical model to estimate manatee abundance. J. Wildl. Manage. 75, 399–412.

Latimer, A.M., Wu, S., Gelfand, A.E., Silander Jr., J.A., 2006. Building statistical models to analyse species distributions. Ecol. Appl. 16, 33–50.

Le Cam, L., 1990. Maximum likelihood — an introduction. ISI Review 58, 153–171.

Lee, Y., Nelder, J.A., 2000. Two ways of modeling overdispersion in non-normal data. App. Stat. 49, 591–598.

Lee, Y., Nelder, J.A., Pawitan, Y., 2006. Generalized Linear Models with Random Effects. Unified Analysis via H-likelihood. Chapman and Hall/CRC, Boca Raton, FL.

Lele, S.R., 2010. Model complexity and information in the data: could it be a house built on sand? Ecology 91, 3493–3496.

Lele, S.R., 2015. Is Non-informative Bayesian Analysis Appropriate for Wildlife Management: Survival of San Joaquin Kit Fox and Declines in Amphibian Populations. arXiv:1502.00483 [q-bio.QM].

Lele, S.R., Dennis, B., 2009. Bayesian methods for hierarchical models: are ecologists making a Faustian bargain? Ecol. Appl. 19, 581–584.

Lele, S.R., Keim, J.L., 2006. Weighted distributions and estimation of resource selection probability functions. Ecology 87, 3021–3028.

Lele, S.R., Moreno, M., Bayne, E., 2012. Dealing with detection error in site occupancy surveys: what can we do with a single survey? J. Plant Ecol. 5, 22–31.

Lewis, T.L., Lindberg, M.S., Schmutz, J.A., Bertram, M.R., Dubour, A.J., 2015. Species richness and distributions of boreal waterbird broods in relation to nesting and brood-rearing habitats. J. Wildl. Manage. 79, 296–310.

Linden, D.W., Roloff, G.J., 2013. Retained structures and bird communities in clearcut forests of the Pacific Northwest, USA. For. Ecol. Manage. 310, 1045–1056.

Link, W.A., 1999. Modeling pattern in collections of parameters. J. Wildl. Manage. 63, 1017–1027.

Link, W.A., 2013. A cautionary note on the discrete uniform prior for the binomial N. Ecology 94, 2173–2179.

Link, W.A., Barker, R.J., 2006. Model weights and the foundations of multimodel inference. Ecology 87, 2626–2635.

Link, W.A., Barker, R.J., 2010. Bayesian Inference with Ecological Applications. Academic Press, London.

Link, W.A., Eaton, M.J., 2012. On thinning of chains in MCMC. Methods Ecol. Evol. 3, 112–115.

Link, W.A., Royle, J.A., Hatfield, J.S., 2003. Demographic analysis from summaries of an age-structured population. Biometrics 59, 778–785.

Link, W.A., Sauer, J.R., 2002. A hierarchical analysis of population change with application to Cerulean warblers. Ecology 83, 2832–2840.

Link, W.A., Yoshizaki, J., Bailey, L.L., Pollock, K.H., 2010. Uncovering a latent multinomial: analysis of mark-recapture data with misidentification. Biometrics 66, 178–185.

Littell, R.C., Milliken, G.A., Stroup, W.W., Wolfinger, R.D., Schabenberger, O., 2008. SAS for Mixed Models, second ed. SAS Institute, Cary, NC.

Little, R.J.A., 2006. Calibrated Bayes: a bayes/frequentist roadmap. Am. Stat. 60, 213–223.

Little, R.J.A., Rubin, D.B., 2002. Statistical Analysis with Missing Data, second ed. Wiley, New York.

Liu, J.S., Wu, Y.N., 1999. Parameter expansion for data augmentation. J. Am. Stat. Assoc. 94, 1264–1274.

Lukacs, P.M., Burnham, K.P., 2005. Estimating population size from DNA-based closed capture–recapture data incorporating genotyping error. J. Wildl. Manage. 69, 396–403.

Lunn, D., Jackson, C., Best, N., Thomas, A., Spiegelhalter, D., 2013. The BUGS Book: A Practical Introduction to Bayesian Analysis. Chapman and Hall/CRC.

Lunn, D.J., Spiegelhalter, D., Thomas, A., Best, N., 2009. The BUGS project: evaluation, critique and future directions. Stat. Med. 28, 3049–3067.

Lunn, D.J., Thomas, A., Best, N., Spiegelhalter, D., 2000. WinBUGS - a Bayesian modelling framework: concepts, structure, and extensibility. Stat. Comput. 10, 325–337.

Lynch, H.J., Thorson, J.T., Shelton, A.O., 2014. Dealing with under- and overdispersed count data in life history, spatial, and community ecology. Ecology 95, 3173–3180.

Lyons, J.E., Royle, J.A., Thomas, S.M., Elliott-Smith, E., Evenson, J.R., Kelly, E.G., Milner, R.L., Nysewander, D.R., Andres, B.A., 2012. Large-scale monitoring of shorebird populations using count data and N-mixture models: Black Oystercatcher (*Haematopus bachmani*) surveys by land and sea. Auk 129, 645–652.

MacEachern, S.N., Berliner, L.M., 1994. Subsampling the Gibbs sampler. Am. Stat. 48, 188–190.

MacKenzie, D.I., 2005. What are the issues with presence-absence data for wildlife managers? J. Wildl. Manage. 69, 849–860.

MacKenzie, D.I., Bailey, L.L., 2004. Assessing the fit of site-occupancy models. J. Agric. Biol. Environ. Stat. 9, 300–318.

MacKenzie, D.I., Nichols, J.D., Hines, J.E., Knutson, M.G., Franklin, A.B., 2003. Estimating site occupancy, colonization, and local extinction when a species is detected imperfectly. Ecology 84, 2200–2207.

MacKenzie, D.I., Nichols, J.D., Lachman, G.B., Droege, S., Royle, J.A., Langtimm, C.A., 2002. Estimating site occupancy rates when detection probabilities are less than one. Ecology 83, 2248–2255.

MacKenzie, D.I., Nichols, J.D., Royle, J.A., Pollock, K.H., Hines, J.E., Bailey, L.L., 2006. Occupancy Estimation and Modeling: Inferring Patterns and Dynamics of Species Occurrence. Elsevier, San Diego.

MacKenzie, D.I., Nichols, J.D., Seamans, M.E., Gutierrez, R.J., 2009. Modeling species occurrence dynamics with multiple states and imperfect detection. Ecology 90, 823–835.

MacKenzie, D.I., Nichols, J.D., Sutton, N., Kawanishi, K., Bailey, L.L., 2005. Improving inferences in population studies of rare species that are detected imperfectly. Ecology 86, 1101–1113.

MacKenzie, D.I., Royle, J.A., 2005. Designing occupancy studies: general advice and allocating survey effort. J. Appl. Ecol. 42, 1105–1114.

Magnusson, W.E., Caughley, G.J., Grigg, G.C., 1978. A double-survey estimate of population size from incomplete counts. J. Wildl. Manage. 42, 174–176.

Marques, F.F.C., Buckland, S.T., 2003. Incorporating covariates into standard line transect analyses. Biometrics 59, 924–935.

Marques, T.A., Buckland, S.T., Bispo, R., Howland, B., 2013. Accounting for animal density gradients using independent information in distance sampling surveys. Stat. Methods Appl. 22, 67–80.

Marques, T.A., Buckland, S.T., Borchers, D.L., Tosh, D., McDonald, R.A., 2010. Point transect sampling along linear features. Biometrics 66, 1247–1255.

Marques, F.F., Buckland, S.T., Goffin, D., Dixon, C.E., Borchers, D.L., Mayle, B.A., Peace, A.J., 2001. Estimating deer abundance from line transect surveys of dung: sika deer in southern Scotland. J. Appl. Ecol. 38, 349–363.

Marques, T.A., Thomas, L., Fancy, S.G., Buckland, S.T., 2007. Improving estimates of bird density using multiple-covariate distance sampling. Auk 124, 1229–1243.

Marsh, H., Sinclair, D.F., 1989. Correcting for visibility bias in strip transect aerial surveys of aquatic fauna. J. Wildl. Manage. 53, 1017–1024.

Martin, T.G., Kuhnert, P.M., Mergersen, K., Possingham, H.P., 2005. The power of expert opinion in ecological models using Bayesian methods: Impact of grazing on birds. Ecol. Appl. 15, 266–280.

Martin, J., Nichols, J.D., McIntyre, C.L., Ferraz, G., Hines, J.E., 2009. Perturbation analysis for patch occupancy dynamics. Ecology 90, 10–16.

Martin, J., Royle, J.A., Mackenzie, D.I., Edwards, H.H., Kéry, M., Gardner, B., 2011. Accounting for non-independent detection when estimating abundance of organisms with a Bayesian approach. Methods Ecol. Evol. 2, 595–601.

Mata, L., Goula, M., Hahs, A.K., 2014. Conserving insect assemblages in urban landscapes: accounting for species-specific responses and imperfect detection. J. Insect Conserv. 18, 885–894.

Matechou, E., Dennis, E.B., Freeman, S.N., Brereton, T., 2014. Monitoring abundance and phenology in (multivoltine) butterfly species: a novel mixture model. J. Appl. Ecol. 51, 766–775.

Mattsson, B.J., Zipkin, E.F., Gardner, B., Blank, P.J., Sauer, J.R., Royle, J.A., 2013. Explaining local-scale species distributions: relative contributions of spatial autocorrelation and landscape heterogeneity for an avian assemblage. PLoS ONE 8, e55097.

Mazerolle, M.J., 2015. AICcmodavg: Model Selection and Multimodel Inference Based on (Q)AIC(c). R package version 2.0-3. http://CRAN.R-project.org/package=AICcmodavg.

McCarthy, M.A., 2007. Bayesian Methods for Ecology. Cambridge University Press, Cambridge.

McCarthy, M.A., Masters, P., 2005. Profiting from prior information in Bayesian analyses of ecological data. J. Appl. Ecol. 42, 1012–1019.

McCarthy, M.A., Moore, J.L., Morris, W.K., Parris, K.M., Garrard, G.E., Vesk, P.A., Rumpff, L., Giljohann, K.M., Camac, J.S., Bau, S.S., Friend, T., Harrison, B., Yue, B., 2013. The influence of abundance on detectabiliy. Oikos 122, 717–726.

McClintock, B.T., Bailey, L.L., Pollock, K.H., Simons, T.R., 2010a. Unmodeled observation error induces bias when inferring patterns and dynamics of species occurrence via aural detections. Ecology 91, 2446–2454.

McClintock, B.T., Nichols, J.D., Bailey, L.L., MacKenzie, D.I., Kendall, W.L., Franklin, A.B., 2010b. Seeking a second opinion: uncertainty in disease ecology. Ecol. Lett. 13, 659–674.

McCoy, E.D., Heck Jr., K.L., 1987. Some observations on the use of taxonomic similarity in large-scale biogeography. J. Biogeogr. 14, 79–87.

McCrea, R.S., Morgan, B.J.T., 2014. Analysis of Capture-recapture Data. Chapman & Hall/CRC Press, Boca Raton, FL, USA.

McCullagh, P., Nelder, J.A., 1989. Generalized Linear Models. Chapman & Hall, London.

McCulloch, C.E., Searle, S.R., 2001. Generalized, Linear, and Mixed Models. Wiley, New York.

McIntyre, A.P., Jones, J.E., Lund, E.M., Waterstrat, F.T., Giovanini, J.N., Duke, S.D., Hayes, M.P., Quinn, T., Kroll, A.J., 2012. Empirical and simulation evaluations of an abundance estimator using unmarked individuals of cryptic forest-dwelling taxa. For. Ecol. Manage. 286, 129–136.

McKann, P.C., Gray, B.R., Thogmartin, W.E., 2013. Small sample bias in dynamic occupancy models. J. Wildl. Manage. 77, 172–180.

McKenny, H.C., Keeton, W.S., Donovan, T.M., 2006. Effects of structural complexity enhancement on eastern red-backed salamander (*Plethodon cinereus*) populations in northern hardwood forests. For. Ecol. Manage. 230, 186–196.

McManamay, R.A., Orth, D.J., Jager, H.I., 2014. Accounting for variation in species detection in fish community monitoring. Fish. Manage. Ecol. 21, 96–112.

McNew, L.B., Handel, C.M., 2015. Evaluating species richness: biased ecological inference results from spatial heterogeneity in detection probabilities. Ecol. Appl. 25, 1669–1680.

Mead, R., 1988. The Design of Experiments: Statistical Principles for Practical Applications. Cambridge University Press, Cambridge UK.

Metropolis, N., Rosenbluth, A.W., Rosenbluth, M.N., Teller, A.H., Teller, E., 1953. Equation of state calculations by fast computing machines. J. Chem. Phys. 21, 1087–1092.

Mihaljevic, J.R., Joseph, M.B., Johnson, P.T.J., 2015. Using multispecies occupancy models to improve the characterization and understanding of metacommunity structure. Ecology 96, 1783–1792.

Millar, R.B., 2009. Comparison of hierarchical Bayesian models for overdispersed count data using DIC and Bayes' factors. Biometrics 65, 962–969.

Miller, A., 2002. Subset Selection in Regression. Chapman & Hall/CRC.

Miller, D.A., Bailey, L.L., Grant, E.H.C., McClintock, B.T., Weir, L., Simons, T.R., 2015. Performance of species occurrence estimators when basic assumptions are not met: a test using field data where true occupancy status is known. Methods Ecol. Evol. 6, 557–565.

Miller, D.A., Nichols, J.D., McClintock, B.T., Grant, E.H.C., Bailey, L.L., Weir, L., 2011. Improving occupancy estimation when two types of observational errors occur: non-detection and species misidentification. Ecology 92, 1422–1428.

Miller, D.A.W., Nichols, J.D., Gude, J.A., Rich, L.N., Podruzny, K.M., Hines, J.E., Mitchell, M.S., 2013b. Determining occurrence dynamics when false positives occur: estimating the range dynamics of wolves from public survey data. PLoS ONE 8, e65808.

Miller, D.L., Burt, M.L., Rexstad, E.A., Thomas, L., 2013a. Spatial models for distance sampling data: recent developments and future directions. Methods Ecol. Evol. 4, 1001–1010.

Miller, D.L., Thomas, L., 2015. Mixture models for distance sampling detection functions. PLoS ONE 10, e0118726. http://dx.doi.org/10.1371/journal.pone.0118726.

Moore, J.E., Barlow, J., 2011. Bayesian state-space model of fin whale abundance trends from a 1991–2008 time series of line-transect surveys in the California Current. J. Appl. Ecol. 48, 1195–1205.

Morales, J.M., Haydon, D.T., Frair, J., Holsinger, K.E., Fryxell, J.M., 2004. Extracting more out of relocation data: Building movement models as mixtures of random walks. Ecology 85, 2436–2445.

Mordecai, R.S., Mattsson, B.J., Tzilkowski, C.J., Cooper, R.J., 2011. Addressing challenges when studying mobile or episodic species: hierarchical Bayes estimation of occupancy and use. J. Appl. Ecol. 48, 56–66.

Murtaugh, P.A., 2007. Simplicity and complexity in ecological data analysis. Ecology 88, 56–62.

Nelder, J.A., 1965a. The analysis of randomized experiments with orthogonal block structure. I. Block structure and the null analysis of variance. Proc. R. Soc. Series A 283, 147–162.

Nelder, J.A., 1965b. The analysis of randomized experiments with orthogonal block structure. II. Treatment structure and the general analysis of variance. Proc. R. Soc. Series A 283, 163–178.

Newman, K.B., Buckland, S.T., Lindley, S.T., Thomas, L., Fernandez, C., 2006. Hidden process models for animal population dynamics. Ecol. Appl. 16, 74–86.

Newman, K., Buckland, S.T., Morgan, B., King, R., Borchers, D.L., Cole, D., Besbeas, P., Gimenez, O., Thomas, L., 2014. Modelling Population Dynamics, Model Formulation, Fitting and Assessment Using State-space Methods. Springer.

Nichols, J.D., Bailey, L.L., O'Connell, A.F., Talancy, N.W., Grant, E.H.C., Gilbert, A.T., Annand, E.M., Husband, T.P., Hines, J.E., 2008. Multi-scale occupancy estimation and modelling using multiple detection methods. J. Appl. Ecol. 45, 1321–1329.

Nichols, J.D., Boulinier, T., Hines, J.E., Pollock, K.H., Sauer, J.R., 1998a. Estimating rates of local species extinction, colonization, and turnover in animal communities. Ecol. Appl. 8, 1213–1225.

Nichols, J.D., Boulinier, T., Hines, J.E., Pollock, K.H., Sauer, J.R., 1998b. Inference methods for spatial variation in species richness and community composition when not all species are detected. Conserv. Biol. 12, 1390–1398.

Nichols, J.D., Hines, J.E., MacKenzie, D.I., Seamans, M.E., Gutierrez, R.J., 2007. Occupancy estimation and modeling with multiple states and state uncertainty. Ecology 88, 1395–1400.

Nichols, J.D., Hines, J.E., Sauer, J.R., Fallon, F.W., Fallon, J.E., Heglund, P.J., 2000. A double-observer approach for estimating detection probability and abundance from point counts. Auk 117, 393–408.

Nichols, J.D., Thomas, L., Conn, P.B., 2009. Inferences about landbird abundance from count data: recent advances and future directions. In: Thomson, D.L., Cooch, E.G., Conroy, M.J. (Eds.), Modeling Demographic Processes in Marked Populations. Springer, New York, pp. 201–235.

Nichols, J.D., McIntyre, C.L., Ferraz, G., Hines, J.E., 2009. Perturbation analysis for patch occupancy dynamics. Ecology 90, 10–16.

Niemi, A., Fernandez, C., 2010. Bayesian spatial point processmodeling of line transect data. J. Agric. Biol. Environ. Stat. 15, 327–345.

NIMBLE Development Team, 2015. NIMBLE: An R Package for Programming with BUGS Models, Version 0.4. http://r-nimble.org.

Norris III, J.L., Pollock, K.H., 1996. Nonparametric MLE under two closed capture-recapture models with heterogeneity. Biometrics 639–649.

Ntzoufras, I., 2009. Bayesian Modeling Using WinBUGS. Wiley, Hoboken, New Jersey.

O'Brien, T.G., Baillie, J.E.M., Krueger, L., Cuke, M., 2010. The Wildlife Picture Index: monitoring top trophic levels. Anim. Conserv. 13, 335–343.

O'Hara, R.B., Sillanpää, M.J., 2009. A review of Bayesian variable selection methods: what, how and which. Bayesian Anal. 4, 85–118.

Oedekoven, C.S., Buckland, S.T., Mackenzie, M.L., Evans, K.O., Burger Jr., L.W., 2013. Improving distance sampling: accounting for covariates and non-independency between sampled sites. J. Appl. Ecol. 50, 786–793.

Oedekoven, C.S., Buckland, S.T., MacKenzie, M.L., King, R., Evans, K., Burger, W., 2014. Bayesian methods for hierarchical distance sampling models. J. Agric. Biol. Environ. Stat. 19, 219–239.

Olea, P.P., Mateo-Tomas, P., 2011. Spatially explicit estimation of occupancy, detection probability and survey effort needed to inform conservation planning. Divers. Distrib. 17, 714–724.

Otis, D.L., Burnham, K.P., White, G.C., Anderson, D.R., 1978. Statistical inference from capture data on closed animal populations. Wildl. Monogr. 62, 1–135.

Ovaskainen, O., Hottola, J., Siitonen, J., 2010. Modeling species co-occurrence by multivariate logistic regression generates new hypotheses on fungal interactions. Ecology 91, 2514–2521.

Ovaskainen, O., Soininen, J., 2011. Making more out of sparse data: hierarchical modeling of species communities. Ecology 92, 289–295.

Pacifici, K., Zipkin, E.F., Collazo, J.A., Irizarry, J.I., DeWan, A., 2014. Guidelines for a priori grouping of species in hierarchical community models. Ecol. Evol. 4, 877–888.

Papaïx, J., Cubaynes, S., Buoro, M., Charmantier, A., Perret, P., Gimenez, O., 2010. Combining capture-recapture data and pedigree information to assess heritability of demographic parameters in the wild. J Evol Biol 23, 2176–2184.

Pardo, M.A., Gerrodette, T., Beier, E., Gendron, D., Forney, K.A., et al., 2015. Inferring cetacean population densities from the absolute dynamic topography of the ocean in a hierarchical Bayesian framework. PLoS ONE 10, e0120727.

Pavlacky, D.C., Blakesley, J.A., White, G.C., Hanni, D.J., Lukacs, P.M., 2012. Hierarchical multi-scale occupancy estimation for monitoring wildlife populations. J. Wildl. Manage. 76, 154–162.

Pearce, J.L., Boyce, M.S., 2006. Modelling distribution and abundance with presence-only data. J. Appl. Ecol. 43, 405–412.

Penteriani, V., 2003. Breeding density affects the honesty of bird vocal displays as possible indicators of male/territory quality. Ibis 145, E127–E135 (on-line).

Phillips, S.J., Dudik, M., 2008. Modeling of species distributions with Maxent: new extensions and a comprehensive evaluation. Ecography 31, 161–175.

Phillips, S.J., Elith, J., 2013. On estimating probability of presence from use-availability or presence-background data. Ecology 94, 1409–1419.

Pinheiro, J.C., Bates, D.M., 2000. Mixed-Effects Models in S and S-Plus. Springer, New York.

Plummer, M., 2003. JAGS: a program for analysis of Bayesian graphical models using Gibbs sampling. In: Hornik, K., Leisch, F., Zeileis, A. (Eds.), Proceedings of the 3rd International Workshop in Distributed Statistical Computing (DSC 2003), March 20–22. Technische Universität, Vienna, Austria, pp. 1–10.

Plummer, M., 2015. rjags: Bayesian Graphical Models Using MCMC. R package version 3-15. http://CRAN.R-project.org/package=rjags.

Pollock, J.F., 2006. Detecting population declines over large areas with presence-absence, time-to-encounter, and count survey methods. Conserv. Biol. 20, 882–892.

Pollock, K.H., 1982. A capture-recapture design robust to unequal probability of capture. J. Wildl. Manage. 46, 752–757.

Pollock, K.H., Nichols, J.D., Brownie, C., Hines, J.E., 1990. Statistical inference for capture-recapture experiments. Wildl. Monogr. 107, 3–97.

Pollock, K.H., Nichols, J.D., Simons, T.R., Farnsworth, G.L., Bailey, L.L., Sauer, J.R., 2002. Large scale wildlife monitoring studies: statistical methods for design and analysis. Environmetrics 13, 105–119.

Pollock, L.J., Tingley, R., Morris, W.K., Golding, N., O'Hara, R.B., Parris, K.M., Vesk, P.A., McCarthy, M.A., 2014. Understanding co-occurrence by modelling species simultaneously with a Joint Species Distribution Model (JSDM). Methods Ecol. Evol. 5, 397–406.

Post van der Burg, M., Bly, B., Vercauteren, T., Tyre, A.J., 2011. Making better use of monitoring data from low density species using a spatially explicit modeling approach. J. Appl. Ecol. 48, 47–55.

Potts, J.M., Elith, J., 2006. Comparing species abundance models. Ecol. Model. 199, 153–163.

Pradel, R., 2005. Multievent: an extension of multistate capture-recapture models to uncertain states. Biometrics 61, 442–447.
Pradel, R., Hines, J.E., Lebreton, J.D., Nichols, J.D., 1997. Capture-recapture survival models taking account of transients. Biometrics 53, 60–72.
Purvis, A., Hector, A., 2000. Getting the measure of biodiversity. Nature 405, 212–219.
Qian, S.S., Shen, Z., 2007. Ecological applications of multilevel analysis of variance. Ecology 88, 2489–2495.
Railsback, S.F., Grimm, V., 2012. Agent-Based and Individual-Based Modeling. Princeton University Press, Princeton, USA.
Ramsey, D.S.L., Caley, P.A., Robley, A., 2015. Estimating population density from presence-absence data using a spatially explicit model. J. Wildl. Manage. 79, 491–499.
Renner, I.W., Elith, J., Baddeley, A., Fithian, W., Hastie, T., Phillips, S.J., Popovic, G., Warton, D.I., 2015. Point process models for presence-only analysis. Methods Ecol. Evol. 6, 366–379.
Renner, I.W., Warton, D.I., 2013. Equivalence of MAXENT and Poisson point process models for species distribution modeling in ecology. Biometrics 69, 274–281.
Riddle, J.D., Mordecai, R.S., Pollock, K.H., Simons, T.R., 2010. Effects of prior detections on estimates of detection probability, abundance and occupancy. Auk 127, 94–99.
Ridout, M.S., Besbeas, P., 2004. An empirical model for underdispersed count data. Stat. Model. 4, 77–89.
Robert, C., Casella, G., 2010. Introducing Monte Carlo Methods with R. Springer Science & Business Media.
Rosenstock, S.S., Anderson, D.R., Giesen, K.M., Leukering, T., Carter, M.F., 2002. Landbird counting techniques: current practices and an alternative. Auk 119, 46–53.
Rota, C.T., Fletcher Jr., R.J., Dorazio, R.M., Betts, M.G., 2009. Occupancy estimation and the closure assumption. J. Appl. Ecol. 46, 1173–1181.
Roth, T., Amrhein, V., 2009. Estimating individual survival using territory occupancy data on unmarked animals. J. Appl. Ecol. 47, 386–392.
Rout, T.M., Moore, J.L., McCarthy, M.A., 2014. Prevent, search or destroy? A partially observable model for invasive species management. J. Appl. Ecol. 51, 804–813.
Royle, J.A., 2004a. Generalized estimators of avian abundance from count survey data. Anim. Biodivers. Conserv. 27, 375–386.
Royle, J.A., 2004b. N-mixture models for estimating population size from spatially replicated counts. Biometrics 60, 108–115.
Royle, J.A., 2006. Site occupancy model with heterogeneous detection probabilities. Biometrics 62, 97–102.
Royle, J.A., 2008. Modeling individual effects in the Cormack-Jolly-Seber model: a state-space formulation. Biometrics 64, 364–370.
Royle, J.A., 2009. Analysis of capture-recapture models with individual covariates using data augmentation. Biometrics 65, 267–274.
Royle, J.A., Chandler, R.B., Sollmann, R., Gardner, B., 2014. Spatial Capture-Recapture. Academic Press.
Royle, J.A., Chandler, R.B., Sun, C.C., Fuller, A.K., 2013. Integrating resource selection information with spatial capture-recapture. Methods Ecol. Evol. 4, 520–530.
Royle, J.A., Chandler, R.B., Yackulic, C., Nichols, J.D., 2012. Likelihood analysis of species occurrence probability from presence-only data for modeling species distributions. Methods Ecol. Evol. 3, 545–554.
Royle, J.A., Converse, S.J., 2014. Hierarchical spatial capture-recapture models: modelling population density in stratified populations. Methods Ecol. Evol. 5, 37–43.
Royle, J.A., Converse, S.J., Link, W.A., 2012. Data augmentation for hierarchical capture-recapture models arXiv preprint arXiv:1211.5706.
Royle, J.A., Dawson, D.K., Bates, S., 2004. Modeling abundance effects in distance sampling. Ecology 85, 1591–1597.

Royle, J.A., Dorazio, R.M., 2006. Hierarchical models of animal abundance and occurrence. J. Agric. Biol. Environ. Stat. 11, 249–263.

Royle, J.A., Dorazio, R.M., 2008. Hierarchical Modeling and Inference in Ecology. The Analysis of Data from Populations, Metapopulations and Communities. Academic Press, New York.

Royle, J.A., Dorazio, R.M., 2012. Parameter-expanded data augmentation for Bayesian analysis of capture-recapture models. J. Ornithol. 152, 521–537.

Royle, J.A., Dorazio, R.M., Link, W.A., 2007a. Analysis of multinomial models with unknown index using data augmentation. J. Comput. Gr. Stat. 16, 67–85.

Royle, J.A., Kéry, M., 2007. A Bayesian state-space formulation of dynamics occupancy models. Ecology 88, 1813–1823.

Royle, J.A., Kéry, M., Gauthier, R., Schmid, H., 2007b. Hierarchical spatial models of abundance and occurrence from imperfect survey data. Ecol. Monogr. 77, 465–481.

Royle, J.A., Kéry, M., Guélat, J., 2011. Spatial capture-recapture models for search-encounter data. Methods Ecol. Evol. 2, 602–611.

Royle, J.A., Link, W.A., 2005. A general class of multinomial mixture models for anuran calling survey data. Ecology 86, 2505–2512.

Royle, J.A., Link, W.A., 2006. Generalized site occupancy models allowing for false positive and false negative errors. Ecology 87, 835–841.

Royle, J.A., Nichols, J.D., 2003. Estimating abundance from repeated presence-absence data or point counts. Ecology 84, 777–790.

Royle, J.A., Nichols, J.D., Kéry, M., 2005. Modelling occurrence and abundance of species when detection is imperfect. Oikos 110, 353–359.

Royle, J.A., Young, K.G., 2008. A hierarchical model for spatial capture-recapture data. Ecology 89, 2281–2289.

Rubin, D.B., 1984. Bayesianly justifiable and relevant frequency calculations for the applied statistician. Ann. Stat. 12, 1151–1172.

Ruiz-Gutierrez, V., Zipkin, E.F., 2011. Detection biases yield misleading patterns of species persistence and colonization in fragmented landscapes. Ecosphere 2, article 61.

Ruiz-Gutierrez, V., Zipkin, E.F., Dhondt, A.A., 2010. Occupancy dynamics in a tropical bird community: unexpectedly high forest use by birds classified as non-forest species. J. Appl. Ecol. 47, 621–630.

Ruppert, D., 2002. Selecting the number of knots for penalized splines. J. Comput. Gr. Stat. 11, 735–757.

Ruppert, D., Wand, M., Carroll, R., 2003. Semiparametric Regression. Cambridge University Press, Cambridge, UK.

Russell, J.C., Stjernman, M., Lindstrom, A., Smith, H.G., 2015. Community occupancy before-after-control-impact (CO-BACI) analysis of Hurricane Gudrun on Swedish forest birds. Ecol. Appl. 25, 685–694.

Russell, R.E., Royle, J.A., Saab, V.A., Lehmkuhl, J.F., Block, W.M., Sauer, J.R., 2009. Modeling the effects of environmental disturbance on wildlife communities: avian responses to prescribed fire. Ecol. Appl. 19, 1253–1263.

Sadoti, G., Zuckerberg, B., Jarzyna, M.A., Porter, W.F., 2013. Applying occupancy estimation and modeling to the analysis of atlas data. Divers. Distrib. 19, 804–814.

Sanderlin, J.S., Block, W.M., Ganey, J.L., 2014. Optimizing study design for multi-species avian monitoring programmes. J. Appl. Ecol. 51, 860–870.

Sauer, J.R., Link, W.A., 2002. Hierarchical modeling of population stability and species group attributes from survey data. Ecology 86, 1743–1751.

Sauer, J.R., Link, W.A., 2011. Analysis of the North American breeding bird survey using hierarchical models. Auk 128, 87–98.

Sauer, J.R., Blank, P.J., Zipkin, E.F., Fallon, J.E., Fallon, F.W., 2013. Using multi-species occupancy models in structured decision making on managed lands. J. Wildl. Manage. 77, 117–127.

Schaub, M., Abadi, F., 2011. Integrated population models: a novel analysis framework for deeper insights into population dynamics. J. Ornithol. 152 (Suppl. 1), 227–237.

Schaub, M., Gimenez, O., Sierro, A., Arlettaz, R., 2007. Use of integrated modeling to enhance estimates of population dynamics obtained from limited data. Conserv. Biol. 21, 945–955.

Schaub, M., Jakober, H., Stauber, W., 2013. Strong contribution of immigration to local population regulation: evidence from a migratory passerine. Ecology 94, 1828–1838.

Schaub, M., Kéry, M., 2012. Combining information in hierarchical models improves inferences in population ecology and demographic population analyses. Anim. Conserv. 15, 125–126.

Schaub, M., Reichlin, T.S., Abadi, F., Kéry, M., Jenni, L., Arlettaz, R., 2012. The demographic drivers of local population dynamics in two rare migratory birds. Oecologia 168, 97–108.

Schmid, H., Zbinden, N., Keller, V., 2004. Überwachung der Bestandsentwicklung häufiger Brutvögel in der Schweiz. Schweizerische Vogelwarte, Sempach.

Schmidt, B.R., Kéry, M., Ursenbacher, S., Hyman, O.J., Collins, J.P., 2013. Site occupancy models in the analysis of environmental DNA presence/absence surveys: a case study of an emerging amphibian pathogen. Methods Ecol. Evol. 4, 646–653.

Schmidt, J.H., Rattenbury, K.L., 2013. Reducing effort while improving inference: estimating Dall's sheep abundance and composition in small areas. J. Wildl. Manage. 77, 1048–1058.

Schmidt, J.H., Rattenbury, K.L., Lawler, J.P., MacCluskie, M.C., 2012. Using distance sampling and hierarchical models to improve estimates of Dall's sheep abundance. J. Wildl. Manage. 76, 317–327.

Schofield, M.R., Barker, R.J., 2008. A unified capture-recapture framework. J. Agric. Biol. Environ. Stat. 13, 458–477.

Schofield, M.R., Barker, R.J., MacKenzie, D.I., 2009. Flexible hierarchical mark-recapture modeling for open populations using WinBUGS. Envir. Ecol. Stat. 16, 369–387.

Seber, G.A.F., 1965. A note on the multiple recapture census. Biometrika 52, 249–259.

Seber, G.A.F., 1982. The Estimation of Animal Abundance and Related Parameters. Charles Griffin & Company Ltd, London.

Shirk, P.L., Linden, D.W., Patrick, D.A., Howell, K.M., Harper, E.B., Vonesh, J.R., 2014. Impact of habitat alteration on endemic Afromontane chameleons: evidence for historical population declines using hierarchical spatial modelling. Divers. Distrib. 20, 1186–1199.

Sillett, T.S., Chandler, R.B., Royle, J.A., Kéry, M., Morrison, S.A., 2012. Hierarchical distance-sampling models to estimate population size and habitat-specific abundance of an island endemic. Ecol. Appl. 22, 1997–2006.

Sollmann, R., Gardner, B., Chandler, R.B., Royle, J.A., Sillett, T.S., 2015. An open population hierarchical distance sampling model. Ecology 96, 325–331.

Sollmann, R., Gardner, B., Williams, K., Gilbert, A., Veit, R., 2016. A hierarchical distance sampling model to estimate abundance and covariate associations of species and communities. Methods Ecol. Evol. in press.

Solymos, P., 2010. dclone: data Cloning in R. R J. 2, 29–37. http://journal.r-project.org/.

Solymos, P., Lele, S.R., Bayne, E., 2012. Conditional likelihood approach for analyzing single visit abundance survey data in the presence of zero inflation and detection error. Environmetrics 23, 197–205.

Solymos, P., Matsuoka, S.M., Bayne, E.M., Lele, S.R., Fontaine, P., Cumming, S.G., Stralberg, D., Schmiegelow, F.K.A., Song, S.J., 2013. Calibrating indices of avian density from non-standardized survey data: making the most of a messy situation. Methods Ecol. Evol. 4, 1047–1058.

Spiegelhalter, D.J., Best, N.G., Carlin, B.P., van der Linde, A., 2002. Bayesian measure of model complexity and fit. J. R. Stat. Soc. Series B 64, 583–639.

Stanley, T.R., Royle, J.A., 2005. Estimating site occupancy and abundance using indirect detection indices. J. Wildl. Manage. 69, 874–883.

Stearns, S.C., 1992. The Evolution of Life Histories. Oxford University Press, Oxford.

Stefanski, L.A., 2000. Measurement error models. J. Am. Stat. Assoc. 95, 1353–1358.

Strebel, N., Kéry, M., Schaub, M., Schmid, H., 2014. Study of phenology by flexible estimation and modeling of seasonal detectability peaks. Methods Ecol. Evol. 5, 483–490.

Stone, M., 1977. An asymptotic equivalence of choice of model by cross-validation and Akaike's criterion. Journal of the Royal Statistical Society. Series B (Methodological) 44–47.

Sturtz, S., Ligges, U., Gelman, A., 2005. R2WinBUGS: a package for running WinBUGS from R. J. Stat. Softw. 12, 1–16.

Su, Y.-S., Yajima, M., 2014. R2jags: A Package for Running Jags from R. R package version 0.04-01. http://CRAN.R-project.org/package=R2jags.

Sutherland, C., Brambilla, M., Pedrini, P., Tenan, S. A multi-region community model for inference about geographic variation in species richness. Methods Ecol. Evol. (in review).

Sutherland, C., Elston, D.A., Lambin, X., 2012. Multi-scale processes in metapopulations: contributions of stage structure, rescue effect, and correlated extinctions. Ecology 93, 2465–2473.

Sutherland, C., Elston, D.A., Lambin, X., 2013. Accounting for false positive detection error induced by transient individuals. Wildl. Res. 40, 490–498.

Sutherland, C., Elston, D.A., Lambin, X., 2014. A demographic, spatially explicit patch occupancy model of metapopulation dynamics and persistence. Ecology 95, 3149–3160.

Tanadini, L., 2010. Heterogeneity Effects in the N-Mixture Model (Unpublished Masters thesis). University of Neuchâtel, Neuchâtel, Switzerland.

Tanner, M.A., Wong, W.H., 1987. The calculation of posterior distributions by data augmentation. J. Am. Stat. Assoc. 82, 528–540.

Tavecchia, G., Besbeas, P., Coulson, T., Morgan, B.J.T., Clutton-Brock, T.H., 2009. Estimating population size and hidden demographic parameters with state-space modeling. Am. Nat. 173, 722–733.

Tenan, S., O'Hara, R.B., Hendriks, I., Tavecchia, G., 2014a. Bayesian model selection: the steepest mountain to climb. Ecol. Model. 283, 62–69.

Tenan, S., Pradel, R., Tavecchia, G., Igual, J.M., Sanz-Aguilar, A., Genovart, M., Oro, D., 2014b. Hierarchical modelling of population growth rate from individual capture–recapture data. Methods Ecol. Evol. 5, 606–614.

Thomas, A., O'Hara, B., Ligges, U., Sturtz, S., 2006. Making BUGS open. R News 6, 12–17.

Thomas, L., Buckland, S.T., Newman, K.B., Harwood, J., 2005. A unified framework for modelling wildlife population dynamics. Aust. N.Z. J. Stat. 47, 19–34.

Thomas, L., Buckland, S.T., Rexstad, E.A., Laake, J.L., Strindberg, S., Hedley, S.L., Bishop, J.R.B., Marques, T.A., Burnham, K.P., 2010. Distance software: design and analysis of distance sampling surveys for estimating population size. J. Appl. Ecol. 47, 5–14.

Tingley, M.W., Beissinger, S.R., 2013. Cryptic loss of montane avian richness and high community turnover over 100 years. Ecology 94, 598–609.

Tobler, M.W., Hartley, A.Z., Carrillo-Percastegui, S.E., Powell, G.V.N., 2015. Spatiotemporal hierarchical modelling of species richness and occupancy using camera trap data. J. Appl. Ecol. 52, 413–421.

Trolle, M., Kéry, M., 2003. Estimation of ocelot density in the Pantanal using capture-recapture analysis of camera-trapping data. J. Mammal. 84, 607–614.

Tyre, A.J., Tenhumberg, B., Field, S.A., Niejalke, D., Parris, K., Possingham, H.P., 2003. Improving precision and reducing bias in biological surveys: estimating false-negative error rates. Ecol. Appl. 13, 1790–1801.

ver Hoef, J.M., Boveng, P.L., 2007. Quasi-Poisson vs. negative binomial regression: how should we model overdispersed count data. Ecology 88, 2766–2772.

ver Hoef, J.M., Cameron, M.F., Boveng, P.L., London, J.M., Moreland, E.E., 2014. A spatial hierarchical model for abundance of three ice-associated seal species in the eastern Bering Sea. Stat. Methodol. 17, 46–66.

ver Hoef, J.M., Frost, K.J., 2003. A Bayesian hierarchical model for monitoring harbor seal changes in Prince William Sound, Alaska. Environ. Ecol. Stat. 10, 201–219.

ver Hoef, J.M., Jansen, J.K., 2007. Space-time zero-inflated count models of harbour seals. Environmetrics 18, 697–712.
ver Hoef, J.M., Boveng, P.L., 2015. Iterating on a single model is a viable alternative to multimodel inference. J. Wildlife Manag. 79, 719–729.
Waldmann, P., 2009. Easy and Flexible Bayesian Inference of Quantitative Genetic Parameters. Evolution 63, 1640–1643.
Warren, C.C., Veech, J.A., Weckerly, F.W., O'Donnell, L., Ott, J.R., 2013. Detection heterogeneity and abundance estimation in populations of Golden-cheeked warblers (*Setophaga chrysoparia*). Auk 130, 677–688.
Warton, D.I., Shepherd, L.C., 2010. Poisson point process models solve the "pseudo-absence problem" for presence-only data in ecology. Ann. Appl. Stat. 4, 1383–1402.
Webster, R.A., Pollock, K.H., Simons, T.R., 2008. Bayesian spatial modeling of data from avian point surveys. J. Agric. Biol. Environ. Stat. 13, 121–139.
Wells, K., Bohm, S.M., Boch, S., Fischer, M., Kalko, E.K.V., 2011. Local and landscape-scale forest attributes differ in their impact on bird assemblages across years in forest production landscapes. Basic Appl. Ecol. 12, 97–106.
Welsh, A.H., Lindenmayer, D.B., Donnelly, C.F., 2013. Fitting and interpreting occupancy models. PLoS ONE 8, e52015.
Wenger, S.J., Freeman, M.C., 2008. Estimating species occurrence, abundance, and detection probability using zero-inflated distributions. Ecology 89, 2953–2959.
White, G.C., 2005. Correcting wildlife counts using detection probabilities. Wildl. Res. 32, 211–216.
White, G.C., Burnham, K.P., 1999. Program MARK: survival estimation from populations of marked animals. Bird Study 46, 120–139.
White, A.M., Zipkin, E.F., Manley, P.N., Schlesinger, M.D., 2013a. Conservation of avian diversity in the Sierra Nevada: moving beyond a single-species management focus. PLoS ONE 8, e63088. http://dx.doi.org/10.1371/journal.pone.0063088.
White, A.M., Zipkin, E.F., Manley, P.N., Schlesinger, M.D., 2013b. Simulating avian species and foraging group responses to fuel reduction treatments in coniferous forests. For. Ecol. Manage. 304, 261–274.
Whittaker, R.J., Willis, K.J., Field, R., 2001. Scale and species richness: towards a general, hierarchical theory of species diversity. J. Biogeogr. 28, 453–470.
Wiegand, T., Moloney, K.A., 2014. A Handbook of Spatial Point Pattern Analysis in Ecology. Chapman and Hall/CRC Press, Boca Raton.
Williams, B.K., Nichols, J.D., Conroy, M.J., 2002. Analysis and Management of Animal Populations. Academic Press, San Diego.
Wintle, B.A., Bardos, D.C., 2006. Modeling species–habitat relationships with spatially autocorrelated observation data. Ecol. Appl. 16, 1945–1958.
Wood, S.N., 2006. Generalized Additive Models. An Introduction with R. Chapman & Hall/CRC, Boca Raton, USA.
Woodworth, G.G., 2004. Biostatistics: A Bayesian Introduction. Wiley.
Woodward, P., 2011. Bayesian Analysis Made Simple: An Excel GUI for WinBUGS. Chapman & Hall/CRC.
Wright, J.A., Barker, R.J., Schofield, M.R., Frantz, A.C., Byrom, A.E., Gleeson, D.M., 2009. Incorporating genotype uncertainty into mark-recapture-type models for estimating abundance using DNA samples. Biometrics 65, 833–840.
Wu, G., Holan, S.H., Wikle, C.K., 2013. Hierarchical bayesian spatio-temporal Conway–Maxwell Poisson models with dynamic dispersion. J. Agric. Biol. Environ. Stat. 18, 335–356.
Wu, G., Holan, S.H., Nilon, C.H., Wikle, C.K., 2015. Bayesian binomial mixture models for estimating abundance in ecological monitoring studies. Ann. Appl. Stat. 9, 1–26.
Yackulic, C.B., Reid, J., Davis, R., Hines, J.E., Nichols, J.D., Forsman, E., 2012. Neighborhood and habitat effects on vital rates: expansion of the Barred Owl in the Oregon coast ranges. Ecology 93, 1953–1966.

Yamaura, Y., 2013. Confronting imperfect detection: behavior of binomial mixture models under varying circumstances of visits, sampling sites, detectability, and abundance, in small-sample situations. Ornithol. Sci. 12, 73–88.

Yamaura, Y., Kéry, M., Royle, J.A. Study of biological communities subject to imperfect detection: bias and precision of multispecies N-mixture abundance models in small-sample situations. Ecol. Res., in press.

Yamaura, Y., Royle, J.A., Kubio, K., Tada, T., Ikeno, S., Makino, S., 2011. Modelling community dynamics based on species-level abundance models from detection/nondetection data. J. Appl. Ecol. 48, 67–75.

Yamaura, Y., Royle, J.A., Shimada, N., Asanuma, S., Sato, T., Taki, H., Makino, S., 2012. Biodiversity of man-made open habitats in an underused country: a class of multispecies abundance models for count data. Biodivers. Conserv. 21, 1365–1380.

Yang, H.C., Chao, A., 2005. Modeling animals' behavioral response by Markov chain models for capture–recapture experiments. Biometrics 61, 1010–1017.

Yoccoz, N.G., Nichols, J.D., Boulinier, T., 2001. Monitoring biological diversity in space and time. Trends Ecol. Evol. 16, 446–453.

Yoshizaki, J., Pollock, K.H., Brownie, C., Webster, R.A., 2009. Modeling misidentification errors in capture–recapture studies using photographic identification of evolving marks. Ecology 90, 3–9.

Zellweger-Fischer, J., Kéry, M., Pasinelli, G., 2011. Population trends of brown hares in Switzerland: the role of land-use and ecological compensation areas. Biol. Conserv. 144, 1364–1373.

Zipkin, E.F., DeWan, A., Royle, J.A., 2009. Impacts of forest fragmentation on species richness: a hierarchical approach to community modelling. J. Appl. Ecol. 46, 815–822.

Zipkin, E.F., Royle, J.A., Dawson, D.K., Bates, S., 2010. Multi-species occurrence models to evaluate the effects of conservation and management actions. Biol. Conserv. 143, 479–484.

Zipkin, E.F., Grant, E.H.C., Fagan, W.F., 2012. Evaluating the predictive abilities of community occupancy models using AUC while accounting for imperfect detection. Ecol. Appl. 22, 1962–1972.

Zipkin, E.F., Sillett, T.S., Grant, E.H.C., Chandler, R.B., Royle, J.A., 2014b. Inferences about population dynamics from count data using multistate models: a comparison to capture-recapture approaches. Ecol. Evol. 4, 417–426.

Zipkin, E.F., Thorson, J.T., See, K., Lynch, H.J., Grant, E.H.C., Kanno, Y., Chandler, R.B., Letcher, B.H., Royle, J.A., 2014a. Modeling structured population dynamics using data from unmarked individuals. Ecology 95, 22–29.

Zippin, C., 1956. An evaluation of the removal method of estimating animal populations. Biometrics 12, 163–189.

Zuur, A.F., Saveliev, A.A., Ieno, E.N., 2012. Zero-inflated Models and Generalized Linear Mixed Models with R. Highlands Statistics.

Author Index

'*Note:* Page numbers followed by "f" indicate figures and "t" indicate tables.'

A

Aarts, G., 220, 554–555
Abadi, F., 12–13, 628
Adams, M.J., 187, 395–396, 422–423, 444, 453, 455, 460, 474, 527–528, 605
Agresti, A., 356–357
Aing, C., 603–604, 614
Aitkin, M., 72, 342
Alldredge, M.W., 351
Almasi, B., 146
Alpizar-Jara, R., 400–401, 479, 608, 621–622, 664–665
Altman, B., 632–633
Altwegg, R., 727–728
Amrhein, V., 628
Amstrup, S.C., 14
Amundson, C.L., 220, 395–396, 464–465, 472–474, 549
Anderson, D.R., 12, 14, 69–70, 157, 213, 221, 243, 314–315, 350, 355–357, 394–396, 398–399, 404, 417–418, 459, 474–475, 577
Anderson, M.D., 121, 622
Andres, B.A., 236–237
Andrewartha, H.G., 220
Annand, E.M., 600–601
Arlettaz, R., 12–13, 548–549
Arnason, A.N., 12, 14
Arnett, E.B., 632–633
Asanuma, S., 631–632, 650, 716–717
Astruc, G., 232, 245, 259, 429–430
Augustin, N.H., 552–553

B

Baddeley, A., 110–112, 220, 469, 554–555
Bailey, L.L., 14, 46, 245, 252, 310, 369–370, 381, 472, 552–553, 558, 560, 569, 584–585, 589, 590t, 594, 600–601, 605, 615, 621–622, 626–627, 727
Baillie, J.E.M., 552, 650
Baillie, S.R., 12–13, 221, 513
Balangtaa, C., 632–633
Baldwin, R.F., 632–633
Balmford, A., 3
Banerjee, S., 115
Banks-Leite, C., 308–309
Barbaro, L., 716–717
Barbraud, C., 121, 184, 622
Bardos, D.C., 552–553, 727–728
Barker, R.J., 12–14, 53, 71–72, 162, 254, 300, 304–305, 341–342, 390–391, 628
Barlow, J., 308–309

Barnagaud, J.Y., 716–717
Barnett, A., 80, 103
Barros, C.S., 308–309
Bates, D.M., 80, 116–117, 121, 205
Bates, S., 20, 29, 46, 106, 141, 314–315, 326, 632–633
Bau, S.S., 615, 626–627
Bayne, E.M., 11–12, 106, 464, 472–473, 556–557, 559, 608
Beesley, L.S., 716–717
Begon, M., 3, 9
Beier, E., 395–396
Beissinger, S.R., 12–13, 221, 632–633, 679
Bekessy, S.A., 187, 615–616
Bell, T., 602
Belmaker, J., 601–602
Benke, H., 464–465
Berggren, P., 464–465
Berliner, L.M., 24, 31, 64, 76, 152
Berman, M., 110–112
Bertolero, A., 232, 245, 259, 429–430
Bertram, M.R., 632–633
Besbeas, P., 12–13, 108, 201–202, 548–549
Besnard, A., 232, 245, 259, 429–430
Best, N.G., 70–71, 79–80, 146, 148, 157, 184
Bettaso, J., 605
Betts, M.G., 46, 586–587, 603–604, 615, 632–633
Bibby, C.J., 254–255, 367, 552
Bidwell, J.K., 314–315, 464–465
Bierregaard, R.O. Jr., 553
Birch, L.C., 220
Bishop, J.R.B., 395–396, 459
Bispo, R., 527–528
Blakesley, J.A., 603
Blank, P.J., 632–633, 727–728
Bled, F., 552–553, 641–643, 684–685, 727–728
Block, W.M., 632–633, 727
Boch, S., 632–633
Bohm, S.M., 632–633
Bolker, B.M., 23, 80, 116–117, 205
Bomberger, B, M., 121, 622
Bonner, S.J., 12
Borchers, D.L., 12–15, 103, 220–221, 279, 358, 394–396, 398–401, 404–405, 417, 459, 464–465, 474–475, 479, 527–528, 555
Bornand, C.N., 615
Boscolo, D., 308–309
Boulinier, T., 15, 608, 664–665, 672
Boveng, P.L., 252, 263–264, 309, 395–396
Boyce, M.S., 555–556
Brashares, J., 632–633

Brauning, D.W., 472
Brereton, T., 245, 259, 429–430, 514, 657, 716–717
Brockerhoff, E., 716–717
Broms, K.M., 727–728
Brooks, M.E., 80
Brooks, S.P., 12–14, 156–157, 221, 243, 548–549
Brown, C.R., 121, 622
Brown, J.H., 632
Brownie, C., 12, 14
Buckland, S.T., 12–15, 106, 146, 148, 221, 394–396, 398–400, 404, 417, 444–445, 459, 464–465, 474–475, 513–514, 527–528, 552–553
Bueche, L., 615
Bull, E.L., 605
Burger, L.W., Jr., 106, 395–396
Burger, W., 474
Burgess, N.D., 254–255, 367, 552
Burnham, K.P., 12, 14, 30, 69–70, 157, 213, 221, 243, 252, 314–315, 350, 355–357, 394–396, 398–399, 404, 417–418, 459, 474–475, 559, 577
Burt, M.L., 395, 527
Burton, A., 632–633
Byrom, A.E., 12–14, 390–391

C

Calder, C.A., 16, 124–125, 222, 307–309
Caley, P.A., 4–6, 553, 600
Cam, E., 115, 552–553, 608, 621–622, 664–665, 668, 727–728
Camac, J.S., 615, 626–627
Cameron, M.F., 395–396
Campomizzi, A.J., 628, 727–728
Cao, Y., 632–633
Cardenas, J.S., 632–633
Carlin, B.P., 70–71, 115, 157, 342
Carlin, J.B., 104, 107–108, 112, 146, 192, 308
Carrillo-Percastegui, S.E., 716–717, 728
Carrillo-Rubio, E., 632–633
Carroll, R.J., 622, 628, 727–728
Carter, M.F., 394
Cassano, C.R., 308–309
Castaneda, P.G.D., 632–633
Castellón, T.D., 224, 306–307
Caswell, H., 12–13
Catchpole, E.A., 12–13, 124–125, 548–549
Caughley, G.J., 318
Chambert, T., 14, 558, 560, 627
Chandler, C.C., 351, 716–717, 726
Chandler, R.B., 14, 23–24, 30, 40–41, 46, 71, 73, 79–80, 103, 220–221, 225, 279, 282, 306–307, 309–311, 315, 325, 330, 342, 351, 356–359, 369–370, 391, 394–396, 399, 405, 430, 444,
447–448, 464–465, 472–473, 483, 485–486, 515–517, 526–528, 532, 549, 553, 555–556, 589, 600, 608, 631–632, 716–717, 726
Chao, A., 355
Chavez, V.J.A., 716–717, 726
Chelgren, N.D., 187, 395–396, 422–423, 444, 453, 455, 460, 474, 527–528, 605
Chen, G., 14, 554–555, 632–633
Cheylan, M., 232, 245, 259, 429–430
Chib, S., 342
Choquet, R., 124–125, 559
Clark, C.J., 80
Clark, J.D., 553, 600
Clark, J.S., 6–7, 16, 124–125, 222, 257, 307–309, 633, 689–690, 725–726
Clobert, J., 184
Clutton-Brock, T.H., 12–13
Cole, D.J., 12–13, 124–125
Cole, R.A., 605
Collazo, J.A., 351, 603–604, 614, 632–633
Collet, A., 464–465
Collier, B.A., 479, 628, 727–728
Collins, J.P., 604–606
Conn, P.B., 110, 395–396, 472, 479, 727–728
Connor, E.F., 554–555, 628, 634–635, 679, 690–691, 725–726
Conquest, L.L., 727–728
Conroy, M.J., 14, 221, 243, 300, 304–305, 350, 394, 560, 628
Converse, S.J., 14, 103, 225, 279, 319, 323–324, 334, 358, 391, 447–448, 527, 542–543, 555–556, 650
Cooch, E.G., 115, 378, 632–633, 668
Cook, R.D., 314–315, 318–319
Cooper, R.J., 600–601, 603
Cormack, R.M., 12
Coull, B.A., 356–357
Coulson, T., 12–13
Couturier, T., 232, 245, 259, 429–430
Covas, R., 121, 622
Crainiceanu, C.M., 102, 121, 622–623
Creel, S., 390–391
Cressie, N., 16, 115, 124–125, 222, 307–309
Crewe, T.L., 514
Cribari-Neto, F., 203
Cuke, M., 552, 650
Cumming, S.G., 11–12, 106, 464, 472–473, 557

D

Dail, D., 46, 307, 310–311, 513–515
Danchin, E., 115, 668
Davis, R., 727–728
Dawson, D.K., 20, 29, 46, 106, 141, 225, 279, 314–315, 326, 553, 600, 603, 632–633

de Valpine, P., 524–525
Deconchat, M., 716–717
Dénes, F.V., 221
Dennis, B., 286
Dennis, E.B., 245, 259, 429–430, 514, 657, 716–717
Denwood, M.J., 147
DeStefano, S., 351, 464–465
DeWan, A.A., 632–633, 667, 725–726
Dhondt, A.A., 632–633
Dice, L., 279, 281, 309–310
Diefenbach, D.R., 472
Dietterich, T.G., 46, 586–587
Dixon, C.E., 14–15, 464–465
Dobrow, R., 603–604, 614
Dobson, A., 80, 103
Dodd, C.K., 222
Doherty, P.F., Jr., 608, 672
Dolan, C.R., 632–633
Donnelly, C.F., 585, 626–627
Donovan, T.M., 252
Dorazio, R.M., 14, 20–21, 42, 44, 46, 71–72, 103, 108, 110–112, 220–222, 224–225, 237, 243, 248, 256, 278–281, 300–301, 304–308, 314–315, 319, 324, 338–339, 350, 356–358, 405, 417–418, 445–447, 465, 548–549, 553–561, 577, 603–604, 615, 621–622, 627–628, 631–635, 642–643, 650, 668, 679, 683–685, 690–691, 709, 712–713, 715, 715f, 724–726, 728
Dorcas, M.E., 632–633
Doyle, T.J., 46, 314–315
Droege, S., 20, 29, 46, 557–558, 584–586, 626
Dubour, A.J., 632–633
Dudik, M., 555–556
Duke, S.D., 632–633
Dunson, D.B., 104, 107–108, 112, 121, 146, 192, 308
Dupuis, J.A., 641–643, 684–685

E

Eaton, M.J., 64, 152
Edwards, D.P., 632–633
Edwards, F.A., 632–633
Edwards, H.H., 225, 308
Efford, M.G., 103, 220, 225, 279, 358, 405, 527, 553, 555, 600, 603
Efron, B., 192
Elith, J., 108, 112, 220, 552–556, 585, 626
Elkin, C., 263–264, 309
Elliott-Smith, E., 236–237
Ellis, M.M., 245, 584–585, 727
Ellis, T.M., 632–633
Ellison, A.M., 690–691, 709, 712–713
Elston, D.A., 10, 14, 17, 310, 558

Emerson, S.C., 46, 586–587
Eskew, E.A., 632–633
Evans, K.O., 106, 395–396, 474
Evenson, J.R., 236–237

F

Fagan, W.F., 632–633
Fallon, F.W., 314–315, 632–633
Fallon, J.E., 314–315, 632–633
Fancy, S.G., 148, 445
Farnsworth, G.L., 314–315, 319, 325, 369–370
Farrell, S.L., 628, 727–728
Fernandez, C., 12–13, 146, 148, 395–396, 513–514, 532–533
Ferraz, G., 552–553
Fewster, R.M., 400–401, 479, 513
Fieberg, J., 220, 554–555, 603–604, 614
Field, R., 709
Field, S.A., 29, 42, 46, 109, 553, 557–558, 584–585, 626
Fischer, M., 615, 632–633
Fisher, N.I., 110–112
Fiske, I., 23, 30, 40–41, 46, 73, 79–80, 315
Fithian, W., 9, 112, 220, 554–556
Flather, C.H., 608, 621–622, 664–665
Fletcher, R.J. Jr., 603–604, 615
Fonnesbeck, C.J., 300, 304–305, 628
Fontaine, P., 11–12, 106, 464, 472–473, 557
Forney, K.A., 395–396
Forsman, E., 727–728
Franklin, A.B., 46, 552, 560, 603–604, 627, 728
Frantz, A.C., 12–14, 390–391
Freeman, M.C., 232–233, 260, 308
Freeman, S.N., 12–13, 245, 259, 429–430, 514, 548–549, 657, 716–717
Friend, T., 615, 626–627
Frost, K.J., 307–309
Fukaya, K., 726, 728
Furnas, B.J., 632–633

G

Gaillard, J.M., 184
Galvan, S., 605
Ganey, J.L., 245, 632–633, 727
Garcia, J.O., 224, 306–307
Gardner, B., 14, 24, 29, 46, 71, 103, 221, 225, 279, 308, 310–311, 358–359, 394, 399, 405, 448, 464–465, 483, 513–517, 526–528, 532, 549, 553–555, 559, 589–591, 621–622, 632–633, 716–717, 726–728
Garrard, G.E., 187, 615–616, 626–627
Gaston, K.J., 4–6, 552
Gauthier, R., 256, 279–281, 314–315, 367

Gawne, B., 716–717
Geange, S.W., 80
Gelfand, A.E., 6–7, 54, 115, 257, 308, 631–633, 650, 689–690, 725–727
Gelman, A., 64, 73, 75, 80, 104, 107–108, 112, 115–116, 121–122, 146–147, 149, 156–157, 192, 207, 250, 308
Gendron, D., 395–396
Genovart, M., 395–396
George, E.I., 342
Gerrodette, T., 395–396
Ghosh, M., 248
Ghosh, S., 6–7, 257, 308
Giesen, K.M., 394
Gilbert, A.T., 600–601, 716–717, 726
Giljohann, K.M., 615, 626–627
Gilks, W.R., 148
Gilroy, J.J., 632–633
Gimenez, O., 12–14, 121, 184, 243, 548–549, 559, 622
Giovanini, J., 632–633
Gleeson, D.M., 12–14, 390–391
Glimskär, A., 642–643, 650, 668, 683–685, 712, 715, 715f, 724–725
Goffin, D., 14–15, 464–465
Golding, N., 633, 725–727
Gopalaswamy, A.M., 52, 552, 603–604, 614
Gordon, A., 553, 585, 590–591, 626
Gotelli, N.J., 690–691, 709, 712–713
Goula, M., 632–633
Graf, R., 272, 275
Grant, E.H.C., 14, 46, 310–311, 558, 560, 600–601, 627, 632–633, 727
Gray, B.R., 576, 586–587, 626–627
Green, R.E., 3
Gregg, K.B., 14, 367
Grigg, G.C., 318
Grimm, V., 141
Groce, J.E., 628, 727–728
Grosbois, V., 184
Gude, J.A., 14, 558, 560, 627
Guélat, J., 146, 483, 527, 532, 628, 727–728
Guillera-Arroita, G., 245, 274, 278, 305–306, 308, 552–553, 555–556, 584–586, 590–591, 608, 614–615, 621, 626–627, 667, 727
Gutierrez, R.J., 726
Guzy, J.C., 632–633
Gwinn, D.C., 716–717

H

Häfliger, G., 553, 559, 590–591
Hahs, A.K., 632–633
Halls, S., 603–604, 614

Halstead, B.J., 632–633
Hammond, P.S., 464–465
Handel, C.M., 220, 395–396, 464–465, 472–474, 549, 632–633
Hanni, D.J., 603
Hanski, I., 10, 17, 552
Hardegen, A., 110–112
Harper, E.B., 395–396, 444, 455, 460
Harper, J.L., 3, 9
Harrison, B., 615, 626–627
Hartley, A.Z., 716–717, 728
Harwood, J., 12–13, 148, 513–514
Hastie, T.J., 9, 102, 112, 121, 220, 552–556
Hastings,W.K., 55
Hatfield, J.S., 12–13
Haugaasen, T., 632–633
Hautier, Y., 602
Hayne, D.W., 319
He, F., 4–6, 552
Hector, A., 3, 552, 602
Hedley, S.L., 106, 395–396, 444, 459, 527
Heglund, P.J., 314–315
Heide-Jørgensen, M.P., 464–465
Heikkinen, J., 552–553
Heimlich, S., 464–465
Henden, J.A., 632–633, 728
Hendrichsen, D.K., 554–555
Hendriks, I., 338–339, 342
Hestbeck, J.B., 12, 14
Hiby, A.R., 464–465
Higa, M., 632–633
Hill, D.A., 254–255, 367, 552
Hill, J., 80, 115–116, 122, 146
Hines, J.E., 12, 14, 30, 46, 245, 281, 314–315, 319, 325, 472, 552–553, 558–560, 569, 584–585, 600–601, 603–604, 608, 614–615, 621–622, 626–627, 632–633, 664–665, 672, 727–728
Hobbs, N.T., 16, 20–21, 23, 108, 201–203, 213, 337–339
Högmander, H., 552–553
Holan, S.H., 201–202, 308
Holder, M., 633, 650
Holderegger, R., 14
Holt, B.G., 632–633
Holt, R.D., 632
Holtrop, A.M., 632–633
Holyoak, M., 632
Homyack, J.A., 632–633
Hooten, M.B., 16, 20–21, 23, 108, 201–203, 213, 337–339, 727–728
Hostetler, J.A., 310–311
Hottola, J., 633–635, 725–727
Howell, K.M., 395–396, 444, 455, 460

Howland, B., 527–528
Hunt, S.D., 632–633
Hurlbert, S.H., 115, 307
Husband, T.P., 600–601
Hutchinson, R.A., 46, 586–587
Hyman, O.J., 604–606

I

Ieno, E.N., 109, 222, 307–309, 623
Igual, J.M., 395–396
Ikeno, S., 298, 301, 304–305, 716–717, 726
Iknayan, K.J., 632–633
Illian, J.B., 3, 14, 220, 397, 554–555
Ims, R.A., 632–633, 728
Irizarry, J.I., 632–633
Isbell, F., 602
Ivan, J.S., 245, 584–585, 727
Ives, A.R., 679

J

Jackman, S., 109, 337
Jackson, C., 146, 148, 184
Jacobson, J.O., 314–315, 318–319
Jager, H.I., 632–633
Jakober, H., 115
Jansen, J.K., 307–309
Jarzyna, M.A., 608
Jelks, H.L., 46, 248, 314–315, 319, 324
Jenkins, M., 3
Jenni, L., 679
Jenouvrier, S., 121, 622
Jetz, W., 601–602
Joachim, J., 641–643, 684–685
Johnson, D.H., 234, 308–309, 472
Johnson, D.S., 107, 110, 252, 264, 381, 395–396, 434–435, 479, 527, 532–533, 549, 727–728
Johnson, F.A., 224, 306–307
Johnson, P.T.J., 632–633
Jolly, G.M., 12
Jones, J.E., 632–633
Jordan, F., 46, 248, 314–315, 319, 324
Joseph, L.N., 263–264, 309
Joseph, M.B., 632–633

K

Kadane, J.B., 213, 341
Kalinowski, S.T., 390–391
Kalko, E.K.V., 632–633
Kanno, Y., 310–311
Karanth, K.K., 632–633
Karanth, K.U., 281, 309–310, 603–604, 614

Karr, J.R., 621–622
Keeton, W.S., 252
Keil, P., 601–602
Keim, J.L., 555–556, 608
Keith, D.A., 112
Keller, V., 254–255, 349–350, 643
Kellner, K., 147, 149, 154, 336
Kelly, E.G., 236–237
Kendall, W.L., 12, 472, 552, 560, 627
Kéry, M., 4–6, 10, 14, 20–21, 29, 80, 84, 103, 115, 117, 121, 124–127, 146–148, 184, 195, 209, 212–213, 225, 236–237, 245, 249–250, 256, 263–264, 274, 278–281, 308–309, 314–315, 317, 319, 333, 350, 357, 367, 378, 390–391, 395–396, 430, 444, 464–465, 483, 515–516, 527, 532–533, 548–549, 552–555, 557–561, 577, 590, 602, 604–606, 608–609, 615, 618, 621–622, 626, 628, 632–633, 661, 667–668, 679, 683, 685, 688–689, 691, 716–717, 725, 727–728
Kgosi, P.M., 124–125
Kidney, D., 527
King, A.J., 716–717
King, D.I., 46, 220, 306–307, 310–311, 325, 330, 351, 369–370, 464, 472–473, 483, 485–486, 716–717, 726
King, R., 12–14, 221, 243, 474, 548–549
Knape, J., 146, 223–224, 244–245, 257, 556–557, 559, 608
Knowles, S.C.L., 52, 552
Knutson, M.G., 46, 552, 603–604, 627, 728
Koehn, J.D., 716–717
Koenen, K.K., 464–465
Koesters, N.B., 12–13, 148, 394–395, 459, 513–514
Koizumi, I., 632–633
Koneff, M.D., 314–315, 464–465
Korner-Nievergelt, F., 146, 223–224, 244–245, 257, 556–557, 559, 608
Krausman, P.R., 464–465
Krebs, C.J., 3, 220, 222
Kroll, A.J., 632–633
Krueger, L., 552, 650
Kubio, K., 298, 301, 304–305, 716–717, 726
Kujala, K., 553, 585, 626
Kumar, N.S., 281
Kuo, L., 71–72, 338–339

L

Laake, J.L., 14, 107, 110, 252, 264, 381, 394–396, 398–399, 404, 417–418, 434–435, 459, 474–475, 479, 527, 532–533, 549
Lachish, S., 52, 552

Lachman, G.B., 20, 29, 46, 557–558, 584–586, 626
Lahoz-Monfort, J.J., 274, 278, 305–306, 308, 553, 555–556, 584–585, 614–615, 621, 626–627, 727
Lambin, X., 10, 14, 17, 310, 558
Langeland, K., 632–633, 728
Langtimm, C.A., 20, 29, 46, 314–315, 557–558, 584–586, 626
Lassoie, J.P., 632–633
Latimer, A., 631–633, 650, 727
Latimer, A.M., 633, 650
Lawler, J.P., 395–396, 464–465, 542–543
Lazar, N.A., 213, 341
Le Cam, L., 162–163, 245, 585
Leathwick, J.R., 552
Lebreton, J.D., 12, 14, 559
Lee, Y., 80, 121–122, 252, 256, 263–264
Lehmkuhl, J.F., 632–633
Leibold, M.A., 632
Lele, S.R., 11–12, 106, 124–125, 286, 464, 472–473, 555–557, 559, 608
Lenormand, T., 121, 622
Lentini, P.E., 553, 585, 590–591, 626
Leopold, M.F., 464–465
Lepage, D., 514
Letcher, B.H., 310–311
Leu, M., 605
Leukering, T., 394
Lewis, P.O., 633, 650
Lewis, T.L., 632–633
Ligges, U., 147, 149
Lindberg, M.S., 632–633
Linden, D.W., 395–396, 444, 455, 460, 632–633
Lindenmayer, D.B., 585, 626–627
Lindley, S.T., 12–13, 146
Lindstrom, A., 632–633
Link, W.A., 4–6, 12–14, 46, 53, 64, 71–72, 115, 152, 162, 238, 254, 278, 310, 319, 341–342, 358, 417, 445–448, 465, 548–549, 555–556, 558, 560, 576, 627, 650, 657, 668, 679, 681–683, 685, 690–691, 726
Linkie, M., 305–306, 308
Littell, R.C., 80, 114, 117
Little, R.J.A., 169
Liu, J.S., 683
London, J.M., 395–396
Louis, T.A., 115
Lovejoy, T.E., 553
Luder, R., 272, 275
Lukacs, P.M., 14, 603

Lunn, D.J., 79–80, 146, 148, 184
Lupton, J., 632–633
Lynch, H.J., 107–108, 201–202, 310–311
Lyons, J.E., 236–237

M

Ma, K., 14, 554–555, 632–633
MacCluskie, M.C., 395–396, 464–465, 542–543
MacEachern, S.N., 64, 152
Mächler, M., 116–117, 205
MacKenzie, D.I., 12–14, 20, 29, 46, 225, 245, 252, 308, 381, 552–553, 557–558, 560, 569, 584–586, 589, 590t, 594, 603–604, 614–615, 621–622, 626–627, 726–728
MacKenzie, M.L., 106, 395–396, 474
MacNeil, M.A., 632–633
Madsen, J., 14
Madsen, L., 46, 307, 310–311, 513–515
Magnusson, W.E., 318
Makino, S., 298, 301, 304–305, 631–632, 650, 716–717, 726
Malecki, R.A., 12, 14
Mallick, B., 71–72, 338–339
Manley, P.N., 632–633
Manly, B.F.J., 14
Marques, F.F.C., 14–15, 464–465
Marques, T.A., 148, 395–396, 445, 459, 527–528
Marsh, H., 306–307
Marshall, M.R., 472
Martin, J., 224–225, 306–308, 552
Martin, T.G., 263–264, 309
Masters, P., 115, 679–680
Mata, L., 632–633
Matechou, E., 514
Mateo-Tomas, P., 274
Mathewson, H.A., 628, 727–728
Matsuoka, S.M., 11–12, 106, 464, 472–473, 557
Matthiopoulos, J., 220, 554–555
Mattice, J.A., 472
Mattsson, B.J., 600–601, 603, 632–633, 727–728
Maurer, B.A., 632
Mayle, B.A., 14–15, 464–465
Mazerolle, M.J., 250, 252, 380–381, 589
McCarthy, M.A., 20–21, 115, 146–147, 187, 552–553, 585, 590–591, 615–616, 626–627, 633, 679–680, 725–727
McClintock, B.T., 14, 46, 310, 552, 558, 560, 627, 727
McCrea, R.S., 14, 221
McCreary, B., 187, 395–396, 422–423, 444, 453, 455, 460, 474, 527–528, 605
McCullagh, P., 79–80, 101–103, 107, 260, 308–309, 339, 589

McCulloch, C.E., 80, 342
McDonald, R.A., 527–528
McDonald, T.L., 14
McIntyre, C.L., 552
McKann, P.C., 576, 586–587, 626–627
McKenny, H.C., 252
McManamay, R.A., 632–633
McNew, L.B., 632–633
Mead, R., 83, 85, 96, 114
Meng, X.-L., 64, 73, 75, 192, 250
Mihaljevic, J.R., 632–633
Millar, R.B., 71
Miller, A., 342
Miller, D.A.W., 14, 46, 310, 558, 560, 627, 727
Miller, D.L., 395, 527, 560–561
Milliken, G.A., 80, 114, 117
Milne, R.K., 110–112
Milner, R.L., 236–237
Mitchell, M.S., 14, 558, 560, 627
Möller, A.P., 184
Moloney, K.A., 3, 220, 397, 554–555
Monnat, J.Y., 115, 668
Monnerat, C., 553, 559, 590–591
Moore, J.L., 552, 615, 626–627
Mordecai, R.S., 560, 577, 600–601, 603
Moreland, E.E., 395–396
Moreno, M., 556–557, 559, 608
Morgan, B.J.T., 12–14, 121, 124–125, 221, 243, 245, 259, 305–306, 308, 429–430, 548–549, 584–586, 622, 626–627, 657, 667, 716–717, 727
Morreale, S.J., 632–633
Morris, W.K., 615, 626–627, 633, 725–727
Morrison, M.L., 479, 628, 727–728
Morrison, S.A., 395–396, 430, 444, 483, 515–516
Mowat, G., 103
Mugglestone, M.A., 552–553
Mukherjee, B., 248
Murtaugh, P.A., 679, 681–682
Mustoe, S., 254–255, 367, 552

N

Naef-Daenzer, B., 272, 275
Nelder, J.A., 79–80, 101–103, 107, 121–122, 252, 256, 260, 263–264, 308–309, 339, 589, 602
Newman, K.B., 12–13, 146, 148, 394–395, 459, 513–514
Newnam, J.C., 628, 727–728
Nichols, J.D., 4–6, 12, 14–15, 20, 29, 46, 221, 224, 243, 245, 281, 300–301, 304–307, 309–310, 314–315, 318–319, 325, 350, 369–370, 394, 447–448, 472, 552–553, 555–558, 560–561, 569, 584–586, 600–601, 603–604, 608, 614–615, 621–622, 626–627, 632–633, 664–665, 672, 726–728

Niejalke, D., 29, 42, 46, 109, 553, 557–558, 584–585, 626
Nielsen, D.L., 716–717
Niemi, A., 395–396, 532–533
Nilon, C.H., 308
Norris, J.L., III, 356–357
Ntzoufras, I., 71–72, 80, 85, 87–88, 103, 146–147
Nysewander, D.R., 236–237

O

O'Brien, T.G., 552, 650
O'Bryan, C.J., 632–633
O'Connell, A.F., 600–601
O'Donnell, L., 225, 310
O'Hara, R.B., 71–72, 147, 213, 338–339, 342, 633, 725–727
Oedekoven, C.S., 106, 395–396, 474
Øien, N., 464–465
Oken, K., 603–604, 614
Olea, P.P., 274
Ono, S., 632–633
Oro, D., 395–396
Orth, D.J., 632–633
Otis, D.L., 14, 221, 243, 314–315, 350, 355–357, 577
Ott, J.R., 225, 310
Otto, M.C., 314–315, 464–465
Ovaskainen, O., 633–635, 725–727

P

Pacifici, K., 632–633
Papaix, J., 716–717
Pardini, R., 308–309
Pardo, M.A., 395–396
Pardoe, I., 207
Parris, K.M., 29, 42, 46, 109, 553, 557–558, 584–585, 615, 626–627, 633, 725–727
Pasinelli, G., 249–250
Patrick, D.A., 395–396, 444, 455, 460
Pavlacky, D.C., 603
Pawitan, Y., 80
Peace, A.J., 14–15, 464–465
Pearce, J.L., 555–556
Pearl, C.A., 605
Penteriani, V., 310
Penttinen, A., 3, 14, 220, 397, 554–555
Perry, R.W., 632–633
Phillips, S.J., 220, 554–556
Pinheiro, J.C., 80, 121
Plattner, M., 14, 554–555, 632–633, 725, 728
Plummer, M., 79–80, 146–147
Podruzny, K.M., 14, 558, 560, 627
Pollock, K.H., 12, 14, 274, 314–315, 319, 325, 351, 356–357, 369–370, 400–401, 472, 479, 509, 553, 558, 560, 569, 577, 608, 615, 621–622, 626, 664–665

Pollock, L.J., 633, 725−727
Pond, B.A., 727−728
Popovic, G., 220, 554−555
Porter, W.F., 608
Possingham, H.P., 29, 42, 46, 109, 263−264, 309, 553, 557−558, 584−585, 626
Potts, J.M., 108
Poulsen, J.R., 80
Powell, G.V.N., 716−717, 728
Pradel, R., 12, 184, 395−396, 559
Prescott, G.W., 632−633
Price, A., 716−717
Price, S.J., 632−633
Purvis, A., 3, 552
Püttker, T., 308−309

Q

Qian, S.S., 602

R

Rachowicz, L.J., 605
Rahbek, C., 632−633
Railsback, S.F., 141
Ramsey, D.S.L., 4−6, 553, 600
Rattenbury, K.L., 395−396, 464−465, 542−543
Raudales, R., 716−717, 726
Ray, J.C., 727−728
Rebelo, A.G., 631−633, 650, 727
Reboulet, A.M., 559
Reich, P.B., 602
Reid, J., 727−728
Reinitz, D., 605
Ren, Y., 632−633
Renner, I.W., 220, 554−556
Rexstad, E.A., 395−396, 459, 527
Rich, L.N., 14, 558, 560, 627
Riddle, J.D., 560, 577
Ridout, M.S., 108, 201−202, 245, 259, 305−306, 308, 429−430, 584−586, 626−627, 657, 667, 716−717, 727
Rioja-Nieto, R., 632−633
Robley, A., 4−6, 553, 600
Robson, D.S., 12
Rodriguez, D.T., 110, 553−557
Rojas-Murcia, L.E., 632−633
Roloff, G.J., 632−633
Rosenstock, S.S., 394
Rota, C.T., 603−604, 615
Roth, T., 146, 628
Rout, T.M., 552
Roy, D.B., 245, 259, 429−430, 657, 716−717

Royle, J.A., 4−6, 10, 12−14, 20−21, 24, 29, 42, 44, 46, 71−72, 103, 106, 110−111, 135, 141, 169, 220−227, 232−234, 236−237, 243, 245, 249−250, 256, 263−264, 278−282, 298, 300−301, 304−311, 314−315, 317, 319, 323−326, 330, 333−334, 338−339, 350, 356−359, 367, 369−370, 390−391, 394−396, 399, 405, 417−418, 430, 444−448, 464−465, 472−474, 483, 485−486, 513−517, 526−528, 532−533, 542−543, 548−549, 552−553, 555−561, 569, 577, 584−586, 589−590, 600, 603−604, 608−609, 614−615, 621−622, 626−627, 631−633, 642−643, 650, 667−668, 672, 679, 683−685, 688−689, 691, 712, 715−717, 715f, 724−728
Rubin, D.B., 64, 104, 107−108, 112, 121, 146, 156−157, 169, 192, 308
Rue, H., 554−555
Ruiz-Gutierrez, V., 632−633
Rumpff, L., 615, 626−627
Runge, J.P., 300, 304−305, 628
Ruppert, D., 102, 622−623
Russell, J.C., 632−633
Russell, R.E., 632−633

S

Saab, V.A., 632−633
Sadoti, G., 608
Sam, M., 632−633
Samba Kumar, N., 603−604, 614
Samora, B., 187, 395−396, 422−423, 444, 453, 455, 460, 474, 527−528
Sanchez, A., 632−633
Sanderlin, J.S., 245, 632−633, 727
Sanz-Aguilar, A., 395−396
Sato, T., 631−632, 650, 716−717
Sauer, J.R., 238, 314−315, 319, 325, 369−370, 576, 608, 621−622, 632−633, 657, 664−665, 681−682, 727−728
Saveliev, A.A., 109, 222, 307−309, 623
Schabenberger, O., 80, 114, 117
Schaub, M., 12−14, 20−21, 80, 103, 115, 121, 124−127, 148, 184, 195, 209, 212−213, 350, 367, 378, 464−465, 548−549, 553, 558−559, 577, 590−591, 618, 626, 628, 668, 679, 685
Schlesinger, M.D., 632−633
Schmid, B., 602
Schmid, H., 121, 237, 254−256, 263−264, 272, 275, 278−281, 309, 314−315, 317, 349−350, 367, 553, 559, 590−591, 626, 643
Schmidt, A.E., 631−633, 650, 727
Schmidt, B.R., 604−606
Schmidt, J.H., 395−396, 464−465, 542−543
Schmiegelow, F.K.A., 11−12, 106, 464, 472−473, 557
Schmutz, J.A., 632−633

Schofield, M.R., 12–14, 300, 304–305, 390–391, 628
Schuhmacher, D., 110–112
Schwartz, M.K., 245, 584–585, 727
Schwarz, C.J., 12
Seamans, M.E., 726
Searle, S.R., 80
Seber, G.A.F., 12, 14, 221
See, K., 310–311
Senzaki, M., 632–633
Shah, R., 110–112
Sheldon, B.C., 52, 552
Shelton, A.O., 107–108, 201–202
Shen, Z., 602
Shepherd, L.C., 554–555
Shimada, N., 631–632, 650, 716–717
Shirk, P.L., 395–396, 444, 455, 460
Shriner, S.A., 351
Sierro, A., 12–13, 548–549
Siitonen, J., 633–635, 725–727
Silander, J.A. Jr., 631–633, 650, 727
Sillanpää, M.J., 71–72, 213, 338–339
Sillett, T.S., 46, 310–311, 395–396, 430, 444, 464, 483, 513–517, 526, 549
Silveira, L.F., 221
Simons, T.R., 274, 314–315, 319, 325, 351, 369–370, 472, 560, 577, 727
Sinclair, D.F., 306–307
Siriwardena, G.M., 513
Smith, A.F.M., 54
Smith, H.G., 632–633
Snelgrove, R.T., 628, 727–728
Söderström, B., 642–643, 650, 668, 683–685, 712, 715, 715f, 724–725
Soininen, J., 633–635, 725–726
Soldaat, L., 237, 308
Sollmann, R., 14, 24, 46, 71, 103, 221, 225, 279, 310–311, 358–359, 394, 399, 405, 448, 464–465, 513–517, 526–528, 532, 549, 589, 716–717, 726
Solymos, P., 11–12, 106, 147, 464, 472–473, 557, 559
Song, S.J., 11–12, 106, 464, 472–473, 557
Sørbye, S.H., 554–555
Sorci, G., 608, 672
Spiegelhalter, D.J., 70–71, 79–80, 146, 148, 157, 184
Spillmann, J.H., 14
Stanley, T.R., 305
Stauber, W., 115
Stearns, S.C., 12
Stefanski, L.A., 15
Stern, H.S., 64, 73, 75, 104, 107–108, 112, 121, 146, 192, 250, 308
Stevens, M.H.H., 80

Stevenson, B.C., 527
Stith, B.M., 46, 314–315
Stjernman, M., 632–633
Stoeckle, T., 29, 559, 621–622
Stouffer, P.C., 553
Stoyan, D., 3, 14, 397, 554–555
Stoyan, H., 3, 14, 220, 397, 554–555
Stralberg, D., 11–12, 106, 464, 472–473, 557
Strebel, N., 121, 626
Strindberg, S., 395–396, 459
Stroup, W.W., 80, 114, 117
Sturtz, S., 147, 149
Su, Y.-S., 147
Sullivan, P.J., 632–633
Sutherland, C., 10, 14, 17, 310, 558

T

Tada, T., 298, 301, 304–305, 716–717, 726
Taki, H., 631–632, 650, 716–717
Talancy, N.W., 600–601
Tanadini, L., 248, 290
Tanner, M.A., 54, 650, 683
Taper, M.L., 390–391
Tavecchia, G., 12–13, 338–339, 342, 395–396
Taylor, P.D., 514
Tenan, S., 338–339, 342, 395–396
Tenhumberg, B., 29, 42, 46, 109, 553, 557–558, 584–585, 626
Thill, R.E., 632–633
Thogmartin, W.E., 576, 586–587, 626–627
Thomas, A., 79–80, 146–148, 184
Thomas, L., 12–14, 146, 148, 221, 394–396, 398–399, 404, 417, 445, 459, 472, 474–475, 513–514, 527, 560–561
Thomas, S.M., 236–237
Thornton, J.E., 632–633
Thorson, J.T., 107–108, 201–202, 310–311
Tibshirani, R.J., 102, 121, 192, 552–553
Tingley, M.W., 632–633, 679
Tingley, R., 553, 585, 590–591, 626, 633, 725–727
Tobler, M.W., 716–717, 728
Tosh, D., 527–528
Townsend, C.R., 3, 9
Toyan, D., 220
Trolle, M., 281
Trubey, R., 716–717, 726
Truong, C., 14
Tucker, J.M., 245, 584–585, 727
Turner, R., 110–112, 469
Tyre, A.J., 29, 42, 46, 109, 553, 557–558, 584–585, 626
Tzilkowski, C.J., 600–601, 603

U

Ullrich, B., 12–13
Unitt, P., 601–602
Uribe, C.A.M., 632–633
Ursenbacher, S., 604–606

V

Valente, J.V., 46, 586–587
van der Linde, A., 70–71, 157
van Ruijven, J., 602
van Strien, A., 237, 308
Veech, J.A., 225, 310
Vehtari, A., 104, 107–108, 112, 121, 146, 192, 308
Veit, R., 716–717, 726
Ver Hoef, J.M., 16, 107, 124–125, 222, 252, 263–264, 307–309, 381, 395–396, 434–435, 527, 532–533, 549
Vesk, P.A., 615, 626–627, 633, 725–727
Viallefont, A., 124–125
Volet, B., 553, 559, 590–591
von Felten, S., 146
Vonesh, J.R., 395–396, 444, 455, 460

W

Walker, S.C., 116–117, 205
Wand, M.P., 102, 622–623
Warren, C.C., 225, 310
Warton, D.I., 220, 554–556
Weber, D., 29, 559, 621–622
Webster, R.A., 14, 274, 314–315, 319, 325, 351
Weckerly, F.W., 225, 310
Weimerskirch, H., 184
Weir, L., 14, 46, 310, 558, 560, 627, 727
Wells, K., 632–633
Welsh, A.H., 585, 626–627
Wenger, S.J., 232–233, 260, 308
Wheeler, C., 632–633
White, A.M., 632–633
White, D., 632–633
White, G.C., 14, 30, 221, 243, 252, 314–315, 323–324, 350, 355–357, 378, 559, 577, 603
White, J.S., 80
Whittaker, R.J., 709
Wiegand, T., 3, 220, 397, 554–555
Wigley, T.B., 632–633
Wikle, C.K., 16, 115, 124–125, 201–202, 222, 307–309
Wilkins, R.N., 479, 628, 727–728
Williams, B.K., 14, 221, 243, 350, 394, 560

Williams, K., 716–717, 726
Williams, N.S.G., 187, 615
Willis, K.J., 709
Wilson, A.M., 601–602
Wilson, J.D., 513
Wintle, B.A., 187, 552–553, 555–556, 585, 590–591, 615–616, 626–627, 727–728
Wolfinger, R.D., 80, 114, 117
Wong, W.H., 54, 650, 683
Wood, S.N., 102, 121
Woodall, C.W., 633, 689–690, 725–726
Woodcock, P., 632–633
Woodward, P., 147
Woodworth, G.G., 147
Wortham, J.S., 314–315, 464–465
Wright, J.A., 12–14, 390–391
Wu, G., 201–202, 308
Wu, S., 631–633, 650, 727
Wu, Y.N., 683

Y

Yabuhara, Y., 632–633
Yackulic, C.B., 447–448, 555–556, 608, 632–633, 727–728
Yajima, M., 147
Yamaura, Y., 245, 298, 301, 304–305, 631–633, 650, 716–717, 726–727
Yang, H.C., 355
Yoccoz, N.G., 15, 632–633, 728
Yoshizaki, J., 14
Young, K.G., 220, 279, 358, 527, 555
Yue, B., 615, 626–627

Z

Zbinden, N., 254–255, 272, 275, 349–350, 553, 559, 590–591, 643
Zeileis, A., 203
Zellweger-Fischer, J., 249–250
Zhang, J., 14, 248
Zhu, J., 679
Zhu, K., 6–7, 257, 308, 633, 689–690, 725–726
Zipkin, E.F., 310–311, 326, 632–633, 667, 725–728
Zippin, C., 319
Zucchini, W., 14, 221, 400–401, 479
Zuckerberg, B., 608
Zuiderwijk, A., 237, 308
Zuur, A.F., 109, 222, 307–309, 623

Subject Index

'*Note*: Page numbers followed by "f" indicate figures and "t" indicate tables.'

A

abstraction, 9, 561
abundance mapping, 254–255, 255f, 277–278
abundance trajectory, 524–525
abundance, 3–9, 220–222, 254–278. *See also* hierarchical distance sampling (HDS)
abundance-basedcommunity model, 716–717
abuse, 29–30, 709
accelerated failure time, 618
acceptance, 55–57, 60, 68
accessible style, 139
activity center, 31, 282
acyclic graphs, 148
adaptive phase (in MCMC), 147, 167
additive model, 102
advanced hierarchical distance sampling, 464
agent-based model, 141
aggregate of point pattern, 6
AHM. *See* applied hierarchical modeling (*AHM*)
AIC. *See* Akaike's information criterion (AIC)
`AICcmodavg`(R package), 250, 264–265, 589, 732
Akaike's information criterion (AIC), 69–70, 495
alder flycatcher (*Empidonaxalnorum*), 351
algebra, 80, 84, 104, 228t, 558, 563, 564t, 571–572, 718
aliased, 96–97
aliasing, 96
alpha diversity, 642–643, 700f, 708
amphibian, 10, 606
analysis of covariance (ANCOVA), 84, 148, 177–183
analysis of results, 264–277
analysis of variance (ANOVA), 602
ANCOVA. *See* analysis of covariance (ANCOVA)
annual site-covariate. *See* yearly site-covariate
annulus, annuli, 409
ANOVA. *See* analysis of variance (ANOVA)
apparent recruitment, 123, 513–515, 734–735
apparent survival, 123, 513–515, 734–735
applied hierarchical modeling (*AHM*), 17, 241, 729–730
approximation, 34–35, 415, 460, 509–510, 540
Aragon, 80–81, 81f, 85, 98
areal-summary, 14
ASE. *See* asymptotic standard error (ASE)
assessment of model fit, 72–76, 496
assumption violation, 248–250
asymptotic variance-covariance matric
asymptotic standard error (ASE), 38
attenuation, 249
audience, x, xxi, 729

augmentation. *See* data augmentation
autocorrelation, 64–65, 156–157, 278, 603–604, 614, 727–728
autologistic formulation of spatial autocorrelation, 727–728, 735
autoregressive, 727–728, 735
availability, 325, 330, 369–370, 429, 472–473, 478, 484f, 485, 487, 494, 496, 498–499, 498f, 513–514, 518, 603, 604f, 611–613
availability model, 325, 369–370, 481
availability bias, 472

B

`backTransform`(R function in package `unmarked`), 496–497
back-transform, 167, 355
backwards elimination, 258–259
balanced design, 559
Bates, Scott, 326
Bayes' rule, 50–51, 569
Bayesian analysis of line transect data, 418–420
Bayesian conventional distance sampling, 417–426
Bayesian inference, 49–54. *See also* classical inference
Bayesian model averaging, 71–72
Bayesian modeling software, 146
Bayesian p-value, 75–76, 195, 337–338
BBS. *See* breeding bird survey (BBS)
beauty (of hierarchical modeling), 24
behavioral response model, 355–356
behavioral response, 355–356, 560, 580
Bernoulli distribution, 6, 13, 16, 36, 109, 308, 557
Bernoulli GLM, 109–113
Bernoulli trial, 315
Bernoulli-Bernoulli (mixture model), 208, 621, 716–717
Bernoulli-normal (mixture model), 208
Bernoulli-Poisson (mixture model), 300–305
best unbiased predictor (BUP), 234, 439, 538–539
beta diversity, 642–643, 709
beta-binomial, 54, 107–108, 225
between-chain variance, 64
bias, 33, 245–247, 584–587
bin, binned, 415–416
binary response, 109–110
binary variable, 13, 340
binned data, 403–408
binned distance data, 405, 412, 428, 515
binomial distribution, 30, 107, 112, 222, 315, 543
binomial GLM, 198–201. *See also* Poisson GLM

771

binomial GLMM, 208—212
binomial integrated likelihood, 405
binomial N-mixture model, 221—225, 729—730
binomial point process, 534
binomial/Poisson (mixture model), 223, 300, 302, 305t, 657
biodiversity, 12, 729
biogeography, 725, 729
bivariate random effects, 667—672
bivariate, 62, 117, 270f
black box, 123
block structure, 602
blue tit, 257, 303
blue-eyed hooktail(*Onychogomphus uncatus*), 80, 81f
book web site, xxi
bootstrap distribution, 40—41, 73, 75, 570f
bootstrap, 40—41, 434—435, 564, 571
bootstrapping, 40—42
borrow strength, 115
bottom-up, 602—603
boundary estimate, 576, 586—587
bounded count, 112—113
breeding bird survey (BBS), 222, 254—255, 255f, 258f, 608, 643—646
Brooks-Gelman-Rubin, 64, 168
BRugs(R package), 147
BUGS(collectively used for the BUGS language and the BUGS engines WinBUGS, OpenBUGS and JAGS), 145—149
bugs(R function in package R2WinBUGS), 71, 147, 149, 152
BUGS language, 85—88, 105, 112, 146, 148, 150, 208, 214, 447—448, 526, 564t, 672, 732—733
BUGS software, 146—149
BUP. *See* best unbiased predictor (BUP)
"burn-in" period, 64, 167

C

Cairngorms, 591f
calibrate derived model parameters, 123
California, 430, 432f, 443f
canonical hierarchical model, 29—30
canonical link function, 103
capture history, 319
capture-mark-recapture model, 252
capture-recapture model, 44, 319, 353. *See also* spatially stratified capture-recapture models
capture-recapture/distance-sampling, 479
CAR. *See* conditional autoregressive (CAR)
Catalonia, 80—81, 82f, 94—95, 97—98
categorical covariate, 448, 472—473, 476, 478—479, 481
categorical distribution, 315
Catoctin Mountain National Park, 326, 327f
CDS. *See* conventional distance sampling (CDS)

censor, 272, 288
censoring, 421, 617
centering, 88
chameleon, 395—396
Chandler, Richard, 46, 347—348
Chandler's flycatcher data, 351—355
Channel Islands, California, 432f
c-hat(overdispersion correction factor, lack-of-fit ratio), 253, 496—497
c-hat, 252, 277, 292t, 434—435
Chihuahuan musk oxen, 401, 403
chi-square, 192—194, 194f, 196, 197f, 251—253, 261, 589
Cholesky decomposition, 534
CI. *See* confidence interval (CI)
circular, 281, 408, 532
circularity, 310
citizen-science, 554—555, 727
CJS. *See* Cormack-Jolly-Seber model (CJS)
classical inference, 20—21, 31—49. *See also* Bayesian inference
cloglog, 110—111
cloglog-Bernoulli, 142
closed model, 12—13, 618
closed population, 279, 309—310, 330, 428, 464
closure assumption, 125—126, 224
closure, 29—30
cluster size (imperfect observation of), 471—472
cluster size (in distance sampling), 464, 471—472
cluster, 464—472
coal tit, 257, 304
colext(R function in package unmarked), 49t
colonization, 6
colonization/extinction, 10, 728
combined protocol models, 549
combined protocol, 479, 549
commonsense goodness-of-fit (CSGoF), 718
community abundance model, 632—633
community assembly, 640
community distance sampling, 717
community model with covariates, 690—708
community model, 651—690
community N-mixture model, 632, 650—651, 716—724
community occupancy model, 662—679, 682—708
community occupancy parameter, 684
community of species, 621—622
community, 631—632
community-level response, 636f—637f
compartmentalization, 16
competition (among species), 725—726
complementary log-log (link function), 110
components of hierarchical models, 79—80
computer, 205, 694

conditional distribution, 26
conditional estimator, 400
conditional likelihood for binned data, 405
conditional likelihood, 400
conditional multinomial (3-part) model, 334–346, 395–396, 452–455
conditional multinomial formulation of multinomial N-mixture model, 455–457
conditional probability distributions, 24
conditional/full likelihood, 417
conditional autoregressive (CAR), 727–728, 735
conditionally-dependent binomial distributions, 320
confidence interval (CI), 161, 171–172
confound, 401, 611–613
consistency, 34
continuous covariate, 84–92, 99–102
continuous/binned distances, 417
continuous-space data, 406
continuous-time, 615
contrast, 85
conventional distance sampling (CDS), 394, 396–417
convergence(of Markov chains), 63–65
convergence, 63–65, 152–153
Conway-Maxwell-Poisson distribution, 308
co-occurrence pattern, 689–690
cookbook recipes, 729
cookbook, 729
Cormack-Jolly-Seber model (CJS), 12, 124–125, 184, 734–735
correlation, 64–65, 657, 668, 726
counts of unmarked individuals, 221
covariable, 15
covariance, 236, 668
covariate(in linear models), 83–102
coverage, 256, 266
crazy, 165f, 240f, 280f
credible interval (CRI), 157
crested tit, 257, 303
CRI. *See* credible interval (CRI)
cross-validation, 69–70
crPiFun(R function in package unmarked), 350, 355
crPiFun pi function, 350, 370
crPiFun.Mb(R function in package unmarked), 355–356
cryptofrequentist, 52
CSGoF. *See* commonsense goodness-of-fit (CSGoF)
custom multinomial models in unmarked, 346–349
custom multinomial observation models, 346–349

D

DA. *See* data augmentation (DA)
DAG. *See* Directed acyclic graphs (DAG)

Dail-Madsen model (DM model), 464
data along transects, 614–621
data augmentation (DA), 358–362, 445–452, 683–690
data augmentation formulation of multinomial N-mixture models, 358
data augmentation variable, 358–359, 417, 683, 695, 712, 718
data simulation, 123–140, 577–581
data.fn(R function in package AHM), 137
data-simulation function for static occupancy models, 577–581
David Parish, 617–618
Deanna Dawson, 326
declarative language, 148
degree-of-belief, 162–163
delta approximation (= delta rule), 34–35
delta rule, 34–35
demographic modeling, 731
density dependence, 515
density, 6, 9
density-dependent detection, 225, 242–243, 310
denstrip(R function), 157, 158f
dependent double observer sampling, 318–319, 474
dependent multiple observer protocol, 318–319
derivation of binomial N-mixture model, 222–225
derivation of site-occupancy model, 557–561
design matrix, 88–90, 92
design-based, 394
design-unbiased, 526
detectability, 395–396, 472–473, 651
detected/non-detected individuals, 536f
detection error, 261
detection function (distance sampling), 448
detection function, 399f, 409, 489–491
detection probability, 13, 16, 126–127, 131, 400–401
detection, 243, 249
detection/nondetection data (also "presence/absence data"), 11–12, 110–112, 142–143, 198, 553–557, 601–603, 726
detection/nondetection, 11–13, 208, 385f
detection-corrected species, 725
detection-function, 428
deterministic, 4, 190, 533, 553, 609
deviance information criterion (DIC), 70–71
diagnostics, 195, 213, 250, 262f
DIC. *See* deviance information criterion (DIC)
diffuse prior, 53
Directed acyclic graphs (DAG), 148
Dirichlet distribution, 103
dirty data, 554–555
disaggregate data, 220, 590t
disassemble, 4
discrepancy measure, 75, 192

discrete-time, 621
discrete-value, 214
discretization, 4–6, 555
dissimilarity, 192, 709
DISTANCE(software program), 476
distance band, 403–404, 407, 423
distance break, 427
distance sampling with clusters, 464–472
distance sampling with location of encounter, 528–532
distance sampling with locations, 432f, 528–532
distance sampling, 394, 396, 464, 485. *See also* hierarchical distance sampling (HDS); spatial distance sampling
distance-based detection model, 473–474
distance sampling, 394–461, 464–550
distribution mapping, 590–600
distribution modeling, 590–600
distribution, 3–4, 16
distsamp(R function in package unmarked), 395, 428–429
DNA, 606
DO. *See* double observer (DO)
Dorazio-Royle community model, 631–632
Dorazio-Royle community occupancy model, 682–708
Dorazio-Royle model (DR model), 685, 690–691, 701f
Dorazio-Royle-Yamaura model (DRY model), 650, 716–724
double observer (DO), 474
double observer distance sampling, 479–483
double observer sampling, 464
double-observer sampling, 321, 350
doublePiFun(R function in package AHM), 347
downscale, 421
DR model. *See* Dorazio-Royle model (DR model)
dragonfly, 84–86, 89–90, 106, 109–110, 115
DRY model. *See* Dorazio-Royle-Yamaura model (DRY model)
Dspat(R package), 395–396
Dutch bird data, 486
Dutch Centre for Field Ornithology (Sovon), 486
Dutch grassland birds, 487f, 496
dynamics, 510–526

E

e2dist(R function in AHM and SCR packages), 534, 537, 545
ecological process, 28, 128–130, 633–634
ecology, 13–16, 128–130
eDNA. *See* environmental DNA (eDNA)
effective sample area, 279–282
effective sample size, 64–65
effective strip half-width, 398–399
effects-parameterization, 85, 87, 92, 105
e-mail list, xx, xxi

encounter probability, 13, 314, 317, 346–347, 355, 375, 490–491, 540
encounter rate variance, 417
endangered species, 3
environmental DNA (eDNA), 606
error propagation. *See* propagation of uncertainty
estimationist models, 734
estimator, 33
European jay (*Garrulus glandarius*), 367f
European Jay, total population size in Switzerland, 386–387
European red squirrel (*Sciurus vulgaris*), 590–591, 591f
evolution
excess-Poisson, 346f
excess-variation, 345
exchangeability
exchangeable, 114, 117
exegesis, 15
exercise in hierarchical modeling, 222–225, 557–561
exhaustive sampling, 609
expectation, 130, 190
experimental treatment, 542–543, 602
experimentation, 478
explicit hierarchical model, 731
exponential distribution, 102–103
exponential family, 103
exponential population model, exponential growth model, 395, 515, 525f
exponential observation model, 614–621
exponential time-to-detection model (exponential TTD model), 615–617
exponential TTD model. *See* exponential time-to-detection model (exponential TTD model)
extinction, 10, 123, 238
extra residual, 344
extra-binomial, 282–283, 296
extra information
extra-Poisson variation, 297, 333, 375
extra-Poisson, 107, 250, 290, 356–357
extrapolate, 292
extrapolation, 293, 712
extra-uncertainty, 277

F

factor(in linear models), 84–99
factor level, 82, 86, 93t, 571
factor, 80–81
Falco peregrinus, 22f
false–negative errors, 13
false-negative detection error, 14
false-negative error, 135, 141, 220–221, 558, 590–591
false-negative measurement error, 551–552, 554–555
false–positive errors, 14

false-positive detection error, 15, 135, 310
false-positive error, 14–15, 141, 220–221, 224, 310, 554–555, 560
false-positive measurement error, 14
finite-sample inference, 570, 573, 576, 636–637
first law of capture-recapture, 246–247, 586
fitstats(R function in package AHM), 333, 375, 434, 496
fitstats2(R function in package AHM), 375–377
fitstats3(R function in package AHM), 375
fixed effect, 115–117, 119
flat model, 20, 29, 648t–649t
flat-tailed horned lizard (*Phrynosomamcallii*), 45f
Flevoland, 486, 487f
four-part hierarchical model, 473–474
Frederick City Watershed Cooperative Wildlife Management Area, 327f
Freeman-Tukey, 76
frequentist inference, 20–21
frequentist, 33
fs.fn(R function in package AHM), 570–571
full conditional distribution, 55, 62–63, 65
full likelihood, 400
fungus, 606

G

gamma diversity, 709
GAMs. *See* generalized additive models (GAMs)
gdist(R function in package unmarked), 49t
gdistOpen(R function in package unmarked), 49t
gdistsamp(R function in package unmarked), 306–307, 369–370, 395–396, 485–498
general data simulation function for *N*-mixture models, 241–245
generalized additive models (GAMs), 556–557
generalized linear mixed model (GLMM), 81, 556–557
generalized linear model (GLM), 79–80, 102–114, 145–146
GENPRES(software program), 584–585
geoBUGS(part of WinBUGS and OpenBUGS), 146
geometry, 9, 528
getN(R function in package AHM), 439, 442
ghost point pattern, 554–555
Gibbs sampling, 62–63
GLM. *See* generalized linear model (GLM)
glmer(R function in package lme4), 117, 120, 205
GLMM. *See* generalized linear mixed model (GLMM)
Glorious Integrated HDS/Dail-Madsen Model, 522–526
gmultmix(R function in package unmarked), 49t, 306–307, 330–333
GoF. *See* goodness-of-fit (GoF)
goodness of fit assessment, 192–198
goodness-of-fit (GoF), 41, 225, 250–254, 337–338, 589–590

gpcount(R function in package unmarked), 49t, 306–307
grain size, 7–8, 553
great tit (*Parus major*), 125–126, 126f
group stratified model, 358–362
groups, 464–472

H

habitat mask, 527–528
half normal detection model, 401–402, 406, 470
half normal, 398, 418, 505–506
half-width, 398–399, 404
hazard rate detection model, 506, 618
hazard, 490–491, 505–506, 618, 620
HDS. *See* hierarchical distance sampling (HDS)
hessian, hessian matrix, 34, 38, 247
heterogeneity model in BUGS, 358–362
heterogeneity model, 322, 358–362, 621–622
heterogeneity, 6, 10, 660–661
heteroskedasticity, 294–295
hierarchical distance sampling (HDS), 394–395, 426–427, 464. *See also* Bayesian HDS; spatial distance sampling
hierarchical model (HM), 12, 28–31, 79–80, 145–146, 631–632, 729–730. *See also* site-occupancy model
hierarchical models for communities, 631–632
highest-posterior density interval (HPDI), 162
HM. *See* hierarchical model (HM)
home range, 279, 281–282, 613
home-range, 282f
homogeneity, 159, 225, 560–561
homogeneous, 634
homoscedasticity, 160
HPDI. *See* highest-posterior density interval (HPDI)
HPDinterval(R function), 162
hurdle model, 108, 308
hyper-coefficients, 675
hyper-hyperparameters, 675
hyper-parameter, 115–116, 635, 675, 682, 696–697, 726–727
hyper-prior, 673, 685, 719

I

Ian Fiske, 46
identifiability, 30, 124–125, 244–245, 559
IM. *See* integrated model (IM)
imbalanced data, 601f
impala data set, 418–419, 445–447
imperfect observation of cluster size, 471–472
implicit dynamics, 510–513. *See also* population dynamics
implicit hierarchical model, 28
improper prior, 53–54
incidence-based modeling, 716–717
inclusion probability, 308

independence model, 518−520
independent multiple observer protocol, 318
index assumption, 107, 318
indicator variable, 71−72, 342
individual covariate model, 400, 464−465
individual covariate, 245, 400, 464−472
individual effect, 243
individual heterogeneity models, 356−357
individual heterogeneity, 225, 249−250
inference for hierarchical models, 31
initial value, 59, 149, 151−152, 477, 531
`inprod`(inner product, R function), 583
`instRemPiFun`(R function in package `unmarked`), 348
integrated analysis, 309−310
integrated likelihood. *See* Marginal likelihood
integrated model (IM), 20, 300, 368, 549, 628
integrated population model (IPM), 549
integration of mathematical function, 421−422
intensity, 3, 7, 369, 532−533
interaction effects, 177, 184, 187, 572
interaction, 92−96, 631−632
interaction-effects ANCOVA, 84, 91f, 236−237
interactive model, 94−95
invalid parent node, 502−503
inverse-logit, 540, 696−697
IPM. *See* integrated population model (IPM)
Island scrub jay (*Aphelocoma insularis*), 430, 431f, 443f
Island scrub jay, global population size of, 430−444
`issj.sim`(R function in package AHM), 517
issue of space, 279−282

J

Jaccard index, 709, 711f
`jags`(R function in package `jagsUI`), 71, 147, 695
JAGS(software program), 146, 186, 193, 450
`jags.basic`(R function in package `jagsUI`), 695
`jagsUI`(R package), 71, 147, 695
Jim Albert, 337
joint density, 23, 310
joint distribution, 24−25, 32, 61, 65−66
joint probability, 36
joint species distribution model (JSDM), 633, 725−726
JSDM. *See* joint species distribution model (JSDM)
Julian date, 487

K

K (integration limit), 259
Ken Kellner, 502−503
knots (in splines), 623

L

L'Equipe de choc helveto-britannique, 617−618
lack of fit ratio, c-hat, 253
lack-of-fit ratio, 253, 378
large-scale modeling, 298, 603
latent random variable, 114
`Lcond`(R function in package AHM), 402
learning from data, 24
least-squares, 84, 90, 96, 178, 183
leave-one-out, 69−70
Leslie matrix, 735
level(of factor), 179−180
`Lfull`(R function in package AHM), 403
life-history, 672
`Lik.binned`(R function in package AHM), 407−408
`Lik.binned.point`(R function in package AHM), 413
`Lik.cond.point`(R function in package AHM), 414
`Lik.cont.point`(R function in package AHM), 413
likelihood analysis of hierarchical model, 391
likelihood inference, 31, 323−325
likelihood ratio test (LRT), 232−233, 572
likelihood, 51, 151
line transect sampling, wiggly lines, 532
line transect survey, 532
line transect, 404, 418−420, 532
linear model with normal response, 149−167
linear model, 83−102, 177−183, 294−298. *See also* generalized linear model (GLM)
linear regression, 117, 190, 675
linear-logistic, 556f, 605
linear-mixed model, 118
`lme4`(R package), 116−117, 205
`lmer`(R function in package `lme4`), 116−118
log link, log link function, 540
logistic regression. *See* binomial GLM
logit link, logit link function, 30, 103, 110, 223
logit-normal, 250, 356−357, 627
log-linear, 35, 294, 336, 448, 540
LRT. *See* likelihood ratio test (LRT)
lumping error, 390−391

M

M_0 model, 351−355
macroscale, 632
magical covariate, 609−611
main effects, 84, 96, 99, 561−564, 571
mapping, 590−600
MAR. *See* missing at random (MAR)
Marc Mazerolle, 252, 732
marginal distribution, 23−26, 74
marginal likelihood, 42, 65
marginal probability, 42, 324

SUBJECT INDEX 777

MARK(software program), 30
Markov chain Monte Carlo (MCMC), 16, 31, 54–69, 84–92
Markov-modulated point process (MMPP), 614–621
mark-recapture distance sampling (MRDS), 479–483, 540–543
mark-recapture, 471, 479–483, 540
mark-recapture/distance-sampling, 471, 479–483, 540
mark-recapture/double observer distance sampling, 314–315, 474, 479–483
marsh tit, 257, 304
matrix model, 91
matrix, 34, 88–89
Maxent(software program), 555–556
maximum likelihood (ML), 148–149, 209
M_b model, 355–356
mb.gof.test(R function in package AICcmodavg), 589
MC error. See Monte Carlo error (MC error)
mean imputation, 174
mean/variance relationship, 337
means parameterization, 87, 90, 92
measurement error model, 13–16, 256, 305, 730–731
measurement error, 13, 220–221
mental removal model, 351
mental removal protocol, 325
meta-analysis, 679–680
metacommunity data, 643–646
metacommunity, 633–643, 646–651
metadata, 47, 428–429
meta-population design, 10–12
metapopulation ecology, 552
metapopulation, 10
meta-population, 10, 221
Metropolis algorithm, 58–60
Metropolis-Hastings algorithm (MH algorithm), 55–56
Metropolis-Hastings, 56–58
Metropolis-within-Gibbs algorithm, 66
S-fold data augmentation, 465
MH algorithm. See Metropolis-Hastings algorithm (MH algorithm)
M_h, 356–357
MHB. See Monitoring Häufige Brutvögel
MhPiFun(R function in package AHM), 357
MhPiFun pi function, 357, 361
Mike Meredith, 532–533
misclassification errors, 390–391
misidentification, 727
missing at random (MAR), 169
missing not at random (MNAR), 169
missing value imputation, 174
missing value, 169–176
mixed model, 114–121

mixing(of Markov chains), 63–65
mixture model. See binomial mixture model; N-mixture model; multinomial mixture model
ML. See maximum likelihood (ML)
MLE (maximum likelihood estimate, maximum likelihood estimator), 20–21, 33–34, 146, 732
MMPP. See Markov-modulated point process (MMPP)
MNAR. See missing not at random (MNAR)
modavgPred(R function in package AICcmodavg), 264–265, 274, 381
model averaging, 69
model checking, 72
model criticism, 261–264
model fitting, 72–76
modelM_h, 314–315
model M_t, 351–355
modelM_x, 356–357
model complexity, 70–71, 731
model selection by AIC, 69–70
model selection by DIC, 70–71
model selection, 69
model.matrix(R function), 89–90, 92, 98, 100, 581–584
model-averaging, 70
model-based, 394
modeling a variance parameter, 121–122
modeling of a variance, 121–122
modeling of abundance, 220–222
modeling species richness in two-step analysis, 679–682
model-selection, 342
modSel(R function in package unmarked), 329, 354, 373, 495
moment matching, 23, 108, 201–203
moment-estimator, 223–224
Monitoring Häufige Brutvögel(MHB, Swiss breeding bird survey), 367, 643
monitoring, 3–4, 367
monograph, 729–731, 733
Monte Carlo error (MC error), 54, 64–65, 155
motorbike, 125
movement, 279, 394, 503, 600, 609
MRDS. See mark-recapture distance sampling (MRDS)
M_t model, 351–355
multinomial cell probabilities, 316–317, 346–348, 350–351, 404–405, 412, 423, 504
multinomial distribution, 315, 316f
multinomial distribution, product of independent Poisson, 342–346
multinomial N-mixture model, 316–325
multinomial observation models, 314–315
multinomial trial, 315
multinomial/Poisson(mixture model), 324, 334, 452–453

multinomPois (R function in package unmarked), 49t, 325, 329–330
multiple linear regression, 149–167
multiple observer protocol, dependent observers
multiple observer protocol, independent observers, 318
multiple observer sampling, 319, 349–350
multiple scale, 290–294
multi-scale N-mixture model, 306–307
multiscale occupancy model, 600–608. *See also* site-occupancy model
multispecies(model), 298, 662, 668, 718, 726
multistate model, 14, 559
M_x model, 351–355
n.eff(effective MCMC sample size), 156–157
National Park Service (NPS), 430
Navarra, 80, 84–85, 94–95, 98
NB distribution. *See* negative binomial distribution (NB distribution)

N

negative binomial abundance model, 330, 377, 395–396, 429, 457
negative binomial distribution (NB distribution), 107, 203, 365, 438
negative exponential detection model, 400–401
negative log-likelihood, 37–38, 69–70, 187
nested ANOVA, 307, 602
nested design, 601–602
n-fold single species occupancy model, 662–667
Nhat(R function in package AHM), 275
Nichols, Jim, 14, 314–315, 726
NIL dereference read (dreaded BUGS error message), 170
NIMBLE(software program), 147
Nmix.gof.test(R function in package AICcmodavg), 250
N-mixture estimator, 246–247
N-mixture model for abundance, 30
N-mixture model with covariates, 229–241
N-mixture model, 203
non-hierarchical model ("flat" model), 29, 225
non-hierarchical, 402
non-independence, 478–479, 658
non-linearmodels, 35
non-missing data, 170, 349
non-model-based approximation, 187
non-normalresponses, 103, 187
non-parallel regression lines, 92
non-parallelism, 166f
nonparametric bootstrapping, 40–41
non-standard likelihood, 184–187
normal data distribution, 117–120
normal distribution, 13, 62, 204, 635–636, 696–697

normal GLM, 149–167, 177–183
normal-normal(mixture model), 117
normal-normal generalized linear mixed model, 117–120
normal-Poisson(mixture model), 324
normal-Poisson generalized linear mixed model, 203–207
NPS. *See* National Park Service (NPS)
numerical integration, 398–399
numerical over/underflow, 459

O

obsCovs. (argument for feeding into an unmarked data frame observational or sampling covariates), 47, 48t
observable random variable, 32
observation model, 28–29, 127
observation process, 28–29, 131–135
observation/measurement error model, 577
observed community size, 651–661
observer effect, 317
obsToY(R function in package unmarked), 348
obtain an estimated Ntotal which is a bit less than the BUP, 459
occasion, 29–30, 349, 370, 614
occu(R function in package unmarked), 30, 43
occuFP(R function in package unmarked), 49t
occupancy model for species distribution, 29–30
occupancy model, 65–66
occupancy modeling of a community of species, 621–622
occupancy models with removal design, 621
occupancy, 552–553
occuPEN(R function in package unmarked), 49t, 586–587
occuRN(R function in package unmarked), 301, 627
occurrence, 9, 14
OD. *See* overdispersion (OD)
offset in Poisson GLM, 105–106
offset, 105
one factor and one continuous covariate (in linear model), 84–96
ones trick, 184–187
one-way ANOVA, 93t
open HDS models, 483–510
open population distance sampling, 12–13, 518–526
OpenBUGS(software program), 146–147
open-population models, 518–526
optim(R function), 37
outlier, 107–108, 377
ovenbird(*Seiurusaurocapillus*), 326, 328f
overdispersed model, 64, 151, 378
overdispersion (OD), 107–108, 242, 252, 264, 292
overdispersion at multiple scales, 290–294
overdispersion correction, 107
overparameterization, 93t

P

parallel regression lines, 91
parallel computing with BUGS, 147, 155, 625
parameter redundancy, 96
parameter-expanded data augmentation (PX-DA). *See* data augmentation (DA)
parametric bootstrap, 41, 192–198, 438
parametric bootstrapping, 40–41, 73–75, 192–198
parasite, 81–82, 82f, 109–110
`parboot`(R function in package `unmarked`), 73, 192–193, 250, 333
passerine, 125–126, 527
`pcount`(R function in package `unmarked`), 30, 227, 232–233, 324, 540–543
`pcount.spHDS`(R function in package AHM), 540
`pcountOpen`(R function in package `unmarked`), 49t
PCR. *See* polymerase chain reaction (PCR)
pD(estimated number of parameters), 70–71
pdf. *See* probability density function (pdf)
penalized likelihood, 559, 586–587
penalized spline, 622–626
percentile-based credible interval, 162f, 277
perceptibility, 472
peregrine falcon (*Falco peregrinus*), 22f
peregrine spring survey, 617–621
peregrine, 21, 24–25
per-individual detection probability, 223, 300–301, 721
permanent emigration, 514
per-quadrat detection probability (also per-site detection probability), 300–301
phylogenetic independence, 679
pi function, 316–317, 325, 346–348, 356–357, 370
`PiFun`(R function in package `unmarked`), 346–347, 350, 355, 363
pixel (also, cell, quadrat), 278, 441–442, 532–533, 535–536, 543
`playRN`(R function in package AHM), 301
`plot.Nmix.resi`(R function in package AHM), 261
pmf. *See* probability mass function (pmf)
point count data set, 125–135
point count, 125
point pattern models (PPMs), 554–555
point pattern, 3, 9
point process model (PPM), 3, 8f, 16, 397, 529–530
point process, 3–9
point transect data, 408–414
point transect HDS, 455–457
point transect sampling, 394
pointtransect, 394
Poisson (Poisson distribution), 222, 317
Poisson data distribution, 120–121
Poisson formulation of multinomial mixture model, 342–346
Poisson GLM, 104–105, 187–192. *See also* binomial GLM
Poisson GLMM, 203–207
Poisson integrated full likelihood, 405
Poisson intensity function, 533, 543
Poisson observation model, 614
Poisson point process model (PPPmodel), 533
Poisson PPM. *See* Poisson point process model (PPPmodel)
Poisson random-effects model, 657–658
Poisson/Bernoulli (mixture model), 734
Poisson/log-normal (mixture model), 336, 344
Poisson/Normal (mixture model), 657
Poisson/Poisson (mixture model), 305–306
Poisson/Poisson *N*-mixture model, 305–306
Poisson-log normal model (PLN model), 237, 459
polymerase chain reaction (PCR), 606
polynomial, 99–102
pond-breeding amphibian, 10
poor man's abundance (occupancy), 4–6, 626–627
population distribution, 398
population dynamics, 513–526. *See also* implicit dynamics
posterior distribution, 53–54
posterior mean, 53–54, 58, 191f, 286f, 293t, 297, 362, 450, 509, 588t, 713
posterior model probabilities, 72, 338–342
posterior predictive check, 192–198
posterior standard deviation, 159, 664–665, 680
`ppc.plot`(R function in package AHM), 253
PPM. *See* point process model (PPM)
PPMs. *See* point pattern models (PPMs)
PPPmodel. *See* Poisson point process model (PPPmodel)
pre–BUGS era of ecological modeling (=stone–age in ecology), xiv
precision, 86, 245–247, 584–587
`predict`(R function in package `unmarked`), 88, 233, 265, 271, 275
prediction, 46, 190, 256, 264
`predictSE`(R function in package AICcmodavg), 264–265
PRESENCE(remarkable software program), 30, 559, 606, 614, 621–622
presence, 552–553, 558–559
presence/absencedata, 11–12, 556–557
presence/background data, 555–556
presence-onlydata, 555–556
prevalence, 35
primary sample, 484
primary period, (primary occasion), 48t, 352–353, 484, 499, 509, 514, 560
prior distribution, 52–53
probability density function (pdf), 21–27
Probability distribution function. *See* probability density function (pdf)

probability games, 163
probability mass function (pmf), 21, 315
probability, 21–27, 552–553
process model, 305
process of hierarchical modeling, 31
process/state, 549
progression(of models), 633, 646–647, 657
properties of MLEs, 33–34
proportion of variance explained (R^2), 183–184
pseudoreplication, 115, 203, 307

Q

Q-Q plot. *See* quantile-quantile plot (Q-Q plot)
Quadrat population size model, 256
quantile-quantile plot (Q-Q plot), 160
quasi-likelihood, 107, 380–381

R

R(software program), 320–323
R function model.matrix with BUGS, 581–584
R package rjags, 167–169
R package unmarked, 46–49
R^2. *See* proportion of variance explained (R^2)
R2jags(R package), 71, 147
R2OpenBUGS(R package), 154–155
R2WinBUGS(R package), 71, 147, 149, 155
random effect, 51, 65, 79–80, 107, 114, 117–121, 145–146, 203, 208
random effects models, 114–121
random variable, 21
random-effects binomial GLM, 208–212
random-effects Poisson GLM, 203–207
Random-effects Poisson model (REP model), 657
randomized-block ANOVA model, 114
raster(R package), 383, 442
rate parameter, 12–13
rbinom(R function), 36
RD. *See* Robust design (RD)
realized abundance (vs. expected abundance), 7, 229–230
recruitment, 123
rectangle, 398, 421
rectangular rule (of integration of mathematical function), 421
reduced-dynamics model, 520–522
reduced-information-content, 552
regression model, 13, 15
"regularizing" effect, 247
relative abundance, 308–309
REML. *See* restricted maximum likelihood (REML)
removal design, 621
removal observation model, 614–621
removal protocol, 326, 363–364
removal sampling, 319, 325–333

removalPiFun(R function in package unmarked), 347
REP model. *See* Random-effects Poisson model (REP model)
reparameterization, 34–35, 478
replicates within and among years, 509–510
residual, 73
restricted maximum likelihood (REML), 117, 141–142
rgdal(R package), 383
Rhat(convergence diagnostic statistic, potential scale reduction factor, Gelman-Rubin statistic), 156–157
R-hat statistic, 64
right-truncated, 690–691
ring-recovery data, 12
rjags(R function), 167–169
RMSE. *See* root mean square error (RMSE)
rmultinom(R function), 320, 322
RNmodel. *See* Royle-Nichols model (RNmodel)
rnorm(R function), 59
Robust design (RD), 484, 509–510, 560
root mean square error (RMSE), 354
Royle-Nichols model (RNmodel), 300–305, 716–717
rpois(R function), 130
runifdisc(R function), 469
runjags(R function), 147

S

Sample size, 127–128
sampling area, 275, 279–281, 311, 369, 379, 380f, 553, 600
Sampling error, 123–124
Sampling variance of estimator, 33
sampling/observation covariate (obsCovs. in unmarked), 368
Santa Cruz Island, California, 430, 432f
SAS(software program), 214
scalar, 208, 572
scale-dependence, 554–555
Schmid, Hans, 229, 254–255, 272, 275
scientific model, 320
SCR models. *See* spatial capture-recapture models (SCR models)
SD. *See* standard deviation (SD)
SDM. *See* species distribution model (SDM)
SE. *See* Standard error (SE)
second law of capture-recapture, 225, 249, 560–561, 647, 690
secondary period. *See* secondary season
secondary sample, 484
secondary season, 509
semi-frequentist, 192
semiparametric, 293
sensitivity to bin width, 415–416
sensitivity to prior issue, 342
sequential analysis, 52
serial correlation, 305–306, 308

set.Seed(R function), 42
sex, 12, 80–82, 86
shape parameter, 491
shifted Poisson distribution, 466
short-term predictive success, 69–70
short—term, 485
shrinkage of Bayesian estimates, 548
shrinkage, 115
significance test, 94–95, 160–162
sim.fn(R function in package AHM), 3–4, 7
sim.ldata(R function in package AHM), 406
sim.pdata(R function in package AHM), 461, 528–529
sim.spatialDS(R function in package AHM), 534
sim.spatialHDS(R function in package AHM), 544
sim3Occ(R function in package AHM), 604, 604f
simComm(R function in package AHM), 634
simHDS(R function in package AHM), 444, 446f–447f
simHDSg(R function in package AHM), 466, 469
simHDSopen(R function in package AHM), 485, 499
simHDStr(R function in package AHM), 474, 479
similarity among sites, 709–712
similarity among species, 709–712
similarity, 192, 679
simNmix(R function in package AHM), 241–245
simOcc(R function in package AHM), 577–581
simOccttd(R function in package AHM), 615–616
simpleNmix(R function in package AHM), 298
simplest DR community model, 683–690
simplest possible N-mixture model, 225–228
simplest possible site-occupancy model, 561–564
simulating binned distance sampling data, 406–408
simulating community data, 633
simulating distance sampling data, 401–403
simulating HDS data with group size, 400
simulating HDS data, 444–445
simulating MRDS data, 479–480
simulating multinomial observations, 320–323
simulating occupancy data, 215
simulating point transect data, 410–412
simulating temporary emigration system, 499–503
simulating time-removal/DS data, 474–479
simulation of metacommunity, 83–102
single-degree-of-freedom, 89
single-season site-occupancy model, 46
single-species occupancy models, 644, 667, 716–717, 727
single-visit, 580
site covariate(siteCovs. in unmarked), 127, 317
site heterogeneity, 248–249
site-by-occasion, 248
site-by-replicate, 644
site-by-survey, 290–291

siteCovs(argument for feeding into an unmarked data frame site covariates), 328
site-occupancy estimator, 584–587
site-occupancy model with covariates, 564–576
site-occupancy model, 557–564. *See also* multiscale occupancy models
site-structured, 221, 447–448, 632
slow mixing, 64
small sample bias, 576
SODA(R package), 584–585
Sovon(Dutch Centre for Field Ornithology), 486
space, 279–282
space-for-time substitution, 608–613
sparrowhawk, 710
spatial capture-recapture models (SCR models), 220, 279, 483, 555. *See also* mark-recapture distance-sampling (MRDS)
spatial covariates, 532–539
spatial distance sampling, 526–548
spatial HDS models in unmarked, 540–543
spatial hierarchical distance sampling, 543
spatial prediction, multinomial N-mixture model, 383–386
spatial sampling, 416–417
spatial temporary emigration, 485
spatial variability, 6, 9, 15, 642–643
spatially explicit models, 14–15, 527, 534
spatially stratified capture-recapture models, 349–366. *See also* multinomial N-mixture models
spatstat(R package), 469
species accumulation curves, 712–716
species distribution model (SDM), 278, 662
species distribution, 551–557
species identity, 650–682
species richness map, 712–716
species richness, 4, 9, 16
spline model, 626
spline.prep(R function in package AHM), 623
split-plot design, 602
splitting error, 391
STAN(software program), 146
standard deviation (SD), 22–23
Standard error (SE), 164
standardization, 64
state parameter, 12–13, 429
state-space, 529–532
static model, 222, 464
statistical inference, 27
statistical model, 23–24
stochastic process, 3, 7
strategy of model building, 212–213
stratified populations (data augmentation for), 465
study design, 245–247

subsampling, 10–11, 306–307, 560
subtracting the intercept, 87
subunit, 10, 307, 601f
success parameter, 6, 184, 222
success probability, 21
succession, 62–63
suitability, 108, 242–243
sum-of-squared errors, 375
sum-to-zero constraints, 87–88
sunflower effect, 279, 280f
survey error, 15
survey, 590–591, 617–621
survival model observation process, 615–617
survival, 12
survival, survival probability, 12, 17
survival/recruitment, 513–516
Swiss breeding bird survey (MHB), 643–646
Swiss Great tits, 254–278, 282–298
Swiss MHB, 279, 305–306
Switzerland, 254–255, 281, 386–387

T

t-distribution, 107–108
temporal autocorrelation, 115, 308
temporary emigration models, 514, 549
temporary emigration processes, 483–510
temporary emigration, 472
temporary removal, 472–473
territory mapping, 255, 329, 367
thinning, 64
3-dimensional array, 644, 662, 709
3-level model, 601–602, 608, 610
3-part model, 334–346, 395–396, 452–455
time of removal models, 464–465, 472–473, 478–479
time to detection (TTD), 615, 617–621
time-for-space substitution, 298–299
time-of-removal, 473–474
time-removal, 472–479
time-to-detection model, 615–617
time-to-event analysis, 615
tits, 131–135, 254–278, 282–298
top-down analysis, 603
tortoises, 472, 485
`traceplot`(R function in package `jagsUI`), 155
transect, 614–621
transformation, 339
transientresponse, 355
trap attraction, 355
trap avoidance, 355
trap happiness. *See* trap attraction
trap response, 355
trap shyness. *See* trap avoidance

treatise, 421–422
treatment structure, 602
treatments, 602
trend, 12, 510, 524–525, 731
triangular distribution, 408, 422–423, 529–532
triangular relationship, 668
triple-Bernoulli, 603–604
true species richness, 7, 651, 683
truncate, 324, 428–429
TTD. *See* time to detection (TTD)
t-test, 83
turnover, 708
two continuous covariates, 99–102
two factors, 96–99
two-level hierarchical model, 557
2-level model, 603, 608, 611
2-dimensional splines, 628
two-step analysis, 679–682
two-way ANOVA with main effects, 96
two-way interaction effects ANOVA, 96
types of multinomial models, 318–319

U

unbiased estimate, 33
unbiased, 13
unbounded count, 104–105
uncertainty, 34, 50, 190, 513–514, 615
uncondition, 405
unconditional, 42, 636–637, 642–643
undefined node (`BUGS` error message), 174
undefined real result (`BUGS` error message), 209, 211
undercount, 226–227
underdispersion, 107–108
uniform distribution, 150–151, 408
uniformity assumption (distance sampling), 526
uniformity assumption, 400, 529–530, 532–533
uninformative prior, 53
unit-subunit, 604
Unix/Linux, 23
`unmarked`(R package), 40–41, 46–49
`unmarkedFrameDS`(R function in package `unmarked`), 49t
`unmarkedFrameGDS`(R function in package `unmarked`), 49t
`unmarkedFrameGDSO`(R function in package `unmarked`), 49t
`unmarkedFrameGMM`(R function in package `unmarked`), 49t
`unmarkedFrameGMM`(R function in package `unmarked`), 49t
`unmarkedFrameMpois`(R function in package `unmarked`), 49t

unmarkedFrameOccu(R function in package
    unmarked), 49t
unmarkedFrameOccuFP(R function in package
    unmarked), 49t
unmarkedFramePCO(R function in package
    unmarked), 49t
unmarkedFramePCount(R function in package
    unmarked), 49t
unmarkedMultFrame(R function in package
    unmarked), 49t
update, 66, 170
use of R function model.matrix with BUGS, 581–584
user group, 147

V

vague prior, 53, 58f, 72, 586–587
variability, 138f, 697
variable, 21
variance, 260, 294–298, 673
variance/mean ratio, 107
variance/mean relationship, 263–264, 309
variance-covariance matrix, 34, 43, 252, 264, 534, 679, 726
vectorized language, 148

W

"wald-type" confidence interval, 34
warm-up period. See "burn-in"; period
Weibull distribution, 618
Weibull observation model, 614–621
Weibull occupancy model, 617–621
wiggly covariate relationship, 278, 552–553, 622–626
wigglyOcc(R function in package AHM), 622, 623f
wildlife biology, 12

willow tit, 302, 305t
WinBUGS(= revolutionary software program), 146–147
within-chain variance, 488
within-unit variation in density, 526–548
wonderful(= BUGS language), xxiii

X

X matrix. See Design matrix

Y

yearly site-covariate, 479
yellow wagtail (*Motacilla flava*), 486f–487f

Z

Z matrix, 709–712
zero inflation, 108–109
zero-centered random effects, 204–205, 209
zero-inflated binomial model (ZIB model), 109
zero-inflated models, 7
zero-inflated Poisson (ZIP), 232–233, 237
zero-inflated Poisson Dorazio-Royle-Yamaura model (ZIP-DRY model), 718
zero-inflation parameter, 257
zero-inflation, 108–109
zeros trick, 148, 184, 421–423, 453, 455
zero-truncated Poisson, 108
ZIB model. See zero-inflated binomial model (ZIB model)
ZIP N-mixture model, 283–289
ZIP. See zero-inflated Poisson (ZIP)
ZIP-DRY model. See zero-inflated Poisson Dorazio-Royle-Yamaura model (ZIP-DRY model)

Printed and bound by CPI Group (UK) Ltd, Croydon, CR0 4YY
08/06/2025
01896870-0016